HERL-0030

UNITED STATES GOVERNMENT PROPERTY
This book is the property of the United
States Government Environmental Protection
Agency

VENOMS:
Chemistry and Molecular Biology

VENOMS:
Chemistry and Molecular Biology

ANTHONY T. TU
Professor of Biochemistry and Anatomy
Colorado State University

A WILEY-INTERSCIENCE PUBLICATION
JOHN WILEY & SONS, New York • London • Sydney • Toronto

Copyright © 1977 by John Wiley & Sons, Inc.

All rights reserved. Published simultaneously in Canada.

No part of this book may be reproduced by any means, nor transmitted, nor translated into a machine language without the written permission of the publisher.

Library of Congress Cataloging in Publication Data:

Tu, Anthony T   1930–
  Venoms: chemistry and molecular biology.

  Includes index.
  1. Snake venom – Physiological effect.   2. Snake venom.   3. Venom.   4. Hydrolases.   I. Title.
QP941.S6T8   591.1'9'245   76–30751
ISBN 0-471-89229-7

Printed in the United States of America

10 9 8 7 6 5 4 3 2 1

To the Memory of

    My Mother, Songsui Lin Tu (1901–1968)

    and

    My Ph.D. Research Adviser, Professor Hubert S. Loring (1908–1974)
    Stanford University

PREFACE

There are numerous books, review articles, and multiauthored monographs on the subject of venoms, which are excellent sources for specialists' research and have their archival values. However, because so many different people have contributed portions of the information, it is difficult to obtain a comprehensive, unified view of venoms. For scientific laymen who want a general idea about "What is venom?" there is a need for a book written by a single author.

When asked by the publisher about the possibility of writing such a book, I knew that it would be a monumental task to assemble all references, to digest the contents and analyze the data, to reconstruct these materials, and to present them in a logical and unified form.

In this book, I have tried to be comprehensive so that specialists can use it as a source of information. At the same time, I have also tried to be selective so as to present general and systematic views to scientists in general, from the vast store of information on venoms. To balance these two basically opposing aims has been difficult.

Overall, materials presented here are expansions of my lecture course, "Chemistry and Pharmacology of Animal Toxins," given at Colorado State University, and my many review articles, lectures at different colleges and universities, and special lectures at scientific meetings. I sincerely hope that this book will stimulate more interest in venoms and increase the understanding of them.

I was fortunate to have Dr. Charlotte L. Ownby, Assistant Professor of Physiology, Oklahoma State University, participate on Chapter 27 on the chemical neutralization of snake venoms. Thanks are also extended to the following persons, who helped in the many phases of the writing of this book: Dr. A. Bieber, Dr. G. Happ, Dr. N. Iritani, Dr. B. Joyce, Dr. D. Leuker, Dr. C. Ownby, Dr. D. Will, Ms. A. Kano, Ms. L. Rimsay, Mr. J. Fox, Mr. J. Pardee, Mr. M. Stringer, and Mr. J. Yadlowski.

I am grateful also to many colleagues — Drs. P. J. Delori, W. B. Elliott, D. M. Fambrough, N. Frontali, E. Heilbronn, J. Ishay, D. Munjal, T. Okonogi, R. Perez-Polo, Y. Setoguchi, J. W. Simpson, and R. Yarom — who supplied valuable photographs; without these, the book could not have been published. I thank Teresa Szidon and Dee Kohut for the final typing of the manuscript.

Finally, my thanks are extended for encouragement by Kazuko Yamomoto Tu, my wife, and Dr. Tsungming Tu, my father, who has done many years of snake venom research.

My deepest regret is that my mother, Songsui Lin Tu, and my Ph.D. research adviser, Professor Hubert S. Loring of Stanford University, cannot see this book.

Anthony T. Tu

Fort Collins, Colorado
January 1977

CONTENTS

I   SNAKE VENOMS: General Background and Composition   1

    1   Composition of Snake Venoms: Nonprotein Components, 5

II   SNAKE VENOMS: Enzymes   21

    2   Phospholipase A_2, 23
    3   Phosphodiesterase, 64
    4   Phosphomonoesterase, 78
    5   L-Amino Acid Oxidase, 85
    6   Acetylcholinesterase, 97
    7   Proteolytic Enzymes, 104
    8   Arginine Ester Hydrolase and Other Esterases, 127
    9   Other Enzymes, 132
   10   Enzyme Inhibitors, 139

III   SNAKE VENOMS: Properties and Actions   149

   11   Venoms of Hydrophiidae (Sea Snakes), 151
   12   Elapidae Venoms, 178
   13   Viperidae Venoms, 201
   14   Venoms of Crotalidae (Crotalids, Pit Vipers), 211
   15   Colubridae Venoms, 234
   16   Distribution of Venoms in Envenomated Animals, 236
   17   Binding of Neurotoxins to Acetylcholine Receptors, 240
   18   Neurotoxins: Chemistry and Structural Aspects, 257
   19   Nonneurotoxic Basic Proteins (Cardiotoxins, Cytotoxins and Others), 301
   20   Hemolysis, 321

21 Blood Coagulation, 329
22 Nerve Growth Factor, 361
23 Hemorrhage, Myonecrosis, and Nephrotoxic Action, 372
24 Autopharmacological Action, 400
25 Metabolic and Teratogenic Effects, 412
26 Immunology, 420
27 Chemical Neutralization of Snake Venoms, 435

IV OTHER VENOMS 457

28 Scorpion Venoms, 459
29 Spider Venoms, 484
30 Venoms of Bees, Hornets, and Wasps, 501
31 Ant Venoms, 527
32 Gila Monster Venom, 531

Appendix, 535

Index, 541

VENOMS:
Chemistry and Molecular Biology

SNAKE VENOMS: General Background and Composition

Human beings have traditionally, and justifiably, been fearful of and puzzled by the violent action of venoms from such small creatures as venomous snakes. The amount of venom injected by these snakes is usually very small, but in some cases the results are fatal. Symptoms arising in the victim result from the combined effects of complex protein components present in the venom.

Venom is not composed of a single substance common to all poisonous snakes, although almost all venoms consist of approximately 90% protein. The proportions of the different substances in venom and their specific characteristics vary among the species. However, usually the closer the phylogenetic relationship of the snakes, the more similar are the venom properties and composition.

Any given snake venom usually contains more than one toxic principle, and these tend to act in combination in an actual poisoning. The overall toxicity is due to enzyme as well as to nonenzymatic proteins. However, the main lethal action, especially in Elapidae and Hydrophiidae snakes, can be attributed to neurotoxins that are not enzymes. This does not mean that enzymes are unrelated to the toxic actions of venoms. Many venom enzymes actively participate in blood coagulation, anticoagulation, hemorrhage, hemolysis, autopharmacological action, and lysis of cell and mitochondrial membranes.

Of the nearly 2000 different types of snakes that exist, about 300 are known to be venomous. The venomous snakes are classified according to morphological characteristics and comprise five families: Crotalidae (crotalids, pit vipers), Viperidae (viperids, vipers), Elapidae (elapids), Hydrophiidae (sea snakes), and Colubridae (colubrids).

Crotalidae comprise six genera: *Crotalus, Sistrurus, Agkistrodon, Bothrops, Lachesis*, and *Trimeresurus*. *Crotalus* and *Sistrurus* are the rattlesnakes and can be found only in North, Central, and South America. *Bothrops* occurs only in Central and South America. *Agkistrodon* includes the copperheads and moccasins. This is the only genus of snake that can be found in both the New and the Old World. *Lachesis* has only one species and is distributed from Central to South America. *Trimeresurus* is the Asiatic pit viper, which lives only in Asia; there are 31 species in this genus.

Viperidae are known commonly as viperids or vipers and can be found in Africa, Europe, and Asia. They are not found in Australia or on the American continent. Viperidae comprise the genera *Vipera, Atractaspis, Bitis, Causus, Cerastes, Echis, Adenorhinos, Atheris, Eristicophis, Pseudoceratstes*, and *Azemiops*. Africa and the Middle East are particularly rich in varieties of Viperidae. In Asia, there is only one genus of *Vipera*.

Elapidae include well-known cobras, mambas, and kraits. All of the poisonous snakes in Australia and New Guinea belong to the family Elapidae. There are only two genera of Elapidae in North America and Central America. In North America, Elapidae are coral snakes, which belong to the genera *Micruroides* and *Micrurus*. In South America, there is *Leptomicrurus* in addition to the other two genera mentioned. These are the only Elapidae that migrated to the New World from Asia through the Bering land bridge many millions of years ago. Australia and New Guinea, on the other hand, are rich in genera belonging to this family: *Acanthophis, Brachyaspis, Demansia, Denisonia, Elapognathus, Glyphodon, Hoplocephalus, Micropechis, Notechis, Oxyuranus, Parademansia, parapistocalamus, Pseudapistocalamus, Pseudechis, Rhinoplocephalus, Toxicolamus*, and *Vermicella*. Other genera of Elapidae are *Elapsoidea, Naja, Walterinnesia, Aspidelaps, Boulengerina, Dendroaspis, Elaps, Elapsoidea, Hemachatus, Paranaja, Pseudohaje, Bungarus, Calliophis, Maticora, Ophiophagus, Apistocalamus, Aspidomorphus, Brachyurophis, Ormodon, Rhynchoelaps, Tropidechis*, and *Ultrocalamus, Ophiophagus*, the king cobra by common name, is the largest poisonous snake in the world, reaching more than 10 ft in length. *Bungarus* (the krait) is an Asiatic poisonous snake.

Hydrophiidae are sea snakes and live in tropical and subtropical sea waters bordering the Indian and Pacific oceans. They are not found in the Atlantic Ocean or the Mediterranean Sea. Only one genus, *Pelamis*, occurs in the coastal waters of Central and South America.

Classification of sea snakes is not complete and is still in a state of confusion. In this book, the classification by Smith, *Monograph of the Sea-Snakes* (British Museum, 1926), is followed. The family Hydrophiidae has two subfamilies, Laticaudinae and Hydrophiinae. Laticaudinae include such genera as *Laticauda, Aipysurus*, and *Emydocephalus*. Hydrophiinae include *Hydrelaps, Kerilia, Thalassophina, Enhydrina, Hydrophis, Acalyptophis, Thalassophis, Kolpophis, Lapemis, Astrotia, Pelamis*, and *Microcephalophis*. There are many species within the genus *Hydrophis*, and species identification is very difficult.

Colubridae constitute by far the largest family of snakes and consist of 250 genera and over 1000 species. But not all of them are poisonous. Poisonous Colubridae include the genera *Dispholidus* and *Thelotornis*, both of which are found in Africa. They are rear-fanged snakes and, because of the awkward position of the fangs, seldom envenomate victims by natural bite.

1 Composition of Snake Venoms: Nonprotein Components

1 INORGANIC CONSTITUENTS	8
1.1 Metal Content, 8	
1.2 Biological Significance, 8	
2 ORGANIC CONSTITUENTS	9
2.1 Amino Acids and Small Peptides, 9	
2.2 Nucleotides and Related Compounds, 10	
2.3 Carbohydrates, 12	
2.4 Lipids, 15	
2.5 Biogenic Amines, 16	
References	17

About 90 to 95% of dry snake venom consists of proteins. Protein fractions are biologically more important than nonprotein ones as most of the biological activities reside in protein fractions. Protein fractions contain major as well as minor toxins, nontoxic proteins, and enzymes. Most enzymes are hydrolytic in nature with the notable exception of L-amino acid oxidase, which causes oxidative deamination of amino acids. Because of the hydrolytic nature of venom enzymes, it is thought that they facilitate tissue damage on the prey to help eventual digestion (exodigestion). In some cases, venom enzymes are considered to play an important role in the self-defense of snakes. It is hard to define the exact role of enzymes in snake venoms, but the fact is that venoms do contain a number of enzymes.

In this chapter, only the nonprotein portions of snake venoms will be discussed. Subsequent chapters will be devoted to the various proteins.

For convenience, the nonprotein components will be divided into inorganic and organic constituents. Organic constituents are further classified into free amino acids and small peptides, nucleotides and related compounds, carbohydrates, lipids, and biogenic amines.

Table 1 Metal Contents (Micrograms Metal per Gram Venom) of Snake Venoms before and after Dialysis, Analyzed by Atomic Absorption

Venom (Origin)	Hr*	Ca	Zn	Mg	Na	K	Cu	Mn	Fe	Other metals†
Elapidae										
Naja naja (India)	0	1000	1600	840	60200	150	0	200	0	0
	48	105	360	650	24800	100		521		
N. naja atra (Formosa)	0	1000	380	650	43600	300	0	13	0	0
	48	138	170	317	25250	109		3		
Bungarus fasciatus (Thailand)	0	1620	196	810	26500	391	0	0	0	0
	48	137	139	500	24700	110				
Viperdae										
Bitis arietans (South Africa)	0	2306	1000	700	41500	500	0	500	0	0
	48	1200	846	274	500	439		52		
B. gabonica (South Africa)	0	2900	690	636	36400	220	0	0	0	0
	48	1080	680	277	750	220				
Vipera russelli siamensis (Thailand)	0	1987	1800	976	34100	760	0	0	0	0
	48	1306	809	306	654	310				
Crotalidae										
Agkistrodon acutus (Formosa)	0	3000	1200	450	36977	1070	175	0	0	0
	48	2668	522	409	12780	965	42			
A. contortrix laticinetus (U.S.A.)	0	2438	964	493	18600	1463	10	49	36	0

Species	Time*								
Crotalus atrox (U.S.A.)	0	4196	1394	701	57300	410	0	0	0
	48	3780	1093	344	24600	320	0	0	0
C. adamanteus (U.S.A.)	0	1610	773	107	42300	750	0	0	0
	48	1604	452	97	8400	750	0	0	0
C. basiliscus (Mexico)	0	1989	1400	376	16800	670	0	0	0
	48	1990	990	310	10200	638	0	0	0
C. durissus (Central America)	0	3003	1203	1470	36700	13500	0	0	0
	48	2968	700	775	12800	3970	0	0	0
C. durissus terrificus (South America)	0	2390	1856	342	45700	1660	0	0	0
	48	2280	1380	204	1780	1440	0	0	0
C. durissus totonacus (Mexico)	0	1633	840	117	28800	590	0	0	0
	48	1590	680	100	1500	550	0	0	0
C. horridus horridus (U.S.A.)	0	4930	980	973	53000	420	0	0	0
	48	3629	800	406	21900	400	0	0	0
C. horridus atricaudatus (U.S.A.)	0	150	680	129	49900	350	0	0	0
	48	97	657	91	10010	240	0	0	0
C. viridis viridis (U.S.A.)	0	4560	1847	240	26400	710	0	0	0
	48	2730	1050	209	1200	600	0	0	0
Sistrurus milarius barbouri (U.S.A.)	0	4000	2010	446	39500	2540	200	0	0
	48	2750	1525	297	1550	2159	90	0	0

* Length of time that crude venoms were dialyzed against distilled water before analysis.

† Mo, Bi, Se, Pt, Pd, Ag and Au.

1 INORGANIC CONSTITUENTS

1.1 Metal Content

Since the major constituents of snake venoms are proteins which are charged macromolecules, it is quite natural that snake venoms contain various cations or anions to neutralize the charges. Some of the metals, especially monovalent cations, serve this purpose. Actually sodium is present in any venom in far greater quantity than other cations (Table 1), but monovalent ions probably have relatively little significance in terms of biological and enzymatic activities. Some divalent metals are required as cofactors for many different enzymatic and biological activities.

The inorganic constituents of *Naja naja atra* from Formosa were relatively well studied by Ueda et al. (1951). By heating the venom in an oven, they obtained 2.16% ash. Chemical analysis indicated that 14.57% Zn, 3.37% Ca, 6.49% Mg, 20.11% K, 10.16% Na, 27.94% SO_4^{2-}, 11.90% Cl^-, and 6.69% P_2O_5 were present in the ash.

The iron contents of *Naja naja* and *Vipera russellii* venoms are 0.028 and 0.016 mg %, respectively (Devi, 1968). Zinc is commonly found in many venoms (Devi, 1968; Friederich and Tu, 1971). The zinc content of dialyzed *Vipera palestinae* venom is 670 ppm (Gitter et al., 1963), a value comparable to the zinc contents of many other dialyzed venoms (Table 1). Copper was also found in high concentration in *V. palestinae* venom after dialysis. A fair amount of zinc remained after dialysis in many snake venoms, but this was not the case for copper (Table 1). The copper contents of various snake venoms were reported by Gitter et al. (1963) and are summarized as follows:

Species	Condition of Venom	Copper Content (ppm)
Vipera palestinae	Dialyzed	800
Naja naja	Dialyzed	1600
	Nondialyzed	1100
Echis carinatus	Dialyzed	3900
Bitis arietans	Dialyzed	3900

These values are 10 to 20 times higher than the ones reported by Friederich and Tu (1971), which are shown in Table 1.

The metal contents of the venoms of *Trimeresurus elegans* and *T. flavoviridis* shown in Table 2 (Hirakawa, 1974) are roughly comparable to those reported by Friederich and Tu (1971).

The inorganic components of *Crotalus durissus cumanensis* venom were investigated by Rodriguez et al. (1974), who found the following nonmetal and metal contents:

0.068% total P, 0.038% inorganic P, 1.66% Na, 0.45% Ca, 0.24% K, 0.12% Mg, 0.078% Zn, 0.049% Fe, less than 0.007% Co, less than 0.001% Mn.

1.2 Biological Significance

Some venoms possess anticholinesterase activity, which requires the Zn(II) ion or, to a lesser degree, the Co(II) ion (Kumar et al., 1973). Some venom proteases are metalloproteins and contain Ca(II) and Zn(II), as summarized in Chapter 7. Moderate

2 Organic Constituents

Table 2 Metal Contents (Micrograms Metal per Gram Venom) of Venoms of Trimeresurus elegans and T. flavoviridis

Venoms	Ca	Zn	Mg	Na	K	Cu	Fe	Ni
Trimeresurus elegans (Japan)	600	1500	100	1200	6400	100	trace	trace
T. flavoviridis (Japan)	700	1300	100	6800	2500	trace	trace	trace

The data are obtained from the paper of Hirakawa, 1974.

amounts of Ca(II) are presents in venoms, and some of this is no doubt used for the activation of phospholipase A_2 (Boffa et al., 1969; Augustyn et al., 1970). To lesser extent, Ca(II) is needed for the activation of the direct hemolytic factor (Mirsalikhova et al., 1975).

2. ORGANIC CONSTITUENTS

2.1 Amino Acids and Small Peptides

Venoms contain only a small amount of free amino acids. Eleven of these amino acids were detected in the venom of *Trimeresurus mucrosquamatus* (Sasaki, 1960): glycine, serine, cysteine, threonine, lysine, alanine, tyrosine, valine, phenylalanine, and leucine. Venom of *Vipera ammodytes* contains histidine, aspartic acid, glycine, glutamic acid, serine, alanine, and spermine (Shipolini et al., 1965). Thirteen different amino acids — arginine, glutamic acid, glycine, leucine, isoleucine, phenylalanine, serine, tyrosine, valine, cystine, lysine, histidine, and proline — were detected in the venoms of *Trimeresurus elegans* and *T. flavoviridis* (Hirakawa, 1974).

Two types of pyroglutamylpeptides were isolated from the venom of *Agkistrodon halys blomboffii*: pGlu–Asn–Trp and pGlu–Glu–Trp (Kato et al., 1966). The occurrence of pyroglutamylpeptides is apparently not restricted to Japanese snake venom. Venom of *Trimeresurus gramineus* contains pGlu–Trp–Lys and pGlu–Gln–Trp, while that of *T. mucrosquamatus* has pGlu–Asn–Trp in addition to the other two peptides mentioned. The venom of *Agkistrodon acutus* contains only one peptide, pGlu–(Glx$_2$ Asx$_2$ Trp)–Trp (Lo et al., 1973).

A number of proline-rich peptides that potentiate bradykinin are present in snake venoms. The following seven such peptides were isolated from the venom of *Bothrops jararaca* and their sequences identified (Bodanszky et al., 1971):

Peptide I   pGlu–Asn–Trp–Pro–His–Pro–Gln–Ile–Pro–Pro
       II pGlu–Ser–Trp–Pro–Gly–Pro–Asn–Ile–Pro–Pro
       III pGlu–Trp–Pro–Arg–Pro–Gln–Ile–Pro–Pro

IV   pGlu–Trp–Pro–Arg–Pro–Thr–Pro–Gln–Ile–Pro–Pro
V   pGlu–Asn–Trp–Pro–Arg–Pro–Gln–Ile–Pro–Pro
VI   pGlu–Trp–Pro–Arg–Pro
VII pGlu–Lys–Phe–Ala–Pro

The sequences of three bradykinin-potentiating peptides from tne venom of *Agkistrodon halys blomhoffii* were identified by Okada et al. (1973). All of these peptides have pyroglutamic acid at the amino terminal and proline at the carboxyl terminal. Potentiator E has the following sequence:

pGlu–Lys–Trp–Asp–Pro–Pro–Pro–Val–Ser–Pro–Pro

The sequences of potentiators B and C are as follows:

Potentiator B: pGlu–Gly–Leu–Pro–Pro–Arg–Pro–Lys–Ile–Pro–Pro
Potentiator C: pGlu–Gly–Leu–Pro–Pro–Gly–Pro–Pro–Ile–Pro–Pro

The amino acid composition of potentiator D for *A. halys blomhoffii* is Arg, Glu, 4 Pro, 2 Gly, Ile, Leu (Kato and Suzuki, 1969). The sequence of potentiator A in the venom of *A. halys blomhoffii* is as follows (Kato et al., 1973):

pGlu–Gly–Arg–Pro–Pro–Gly–Pro–Pro–Ile–Pro

A tripeptide with unknown biological activity is also present in the same venom and has the sequence pGlu–Lys–Ser (Okada et al., 1974). A small peptide with bradykinin-potentiating activity was isolated from the venom of *Bothrops jararaca* (Ferreira et al., 1970a, b).

Pipecolic acid,

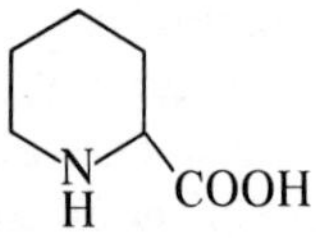

was found in the hydrolysate of *B. jararaca* venom (Michl, 1957). Since pipecolic acid was obtained only after hydrolysis, the conclusion was that it is not present as a free molecule in whole venom. How the pipecolic acid is incorporated into venom protein molecules awaits clarification.

2.2 Nucleotides and Related Compounds

Study on the contents of venom nucleotide (and related compounds) has been very limited. Therefore we cannot really generalize that all snake venoms contain the compounds listed in Table 3, which summarizes all venom nucleotides and related compounds. There is one common finding from analyzing the table: all the compounds listed are derivatives of purine. Doery (1957) considers that adenosine and guanosine are natural constituents of venoms as they are found in freshly milked venoms. Inosine and hypoxanthine could arise from adenosine by enzymatic action during the drying process.

The total content of purine compounds amounts to 1.4 to 4.3% of venom dry weight. Thus the purine content is not insignificant. However, the role of these compounds in snake venom action has not been studied and is not clear as yet.

Table 3 Presence of Nucleotides and Related Compounds in Snake Venoms

Family Genus Species Subspecies	Elapidae Acanthophis antarctica	Denisonia superba	Notechis scutatus	Bungarus multicinctus	Naja naja atra	Dendraspis angusticeps	Viperidae Bitis arietans	Crotalidae Agkistrodon acutus
Origin	Australia	Australia	Australia	Formosa	Formosa	Africa	Africa	Formosa
Compounds								
Base								
hypoxanthine			+					
Nucleoside								
adenosine	+	trace	+		+	+	+	
guanosine			+	+	+			+
inocine			+		+			
Nucleotide								
AMP-3'	trace	+	+			–	–	
Unidentified								
guanine compounds	trace	–	trace			–	trace	
% weight	2.4	2.0	1.4	1.1		4.3	3.8	
Reference	Dorey, 1956	Dorey, 1956	Dorey, 1957	Wei and Lee, 1965; Lo, 1972	Lo and Chen, 1972	Dorey, 1957	Fischer, F.G. and H. Dörfel, 1954	Lo, 1972

2.3 Carbohydrates

In snake venoms carbohydrates are in the form of glycoproteins rather than free sugars, and they are found in many venoms (Basu et al., 1970).

Sialic acid was detected in the venom of *Naja naja*, but treatment with neuraminidase did not affect the toxicity of the venom (Braganca and Patel, 1965).

Various types of carbohydrates present in snake venoms were systematically investigated by Oshima and Iwanaga (1969). Most sugars were either neutral sugars, amino sugars, or sialic acid; no mucopolysaccharides were found.

The venom glands of *Agkistrodon piscivorus piscivorus, Vipera lebetina, Echis carinatus*, and *Vipera berus* contain large amounts of mucopolysaccharides (Rhoades et al., 1967; Zakharov, 1966). Since a venom gland is a site of venom biosynthesis, and

Table 4   Carbohydrate Contents in Snake Venoms

Snake venom	Neutral sugars	Amino sugar (µg/mg of venom)	Sialic acid	% content
Elapidae:				
N. naja atra	0.9	4.7	3.1	1.3
N. melanoleuca	17.4	22.7	9.4	4.6
Sepedon haemachates	12.1	13.1	8.4	3.6
Bungarus fasciatus	6.0	8.7	3.4	1.9
Viperidae:				
V. russelli	16.0	25.8	8.4	5.0
Bitis gabonica	56.5	48.4	13.3	11.8
Echis carinatus	15.7	38.0	14.6	6.8
Crotalidae:				
A. halys blomhoffii	25.2	34.0	11.0	7.0
A. piscivorus piscivorus	15.0	28.2	8.5	5.2
A. contortrix contortrix	10.4	22.1	5.4	3.8
C. adamanteus	25.7	43.4	10.0	7.9
C. atrox	10.8	23.5	6.0	4.0
B. jajaraca	9.4	18.6	8.4	3.6
B. atrox	6.0	119.1	5.2	2.2
T. flavoviridis	8.2	12.7	4.0	2.5

The data were obtained from the paper of Oshima and Iwanga, 1969.

2 Organic Constituents

many substances are secreted into the cavity of the poison sac as venom, it is interesting to note that mucopolysaccharides are not released as a constituent of venom.

The carbohydrate content is as high as 11.8% in the case of *Bitis gabonica* venom (Oshima and Iwanage, 1969). Contents of neutral sugars, amino sugars, and sialic acid are summarized in Table 4. Neutral sugars include galactose, mannose, and fucose. Only glucosamine is present as an amino sugar; galactosamine is lacking. The ratios of D-galactose, D-mannose, and L-fucose are summarized in Table 5. The chemical formulas of sugars found in snake venoms are summarized in Fig. 1.

It is of great interest and also of great importance to know whether major toxins are glycoproteins. There were earlier reports that cobra neurotoxins were glycoproteins (Braganca and Patel, 1965; Kabara, 1971). However, the phenol–sulfuric acid sugar test applied to *Naja naja atra* neurotoxin and α- and β-bungarotoxins from *Bungarus multicinctus* venom showed negative reults (Lin and Lee, 1971). Pure rattlesnake toxin from the venom of *Crotalus scutulatus* and *Pelamis platurus* sea snake toxin also gave

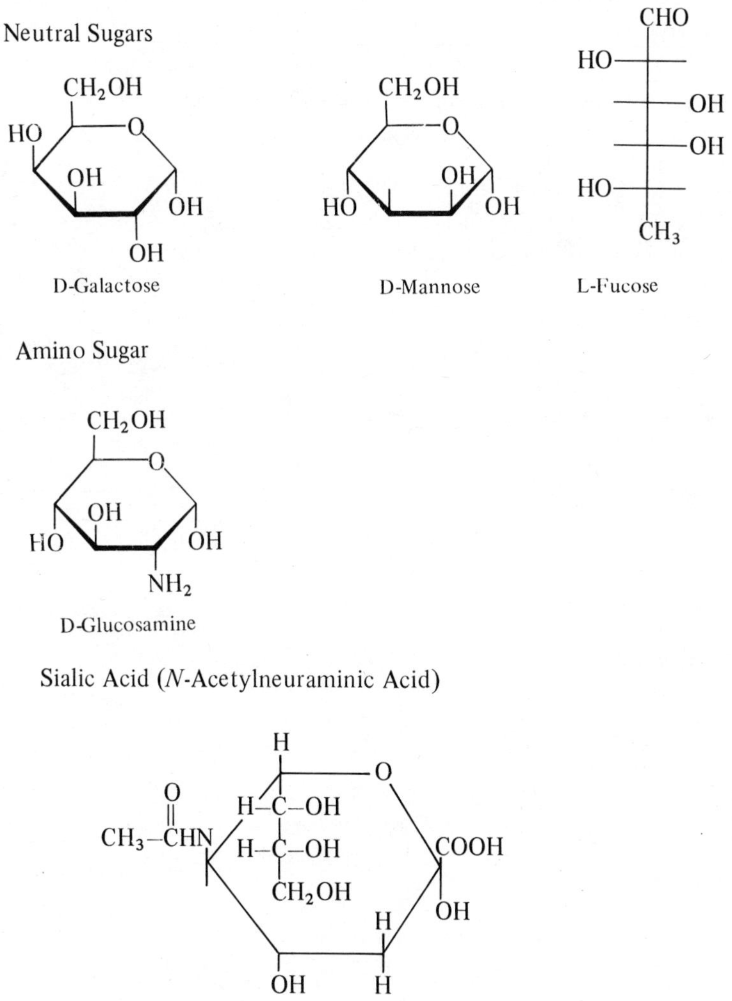

Figure 1.1. Structure of carbohydrates found in snake venoms.

Table 5 Ratio of Galactose, Mannose, and Fucose in Neutral Sugars

	D-galactose	D-mannose	L-fucose
A. halys blomhoffii	1.0	0.29	0.14
A. piscivorus piscivorus	1.0	0.81	0.72
C. adamanteus	1.0	0.31	0.12
B. jajaraca	1.0	0.69	0.34
T. flavoviridis	1.0	1.15	0.41
V. russelli	1.0	0.86	0.17
Bitis gabonica	1.0	1.18	0.60—0.90
Echis carinatus	1.0	0.84	0.18

The data were obtained from the paper of Oshima and Iwanga, 1969.

negative results when subjected to orcinol test for carbohydrate (Bieber et al., 1975; Tu et al., 1975). Thus it is clear that major toxins are not glycoproteins. The exact biological functions of venom carbohydrates have not yet been determined.

Many important venom components are glycoproteins. For instance, Arvin, a blood coagulant isolated from the venom of *Agkistrodon rhodostoma*, is a glycoprotein (Hatton, 1973).

Table 6 Lipids Found in the Venom of Naja naja

Compounds	mg/g	% of total lipid
Hydrocarbon	0	0
Cholesterol esters	0.016	0.4
Triglycerides	0.11	2.5
Cholesterol	0.43	10.0
Monoglycerides	0.10	2.3
Diglycerides	0.08	1.8
Free fatty acids	0	0
Phospholipids	3.6	83.0
Total	4.34	100

This table is taken from the article of Kabara and Fischer, 1969.

2 Organic Constituents

2.4 Lipids

The total content of lipids in *Naja naja* venom is very small and accounts for only 0.43% of dry weight (Kabara and Fischer, 1969). The major component is a phospholipid, phosphatidylcholine. Because of the small quantity of venom lipids and also the absence of unusual lipids in the venom, it is unlikely that these substances play any significant role in venom action. Table 6 summarizes the contents of different lipids found in *N. naja* venom. All the fatty acids are present in the neutral and phospholipid fractions. The types of fatty acids are shown in Table 7.

Many venoms exhibit a yellow color. This is due to the presence of L-amino acid oxidase, which contains riboflavin as a prosthetic group. The prosthetic group FAD is

Table 7   Fatty Acid Compositions of Lipids

Fatty Acid	Neutral lipids (%)	Phospholipids (%)
C6	Trace	Trace
C8	Trace	Trace
C10	Trace	Trace
C12	1	Trace
C12-1	—	—
C14	3	3
C14-1	Trace	2
C16	11	0.2
C16-1	2	0.2
C18	21	4
C18-1	9	7
C18-2	1	5
C18-3	—	1
C20	24	0.4
C20-4	7	59
C22	13	4
C22-4	Trace	0.2
C24	7	3
C24-4	Trace	3
C26	—	3
C26-4	—	1
Unidentified	1 (two peaks)	4 (five peaks)

This table was obtained from the paper of Kabara and Fisher, 1969.

very tightly bound to the apoenzyme moiety (Wellner, 1971) and has been found in all snake venoms investigated. The only FMN-containing L-amino acid oxidase reported was the one obtained from *Trimeresurus flavoviridis* venom (Inamasu et al., 1974). Dimitrov and Kankonkar (1968) demonstrated that L-amino acid oxidase was present in a lesser amount in the white venom than in the yellow venom of *Vipera russellii*. Kornalik and Master (1964) examined the L-amino acid oxidase content of yellow and white venoms of *Vipera ammodytes* and found that the yellow venom contained 200 times more L-amino acid oxidase than did the white venom. Thus it is clear that the more yellow the venom, the more riboflavin it contains. The chemical nature and biological importance of L-amino acid oxidase are discussed in detail in Chapter 5.

2.5 Biogenic Amines

From clinical cases, it has been observed that pain production at the site of snakebite appears to be more common for Viperidae (vipers) and Crotalidae (pit vipers) than for Elapidae (elapids) and Hydrophiidae (sea snakes). This difference is believed to be due to the compositional differences of the venoms. Several biogenic amines are potent pain producers. They include such compounds as bradykinin, histamine, 4-hydroxytryptamine, *N*-methyltryptamine, *N,N'*-dimethyl-5-hydroxytryptamine (bufotenine), *N*-methyl-5-hydroxytryptamine, and serotonin (5-hydroxytryptamine). Some of these amines, such as histamine, are present in the venoms of wasps (Jaques and Schachter, 1954) and hornets (Bhoola et al., 1961).

Histamine and spermine were detected in the venom of *Trimeresurus mucrosquatnatus* by paper chromatography (Sasaki, 1959, 1960). The content of spermine in the venom is 2.1%. Serotonin is present in the venoms of the scorpion (Adam and Weiss, 1956), Gila monster (Zarafonetis and Kalas, 1960), spider (Welsh and Batty, 1963), hornet (Bhoola et al., 1961), and wasp (Welsh and Batty, 1963), as well as in frog toxin (Welsh, 1964) and other amphibian toxins (Erspamer, 1966).

The amount of serotonin in snake venom is usually very small. Zarafonetis and Kalas (1960) observed that the serotonin content in the venom was 0.15 to 0.3 $\mu g\ ml^{-1}$ for *Crotalus atrox*; 0.1 $\mu g\ ml^{-1}$ for *C. adamanteus*; and 0.35 $\mu g\ ml^{-1}$ for *Agkistrodon piscivorus*. The presence of serotonin in snake venom was systematically studied by Welsh (1966a). Elapidae venoms such as those of *Bungarus fasciatus, Hemachatus hemachatus, Naja naja, Pseudechis mortonensis, Dendroaspis polylepis,* and *D. angusticeps* do not contain serotonin. However, venoms of Viperidae (*Vipera russellii, Bitis gabonica*) and Crotalidae (*Sistrurus miliarius barbouri, Agkistrodon contortrix, A. piscivorus*) contain serotonin which is detected by both bioassay and spectrofluorometric methods. Norepinephrine is absent in the venoms of *Crotalus adamanteus* and *Agkistrodon piscivorus* (Anton and Gennero, 1965).

An acetylcholinelike substance is present in some snake venoms. Welsh (1966b), using bioassay as well as paper chromatographic techniques, concluded that venoms of *Dendroaspis jamesoni, D. polylepis,* and *D. angusticeps* contain acetylcholine. There was no indication of the presence of an acetylcholinelike substance in the venoms of *Bungarus bungarus, Hemachatus haemachatus, Naja naja, Pseudechis mortonensis, Vipera russellii, Bitis gabonica, Crotalus horridus, Sistrurus miliarius barbouri, Agkistrodon contortrix, A. piscivorus,* and *Bothrops atrox*. However, more solid chemical evidence is required to confirm the presence or absence of acetylcholine in certain venoms.

REFERENCES

Adam, K. R. and Weiss, C. (1956). 5-Hydroxytryptamine in scorpion venom, *Nature,* 178, 421.

Anton, A. H. and Gennaro, J. F., Jr. (1965). Norepinephrine and serotonin in the tissues and venoms of two pit vipers, *Nature,* 208, 1174

Augustyn, J. M., Parsa, B., and Elliott, W. B. (1970). Structural and respiratory effects of *Agkistrodon piscivorus* phospholipase A on rat liver mitochondria, *Biochim. Biophys. Acta,* 197, 185.

Basu, A. S., Parker, R., and O'Connor, R. (1970). Disc electrophoresis of glycoproteins in snake venoms, *Toxicon,* 8, 279.

Bhoola, K. D., Calle, J. D., and Schachter, M. (1961). Identification of acetylcholine, 5-hydroxytryptamine, histamine, and a new kinin in hornet venom (*V. crabro*), *J. Physiol.,* 159, 167.

Bieber, A. L., Tu, T., and Tu, A. T. (1975). Studies of an acidic cardiotoxin isolated from the venom of Mojave rattlesnake (*Crotalus scutulatus*), *Biochim. Biophys. Acta,* 400, 178.

Bodanszky, A., Ondetti, M. A., Ralofsky, C. A., and Bodanszky, M. (1971). Optical rotatory dispersion of the proline rich peptides from the venom of *Bothrops jararaca*, *Experientia,* 27, 1269.

Boffa, M. C., Josso, F., and Boffa, G. A. (1969). Relation between the hemolytic action and lecithinase activity of cobra (*Naja nigricollis*) venom, *C. R. Acad. Sci. Paris,* Ser. D, 269, 2036.

Braganca, B. M. and Patel, N. T. (1965). Glycoproteins as components of the lethal factors in cobra venom. (*Naja naja*), *Can. J. Biochem.,* 43, 915.

Devi, A. (1968). "The protein and nonprotein constituents of snake venoms," in W. Bucherl, E. E. Buckley, and V. Deulofeu, Eds., *Venomous Animals and Their Venoms,* Academic, New York.

Dimitrov, G. D. and Kankonkar, R. C. (1968). Fractionation of *Vipera russellii* by gel filtration. II. Comparative study of yellow and white venoms of *Vipera russellii* with special reference to the local necrotizing and lethal actions, *Toxicon,* 5, 283.

Doery, H. M. (1956). Purine compounds in snake venoms, *Nature,* 177, 381.

Doery, H. M. (1957). Additional purine compounds in the venom of the tiger snake (Notechis scutatus), *Nature,* 180, 799.

Erspamer, V. (1966). *Handbook of Experimental Pharmacology,* Vol. XIX, Springer-Verlag.

Ferreira, S. H., Bartlet, D. C., and Greene, L. J. (1970a). Isolation of bradykinin-potentiating peptides from *Bothrops jararaca* venom, *Biochemistry,* 9, 2583.

Ferreira, S. H., Greene, L. J., Alabaster, V. A., Bakhle, Y. S., and Vane, J. R. (1970b). Activity of various fractions of bradykinin-potentiating factor against angiotensin I-converting enzyme, *Nature,* 225, 379.

Fischer, F. G. and Dörfel, H. (1954). Die Aminosäuren-Zusammensetzung von Crotoxin, *Z. Physiol. Chem.,* 297, 278.

Friederich, C. and Tu, A. T. (1971). Role of metals in snake venoms for hemarrhagic, esterase and proteolytic activities, *Biochem. Pharmacol.,* 20, 1549.

Gitter, S., Amiel, S., Gilat, G., Sonnino, T., and Welwart, Y. (1963). Neutron activation analysis of snake venoms: Presence of copper, *Nature,* 197, 383.

Hatton, M. W. C. (1973). Studies on the coagulant enzyme from *Agkistrodon rhodostoma* venom: Isolation and some properties of the enzyme, *Biochem. J.,* 131, 799.

Hirakawa, Y. (1974). Venom of *Trimeresurus elegans* and *Trimeresurus flavoviridis, Kagoshima. Daigaku Igaku Zasshi,* 26, 611.

Inamasu, Y., Nakano, K., Kobayashi, M., Sameshima, Y., and Obo, F. (1974). On the nature of the prosthetic group of the L-amino acid oxidase from habu snake (*Trimeresurus flavoviridis*) venom, *Acta Med. Univ. Kagoshima,* 16, 23.

Jaques, R. and Schachter, M. (1954). The presence of histamine, 5-hydroxytryptamine and a potent, slow-contracting substance in wasp venom, *Brit. J. Pharmacol.,* 9, 53.

Kabara, J. J. (1971). "The Chemical Composition of *Naja naja* venom: Isolation of two toxins," in A. de Vries and E, Kochhva, Eds., *Toxins of Animal and Plant Origin,* Vol I, Gordon and Breach, New York, p. 293.

Kabara, J. J. and Fischer, G. H. (1969). Chemical composition of *Naja naja:* Extractable lipids, *Toxicon,* 7, 223.

References

Kato, H. and Suzuki, T. (1969). Bradykinin-potentiating peptides from the venom of *Agkistrodon halys blomhoffii, Experientia,* **25,** 694.

Kato, H., Iwanaga, S., and Suzuki, T. (1966). The isolation and amino acid sequences of new pyroglutamyl-peptides from snake venoms, *Experientia,* **22,** 49.

Kato, H., Suzuki, T., Okada, K., Kimura, T., and Sakakibara, S. (1973). Structure of potentiator A, one of the five bradykinin potentiating peptides from the venom of *Agkistrodon halys blomhoffii, Experientia,* **29,** 574.

Kornalik, F. and Master, R. W. P., (1964). A comparative examination of yellow and white venoms of *Vipera ammodytes, Toxicon,* **2,** 109.

Kumar, V., Rejent, T. A., and Elliott, W. B. (1973). Anticholinesterase activity of elapid venoms. *Toxicon,* **11,** 131.

Lin, S. S. and Lee, C. Y. (1971). Are neuroproteins from elapid venoms glycoproteins? *Toxicon,* **9,** 295.

Lo, K. M., Chen, S. W., and Lo, T. B. (1973). Isolation and chemical characterization of small peptides from Formosan snake venoms, *J. Chin. Biochem. Soc.,* **2,** 33.

Lo, T. B. (1972). Chemical studies of Formosan snake venoms, *J. Chin. Biochem. Soc.,* **1,** 39.

Lo, T. B. and Chen, Y. H. (1966). Chemical studies of Formosan cobra (*Naja naja atra*) venom. II. Isolation and characterization of guanosine, adenosine, and inosine from cobra venom, *J. Chin. Chem Soc.,* **II,** *13,* 195.

Michl, H. (1957). Presence of pipecolic acid in venoms (from snakes and wasps), *Monatsh. Chem.,* **88** 701.

Mirsalikhova, N. M., Yukel'son, L. Ya., and Ziyamukhamedov, R. (1975). Role of calcium ions and fatty acids in inhibition of (magnesium ion) and (sodium, potassium ion) dependent ATPase by direct hemolysin and phospholipase A of cobra venom, *Ukr. Biokhim. Zh.,* **47,** 61.

Okada, K., Uyehara, T., Hiramoto, M., Kato, H., and Suzuki, T. (1973). Application of mass spectrometry to sequence analysis of pyroglutanyl peptides from snake venoms: Contribution to the confirmation of the amino acid sequence of bradykinin-potentiating peptides B, C, and E isolated from the venom of *Agkistrodon halys blomhoffii, Chem. Pharm. Bull.,* **21,** 2217.

Okada, K., Nagai, S., and Kato, H. (1974). A new pyroglutamyl peptide (Pyr–Lys–Ser) isolated from the venom of *Agkistrodon halys blomhoffii, Experientia,* **30,** 459.

Oshima, G. and Iwanaga, S. (1969). Occurrence of glycoproteins in various snake venoms, *Toxicon,* 7,235.

Rhoades, R., Lorincz, A. E., and Gennaro, J. F., Jr. (1967). Polysaccharide content of the poison apparatus of the cottonmouth moccasin *Agkistrodon piscivorus piscovorus, Toxicon,* **5,** 125.

Rodriguez, O. G., Scannone, H. R., and Parra, N. D. (1974). Enzymatic activities and other characteristics of *Crotalus durissus cumanensis* venom, *Toxicon,* **12,** 297.

Sasaki, T. (1959). Chemical Studies on the venom of Formosan habu (*Trimeresurus mucrosquamatus* Cantor). I. Basic substance in the venom, *Yakugaku Zasshi,* **79,** 221.

Sasaki, T. (1960). Chemical studies on the venom of Formosan habu (*Trimeresurus mucrosquamatus* Cantor). III. On the dialyzable substances in the venom, *Yakugaku Zasshi,* **80,** 844.

Shipolini, R., Ivanov, C. P., Dimitrov, G., and Alexiev, B. V. (1965). Composition of the low molecular fraction of the Bulgarian Viper venom, *Biochim. Biophys. Acta,* **104,** 292.

Tu, A. T., Lin, T. S., and Bieber, A. L. (1975). Purification and chemical characterization of the major neurotoxin from the venom of *Pelamis platurus, Biochemistry,* **14,** 3408.

Ueda, E., Sasaki, T., and Peng, M. T. (1951). A chemical study of Formosan cobra venom, *Mem. Fac. Med. Nat. Taiwan Univ.,* **1,** 194.

Wei, A. L. and Lee, C. Y. (1965). A nucleoside isolated from the venom of *Bungarus multicinctus, Toxicon,* **3,** 1.

Wellner, D. (1971). L-Amino acid oxidase (snake venom), *Methods Enzymol.,* **17,** Part B, 597.

Welsh, J. H. (1964). Composition and mode of action of some invertebrate venoms, *Ann. Rev. Pharmacol.,* **4,** 293.

Welsh, J. H. (1966a). Serotonin and related tryptamine derivatives in snake venoms, *Mem. Inst. Butantan,* **33,** 509.

Welsh, J. H. (1966b). "Acetylcholine in snake venoms," in F. E. Russell and B. P. R. Saunders, Eds., *Animal Toxins,* Pergamon, New York, pp. 363-368.

References

Welsh, J. H. and Batty, C. S. (1963). 5-Hydroxytryptamine content of some arthropod venoms and venom-containing parts, *Toxicon,* 1, 165.

Zakharov, A. M. (1966). Histochemical study of polysaccharides in the venom secreting glands of Viperidae, *Arkh. Anat. Gistol. Embriol.,* 51, 118.

Zarafonetis, C. J. and Kalas, J. P. (1960). Serotonin, catecholamine, and amine oxidase activity in the venoms of certain reptiles, *Am. J. Med. Sci.,* 240, 764.

II SNAKE VENOMS: Enzymes

2 Phospholipase A_2

Phosphatide Acylhydrolase   EC 3.1.1.4

1 CHEMISTRY   26

    1.1 Isolation, 26
    1.2 Sequence, 27
    1.3 Site of Hydrolysis, 32
    1.4 Mechanism of Enzyme Action, 36
    1.5 Substrate Specificity, 39
    1.6 Lysophospholipase Activity of Phospholipase A_2, 43
    1.7 Conformation, 43
    1.8 Isozymes, 44
    1.9 Effects of pH and Temperature, 46

2 BIOLOGICAL EFFECTS   46

    2.1 Effect on Mitochondria, 46
    2.2 Effect on Nerves and Muscles, 48
    2.3 Effect on Cell Membranes, 52
    2.4 Toxic Action of Phospholipase A_2, 56
          Lethality, 56
          Other Toxic Action, 56

3 OCCURRENCE   57

    References   57

Phospholipase A_2 is one of the venom enzymes extensively studied by many investigators for its chemical properties as well as for its biological effects. In the past, the enzyme was known as lecithinase, but now it is called phospholipase A or A_2.

It has been known for many years that this enzyme is extremely stable. Therefore many investigators simply used heat-treated venoms and considered any effects arising from heat-stable components in the venom to be actions of phospholipase A_2. This is, of course, not a very good way to study the biological and biochemical effects of phospholipase A_2 actions. For instance, we know now that neurotoxins present in the venoms of Hydrophiidae and Elapidae are also relatively heat stable.

Recently, phospholipase A_2 has been isolated from a variety of venoms; all the venom

Table 1 Isolation of Phospholipase A_2 from Various Snake Venoms

Venoms	Origin	Name	Molecular Weight	Reference
Hydrophiidae				
Laticauda semifasciata	E. China Sea	phospholipase A	9,000~11,000	Uwatoko-Setoguchi and Obo, 1969
	E. China Sea	phospholipase A	11,000	Tu et al, 1970
Elapidae				
Haemachatus haemachatus	S. Africa	phospholipase A De-I	13,500	Joubert, 1975c
Naja melanoleuca	S. Africa	phospholipase A DE-III	14,300[a] 27,200[b]	Joubert, Van der Walt, 1975
	[a, I=0.1, pH=2.1; b, I=0.3, pH=7.2]			
N. naja	India	phospholipase A isozyme	8,500~23,800	Salach et al, 1971a
	India	phospholipase A_2	—	Smith et al, 1972
		peak 1	24,000	Currie et al, 1968
		peak 2	24,000	Currie et al, 1968
		phospholipase A_2	11,000	Deems and Dennis, 1975
N. naja atra	Formosa	phospholipase A	12,500	Chang and Lo, 1972
			28,655	Sasaki, 1958
N. nigricollis	Africa	Basic phospholipase	13,500	Dumarey et al, 1975
		phospholipase A A-I	13,000	Wahlström, 1971
		A-II	14,600	Wahlström, 1971
N. oxiana	Russia	phospholipase A	14,000~15,000	Sakhibov et al, 1970
Viperidae				
Bitis arietans	S. Africa	phospholipase A_2	14,000	Howard, 1975
B. gabonica	S. Africa	phospholipase A	13,347	Botes and Viljoen, 1974a
Vipera ammodytes	Yugoslavia	phospholipase A protein i protein k		Gubensek and Lapanje, 1974
V. berus	France	phospholipase A_2		Delori, 1971
	France	Phospholipase A		
		P_1	36,000	Delori, 1973
		P_2	26,000	"
		P_3	11,000	"
		P_4	11,000	"

Venoms	Origin	Name	Molecular Weight	Reference
V. russellii	Asia	phospholipase A isozymes	15,900~23,800	Salach et al, 1971a
V. palestinae	Palestine	phospholipase A		
		F_1	16,000	Shiloah et al, 1973
		F_2	16,000	"
Crotalidae				
Agkistrodon halys blomhoffii	Japan	phospholipase, A-I	13,130~14,600	Kawauchi et al, 1971
		A-II	12,000~15,600	"
A. piscivorus	U.S.A.	phospholipase A	14,000	Augustyn and Elliott, 1970
Bothrops atrox	Argentina	phospholipase A		Vidal and Stoppani, 1971
		p-1		
		p-2		"
B. jararaca	Argentine	phospholipase A		
		p-1		"
		p-2		"
B. jararacussu	Argentina	phospholipase A		
		p-1		"
		p-2		"
B. neuwiedii	Argentina	phospholipase A		
		p-1	9,500	Vidal et al, 1972
		p-2	9,400	"
		p-1 dimer	20,000	"
		p-2 dimer	20,000	"
Crotalus atrox	U.S.A.	phospholipase A_2	14,500	Wu and Tinker, 1969
	U.S.A.	phospholipase A		Woelk and Debuch, 1971
	U.S.A.	phospholipase A_2	29,500	Hachimori et al, 1971
C. adamanteus	U.S.A.	phospholipase A		Wells and Hanahan, 1969
		α	29,500~29,900	
		β	29,800~31,900	"
	U.S.A.	phospholipase A_2-α		Tsao et al, 1975
C. durissus terrificus	S. America	phospholipase A	16,200	Omori-Satoh et al, 1975

phospholipase A_2 isolated in pure form are summarized in Table 1. Therefore, research done on the pure enzyme is stressed in this chapter rather than studies on the ill-defined phopholipase A_2.

1. CHEMISTRY

1.1 Isolation

Phospholipase A_2 is present in venoms of all families of snakes, and in many of them this enzyme has been purified. Although phospholipases A_2 from various venoms exhibit more or less similar enzymatic properties, they are not identical immunologically (Fig. 1). Specific antiserum against the venom of a particular species inhibits the phospholipase A_2 activity of the venoms of closely related species, but in most cases fails to suppress the phospholipase A_2 activity of venoms from other families (Nair et al., 1975).

The enzyme from Hydrophiidae (sea snake) venom was investigated by two groups, and both showed that the molecular weight was about 11,000 (Tu et al., 1970; Uwatoko-Setoguchi, 1970). For the enzymes from the venoms of Elapidae, Viperidae, and Crotalidae, the molecular weight ranges from 12,000 to 15,000 for the monomer and is 30,000 for the dimer. Joubert and Van der Walt (1975) showed that, by changing the pH and ionic strength, phospholipase A_2 can be transformed to the dimer. The enzyme isolated from the venom of *Crotalus adamanteus* has a total of 133 amino acid residues (Tsao et al., 1975). The number obtained by Wells and Hanahan (1969) for the enzyme

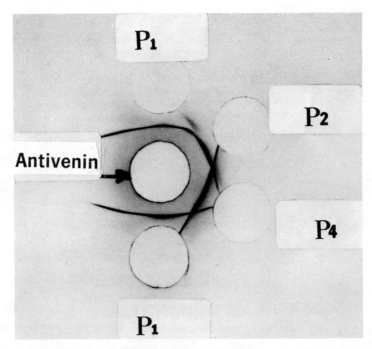

Figure 2.1 Immunological relationship of phospholipase A_2 isozymes, P1, P2, and P4 isolated by Delori (1973). (Photograph kindly supplied by Dr. P. J. Delori and originally published in Delori, 1973.)

1 Chemistry

from the same venom was 266, which is exactly twice 133. Thus the high numbers of the residues and molecular weight are for the dimer.

The amino acid compositions of phospholipase A_2 obtained from a variety of venoms are summarized in Table 2. Since the enzyme does not contain any free sulfhydryl group (Wells and Hanahan, 1969; Kawauchi et al., 1971a, b; Joubert and Van der Walt, 1975; Sakhibov et al., 1975), the total half-cystine content of 12 to 14 represents six to seven disulfide bonds per molecule.

Conversion of dimer to monomer is accompanied by a quenching tryptophan fluorescence. The pH dependence of this quenching suggests that two protons per monomer are involved in the dimer-to-monomer conversion (Wells, 1971a).

In the past, the enzyme was isolated by conventional column chromatography using Sephadex and/or ion-exchange resins. More recently, however, affinity chromatography has been used to isolate phospholipase A_2. Rock and Snyder (1975) used alkyl ether anologs of ethanolamine and choline phospholipids as ligands to purify the enzyme from *C. adamanteus* venom by affinity chromatography.

The phospholipase A_2 of *Vipera ammodytes* immobilized on CNBr-activated Sepharose 4B retains enzyme activity and toxicity (Gubensek, 1976).

1.2 Sequence

Amino acid sequences of phospholipase A_2 from a number of different venoms have been identified by several investigators and are summarized in Table 3. There is a high degree of similarity, especially in regard to the positions of 14 sulfur atoms. There is some similarity between phospholipase A_2 from venom and that of porcine pancreatic origin (De Haas et al., 1970b), although the latter lacks sulfur atoms in two positions that normally would be occupied by the venom enzyme. The relative positions of the rest of the 12 sulfur atoms are identical for pancreatic phospholipase A_2 and the venom enzymes. The exact positions of disulfide linkages in porcine pancreatic phospholipase are known (De Haas et al., 1970a; Volwerk et al., 1974).

The characteristics of the different loops of venom phospholipase A_2 shown in Table 3 can be summarized here as follows:

LOOP
1. Consists of 10 amino acid residues, 9 of which are almost homologous in all enzymes from different snakes.
2. Is a relatively constant region.
3. Contains only 1 residue, which is always tyrosine regardless of the species of snake or source of nonsnake origin.
4. Is fairly constant.
5. Consists of 5 residues for enzymes of porcine pancreas and Elapidae origin, and 4 residues for those from Viperidae and Crotalidae.
6. Is fairly constant.
7. Consists of 14 residues in the porcine pancreas enzyme, but has fewer in venom enzymes.
8. Contains 7 residues in porcine pancreas enzyme but fewer in venom enzymes.
9. Is the most variable region. Enzymes from different sources exhibit different sequences.
10. Has only 3 or 4 residues.
11. Has only 1 residue.

Table 2 Amino Acid Composition of Phospholipase A_2

Family	Hydrophiidae	Elapidae						Viperidae			
Genus Species Subspecies	Laticanda semifasciata	Naja melanoleuca			Naja naja			Naja naja atra	Naja naja atra	Naja nigricollis		Bitis gabonica
Origin	E. China Sea	S. Africa			India			Formosa	Formosa	Africa		S. Africa
Name	phospholipase A	phospholipase A			peak 1	peak 2	phospholipase A_2	phospholipase A	phospholipase A	X	A_1	phospholipase A
Amino Acid		DE-I	DE-II	DE-III								
Lysine	7	8	5	4	10	10	5	11	5	10	10	8
Histidine	2	3	2	3	2	2	1	11	1	2-3	3	2
Arginine	4	6	6	6	10	12	4	11	5	5	5	3
Aspartic acid	11	19	20	20	42	40	19	42	20	17	17	19
Threonine	6	7	7	6	8	8	4	10	5	5	5	8
Serine	7	3	6	6	10	9	7	7	6	3	2-3	6
Glutamic acid	9	7	8	6	15	15	6	20	8	4	4-5	10
Proline	5	4	3	4	8	6	5	13	5	7	6	2
Glycine	10	9	9	9	19	17	9	13	10	11	11	13
Alanine	8	9	10	11	23	22	8	12	11	10	10	4
Valine	4	4	4	4	8	8	4	10	4	4	4	3
Methionine	1	1	1	1	0	0	1	0	1	2	2	3
Isoleucine	3	5	5	6	8	8	3	—	5	4	4	6
Leucine	5	3	3	3	10	10	5	20	5	5	5	2
Tyrosine	10	9	9	9	14	14	6	14	9	8-9	9	10
Phenylalanine	3	4	4	4	8	8	3	12	4	5	5	5
Half-cystine	12	14	14	14	19	22	11	22	12	14	13	12
Tryptophan	1	3	3	3	7	6	1	7	3	2-3	2	2
Total residue	108	118	118	119	221	217	102		119	118-121	116-118	118
References	Tu et al, 1970	Joubert and Van der Walt, 1975			Currie et al, 1968		Deems and Dennis, 1975	Sasaki, 1958	Chang and Lo, 1972	Dumarey, et al, 1975		Botes and Viljoen, 1974

Family	Viperidae					Crotalidae					
Genus Species Subspecies	Bitis arietans	Vipera berus			Vipera palestinae		Agkistrodon halys blomhoffii		Crotalus adamanteus			C. atrox
Origin	S. Africa	France			Palestine		Japan		U.S.A.		U.S.A.	U.S.A.
Name	phospholipase A_2	phospholipase A			phospholipase A		phospholipase		phospholipase		phospholipase $A_{\alpha-\alpha}$	phospholipase A_2
		P_1	P_2	P_4	F_1	F_2	A-I	A-II	A_α	A_β		
Amino Acid												
Lysine	12	19	13	8	10	9	17	8	16	16	7	14
Histidine	2	4	5	2	3	3	2	1	5	5	2	4
Arginine	4	10	10	4	3	3	6	4	12	12	5	10
Aspartic acid	14	40	34	11	18	18	14	17	30	30	16	34
Threonine	6	12	13	5	7	7	6	5	13	13	7	16
Serine	5	17	16	4	7	7	5	4	13	13	7	16
Glutamic acid	10	22	19	6	12	12	6	13	24	24	13	28
Proline	5	12	14	4	4	4	5	5	16	16	9	18
Glycine	11	35	27	8	14	15	10	13	24	24	13	30
Alanine	6	13	13	4	7	7	5	7	15	15	8	18
Valine	3	14	14	2	7	7	4	4	11	11	5	8
Methionine	2	11	8	2	4	4	3	2	2	2	1	4
Isoleucine	5	16	16	4	4	4	7	7	11	11	5	14
Leucine	4	15	15	5	9	9	5	5	11	11	5	14
Tyrosine	8	21	14	7	9	9	10	10	16	16	8	14
Phenylalanine	5	14	10	4	8	8	5	5	10	10	5	8
Half-cystine	14	27	17	10	12	12	14	14	30	30	14	28
Tryptophan	1	4	5	1	2	2	2	2	7	7	3	2
Total residue	117	298	259	91	140	140	126	126	266	266	133	280
References	Howard, 1975	Delori, 1973			Shiloah et al, 1973		Kawanchi et al, 1971		Wells and Hanahan, 1969		Tsao et al, 1975	Hachimori et al, 1971

Table 3 Amino Acid Sequences of Snake Venom Phospholipase A$_2$

	Source	Loop 1		Loop 2	
Elapidae	Porcine pancreas	H$_2$N-Ala-Leu-Trp-Gln-Phe-Arg-Ser-Met-Ile-Lys-	-Pro-Gly-Ser-His-Pro-Leu-Met-Asp-	Phe-Asn-Asn- - - - -Tyr-Gly
	Naja melanoleuca (DE-I)	H$_2$N-Asn-Leu-Tyr-Gln-Phe-Lys-Asn-Met-Ile-His-Cys-	- -Pro-Trp-Trp-His-Pro-Ala-Asn-	- - -Tyr-Gly
	Naja melanoleuca (DE-II)	H$_2$N-Asn-Leu-Tyr-Gln-Phe-Lys-Asn-Met-Ile-Gln-Cys-	-Thr-Val-Pro-Asn-Arg- - -Ser-Trp-Trp-His-Pro-Ala-Asn-	- - -Tyr-Gly
	Naja melanoleuca (DE-III)	H$_2$N-Asn-Leu-Tyr-Gln-Phe-Lys-Asn-Met-Ile-His-Cys-	-Thr-Val-Pro-Asn-Arg- - -Ser-Trp-Trp-His-Pro-Ala-Asn-	- - -Tyr-Gly
	Hemachatus haemachatus (DE-I)	H$_2$N-Asn-Leu-Try-Gln-Phe-Lys-Asn-Met-Ile-Cys-	-Thr-Val-Pro-Arg- - - - -Ser-Trp-Trp-His-Pro-Ala-Asn-	- - -Tyr-Gly
	Naja naja atra	H$_2$N-Asn-Leu-Try-Gln-Phe-Lys-Asn-Met-Ile	-Ser-Arg- - - - - - - -	
Viperidae	Bitis gabonica	H$_2$N-Asp-Leu-Thr-Gln-Phe-Gly-Asn-Met-Ile-Asn-	- - -Lys-Met-Gly-Glu-Ser-	Val-Phe-Asp-Tyr-Ile-Tyr Tyr-Gly
Crotalidae	Crotalus adamanteus	H$_2$N-Ser-Leu-Val-Gln-Phe-Glu-Thr-Leu-Ile-Met-	- - -Lys-Val-Ala-Lys-Arg-Ser-Gly-Leu-Trp-	Tyr-Ser-Ala-Tyr-Gly

	Source	Loop 3		Loop 4	Loop 5
Elapidae	Porcine pancreas	Cys-Tyr-Cys-Gly-Leu-Gly-Gly-	-Ser-Gly-Thr-Pro-Val-Asn-Glu-Leu-Asn-Arg-Cys-	-Gly-His-Thr-Asp-Asn-
	Naja melanoleuca (DE-I)	Cys-Tyr-Cys-Gly-Arg-Gly-Gly-Lys-	-Gly-Thr-Pro-Val-Asp-Asp-Asp-Leu-Asp-Arg-Cys-	-Cys-Gln-Ile-His-Asp-Lys-
	Naja melanoleuca (DE-II)	Cys-Tyr-Cys-Gly-Arg-Gly-Gly-	-Ser-Gly-Thr-Pro-Val-Asp-Asp-Asp-Leu-Asp-Arg-Cys-Cys-Gln-Ile-His-Asp-Asn-
	Naja melanoleuca (DE-III)	Cys-Tyr-Cys-Gly-Arg-Gly-Gly-	-Ser-Gly-Thr-Pro-Val-Asp-Asp-Asp-Leu-Asp-Arg-Cys-Cys-Gln-Ile-His-Asp-Asn-
	Hemachatus haemachatus (DE-I)	Cys-Tyr-Cys-Gly-Arg-Gly-Gly-	-Ser-Gly-Thr-Pro-Val-Asp-Asp-Asp-Leu-Asp-Arg-Cys-Cys-Gln-Thr-His-Asp-Asn-
	Naja naja atra	Cys-Tyr-Cys-Gly-	-Gly-Lys-Pro-Ile-Asp-Ala-Thr-Asp-Arg-Cys-Cys-Phe-Val-His-Asp-
Viperidae	Bitis gabonica	Cys-Tyr-Cys-	-Gly-Gly-Arg-Pro-Gln-Asx-Ala-Thr-Ser-Arg-Cys-Cys-Phe-Val-His-Asx-
Crotalidae	Crotalus adamanteus	Cys-Tyr-Cys-		

	Source	Loop 6	Loop 7	
Elapidae	Porcine pancreas	-Tyr-Arg-Asp-Ala-Lys-Asn-Leu-Asp-Ser-Cys-Lys-Phe-Leu-Val-Asp-Asn-Pro-Tyr-Thr-	Glu-Ser-Tyr-	
	Naja melanoleuca (DE-I)	-Tyr-Asp-Glu-Ala-Glu- -Lys-Ile-Ser-Gly-Cys-Trp-Pro-Tyr-Ile-Lys-Thr-	-Tyr-Thr-	-Tyr-Glu-Ser-
	Naja melanoleuca (DE-II)	-Tyr-Gly-Glu-Ala-Glu- -Lys-Ile-Ser-Gly-Cys-Trp-Pro-Tyr-Ile-Lys-Thr-	-Tyr-Thr-	-Tyr-Glu-Ser-
	Naja melanoleuca (DE-III)	-Tyr-Gly-Glu-Ala-Glu- -Lys-Ile-Ser-Gly-Cys-Trp-Pro-Tyr-Ile-Lys-Thr-	-Tyr-Thr-	-Tyr-Asp-Ser-
	Hemachatus haemachatus (DE-I)	-Tyr-Ser-Asp-Ala-Glu- -Lys-Ile-Ser-Gly-Cys-Arg-Pro-Tyr-Phe-Lys-Thr-	-Tyr-Ser-	-Tyr-Asp-
	Naja naja atra			
Viperidae	Bitis gabonica	Cys- - -Tyr-Gly-Lys-Met-Gly- - -Thr-Tyr-Asp-Thr- -Lys-Trp-Thr-Ser-Tyr-Asn-	-Tyr- -Glu-Ile-	
Crotalidae	Crotalus adamanteus	Cys- -Tyr-Gly-Lys-Ala-		

30

	Source		Loop 8	Loop 9	Loop 10	Loop 11
Elapidae	Porcine pancreas		Cys-Ser-Ser-Asn-Thr-Glu-Ile-Thr-Cys-Asn-Ser-Lys-Asn-Ala-Cys-Glu-Ala-Phe-Ile-Val-Cys-Asn-			
	Naja melanoleuca (DE-I)		Cys-Gln-Gly- - -Thr-Leu-Thr-Cys-Lys-Asp-Gly-Gly-Lys- -Ala-Ala-Ser-Val-Cys-Asp-Cys-
	Naja melanoleuca (DE-II)		Cys-Gln-Gly- - -Thr-Leu-Thr-Ser-Cys-Gly-Ala-Asn-Lys- -Ala-Ala-Ser-Val-Cys-Asp-Cys-
	Naja melanoleuca (DE-III)		Cys-Gln-Gly- - -Thr-Leu-Thr-Ser-Cys-Gly-Ala-Ala-Asn-Asn- -Ala-Ala-Ser-Val-Cys-Asp-Cys-
	Hemachatus haemachatus (DE-I)		Cys-Thr-Lys-Gly-Lys-Leu-Thr- - -Cys-Lys-Glu-Gly-Asn-Asn-Glu-Cys-Val-Ala-Phe-Leu- -Ala-Arg-His-
Viperidae	Naja naja atra					
Crotalidae	Bitis gabonica		- -Gln- - -Asn-Gly-Gly-Ile-Asp-Cys-Asp-Glu-Asp-Pro-Gln-Lys- -Lys-Glu-Leu- -Cys-Glu-Cys-
	Crotalus adamanteus					

	Source	Loop 12	Loop 13
Elapidae	Porcine pancreas	Asp-Arg-Asn-Ala-Ala-Ile-Cys-Phe-Ser-Lys-Ala-Pro-Tyr-Asn-Lys-Glu-His-Lys-Asn- - -Leu-Asn-Thr-Lys-Lys-Tyr- - -
	Naja melanoleuca (DE-I)	Asp-Arg-Val-Ala-Ala-Asn-Cys-Phe-Ala-Arg-Ala-Thr-Tyr-Asn-Asp-Lys-Asn- - -Tyr-Asn-Ile-Asp-Phe-Asn-Ala-Arg-
	Naja melanoleuca (DE-II)	Asp-Arg-Val-Ala-Ala-Asn-Cys-Phe-Ala-Arg-Ala-Thr-Tyr-Asn-Asp-Lys-Asn- - -Tyr-Asn-Ile-Asp-Phe-Asn-Ala-Arg-
	Naja melanoleuca (DE-III)	Asp-Arg-Val-Ala-Ala-Asn-Cys-Phe-Ala-Arg-Ala-Ala-Pro-Tyr-Ile-Asp-Lys-Asn- - -Tyr-Asn-Ile-Asp-Phe-Asn-Ala-Arg-
	Hemachatus haemachatus (DE-I)	Asp-Arg-Leu-Ala-Ala-Ile-Cys-Phe-Ala-Gly-Ala-His-Tyr-Asn-Asp-Asn-Asn-Tyr- - -Ile-Asp-Phe-Asn-Ala-Arg-
Viperidae	Naja naja atra		
Crotalidae	Bitis gabonica	Asp-Arg-Val-Ala-Ala-Ala-Cys-Phe-Ala-Asn-Asn-Arg-Asn-Thr-Tyr-Asn-Ser-Asn-Tyr- - -Phe-Gly-His-Ser-Ser-Lys-
	Crotalus adamanteus		

	Source	Loop 14	References
Elapidae	Porcine pancreas	Cys- - - - - - - - -OH	De Haas et al. (1970)
	Naja melanoleuca (DE-I)	Cys-Gln- - - - - -OH	Joubert (1975a)
	Naja melanoleuca (DE-II)	Cys-Gln- - - - - -OH	Joubert (1975a)
	Naja melanoleuca (DE-III)	Cys-Gln- - - - - -OH	Joubert (1975b)
	Hemachatus haemachatus (DE-I)	Cys-Gln- - - - - -OH	Joubert (1975c)
Viperidae	Naja naja atra		Chang and Lo (1972)
Crotalidae	Bitis gabonica	Cys-Thr-Gly-Thr-Glu-Gln-Cys-OH	Botes and Viljoen (1974b)
	Crotalus adamanteus		Tsao et al. (1975)

12 Always has 6 residues and shows a high degree of homologous relationships.
13 Is the longest loop.
14 Consists of 1 to 6 residues. There is no loop 14 in porcine pancreas phospholipase A_2.

Bitis gabonica phospholipase A_2 has a longer peptide chain than the enzyme of Elapidae origin. The enzymes obtained from the venoms of Elapidae have more similarity within themselves than to those of another family, Viperidae. Again, the closer the relationship in phylogenecity of snakes, the more similar is the chemical structure of a venom component.

The sequence of *Agkistrodon halys blomhoffii* phospholipase A_2 was also determined (Samejima et al., 1974) but homologies to other venom phospholipases A_2 are not great. Its sequence is as follows:

```
                                  10
⌐Glu–Phe–Glu–Thr–Leu–Ile–Met–Ser–Leu–Met–Lys–Ile–Ala–Gly–Arg–
          20                                          30
Ser–Gly–Ile–Tyr–Tyr–Gly–Ser–Tyr–Cys–Gly–Cys–Tyr–Gln–Gly–Ala–Gly–Gly–
                              40
    Gln–Gly–Cys–Pro–Arg–Ser–Ala–Asp–Asp–Arg–Cys–Cys–Phe–Ile–Cys–
          50                                          60
    Glu–Cys–Asp–Lys–Asp–Ala–Ala–Arg–Asp–Asn–Ile–Asp–Thr–Tyr–Asp–
                              70
    Asn–Lys–Val–Thr–Gly–Cys–Asx–Pro–Lys–Leu–Asp–Val–Tyr–Thr–Tyr–
          80                                          90
    Thr–Glu–Glu–Asp–Gly–Ala–Ile–Val–Cys–Gly–Gly–Asx–Asx–Asx–Ala–
                              100
    Glx–Pro–Lys–Lys–Cys–Cys–Phe–Val–His–Asp–Cys–Cys–Tyr–Gly–Lys–
          110                                         120
    Ile–Tyr–Trp–Trp–Phe–Pro–Phe–Ala–Lys–Asn–Cys–Gln–Cys–Glu–Ser–
                              126
    Pro–Gly–Glu–Cys–OH.
```

1.3 Site of Hydrolysis

The compositions of released fatty acids have been investigated by a number of workers, and their results are summarized in Tables 4, 5, and 6. Regardless of the origin of the venoms, all purified phospholipases A_2 liberated predominantly polyenic fatty acids. It has been shown that phosphatidylcholine contains predominantly saturated fatty acid esterified at the 1 position, and unsaturated fatty acids at the 2 position (Menzel and Olcott, 1964). This indicates that the site of hydrolysis was at the β position (C_2) of phosphatidylcholine, as shown in the following diagram:

$$\begin{array}{c}
\downarrow A_1 \\
O1\ CH_2-O-C-R \\
\|\| \\
\|O \\
R-C-O-CH_2\ 2 \\
A_2\uparrow|O \\
|\| \\
3\ CH_2-O-P-O-CH_2CH_2N^+(CH_3)_3 \\
| \\
OH
\end{array}$$

1 Chemistry

Since phospholipase hydrolyzes the acyl bond at C_2 (β-carbon), it is frequently called phospholipase A_2. The enzyme that hydrolyzes C_1 (α-carbon) is sometimes termed phospholipase A_1. The phospholipase found in snake venom is the A_2 type and in this book will simply be called phospholipase A_2.

The specificity of the enzyme is directed toward "the site of fatty acids" rather than "the type of fatty acids." Thus α-oleolyl-β-stearoylphosphatidylcholine yields only stearic acid (De Haas and Van Deenen, 1961). This positional specificity of venom enzyme is further confirmed by unequivocal evidence obtained by using the synthetic substrates of known fatty acids at the known position. De Haas et al. (1962) used (α-palmitoyl-β-linolenoyl)-L-α-phosphatidylethanolamine and (α-linolenoyl-β-palmitoyl)-L-α-phosphatidylethanolamine, and analyzed the liberated fatty acids as well as the fatty acids in the resultant lysophospholipids. The results are illustrated in Scheme 1.

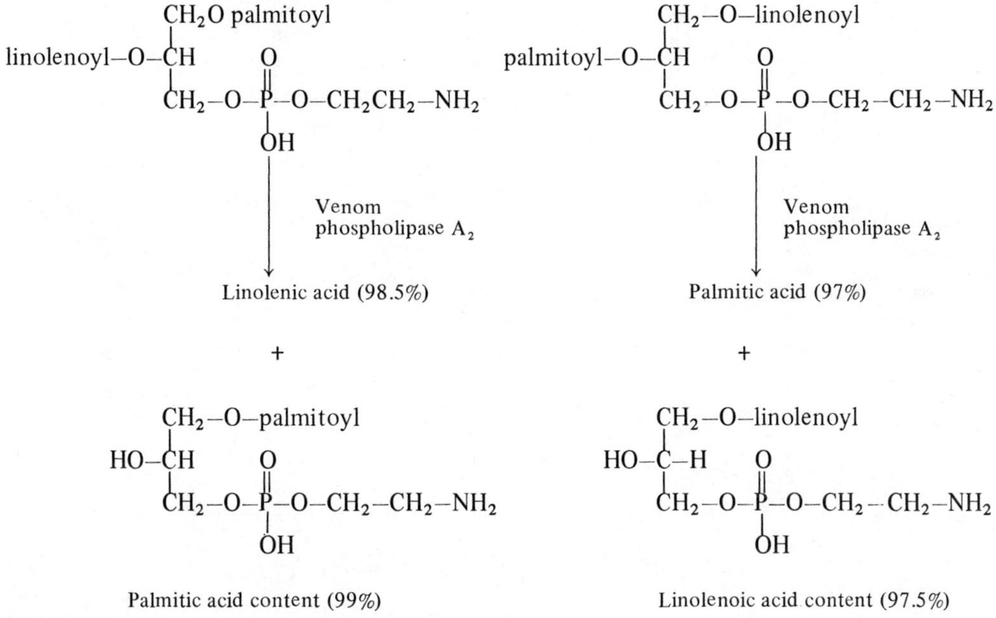

Scheme 1

Cleavage at the C_2 position by *Naja naja atra* phospholipase A_2 was also unequivocally determined by the nuclear magnetic resonance spectroscopic method (Chang and Lo, 1975). Moore and Williams (1963–1964) found, however, that the type of fatty acids in α and β positions exerts a considerable influence on the rate of hydrolysis at the β-position. They concluded that the rate of hydrolysis of the various types of phosphatidylcholine molecules by *Crotalus adamanteus* phospholipase A_2 decreased in the following order: α-unsaturated β-saturated, α-unsaturated β-unsaturated, α-saturated β-polyunsaturated, α-saturated β-monounsaturated, α-saturated β-saturated.

In the hydrolysis of a phosphatidylcholine by venom phospholipase A_2 two sites of cleavages are possible (Fig. 2). To clarify the exact site of hydrolysis, Wells (1971b) used isotopic ^{18}O-labeled water and found that ^{18}O was incorporated into fatty acid. This clearly indicates that the hydrolysis occurred by O-acyl cleavage via the mechanism shown in Scheme 2.

Table 5 Fatty Acid Compositions of Products from Viperidae Phospholipase A_2 Hydrolysis of Phosphatidylcholine

Sample	Origin of Venom	Fatty acid (mole %)			
			phosphatidyl choline (egg)	Free fatty acid	non-hydrolyzed fatty acid
Laticauda semifasciata venom	E. China Sea	14:0	Tr	Tr	Tr
		16:0	36.2	10.7	52.9
		18:0	13.8	4.1	21.7
		14:1	Tr	Tr	Tr
		16:1	1.4	1.6	2.8
		18:1	34.0	58.8	18.8
		18:2	12.9	22.4	3.6
		20:4	1.6	2.4	-0-
L. semifasciata phospholipase A	E. China Sea	14:0	Tr	Tr	Tr
		16:0	36.2	7.8	59.3
		18:0	13.8	2.6	23.5
		14:1	Tr	Tr	Tr
		16:1	1.4	1.5	2.8
		18:1	34.0	60.2	12.6
		18:2	12.9	25.8	1.6
		20:4	1.6	2.0	-0-

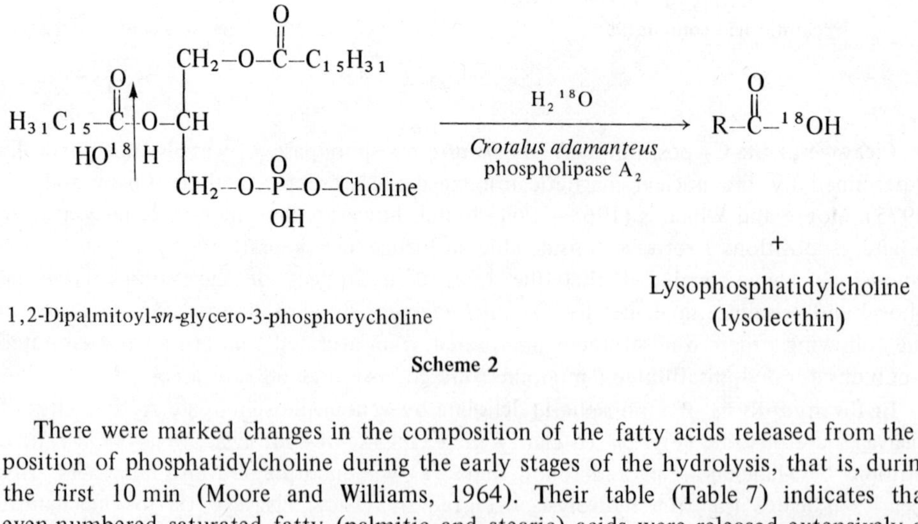

$$\underset{\text{1,2-Dipalmitoyl-}sn\text{-glycero-3-phosphorycholine}}{\begin{array}{c} \text{O} \\ \text{H}_{31}\text{C}_{15}-\overset{\|}{\text{C}}\text{-O-CH} \\ \text{HO}^{18}\text{H} \end{array} \begin{array}{c} \text{O} \\ \text{CH}_2\text{-O-}\overset{\|}{\text{C}}\text{-C}_{15}\text{H}_{31} \\ | \\ \text{CH}_2\text{-O-P-O-Choline} \\ \text{OH} \end{array}} \xrightarrow[\text{phospholipase } A_2]{\underset{\textit{Crotalus adamanteus}}{H_2^{18}O}} \text{R-}\overset{\overset{\text{O}}{\|}}{\text{C}}\text{-}^{18}\text{OH}$$

+

Lysophosphatidylcholine (lysolecithin)

Scheme 2

There were marked changes in the composition of the fatty acids released from the β position of phosphatidylcholine during the early stages of the hydrolysis, that is, during the first 10 min (Moore and Williams, 1964). Their table (Table 7) indicates that even-numbered saturated fatty (palmitic and stearic) acids were released extensively at

1 Chemistry

Table 5 Fatty Acid Compositions of Products from Viperidae Phospholipase A_2 Hydrolysis of Phosphatidylcholine

Sample	Origin of Venom	Fatty acid (mole %)			Reference
		phosphatidyl choline (egg)	Free fatty acid	non-hydrolyzed fatty acid	
Bitis gabonica phospholipase A_2	S. Africa	10:0 3.1 12:0 7.1 14:0 0.2 15:0 Tr 16:0 (iso) Tr 16:0 32.2 17:0 Tr 18:0 (iso) Tr 18:0 11.9 20:0 Tr 16:1 1.6 18:1 28.1 18:2 14.3 18:3 Tr 20:2 Tr 20:3 1.3	— Tr Tr Tr Tr 8.7 Tr Tr 4.4 Tr 1.4 56.0 27.7 Tr Tr 1.6	Tr Tr 0.4 Tr Tr 52.6 Tr Tr 18.8 Tr 1.9 15.8 8.1 Tr Tr 1.3	Botes and Viljoen, 1974a
B. arietans	S. Africa	16:0 18:0 16:1 18:1 18:2 20:3	3.1 1.0 1.4 55.5 28.2 10.8	— — — — — —	Howard, 1975
Vipera berus Venom	France	14:0 16:0 18:0 16:1 18:1 18:2 18:3 20:2 20:3 20:4	3.2 5.8 3.3 1.0 53.1 32.2 0 0 0 1.2	— — — — — — — — — —	Delori, 1973
V. berus phospholipase A_2 P_4	France	14:0 16:0 18:0 16:1 18:1 18:2 18:3 20:2 20:3 20:4	3.2 3.6 2.2 0.9 53.7 32.1 0.5 0.6 0.5 6.5	— — — — — — — — — —	

```
              O                                            O
              ‖                                            ‖
           CH₂—OC—R                                   CH₂—O—C—R
    O      |                              O           |
    ‖      |                              ‖           |
R—C—┼—O—CH                            R—C—O—┼—CH
   HO   H  |                              H   OH     |
           |   O                                     |   O
           CH₂—O—P—O—Choline                         CH₂—O—P—O—Choline
               |                                         |
               OH                                        OH

              A                                            B
```

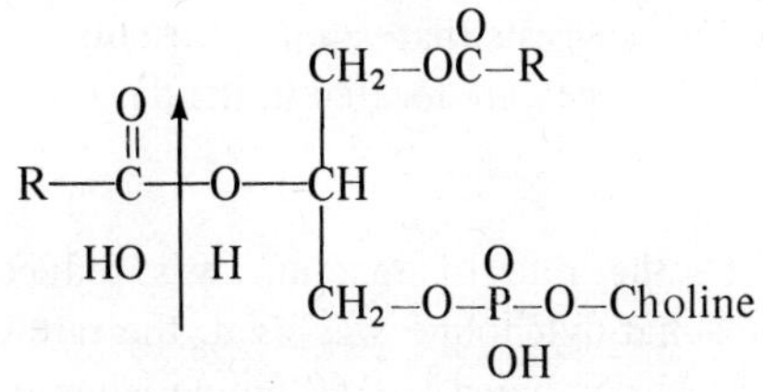

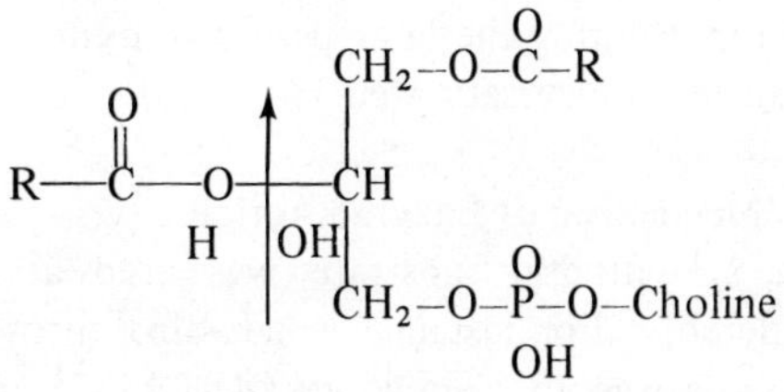

Figure 2.2 There are two mechanisms possible, A or B, for the hydrolysis of the fatty acid ester bond at the C-2 position.

Table 6 Fatty Acid Compositions of Products from Crotalidae Phospholipase A_2 Hydrolysis of Phosphatidylcholine

Sample	Origin of Venom	Origin of phosphatidyl choline	Fatty acid (mole %)				Reference
				phosphatidyl choline	Free fatty acid	non-hydrolyzed fatty acid	
Bothrops neuwiedii phospholipase A_2, P-1	Argentina	egg	16:0	28.0	6.5	59.2	Vidal et al, 1972
			17:0	0.2	Tr	0.4	
			18:0	16.1	6.5	34.0	
			19:0	0.1	Tr	Tr	
			20:0	0.2	0.2	Tr	
			22:0	0.05	Tr	0.2	
			16:1	0.2	0.4	0	
			17:1	Tr	Tr	0	
			18:1	39.5	59.0	1.9	
			18:2	9.5	16.6	0.1	
			18:3	0.1	0.2	0	
			20:U	4.0	5.9	0.8	
			22:U	1.9	4.2	0	
B. neuwiedii phospholipase A_2, P-2	Argentina	egg	16:0	28.0	5.8	49.0	Vidal et al, 1972
			17:0	0.2	Tr	0.6	
			18:0	16.1	4.4	32.0	
			19:0	0.1	Tr	Tr	
			20:0	0.2	0.2	Tr	
			22:0	0.05	Tr	0.7	
			16:1	0.2	0.4	0	
			17:1	Tr	Tr	0	
			18:1	39.5	61.4	4.9	
			18:2	9.5	17.0	0.9	
			18:3	0.1	0.2	0	
			20:U	4.0	6.4	1.9	
			22:U	1.9	4.1	0	

the early stage of the hydrolysis. Release of unsaturated (oleic and linoleic) acids was more rapid during the later period of hydrolysis. This suggests that venom phospholipase A_2 can hydrolyze saturated fatty acid esters as long as they are located in the β position.

1.4 Mechanism of Enzyme Action

When a synthetic substrate was used at $23°C$, the rate of reaction was reduced considerably. For instance, when dipalmitoylphosphatidylcholine was used, the rate of hydrolysis was 13.5 μmole RCOOH $\min^{-1}$ mg^{-1}, as compared to 160 μmole when egg phosphatidylcholine served as a substrate. Moreover, a lag period was observed with a synthetic substrate (Tu et al., 1970).

Table 6 *Continued*

Sample	Origin of Venom	Origin of phosphatidyl choline	Fatty acid (mole %)			Reference
			phosphatidyl choline	Free fatty acid	non-hydrolyzed fatty acid	
Crotalus atrox phospholipase A_2	U.S.A.	liver	14:0 0.5	1.0	0.8	Wu and Tinker, 1969
			15:0 0.2	0.1	0.4	
			16:0 17.5	4.1	30.5	
			18:0 18.8	2.5	47.8	
			22:0 0.7	1.4		
			16:1 1.4	1.5	1.2	
			18:1 13.3	12.5	13.9	
			18:2 15.6	27.3	2.2	
			20:1 0.1	0.2	0.2	
			20:2 0.5	0.8	2.9	
			20:3 2.4	4.4		
			20:4 24.2	39.9		
			20:5 Tr	0.1		
			22:3 0.1	0.5		
			22:4 0.7	0.4		
			22:5 0.4	0.7		
			22:6 3.5	5.0		
C. atrox phospholipase A_2	U.S.A.	egg	16:0 42.0	6.6		Hachimori et al, 1971
			18:0 10.3	2.0		
			16:1 1.7	1.3		
			16:2 0.3	0.2		
			18:1 32.4	60.0		
			18:2 11.0	24.3		
			18:3 0.4	1.4		
			20:4 1.9	4.2		

Ribeiro and Dennis (1973) studied the colloidal states of dipalmitoylphosphatidylcholine at different temperatures in the presence of Triton X-100. They found that at a higher temperature the synthetic phospholipid in aqueous dispersion undergoes a thermotropic phase transition from bilayer to stable mixed micelles. The activity of phospholipase A_2 toward dipalmitoylphosphatidylcholine and dimyristoylphosphatidylcholine at different temperatures showed a direct effect of the thermotropic phase transition of dipalmitoylphosphatidylcholine on enzyme activity (Dennis, 1973a). This indicates that substrates must be in micellar form rather than either monomers or bilayers (Dennis, 1973b; Deems et al., 1975).

For phospholipase A_2 action, Ca(II) is required. The calcium ion binds to the polar

Table 7 Compositions of the Fatty Acids (Molar Percentage of the Total) Produced after Various Time Intervals during the Hydrolysis of Phosphatidylcholine by Snake Venom Phospholipase A_2

Time	Fatty Acid						
	16:0	18:0	16:1	18:1	18:2	20:4	22:6
15 sec	39.1	16.3	2.1	29.3	13.5	—	—
30 "	23.9	12.6	2.2	34.5	16.9	2.9	5.5
45 "	16.1	9.1	1.1	39.5	19.6	4.5	8.0
60 "	13.3	6.1	1.0	41.2	21.1	5.2	10.1
75 "	11.2	4.5	1.1	43.8	23.6	5.0	10.0
90 "	8.4	3.7	1.1	45.4	24.3	5.4	10.5
105 "	5.7	3.1	1.0	44.8	26.6	5.4	10.8
4 min	6.1	1.3	0.9	44.8	29.1	5.8	11.4
6 "	2.8	0.9	0.9	47.4	27.6	6.6	11.7
8 "	1.8	1.0	0.8	49.1	27.2	7.0	11.9
10 "	1.5	0.8	0.8	50.7	27.2	6.4	11.1
20 "	1.0	0.2	0.8	55.7	27.2	5.8	9.2
30 "	1.3	0.5	0.7	55.9	26.1	5.4	9.1
40 "	1.0	0.3	0.8	55.8	26.4	5.7	8.5
50 "	1.0	0.4	0.8	55.8	26.4	5.5	8.5
60 "	1.0	0.4	0.9	55.4	26.6	5.3	8.7
70 "	1.0	0.3	0.7	55.6	26.7	5.7	9.1
80 "	1.1	0.2	0.9	53.9	27.8	6.1	9.4

Obtained from Moore and Williams, 1964.

part of the phospholipid molecule. The binding of Ca(II) produces a partial positive charge on the carbonyl carbon and promotes nucleophilic attack (Wells, 1972).

The Ca(II) adds to the enzyme before the substrate attaches to the enzyme (Viljoen et al., 1974). The attachment of Ca(II) induces a conformational change, as can be seen in the spectral charge (Viljoen et al., 1975). After the enzyme–Ca(II)–substrate complex is formed, the fatty acid is released from the enzyme, followed by the other product, lysophosphatidylcholine (lysolecithin) (Wells, 1972; Viljoen et al., 1974). The stepwise reactions are summarized in Scheme 3, where E represents phospholipase A_2, S is a substrate such as phosphatidylcholine, and RCOOH is a fatty acid.

The binding of certain cations such as Ca(II), Ba(II), and Sr(II) to phospholipase A_2 obtained from the venom of *Crotalus adamanteus* causes spectral perturbations of tryptophan residues in the protein. These are believed to be due to the removal of a charged group from the vicinity of the tryptophan (Wells, 1973a).

1 Chemistry

$$E + Ca(II) \rightleftharpoons E-Ca(II) \xrightleftharpoons{S} E-Ca(II)-S \rightleftharpoons$$

$$E-Ca(II)-\text{lysolecithin}-RCOOH \xrightleftharpoons{-RCOOH} E-Ca(II)-\text{lysoleithin} \rightleftharpoons$$

$$\text{Lysolecithin} + E-Ca(II)$$
$$\downarrow$$
$$E + Ca(II)$$

Scheme 3

Purified phospholipase A_2 obtained from the venom of *Naja naja atra* loses its enzyme activity when it is treated with nitrous acid, and the loss of enzyme activity cannot be reversed by the application of hydrogen sulfide. These facts suggest that phospholipase A_2 activity is dependent on the tyrosine residue and not on the amino or sulfhydryl groups (Sasaki, 1959).

In the dimer state, tryptophan and tyrosine residues in phospholipase of *Crotalus adamanteus* venom are in an unusual environment, which can be detected by an unusual absorption spectrum (Wells, 1971a).

When two tryptophan residues of *C. adamanteus* phospholipase A_2 were modified with *N*-bromosuccinimide, the enzyme activity was lost (Wells, 1973b). These tryptophans are probably located close to the active site of the enzyme. When one lysine per dimer was modified with ethoxyformic acid, the enzyme could also be inactivated, although one cation binding site was normal. The acylation of one lysine per dimer reduced the stability of the dimer and caused dissociation into subunits at pH 5.0. One subunit was normal and could reassociate to form a native, active dimer, whereas the modified subunit remained in the monomeric form.

Laticauda semifasciata phospholipase A_2 was activated by Ca(II) and to a lesser degree by Mg(II). The cations Zn(II) and Cd(II) both inhibited the enzyme completely, even in the presence of Ca(II) (Tu et al., 1970).

1.5 Substrate Specificity

Phosphatidylcholine (lecithin) is the most common substrate for phospholipase A_2. The hydrolysis of this substrate can take place in free substrate form, in egg yolk (Tu et al., 1970), and in serum (Marinetti, 1961; Chinen, 1972). Phospholipids of intact human serum high-density lipoprotein can be hydrolyzed by phospholipase A_2 isolated from *Crotalus adamanteus* venom (Pattanik et al., 1976).

In addition to phosphatidylcholine, a number of its natural and synthetic derivatives can be used as substrates. Phospholipase A_2 obtained from the venom of *Bothrops neuwiedii* hydrolyzed the substrates phosphatidylethanolamine, phosphatidylcholine, phosphatidylinositol, phosphatidylserine, and cardiolipin in the preceding order of activity (Vidal et al., 1972). The enzyme from the venom of *Agkistrodon piscivorus* hydrolyzed phosphatidylcholine, phosphatidylethanolamine, and dipalmitoylphosphatidylcholine (Augustyn and Elliott, 1970). Three synthetic substrates – dipalmitoyl-, dioleoyl-, and dilinoleoylphosphatidylcholines – were hydrolyzed by *Naja naja* phospholipase A_2 (Smith et al., 1972). Phospholipase A_2 from sea snake venom did not hydrolyze phosphatidylethanolamine, phosphatidyl-L-serine, phosphatidyl inositide, phosphatidic acid, lysolecithin, sphingomyelin, cerebroside, and cardiolipin (Tu et al., 1970).

The hydrolytic activity of the *Crotalus atrox* enzyme on different substrates was investigated by Coles et al. (1974). In ether, the relative rates of hydrolysis are phosphatidylcholine > phosphatidalcholine ≫ phosphatidylserine = phosphatidalethanolamine > phosphatidylethanolamine. In water, the rates are phosphatidylcholine > phosphatidylethanolamine > phosphatidylserine = phosphatidalethanolamine = phosphatidalcholine.

Natural phosphatidylcholine contains a high proportion of unsaturated fatty acid in β-position. When stearic acid, a saturated fatty acid, is placed in β-position, venom phospholipase A_2 can release stearic acid. A synthetic substrate, γ-elaidyl-β-(9,10-tritium)stearyl-L-α-glycerlophosphorylcholine, was hydrolyzed with *Crotalus adamanteus* phospholipase A_2, and stearic acid was the only fatty acid observed (Danvillier et al., 1964). The rate of hydrolysis of phosphatidylcholine was independent of the degree of unsaturation of the fatty acids (Woelk and Debuch, 1971). Compared to 1,2-diacyl-*sn*-glycerol-3-phosphorylcholine and ethanolamine, the corresponding 1-alk-1'-enyl-2-acyl compounds (plasmalogens) were cleaved more slowly by the purified phospholipase A_2 of *Crotalus atrox* venom:

$$\begin{array}{c} \text{CH}_2\text{OCH=CHR} \\ \text{R}'-\overset{O}{\overset{\|}{C}}-\text{CH} \\ \overset{|}{\text{CH}_2}-\text{O}-\overset{O}{\overset{\|}{\underset{|}{P}}}-\text{O}-\text{CH}_2-\text{CH}_2-\overset{+}{\text{NH}_3} \\ \text{O}_- \end{array}$$

Plasmalogen-type compounds served as competitive inhibitors for the hydrolysis of 1,2-diacyl-*sn*-glycero-3-phosphorylcholine or ethanolamine (Woelk and Debuch, 1971). Both acylalkenyl and acylalkyl-*sn*-glycero-3-phosphorylcholine are cleaved approximately 20% as much as is phosphatidylcholine. The hydrolysis rate depends on the fatty acid moiety at the 2-position of the phosphoglycerides (Woelk and Peiler-Ichikawa, 1974). When various compounds were tested by Woelk and Peiler-Ichikawa, the rates of hydrolysis were found to be as follows:

Compound	Specific activity (units/mg^{-1} of protein)
1-Acyl-2-[^{14}C] linoleoyl-*sn*-glycero-3-phosphorylcholine	151.4 ± 10.2
1-Alk-1'-enyl-2-[^{14}C] linoleoyl-*sn*-glycero-3-phosphorylcholine	25.6 ± 1.8
1-Alk-1'-enyl-2-[^{14}C] linolenoyl-*sn*-glycero-3-phosphorylcholine	34.8 ± 2.0
1-Alk-1'-enyl-2-[^{14}C] arachidonoyl-*sn*-glycero-3-phosphorylcholine	19.8 ± 1.5
1-Alkyl-2-[^{14}C] linoleoyl-*sn*-glycero-3-phosphorylcholine	32.4 ± 1.,
1-Alkyl-2-[^{14}C] linolenoyl-*sn*-glycero-3-phosphorylcholine	26.1 ± 2.0
1-Alkyl-2-[^{14}C] arachidonoyl-*sn*-glycero-3-phosphorylcholine	18.2 ± 1.4

1 Chemistry

The hydrolysis of cardiolipin (diphosphatidylglycerol) by snake venom phospholipase A_2 has also been well studied (Marinetti, 1964; Okuyama and Najima, 1965). Cardiolipin is the only phosphatide with known immunological properties. The enzyme reaction with cardiolipin and *Naja naja* venom phospholipase A_2 is slow, but is greatly stimulated by the addition of beef heart phosphatidylcholine (Marinetti, 1964).

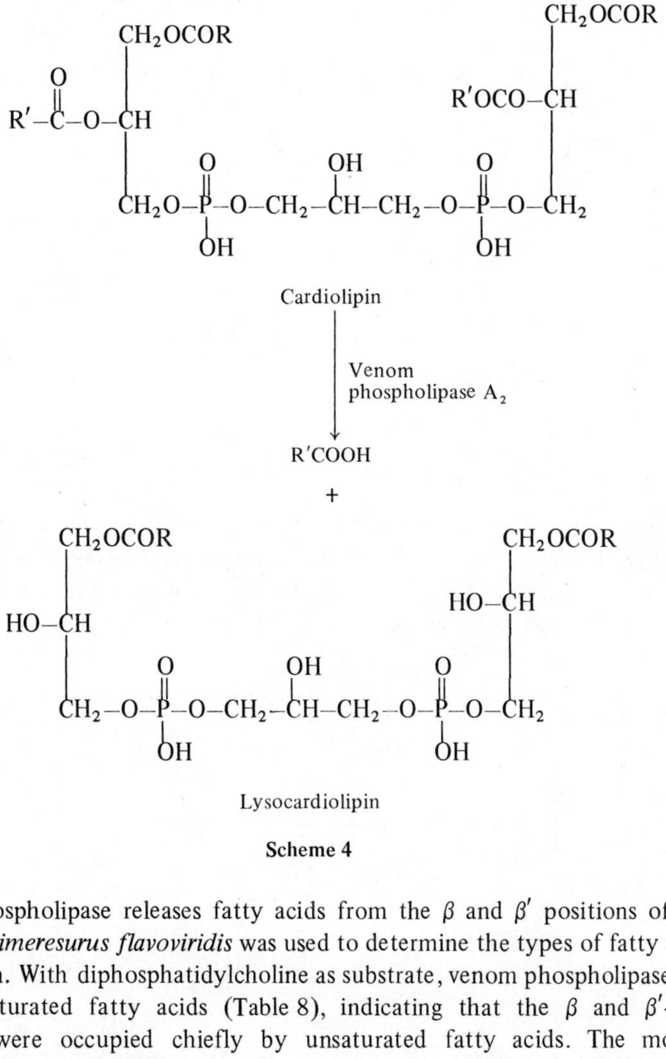

Cardiolipin

↓ Venom phospholipase A_2

R′COOH +

Lysocardiolipin

Scheme 4

Since phospholipase releases fatty acids from the β and β' positions of cardiolipin, venom of *Trimeresurus flavoviridis* was used to determine the types of fatty acids present in cardiolipin. With diphosphatidylcholine as substrate, venom phospholipase A_2 released mainly unsaturated fatty acids (Table 8), indicating that the β and β'-positions of cardiolipin were occupied chiefly by unsaturated fatty acids. The most common unsaturated fatty acid released was linoleic acid; it accounted for 80% of all the fatty acids released. The next most common was oleic acid, which amounted to about 12 to 14% (Okuyama and Najima, 1965).

Phospholipase A_2 catalyses the hydrolysis of the fatty acid ester linkage located adjacent to the glycerophosphoric acid ester bond only when the fatty acid to be released is allowed to occupy a certain steric position during the substrate–enzyme interaction. In a very interesting experiment De Haas and Van Deenen (1963) treated β-phosphatidylcholine (I, Scheme 5) with *Crotalus adamanteus* venom. They found that α,γ-distearoyl-β-glycerylphophorylcholine was hydrolyzed to produce β-lysophosphatidylcholine.

Table 8 Fatty Acid Compositions of Fatty Acids Released from the Snake Venom Hydrolysis of Cardiolipin

Fatty acids	From beef heart Cardiolipin	From human heart Cardiolipin	Reference
Less than 16	—	0.3	Okuyama
16:0	0.3-0.6	11.9	and Nojima,
18:0	—	—	1965
Larger than 18	Tr	Tr	
16:1	3.2-3.5	2.8	
17:	1.0-1.5	—	
18:1	11.8-14.0	9.8	
18:2	78.6-81.3	74.7	
18:3	1.6-2.6	0.5	

Further incubation of β-lysophosphotidylcholine did not produce any glycerly-phosphorylcholine. Therefore the venom phospholipase A_2 catalyzes the liberation of one fatty acid only from the β-phosphatidylcholine molecule. However, there are two possible mechanisms for such reactions (see Scheme 5). De Haas and Van Deenen tentatively concluded that product II is more likely from the theoretical viewpoint.

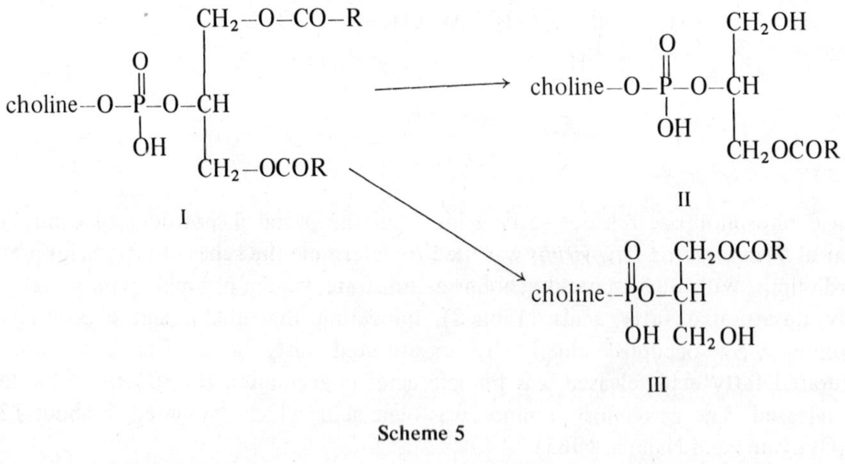

Scheme 5

Venom and certain bacterial toxin phospholipase A_2 can also release fatty acids from lipoprotein-bound phosphatidylcholine (Marinetti, 1961; Matsumoto, 1961). The rate of hydrolysis in lipoprotein-bound phospholipids is 10 to 20 times higher than that in isolated free phospholipids (Condrea et al., 1962; Slotta et al., 1971). Therefore snake venom phospholipase A_2 has a similarity to lipoprotein lipase, which liberates fatty acids from lipoproteins.

1 Chemistry

1.6 Lysophospholipase Activity of Phospholipase A_2

This is the only enzyme with dual activities (Shiloah et al., 1973b). It may be that this is true for all phospholipase A_2, but no one has examined the phospholipase B activity of phospholipase A_2 at pH 10. Purified phospholipase A_2 from *Naja naja* venom also has lysophospholipase (phospholipase B) activity. Phospholipase B activity is low at pH 9 but increases steadily beyond pH 10.5.

In addition to phospholipase A_2 action, the enzyme obtained from *Vipera palestinae* venom also exhibited phospholipase B (lysophospholipase) activity at pH 10. The hydrolysis of the acyl group from lysophosphatidylcholine (lysolecithin) was not due to high basicity of the solution since the activity was proportional to enzyme concentration. At the proper pH the phospholipase A_2 hydrolyzed lysophophatidylcholine, producing fatty acids and glycerophosphorylcholine (see Scheme 6).

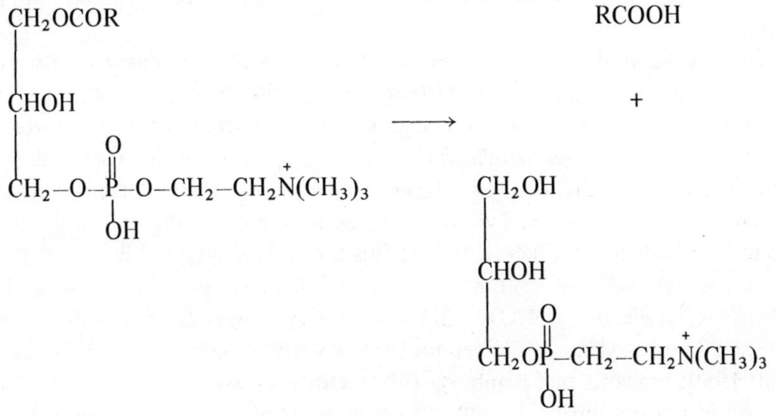

$$\begin{array}{c} CH_2OCOR \\ | \\ CHOH \\ | \\ CH_2-O-\overset{O}{\underset{OH}{P}}-O-CH_2-CH_2\overset{+}{N}(CH_3)_3 \end{array} \longrightarrow RCOOH + \begin{array}{c} CH_2OH \\ | \\ CHOH \\ | \\ CH_2O\overset{O}{\underset{OH}{P}}-CH_2-CH_2\overset{+}{N}(CH_3)_3 \end{array}$$

Scheme 6

1.7 Conformation

Circular dichroism (CD) studies of phospholipase A_2 dimer obtained from the venom of *Crotalus adamanteus* indicated that it had α-helix content of nearly 70% (Wells, 1971a). Similar results were obtained for two phospholipases A_2 obtained from the venom of *Agkistrodon halys blomhoffii* (Kawauchi et al., 1971a). A negative trough at 233 nm in CD spectra suggested that the enzyme consisted of a right-handed α-helical structure. The α-helix contents of the two phospholipases A_2 from *Vipera ammodytes* venom were estimated to be 24 and 23%, respectively. The nonhelical portions of the molecules were considered to be of random structure (Gubensek and Lapanje, 1974).

Circular dichroism spectra of three phospholipases A_2 obtained from the venom of *Naja melanoleuca* are very similar. The spectra show two negative maxima in the regions of 210 and 233 nm. The amounts of α-helix, calculated by the method of Bannister et al. (1973), were 28, 27, and 22% for DE-I, DE-II, and DE-III, respectively. The β-structure content was considered to be very small or non-existent in phospholipase A_2 (Joubert and Van der Walt, 1975).

Crystals of two venom phospholipases A_2 were subjected to X-ray diffraction analysis (Pasek et al., 1975). The space group for *Crotalus adamanteus* enzyme is $C222_1$, and $a = 108.1$ Å, $b = 79.4$ Å, $c = 63.9$ Å, with one dimer per asymmetric unit. Phospholipase A_2 from *C. atrox* has a space group of $P2_12_12_1$, and $a = 53.8$ Å, $b = 99.0$ Å, $c = 49.5$ Å, with one dimer per asymmetric unit.

1.8 Isozymes

Multiple forms of phospholipase A_2 of venom origin have been observed by a number of investigators. Multiple components with phospholipase A_2 activity from *Naja naja* venom were observed by Braganca and Sambray (1967). The isoelectric points of these components were 4.64, 4.90, 4.94, 5.02, 5.51, and 5.56 (Salach et al., 1971a). The molecular weights of the catalytically active multiple components observed ranged from 8500 to 22,000. In view of the difference in molecular weight, there is some question whether these are isozymes in the traditional sense. The most abundant form of isozyme from *N. naja* venom was the one with a pI of 4.95 (Salach et al. 197, a). Six phospholipase A_2 bands are shown in the venom of *N naja* (Shiloah et al., 1973a; Shiloah, 1974). Each of the purified phospholipase A_2-preparations gives only one distinct band on SDS acrylamide gel electrophoresis after incubation with β-mercaptoethanol. These preparations are antigenically identical and are also homogeneous in immunoelectrophoresis.

The enzyme isolated from the venom of *Vipera russellii* contains seven isozymes with molecular weights ranging from 15,000 to 23,800 (Salach et al., 1971a). Two phospholipase A_2 having similar molecular weight of approximately 9500 were isolated from the venom of *Bothrops neuwiedii* (Vidal et al., 1972). The enzymes readily dimerize to higher molecular weights of about 20,000. There are three different phospholipases A_2 in the venom of *Vipera berus*. Two of them are immunologically identical, whereas the other is distinctly different (Delori, 1973). This is clearly shown in Fig. 2.

The two isozymes in the venom of *Naja nigricollis* have isoelectric points of 7.8 and 5.5 and molecular weights of 13,000 and 14,667, respectively (Dumarey et al., 1975).

There are two isozymes in the venom of *Agkistrodon halys blomhoffii* (Iwanaga and Kawachi, 1959; Braganca and Sambray, 1967). One is a basic protein, the other, a neutral protein. Basic phospholipase A_2 was also isolated from the same venom (Martin et al., 1975). The crotoxin complex isolated from the venom of *Crotalus durissus terrificus* contains three phospholipases A_2 with pI values of 9.7, 9.7, and 4.8 (Breithaupt et al., 1975). The fact that proteins with such different isoelectric points show the same enzyme activity is rather fascinating and may suggest that hydrophilic groups are not essential in the enzyme action of phospholipase A_2.

Lecithinase A-II, as it is called by the author hydrolyzes both dipalmitoyl-L-α-phosphatidylcholine and dipalmitoleyl-L-α-phosphatidylcholine. Lecithinase A-1, however, hydrolyzes the second substrate and has very low hydrolytic action on the first substrate with saturated fatty acids as the substituents (Wakui and Kawachi, 1961).

Not all snake venoms contain multiple forms of phospholipase A_2. For instance, venoms of *Bothrops alternata* and *B. jararaca* show only a single band with the enzyme activity after electrophoresis (Braganca and Sambray, 1967). Sea snake venom from *Laticauda semifasciata* has only one phospholipase, which has been isolated in two laboratories and well characterized (Tu et al., 1970; Uwatoko-Setoguchi, 1970). Venom of *L. semifasciata* was fractionated in an isoelectric focusing column, using the amphorline range of pH 3 to 10 (Fig. 3A). The enzyme activity of each fraction was determined, and only the fraction with a pI of 6.0 was found to have phospholipase A_2 activity (Fig. 3A). After pure phospholipase A_2 was isolated from the venom by two-step CM-cellulose column chromatography, the enzyme was placed in an isoelectric focusing column. The pI of the pure enzyme focused at 6.5 (Fig. 3B). Thus it is clearly demonstrated that the venom of *L. semifasciata* contains only one type of phospholipase A_2 without isozymes.

1 Chemistry

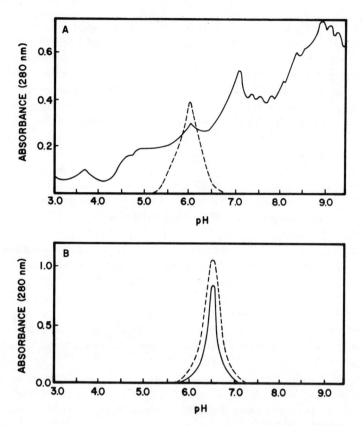

Figure 2.3 There is no evidence of an isozyme of phospholipase A_2 in the venom of *Laticauda semifasciata* (Tu and Passey, 1971). (A) Distribution of p*I*s (isoelectric points) of *L. semifasciata* venom. The dotted line indicates the phospholipase A_2 line. (B) Isoelectric focusing pattern of the purified enzyme, indicating the p*I* value of 6.5, which is similar to the 6.0 shown for the original venom.

Two immunologically identical phospholipases A_2 isolated from the venom of *Vipera palestinae* (Shiloah et al., 1973a) have identical molecular weights of 16,000.

All these multiple forms of the enzyme can be explained in many different ways.

1. Phospholipase A_2 can have isozymes in the traditional sense.
2. Many forms of phospholipase A_2 are actually polymerized forms of the enzyme.
3. Some forms are artifacts produced during the purification process. For instance, amide groups of asparagine and glutamine may be hydrolyzed to aspartic and glutamic acids, respectively.
4. Some of the multiple forms of the enzymes are products of the proteolytic degradation of the true enzyme.
5. Some forms are intermediate products produced during the synthesis of the true enzyme.

The second case should be considered seriously, as phospholipase A_2 tends to dimerize under certain conditions of pH and ionic strength. There is also some evidence to support case 4. Actually phospholipase A_2 activity was enhanced when the enzyme was subjected to trypsin or mild HC1 treatment (Marinetti, 1965). This suggest that removal of some portion of the enzyme molecule does not destroy the biological activity. However, this possibility may be excluded in the case of Elapidae venoms. Usually Elapidae venoms do not exhibit endopeptidase activity, yet they contain multiple forms of phospholipase A_2.

Case 5 may be a contributory factor for the multiplicity of the enzyme. This will be clarified by determination of the amino acid sequences of all the multiple-form enzymes in the future and also by study of the biosynthesis of phospholipase A_2.

1.9 Effects of pH and Temperature

Phospholipase A_2 from the venom of *Laticauda semifasciata* exhibits a rather narrow pH optimum, with maximum activity at pH 8.0. No enzymatic activity was detected at pH values below 6.0 or above 10.0 (Tu et al., 1970, Uwatoko-Setoguchi, 1970). The optimum pH for phospholipase A_2, DE-II, and DE-III was around 8.5 to 9 with a rather broad range. The pH optimum for DE-I was somewhat lower, at pH 7.5 to 8.5, and also extended over a broad range (Joubert and Van der Walt, 1975). For phospholipase A_2 obtained from the venom of *Bothrops neuwiedii*, the optimum pH was around 7.4 to 7.6, using egg yolk lipoprotein as a substrate (Vidal et al., 1972).

Laticauda semifasciata phospholipase A_2 has an optimum temperature of 35 to 40°C. The enzyme activity was destroyed at 75°C. The enzyme was still quite active, however at 1°C. The activation energy calculated from the Arrhenius plot is 6900 cal mole^{-1} (Tu et al., 1970). *Naja naja atra* venom phospholipase A_2 was also quite stable and retained its full activity at 90°C after 5 min (Chiang et al., 1973).

Phospholipase A_2 obtained from the venom of *Naja nigricollis* has an optimum pH between 7.7 and 8.3. When the pH is maintained at 8.0, the optimum pH between 7.7 and 8.3. When the pH is maintained at 8.0, the optimum temperature lies between 60 and 70°C (Dumarey et al., 1975).

2 BIOLOGICAL EFFECTS

2.1 Effect on Mitochondria

Most studies on the effect of phospholipase A_2 on mitochondrial morphology and electron transport systems were done using heat-treated crude venoms. The investigators simply assumed that the effect obtained from a heat-treated venom was due solely to the phospholipase A_2 present in the venom. This assumption will confuse and mislead investigators who are attempting to determine the real effects of phospholipase A_2. Because of the vast amount of research done with crude venoms in earlier days, these findings are also included in this book. However, more recent studies using isolated pure phospholipase A_2 are stressed in this section.

It is well established that cobra venom inhibits a number of dehydrogenase and oxidase activities. Succinate-cytochrome *c* reductase of mouse heart origin and cytochrome *c* oxidase were inhibited by *Naja naja atra* venom (Huang, 1954; Lin and Chang, 1957). NADH-cytochrome *c* reductase of pigeon breast muscle and rabbit kidney were also inhibited by the same venom (Lee, 1954). Ghosh and Sarka (1956) postulated that inhibition of glycolysis by *N. naja* venom was due to the inhibition of cytochrome *c* oxidase activity. These earlier works did not indicate that the inhibition of oxidoreductase systems was due to venom phospholipase A_2.

Venom phospholipase A_2 is considered to be a prime factor for the disruption of the electron transport chain and the integrity of mitochondrial structure. The idea originated mainly from the fact that lipolytic agents can uncouple electron transport chains as well as cause disintegration of mitochondrial structure. As early as 1959, Petrushka et al. (1959) used heat-treated venoms of *Agkistrodon piscivorus* and *Naja naja* as lipolytic agents to study the role of phospholipids in mitochondrial oxidative phosphorylation and structure. Since then, numerous investigations have been made of the biochemical and biological effects of venom phospholipase A_2. As can be seen from Table 1, a large number of pure phospholipases A_2 have been isolated from many different venoms and by many workers.

Aravindakshan and Braganca (1959) isolated partially uncoupled rat liver and brain mitochondria after *Naja naja* venom was injected into an animal. Inactivation of NADH oxidase, succinate oxidase, and cytochrome c oxidase in a submitochondrial particle system by heat-treated venom was attributed to the venom phospholipase A_2 (Luzikov and Romashina, 1972). The components obtained from the venoms of *N. naja atra* and *Bungarus multicinctus* were used to investigate the effects on mitochondrial respiratory enzymes. As the phospholipase A_2 activity of a given fraction increased, the inhibitory effect on succinoxidase and succinate-cytochrome c reductase also increased (Lin and Lee, 1974). Neither cobra neurotoxin nor α-bungarotoxin produces such inhibitory effects. It is, clear, therefore, that the effect was a phospholipase A_2-induced inhibition.

It has been shown that exposure of rat-liver mitochondria to snake venom phospholipase A_2 leads initially to an uncoupling of oxidative phosphorylation and ultimately to the cessation of oxidation of various substrates with a concomitant disarrangement of mitochondrial structure. The incubation of submitochondrial heart muscle particles (Keilin–Hartree preparation) with purified phospholipase A_2 from *Bothrops neuwiedi* venom inhibited the electron transport activity (Vidal et al., 1966). The segment coenzyme Q-cytochrome c is most sensitive to phospholipase A_2 action. The fact that quinone reductase activity was affected, whereas NADH-dehydrogenase activity was not, suggested that there must be another site of inhibition between the flavoprotein (NADH-dehydrogenase) and coenzyme Q. The third point of inhibition is near the cytochrome oxidase.

An oxygen electrode study (Fig. 4) showed that isolated *Agkistrodon piscivorus* phospholipase A_2 altered mitochondrial respiration and phosphorylation in a manner identical to that of whole venom (Augustyn et al., 1971). At low concentrations it increased mitochondrial respiration in the absence of phosphate acceptor. At high concentrations it caused severe inhibition of electron transport. At intermediate concentrations it produced a stage of respiratory decline in which ADP acted as an inhibitor (phosphate acceptor inhibition).

Electron microscope studies confirmed that whole *A. piscivorus* venom and isolated phospholipase A_2 produced identical morphological alterations of mitochondria and that structural disruption accompanied respiratory decline (Figs. 5, 6, and 7). Serum albumin reversed the uncoupling caused by whole *A. piscivorus* venom or its purified phospholipase A_2 (Augustyn et al., 1970). In the submitochondrial system involving NADH-cytochrome c oxidoreductase-cytochrome oxidase, cobra venom phospholipase A_2 inhibited the electron transfer at the level of cytochrome c first. Other sites of the respiratory chain were affected more slowly (Romashina et al., 1972).

Phospholipase A_2 caused solubilization of mitochondrial membrane-bound enzymes.

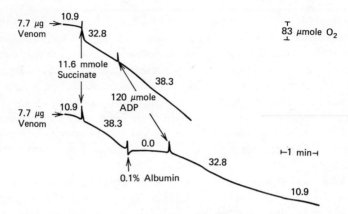

Figure 2.4 Inhibition of mitochondrial respiration by *Agkistrodon piscivorus* venom and phospholipase A_2. (A) Control, showing a normal curve. (B) Addition of 10.4 µg venom or 1.2 µg phospholipase A_2. (C) Addition of 16.7 µg venom or 2.4 µg phospholipase A_2. (D) Addition of 41.9 µg venom or 3.6 µg phospholipase A_2. Note severe inhibition. The oxygen uptake was measured with an oxygen electrode. The activity is expressed as micromoles O_2 per second. (Reproduced from Augustyn et al., 1970, by permission of the copyright owner, Elsevier Scientific Publishing Company.)

Reduced nicotinamide adenine dinucleotide dehydrogenase from inner membranes of heart mitochondria and α-glycerophosphate dehydrogenase from brain mitochondria were released by the action of venom phospholipase A_2 (Salach et al., 1971b).

Hydrolysis of membrane phosphatidylcholine was much faster than hydrolysis of phosphatidylethanolamine in the mitochondrial system. *Vipera palestinae* venom requires another component, direct lytic factor (DLF), for the hydrolysis of phospholipids in mitochondria (Condrea et al., 1965) because this venom lacks DLF, which is present in other venoms, such as those of cobras. Venom phospholipase A_2 can be used for solubilizing membrane-bound enzymes. For instance, NAD(P) transhydrogenase (NADPH:NAD$^+$ oxidoreductase, EC 1.6.1.1) of rat liver mitochondria can be isolated from a true solution after treatment with phospholipase A_2 from the venom of *Agkistrodon piscivorus* (Salvenmoser and Kramar, 1970).

2.2 Effect on Nerves and Muscles

It is known that low concentration of certain venoms markedly increase membrane permeabilities of lobster leg axon and giant axon of squid. At higher venom concentrations, axonal conduction is blocked (Rosenberg and Ehrenpreis, 1961; Rosenberg and Dettbarn, 1964). Such permeability increase or axonal conduction block was attributed to phospholipid splitting action of the venom phospholipase A_2 (Condrea and Rosenberg, 1968; Rosenberg, 1975).

Neither phospholipase A_2 obtained from *Naja naja atra* venom nor lysolecithin had any effect on the axonal conduction. The blocking action of cardiotoxin, however, was markedly accelerated by the addition of phospholipase A_2. Cardiotoxin depolarized the membrane of superficial muscle fibers, while the phospholipase required a dose 30 times higher and a prolonged period of incubation to induce depolariziation of similar extent. The slow depolarizing effect of phospholipase A_2 was enhanced by high concentrations of Ca(II). The synergistic effect of phospholipase A_2 on cardiotoxin is due to accelaration rather than potentiation of its action (Chang et al., 1972).

The acidic component isolated from crotoxin had no stimulatory effect on isolated

2 Biological Effects

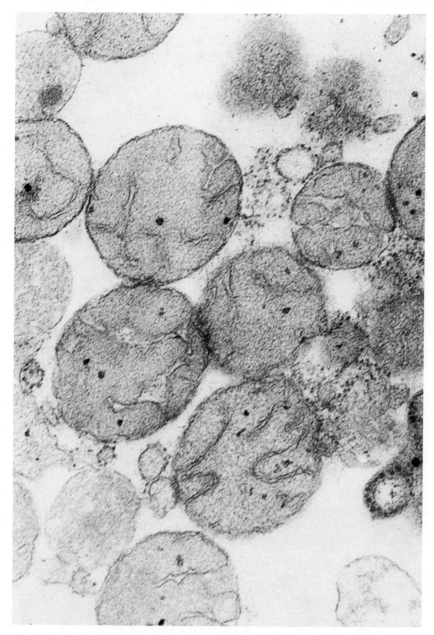

Figure 2.5 Control mitochondria. (Electron micrograph kindly supplied by Dr. W. B. Elliott and originally published in Augustyn et al., 1970.)

frog sartorius nerve–muscle preparation. However, the activity was enhanced by phospholipase A_2 (Brazil et al., 1973).

In addition to occurring in snake venoms, phospholipase A_2 is commonly found in the pancreas. There is some difference in the actions of the enzymes from the two sources. The enzyme from pancreas requires deoxycholate for activation, whereas venom enzyme did not require an activator for its action. *Naja naja* venom phospholipase A_2 split

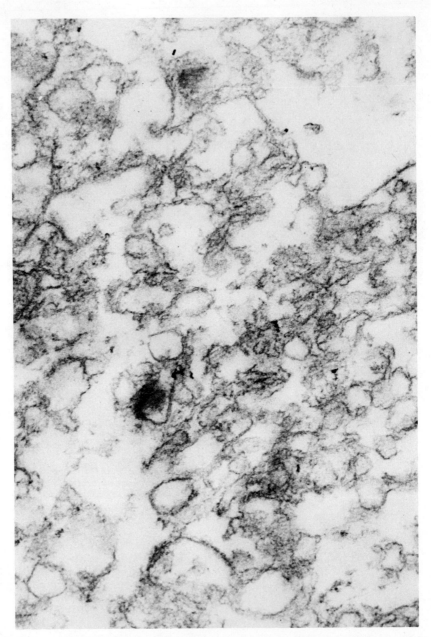

Figure 2.6 Destruction of mitochondria by phospholipase A_2 isolated from *Agkistrodon contortrix* venom. The enzyme also inhibited mitochondrial respiration and oxidative phosphorylation. (Kindly supplied by Dr. W. B. Elliott and originally published in Augustyn et al., 1970.)

phospholipids intact muscle cells, but the pancreatic enzyme attacked these only slowly, if at all (Ibrahim et al., 1964).

Venoms of *Haemachatus haemachatus* and *Agkistrodon piscivorus piscivorus* and phospholipase A_2 fraction (by heat treatment) blocked the electrical activity and facilitated the penetration of curare and acetylcholine into axons from the walking leg of

2   Biological Effects   51

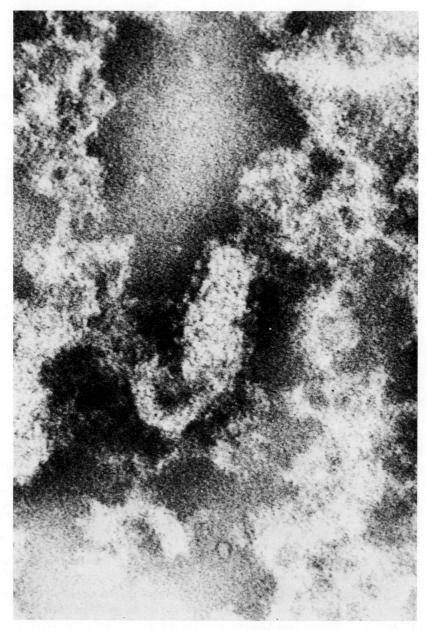

Figure 2.7   Negatively stained mitochondria destroyed after treatment by phospholipase A_2. Kindly supplied by Dr. W. B. Elliott and originally published in Augustyn et al., 1970.)

lobster (Condrea et al., 1967). These were the results of phospholipid hydrolysis in the axons. Venoms of *Vipera palestine* and *Crotalus admanteus* do not have such a blocking action, nor do they split axonal phospholipids.

The same results were observed for squid giant axon. The block and the permeability increase were attributed to lysolecithin (lysophosphatides) produced by the action of phospholipase A_2 on axonal phospholipids. Evidence showed that lysolecithin itself produced the same result. A second line of evidence is that phospholipase C (obtained from nonvemon source) caused extensive splitting of axonal phospholipids without producing such blocking effects. Unlike phospholipase A_2, phospholipase C produces diglycerides and phosphorylated bases; none of these products has detergent property (Rosenberg and Condrea, 1968).

The postsynaptic potential and action potential of isolated single electroplax from the electric eel are blocked when more than one-third of the phosphatidylcholine of the cell is hydrolyzed by *Agkistrodon piscivorus piscivorus* venom or phospholipase A_2 (Bartels and Rosenberg, 1972). Almost all of the phospholipids of the eel electroplax can be split with only a small alteration in gross membrane permeability being observed. Thus the presence of intact phospholipids does not appear essential to the maintenance of normal membrane permeability. The venom increases the permeability of isolated single electroplax (Rosenberg, 1973). The component responsible for this increase has not yet been identified. Phospholipase A_2 alone is not responsible for the marked increase in permeability produced by the venom as the acid-boiled venom has almost the same phospholipase A_2 activity as the unboiled venom.

Synaptic vesicles from rat brain cortex and cholinergic vesicles isolated from the electric organ of *Torpedo nobiliana* also served as substrates for purified phospholipase A_2 obtained from the venom of *Naja naja siamensis* (Heilbronn, 1972). Apparently, lysolecithin formed by the action of venom phospholipase A_2 weakens and breaks down synaptic vesicles. Eventually, vesicle membranes break down completely and form the open membrane pieces that indicate spots in the original membrane particularly sensitive to the enzyme or to the lysophosphatides formed by its action. Vesicle-bound acetylcholine is released as a consequence of the enzyme treatment. Electron micrographs of synaptic vesicles from rat brain cortex before and after phospholipase A_2 treatment are shown in Figs. 8 and 9. Phospholipase C is less effective than phospholipase A_2 in decreasing the acetylcholine content of brain cortex (Heilbronn, 1969).

In its action on central nervous system myelin, phospholipase A_2 showed a preference for phosphatidylethanolamine over phosphatidalethanolamine (Coles et al., 1974). The relative rates of hydrolysis of phosphoglycerides in the membrane were as follows: phosphotidylserine > phosphatidylcholine > ethanolamine phospholgycerides. The products (fatty acid and lysophosphatide) remained associated with the membrane after enzyme action.

Studies in which iodine-labeled phospholipase A_2 was used indicated that the enzyme was enriched in the liver after i.v. injection into mice. Only a small percentage of the injected enzyme was found in the brain, indicating that the enzyme did not pass significantly through the blood—brain barrier. The diaphragm contained about twice the amount of phospholipase A_2 that was observed in skeletal muscle (Habermann et al., 1972).

Pure phospholipase A_2 isolated from *Laticauda semifasciata* venom caused mild myonercrosis in mice (Tu et al., 1970; Tu and Passey, 1971). An example of myonecrosis is shown in Fig. 10.

2.3 Effect on Cell Membranes

Phospholipase A_2 of *Naja naja* venom increased the permeability of rat liver cells (Gallai-Hatchard and Gray, 1968), as measured by the release of an intracellular enzyme, glutamic oxaloacetic transaminase. There was only a small leakage of the intracellular enzyme from the cell until 25 to 35% of the phosphatidylethanolamine had been hydro-

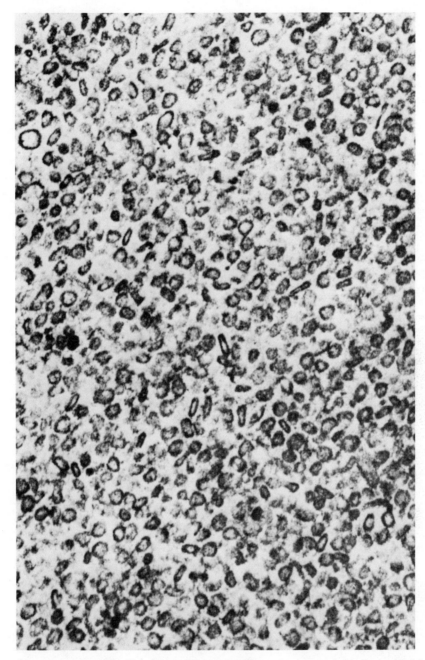

Figure 2.8 Synaptic vesicles from rat brain cortex. (Electron micrograph kindly supplied by Dr. Edith Heilbronn and originally published in Heilbronn, 1972.)

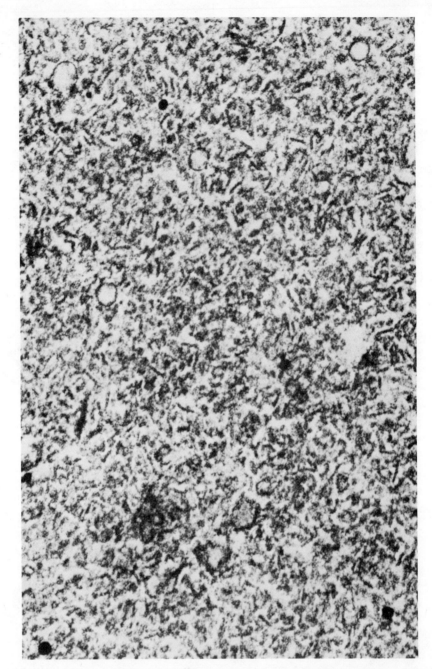

Figure 2.9 Synaptic vesicles from rat brain cortex in the presence of phospholipase A_2 isolated from the venom of *Naja naja siamensis*. (Electron micrograph kindly supplied by Dr. Edith Heilbronn and originally published in Heilbronn, 1972.)

2   Biological Effects

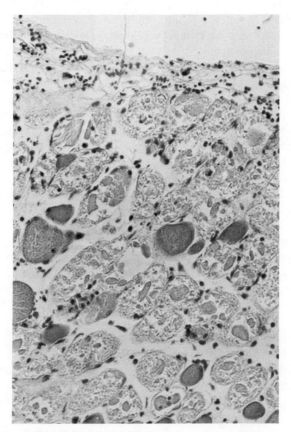

Figure 2.10 Mild myonecrotic action of phospholipase A_2 (100 μg) isolated from the venom of *Laticauda semifasciata* venom. Photograph by Mr. R. B. Passey of author's laboratory.)

lyzed. Further hydrolysis, however, caused a rapid increase in the release of the enzyme. The gross morphology of the cell was unaltered by hydrolysis of up to 50 to 60% of the phosphatidylethanolamine. Phosphatidylcholine in the cell membrane was not hydrolyzed with the phospholipase A_2 concentration used. This result suggests that phosphatidylethanolamine is relatively more important than phosphatidylcholine, in terms of permeability to a protein molecule and the structural integrity of the liver cell plasma membrane.

Certain substances such as Gramicidin S, synthetic basic α-amino acid copolymers, poly(Orn–Leu), poly(Orn–Leu–Ala), and poly(Lys–Leu) and the direct lytic factor (DLF) facilitate the action of phospholipase A_2 on red cell membranes (Kilbansky et al., 1968; Condrea et al., 1970). These basic compounds attach to membranes through electrostatic attraction and expose the cell membrane phospholipids to the phospholipase A_2. In the presence of DLF, EDTA inhibits the splitting of red cell membrane phospholipids by phospholipase A_2 by removing Ca(II), which is required for the enzyme action (Condrea et al., 1970).

Phospholipase A_2 isolated from *Laticauda semifasciata* venom causes no lytic effect on mouse embryo cells in tissue culture even when incubated with ovolecithin for 18 hr (Tu and Passey, 1971).

Platelet membranes are more susceptible to *Vipera russellii* venom than are erythrocyte membranes, and the membrane phospholipids were hydrolyzed (Bradlow and Marcus, 1966). The high susceptibility of platelet membrane phospholipids in comparison to erythrocytes suggests that the lipid arrangement in platelets is different from that in erythrocytes. Platelet and leucocyte phospholipids can be hydrolyzed *in vivo* by the venom of *Naja naja* but not by *Vipera palestinae* venom (Kirschmann et al., 1964; Klibansky et al., 1967) because *N. naja* venom contains both phospholipase A_2 and DLF, whereas *V. palesinae* venom contains only phospholipase A_2. Purified phospholipase A_2 itself does not hydrolyze leucocyte phospholipids (Klibansky et al., 1967).

Whether the intact cell membranes will be ruptured depends on many factors such as the type of cells employed and the presence or absence of phosphatidylcholine and DLF.

Phospholipase A_2 also plays a very important role in the lysis of red cell membranes, that is, in hemolysis. This is discussed in Chapter 20.

2.4 Toxic Action of Phospholipase A_2

Lethality. Phospholipase A_2 can be separated from the major neurotoxic fraction by fractionation of *Laticauda semifasciata* venom (Tu et al., 1970; Uwatoko-Setoguchi, 1970). At 5 µg enzyme per gram body weight of mouse, the enzyme is non toxic, whereas the purified toxins of *L. semi fasciata* venom have LD_{50} values of 0.05 to 0.06 µg g^{-1} (Tu and Passey, 1971).

The LD_{50} of pure *Naja naja atra* phospholipase A_2 is 8 µg g^{-1} in mice (Lo et al., 1975). Since the LD_{50} of crude *N. naja atra* is 0.29 µg g^{-1} (Tu, 1974), phospholipase A_2 is 28 times less toxic than the original venom. In comparison to the purfied cobra toxin, the toxicity of the enzyme is even weaker.

In *Agkistrodon acutus* venom, the main toxicity can be separated from the phospholipase A_2 activity, (Yang et al., 1959). Chiu (1960) investigated phospholipase A_2 activity in six venoms of Formosan snakes and found no correlation between venom enzyme activity and venom potency. Again, this is evidence that enzymes are not responsible for the major lethal action of snake venoms.

Crotoxin isolated from the venom of *Crotalus durissus terrificus* is a mixture of acidic and basic proteins. One of the phospholipases A_2 is an acidic protein and is nontoxic (Breithaupt et al., 1975).

In the fractionation of *Naja flava* venom, Radomski and Deichman (1958) observed that enzymes such as phospholipase A_2, ATPase, diphosphopyridine nucleotidase, and RNase were not of prime importance toxicologically. Entirely similar results were obtained by Yang et al. (1959). They also found that phosphodiesterase (exonuclease), phosphomonoesterase, 5'-nucleotidase, ATPase, proteinase, and phospholipase A_2 were clearly separated from three toxic principles of *Naja naja atra* venoms. Similarly, toxic fractions of *Bungarus multicinctus* venom were separated from many enzymes such as phosphomonoesterase, phospholipase A_2, cholinesterase, L-amino acid oxidase, and hyaluronidase (Yang et al., 1960). In a gel electrophoretic study of *Naja nigricollis* venom, two toxic fractions were separated from phospholipase A_2 hemolytic activity, ATPase and proteinase (Mohamed et al., 1972).

Other Toxic Action. It was thought that the histamine-releasing action of rattlesnake venom was due to the phospholipase A_2 present in the venom (Feldberg and Kellaway, 1937; Trethewie, 1939). However, these two activities are clearly separated by Amberlite CG-50 ion-exchange chromatography. Thus phospholipase A_2 is not responsible for histamine release.

Phospholipase A_2 was considered to be responsible for convulsions induced by intraventicular injection of *Naja naja* venom. Partially purified phospholipase A_2 produced convulsions even at very low doses of 2.5 μg (Lysz and Rosenberg, 1974). High doses of purified cobratoxin or cardiotoxin did not produce hyperexcitability when injected similarly.

Pure phospholipase A_2 of *Laticauda semifasciata* venom produced no hemorrhagic activity and caused only very weak myonecrosis in skeletal muscle (Tu et al., 1970; Tu and Passey, 1971). The enzyme was also separated from main lethal fractions. The nonhemorrhagic nature of phospholipase is further confirmed by a study of phospholipase A_2 obtained from the venom of *Vipera palestinae*. Subcutaneous and systemic administrations of phospholipase A_2 at a concentration of 0.003 mg per 0.1 ml did not produce lysis of capillary endothelial cells or massive local hemorrhage (McKay et al., 1970).

Direct lytic factor isolated from *Naja naja* venom was cytotoxic, but phospholipase A_2 itself caused no degranulation of mast cells (Damerau et al., 1975). In comparison to the toxic action of a neutrotoxin, the toxic effects of venom phospholipase A_2 are much weaker. Slotta et al. (1971) investigated the effect of *Naja naja* phospholipase A_2 on blood pressure, electrocardiogram, and electroencephalogram. A precipitous fall in arterial pressure and a small increase in central venous pressure were noted concomitantly with a short period of apnea and a slight bradycardia. All parameters, including the loss of cortical activity, returned to control or near-control values within 15 to 30 min postinjection. No notable change was shown in EKG pattern.

3 OCCURRENCE

Phospholipase A_2 is one of the most common enzymes found in snake venoms. All venoms of Hydrophiidae (sea snakes), Elapidae, Viperidae, and Crotalidae so far investigated contain this enzyme. The only venom from Colubridae (opisthoglyphous) that has been investigated is that of *Leptodiera annulata*. Mebs (1968) found that there was phospholipase A_2 activity in the venom of *L. annulata*, but the activity was relatively weak.

REFERENCES

Aravidakshan, I. and Braganca, B. M. (1959). Oxidative phosphorylation in brain and liver mitochondria of animals injected with cobra venom, *Biochim. Biophys. Acta,* **31**, 463.

Augustyn, J. M. and Elliott, W. B. (1970). Isolation of a phospholipase A from *Agkistrodon piscivorus* venom, *Biochim. Biophys. Acta,* **206**, 98.

Augustyn, J. M., Parsa, B., and Elliott, W. B. (1970). Structural and respiratory effects of *Agkistrodon piscivorus* phospholipase A on rat liver mitochondria, *Biochim. Biophys. Acta,* **197**, 185.

Bannister, W. H., Bannister, J. V., and Camilleri, P. (1973). Conformational analysis of bovine erythrocuprein from circular dichroism and infra-red spectra, *Int. J. Biochem.,* **4**, 365.

Bartels, E. and Rosenberg, P. (1972). Correlation between electrical activity and splitting of phospholipids by snake venom in the single electroplax, *J. Neurochem.,* **19**, 1251.

Bethell, F. H. and Bleyl, K. (1942). The production of micropherocytosis of red cells and hemolytic anemia by the injection of rattlesnake (*Crotalus atrox*) venom, *J. Clin. Invest.,* **21**, 641.

Botes, D. P. and Viljoen, C. C. (1974a). Purification of phospholipase A from *Bitis gabonica* venom, *Toxicon,* **12**, 611.

Botes, D. P. and Viljoen, C. C. (1974b). *Bitis gabonica* venom: The amino acid sequences of phospholipase A, *J. Biol. Chem.,* **249**, 3827.

Bradlow, B. A. and Marcus, A. J. (1966). Action of snake venom phospholipase A on isolated platelet membranes, *Proc. Soc. Exp. Biol. Med.,* 123, 889.

Braganca, B. A. and Sambray, Y. M. (1967). Multiple forms of cobra venom phospholipase A, *Nature,* 216, 1210.

Brazil, O. V., Excell, B. J., and Santana de Sa, S. (1973). The importance of phospholipase A in the action of the crotoxin complex at the frog neuromuscular junction, *J. Physiol.,* 234, 63.

Breithaupt, H., Omori-Satoh, T., and Lang, J. (1975). Isolation and characterization of three phospholipases A from the crotoxin complex, *Biochim. Biophys. Acta,* 403, 355.

Byrd, F. and Johnson, B. D. (1970). Some effects of heat-labile venom components on indirect hemolysis by crotalid venoms, *Am. J. Trop. Med. Hyg.,* 19, 724.

Chang, W. C. and Lo, T. B. (1972). Amino terminal sequence of phospholipase A from Formosan cobra (*Naja naja atra*) venom, *J. Chin. Biochem. Soc.,* 1, 77.

Chang, W. C. and Lo, T. B. (1975). The differentiation of phospholipase A_1 and A_2 by nuclear magnetic resonance spectroscopy, *J. Chin. Biochem. Soc.,* 4, 43.

Chang C. C., Chuang, S., Lee, C. Y., and Wei, J. W. (1972). Role of cardiotoxin and phospholipase A in the blockade of nerve conduction and depolarization of skeletal muscle induced by cobra venom, *Br. J. Pharmacol.,* 44, 752.

Chiang, H. C., Chang, W. C., and Lo, T. B. (1973). Enzymatic properties of phospholipase A from Formosan cobra (*Naja naja atra*) venom, *J. Chin. Chem. Soc.,* 2, 16.

Chinen, I. (1972). Phospholipase A activity of habu snake venom by using chicken serum as a substrate, *Ryukyu Daigaku Nogakubu Gakujutsu Hokoku,* 19, 259.

Chiu, W. C. (1960). Studies on the snake venom enzyme. IX. Electrophoresis studies of Russell's viper (*Vipera russellii formosensis*) venom and the relation of toxicity with enzyme activities, *J. Yamaguchi Med. Assoc.,* 9, 1361.

Coles, E., McIlwain, D. L., and Rapport, M. M. (1974). The activity of pure phospholipase A_2 from *Crotalus atrox* venom on myelin and on pure phospholipids, *Biochim. Biophys. Acta,* 337, 68.

Condrea, E. and Rosenberg, P. (1968). Demonstration of phospholipid as the factor responsible for increased permeability and block of axonal conduction by snake venom, *Biochim. Biophys. Acta,* 150, 271.

Condrea, E., De Vries, A., and Mager, J. (1962). Action of snake-venom phospholipase A on free and lipoprotein bound phospholipids, *Biochim. Biophys. Acta.* 58, 389.

Condrea, E., De. Vries, A., and Mager, J. (1964). Hemolysis and splitting of human erythrocyte phospholipids by snake venom, *Biochim. Biophys. Acta,* 84, 60.

Condrea, E., Mager, J., and De Vries, A. (1965). Action of snake venom on phospholipids in cellular and subcellular membranes, *Biol. Conf. "Oholo",* 10, 56.

Condrea, E., Rosenberg, P., and Dettbarn, W. D. (1967). Demonstration of phospholipid splitting as the factor responsible for increased permeability and block of axonal conduction induced by snake venom, *Biochim. Biophys. Acta,* 135, 669.

Condrea, E., Barzilay, M., and Mager, J. (1970). Role of cobra venom direct lytic factor and Ca^{2+} in promoting the activity of snake venom phospholipase A. *Biochim. Biophys. Acta,* 210, 65.

Currie, B. T., Oakley, D. E. and Broomfield, C. A. (1968). Crystalline phospholipase A associated with a cobra venom toxin, *Nature,* 220, 371.

Damerau, B., Lege, L., Oldigs, H. D., and Vogt, W. (1975). Histamine release, formation of prostaglandin-like activity (SRS–C) and mast cell degranulation by the direct lytic factor (DLF) and phospholipase A of cobra venom, *Naunyn–Schmiedebergs Arch. Exp. Pathol. Pharmakol.,* 287, 141.

Danvillier, P., Deltaas, G. H., Van Deenen, L. L. M., and Raulin, J. (1964). Mode d'action de la phospholipase A du venin de serpent sur les substrats élaïdisés, *C. R. Acad. Sci. Paris,* 259, 4865.

Deems, R. A. and Dennis, E. A. (1975). Characterization and physical properties of the major form of phospholipase A_2 from cobra venom (*Naja naja naja*) that has a molecular weight of 11,000, *J. Biol. Chem.,* 250, 9008.

Deems, R. A., Eaton, B. R., and Dennis, E. A. (1975). Kinetic analysis of phospholipase A_2 activity toward mixed micelles and its implications for the study of lipolytic enzymes, *J. Biol. Chem.,* 250, 9013.

De Haas, H. and Van Deenen, L. L. M. (1961). "The site of action of phospholipase A on synthetic 'mixed-acid' lecithins," in P. Desnuelle, Ed., *The Enzymes in Lipid Metabolism,* Pergamon, London, pp. 53–59.

References

De Haas, G. H. and Van Deenen, L. L. M. (1963). The stereospecific action of phospholipase A on β lecithins, *Biochim. Biophys. Acta,* 70, 469.

De Haas, G. H., Daemen, F. J. M., and Van Deenen, L. L. M. (1962). Positional specificity of phosphatide acyl hydrolase (phospholipase A), *Nature,* 196, 68.

De Haas, G. H., Slotboom, A. J., Bonsen, P. P. J., Nieuwenhuizen, W., and Van Deenen, L. L. M. (1970a). Studies on phospholipase A and its zymogen from porcine pancreas. II. The assignment of the position of the six disulfide bridges, *Biochim. Biophys. Acta,* 221, 54.

De Haas, G. H., Slotboom, A. J., Bonsen, P. P. M., Van Deenen, L. L. M., Maroux, S., Puigserver, A., and Desnuelle, P. (1970b). Studies on phospholipase A and its zymogen from porcine pancreas. I. The complete amino acid sequence, *Biochim. Biophys. Acta,* 221, 31.

Delori, P. J. (1971). Isolement, purification et étude d'une phospholipase A_2 toxique du venin de *Vipera berus, Biochimie,* 53, 941.

Delori, P. J. (1973). Purification et propriétés physico-chimiques, chimiques et biologiques d'une phospholipase A_2 toxique isolée d'un venin de serpent Viperidae: *Vipera Berus, Biochimie,* 55, 1031.

Dennis, E. A. (1973a). Phospholipase A_2 activity towards phosphatidylcholine in mixed micelles: Surface dilution kinetics and the effect of thermotropic phase transitions, *Arch. Biochem. Biophys.,* 158, 485.

Dennis, E. A. (1973b). Kinetic dependence of phospholipase A_2 activity on the detergent Triton X-100, *J. Lipid Res.,* 14, 152.

Dumarey, C., Sket, M. D., Joseph, D., and Boquet, P. (1975). Etude d'une phospholipase basique du venin de *Naja nigricollis, C. R. Acad. Sci. Paris,* Ser. D, 280, 1633.

Feldberg, W. and Lellaway, C. H. (1937). Liberation of histamine and its role in symptomatology of bee venom poisoning, *Aust. J. Exp. Biol. Med. Sci.,* 15, 461.

Gallai-Hatchard, J. and Gray, G. M. (1968). The action of phospholipase A on the plasma membranes of rat liver cells, *Eur. J. Biochem.,* 4, 35.

Ghosh, B. N. and Sarka, N. K. (1956). "Active principles of snake venoms," in E. Buckley and N. Porges, Ed., *Venoms,* American Association for the Advancement of Science, Washington, D.C., pp. 189-196.

Gubensek, F. (1976). Sepharose bound toxic phospholipase A, *Bull. Inst. Pasteur Paris,* 74, 47.

Gubensek, F. and Lapanje, S. (1974). Circular dichroism of two phospholipase A from *Vipera ammodytes* venom, *FEBS Lett.,* 44, 182.

Gul, S. and Smith, A. D. (1972). Haemolysis of washed human red cells by the combined action of *Naja naja* phospholipase A_2 and albumin, *Biochim. Biophys. Acta,* 288, 237.

Habermann, E., Walsch, P., and Breithaupt, H. (1972). Biochemistry and pharmacology of the crotoxin complex. II. Possible interrelationships between toxicity and organ distribution of phospholipase A, crotapotin and their combination, *Naunyn-Schmiedebergs Arch. Exp. Pathol. Pharmakol.,* 273, 313.

Hachimori, Y., Wells, M., and Hanahan, D. J. (1971). Observations on the phospholipase A_2 of *Crotalus atrox*: Molecular weight and other properties, *Biochemistry,* 10, 4084.

Heilbronn, E. (1969). The effect of phospholipases on the uptake of atropine and acetylcholine by slices of mouse brain cortex, *J. Neurochem.,* 16, 627.

Heilbronn, E. (1972). Action of phospholipase A on synaptic vesicles: A model for transmitter release? *Prog. Brain Res.* 36, 38.

Hölzl, J. and Wagner, H. (1968). Über die Hämolyseaktivität von Lysolecithin, das durch Lecithinspaltung mit Phospholipase A in wässrigem Milieu erhalten wird, *Z. Naturforsch,* B, 23, 449.

Howard, N. L. (1975). Phospholipase A_2 from puff adder (*Bitis arietans*) venom, *Toxicon,* 13, 21.

Huang, P. O. (1954). Prevention and reversal of succinate-cytochrome c reductase inhibition caused by cobra venom, *J. Formosan Med. Assoc.,* 53, 353.

Ibrahim, S. A., Sanders, H., and Thompson, R. H. S. (1964). The action of phospholipase A on purified phospholipids, plasma and tissue preparations, *Biochem. J.,* 93, 588.

Iwanaga, S. and Kawachi, S. (1959). Studies on snake venoms. V. column chromatography of lecithinase A in Japanese Mamushi venom (*Agkistrodon halys blomhoffii* Boie), *Yakugaku Zasshi,* 79, 582.

Joubert, F. J. (1975a). *Naja melanoleuca* (forest cobra) venom: The amino acid sequence of phospholipase A, fractions DE-I and DE-II, *Biochim. Biophys. Acta,* 379, 345.

Joubert, F. J. (1975b). *Naja melanoleuca* (forest cobra) venom: The amino acid sequence of phospholipase A, Fraction DE-III, *Biochim. Biophys. Acta,* 379, 329.

Joubert, F. J. (1975c). *Hemachatus haemachatus* (ringhals) venom: Purification, some properties and amino acid sequence of phospholipase A (fraction DE-I), *Eur. J. Biochem.,* 52, 539.

Joubert, F. J. and Van der Walt, S. J. (1975). *Naja melanoleuca* (forest cobra) venom: Purification and some properties of phospholipase A, *Biochim. Biophys. Acta,* 379, 317.

Kawauchi, S., Iwanaga, S., Samejima, Y., and Suzuki, T. (1971a). Isolation and characterization of two phospholipase A's from the venom of *Agkistrodon halys blomhoffii, Biochim. Biophys. Acta,* 236, 142.

Kawauchi, S., Samejima, Y., Iwanaga, S., and Suzuki, T. (1971b). Amino acid compositions of snake venom phospholipase A's, *J. Biochem.,* 69, 433.

Kirschmann, C., Condrea, E., Moav, N., Aloof, S., and De Vries, A. (1964). Action of snake venom on human platelet phospholipids, *Arch. Int. Pharmacodyn.,* 150, 372.

Klibansky, C., Shiloas, J., and De Vries, A. (1967). Hydrolysis of human leucocyte phospholipids by snake venoms, *Experientia,* 23, 333.

Klibansky, C., London, Y., Frenkel, A., and De Vries, A. (1968). Enhancing action of synthetic and natural basic polypeptides on erythrocyte-ghost phospholipid hydrolysis by phospholipase A, *Biochim. Biophys. Acta,* 150, 15.

Lee, Y. P. (1954). Effects of snake venoms on diphosphopyridine nucleotide cytochrome c reductase and cytochrome c oxidase, *J. Formosan Med. Assoc.,* 53, 361.

Lin, Shiau S. Y. and Lee, C. Y. (1974). Studies on the respiratory-enzyme inhibitors in Formosan elapid venoms, *J. Chin. Biochem. Soc.,* 3, 77.

Lin, Y. C. and Chang, L. T. (1957). Effect of heat treatment on the inhibition of succinate-cytochrome c reductase by cobra venom and its lecithinase A, *J. Formosan Med. Assoc.,* 56, 38.

Lo, T. B., Chang, W. C., Chiang, H. C., and Chang, C. S. (1975). "Phospholipase A from Formosan cobra (*Naja naja atra*) venom," in T. A. Bewley, M. Lin, and J. Ramachandran, Eds., *Proceedings of the International Workshop on Hormones and Proteins,* pp. 79-90.

Luzikov, V. N. and Romashina, L. V. (1972). Studies on stabilization on the oxidative phosphorylation system. II. Electron transfer-dependent resistance of succinate oxidase and NADH oxidase system of submitochondrial particles to proteinase and cobra venom phospholipase, *Biochim. Biophys. Acta,* 267, 37.

Lysz, T. W. and Rosenberg, P. (1974). Convulsant activity of *Naja naja* venom and its phospholipase A component, *Toxicon,* 12, 253.

Marinetti, G. V. (1961). *In vitro* lipid transformations in serum, *Biochim. Biophys. Acta,* 46, 468.

Marinetti, G. V. (1964). Hydrolysis of cardiolipin by snake venom phospholipase A, *Biochim. Biophys. Acta,* 84, 55.

Marinetti, G. V. (1965). The action of phospholipase A on lipoproteins, *Biochim. Biophys. Acta,* 98, 554.

Martin, J. K., Luthra, M. A., Wells, M. A., Watts, R. P., and Hanahan, D. J. (1975). Phospholipase A_2 as a probe of phospholipid distribution in erythrocyte membranes: Factors influencing the apparent specificity of the reaction, *Biochemistry,* 14, 5400.

Matsumoto, M. (1961). Studies on phospholipids. II. Phospholipase activity of *Clostridium perfringens* toxin, *J. Biochem.,* 49, 23.

McKay, D. G.. Moroz, C., De Vries, A., Csavossy, I., and Cruse, V. (1970). Action of hemorrhagin and phospholipase derived from *Vipera palestinae* venom on the microcirculation, *Lab. Invest.,* 22, 387.

Mebs, V. D. (1968). Vergleichende Enzymutersuchungen an Schlangengiften unter besonderer Berücksichtingung ihrer Casein-spaltenden Proteasen, *Hoppe-Seylers Z. Physiol. Chem.,* 349, 1115.

Menzel, D. B. and Olcott, H. S. (1964). Positional distribution of fatty acids in fish and other animal lecithins, *Biochim. Biophys. Acta,* 84, 133.

Mohamed, A. H., Kamel, A., Selim, R., and Hani-Ayobe, M. (1972). Starch gel electrophoresis of *Naja nigricollis* venom: Enzymatic and toxicity studies, *Toxicon,* 10, 7.

Moore, J. H. and Williams, D. L. (1963). A time study of the hydrolysis of lecithin by snake-venom phospholipase A, *Biochim. Biophys. Acta,* 70, 348.

References

Moore, J. H. and Williams, D. L. (1964). Some observations on the specificity of phospholipase A, *Biochim. Biophys. Acta*, 84, 41.

Nair, B. C., Nair, C., and Elliott, W. B. (1975). Action of antisera against homologous and heterologous snake venom phospholipase A_2, *Toxicon*, 13, 453.

Okuyama, H. and Najima, S. (1965). Studies on hydrolysis of cardiolipin by snake venom phospholipase A, *J. Biochem. (Tokyo)*, 57, 529.

Omori-Satoh, T., Lang, J., Breithaupt, H., and Habermann, E. (1975). Partial amino acid sequence of the basic *Crotalus* phospholipase A, *Toxicon*, 13, 69.

Pasek, M., Keith, C., Feldman, D., and Singler, P. B. (1975). Characterization of crystals of two venom phospholipase A_2, *J. Mol. Biol.*, 97, 395.

Pattnaik, M. M., Kezdy, F. J., and Scanu, A. M. (1976). Kinetic study of the action of snake venom phospholipase A_2 on human serum high density lipoprotein 3, *J. Biol. Chem.*, 251, 1984.

Petrushka, E., Quastel, J. H., and Scholefield, P. G. (1959). Role of phospholipids in oxidative phosphorylation and mitochondrial structure, *Can. J. Biochem. Physiol.*, 37, 989.

Radomski, J. L. and Deichmann, W. B. (1958). The relationship of certain enzymes in cobra and rattlesnake venoms to the mechanism of action of these venoms, *Biochem. J.*, 70, 293.

Ribeiro, A. A. and Dennis, E. A. (1973). Effect of thermotropic phase transitions of dipalmitoyl phosphatidylcholine on the formation of mixed micelles with Triton X–100, *Biochim. Biophys. Acta*, 332, 26.

Rock, C. O., and Snyder, F. (1975). Rapid purification of phospholipase A_2 from *Crotalus adamanteus* venom by affinity chromatography, *J. Biol. Chem.*, 250, 6564.

Romashina, L. V., Voznaya, N. M., Grosse, R., Rakhrmov, M. M., and Luzikov, V. N. (1972). Mechanism of the respiratory chain inactivation of cobra venom phospholipase, *Biokhimiya*, 37, 1204.

Rosenberg, P. (1973). Venoms and enzymes: Effects on permeability of isolated single electroplax, *Toxicon*, 11, 149.

Rosenberg, P. (1975). Penetration of phospholipase A_2 and C into the squid (*Loligo pealii*) giant axon, *Experientia*, 31, 1401.

Rosenberg, P. and Condrea, E. (1968). Maintenance of axonal conduction and membrane permeability in presence of extensive phospholipid splitting, *Biochem. Pharmacol.*, 17, 2033.

Rosenberg, P. and Dettbarn, W. D. (1964). Increased cholinesterase activity of intact cells caused by snake venoms, *Biochem. Pharmacol.*, 13, 1157.

Rosenberg, P. and Ehrenpreis, S. (1961). Reversible block of axonal conduction of curare after treatment with cobra venom, *Biochem. Pharmacol.*, 8, 192.

Rosenberg, P. and Hoskin, F. C. G. (1965). Penetration of acetylcholine into giant squid axons, *Biochem. Pharmacol.*, 14, 1765.

Rosenberg, P. and Ng, K. Y. (1963). Factors in venoms leading to block of axonal conduction by curare, *Biochim. Biophys. Acta*, 75, 116.

Rosenberg, P. and Podelski, T. R. (1963). Ability of venoms to render squid axons sensitive to curare and acetylcholine, *Biochim. Biophys. Acta*, 75, 105.

Rothschild, A. M. (1967). Chromatographic separation of phospholipase A from a histamine releasing component of Brazilian rattlesnake venom (*Crotalus durissus terrificus*), *Experientia*, 23, 741.

Sakhibov, D. N., Sorokin, V. M., and Yukel'son, L. Ya. (1970). Isolation of phospholipase A from the venom of the Central-Asian cobra, *Biokhimiya*, 35, 13.

Sakhibov, D. N., Yukel'son, L. Y., and Salikhov, R. (1975). Amino acid compositions of phospholipases A_2 from Central-Asian cobra venom, *Khim. Prir. Soedin.*, 11, 223.

Salach, J., Turini, P., Seng, R., Hauber, J., and Singer, T. P. (1971a). Phospholipase A of snake I. Isolation and molecular properties of isozymes from *Naja naja* and *Vipera russellii* venoms, *J. Biol. Chem.*, 246, 331.

Salach, J. I., Seng, R., Tisdale, H., and Singer, T. P. (1971b). Phospholipase A of snake venoms. II. Catalytic properties of the enzyme from *Naja naja*, *J. Biol. Chem.*, 246, 340.

Salvenmoser, F. and Kramar, R. (1970). Versuche zur Solubilisierung von Mitochondrialer Nad(P)-Transhydrogenase Mittles Schlangengift, *Enzymologia*, 40, 322.

Samejima, Y., Iwanaga, S., and Suzuki, T. (1974). Complete amino acid sequence of phospholipase A_2-II isolated from *Agkistrodon halys blomhoffii* venom, *FEBS Lett.*, 47, 348.

Sasaki, T. (1958). Chemical Studies on the poison of Formosan cobra. IV. Terminal amino acid residues, molecular weight and amino acid composition of purified lecithinase, *Yakugaku Zasshi,* 78, 516.

Sasaki, T. (1959). Chemical studies on the poison of Formosan cobra. VI. Essential group of lecithinase, *Yakugaku Zasshi,* 79, 35.

Shiloah, J. (1974). Phospholipase B activity of purified phospholipase A from *Vipera palestinae* and *Naja naja* snake venoms, *Israel J. Chem.,* 12, 605.

Shiloah, J., Klibansky, C., and De Vries, A. (1973a). Phospholipase isoenzymes from *Naja naja* venom. I. Purification and partial characterization, *Toxicon,* 11, 481.

Shiloah, J., Klibansky, C., De Vries, A., and Berger, A. (1973b). Phospholipase B activity of a purified phospholipase A from *Vipera palestinae* venom, *J. Lipid Res.,* 14, 267.

Slotta, K. H., Vick, J. A., and Ginsberg, N. J. (1971). "Enzymatic and toxic activity of phospholipase A," in A. De Vries and E. Kochva, Eds., *Toxins of Animal and Plant Origin,* Vol. I, Gordon and Breach, New York, pp. 401-418.

Smith, A. D., Gul, S., and Thompson, R. H. S. (1972). The effect of fatty acids and of albumin on the action of a purified phospholipase A_2 from cobra venom on synthetic lecithins, *Biochim. Biophys. Acta,* 289, 147.

Trethewie, E. R. (1939). Comparison of haemolysis and liberation of histamine by two Australian snake venoms, *Aust. J. Exp. Biol. Med. Sci.,* 17, 145.

Tsao, F. H. C., Keim, P. S., and Heinrikson, R. L. (1975). *Crotalus adamanteus* phospholipase A_2-α: Subunit structure, NH_2-terminal sequence, and homology with other phospholipases, *Arch. Biochim. Biophys.,* 167, 706.

Tu, A. T. (1974). Sea snake venoms and neurotoxins, *J. Agr. Food Chem.,* 22, 36.

Tu, A. T. and Passey, R. B. (1971). "Phospholipase A from sea snake venom and its biological properties," in A. De Vries and E. Kochva, *Toxins of Animal and Plant Origin,* Vol. I, Gordon and Breach, New York, pp. 419-436.

Tu, A. T., Passey, R. B., and Toom, P. M. (1970). Isolation and characterization of phospholipase A from sea snake, *Laticauda semifasciata,* venom, *Arch. Biochem. Biophys.,* 140, 96.

Uwatoko-Setoguchi, Y. (1970). Studies on sea snake venom. VI. Pharmacological properties of *Laticauda semifasciata* venom and purification of toxic components, acid phosphomonoesterase and phospholipase A in the venom, *Acta Med. Univ. Kagoshima.,* 12, 73.

Uwatoko-Setoguchi, Y. and Ohbo, F. (1969). Studies on sea snake venom. V. Some properties of phospholipase A in *Laticauda semifasciata* venom, *Acta. Med. Univ. Kagoshima.,* 11, 139.

Vidal, J. C. and Stoppani, A. O. M. (1971). Isolation and purification of two phospholipase A from *Bothrops* venoms, *Arch. Biochem. Biophys.,* 145, 543.

Vidal, J. C., Badano, B. N., Stoppani, A. O. M., and Boveris, A. (1966). Inhibition of electron transport chain by purified phospholipase A from *Bothrops neuwiedii* venom, *Mem. Inst. Butantan,* 33, 913.

Vidal, J. C., Cattaneo, P., and Stoppani, A. O. M. (1972). Some characteristic properties of phospholipase A_2 from *Bothrops neuwiedii* venom, *Arch. Biochem. Biophys.,* 151, 168.

Viljoen, C. C., Schabort, J. C., and Botes, D. P. (1974). *Bitis gabonica* venom: A kinetic analysis of the hydrolysis by phospholipase A_2 of 1,2-dipalmitoyl-sn-glycero-3-phosphoryl-choline, *Biochim. Biophys. Acta,* 360, 156.

Viljoen, C. C., Botes, D. P., and Schabort, J. C. (1975). Spectral properties of *Bitis gabonica* venom phospholipase A_2 in the presence of divalent metal ion, substrate and hydrolysis products, *Toxicon,* 13, 343.

Volwerk, J. J., Pieterson, W. A., and De Haas, G. H. (1974). Histidine at the active site of phospholipase A_2, *Biochemistry,* 13, 1446.

Wahlström, A. (1971). Purification and characterization of phospholipase A from the venom of *Naja nigricollis, Toxicon,* 9, 45.

Wakui, K. and Kawachi, S. (1961). Properties of the two lecithinase A in snake venom, *J. Pharm. Soc. Jap.,* 81, 1394.

Wells, M. A. (1971a). Spectral peculiarities of the monomer-dimer transition of the phospholipase A_2 of *Crotalus adamanteus* venom, *Biochemistry,* 10, 4078.

References

Wells, M. A. (1971b). Evidence of O-acyl cleavage during hydrolysis of 1,2-diacyl-sn-glycero-3-phosphorylcholine by the phospholipase A_2 of *Crotalus adamanteus* venom, *Biochim. Biophys. Acta*, 248, 80.

Wells, M. A. (1972). A kinetic study of the phospholipase A_2 (*Crotalus adamanteus*) catalyzed hydrolysis of 1,2-dibutyryl-sn-glycero-3-phosphorylcholine, *Biochemistry*, 11, 1030.

Wells, M. A. (1973a). Spectral perturbations of *Crotalus adamanteus* phospholipase A_2 induced by divalent cation binding, *Biochemistry*, 12, 1080.

Wells, M. A. (1973b). Effects of chemical modification on the activity of *Crotalus adamanteus* phospholipase A_2: Evidence for an essential amino group, *Biochemistry*, 12, 1086.

Wells, M. A. and Hanahan, D. J. (1969). Studies on phospholipase A. I. Isolation and characterization of two enzymes from *Crotalus adamanteus* venom, *Biochemistry*, 8, 414.

Woelk, H. and Debuch, H. (1971). Die Wirkung der Phospholipase A aus Schlangengift auf Phosphatidylcholin, Phosphatidyläthanolamin und auf die entsprechenden Plasmalogene, *Hoppe-Seylers Z. Physiol. Chem.*, 352, 1275.

Woelk, H. and Peiler-Ichikawa, K. (1974). The action of phospholipase A_2 purified from *Crotalus atrox* venom on specifically labeled 2-acyl-l-alk-l'-enyl- and 2-acyl-l-alkyl-sn-glycero-3-phosphorylcholine, *FEBS. Lett.*, 45, 75.

Wu, T. W. and Tinker, D. O. (1969). Phospholipase A_2 from *Crotalus atrox* venom. I. Purification and some properties, *Biochemistry*, 8, 1558.

Yang, C. C., Su, C. C., and Chen, C. J. (1959). Biochemical studies on the Formosan snake venoms. V. The toxicity of Hyappoda (*Agkistrodon acutus*) venom, *J. Biochem. (Tokyo)*, 46, 1209.

Yang, C. C., Kao, K. C., and Chiu, W. C. (1960). Biochemical studies on the snake venoms. VIII. Electrophoretic studies of banded krait (*Bungarus multicinctus*) venom and the relation of toxicity with enzyme activities, *J. Biochem. (Tokyo)*, 48, 714.

3 Phosphodiesterase

Orthophosphoric Diester Phosphohydrolase   EC 3.1.4.1

1 EXONUCLEASE   64

   1.1 Types of Nucleotides and Chain Length, 65
   1.2 Position of Phosphates, 67
   1.3 Types of Bases, 67
   1.4 Types of Sugars, 68
   1.5 Synthetic Substrates, 69
   1.6 Endonucleolytic Activity of Exonuclease, 69
   1.7 Isolation, 70
   1.8 Chemical Properties, 71
   1.9 Occurrence, 71

2 HYDROLYSIS OF ATP   72

3 ENDOPOLYNUCLEOTIDASES (RNase and DNase)   73

   3.1 RNase, 73
   3.2 DNase, 74

   References   74

Snake venoms contain many enzymes that hydrolyze phosphomonoester and phosphodiester bonds. Laskowski (1971) stated in his review article that so-called venom phosphodiesterase should be termed "venom exonuclease," and "phosphodiesterase" should be used as a general name for all enzymes attacking diesterified phosphate. I am in complete agreement with Dr. Laskowski's view and will follow his suggestion in this chapter.

1 EXONUCLEASE

Among venom phosphodiesterases, exonuclease is one of the most extensively studied enzymes and is used commonly for the degradation of nucleic acid. Substrates for this enzyme are DNA (native double strands or denatured), RNA, or any polynucleotide

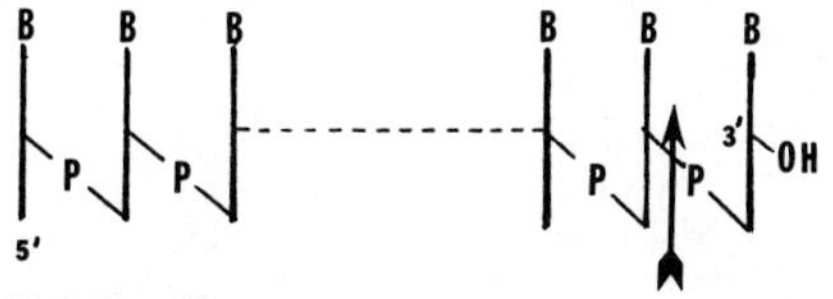

1 Exonuclease

Figure 3.1 Specificity of snake venom exonuclease. The enzyme hydrolyzes phosphodiester linkage from the 3' end, yielding nucleoside-5'-monophosphate.

chains. Exonuclease removes successive mononucleotide units from the polynucleotide chain in stepwise fashion, starting from the end that bears a free 3'-hydroxyl group. The specificity is illustrated in Fig. 1.

Exonuclease yields pG, pC, pA, pU, and pT. Variations in the types of bases or sugars have no effect on the reaction rate (Razzell and Khorana, 1958, 1959a, b). The enzyme can hydrolyze single-stranded as well as double-stranded nucleic acids. Indeed, there is a report that native thymus DNA is initially degraded more rapidly than heat-denatured DNA (Björk, 1967).

1.1 Types of Nucleotides and Chain Length

Snake venom exonucleases can hydrolyze almost any type of polynucleotide of any chain length. However, there are small differences, depending on type of nucleotide, chain length, and state (native vs. denatured).

Björk (1967) observed that *Hemachatus haemachatus* exonuclease exhibited the same activity toward tri, tetra-, penta-, and hexanucleotides at high ionic strength and at high substrate concentration. However, when the Mg(II) concentration was lowered, enzymic activity was dependent on the chain length of the substrate.

Venom enzyme can hydrolyze liver ribosomal RNA completely to yield pG, pC, pA, and pU (Hadjiolov et al., 1967). In addition, cyclic oligonucleotides that do not have a terminal group can be hydrolyzed slowly by venom exonuclease (Razzell and Khorana, 1959a).

The exonuclease can also hydrolyze cyclic nucleoside monophosphate. When cyclic 3', 5'-AMP is used, the product is 5'-AMP only. Thus exonuclease hydrolyzes the 3' linkage as follows:

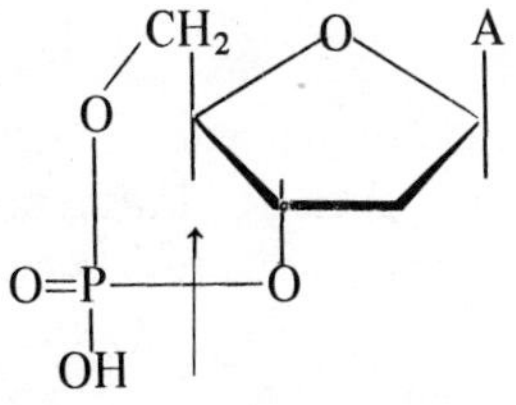

When cyclic 2', 3'-AMP is used, the product is 3'-AMP only (Suzuki et al., 1960d).

Exonuclease obtained from the venoms of *Crotalus adamanteus* and *C. durissus terrificus* hydrolyze 3',5'-AMPs (adenosine 3,5'-cyclic phosphoriothioate) to 5'-AMP, with adenosine 5'-phosphorothioate (AMPs) as an intermediate (Eckstein and Bär, 1969). This is illustrated in Scheme 7.

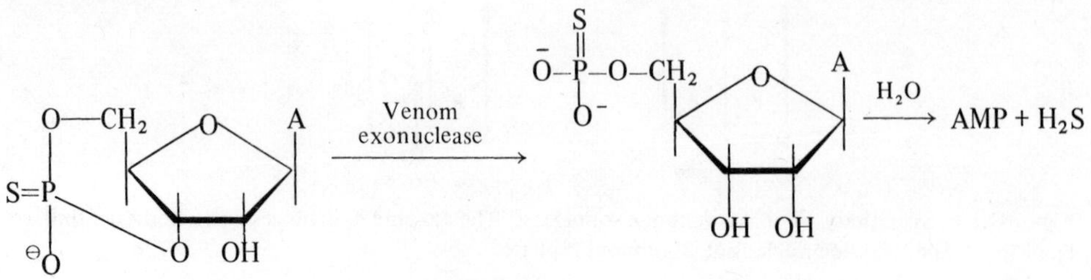

Scheme 7

It is not surprising that venom exonuclease hydrolyzes tRNA with a free hydroxyl group at the 3' end. However, it will be extremely interesting to determine whether the enzyme can hydrolyze tRNA esterized at the 3' end. The function of tRNA is to pick up an amino acid at the 3' end. Yot et al. (1968) studied aminoacyl-tRNA using exonuclease of *Crotalus adamanteus* venom. Their interesting results are summarized in Scheme 8.

$$\text{tRNA--CpCpA--}O\text{--Val--NHCOCH}_3 \xrightarrow{\text{Venom exonuclease}} \text{--CpC}$$

Partial structure of
3'(2')-*O*-(*N*-acetyl-L-Val)-tRNA

$+$

3'(2')-*O*-(*N*-acetyl-L-Val)-AMP
$\downarrow$ Alkaline phosphatase
P_i + 3'(2')-*O*-(*N*-acetyl-L-Val)-adenosine

$$3'(2')\text{-}O\text{-L-Glu-tRNA} \xrightarrow{\text{Venom exonuclease}} \text{--CpC} + 3'(2')\text{-}O\text{-Glu-AMP}$$

$$3'(2')\text{-}O\text{-L-Lys-tRNA} \xrightarrow{\text{Venom exonuclease}} 3'(2')\text{-}O\text{-Lys-AMP} + \text{--CpC}$$

$$3'(2')\text{-}O\text{-(}N\text{-acetyl-Phe-Val)-tRNA} \xrightarrow{\text{Venom exonuclease}} \text{--CpC}$$

$+$

3'(2')-*O*-(*N*-acetyl-Phe-Val)-AMP

Scheme 8

It is clear that the esterification of the 3' end of tRNA does not have any effect on the hydrolysis of tRMA by venom exonuclease. Furthermore, side chains of the esterified amino acids have no significant effect, whether they are acidic peptides, basic peptides, or oligo peptides.

However, the presence of an aminoacyl group at the 3'(2') terminus of tRNA decreases the rate of hydrolysis of the terminal AMP by a factor of 5. The product of this reaction is a 3'(2')-*O*-aminoacyl-AMP. Esterification of the 2'(3')-OH of the terminal adenosyl residue is the primary reason for the decrease in rate of hydrolysis (Miller et al., 1970).

Exonuclease immobilized on Concanavalin-A–Sepharose can hydrolyze the substrate (Sulkowski and Laskowski, 1974). The immobilized enzyme hydrolyzed polyA completely. Conversely, immobilized DNA can be hydrolyzed with venom exonuclease (Pritchard and Eichinger, 1974).

1 Exonuclease

It is known that a number of snake venoms hydrolyze NAD into nicotinamide mononucleotide and 5'-AMP. This may be due to the action of venom exonuclease rather than of nucleotide pyrophosphatase (Suzuki et al., 1960a; Rakitzis, 1972). However, the splitting of NAD to nicotinamide and adenosine diphosphate ribose is believed to be due to an independent enzyme, nicotinamide adenine nucleotidase (Boman, 1959; Suzuki et al., 1960b).

The snake venom enzyme cannot recognize nucleotide units in the syn conformation. For this study Ogilvie and Hruska (1976) synthesized dinucleotides containing 6-methyldeoxyurdine and inosine. Spleen exonuclease degraded all nucleotides tested, whereas venom caused very little degradation of nucleotides having 6-methyldeoxyuridine in the 3'-terminal position.

1.2 Position of Phosphates

The presence of the phosphoryl group in the 5' position exerts a favorable effect on the action of rattlesnake venom phosphodiesterase. The removal of the terminal 5'-phosphoryl group produces a striking decrease in the rate of hydrolysis (Privat de Garilhe and Laskowski, 1956). Nikol'skaya et al. (1965), however, concluded that the 5'-terminal phosphate oligonucleotide is hydrolyzed at a much slower rate than the substrate containing a 5'-terminal hydroxyl.

A study using synthetic nucleotides $(Ap)_n Cp$ and $(Ap)_n C(N_{ave} = 15)$ and the enzyme isolated from *Crotalus adamanteus* venom indicated that a free terminal 3'-OH is indeed essential for enzyme activity (Hadjiolov et al., 1966). Richards and Laskowski (1969a) further investigated the role of the 3' end in the hydrolysis of polynucleotide by venom exonuclease. The resistance to hydrolysis conferred by a 3'-phosphoryl group is due to negative charge, particularly a double charge. They demonstrated that an oligonucleotide with a 3'-phosphate can be hydrolyzed completely at a pH between 5 and 8. By using this property, they were able to prepare mononucleotide diphosphates (pNp). At lower pH, the 3'-phosphate becomes protonated, thereby allowing enzymatic attacks and subsequent hydrolysis of the nucleotide (Richards and Laskowski, 1969b). This finding led to the conclusion by Lakowski (1971) that earlier experimentation indicating that nucleotides bearing a 3'-monophosphoryl group were totally resistant to hydrolysis (Felix et al., 1960) was erroneous.

1.3 Types of Bases

The types of normal bases (A, G, U, C, T) in RNA or DNA usually have little effect on the rate of hydrolysis. However, the enzyme cannot hydrolyze the internucleotide bond adjacent to abnormal nucleotides such as hydroxymethylcytidylic acid or glucosylated derivatives present in phage T_2 DNA (Nikol'skaya et al., 1965). Venom exonuclease cannot hydrolyze the linkage between TMP and pseudouridylic acid, as studied with TpψCpG (Venkstern, 1966).

Since tRNA can be hydrolyzed completely with *Crotalus adamanteus* exonuclease (Keller, 1964), the presence of rare bases does not affect the exonuclease activity. Normally tRNA contains such rare bases as 7-methylguanine, N^2-dimethylguanine, 5,6-dihydrouridine, and pseudouridine. The tRNA from *Escherichia coli* can be hydrolyzed by *C. adamantues* exonuclease (Miller et al., 1970). The terminal AMP is removed 4 times faster than the first CMP and 65 times faster than the second CMP.

However, Vasilenko (1964) reported a different result: that tRNA was hydrolyzed approximately 25% by cobra venom exonuclease. The rate of release of nucleotides depends on the types of bases present. When DNA is irradiated with ultraviolet light, thymine bases form dimers. When DNA is used for enzymatic hydrolysis with venom exonuclease, there are oligonucleotide sequences that are resistant to enzymatic degradation. Most of the resistant oligonucleotides contain thymine dimers in sequences whose general structure is pXp$\overline{\text{TpT}}$ (Setlow et al., 1964).

1.4 Types of Sugars

All three linkages, $2'-5'$, $3'-5'$, and $5'-5'$, in dinucleotides are hydrolyzed by exonuclease purified from the venom of *Crotalus adamanteus* (Richards et al., 1967). The maximum velocity (V_{max}) decreases in this order: $3'-5' > 5'-5' > 2'-5'$. Moreover, substrates containing arabinose are equally as good as those containing ribose or deoxyribose. Hydrolysis is slowest with compounds that lead to the formation of $5'$-arabinose mononucleotides. Compounds that lead to the formation of arabinose nucleosides are hydrolyzed relatively faster. The types of substrates and their hydrolysis products are listed in Scheme 9, where r = ribose, d = deoxyribose, and a = arabinose.

$$aC2'p5'rA \longrightarrow aC + prA$$
$$aC3'p5'rA \longrightarrow aC + prA$$
$$aC5'p5'rA \longrightarrow aC + prA$$
$$rA2'p5'rA \longrightarrow rA + paC$$
$$rA3'p5'aC \longrightarrow rA + paC$$
$$rC3'p5'rC \longrightarrow rC + prC$$
$$rC5'p5'rC \longrightarrow rC + prC$$
$$aC3'p5'aC \longrightarrow aC + paC$$
$$aC5'p5'aC \longrightarrow aC + paC$$
$$rC5'p5'dC \longrightarrow \begin{cases} rC + pdC \\ dC + prC \end{cases}$$
$$aC5'p5'rC \longrightarrow aC + prC$$
$$aC5'p5'dC \longrightarrow aC + pdC$$
$$\text{Cytosine arabinoside } 5'\text{-}O\text{-phenylphosphate} \longrightarrow C_6H_5 + paC$$
$$\text{Cytosine arabinoside } 5'\text{-}O\text{-methylphosphate} \longrightarrow CH_3 + paC$$

Scheme 9

For further study of the specificity of exonuclease for the substrate sugar, the trinucleoside diphosphate containing arabinose, a-CpCpC (all sugars are arabinose), was synthesized by Wechter et al. (1968). Venom exonuclease hydrolyzed the substrate completely. The reaction proceeds in two steps:

$$\text{a-CpCpC} \longrightarrow \text{a-CpC + a-pC}$$
$$\text{a-CpC} \longrightarrow \text{a-C + a-pC}$$

Apparently, venom exonuclease is "blind to sugar" for its enzymatic hydrolysis. This is quite a contrast to micrococcal nuclease, which does not hydrolyze phosphodiester linkages involving arabinose.

1 Exonuclease

The 2'→5'-linked dinucleoside phosphate is hydrolyzed very slowly with venom exonuclease. For example, CapCa and CapdU cannot be hydrolyzed in 2 hr; these compounds require more than 24 hr for complete hydrolysis (Wechter, 1967).

1.5 Synthetic Substrates

A number of synthetic substrates have been employed for venom phosphodiesterase activity assays. Chromophoric synthetic substrates are conveniently employed to measure enzymatic reaction rates directly by optical methods. p-Nitrophenyl phosphate diesters and 3-pyridyl diesters are frequently used for this purpose (Razzell and Khorana, 1959a; Wigler, 1963). All of these synthetic substrates have similarities in structure (see Scheme 10).

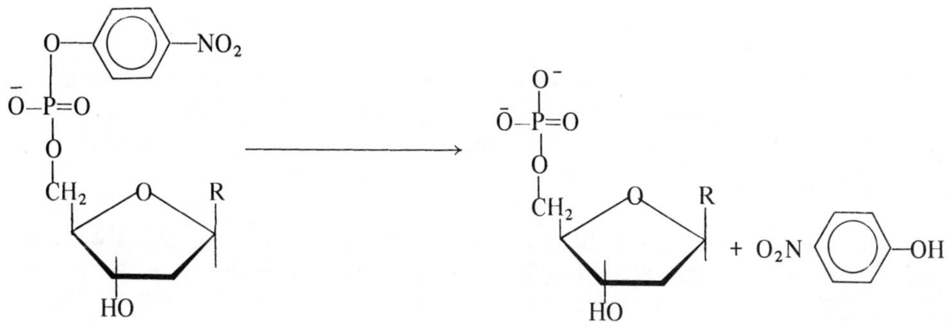

Scheme 10

1.6 Endonucleolytic Activity of Exonuclease

It is well documented that venom exonuclease hydrolyzes a polynucleotide chain beginning from the 3'-OH terminal and liberates 5'-nucleotide. However, an important question is whether or not exonuclease also has some endonucleolytic activity. As early as 1958, Razzell and Khorana (1958, 1959b) observed that venom exonuclease is capable of hydrolyzing cyclic oligonucleotides in which the 3'-hydroxyl group is in an ester linkage with the 5'-phosphate group at the other end of the chain. The compounds used were cyclo-pTpT, cyclo-pTpTpT, and cyclo-pTpTpTpT. These oligonucleotides, which do not have any terminal group, were hydrolyzed by venom exonuclease. No intermediates could be detected during their degradation, and pT was the only product. The endonucleolytic activity of venom exonuclease was also recognized by Björk (1967), using phage T_2 DNA.

Endonucleolytic activity by exonuclease isolated from *Vipera russellii* venom was also observed on wheat germ and RNA by McLennan and Lane (1968a,b). Wheat embryo ribosomal RNA contains about 1300 nucleotide residues. About 5 to 10 endonucleolytic breaks occurred during the first hour of hydrolysis. As the chain length decreased, the rate of endonucleolytic activity also dropped. It appears that endonucleolysis by venom exonuclease becomes more active with increasing polynucleotide chain length. Another very interesting finding is that the proportion of the minor component, pseudouridylate, is relatively high among the 5'-linked termini of the fragments produced by venom exonuclease. This may suggest that endonucleolytic scission occurs preferentially in regions of the RNA molecules rich in pseudouridylate. Laskowski (1971) stated in a review article that it is not clear whether the endonucleolysis is due to an intrinsic catalytic property of pure exonuclease or to the presence of other enzymes in venom exonuclease preparations. When RNA is digested with pancreatic RNase, the "RNA core" left behind is resistant to further RNase action. However, when purified exonuclease of *Crotalus adamanteus* is used, complete hydrolysis of tRNA occurs (Keller, 1964). Digestion of such an oligonucleotide core has been attributed to the endonucleolytic action of venom exonuclease (Felix et al., 1960).

The endonuclease activity of cobra exonuclease can be inhibited in the presence of Zn(II) without affecting the exonuclease activity (Vasilenko, 1964). When phage MS_2 RNA is hydrolyzed with *Vipera lebetina* exonuclease for a limited period, the composition of liberated 5'-nucleotides does not correspond with the theoretical composition of nucleotides that should have been released from the 3'-terminal. This may be due to the endonucleolytic activity of venom exonuclease, liberating nucleotides from the center portion of RNA by an endonucleolytic split (Jansone et al., 1975).

1.7 Isolation

Snake venom contains more than one exonuclease. Venom of *Crotalus adamanteus* has three different exonucleases whose activities are parallel to those of DNase (Boman and Kaletta, 1957). Two different types of exonucleases were observed in the venom of *Naja naja atra* (Suzuki and Iwanaga, 1958a, b). Suziki et al. (1960a, b) isolated three different fractions of exonuclease from the venom of *Agkistrodon halys blomhoffi*.

Snake venom exonuclease has been isolated by many investigators, notably Laskowski. Normally isolation is achieved through the use of ion-exchange and Sephadex column chromatography (Privat de Garilhe and Laskowski, 1955; Suzuki et al., 1960a, b; Laskowski, 1966). The isolation and enzymatic properties of venom exonuclease have been reviewed by Heppel and Rabinowitz (1958); therefore no attempt will be made here to describe these techniques. More recently Laskowski (1966, 1967, 1971) and Razzell (1961) have summarized the methods of assay and preparation of exonuclease from snake venoms. In this section, I would like to describe a more unconventional method of preparation.

Recently, an affinity chromatography technique was used for purification. The enzyme was isolated from the venom of *Bothrops atrox* by the use of O-(4-nitrophenyl)-O'-phenylthiophosphate ester coupled to activated Sepharose (Frischauf and Eckstein, 1973). 3',5'-Thiophosphates are resistant to action by snake venom. The compound O-(4-nitrophenyl)-O'-phenylthiophosphate

is also resistant to snake venom exonuclease and is a competitive inhibitor with K_i = 10 μM for this enzyme.

However, affinity chromatography alone does not give complete purification. Eckstein and Frischauf (1974) recommend using a combination of phosphocellulose chromatography followed by affinity chromatography.

Immobilized phosphodiesterase in a Concanavalin-A—Sepharose column was used for purification of the enzyme because endonuclease and nonspecific phosphatase could be eluted without adsorption (Dolaphchiev et al., 1974). The NAD can be immobilized on a Sepharose 4B column. Exonuclease can be eluted without adsorption, whereas 5'-nucleotidase is strongly adsorbed (Tatsuki et al., 1975).

1 Exonuclease

Venom exonuclease can be obtained commercially. The preparation of the commercially available enzyme follows the method of Sinsheimer and Koerner (1952) and Koerner and Sinsheimer (1957). The enzyme is 99.9% free of 5'-nucleotidase activity, but for some applications the remaining 0.1% can be a source of trouble.

Several methods can be used to inactivate the contaminating 5'-nucleotidase. For example, it can be selectively inactivated with $ZnCl_2$, a method used for purification of exonuclease from viper venom (Nikol'skaya et al., 1963). A method using a Dowex 50 column was proposed by Keller (1964), and a simple method was devised by Sulkowski and Laskowski (1971). They recommend incubating a commerical preparation of venom exonuclease at 37° and pH 3.6 for 3 hr. The contaminating 5'-nucleotidase activity is reduced by 100- to 1000-fold, without loss of exonuclease.

The exonuclease was isolated from *Crotalus* spp. venom by Philipps (1975). The molecular weight is 115,000. The enzyme is relatively stable and can be stored at 4° in the presence of Mg(II) and serum albumin for years. However, at 75°C inactivation of the enzyme takes place in 4 min. A relatively stable exonuclease was also isolated from the venom of *Bothrops atrox* (Philipps, 1976).

1.8 Chemical Properties

Despite the extensive use of venom exonuclease in nucleic acid research, very little is known about its chemical properties because of a scarcity of the enzyme preparation due to the high price of snake venoms. Venom exonuclease is heat labile. The enzyme isolated from *Bothrops atrox* venom is inactivated at temperatures higher than 50°C. The enzyme is stable in the pH range of 6 to 9, but outside these pH limits it is rapidly inactivated (Björk, 1963).

1.9 Occurrence

Exonuclease (phosphodiesterase) is widely distributed in a variety of venoms from all five families of venomous snakes. Venoms from the following species have been found to contain exonuclease:

Hydrophiidae

Enhydrina schistosa	Tu and Toom, 1971
Laticauda semifasciata	Setoguchi et al., 1968

Elapidae

Bungarus fasciatus	Mebs, 1968; Lu and Lo, 1974
Dendroaspis angusticeps	Mebs, 1968
Hemachatus hemachatus	Björk and Boman, 1959
Naja haje	Mebs, 1968
N. naja	Mebs, 1968
N. naja atra	Suzuki and Iwanaga, 1958b; Chiu, 1960; Brisbois et al., 1968
N. naja siamensis	Suzuki and Iwanaga, 1958b
N. nigricollis	Richards et al., 1965; Mebs, 1968
N. nivea	Pfleiderer and Ortanderl, 1963; Mebs, 1968
N. oxiana	Nigmatov et al., 1972
Ophiophagus hannah	Suzuki and Iwanaga, 1958; Mebs, 1968
Pseudechis collettii	Mebs, 1968

Viperidae

Bitis gabonica	Mebs, 1968
Vipera russellii	Mebs, 1968

Crotalidae

Agkistrodon acutus	Suzuki and Iwanaga, 1958a; Yang et al., 1959
A. bilineatus	Kocholaty et al., 1971
A. halys blomhoffii	Suzuki and Iwanaga, 1958a; Suzuki et al., 1960a; Gafurov and Rasskazov, 1972
A. piscivorus	Richards et al., 1965; Kocholaty et al., 1971
A. rhodostoma	Kocholaty et al., 1971
Bothrops atrox	Richards et al., 1965
B. atiox asper	Kocholaty et al., 1971
B. jararaca	Mebs, 1968
Crotalus adamanteus	Richards et al., 1965; Kocholaty et al., 1971
C. atrox	Richards et al., 1965; Brown, 1966; Mebs, 1968; Kocholaty et al., 1971
C. durissus cumanensis	Rodriguez et al., 1974
C. durissus durissus	Kocholaty et al., 1971
C. durissus terrificus	Mebs, 1968; Kocholaty et al., 1971
C. horridus horridus	Kocholaty et al., 1971
C. scutulatus	Kocholaty et al., 1971
C. viridis helleri	Kocholaty et al., 1971
C. viridis viridis	Kocholaty et al., 1971
Lachesis muta	Yarleque and Campås, 1973
Sistrurus miliarius barbouri	Kocholaty et al., 1971
Trimeresurus elegans	Kimura and Nakagawa, 1965
T. flavoviridis	Kimura and Nakagawa, 1965; Kocholaty et al., 1971
T. gramineus	Suzuki and Iwanaga, 1958a
T. mucrosquamatus	Suzuki and Iwanaga, 1958a

Colubridae

Leptodeira annulata	Mebs, 1968

2 HYDROLYSIS OF ATP

When ATP is incubated with snake venoms, adenosine, P_i (orthophosphate), and PP (pyrophosphate) can be obtained. This is due to the actions of two enzymes in snake venom (Johnson et al., 1953) as shown in Scheme 11.

$$\text{ATP} \longrightarrow \text{PP} + \text{AMP} \qquad (1)$$
$$\text{AMP} \longrightarrow P_i + \text{Adenosine} \qquad (2)$$

Scheme 11

There are important questions concerning the identity of the enzyme that hydrolyzes ATP to PP and AMP (reaction 1). Is this hydrolysis due to an ATPase or to venom exonuclease activity? The ATPase of most biological systems hydrolyzes the γ-phosphate, producing ADP + P_i. If the ATP-cleaving enzyme in snake venoms is an ATPase, then venom ATPase is unique. Fellig (1955) confirmed that pyrophophosphate was released by incubating ATP with *Crotalus adamanteus* venom and concluded that the responsible enzyme is an ATPase. However, he observed that UTP also generated PP in the same system.

Some investigators have expressed doubts about the presence of an ATPase in venoms (Björk and Boman, 1959; Boman, 1959; Pfleiderer and Ortanderl, 1963; Suzuki, 1966). They propose that the hydrolysis of ATP is due to an exonuclease present in snake venoms rather than to a specific ATPase.

On the other hand, venom ATPase and exonuclease activities respond differently toward Co(II), Mg(II), Zn(II), pH, and ionic strength (Pereira Lima et al., 1971). This constitutes only indirect evidence, however, and the differentiation of these enzymes should eventually be based on their isolations.

Dolapchiev et al. (1974) observed that during the preparation of exonuclease ATPase could not be freed from exonuclease activity. Moreover, the two activities paralleled each other during the purification steps. They concluded that both activities are the intrinsic property of the same enzyme molecule.

Venom exonuclease has also been demonstrated to hydrolyze GTP and produce 5'-GMP and pyrophosphate. From all this evidence it seems reasonable to conclude that there is no separate enzyme ATPase in snake venoms. Rather, the hydrolysis of ATP is associated with the intrinsic nature of venom exonuclease.

3 ENDOPOLYNUCLEOTIDASES (RNase and DNase)

There has been some question whether snake venoms contain true RNase or DNase. Privat de Garilhe and Laskowski (1955) tested their purified exonuclease from *Crotalus adamanteus* venom on DNA. They found that the purified enzyme hydrolyzed both calcium [bis(*p*-nitrophenyl)phosphate] and oligonucleotides, but that no action could be observed toward a highly polymerized DNA. Thus it was believed that venom exonuclease is not a DNase. However, this conclusion is not above challenge. Boman and Kaletta (1956) partially purified two fractions of exonucleases from *C. adamanteus* venom and found that exonuclease activity assayed on calcium [bis(*p*-nitrophenyl)phosphate] and DNase activities assayed on DNA at the same fractions. The striking parallelism even after the fractions were subjected to heat denaturation led Boman and Kaletta (1956) to conclude that the same enzyme is responsible for the two types of enzymatic reactions.

Some investigators use whole venoms for the hydrolysis of RNA and DNA. In such cases, it is not possible to determine whether the enzyme is RNase or DNase. The result simply indicates that snake venoms contain endonuclease.

Since it is not settled yet whether the hydrolysis of RNA or DNA is due to the specific RNase or DNase or due to non-specific venom endonuclease, the data were summarized and are presented here without the author's comment.

3.1 RNase

An endopolynucleotidase, RNase, is present in snake venoms but in very small amounts. Venom RNase (*Vipera russellii*) has preferential specificity toward pyrinidine containing

PypA bonds in ribonucleate chains (McLennan and Lane, 1968a, b). The RNase has a pH optimum of 7 to 9 when ribosomal RNA is used as a substrate.

Chemically modified tRNA can still be hydrolyzed by *Naja oxiana* RNase. Babkina et al. (1971) modified tRNA with N-cyclohexyl-N-[β-(4-methylmorpholinium)-ethyl]-carbodiimide p-toluenesulfonate and found that the bonds adjoining modified guanosine residues showed relatively increased stability but did not prevent hydrolysis.

RNase was purified from *N. oxiana* venom (Vasilenko and Rait, 1975). The enzyme has a molecular weight of 15,900 and is homogeneous on polyacrylamide gels.

3.2 DNase

The venom of *Crotalus adamanteus* contains potent DNase II with a pH optimum of 5 (Haessler and Cunningham, 1957). Whereas DNase II requires no divalent metal ions for its activation, DNase I, such as that of bovine pancreas, requires Mg(II).

Endonuclease was purified approximately 1000-fold from the venom of *Bothrops atrox* (Georgatsos and Laskowski, 1962). The enzyme could hydrolyze both RNA and DNA at similar rates. It has an optimal activity at pH 5.0 and requires no Mg(II). It acts on DNA as an endonuclease and produces predominantly tri- or higher oligonucleotides, all of which terminate in 3'-monoesterified phosphate. At the early stages of digestion d-GpGp is the most susceptible bond. As digestion progresses, the specificity in respect to the adjacent bases decreases, and the length of the substrate becomes more significant.

Venom of *Crotalus adamanteus* contains two DNases (Laskowski et al., 1957). There are two pH optima (5 and 9) for the action of DNase. The endonuclease with an optimum at pH 5 was observed by Haessler and Cunningham (1957) and Georgatsos and Laskowski (1962). As the purification proceeds, the optimum at pH 5 disappears.

REFERENCES

Babkina, G. T., Knorre, D. G., and Malygin, E. G. (1971). Effects of cobra venom RNase on tRNA, modified with water-soluble carbodiimide, *Mol. Biol.*, **5**, 102.

Bhattacharya, K. L. (1953). Effect of snake venoms on coenzyme I, *J. Indian Chem. Soc.*, **30**, 685.

Björk, W. (1963). Purification of phosphodiesterase from *Bothrops atrox* venom, with special consideration of the elimination of monophosphatases, *J. Biol. Chem.*, **238**, 2487.

Björk, W. (1967). Interactions of snake venom phosphodiesterase with DNA and DNA fragments, *Ark. Kemi*, **27**, 515.

Björk, W. and Boman, H. G. (1959). Fractionation of venom from the ringhals cobra, *Biochim. Biophys. Acta*, **34**, 503.

Boman, H. G. (1959). On the specificity of the snake venom phosphodiesterase, *Ann. N.Y. Acad. Sci.*, **81**, 800.

Boman, H. G. and Kaletta, U. (1956). Identity of the phosphodiesterase and deoxyribonuclease in rattlesnake venom, *Nature*, **178**, 1394.

Boman, H. G. and Kaletta, U. (1957). Chromatography of rattlesnake venom; A separation of three phosphodiesterases, *Biochim. Biophys. Acta*, **24**, 619.

Brisbois, L., Delori, P., and Gillo, L. (1968). Venins de serpents: Fractionnement d'un venin de cobra (*Naja naja atra*) par chromatographies sur gels de detrane, *L'Ing. Chim.*, **50**, 45.

Brown, J. (1966). Effect of pH, termperature, antivenin and functional group inhibitors on the toxicity and enzymatic activities of *C. atrox* venom, *Toxicon*, **4**, 99.

Chiu, W. C. (1960). Studies on the snake venom enzyme. IX. Electrophoresis studies of Russell's viper (*Vipera russellii formosensis*) venom and the relation of toxicity with enzyme activities, *J. Yamaguchi Med. Assoc.*, **9**, 1361.

Dolapchiev, L. B., Sulkowski, E., and Laskowski, M., Sr. (1974). Purification of exonuclease (phosphodiesterase) from the venom of *Crotalus adamanteus, Biochem. Biophys. Res. Commun*, **61**, 273.

References

Eckstein, F. and Bär, H. P. (1969). Enzymatic hydrolysis of adenosine 3′,5′-cyclic phosphorothioate, *Biochim. Biophys. Acta,* **191,** 316.

Eckstein, F. and Frischauf, A. M. (1974). Phosphodiesterase from snake venom, *Methods Enzymol.,* **35,** 605.

Felix, F., Potter, J. L., and Laskowski, M. (1960). Action of venom phosphodiesterase on deoxyribooligonucleotides carrying a monoesterified phosphate on carbon 3′, *J. Biol. Chem.,* **235,** 1150.

Fellig, J. (1955). Specificity and mode of action of rattlesnake venom adenosinetriphosphatase, *J. Am. Chem. Soc.,* **77,** 4419.

Frischauf, A. M. and Eckstein, F. (1973). Purification of a phosphodiesterase from *Bothrops atrox* venom by affinity chromatography, *Eur. J. Biochem.,* **32,** 479.

Gafurov, N. N. and Rasskazov, V. A. (1972). Characteristics of 5′-nucleotidase from *Ussurian mamushi* venom, *Biokhimiya,* **37,** 184.

Garlike, M. P. and Laskowski, M. (1955). Studies of the phosphodiesterase from rattlesnake venom, *Biochim. Biophys. Acta,* **18,** 370.

Georgatsos, J. G. and Laskowski, M., Sr. (1962). Purification of an endonuclease from the venom of *Bothrops atrox, Biochemistry,* **1,** 288.

Hadjiolov, A. A., Dolapchiev, L. B., and Milchev, G. I. (1966). Hydrolysis of synthetic polynucleotides with snake venom phosphodiesterase, *C. R. Acad. Bulg. Sci.,* **19,** 949.

Hadjiolov, A. A., Venkov, P. V., Dolapchiev, L. B., and Genchev, D. D. (1967). The action of snake venom phosphodiesterase on liver ribosomal ribonucleic acid, *Biochim. Biophys. Acta,* **142,** 111.

Haessler, H. A. and Cunningham, L. (1957). A comparison of several deoxyribonucleases of type II, *Exp. Cell Res.,* **13,** 304.

Heppel, L. A. and Rabinowitz, J. C. (1958). Enzymology of nucleic acids, purines, and pyrimidines, *Ann. Rev. Biochem.,* **27,** 613.

Jansone, I., Rozentals, G., Renhofs, R., and Grens, E. (1975). Phage MS2 hydrolysis by snake venom phosphodiesterase and template properties of the RNA produced by limited exonucleolytic action, *Mol. Biol. (Moscow),* **9,** 524.

Johnson, M., Kaye, M. A. G., Hems, R., and Krebs, H. A. (1953). Enzymic hydrolysis of adenosine phosphates by cobra venom, *Biochem. J.,* **54,** 625.

Keller, E. B. (1964). The hydrolysis of "soluble" ribonucleic acid by snake venom phosphodiesterase, *Biochem. Biophys. Res. Commun.,* **17,** 412.

Kimura, S. and Nakagawa, Y. (1965). Investigation of snake venoms. 4. Studies on the phosphodiesterase in venoms of Habu (*Trimeresurus flavoviridis flavoviridis* Hallowell) and Sakishima Habu (*Trimeresurus elegans* Grey), *Acta Med. Univ. Kagoshima.,* **17,** 36.

Kocholaty, W. F., Ledford, E. B., Daly, J. G., and Billings, T. A. (1971). Toxicity and some enzymatic properties and activities in the venoms of Crotalidae, Elapidae and Viperidae, *Toxicon,* **9,** 131.

Koerner, J. F. and Sinsheimer, R. L. (1957). A deoxyribonuclease from calf spleen, *J. Biol. Chem.,* **228,** 1049.

Laskowski, M., Sr. (1966). Exonuclease (phosphodiesterase) and other nucleolytic enzymes from venom, *Proc. Nucleic Acid Res.,* **1,** 154–187.

Laskowski, M., Sr. (1967). DNases and their use in the studies of primary structure of nucleic acids, *Adv. Enzymol.,* **29,** 165.

Laskowski, M., Sr. (1971). "Venom exonuclease," in P. D. Boyer, Ed., *Enzymes,* 3rd ed., Academic, New York, pp. 313–328.

Laskowski, M., Hargerty, G., and Laurila, U. (1957). Phosphodiesterase from rattlesnake venom, *Nature,* **180,** 1181.

Lu, M. S. and Lo, T. B. (1974). Chromatographic separation of *Bungarus fasciatus* venom and preliminary characterization of its components, *J. Chin. Biochem. Soc.,* **3,** 57.

McLennan, B. D. and Lane, B. G. (1968a). The chain termini of polynucleotides formed by limited enzymic fragmentation of wheat embryo ribosomal RNA. I. Studies of snake venom phosphodiesterase, *Can. J. Biochem.,* **46,** 81.

McLennan, B. D. and Lane, B. G. (1968b). The chain termini of polynucleotides formed by limited enzymic fragmentation of wheat embryo ribosomal RNA. II. Studies of a snake venom ribonuclease and pancreas ribonuclease, *Can. J. Biochem.,* **46,** 93.

Mebs, V. D. (1968). Vergleichende Enzymuntersuchungen an Schlangengiften unter besonderer Berücksicktigung ihrer Casein-spaltenden Proteasen, *Hoppe-Seylers Z. Physiol. Chem.,* **349,** 1115.

Miller, J. P., Hirst-Bruns, M. E., and Philipps, G. R. (1970). Action of venom phosphodiesterase on transfer RNA from *Escherichia coli, Biochim. Biophys. Acta,* **217,** 176.

Nigmatov, Z. N., Sorokin, V. M., and Yukel'son, L. Ya. (1972). Phosphodiesterase in Central Asian cobra venom, *Khim. Prir. Soedin.,* **5,** 688.

Nikol'skaya, I. I., Shaline, N. M., and Budovskii, E. I. (1963). Method for phosphodiesterase isolation from viper venom, *Biokhimiya,* **28,** 759.

Nikol'skaya, I. I., Kislina, O. S., and Tikhonenko, T. I. (1964). Separation of 5'-nucleotidase of *Vipera levetina* venom from interfering enzymes, *Dokl. Akad. Nauk SSSR,* **157,** 475.

Nikol'skaya, I. I., Kislina, O. S., Shalina, N. M., and Tikhonenko, T. I. (1965). The substrate specificity of *Vipera lebetina* venom phosphodiesterase, *Biokhimiya,* **30,** 1236.

Ogilvie, K. K. and Hruska, F. H. (1976). Effect of spleen and snake venom phosphodiesterases on nucleotides containing nucleosides in the *syn* conformation, *Biochem. Biophys. Res. Commun.,* **68,** 375.

Pereira Lima, R. A., Schenberg, S., Schiripa, L. N., and Nagamori, A. (1971). "ATPase and phosphodiesterase differentiation in snake venoms," in A. De Vries and E. Kochva, Eds., *Toxins of Animal and Plant Origin,* Gordon and Breach, New York, pp. 464–470.

Pfleiderer, G. and Ortanderl, F. (1963). Identität von Phosphodiesterase und ATP-Pyrophosphatase aus Schlangengift, *Biochem. Z.,* **337,** 431.

Philipps, G. R. (1975). Purification and characterization of phosphodiesterase from *Crotalus* venom, *Hoppe-Seylers Z. Physiol. Chem.,* **356,** 1085.

Philipps, G. R. (1976). Purification and characterization of phosphodiesterase I from *Bothrops atrox, Biochim. Biophys. Acta,* **432,** 237.

Pritchard, A. E. and Eichinger, B. E. (1974). The isolation of terminally cross-linked DNA and kinetics of venom phosphodiesterase, *Biochemistry,* **13,** 4455.

Privat de Garilhe, M. and Laskowski, M. (1955). Studies of the phosphodiesterase from rattlesnake venom, *Biochim. Biophys. Acta.,* **18,** 370.

Privat de Garilhe, M. and Laskowski, M. (1956). Optical changes occurring during the action of phosphodiesterase on oligonucleotides derived from deoxyribonucleic acid, *J. Biol. Chem.,* **223,** 661.

Rakitzis, E. T. (1972). Nucleotide pyrophosphatase activity of *Crotalus adamanteus* venom phosphodiesterase, *Folia Biochim. Biol. Graeca,* **9,** 100.

Razzell, W. E. (1961). Phosphodiesterase, *Methode Enzymol.,* **6,** 236.

Razzell, W. E. and Khorana, H. G. (1958). The stepwise degradation of thymidine oligonucleotides by snake venom and spleen phosphodiesterases, *J. Am. Chem. Soc.,* **80,** 1770.

Razzell, W. E. and Khorana, H. G. (1959a). Studies on polynucleotides. III. Enzymic degradation, substrate specificity and properties of snake venom phosphodiesterase, *J. Biol. Chem.,* **234,** 2105.

Razzell, W. E. and Khorana, H. G. (1959b). Studies on polynucleotides. IV. Enzymic degradation: The stepwise action of venom phosphodiesterase on deoxyribo-oligo-nucleotides, *J. Biol. Chem.,* **235,** 2114.

Richards, G. M. and Laskowski, M., Sr. (1969a). Negative charge at the 3' terminus of oligonucleotides and resistance to venom exonuclease, *Biochemistry,* **8,** 1786.

Richards, G. M. and Laskowski, M., Sr. (1969b). Use of venom exonuclease at low pH for preparation of mononucleoside diphosphates, *Biochemistry,* **8,** 4858.

Richards, G. M., de Vair, G., and Laskowski, M., Sr. (1965). Comparison of the levels of phosphodiesterase, endonuclease, and monophosphatases in several snake venoms, *Biochemistry,* **4,** 501.

Richards, G. M., Tutas, D. J., Wechter, W. J., and Laskowski, M., Sr. (1967). Hydrolysis of dinucleoside monophosphates containing arabinose in various internucleotide linkages by exonuclease from the venom of *Crotalus adamanteus, Biochemistry,* **6,** 2908.

Rodriguez, O. G., Scannone, H. R., and Parra, N. D. (1974). Enzymatic activities and other characteristics of *Crotalus durissus cumanensis* venom, *Toxicon,* **12,** 297.

Setlow, R. B., Carrier, W. L., and Bollum, R. J. (1964). Nuclease-resistant sequences in ultraviolet-irradiated deoxyribonucleic acid, *Biochim. Biophys. Acta,* **91,** 446.

Setoguchi, Y., Morisawa, S., and Obo, F. (1968). Investigation of sea snake venom. III. Acid and alkaline phosphatases (phosphodiesterase, phosphomonoesterase, 5'-nucleotidase and ATPase) in sea snake (*Laticauda semifasciata*) venom, *Acta Med. Univ. Kagoshima.*, **10**, 53.

Shipolini, R., Ivanov, C., and Dimitrov, G. (1964). Venom of the Bulgarian viper. II. The proteolytic, hemolytic, and toxic activity of fractions separated by paper electrophoresis, *God. Kim-Teknol. Inst.*, **11**, 87.

Sinsheimer, R. L. and Koerner, J. F. (1952). A purification of venom phosphodiesterase, *J. Biol. Chem.*, **198**, 293.

Sulkowski, E. and Laskowski, M., Sr. (1971). Inactivation of 5'-nucleotidase in commercial preparations of venom exonuclease (phosphodiesterase), *Biochim. Biophys. Acta*, **240**, 443.

Sulkowski, E. and Laskowski, M., Sr. (1974). Venom exonuclease (phosphodiesterase) immobilized on concanavalin-A-Sepharose, *Biochem. Biophys. Res. Commun.*, **57**, 463.

Suzuki, T. (1966). Separation methods of animal venom constituents, *Mem. Inst. Butantan*, **33**, 389.

Suzuki, T. and Iwanaga, S. (1958a). Studies on snake venoms. II. Some observations on the alkaline phosphatases of Japanese and Formosan snake venoms, *Yakugaku Zasshi*, **78**, 354.

Suzuki, T. and Iwanaga, S. (1958b). Studies on snake venoms. IV. Purification of alkaline phosphatases in cobra venoms, *Yakugaku Zasshi*, **78**, 368.

Suzuki, T., Iwanaga, S., and Satake, M. (1960a). Studies on snake venom. VI. Fractionation of three phosphodiesterases from venom of Mamushi (*Agkistrodon halys blomhoffi* Boie), *Yakugaku Zasshi*, **80**, 857.

Suzuki, T., Iwanaga, S., and Satake, M. (1960b). Studies on snake venom. VII. On the properties of three phosphodiesterases obtained from Mamushi venom (*Agkistrodon halys blomhoffi* Boie), *Yakugaku Zasshi*, **80**, 861.

Suzuki, T., Iizuka, K., and Murata, Y. (1960c). Studies on snake venom. IX. Studies on snake nucleotidase in snake venom, *Yakugaku Zasshi*, **80**, 868.

Suzuki, T., Iwanaga, S., and Nitta, K. (1960d). Studies on snake venom. XI. On the hydrolysis of cyclic mononucleotides by snake venom, *Yakugaku Zasshi*, **80**, 1040.

Tatsuki, T., Iwanaga, S., and Suzuki, T. (1975). Simple method for preparation of snake venom phosphodiesterase almost free from 5'-nucleotidase, *J. Biochem.*, **77**, 831.

Tu, A. T. and Toom, P. M. (1971). Isolation and characterization of the toxic component of *Enhydrina schistosa* (common sea snake) venom, *J. Biol. Chem.*, **246**, 1012.

Vasilenko, S. K. (1964). Kinetics of transfer-RNA hydrolysis by cobra venom phosphodiesterase, *Biokhimiya*, **29**, 1190.

Vasilenko, S. K. and Rait, V. K. (1975). Isolation of highly purified ribonuclease from cobra (*Naja oxiana*) venom, *Biokhimiya*, **40**, 578.

Vasilenko, S. K., Serbo, N. A., Ven'yaminova, A. G., Boldyreva, L. G., Budker, V. G., and Kobets, N. D. (1976). Preparative isolation of 5'-oligoribonucleotides using cobra venom ribonuclease, *Biokhimiya*, **41**, 260.

Venkstern, T. V. (1966). Specificity of snake venom phosphodiesterase towards the phosphodiesteric linkages of minor ribonucleotides, *Dokl. Akad. Nauk SSSR*, **170**, 718.

Wechter, W. J. (1967). Nucleic acids. I. The synthesis of nucleotides and dinucleoside phosphates containing ara-cytidine, *J. Med. Chem.*, **10**, 762.

Wechter, W. J., Mikulski, A. J., and Laskowski, M., Sr. (1968). Gradation of specificity with regard to sugar among nucleases, *Biochem. Biophys. Res. Commun.*, **30**, 318.

Wigler, P. W. (1963). The kinetics of snake venom phosphodiesterase, with a new type of substrate, 3-pyridyl thymidine 5'-phosphate, *J. Biol. Chem.*, **238**, 1767.

Yang, C. C., Su, C. C., and Chen, C. J. (1959). Biochemical studies on the Formosan snake venoms. V. The toxicity of Hyappoda (*Agkistrodon acutus*) venom, *J. Biochem. (Tokyo)* **46**, 1209.

Yarleque, A. and Campos, S. (1973). Phosphodiesterase activity in the venom of the snake *Lachesis muta, Bol. Soc. Quim. Peru*, **39**, 141.

Yot, P., Gueguen, P., and Chapeville, F. (1968). Hydrolyse des aminoacyl-tARN par la phosphodiesterase du venin, *FEBS Lett.*, **1**, 156.

4 Phosphomonoesterase

Orthophosphoric Monoester Phosphohydrolase   EC 3.1.3.2

1 NONSPECIFIC PHOSPHATASE   78
   1.1 Chemistry, 78
   1.2 Occurrence, 80

2 SPECIFIC PHOSPHATASE   81
   2.1 5'-Nucleotidase, 81
   2.2 Occurrence, 82

   References   83

Snake venoms contain nonspecific as well as specific phosphomonoesterase. Phosphomonoesterase is also commonly called phosphatase. Depending on the optimum pH of the phosphomonoesterase actions, the enzyme is also termed acid phosphatase or alkaline phosphatase.

The presence of phosphomonoesterase seems fairly common. Some venoms contain both acid and alkaline phosphatase, whereas others contain only one type (Tu and Chua, 1966). For instance, venom of *Agkistrodon acutus* shows only acid phosphatase activity; venom of *Ophiophagus hannah*, only alkaline. Many venoms, such as *Naja naja atra, N. naja samarensis, N. haje,* and *N. melanoleuca*, show both acid and alkaline phosphatase activity. Examples of each case are given in Fig. 1.

1 NONSPECIFIC PHOSPHATASE

1.1 Chemistry

The optimum pH for most venom acid phosphatases is around 4 to 5 (Tu and Chua, 1966). Acid phosphatase was isolated from venom of the sea snake, *Laticauda semifasciata* (Uwatoko-Setoguchi, 1970). The purified enzyme hydrolyzes *p*-nitrophenyl phosphate, phenyl phosphate, 2'-AMP, 3'-AMP, 5'-AMP, and ATP. Glucose-1-phosphate, glucose-6-phosphate, and glycerophosphoric acid are resistant to hydrolysis and liberate

1 Nonspecific Phosphatase 79

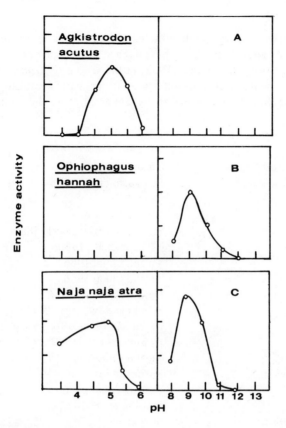

Figure 4.1 Phosphomonoesterase activity in snake venoms. (A) Acid phosphatase (*Agkistrodon acutus* venom). (B) Alkaline phosphatase (*Ophiophagus hannah* venom). (C) Both acid and alkaline phosphatases (*Naja naja atra* venom). (Based on data of Tu and Chua, 1966.)

phosphate only after 24 hr exposure to the enzyme. Calcium di-*p*-nitrophenyl phosphate is not hydrolyzed by the purified enzyme. The enzyme has an optimum pH of 5 and a K_m of 0.59 mM when *p*-nitrophenyl phosphate is used as a substrate. Enzyme activity is rapidly destroyed at 50°C.

A purified (200-fold increase in the specific activity) alkaline phosphatase obtained from *Bothrops atrox* venom is activated by Ca(II) or Mg(II) and has a pH optimum of approximately 9.5. The enzyme hydrolyzes 5′-AMP, 5′-dAMP, 3′-AMP, ribose-5-P, ATP, d-XpYp, d-pGp, flavin mononucleotide, nicotinamide mononucleotide, and 5-phosphorylribose 1-pyrophosphate, but no activity was exhibited toward pyrophosphate and 2′,3′-cyclic phosphate (Sulkowski et al., 1963).

In spite of the common occurrence of the enzyme in snake venoms, complete isolation has not been achieved. Therefore we know very little about the chemical and physical properties of the enzyme. Perhaps one reason for the lack of concern by investigators toward isolating the protein is that there has been no report on any important biological effect of phosphatase associated with snake venom toxic actions. It has been proved that phosphomonoesterase can be separated from three toxic principles of *Naja naja atra* venom (Yang et al., 1959a).

1.2 Occurrence

Phosphomonoesterase is present in the venoms of four families of snakes: Hydrophiidae, Elapidae, Viperidae, and Crotalidae. The presence of phosphomonoesterase in snake venoms was first reported by Uzawa (1932) in the venom of *Trimeresurus flavoviridis*. Since then the enzyme has been detected in venoms by many investigators. Venoms from the following species have been reported to contain nonspecific phosphomonoesterase:

Hydrophiidae

Enhydrina schistosa	Tu and Toom, 1971
Laticauda semifasciata	Setoguchi et al., 1968

Elapidae

Bungarus multicinctus	Suzuki and Iwanaga, 1958a; Chen and Su, 1959; Yang et al., 1960b
Naja haje	Tu and Chua, 1966
N. melanoleuca	Tu and Chua, 1966
N. naja	Devi and Sarkar, 1966
N. naja atra	Suzuki and Iwanaga, 1958b; Chen and Su, 1959; Tu and Chua, 1966; Yang et al., 1960a; Simon et al., 1969
N. naja samarensis	Tu and Chua, 1966
N. naja siamensis	Suzuki and Iwanaga, 1958a
N. nigricollis	Richards et al., 1965
N. oxiana	Sorokin et al., 1972
Ophiophagus hannah	Suzuki and Iwanaga, 1958a; Tu and Chua, 1966

Viperidae

Vipera russellii	Devi and Sarkar, 1966
V. russellii formosensis	Chen and Su, 1959; Yang et al., 1960a

Crotalidae

Agkistrodon acutus	Suzuki and Iwanaga, 1958a; Chen and Su, 1959; Yang et al., 1960a
A. halys blomhoffii	Suzuki and Iwanaga, 1958a; Gafurov and Rasskazov, 1972
A. piscivorus	Richards et al., 1965
Bothrops atrox	Sulkowski et al., 1963; Richards et al., 1965
Crotalus adamanteus	Richards et al., 1965; Devi and Sarkar, 1966
C. atrox	Brown and Bowles, 1965; Richards et al., 1965
Trimeresurus elegans	Kimura et al., 1965
T. flavoviridis	Uzawa, 1932; Suzuki and Iwanaga, 1958a; Kimura et al., 1965
T. gramineus	Suzuki and Iwanaga, 1958a; Chen and Su, 1959; Yang et al., 1960a
T. mucrosquamatus	Suzuki and Iwanaga, 1958a; Chen and Su, 1959; Yang et al., 1960b; Tu and Chua, 1966
T. okinavensis	Suzuki and Iwanaga, 1958a

2 SPECIFIC PHOSPHATASE

2.1 5'-Nucleotidase (5'-Ribonucleotide Phosphohydrolase   EC 3.1.3.5)

A very specific phosphomonoesterase, 5'-nucleotidase, is widely distributed in snake venoms.

A purified 5'-nucleotidase from the venom of *Bothrops atrox* hydrolyzes only 5'-mononucleotides. It is inactive against ribose 5'-phosphate, mononucleoside 3',5'-diphosphates, or higher oligonucleotides. It hydrolyses ribo- and deoxyribonucleotides, the rate being dependent on base composition.

Crude venoms from different species (*Hemachatus haemachatus, Vipera russelli, Crotalus adamanteus, Bothrops atrox*) exhibit different pH optima for 5'-nucleotidase activity, but these differences disappear during purification. The enzymatic activity is not influenced by mild oxidizing or reducing agents or by sulfhydryl reagents, indicating that disulfide bonds and free sulfhydryl groups are not involved in enzyme activity (Björk, 1964).

The enzyme has been purified by ammonium sulfate fractionation from the venom of *Vipera lebetina*. Neither the enzyme nor the crude venom hydrolyzes 2'- or 3'-phosphates of adenosine, guanosine, cytidine, uridine, or adenosine 2',3'-cyclic phosphates. The enzyme hydrolyzes 5'-phosphates of deoxynucleosides and ribonucleosides at similar rates. Product inhibition is more pronounced with purines than with pyrimidines. Inorganic phosphate at high concentration completely inhibits enzyme activity. Cysteine and EDTA are inhibitors, whereas Mg(II), Na(I), and K(I) are activators (Nikol'skaya et al., 1965).

5'-Nucleotidase has also been isolated from the venom of *Naja naja atra* (Chen and Lo, 1968). The molecular weight of the enzyme was estimated from gel filtration and total amino acid composition to be approximately 10,000. The optimum pH is at 6.5 to 7.0, and the enzyme is activated by Mg(II) and Mn(II); Ni(II) and Zn(II) are inhibitors. The enzyme is heat labile and has no activity toward the 2'- and 3'-phosphates of nucleosides.

A number of substrates have been tested using purified 5'-nucleotidase. Their relative activities, based on 5'-AMP as 100, are as follows:

Substrate	Activity
5'-GMP	56.2
5'-CMP	91.0
5'-UMP	91.2
2'(3')-AMP	1.6
d-5'-GMP	8.2
d-5'-CMP	19.8
GDP	5.1
GTP	0

From the above data, one can conclude that 5'-nucleotidase is rather specific for 5'-ribonucleoside monophosphate.

The enzyme was also isolated from the venom of *Agkistrodon halys blomhoffii* (Gafurov and Rasskazov, 1972). The enzyme preparation was free from 3'-nucleotidase, alkaline phosphatase, and phosphodiesterase. The pH optimum is 6.8 to 6.9, and enzyme activity is enhanced by Mg(II) and Mn(II); Zn(II) is an inhibitor. The optimum pH of the purified enzyme from *A. halys blomhoffii* from central Asia is remarkably similar to that from *N. naja atra* from Formosa.

Lachesis muta venom exhibits 5'-nucleotidase activity and hydrolyzes 5'-AMP and 5'-GMP, but not pyridoxal 5-phosphate, d-2'-CMP, naphthyl phosphate and β-pyrophosphate (Campos and Yarleque, 1974). The enzyme is active in the pH range of 5 to 10.2 with peaks at 7.8 and 9.8. At pH 9.8, Mg(II) strongly activates the enzyme whereas (Ca(II) inhibits it.

The fact that the enzyme was separated from cardiotoxic and neurotoxic fractions suggests that it does not participate in a major toxic effect of *Bungarus fasciatus* venom (Lu and Lo, 1974). This is further confirmed by the toxicity of 5'-nucleotidase, which has a LD_{50} value greater than 50 $\mu g\,g^{-1}$ in mice. Neurotoxin isolated from the same venom has a very low LD_{50} value, 0.04 $\mu g\,g^{-1}$.

Similarly, 5'-nucleotidase present in the venom of *Naja naja atra* was separated from the three toxic principles (Yang et al., 1959a). In *N. oxiana* venom, toxic fraction was also separated from 5'-nucleotidase (Turakulov et al., 1969).

2.2 Occurrence

The occurrence of 5'-nucleotidase in the venoms of Hydrophiidae has not been investigated, and therefore we do not know whether this enzyme is present in sea snake venoms. For Colubridae, only one venom, from *Leptodeira annulata*, has been investigated. It was found that this rear-fanged snake does not show 5'-nucleotidase activity (Mebs, 1968a, b).

The presence of 5'-nucleotidase in the venoms of Elapidae, Viperidae, and Crotalidae is common. Venoms from the following species are reported to have this enzyme:

Elapidae

Bungarus fasciatus	Mebs, 1968a; Lu and Lo, 1974
B. multicinctus	Suzuki and Iwanaga, 1958a; Yang et al., 1959a; Chen and Su, 1959
Dentroaspis angusticeps	Mebs, 1968a
Haemachatus hemachatus	Björk, 1964; Mebs, 1968a
Naja haje	Mebs, 1968a
N. naja	Devi and Sarkar, 1966
N. naja siamensis	Suzuki and Iwanaga, 1958a, b
N. naja atra	Suzuki and Iwanaga, 1958a; Yang et al., 1959a; Chen and Su, 1959; Chiu, 1960; Mebs, 1968a
N. nigricollis	Richards et al., 1965; Mebs, 1968a
N. nivea	Mebs, 1968a
N. oxiana	Turakulov et al., 1969; Sakhibov et al., 1968; Sorokin et al., 1972
Ophiophagus hannah	Suzuki and Iwanaga, 1958a; Mebs, 1968a
Pseudechis collettii	Mebs, 1968a

Viperidae

Bitis gabonica	Mebs, 1968a
Vipera lebetina	Nikol'skaya et al., 1964
V. russellii	Björk, 1964; Devi et al., 1966; Mebs, 1968a; Kucerova et al., 1968
V. russelli formosensis	Yang et al., 1959a; Chen and Su, 1959

Crotalidae

Agkistrodon acutus	Suzuki and Iwanaga, 1958a; Yang et al., 1959b; Chen and Su, 1959
A. halys blomhoffi	Suzuki and Iwanaga, 1958a; Gafurov and Rasskazov, 1972
A. piscivorus	Richards et al., 1965
Bothrops atrox	Sulkowski et al., 1963; Björk, 1964
B. jararaca	Devi et al., 1966; Mebs, 1968a
Crotalus adamanteus	Björk, 1964; Richards et al., 1965; Devi et al., 1966
C. atrox	Richards et al., 1965; Devi and Sarkar, 1966; Mebs, 1968a
C. durissus cumanesis	Rodriguez et al., 1974
C. durissus terrificus	Mebs, 1968a
Lachesis muta	Campos and Yarleque, 1974
Trimeresurus flavoviridis	Suzuki and Iwanaga, 1958a
T. gramineus	Suzuki and Iwanaga, 1958a; Yang et al., 1959a; Chen and Su, 1959
T. mucrosquamatus	Suzuki and Iwanaga, 1958a; Yang et al., 1959a; Chen and Su, 1959
T. okinavensis	Suzuki and Iwanaga, 1958

REFERENCES

Björk, W. (1964). Activation and stabilization of snake venom 5′-nucleotidase, *Biochim. Biophys. Acta*, **89**, 483.

Brown, J. H. and Bowles, M. E. (1965). Effect of enzymic group inhibitors on the esterases of *Crotalus atrox* venom, *U.S. Army Med. Res. Lab., Fort Knox, Ky. Rept.* No. 627.

Campos, S. and Yarleque, A. (1974). 5′-Nucleotidases in the venom of the snake *Lachesis muta*, *Bol. Soc. Quim. Peru*, **40**, 202.

Chen, C. J. and Su, C. C. (1959). Studies on the snake venom enzyme. I. On the phosphatase and lecithinase activities of Formosan snake venoms and their toxicities, *J. Yamaguchi Med. Assoc.*, **8**, 570.

Chen, Y. and Lo, T. B. (1968). Chemical studies of Formosan cobra (*Naja naja atra*) venom. V. Properties of 5′-nucleotidase, *J. Chin. Chem. Soc.*, **15**, 84.

Chiu, W. C. (1960). Studies on the snake venom enzyme. IX. Electrophoresis studies of Russell's viper (*Vipera russellii formosensis*) venom and the relation of toxicity with enzyme activities, *J. Yamaguchi Med. Assoc.*, **9**, 1361.

Devi, A. and Sarkar, N. K. (1966). Cardiotoxic and cardiostimulating factors in cobra venom, *Mem. Inst. Butantan*, **33**, 573.

Devi, A., Ashgar, S. S., and Sarkar, N. K. (1966). 5′-Nucleotidase activity in snake venoms, *Mem. Inst. Butantan*, **33**, 943.

Gafurov, N. N. and Rasskazov, V. A. (1972). Characteristics of 5′-nucleotidase from Ussurian mamushi venom, *Biokhimiya*, **37**, 184.

Kimura, Y., Nakagawa, Y., and Yoshihira, A. (1965). Investigation of snake venoms. 3. Studies on the alkaline monophosphatase in venom of Habu (*Trimeresurus flavoviridis flavoviridis* Hallowell) and Sakishima Habu (*T. elegans* Gray), *Acta Med. Univ. Kagoshima.*, **17**, 33.

Kucerova, Z., Skoda, J., Holy, A., and Sorm, F. (1968). Factors affecting the cleavage of adenosine and unidine 5′-phosphites by the venom of *Vipera russelli*, *Collect. Czech. Chem. Commun.*, **33**, 4350.

Lu, M. S. and Lo, T. B. (1974). Chromatographic separation of *Bungarus fasciatus* venom and preliminary characterization of its components, *J. Chin. Biochem. Soc.*, **3**, 57.

Mebs, D. (1968a). Vergleichende Enzymuntersuchungen an Schlangengiften unter besonderer Berücksichtigung ihrer Casein-spaltenden Proteasen, *Hoppe-Seylers Z. Physiol. Chem.*, **349**, 1115.

Mebs, D. (1968b). Analysis of *Leptodeira annulata* venom, *Herpetologica*, **24**, 338.

Nikol'skaya, I. I., Kislina, O. S., and Tikhonenko, T. I. (1964). Separation of 5'-nucleotidase of *Vipera lebetina* venom from interfering enzymes, *Dokl. Akad. Nauk SSSR*, **157**, 475.

Nikol'skaya, I. I., Kislina, O. S., and Tikhonenko, T. I. (1965). Properties of 5'-nucleotidase of viper venom, *Biokhimiya*, **30**, 107.

Richards, G. M., du Vair, G., and Laskowski, M., Sr. (1965). Comparison of the levels of phosphodiesterase, endonuclease, and monophosphatases in several snake venoms, *Biochemistry*, **4**, 501.

Rodriguez, O. G., Scannone, H. R., and Parra, N. D. (1974). Enzymatic activities and other characteristics of *Crotalus durissus cumanensis* venom, *Toxicon*, **12**, 297.

Sakhibov, D. N., Sorokin, V. M., and Yukel'son, L. Y. (1968). Purification and extraction of biologically active substances from cobra venom, *Uzb. Biol. Zh.*, **12**, 65.

Setoguchi, Y., Morisawa, S., and Obo, F. (1968). Investigation of sea snake venom. III. Acid and alkaline phosphatases (phosphodiesterase, phosphomonoesterase, 5'-nucleotidase, and ATPase) in sea snake (*Laticauda semifasciata*) venom, *Acta Med. Univ. Kagoshima.*, **10**, 53.

Simon, J., Brisbois, L., and Gillo, L. (1969). Fractionation of cobra venom by electrofocusing, *J. Chromatogr.*, **44**, 209.

Sorokin, V. M., Nigmativ, Z., and Yukelson, L. Ya (1972). Cholinesterase of central Asian cobra venom, *Khim. Prir. Soedin.*, **6**, 783.

Sulkowski, E., Björk, W., and Laskowski, M., Sr. (1963). A specific and nonspecific alkaline monophosphatase in the venom of *Bothrops atrox* and their occurrence in the purified venom phosphodiesterase, *J. Biol. Chem.*, **238**, 2477.

Suzuki, T. and Iwanaga, S. (1958a). Studies on snake venoms. II. Some observations on the alkaline phosphatases of Japanese and Formosan snake venoms, *Yakugaku Zasshi*, **78**, 354.

Suzuki, T. and Iwanaga, S. (1958b). Studies on snake venoms. IV. Purification of alkaline phosphatases in cobra venoms, *Yakugaku Zasshi*, **78**, 368.

Tu, A. T. and Chua, A. (1966). Acid and alkaline phosphomonoesterase activities in snake venoms, *Comp. Biochem. Physiol.*, **17**, 297.

Tu, A. T. and Toom, P. M. (1971). Isolation and characterization of the toxic component of *Enhydrina schistosa* (common sea snake) venom, *J. Biol. Chem.*, **246**, 1012.

Turakulov, Ya. Kh., Sakhibov, D. N., Sorokin, V. M., and Yukel'son, L. Ya. (1969). Separation of central-Asian cobra venom by means of gel filtration through Sephadex and determination of biological activity of the resulting fractions, *Biokhimiya*, **34**, 1119.

Uwatoko-Setoguchi, Y. (1970). Studies on sea snake venom. VI. Pharmacological properties of *Laticauda semifasciata* venom and purification of toxic components, acid phosphomonoesterase and phospholipase A in the venom, *Acta Med. Univ. Kagoshima.*, **12**, 73.

Uzawa, S. (1932). Über die phosphomonoesterase und die phosphodiesterase, *J. Biochem. (Tokyo)*, **15**, 19.

Yang, C. C., Chen, C. J., and Su, C. C. (1959a). Biochemical studies on the Formosan snake venoms. IV. The toxicity of Formosan cobra venom and enzyme activities, *J. Biochem. (Tokyo)*, **46**, 1201.

Yang, C. C., Su, C. C., and Chen, C. J. (1959b). Biochemical studies on the Formosan snake venoms. V. The toxicity of Hyappoda (*Agkistrodon acutus*) venom, *J. Biochem. (Tokyo)*, **46**, 1209.

Yang, C. C., Chiu, W. C., and Kao, K. C. (1960a). Biochemical studies on the snake venoms. VII. Isolation of venom cholinesterase by zone electrophoresis, *J. Biochem. (Tokyo)*, **48**, 706.

Yang, C. C., Kao, K. C., and Chiu, W. C. (1960b). Biochemical studies on the snake venoms. VIII. Electrophoretic studies of banded krait (*Bungarus multicinctus*) venom and the relation of toxicity with enzyme activities, *J. Biochem. (Tokyo)*, **48**, 714.

5 L-Amino Acid Oxidase

L-Amino acid: Oxygen
Oxidoreductase (Deaminating)   EC 1.4.3.2

1 CHEMISTRY 85
  1.1 Physical and Chemical Properties, 85
  1.2 Isozymes, 89
  1.3 Enzymatic properties, 89

2 BIOLOGICAL SIGNIFICANCE 93

3 OCCURRENCE 93

  References 95

Unlike most snake venom enzymes, which are hydrolytic, L-amino acid oxidase is a nonhydrolytic enzyme. It converts free amino acid into an α-keto acid. L-Amino acid oxidase is one of the venom enzymes that have been extensively characterized by many investigators; it is used to make special α-keto acid.

1 CHEMISTRY

1.1 Physical and Chemical Properties

L-Amino acid oxidase was first purified from *Crotalus adamanteus* venom by Wellner and Meister (1960). The preparation and the assay method for L-amino acid oxidase have been described by Wellner (1971).

The enzyme exhibits absorption maxima at 275, 390, and 462 nm (Wellner, 1971) or 280, 390, and 460 nm (Kurth and Aurich, 1973). The absorption spectra of L-amino acid oxidase (*C. adamanteus* venom origin) under different conditions are shown in Fig. 1, and the fluorescence spectra of L-amino acid oxidase purified from the venom of *Vipera ammodytes* in Fig. 2. The fluorescence spectra, which originate from the prosthetic group, FAD, present in the enzyme, can be obtained by an excitation at 360 nm, the wavelength of maximum absorption.

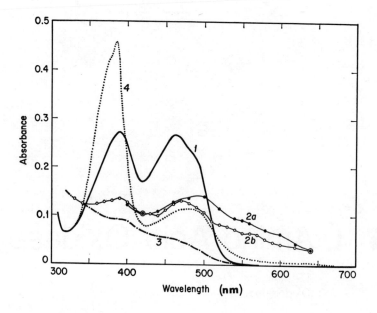

Figure 5.1 Spectra of L-amino acid oxidase (*Crotalus adamanteus* venom). (1) Oxidized enzyme. (2a) Intermediate produced in stopped-flow experiment with L-phenylalanine. (2b) Intermediate produced in stopped-flow experiment with L-leucine. (3) Reduced enzyme. (4) Semiquinonoid enzyme. (Reproduced from Massey and Curti, 1967, by permission of the copyright owner, *The Journal of Biological Chemistry*.)

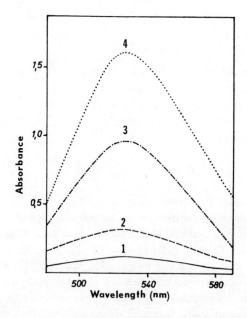

Figure 5.2 Fluorescence emission spectra of L-amino acid oxidase isolated from the venom of *Vipera ammodytes*. The enzyme is excited at 360 nm. (1) In 0 to $1M$ phosphate buffer, pH 7.5. (2) After being kept at 35°C. (3) After the addition of urea to the final concentration of $8M$. (4) On maintaining pH at 1.0 with $6N$ HCl. (Reproduced from Kurth and Aurich, 1973.)

1 Chemistry

Table 1   Molecular Weights and FAD Contents of L-Amino Acid Oxidases Isolated from Snake Venoms

Molecular Weight	mole FAD	Species	Reference
97,000	2	V. ammodytes	Zwisler, 1965
128,000 to 153,000	2	C. adamanteus	Meister and Wellner, 1966
130,000	2	C. adamanteus	Wellner, 1971
130,000	—	V. palaestinae	Shaham and Bdolah, 1973
100,000	—	V. lebetina turanica	Sakhibov et al, 1973
135,000	2.2	V. ammodytes	Kurth and Aurich, 1973
85,000	—	Agkistrodon caliginosus	Sugiura et al, 1975

The molecular weights and numbers of moles of FAD reported for various venoms are shown in Table 1. A molecular weight of about 130,000 daltons is most consistently reported. The enzyme has a sedimentation velocity of 6.35 at 0.1% protein concentration (Radda, 1964) and of 6.63 at 2.3 mg ml^{-1} (Wellner, 1971). The diffusion coefficient, D_{20}, is 4.6×10^{-7} cm^2 sec^{-1} (Wellner, 1971). The enzyme is a glycoprotein containing about 2 to 5% carbohydrate, including sialic acid (de Kok and Rawitch, 1969).

The L-amino acid oxidase of *C. adamanteus* is inactivated by storage in the frozen state at temperatures between $-5°$ and $-60°$, with maximal inactivation being observed at approximately $-20°$C (Curti et al., 1968). The rate of inactivation is dependent on the pH during storage and on the ionic composition of the medium. The inactivation is due to a limited conformational change in the enzyme structure, particularly in the vicinity of the flavin prosthetic group. A similar instability of the enzyme (from *V. ammodytes* venom) was also reported by Kurth and Aurich (1973, 1976). They observed that stability depended on the temperature of storage and the presence of salts; the half-life with Tris–HCl buffer was 13 days at room temperature, and 130 days at 7°. Shaham and Bdolah (1973) reported that the optimal condition for storage of the enzyme was either at $-15°$ with 0.2M acetate or under nitrogen at 0°.

Inactivation and reactivation of the *C. adamanteus* and *Agkistrodon piscivorus*

piscivorus enzymes are associated with conformational changes at the active center of the enzyme and changes in the mode of binding of the flavin-adenine dinucleotide to the protein (Wellner, 1966). Inactivation does not involve aggregation, dissociation into subunits, dissociation of the enzyme, or changes in antibody binding. There are shifts in the absorption spectrum and optical rotatory dispersion spectrum upon denaturation. Renaturation of the enzyme reverses the spectra completely.

Amino acid compositions of L-amino acid oxidases from the venoms of *Vipera ammodytes* and *C. adamanteus* are summarized in Table 2. Both preparations show high acidic amino acid contents.

Table 2 Amino Acid Composition of Snake Venom L-Amino Acid Oxidase

Genus Species	Vipera ammodytes	Crotalus adamanteus
Origin	Europe	U.S.A.
Amino Acid		
Lysine	85	82
Histidine	28	30
Arginine	34	78
Aspartic acid	133	116
Threonine	53	67
Serine	87	70
Glutamic acid	101	116
Proline	51	41
Glycine	187	73
Alanine	108	83
Valine	65	67
Methionine	7	19
Isoleucine	55	70
Leucine	66	78
Tyrosine	26	58
Phenylalanine	34	55
Half-cystine	—	—
Tryptophan	—	—
References	Kurth and Aurich, 1973	Hayes and Wellner, 1969

1 Chemistry

The isoelectric point of the enzyme isolated from the venom of *Agkistrodon caliginosus* is 4.9 (Sugiura et al., 1975), and that of *Trimeresurus flavoviridis* is 8.4 (Nakano et al., 1972).

1.2 Isozymes

L-Amino acid oxidase obtained from *Crotalus adamanteus* consists of three isozymes that demonstrate equal specific activities (Hayes and Wellner, 1969; Wellner, 1971). Three bands were also observed by Shaham and Bdolah (1973) for the enzyme obtained from *Vipera palaestinae*. When sialic acid residues were removed from the enzyme (*C. adamanteus*), the electrophoretic mobility was altered but three isozyme bands were still evident (de Kok and Rawitch, 1969). The native enzyme is a noncovalent dimer consisting of two subunits of molecular weight near 70,000. Two types of polypeptide subunits occur in the enzyme but in unequal amounts, with a ratio of 2.5 to 1. The three isozymes appear to be the result of the various combinations of these different subunits to give the native dimer. The subunits differ in electrophoretic mobility as well as amino terminal residues.

When three crystalline L-amino acid oxidases (*C. adamanteus* venom origin) are placed on an isoelectric focusing column, at least 18 fractions with amino acid oxidase activity can be obtained. The isoelectric points of these isozymes are 5.20, 5.35, 5.45, 5.60, 5.71, 5.87, 6.18, 6.41, 6.64, 6.89, 7.09, 7.25, 7.45, 7.68, 7.87, 8.05, 8.25, and 8.40, respectively (Hayes and Wellner, 1969). Changes in substrate specificity during purification also indicate that there may be several different L-amino acid oxidases in the venom of *Trimeresurus flavoviridis* (Nakano et al., 1972).

1.3 Enzymatic Properties

The optimum pH of L-amino acid oxidase usually lies between 7 and 9. Paik and Kim (1965) extensively studied the pH–substrate relationship and found that the enzyme isolated from *Crotalus adamanteus* venom exhibits six different shapes of pH curves, depending on the amino acid used as substrate. The enzyme required Mg(II) and was inhibited by Ca(II) and PO_4^{3-} as well as *p*-chloromercuribenzoate (Paik and Kim, 1967).

As can be seen from Table 1, the prosthetic group of L-amino acid oxidase is FAD. However, Inamasu et al. (1974) reported that the enzyme from *Trimeresurus flavoviridis* venom contained FMN instead of FAD.

The enzyme isolated from the venom of *Agkistrodon caliginosus* was most active at pH 8.0 to 8.5, with L-leucine used as substrate (Sugiura et al., 1975).

The enzyme catalyzes the oxidative deamination of L-amino acid. The reaction takes place in two steps (see Schemes 12 and 13), the step in Scheme 12 being rate controlling. The intermediate imino acid is readily converted to α-keto acid in the presence of water.

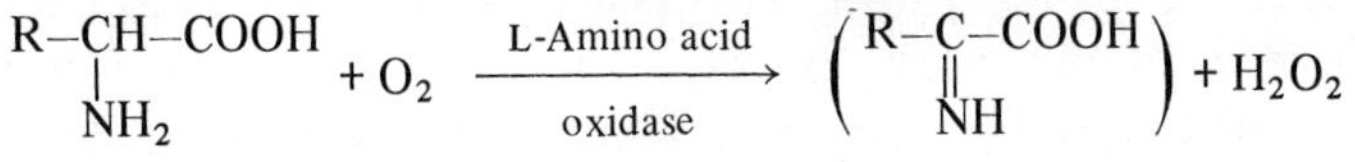

$$\underset{NH_2}{R-CH-COOH} + O_2 \xrightarrow{\text{L-Amino acid oxidase}} \left(\underset{NH}{R-C-COOH} \right) + H_2O_2$$

Scheme 12

$$\left(\underset{NH}{R-C-COOH} \right) + H_2O \longrightarrow \underset{O}{R-C-COOH}$$

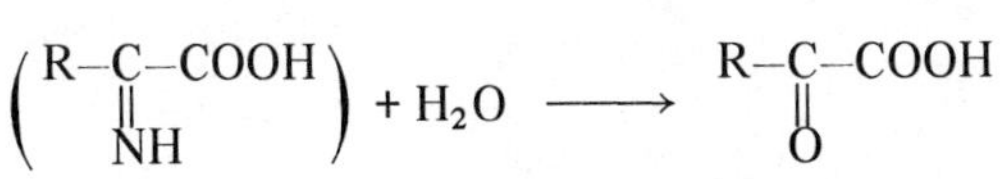

Scheme 13

Without the presence of catalase, an α-keto acid is actually the product of an L-amino acid treated with the enzyme. In the presence of catalase, however, α-keto acids are decarboxylated and oxidized to carboxylic acids (see Scheme 14).

$$\underset{O}{R-\overset{\|}{C}-COOH} + H_2O_2 \xrightarrow{\text{Catalase}} RCOOH + CO_2 + H_2O$$

Scheme 14

In Schemes 12 and 13 the parentheses indicate that the imino acids $R-\underset{NH}{\overset{\|}{C}}-COOH$ are intermediate products. It had long been thought that the formation of imino acid could not be demonstrated, but this was achieved by allowing the amino acid oxidase reaction to proceed in the presence of $NaBH_4$. Under this condition the imino acid was reduced to the racemic amino acid instead of being oxidized to the α-keto acid (Hafner and Wellner, 1971).

Many α-keto acids have been prepared by using the enzyme and α-amino acid analogs. For instance, Meister (1956) prepared the following α-keto acids:

α-ketoadipamic, α-keto-N-methyladipamic, α-ketobutyric, α-ketoheptylic, α-keto-ε-hydroxycaproic, α-ketophenulacetic, α-keto-δ-guanidinovaleic, α-ketosuccinamic, α-keto-δ-carbamidovaleric, β-cyclohexylpyruvic, β-sulfopyruvic, α-keto-γ-ethiolbutyric, α-ketoglutaramic, α-keto-γ-methylglutaramic, α-keto-N-methylglutaramic, α-keto-N--dimethylglutatamic, d-α-keto-β-methylvaleric, l-α-keto-β-methylvaleric, α-ketoisocaproic, α-keto-ε-aminocaproic, α-keto-ε-N-chloroacetyl-ε-aminocaproic, α-keto-γ-methiolbutyric, α-keto-δ-nitroguanidinovaleric, α-ketocaproic, α-keto-δ-aminovaleric, phenylpyruvic, S-benzyl-β-mercaptopyruvic, p-hydroxyphenylpyruvic, α-ketoisovaleric, and α-ketoglutaric acid-γ-ethyl ester.

The oxidation of L-lysine and L-ornithine derivatives was extensively studied by Paik and Kim (1964), who found that the following compounds were oxidized by the action of the enzyme:

ε-N-methyl-L-lysine, ε-N-ethyl-L-lysine, ε-N-dimethyl-L-lysine, ε-N-glycyl-L-lysine, ε-N-acetyl-L-lysine, ε-N-chloroacetyl-L-lysine, ε-N-formyl-L-lysine, ε-N-propionyl-L-lysine, ε-N-(acetylglycyl)-L-lysine, ε-N-benzoyl-L-lysine, ε-N-carbobenzoxy-L-lysine, ε-N-toluenesulfonyl-L-lysine, L-ornithine, δ-N-methyl-L-ornithine, δ-N-acetyl-L-ornithine.

The compounds not oxidized with snake venom L-amino acid oxidase are ε-N-acetyl-L-lysine amide and ε-N-acetyl-L-lysylglycine.

Not every amino acid can be oxidized readily by the enzyme. Some amino acids such as glutamate, aspartate, glutamine, and asparagine are resistant to oxidation by the enzyme (Nakano et al., 1972).

The oxidative deamination of sulfur amino acids by *Crotalus durissus terrificus* L-amino acid oxidase was extensively studied by Chen et al. (1971). They found no measurable activity with any of the sulfoxides, S-adenosyl-L-methionine, cysteic acid, homocysteic acid, lanthionine, or cystathionine. The affinity for the various substrates at pH 7.5 is, in decreasing order, as follows: L-methionine, L-homocystine, L-homocysteine,

1 Chemistry

S-adenosyl-L-homocysteine, *S*-ribosyl-L-homocysteine, L-djenkolic acid, L-cysteine, and L-cystine. Chen et al. also observed that prolonged incubation of cysteine and homocysteine allowed spontaneous decomposition to H_2S and α-keto acids. The mechanism is illustrated in Scheme 15.

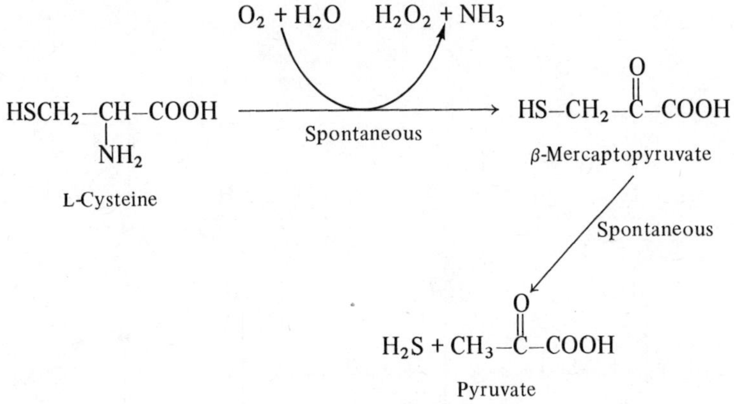

$$\text{HSCH}_2\text{-CH(NH}_2\text{)-COOH} \xrightarrow[\text{Spontaneous}]{O_2 + H_2O \rightarrow H_2O_2 + NH_3} \text{HS-CH}_2\text{-CO-COOH}$$

L-Cysteine → β-Mercaptopyruvate

$$\text{β-Mercaptopyruvate} \xrightarrow{\text{Spontaneous}} H_2S + CH_3\text{-CO-COOH (Pyruvate)}$$

Scheme 15

Cysteine is the least active substrate in comparison to other amino acids.

Islamova (1970) observed that L-amino acid oxidase oxidized L-arginine and L-asparagine faster than L-glutamic acid and L-cysteine. Sakhibov et al. (1973) obtained the same result and also reported that the enzyme was active against aromatic amino acids, valine, arginine, histidine, and, to a lesser extent, glutamic acid and cysteine.

The reaction of L-cystine with L-amino acid oxidase (from *C. durissus terrificus* venom) in the absence of catalase was extensively studied by Ubuka and Yao (1973). Their findings are briefly summarized in Scheme 16, where OCETC represents *S*-(2-oxo-2-carboxymethylthio)cysteine and CMTC denotes *S*-(carboxymethylthio)cysteine.

L-Cysteine $\xrightarrow[O_2 + H_2O \rightarrow NH_3 + H_2O_2]{\text{L-Amino acid oxidase}}$ S–CH$_2$–CO–COOH / S–CH$_2$–CH(NH$_2$)–COOH (OCETC)

OCETC $\xrightarrow[NH_3 + H_2O_2]{O_2 + H_2O, \text{L-Amino acid oxidase}}$ S–CH$_2$–CO–COOH / S–CH$_2$–CO–COOH

OCETC $\xrightarrow[CO_2 + H_2O]{H_2O_2}$ S–CH$_2$–COOH / S–CH$_2$–CH(NH$_2$)–COOH (CMTC)

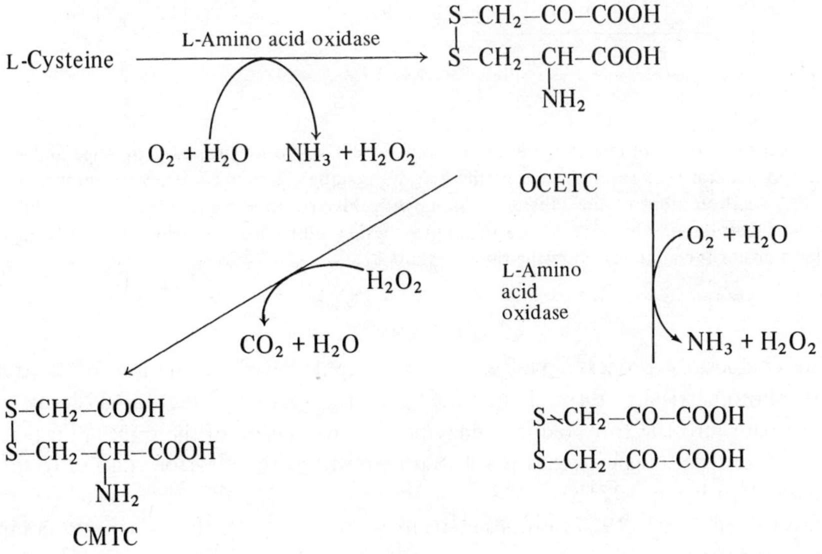

Scheme 16

As the L-amino acids are oxidized by the enzyme, the FAD in the enzyme is reduced. Page and Van Etten (1969, 1971a, b) studied the mechanism of this FAD reduction and concluded that a proton transfer step is mediated by a basic group on the enzyme during the reduction reaction. The basic group is probably a histidyl residue. The proposed mechanism for the reduction of L-amino acid oxidase by substrate is illustrated in Scheme 17.

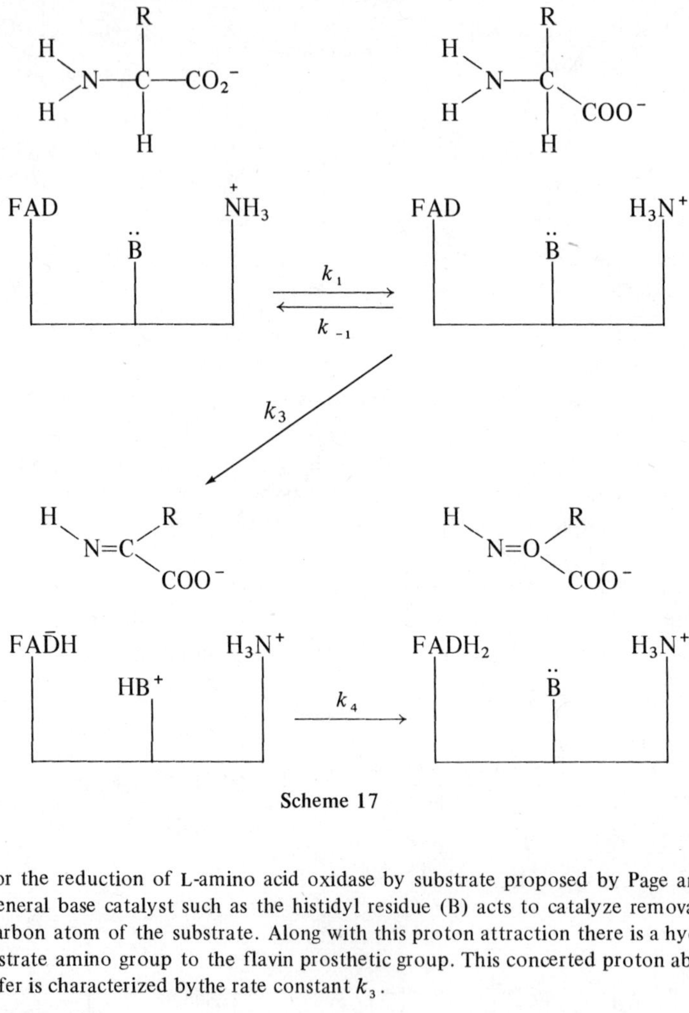

Scheme 17

Mechanism for the reduction of L-amino acid oxidase by substrate proposed by Page and Van Etten (1971a). A general base catalyst such as the histidyl residue (B) acts to catalyze removal of a proton from the α-carbon atom of the substrate. Along with this proton attraction there is a hydride transfer from the substrate amino group to the flavin prosthetic group. This concerted proton abstraction and hydride transfer is characterized by the rate constant k_3.

The reduction process involves a very rapid initial formation of a transient, spectrophotometrically distinct intermediate, followed by the rapid decay of this intermediate into the fully reduced enzyme. The formation of the transient intermediate involves a proton transfer from the substrate α position and electron transfer to the flavin (Page and Van Etten, 1971b).

Massey and Curti (1967) proposed from a kinetic study that a ternary complex of enzyme, substrate, and molecular oxygen is involved in the overall catalytic reaction.

The activity of snake venom L-amino acid oxidase is very labile at alkaline pH. Some amino acids such as L-leucine offer complete protection from pH inactivation. Aliphatic and aromatic monocarboxylic acids protect the enzyme even in the absence of substrate (Paik and Kim, 1967, 1975).

Zeller et al. (1974) investigated the effect of meta- and parasubstituted phenylalanine in the enzyme reaction. The results indicated that there was a linear Hammett plot of positive slope, which correlated with increasing Van der Waals radii of the substituents. This suggested that large substituents cause the substrate molecules to be displaced from their normal (eutopic, productive) positions at the active site, resulting in lower velocities.

2 BIOLOGICAL SIGNIFICANCE

Zeller (1948, 1951) proposed that snake venom L-amino acid oxidase activates tissue peptidases and proteolytic enzymes. However, activation of proteolytic enzymes was not experimentally demonstrated. Zwisler (1965) investigated the biological effect of the enzyme and concluded that L-amino acid oxidase from *Vipera ammodytes* venom did not activate the leucine aminopeptidase of human serum. He also found that the arterial blood pressure in rabbits was not decreased, and that necrosis was produced by the enzyme.

The LD_{50} of L-amino acid oxidase obtained from *Crotalus adamanteus* venom is 9.13 mg kg^{-1} in mice (Russell et al., 1963), approximately half the lethal value of the original venom. However, because of the low content of the enzyme in the whole venom, L-amino acid oxidase contributes less than 1% of the total lethality of the venom. Russell et al. also found that the enzyme does not affect nerve, muscle, or neuromuscular transmission.

In the fractionation of *Naja naja atra* venom Yang et al. (1959) found that the L-amino acid oxidase fraction can clearly be separated from the neurotoxic fraction and concluded that the enzyme does not contribute to the main toxic effect of cobra venom.

L-Amino acid oxidase is responsible for the yellow color in snake venoms. Chemical differences between the yellow and white venoms from snakes of some species were studied by various investigators. Zeller (1948) observed that white venoms of *Denisonia textilis* and *Bothrops itapetiningae* have no traces of L-amino acid oxidase activity. Kornalik and Master (1964) found that the yellow venom of *Vipera ammodytes* contains 200 times more enzyme than the white venom of the same snake.

3 OCCURRENCE

Many snake venoms contain L-amino acid oxidase, and frequently venoms are used as the source of the enzyme. There is no correlation between enzyme occurrence and taxonomy of venomous snakes. Venoms of the following snakes are known to contain L-amino acid oxidase activity:

Elapidae

Bungarus caeruleus (low activity)	Kocholaty et al., 1971
B. fasciatus (low activity)	Kocholaty et al., 1971
B. multicinctus (low activity)	Yang and Chang, 1960; Kocholaty et al., 1971

Micrurus fulvis fulvis (low activity)	Kocholaty et al., 1971
Naja melanoleuca (low activity)	Kocholaty et al., 1971
N. naja atra	Yang and Chang, 1960; Brisbois et al., 1968a,b; Simon et al., 1969
N. naja naja (low activity)	Kocholaty et al., 1971
N. naja kaouthia (low activity)	Kocholaty et al., 1971
Ophiophagus hannah (low activity)	Kocholaty et al., 1971

Viperidae

Vipera lebtina	Islamova, 1970
V. lebtina turanica	Sakhibov et al., 1973
V. ammodytes	Master and Kornalik 1965; Kurth and Aurich, 1973; Zwisler, 1965
V. palaestinae	Shaham and Bdolah, 1973
V. ursini	Islamova, 1970
V. aspis	Zeller, 1951
V. russellii	Master and Kornalik, 1965
V. russellii formosensis	Chiu, 1960

Crotalidae

Agkistrodon bilineatus	Kocholaty et al., 1971
A. caliginosus	Sugiura et al., 1975
A. contortrix mokesen	Kocholaty et al., 1971
A. halys	Islamova, 1970
A. halys blomhoffii	Suzuki and Iwanaga, 1960
A. piscivorus piscivorus	Kocholaty et al., 1971
A. rhodostoma	Radda, 1964; Kocholaty et al., 1971
Bothrops atrox	Jiménez-Porras, 1964; Kocholaty et al., 1971
Crotalus adamanteus	Boman and Kaletta, 1957; Paik and Kim, 1965, 1967; Curti et al., 1968; Meister and Daniel, 1966; Wellner, 1971; Kocholaty et al., 1971
C. atrox	Kocholaty et al., 1971
C. durissus cumanensis	Rodriguez et al., 1974
C. durissus durissus	Kocholaty et al., 1971
C. durissus terrificus	Chen et al., 1971; Ubuka and Yao, 1973; Kocholaty et al., 1971
C. horridus horridus	Kocholaty et al., 1971
C. scutulatus	Kocholaty et al., 1971
C. viridis helleri	Kocholaty et al., 1971
C. viridis viridis	Kocholaty et al., 1971
Sistrurus miliarius barbouri	Kocholaty et al., 1971
Trimeresurus flavoviridis	Kocholaty et al., 1971; Sato and Takaki, 1972; Nakano et al., 1972
T. gramineus formosensis	Yang and Chang, 1960
T. okinavensis	Suzuki and Iwanaga, 1960

Venoms that do not contain the enzyme are those of *Agkistrodon acutus* and *Vipera russellii formosensis* (Yang and Chang, 1960). The presence of L-amino acid oxidase has also not been reported for venoms of the sea snakes (Hydrophiidae). This does not necessarily mean, however, that the enzyme is absent in sea snake venoms; rather it indicates a lack of investigation.

REFERENCES

Boman, H. G. and Kaletta, U. (1957). Chromatography of rattlesnake venom: A separation of three phosphodiesterases, *Biochim. Biophys. Acta,* **24,** 619.

Brisbois, L., Delori, P., and Gillo, L. (1968a). Venins de serpents: Fractionnement d'un venin de cobra (*Naja naja atra*) par chromatographies sur gels de detrane, *L'Ing. Chim.,* **50,** 45.

Brisbois, L., Rabinovitch-Mahler, N., Delori, P., and Gillo, L. (1968b). Etude des fractions obtenues par chromatographie du venin de *Naja naja atra* sur sulphoéthyl-Sephadex, *J. Chromatogr.,* **37,** 463.

Chen, S. S., Walgate, J. H., and Duerre, J. A. (1971). Oxidative deamination of sulfur amino acids by bacterial and snake venom L-amino acid oxidase, *Arch. Biochem. Biophys.,* **146,** 54.

Chiu, W. C. (1960). Studies on the snake venom enzyme. IX. Electrophoresis studies of Russell's viper (*Vipera russellii formosensis*) venom and the relation of toxicity with enzyme activities, *J. Yamaguchi Med. Assoc.,* **9,** 1361.

Curti, B., Massey, V., and Zmudka, M. (1968). Inactivation of snake venom L-amino acid oxidase by freezing, *J. Biol. Chem.,* **243,** 2306.

De Kok, A. and Rawitch, A. B. (1969). Studies on L-amino acid oxidase. II. Dissociation and characterization of its subunits, *Biochemistry,* **8,** 1405.

Hafner, E. W. and Wellner, D. (1971). Demonstration of amino acids as products of the reactions catalyzed by D- and L-amino acid oxidase, *Proc. Natl. Acad. Sci.,* **68,** 987.

Hayes, M. B. and Wellner, D. (1969). Microheterogeneity of L-amino acid oxidase: Separation of multiple components by polyacrylamide gel electrofocusing, *J. Biol. Chem.,* **244,** 6636.

Inamasu, Y., Nakano, K., Kobayashi, M., Sameshima, Y., and Obo, F. (1974). On the nature of the prosthetic group of the L-amino acid oxidase from habu snake (*Trimeresurus flavoviridis*) venom, *Acta Med. Univ. Kagoshima.,* **16,** 23.

Islamova, G. (1970). Comparative studies concerned with L-amino acid oxidases in poisons of five species of central Asiatic snakes, *Uzb. Biol. Zh.,* **14,** 3.

Jiménez-Porras, J. M. (1964). Venom proteins of the fer-de-lance, *Bothrops atrox,* from Costa Rica, *Toxicon,* **2,** 155.

Kocholaty, W. F., Ledford, E. B., Daly, J. G., and Billings, T. A. (1971). Toxicity and some enzymatic properties and activities in the venoms of Crotalidae, Elapidae and Viperidae, *Toxicon,* **9,** 131.

Kornalik, F. and Master, R. W. P. (1964). A comparative examination of yellow and white venoms of *Vipera ammodytes, Toxicon,* **2,** 109.

Kurth, J. and Aurich, H. (1973). Purification and some properties of L-amino acid oxidase from the venom of sand viper (*Vipera ammodytes*), *Acta Biol. Med. Ger.,* **31,** 641.

Kurth, J. and Aurich, H. (1976) Einfluss von pH-Wert und Tempertur auf die Stabilität der L-Aminosäureoxydase aus dem Gift der Sandotter, *Acta Biol. Med. Germ.,* **35,** 175.

Massey, V. and Curti, B. (1967). On the reaction mechanism of *Crotalus adamanteus* L-amino acid oxidase, *J. Biol. Chem.,* **242,** 1259.

Master, R. W. P. and Kornalik, F. (1965). Biochemical differences in yellow and white venoms of *Vipera ammodytes* and Russell's viper, *J. Biol. Chem.,* **240,** 139.

Meister, A. (1956). "The use of snake venom L-amino acid oxidase for the preparation of α-keto acid," in E. E. Buckley and N. Porges, *Venoms,* American Association of the Advancement of Science, Washington, D.C. pp. 295–302.

Meister, A. and Wellner, D. (1966). "L-amino acid oxidase," in E. C. Slater, Ed., *Flavins and Flavoproteins,* Elsevier, Amsterdam, pp. 226–241.

Nakano, K., Inamasu, Y., Hagihara, S., and Obo, F. (1972). Isolation and properties of L-amino acid oxidase in the Habu snake (*Trimeresurus flavoviridis*) venom, *Acta Med. Univ. Kagoshima.,* **14,** 229.

Page, D. S. and Van Etten, R. L. (1969). L-Amino-Acid Oxidase. I. Effect of pH, *Biochim. Biophys. Acta,* **191,** 38.

Page, D. S. and Van Etten, R. L. (1971a). L-Amino acid oxidase. II. Deuterium isotope effects and the action mechanism for the reduction of L-amino acid oxidase by L-leucine, *Biochim. Biophys. Acta,* **227,** 16.

Page, D. S. and Van Etten, R. L. (1971b). L-Amino acid oxidase. III. Substrate substituent effect upon the reaction of L-amino acid oxidase with phenylalanine, *Bioorg. Chem.,* **1,** 36.

Paik, W. K. and Kim, S. (1964). Enzymic synthesis of ϵ-*N*-acetyl-L-lysine, *Arch. Biochem. Biophys.,* **108,** 221.

Paik, W. K. and Kim, S. (1965). pH-Substrate relation of L-amino acid oxidases from snake venom and rat kidney, *Biochim. Biophys. Acta,* **96,** 66.

Paik, W. K. and Kim, S. (1967). Studies on the stability of L-amino acid oxidase of snake venom, *Biochim. Biophys. Acta,* **139,** 49.

Paik, W. K. and Kim, S. (1975). A factor which counteracts the stabilizing activity of acetate ion on ophioxidase at alkaline pH, *Experientia,* **31,** 150.

Radda, G. K. (1964). Electronic effects and L-amino acid oxidase specificity, *Nature,* **203,** 936.

Rodriguez, O. G., Scannone, H. R., and Parra, N. D. (1974). Enzymatic activities and other characteristics of *Crotalus durissus cumanensis* venom, *Toxicon,* **12,** 297.

Russell, F. E., Buess, F. W., Woo, M. Y., and Eventov, R. (1963). Zootoxicological properties of venom L-amino acid oxidase, *Toxicon,* **1,** 229.

Sakhibov, D. N., Islamova, G., and Yukelson, L. Ya (1973). Isolation and characteristics of L-amino acid oxidase from venom of the Central Asian gursa, *Biokhimiya,* **38,** 216.

Sato, H. and Takaki, S. (1972). Enzymogram of habu venom, *Acta Med. Univ. Kagoshima.,* **14,** 1.

Shaham, N. and Bdolah, A. (1973). L-Amino acid oxidase from *Vipera palaestinae* venom: Purification and assay, *Comp. Biochem. Physiol.,* B, **46,** 691.

Simon, J., Brisbois, L., and Gillo, L. (1969). Fractionation of cobra venom by electrofocusing, *J. Chromatogr.,* **44,** 209.

Sugiura, M., Sasaki, M., Ito, Y., Akatsuka, M., Oikawa, T., and Kakino, M. (1975). Purification and properties of L-amino acid oxidase from the venom of Kankokumamushi (*Agkistrodon caliginosus*), *Snake,* **7,** 83.

Suzuki, T. and Iwanaga, S. (1960). Studies on snake venom. VIII. Substrate specificity of L-amino acid oxidase in Mamushi (*Agkistrodon halys blomhoffi* Boie) and Habu (*Trimeresurus okinavensis* Boulenger) venoms, *Yakugaku Zasshi,* **80,** 1002.

Ubuka, T. and Yao, K., (1973). Oxidative deamination of L-cystine by L-amino acid oxidase from snake venom: Formation of *S*-(2-oxo-2-carboxyethylthio)cysteine and *S*-(carboxymethylthio)cysteine, *Biochem. Biophys. Res. Commun.,* **55,** 1305.

Wellner, D. (1966). Evidence for conformational changes in L-amino acid oxidase associated with reversible inactivation, *Biochemistry,* **5,** 1585.

Wellner, D. (1971). L-Amino acid oxidase (snake venom), *Methods Enzymol.,* **17,** Part B, 597.

Wellner, D. and Meister, A. (1960). Crystalline L-amino acid oxidase of *Crotalus adamanteus, J. Biol. Chem.,* **235,** 2013.

Yang, C. C. and Chang, C. C. (1960). Biochemical studies on the snake venoms. VI. On the ophio-L-amino acid oxidase activity of Formosan snake venoms, *J. Formosan Med. Assoc.,* **59,** 187.

Yang, C. C., Su, C. C., and Chen, C. J. (1959). Biochemical studies on the Formosan snake venoms. V. The toxicity of Hyappoda (*Agkistrodon acutus*) venom, *J. Biochem. (Tokyo)* **46,** 1209.

Zeller, E. A. (1948). Enzymes of snake venoms and their biological significance, *Adv. Enzymol.,* **3,** 459.

Zeller, E. A. (1951). "Enzymes as essential components of bacterial and animal toxins," in J. F. Sumner and K. Myback, Eds., *The Enzymes,* Vol. 2, Part 2, Academic, New York, p. 986.

Zeller, E. A. and Maritz, A. (1944). Über eine neue L-Amino säureoxydase, *Helv. Chim. Acta,* **27,** 1888.

Zeller, E. A., Clauss, L. M., and Ohlsson, J. T. (1974). Interaction of ophidian L-amino acid oxidase with its substrates and inhibitors: Role of molecular geometry and electron distribution, *Helv. Chim. Acta,* **57,** 261.

Zwisler, O. (1965). L-Amino acid oxidase from the venom of *Vipera ammodytes, Z. Physiol. Chem.,* **343,** 178.

6 Acetylcholinesterase

Acetylcholine Acetyl-Hydrolase   EC 3.1.1.7

1 CHEMISTRY	97
2 OCCURRENCE	100
References	107

Acetylcholinesterase commonly occurs in nerves, serum, erythrocytes, and the retina. This enzyme is also commonly present in the venoms of Elapidae and Hydrophiidae but not in those of Viperidae and Crotalidae. Since acetylcholinesterase is involved in nerve transmission, it was thought at one time that venom acetylcholinesterase was responsible for the neurotoxic action of Elapidae venoms (Zeller, 1951). However, by employing electrophoretic methods, this enzyme was clearly separated from the toxic fraction in venoms of *Naja naja atra* and *Bungarus multicinctus* (Yang et al., 1960). This separation provided a positive statement that there is no cause and effect relationship between cholinesterase activity and venom toxicity.

This finding was important and had a considerable influence on the concept of the action of snake venoms. Biochemists tend to explain every biological activity in terms of enzyme action. There was no exception in the field of venom research, but the separation of venom enzymes from the major neurotoxic action led eventually to the discovery that snake neurotoxins bind to aceylcholine receptors, and that the neurotoxic action has nothing to do with acetylcholinesterase. (See Chapter 17, "Binding of Neurotoxins to Acetylcholine Receptors.")

1 CHEMISTRY

Acetylcholinesterase is quickly destroyed by heat treatment (Yukel'son and Malikov, 1970). Maximum activity was obtained at 37 to 38°C and at pH 8 to 8.5 for the purified enzyme from the venom of *Naja oxiana* (Sorokin et al., 1972). In addition to acetylcholine, acetylthiocholine can serve as a substrate for *N. oxiana* acetylcholinesterase (Sorokin et al., 1972). The enzyme, however, does not hydrolyze butyrylthiocholine.

Our information on the chemical and physical properties of purified venom acetylcholine sterase is due largely to the work of Kumar and Elliott (1973a, b). They isolated the enzyme from *Bungarus fasciatus* venom by affinity chromatography, using the enzyme inhibitor trimethyl-(*m*-aminophenyl)ammonium chloride hydrochloride coupled to Sepharose 4B. The molecular weight of acetylcholinesterase is 130,000 or 126,000, and the enzyme is composed of subunits of 62,700 daltons. The $S_{20,w}$ is 5.82, and $D_{20,w}$ is 4.5×10^{-7}. Combination of two monomers gives a molecular weight of 126,000, which is in agreement with the value of 130,000 obtained by gel filtration.

Table 1 Amino Acid Composition of Acetylcholinesterase from *Bungarus fasciatus* Venom and Electric Eel

Amino Acid	Residues Assuming four histidine residues	
	B. fasciatus [a]	electric eel [b]
Lysine	6	8
Histidine	4	4
Arginine	12	10
Aspartic acid	24	20
Threonine	10	8
Serine	14	12
Glutamic acid	20	16
Proline	18	14
Glycine	18	14
Alanine	16	10
Valine	14	12
Methionine	4	5
Isoleucine	6	6
Leucine	22	16
Tyrosine	8	7
Phenylalanine	12	10
½ Cystine	2	2

[a] From Kumar and Elliott (1973a)

[b] From Leuzinger and Baker (1968)

1 Chemistry

Acetylcholinesterase from *B. fasciatus* has been viewed by electron microscopy, employing negative staining (Nickerson and Kumar, 1974). The basic monomeric unit is circular and has a diameter of 50Å. Associations of subunits can form tetramers, but dimers are most frequently observed.

The molecular weight of cholinesterase obtained from the electric eel is about twice that of the corresponding enzyme from snake venom and is about 260,000. Immunodiffusion studies indicate that acetylcholinesterase isolated from electric eel or bovine erythrocyte does not respond to antibodies produced against the enzyme of *B. fasciatus*

Table 2 Values of V_{max} and K_m for Various Substrates with Acetylcholinesterase from *Bungarus fasciatus* Venom

Substrate	Structural Formulae	V_{max} $\times 10^{-2}$	V_{max}^a	$K_m \times 10^{-5}$
Acetylcholine	$(CH_3)_3\text{-}\overset{+}{N}\text{-}CH_2CH_2\text{-}O\text{-}\overset{O}{\underset{\|}{C}}\text{-}CH_3$	5.14	1.00	4.17
Acetylthiocholine	$(CH_3)_3\text{-}\overset{+}{N}\text{-}CH_2CH_2\text{-}S\text{-}\overset{O}{\underset{\|}{C}}\text{-}CH_3$	8.14	1.59	1.66
Acetyl-β-methylcholine	$(CH_3)_3\text{-}\overset{+}{N}\text{-}CH_2\text{-}\overset{CH_3}{\underset{\|}{CH}}\text{-}O\text{-}\overset{O}{\underset{\|}{C}}\text{-}CH_3$	1.814	0.35	51.66
Acetyl-β-methylthiocholine	$(CH_3)_3\text{-}\overset{+}{N}\text{-}CH_2\text{-}\overset{CH_3}{\underset{\|}{CH}}\text{-}S\text{-}\overset{O}{\underset{\|}{C}}\text{-}CH_3$	4.58	0.89	1.04
Propionylcholine	$(CH_3)_3\ \overset{+}{N}\text{-}CH_2\text{-}CH_2\text{-}O\text{-}\overset{O}{\underset{\|}{C}}\text{-}CH_2\text{-}CH_3$	3.32	0.63	3.31
Propionylthiocholine	$(CH_3)_3\ \overset{+}{N}\text{-}CH_2\text{-}CH_2\text{-}S\text{-}\overset{O}{\underset{\|}{C}}\text{-}CH_2\text{-}CH_3$	4.24	0.83	2.5
Propionyl-β-methylthiocholine	$(CH_3)_3\ \overset{+}{N}\text{-}CH_2\text{-}\overset{CH_3}{\underset{\|}{CH}}\text{-}S\text{-}\overset{O}{\underset{\|}{C}}\text{-}CH_2\text{-}CH_3$	2.97	0.58	3.07
Acetylcholine +2% EtOH	$(CH_3)_3\ N\text{-}CH_2\text{-}CH_2\text{-}O\text{-}\overset{O}{\underset{\|}{C}}\text{-}CH_3$	6.51	1.27	13.18
Phenylacetate +2% EtOH	$\text{C}_6\text{H}_5\text{-}O\text{-}\overset{O}{\underset{\|}{C}}\text{-}CH_3$	5.76	1.12	114.2
Indophenylacetate +2% EtOH	$CH_3\text{-}\overset{O}{\underset{\|}{C}}\text{-}O\text{-}C_6H_4\text{-}N\text{=}C_6H_4\text{=}O$	16.13[b]	—	141.18

[a] V_{max} with acetylcholine is arbitrarily set at unity.
[b] V given as O.D. change/min per mg protein.

This table is reproduced from Kumar and Elliott, 1975a.

venom. This implies that there are no common antigenic sites between the venom enzyme and the enzyme from other origins (Kumar and Elliot, 1973b). However, comparison of the amino acid compositions of the enzymes from *B. fasciatus* and electric eel origins indicates that they are similar (Table 1).

The purified venom enzyme also hydrolyzes a number of substrates other than acetylcholine (see Table 2). Esters of thiocholine have lower K_m and higher V_{max} values than their oxygen analogs. Enzyme behavior in the presence of different inhibitors was also studied by Kumar and Elliott (1975a), as shown in Table 3.

Table 3 Competitive and Noncompetitive Constants for Various Inhibitors of the Hydrolysis of Acetylcholine at pH 7 to 5

Inhibitor	K_i	K_i^1
Choline	7.0×10^{-4}	3.24×10^{-3}
Butrylcholine	7.12×10^{-5}	1.02×10^{-4}
Succinylcholine	2.51×10^{-5}	2.40×10^{-4}
Trimethylamine	5.53×10^{-3}	6.3×10^{-3}
Tetrapropylammonium	1.09×10^{-4}	1.54×10^{-4}
Trimethylphenylammonium	4.10×10^{-5}	5.27×10^{-4}
Fructose-1-6-diphosphate-Tetracyclohexyl ammonium salt	3.77×10^{-5}	7.14×10^{-4}
d-Tubocurarine	1.04×10^{-4}	1.28×10^{-4}
Neostigmine	7.4×10^{-8}	7.1×10^{-7}
Eserine	1.89×10^{-7}	1.4×10^{-6}

The table is reproduced from Kumar and Elliott (1975a).

Acetylcholinesterase of *B. fasciatus* venom shows enzyme activity from pH 5 to 9 with maximum activity at pH 8.0. The optimum temperature is 40°C, indicating instability at high temperature. The enzyme has a pK of 6.52 at 37°C and a heat of ionization of 7310 cal mole^{-1}; the energy of activation is 5700 cal mole^{-1}. Since histidine has a pK value close to 6.52, it is suspected that histidine is important for enzyme action (Kumar and Elliott, 1975b).

Like erythrocyte acetylcholine sterase, the enzyme obtained from *Naja naja oxiana* is inhibited by thiophosphonate compounds (Siigur et al., 1975).

2 OCCURRENCE

Cholinesterase is present in the venoms of Elapidae and Hydrophiidae but not in those of Viperidae and Crotalidae (Zeller, 1948; Chang and Lee, 1955). The following species have

2 Occurrence

been found to contain the enzyme in their venoms:

Elapidae

Acanthophis antarcticus	Zeller, 1947
Bungarus caeruleus	Zeller, 1947; McLean et al., 1971
B. fasciatus	Chowdhury, 1944; Ghosh, 1940, Zeller, 1947; McLean et al., 1971
B. multicinctus	Yang et al., 1960
Demansia textilis	Zeller, 1947
Dendraspis angusticips	Zeller, 1948
Denisonia superba	Zeller, 1947
Elaps corallinus	Zeller, 1947
Naja bungarus	Zeller, 1947
N. flava	Zeller, 1947
N. haje	Zeller, 1947; Mohamed et al., 1969; McLean et al., 1971
N. melanoleuca	Zeller, 1947; McLean et al., 1971
N. naja	Bovet-Nitti, 1947; Bovet-Nitti and Bovet, 1943; Chowdhury, 1944; Ghosh, 1940; Sarkar et al., 1942; Master and Savur, 1959
N. naja atra	Yang et al., 1960; Simon et al., 1969
N. naja oxiana	Yukel'son and Malikov, 1970; Siigur et al., 1974; Sorokin et al., 1972
N. nigricollis	Zeller, 1947
N. nivea	McLean et al., 1971
N. tripudians	Chowdhury, 1942
Notechis scutatus	Zeller, 1947
Pseudechis australis	Zeller, 1947
P. porphyriacus	Zeller, 1947
Sepedon haemachates	Zeller, 1947
Walterinessia aegyptia	Mohamed et al., 1969

Hydrophiidae

Enhydrina schistosa	Tu and Toom, 1971

Zeller lists as follows, the species whose venoms do not contain acetylcholinesterase:

Agkistrodon mokasen
A. piscivorus
Bitis arietans
B. gabonica
Bothrops alternatus
B. atrox
B. cotiara
B. jararaca
B. jararacussu

B. itapetiningae
B. neuwiedii
B. nummifera
Crotalus adamanteus
C. cinereus
C. horridus
C. lucasensis
C. ruber
C. terrificus
C. terrificus basilicus
C. viridis
C. viridis oreganus
Echis carinatus
Sistrurus catenatus
Trimeresurus gramineus
Vipera ammodytes
V. aspis
V. aspis hugyi
V. lebetina
V. russellii

The snakes whose venoms do not show enzyme activity all belong to the families Viperidae and Crotalidae.

The absence of cholinesterase in venoms of *Naja nigricollis, Echis carinatus, E. coloratus, Cerastes cornutus*, and *C. vipera* was reported by Mohamed et al. (1969).

REFERENCES

Bovet-Nitti, F. (1947). Sur la nature de l'éstérase contenue dans le venin de cobra, *Experientia,* **3**, 283.

Bovet-Nitti, F. and Bovet, D. (1943). Application de la méthode de Warburg à l'etude de l'action éstérasique de venin de cobra, *Ann. Inst. Pasteur,* **69**, 309.

Chang, C. C. and Lee, C. Y. (1955). Cholinesterase and anticholinesterase activities in snake venoms, *J. Formosan Med. Assoc.,* **54**, 103.

Chowdhury, D. K. (1942). Isolation of cholinesterase from cobra venom (*Naja tripudians*), *Sci. Cult.,* **8**, 238.

Chowdhury, D. K. (1944). Cholinesterase, *Ann. Biochem. Exp. Med. India,* **4**, 77.

Ghosh, B. N. (1940). Die Enzyme der Schlangengifte, *Osterr. Chem. Stg.,* **43**, 158.

Kumar, V. and Elliott, W. B. (1973a). Acetylcholinesterase from a non-membrane source (*Bungarus fasciatus* venom), *Prep. Biochem.,* **3**, 569.

Kumar, V. and Elliott, W. B. (1973b). The acetylcholinesterase of *Bungarus fasciatus* venom, *Eur. J. Biochem.,* **34**, 586.

Kumar, V. and Elliott, W. B. (1975a). Acetylcholinesterase from *Bungarus fasciatus* venom. I. Substrate specificity, *Comp. Biochem. Physiol.,* **51C**, 249.

Kumar, V. and Elliott, W. B. (1975b). Acetylcholinesterase from *Bungarus fasciatus* venom. II. Effects of pH and temperature, *Comp. Biochem. Physiol.,* **51C**, 255.

Leuzinger, W. and Baker, A. L. (1968). Acetylcholinesterase. I. Large-scale purification, homogeneity, and amino acid analysis, *Proc. Natl. Acad. Sci. U.S.A.,* **57**, 446.

Master, R. W. P. and Savur, S. R. (1959). Identification of cholinesterase in cobra venom after electrophoretic separation on starch gel, *Curr. Sci.,* **28**, 112.

References

McLean, R. L., Massaro, E. J., and Elliott, W. B. (1971). A comparative study of the homology of certain enzymes in elapid venoms. *Comp. Biochem. Physiol.,* **39B,** 1023.

Mohamed, A. H., Kamel, A., and Ayobe, M. M. (1969). Some enzymatic activities of Egyptian snake venoms and a scorpion venom, *Toxicon,* **7,** 185.

Nickerson, P. A. and Kumar, V. (1974). Electron microscopic studies of acetylcholinesterase from *Bungarus fasciatus* venom, *Toxicon,* **12,** 83.

Sarkar, B. B., Maitra, S. R., and Ghosh, B. N. (1942). The effect of neurotoxin, haemolysin and cholinesterase isolated from cobra venom on heart, blood pressure and respiration, *Indian J. Med. Res.,* **30,** 453.

Siigur, E., Ilomets, T., Sarapuu, T., Lendla, M., and Nommeots, M. (1974). Separation of acetylcholinesterase from cobra (*Naja naja oxiana*) venom by affinity chromatography, *Uch. Zap. Tartusk. Gov. Univ.,* **332,** 198.

Siigur, E., Abduvakhabov, A. A., Asviksaar, A., and Ilomets, T. (1975). Specificity of cobra venom cholinesterase in a reaction with thiophosphonates, *Reakts. Sposobn. Org. Soedin.,* **11,** 861.

Simon, J., Brisbois, L., and Gillo, L. (1969). Fractionation of cobra venom by electrofocusing, *J. Chromatogr.,* **44,** 209.

Sorokin, V. M., Nigmatov, Z., and Yukel'son, L. Ya (1972). Fractionation of middle-Asian cobra venom on sulfoethyl-Sephadex and investigation of the biological activity of fractions obtained, *Biokhimiya,* **37,** 112.

Tu, A. T. and Toom, P. M. (1971). Isolation and characterization of the toxic component of *Enhydrina schistosa* (common sea snake) venom, *J. Biol. Chem.,* **246,** 1012.

Yang, C. C., Chiu, W. C., and Kao, K. C. (1960). Biochemical studies on the snake venoms. VII. Isolation of venom cholinesterase by zone electrophoresis, *J. Biochem. (Tokyo),* **48,** 706.

Yukel'son, L. Y. and Malikov, M. (1970). Cholinesterase of snake venoms, *Vop. Med. Khim. Biokhim. Gorm., Deistviya Fiziol. Aktiv. Veshchestv. Radiats.,* **76.**

Zeller, E. A. (1947). Über das Vorkommen und die Natur der Cholinesterase der Schlangengifte, *Experientia,* **3,** 375.

Zeller, E. A. (1948). Enzymes of snake venoms and their biological significance, *Adv. Enzymol.,* **8,** 459.

Zeller, E. A. (1951). Enzymes as essential components of toxins, *Enzymes,* **1,** 986.

7 Proteolytic Enzymes

Peptide Hydrolases

1 ENDOPEPTIDASE 104
   1.1 Absence of Trypsin: Separation of Proteases from Arginine Ester Hydrolases, 105
   1.2 Physical and Chemical Properties, 106
   1.3 Presence of Metals, 107
   1.4 Carbohydrate Content, 110
   1.5 Site of Hydrolysis, 110
   1.6 Biological Activities, 116
       Lethality, 116
       Hemorrhage, 116
   1.7 Occurrence 117

2 EXOPEPTIDASE 119

3 SPECIFIC PROTEASES 119
   3.1 Collagenase, 120
   3.2 Elastase, 122

   References 123

1 ENDOPEPTIDASE

The presence of proteolytic enzymes in various snake venoms has been recognized for many years. Probably the first recorded observation of snake venom protease was that of Lacerda in 1884. Venoms usually contain more than one type of protease. At least five electrophoretically distinct proteases have been observed (Ohsaka, 1960). Three different proteases were isolated from *Crotalus atrox* venom (Pfleiderer and Sumyk, 1961), and three proteinases were separated, using casein as a substrate, from the venom of *Agkistrodon halys blomhoffii* (Satake et al., 1963b). However, many proteolytic enzymes from a given venom are still uncharacterized, so that, our knowledge of venom proteases remain fragmental.

It is of interest that the occurrence of endopeptidases shows some correlation to the family of snakes. Venoms of Crotalidae possess very strong proteolytic activity, whereas

1 Endopeptidase

those of Viperidae are weak. Venoms of Elapidae and Hydrophiidae (sea snakes) show very weak activity or no activity when casein or hemoglobin is used as a substrate. Although Elapidae venoms show very little endopeptidase activity, the venoms of all families exhibit marked exopeptidase activity.

1.1 Absence of Trypsin: Separation of Proteases from Arginine Ester Hydrolase

At one time venom endopeptidase was considered to be identical to trypsin. However, several lines of evidence indicated that the two enzymes were similar rather than identical. After the introduction of synthetic substrates into the assay for trypsin and chymotrypsin, these substrates were used by many investigators to assay venom proteolytic activity. It was observed that venoms of Crotalidae and Viperidae hydrolyzed arginine esters such as BAEE (*N*-benzoyl-L-arginine ethyl ester) and TAME (*p*-toluenesulfonyl-L-arginine methyl ester), which are well-known synthetic substrates for trypsin (Deutsch and Diniz, 1955, 1956; Henriques et al., 1958; Tu et al., 1965b). Since synthetic substrates for chymotrypsin such as BETT (*N*-benzoyl-L-tyrosine ethyl ester) and ATEE (*N*-acetyl-L-tyrosine ethyl ester) were not hydrolyzed, the presence of chymotrypsin in snake venoms was not indicated (Tu et al., 1966a).

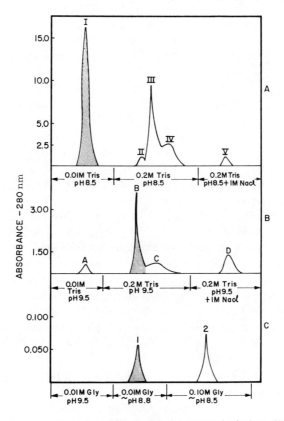

Figure 7.1 Separation of proteolytic enzyme and arginine ester hydrolase. (A) Both arginine ester hydrolase and proteolytic activities are concentrated in fractions I and III. (B) Fractionation of fraction I gives fraction B with high esterase activity and some proteolytic activity. (C) Complete separation of esterase and proteolytic activities. Both fractions originated from fraction B. Fraction 1 shows only arginine ester hydrolase activity; fraction 2, only proteolytic activity. Based on data of Toom et al. (1970).

Although synthetic arginine esters were hydrolyzed by various venoms, it was also observed that hydrolysis was not decreased by soybean trypsin inhibitor. Consequently, venom protease was assumed to be nonidentical to trypsin (Deutsch and Diniz, 1955; Tu et al, 1966a, b). As early as 1958, Henriques et al. observed that, of two venom proteolytic fractions, one had a strong caseinase activity whereas the other showed marked benzoyl-L-arginine amide activity. Apparently, these workers were not aware at that time that they had separated two distinct enzymes. In 1963, two enzymes, one responsible for protein hydrolysis and the other to hydrolyze arginine esters, were isolated (Satake et al., 1963a; Wagner et al., 1968; Delpierre, 1968a, b; Toom et al., 1970; Delpierre et al., 1971).

An example of the separation of the proteolytic enzyme from the arginine esterase is shown in Fig. 1.

1.2 Physical and Chemical Properties

The sedimentation velocities of several venom proteases have been determined by a number of workers. Some reported values are as follows:

Species	Protease	Sedimentation Velocity	Reference
Agkistrodon halys blomhoffii	Proteinase a	3.59	Oshima et al., 1968b
	Proteinase b	5.54	Oshima et al., 1968a
	Proteinase c	4.94	Oshima et al., 1968b
A. piscivorus leucostoma	Leucostoma peptidase A	2.8	Wagner et al., 1968
Trimeresurus flavoviridis	H_2-proteinase	2.43	Takahashi and Ohsaka, 1970

For Viperidae venom, protease A obtained from *Bitis arietans* had a value of 2.5 to 2.6 (Van der Walt and Joubert, 1972a, b).

Diffusion coefficients of 5.26 and 5.88 for proteinases b and c from *A. halys blomhoffii* venom, 10.66 for from *A. piscivorus leucostoma, leucostoma* peptidase A and 9.3 for protease A from *B. arietans* venoms are known (Van der Walt and Joubert, 1971). Protease A tends to aggregate to higher weight species at low pH (Van der Walt and Joubert, 1972a). The oligomer formation of protease also depends on ionic strength (Van der Walt, 1972). The enzyme can form monomers, dimers, trimers, and tetramers. Thus the high molecular weights reported for proteinases a, b, and c from the venom of *A. halys blomhoffii* may be aggregates of dimers, tetramers, and trimers, respectively (Table 1). All other proteases reported are in the molecular weight range of 21,000 to 24,000 (Table 1).

Proteinases I and II obtained from *Vipera russellii* tend to form aggregates with other snake venom components (Dimitrov, 1971).

Proteinases a, b, and c from the venom of *Agkistrodon halys blomhoffii* exhibit pH optima of 10.5, 9.8, and 8.9, respectively. They also react differently toward heat treatment and in the presence of an inhibitor, cystein. However, they are all similar with respect to inhibition by EDTA (Satake et al., 1963a); most snake venom proteolytic activities are inhibited by EDTA (Wagner and Prescott, 1966; Delpierre, 1968b; Friederich and Tu, 1971).

1 Endopeptidase

Table 1   Purified Proteolytic Enzymes from Snake Venoms

Venoms	Name	M.W.	References
Crotalidae			
Agkistrodon halys blomhoffii	proteinase a	50,000	Ohshima et al, 1968b
	proteinase b	95,000	Ohshima et al, 1968a
	proteinase c	70,000	Ohshima et al, 1968b
A. piscivorus leucostoma	leucostoma peptidase A	22,500	Wagner et al, 1968
Bothrops jararaca	protease A		Mandelbaum et al, 1967
Crotalus atrox	α-protease	23,000	Zwilling and Pfleiderer, 1967
Trimeresurus flavoviridis	H_2-proteinase	24,000	Takahashi and Ohsaka, 1970
	proteinase	22,000	Hagihara, 1974
Viperidae			
Bitis arietans	protease A	21,400	Van der Walt and Joubert, 1971
Vipera russelli	proteinase I		Dimitrov, 1971
	proteinase II		Dimitrov, 1971

The amino acid compositions of purified proteolytic enzymes from different snake venoms are summarized in Table 2.

Several proteolytic fractions were obtained from the venom of *Bothrops jararaca*. However, they were immunologically related, suggesting that they were different molecular forms of the same enzyme (Uriel et al., 1968).

1.3   Presence of Metals

Metals are intrinsically involved in venom proteases. The fact that treatment of venoms with EDTA effectively eliminates proteolytic activity suggests that venom proteases are metalloproteins or are activated by certain metals (Friederich and Tu, 1971). Proteinase b from *Agkistrodon halys blomhoffii* venom contains 2 moles of calcium per mole of enzyme. When calcium was removed with EDTA, loss of proteolytic activity was observed, with a concomitant change in the ultraviolet absorption spectrum (Oshima et al., 1971).

The metal contents of various purified proteases are summarized in Table 3. All proteases isolated from Crotalidae venoms contain calcium (as an integrating component of the molecule), although some are reported to contain 2 moles of Ca^{2+} per mole

Table 2 Amino Acid Compositions of Snake Venom Proteolytic Enzymes

Family	Viperidae	Crotalidae			
Genus Species Subspecies	Bitis arietans	Agkistrodon halys blomhoffii	A. halys blomhoffii	A. halys blomhoffii	A. piscivorus leucostoma
Origin	S. Africa	Japan	Japan	Japan	U.S.A.
Name	protease A	proteinase a residue/100g protein	proteinase b residue/100g protein	proteinase c residue/100g protein	leucostoma peptidase A
Amino Acid					
Lysine	8	6.36	4.10	5.01	10
Histidine	8	2.01	2.35	3.58	9
Arginine	10	4.73	4.14	2.34	6
Aspartic acid	26	12.00	11.91	11.11	29
Threonine	8	4.82	3.92	3.10	15
Serine	10	4.21	2.84	3.41	18
Glutamic acid	13	8.32	8.43	8.67	19

	Van der Walt and Joubert, 1971	Oshima et al, 1968b	Oshima et al, 1968a	Oshima et al, 1968b	Wagner et al, 1968
Proline	8	2.77	3.16	3.36	6
Glycine	8	2.88	2.55	3.16	12
Alanine	10	3.15	2.73	2.83	11
Valine	11	3.91	3.66	3.75	15
Methionine	7	1.62	2.26	2.39	5
Isoleucine	12	3.96	4.47	3.62	10
Leucine	8	6.07	5.58	4.21	19
Tyrosine	10	4.05	3.70	6.06	5
Phenylalanine	7	3.18	2.51	2.59	6
Half-cystine	6	5.09	4.20	5.38	6
Tryptophan	6	1.84	1.68	1.99	4
Total residue	176	82.54	76.07	77.79	205
References	Van der Walt and Joubert, 1971	Oshima et al, 1968b	Oshima et al, 1968a	Oshima et al, 1968b	Wagner et al, 1968

Table 3 Presence of Metals in Purified Proteases from Crotalidae and Viperidae Venoms

Venoms	Name	Content	Metals not present or in negligible amounts	Reference
Crotalidae				
Agkistrotrodon halys blomhoffii	proteinase b	2 Ca/mole	Mg, Zn	Oshima et al, 1971
A. piscivorus leucostoma	leucostoma peptidase A	2 Ca/mole 1 Zn/mole	—	Spiekerman et al, 1973
Trimeresurus flavoviridis	proteinase	1 Ca/mole 1 Zn/mole	Pb, Cd, Mg, Fe, Cu, Ni	Hagihara, 1974
Viperidae				
Vipera russelli	proteinase I	0.9% Fe	—	Dimitrov, 1971

protein, whereas others contain only 1 mole. Differences may be due to the method of analysis. Since only three purified proteases from Crotalidae venoms have been subjected to metal analysis, it is premature to judge which values are correct. One mole of zinc per mole of enzyme was reported for *leucostoma* peptidase A from the venom of *A. piscivorus* and the proteinase from *Trimeresurus flavoviridis*, while no zinc was found in proteinase b. The only proteinase isolated from the venom of Viperidae is proteinase I, which contains 0.9% iron. Zinc and calcium were not analyzed, thus making it difficult to judge the similarity of Viperdae protease to the proteases of other venoms.

Additional cations of various types did not increase the proteolytic activities of purified or semipurified enzymes (Ohsaka, 1960; Pfleiderer and Sumyk, 1961; Takahashi and Ohsaka, 1970; Hagihara, 1974).

1.4 Carbohydrate Content

Proteinases a, b, and c contain carbohydrates. The carbohydrate, neutral sugar, amino sugar, and sialic acid contents are summarized in Table 4. These are the only reported proteases of a glycoprotein nature, but carbohydrate content has not been determined for other venom proteases. However, *leucostoma* peptidase A yielded amino acid recoveries of 99.88 and 102.87% when subjected to hydrolysis (Wagner et al., 1968). Therefore it is unlikely that *leucostoma* peptidase A is a glycoprotein. Further investigation of this matter is required.

1.5 Site of Hydrolysis

To determine the specificity of venom protease, the oxidized B chain of insulin is frequently used as a substrate. The hydrolysis sites are summarized in Table 5.

Among the five venom proteases, protease A from the venom of *Bothrops jararaca* is notably different in specificity. The sites of hydrolysis of protease A are completely different from those of the other four proteases. There are considerable similarities in

1   Endopeptidase

Table 4   Carbohydrate Contents of Proteases Obtained from the Venom of *Agkistrodon halys blomhoffii*

	proteinase		
	a	b	c
Total sugar content	13.6	17.5	8.46
Neutral sugars	8.15	6.8-8.2	3.91
Amino sugars	4.72	6.5	3.34
Sialic acid	0.81	3.0-3.2	1.21
Reference	Oshima et al, 1968b	Oshima et al, 1968a	Oshima et al, 1968b

substrate specificity among protease C, α-protease, H_2-protease, and *leucostoma* peptidase A.

The specificity of protease A from the venom of *Bitis arietans* has been extensively studied by Van der Walt and Joubert (1972a). They used toxin α of *Naja nivea,* RNase, and cytochrome *c* as substrates and obtained the results illustrated in Figs. 2, 3, and 4.

The oxidized chain of insulin was also used as a substrate for α-protease from the venom of *Crotalus atrox* (Pflederer and Krauss, 1965). The sites of hydrolysis are as follows:

```
 1    2    3    4    5     6           7       8    9    10    11     12   13
Gly–Ile–Val–Gln–Glu–Cys·SO₃H–Cys·SO₃H–Ala–Ser–Val–Cys·SO₃H–Ser–Leu–
                                      ↑                            ↑
                                     14   15   16   17   18   19    20     21
                                    Tyr–Gln–Leu–Glu–Asn–Tyr–Cys·SO₃H–Asn
                                     ↑    ↑
```

When native insulin is used, the sites of hydrolysis are similar to those obtained with oxidized A and B chains except that hydrolysis does not take place at the minor cleavage sites (Jentsch, 1969b):

```
A-Chain                    S——————————S
H Gly–Ile–Val–Glu(NH₂)–Glu–Cys–Cys–Ala–Ser–Val–Cys–Ser–Leu–Tyr–Glu(NH₂)–Leu
                                    ↑                          ↑
   1    2    3      4     5   6  S 7   8    9   10   11   12   13    14      15    16
                                  S
B-Chain                          
H Phe–Val–Asp(NH₂)–Glu(NH₂)–His–Leu–Cys–Gly–Ser–His–Leu–Val–Glu–Ala–Leu–Tyr–
                               ↑                 ↑                    ↑
   1    2     3         4     5   6    7   8    9   10   11   12   13   14   15   16
```

Table 5 Sites of Hydrolysis of Oxidized B Chain for Various Snake Venom Proteases

Sequence (positions 1–30):
Phe-Val-Asn-Gln-His-Leu-CySO$_3$H-Gly-Ser-His-Leu-Val-Glu-Ala-Leu-Tyr-Leu-Val-CySO$_3$H-Gly-Glu-Arg-Gly-Phe-Phe-Tyr-Thr-Pro-Lys-Ala

Enzyme	Approximate cleavage sites (↑)	Reference
protease c (A. halys blomhoffii)	↑ after position 10; ↑ after 14; ↑ after 21	Oshima et al, 1968b
α-protease (C. atrox)	↑ after 6; ↑ after 10; ↑ after 14; ↑ after 16; ↑ after 24	Pfleiderer and Krauss, 1965
protease A (B. jararaca)	↑ after 11; ↑ after 15; ↑ after 23; ↑↑ after 25	Mandelbaum et al, 1967
H$_2$-protease (T. flavoviridis)	↑ after 10; ↑ after 14; ↑ after 22	Takahashi and Ohsaka, 1970
leucostoma peptidase A (A. piscivorus leucostoma)	↑ after 1; ↑ after 6; ↑ after 10; ↑ after 14; ↑↑↑ around 24–26	Wagner et al, 1968

1 Endopeptidase

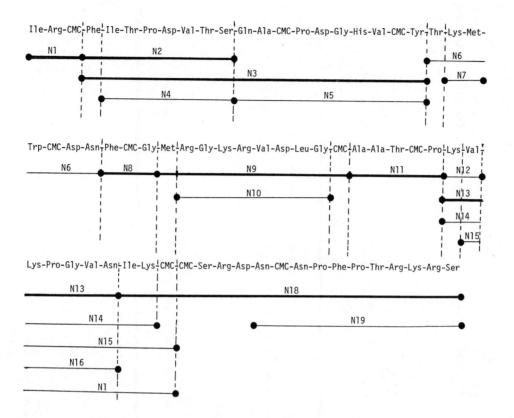

Figure 7.2 The sites of hydrolysis of RSCM-toxin by protease A. Heavy lines indicate a peptide yield greater than 5%. (Reproduced from Van der Walt and Joubert, 1972a.)

Jentsch (1969a) compared the hydrolysis of native insulin by α-protease of *Crotalus atrox* and thermolysin and found that the cleavage sites of the two enzymes were different.

The substrate specificity of α-protease was also tested (Jentsch, 1969b) on melittin, which is the primary toxic peptide from bee venom. Hydrolysis occurred mainly at the peptide bonds whose amino groups were linked to hydrophobic side chains such as valine, leucine, and isoleucine. However, one glutamine bond was cleaved. The hydrolysis sites

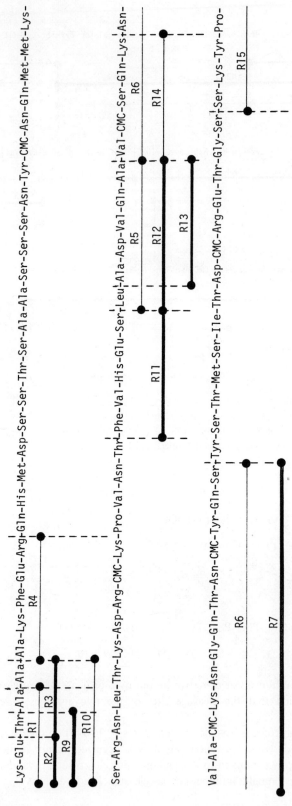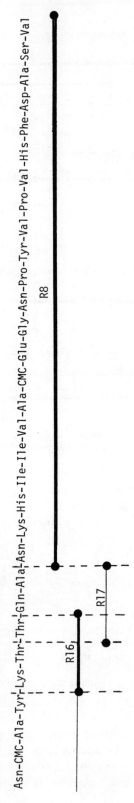

Figure 7.3 The sites of hydrolysis of RSCM-ribonuclease by protease A. Peptides R1 to R8 – 12 min of digestion; R9 to R17 – 38 min digestion. Heavy lines indicate a peptide yield of greater than 1%. (Reproduced from the Van der Walt and Joubert, 1972a.)

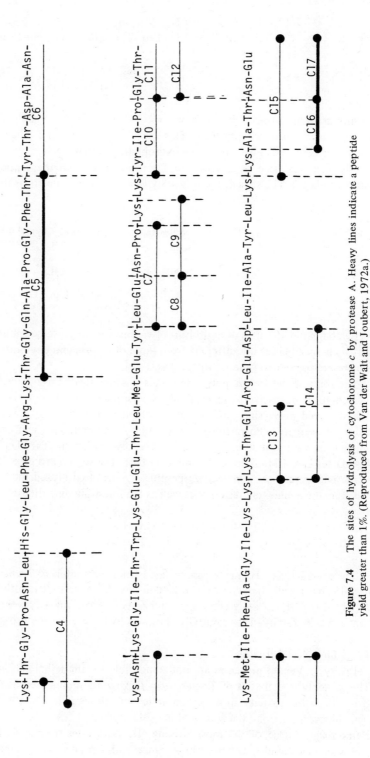

Figure 7.4 The sites of hydrolysis of cytochrome *c* by protease A. Heavy lines indicate a peptide yield greater than 1%. (Reproduced from Van der Walt and Joubert, 1972a.)

are as follows:

```
 1    2    3    4    5    6    7    8    9   10   11   12   13   14   15   16
Gly—Ile—Gly—Ala—Val—Leu—Lys—Val—Leu—Thr—Thr—Gly—Leu—Pro—Ala—Leu—
         ↑              ↑                             ↑
                                  17   18   19   20   21   22   23   24   25   26
                                 Ile—Ser—Trp—Ile—Lys—Arg—Lys—Arg—Gln—Gln
                                           ↑                        ↑
```

Bitis arietans protease has a broad specificity, attacking mainly the amino end of hydrophobic amino acid residues. This specificity is enhanced when the hydrophobic amino acid is preceded by one or two hydrophilic residues.

Glucagon was used as a substrate for protease c from the venom of *Agkistrodon halys blomhoffii*. The sites of hydrolysis by this venom enzyme are as follows:

$$\text{His—Ser—}\overset{NH_2}{\underset{|}{Glu}}\text{—Gly—Thr—Phe—Thr—Ser—Asp—Tyr—Ser—Lys—Tyr—Leu—Asp—Ser—Arg—}$$

```
 1              5 ↑ 6    7 ↑ 8                              15 ↑ 16
```

$$\text{Arg—Ala—}\overset{NH_2}{\underset{|}{Glu}}\text{—Asp—Phe—Val—}\overset{NH_2}{\underset{|}{Glu}}\text{—Trp—Leu—Met—}\overset{NH_2}{\underset{|}{Asp}}\text{—Thr}$$

```
              21 ↑ 22            25 ↑ 26              29
```

By comparing the sites of hydrolysis of various venom proteases, it has become evident that their specificities are different from those of trypsin and chymotrypsin. Therefore it is incorrect to refer to venom proteases as trypsin.

The ability to hydrolyze polyamino acids was investigated using purified protease from the venom of *A. piscivorus leucostoma* (Spiekerman et al., 1973). *Leucostoma* peptidase A hydrolyzed polylysine, polytyrosine, polyphenylalanine, and polytryptophan in that order of activity. Polyalanine, polyglycine, polyglutamic acid, polyhistidine, and polyleucine were not hydrolyzed, nor were Gly—Tyr, Leu—Tyr, Leu—NH_2 Lys—NH_2, Arg—OMe, Lys—OMe, Phe—OEt, and Tyr—OEt. The venom protease thus appeared to be a true endopeptidase. *Leucostoma* peptidase A also hydrolyzed the amino end of basic and aromatic amino acid residues, as well as the following dipeptides:

Z—Gly—Arg, Ac—Gly—Lys—OMe, Ac—Gly—Lys, Z—Gly—Phe—OMe, Z—Gly—Phe—NH_2, Z—Gly—Tyr, Z—Leu—Tyr, and Z—Trp—Tyr.

The specificity of H_2 proteinase isolated from the venom of *Trimeresurus flavoviridis* has also been well characterized (Takahashi and Ohsaka, 1970). Although Z—Gly—Leu, Z—Gly—Leu—NH_2, Z—Gly—Pro—Leu, and Z—Gly—Pro—Leu—Gly were not hydrolyzed, a pentapeptide, Z—Gly—Pro—Leu—Gly—Pro, was hydrolyzed at the Pro—Leu bond.

1.6 Biological Activities

Lethality. Venom proteases are not responsible for the lethal action of snake venoms. The proteolytic activities of *Trimeresurus flavoviridis* venom were completely separated from the lethal toxicity in electrophoretic separations (Ohsaka, 1960) and in chromatographic fractionation (Maeno and Mitsuhashi, 1961).

Hemorrhage. One of the most striking effects of bites from snakes of the Crotalidae and Viperidae families is hemorrhage. Since both Crotalidae and Viperidae venoms also

1 Endopeptidase

contain proteolytic enzymes, it was thought as early as 1930 by Houssay that venom proteases were responsible for inducing these hemorrhages. This question has not been resolved. On many occasions proteases and hemorrhagic factors are separable (Ohsaka et al., 1960; Takahaski and Ohsaka, 1970; Omori-Satoh and Ohsaka, 1970; Toom et al., 1969). However, Iwanaga et al. (1965) showed that a hemorrhagic factor is a proteolytic enzyme. This topic is discussed in further detail in Chapter 23, "Hemorrhage, Myonecrosis, and Nephrotoxic Action."

1.7 Occurrence

The venoms of Crotalidae and Viperida are good sources of proteolytic enzymes. Most Elapidae venoms either contain no proteolytic enzymes (endopeptidases) or demonstrate very weak activity. Sea snakes (family hydrophiidae) show no proteolytic activity. Tu and Toom (1971) found that the venom of *Enhydrina schistosa* hydrolyzed neither casein nor hemoglobin. The only Colubridae venom that has been investigated is that of *Leptodeira annulata*, which Mebs (1968), found to have very strong proteolytic activity.

Proteolytic enzymes have been reported in the venoms of the following snakes:

Crotalidae

Agkistrodon acutus	Murata et al., 1963; Oshima et al., 1969
A. bilineatus	Kocholaty et al., 1971
A. contortrix contortrix	Wagner and Prescott, 1966; Oshima et al., 1969
A. contortrix mokeson	Kocholaty et al., 1971; Oshima et al., 1969
A. halys blomhoffi	Murata et al., 1963; Henriques and Evseeva, 1969
A. halys halys	Henriques and Evseeva, 1969
A. piscivorus leucostoma	Wagner and Prescott, 1966
A. piscivorus piscivorus	Wagner and Prescott, 1966; Kaiser and Raab, 1967; Oshima et al., 1969; Kocholaty et al., 1971
A. rhodostoma	Kocholaty et al., 1971; Soh and Chan, 1974
Bothrops atrox	Kaiser and Raab, 1967
B. atrox asper	Kocholaty et al., 1971
B. jararaca	Henriques et al., 1958; Kaiser and Raab, 1967; Mebs 1968a
B. venezuelae	Drujan et al., 1963
Crotalus adamanteus	Oshima et al., 1969; Kocholaty et al., 1971
C. atrox	Wagner and Prescott, 1966; Brown, 1966; Mebs, 1968a; Kocholaty et al., 1971
C. basciliscuss	Oshima et al., 1969
C. durissus durissus	Kocholaty et al., 1971
C. durissus terrificus	Kaiser and Raab, 1967; Mebs, 1968a; Lodi, 1969; Kocholaty et al., 1971
C. horridus horridus	Kocholaty et al., 1971
C. scutulatus	Kocholaty et al., 1971
C. viridis helleri	Kocholaty et al., 1971
C. viridis viridis	Kocholaty et al., 1971; Oshima et al., 1969
Sistrurus miliarius barbouri	Kocholaty et al., 1971

Trimeresurus flavoviridis flavoviridis	Kocholaty et al., 1971; Ohsaka, 1960; Murata et al., 1963
T. gramineus	Murata et al., 1963
T. mucrosquamatus	Murata et al., 1963

Viperidae

Bitis arietans	Delpierre, 1968a, b; Tu et al., 1969; Oshima et al., 1969; Kocholaty et al., 1969; Van der Walt and Joubert, 1971
B. gabonica	Mebs, 1968a; Delpierre, 1968a, b; Tu et al., 1969; Oshima et al., 1969; Kocholaty et al., 1971
B. nasicornis	Tu et al., 1969
Causus rhombeatus	Delpierre, 1968a, b; Oshima et al., 1969
Echis carinatus	Henriques and Evseeva, 1969; Oshima et al., 1969; Kocholaty et al., 1971
Vipera ammodytes	Shipolini et al., 1964; Oshima et al., 1969
V. aspis	Tu et al., 1969
V. lebetina	Henriques and Evseeva, 1969
V. palaestinae	Oshima et al., 1969
V. russellii	Mebs, 1968a; Oshima et al., 1969; Kocholaty, 1971
V. russellii formosensis	Tu et al., 1969
V. ursini	Henriques and Evseeva, 1969

Elapidae

Haemachatus haemachatus	Mebs, 1968a
Ophiophagus hannah	Mebs, 1968; Oshima et al., 1969; Kocholaty et al., 1971
Pseudechis collettii	Mebs, 1968a

Colubridae

Leptodeira annulata	Mebs, 1968b

Venoms of the following species have been reported to lack proteolytic enzymes or to show very weak activity:

Bungarus fasciatus	Mebs, 1968; Oshima et al., 1969
B. multicinctus	Murata et al., 1963; Oshima et al., 1969
Dendroaspis angusticeps	Mebs, 1968a; Oshima et al., 1969; Robertson et al., 1969
D. polylepis	Robertson et al., 1969
Haemachatus haemachatus	Oshima et al., 1969; Robertson et al., 1969
Naja haje	Mebs, 1968a; Robertson et al., 1969
N. melanolenca	Robertson et al., 1969; Oshima et al., 1969
N. naja atra	Murata et al., 1963; Mebs, 1968a; Oshima et al., 1969
N. naja naja	Murata et al., 1963; Mebs, 1968a

N. naja oxianna	Henriques and Evseeva, 1969
N. naja samarensis	Oshima et al., 1969
N. nigricollis	Mebs, 1968a; Oshima et al., 1969
N. nivea	Mebs, 1968a; Robertson et al., 1969

2 EXOPEPTIDASE

Snake venoms do not contain carboxypeptidase or aminopeptidase, but they hydrolyse many dipeptides and tripeptides. One interesting observation is that venoms of Elapidae, which usually do not have endopeptidase activity, actively hydrolyze di- and tripeptides.

Ghosh et al. (1939) and Ghosh and Chowdbury (1941) observed that the venoms of *Naja naja, Bungarus fasciatus, Echis carinatus,* and *Vipera russellii* hydrolyzed Gly–Gly, Leu–Gly, and Leu–Gly–Gly. The hydrolysis of dipeptides by various snake venoms was extensively investigated by Tu et al., (1965a, 1967) and Tu and Toom (1967, 1968). The following peptides are readily hydrolyzed by different snake venoms.

Ala–Gly, Ala–Leu, Ala–Phe, Ala–Val, Gly–Ala, Gly–Gly, Gly–Ile, Gly–Phe, Gly–Leu, Gly–Met, Gly–Trp, Gly–Tyr, Gly–Val, Gly–Thr, Leu–Ala, Leu–Gly, Met–Met, Pro–Phe, Phe–Gly, Ser–Gly, Tyr–Gly, Val–Gly.

Some dipeptides that are hydrolyzed or not, depending on the venom, are the following:

Leu–Phe, Leu–Tyr, Phe–Phe, Pro–Gly.

The following dipeptides resist hydrolysis:

Gly–Asp, Gly–Glu, Gly–Pro, Leu–Val, Lys–Gly.

Some tripeptides, for instance, Ala–Gly–Gly, Gly–Phe–Phe, Gly–Leu–Tyr, Gly–Gly–Ala, Leu–Gly–Gly, and Leu–Gly–Phe, can be hydrolyzed by snake venoms (Tu et al., 1965a, 1967; Tu and Toom, 1967, 1968). Tetraglycine is not hydrolyzed by any snake venom, although the Gly–Gly bond in Ala–Gly–Gly, Gly–Gly–Ala, and Leu–Gly–Gly can be hydrolyzed. This indicates that the cleavage of glycylglycine linkage is due to nonspecific exopeptidase. It was observed that the hydrolysis of tripeptides always starts at the first bond from the N-terminal side.

3 SPECIFIC PROTEASES

The venom proteases discussed in the previous section are rather nonspecific enzymes that can hydrolyze a number of different proteins, but snake venom can also be a rich source of very specific protease. For instance, the bradykin-releasing factor (kininogenase), which hydrolyzes two specific peptide bonds in bradykininogen to release bradykinin, is an important component of venom, as is discussed in detail in Chapter 24, "Autopharmacological Action." Snake venoms are also known to have a pronounced effect on the blood coagulation system. Venoms contain coagulants as well as anticoagulants, and many of these are also very specific proteases. This topic is discussed in Chapter 2.1, "Blood Coagulation."

3.1 Collagenase

Collagenase is widely found in many normal animal tissues. The presence of collagenolytic activity has been observed in the venoms of *Vipera aspis* (Delaunay et al., 1949), *Agkistrodon piscivorus* (Hadidian, 1956), *A. contortrix contortrix* (Simpson et al., 1971), *A. piscivorus leucostoma* (Simpson et al., 1971), *Bitis nasicornis* (Simpson et al., 1971), and *Crotalus atrox* (Simpson and Rider, 1971; Simpson, 1972). Collagenase activity can be detected by observing decreases in the viscosity of collagen solution.

Collagen, one of the fibrous proteins commonly present in animal tissues, especially connective tissue, is an elongated molecule that can be readily seen under the electron microscope (Fig. 5). Injection of *Agkistrodon contortrix laticinctus* venom caused hydrolysis of collagen, and collagens present in the endoneurial space disappeared completely (Fig. 7 of Chapter 14).

Since *Crotalus atrox* venom attacks mesenteric collagen fibers but not protein, venom collagenolytic enzyme is a true collagenase (Simpson et al., 1971) (Fig. 6). Moreover,

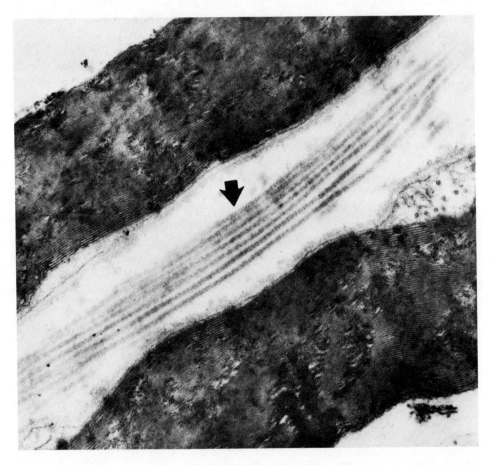

Figure 7.5 Collagen (arrow) in the endoneurial space. (Electron micrograph by Mr. M. Schmid of author's laboratory.)

3 Specific Proteases

EDTA inhibits collagenolytic activity but has no effect on arginine esterase (Simpson, 1972). Venom protease (with casein as a substrate) can be inhibited by the Hg(II) ion, but collagenolytic activity is not affected. These items of evidence indicate that venom collagenase is different from nonspecific protease and arginine esterase. Furthermore, the fact that collagenase and trypsinlike protease can be separated from *C. atrox* venom suggests that venom collagenase is a very specific enzyme (Simpson, 1971).

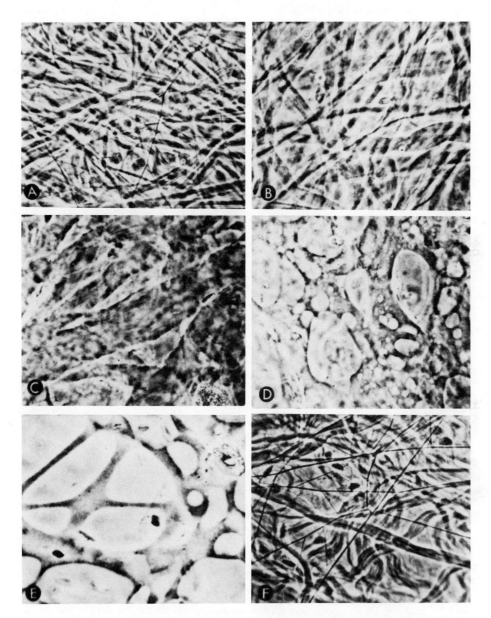

Figure 7.6 Effect of *Crotalus atrox* venom on native collagen fibers of rat mesentery in tissue cultures. (A) Freshly mounted preparation. (B) After 15-min incubation. (C) After 1-hr incubation. (D) After 4-hr incubation. (E) After 24-hr incubation. (F) After 24-hr incubation with boiled venom. (Kindly supplied by Dr. John W. Simpson and originally published in Simpson et al., 1971.)

Collagenase activity is strongest in the venoms of Crotalidae, being less potent in those of Viperidae. Most Elapidae venoms contain no or minimal activity (Simpson, 1975). Therefore the distribution of collagenolytic activity correlates well with the taxonomy of poisonous snakes at the family level.

3.2 Elastase

Elastin is a yellow scleroprotein present in elastic tissues. It is known that rattlesnake venom degrades elastin as well as collagen fibers of the mesentery. Simpson and Taylor (1973) showed histologically that native elastic fibers of rat aorta were degraded by *Crotalus atrox* venom (Fig. 7). Moreover, they found that the venom hydrolyzed the synthetic elastase substrate t-BOC-L-alanine-p-nitrophenol and solubilized Congo red–elastin. Thus they concluded that an elastaselike activity was present. However, designating the enzyme as elastase must await isolation and complete characterization of the protein. Since purified thrombinlike enzyme from the venom of *Crotalus adamanteus* (Markland and Damus, 1971) and Arvin from *Agkistrodon rhodostoma* (Esnouf and Tunnah, 1967) can also hydrolyze t-BOC-L-alanine-p-nitrophenyl ester, isolation of this enzyme is necessary to establish whether it is a true elastase.

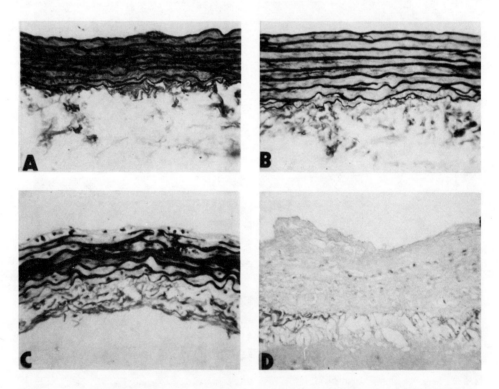

Figure 7.7 Effect of *Crotalus atrox* venom on native elastic fibers of rat aorta. Aorta segments were incubated with 0.5% venom at 37°. (A) After 4-hr incubation without venom. (B) After 20-hr incubation without venom. (C) After 4-hr incubation with venom. (D) After 20-hr incubation with venom. (Kindly supplied by Dr. John W. Simpsoe and orginally published in Simpson and Taylor, 1973.)

Elastase is present in the venoms of Crotalidae and Viperidae, but not in those of Elapidae. The distribution of elastase, like that of collagenase, has a biotaxonomic significance at the family level (Bernick and Simpson, 1976).

REFERENCES

Bernick, J. J. and Simpson, J. W. (1976). Distribution of elastaselike enzyme activity among snake venoms, *Comp. Biochem. Physsiol.,* **54B,** 51.

Brown, J. (1966). Effect of pH, temperature, antivenin and functional group inhibitors on the toxicity and enzymatic activities of *C. atrox* venom, *Toxicon,* **4,** 99.

Delaunay, M., Guillaumie, M., and Delaunay, A. (1949). Études sur le collagène. I. A propos des collagènases bactériennes, *Ann. Inst. Pasteur,* **76,** 16.

Delpierre, G. R. (1968a). African snake venoms. II. Differentiation between proteinase and amino-acid esterase activities of some African Viperidae venoms, *Toxicon,* **6,** 103.

Delpierre, G. R. (1968b). Studies on African snake venoms. I. The proteolytic activities of some African Viperidae venoms, *Toxicon,* **5,** 233.

Delpierre, G. R., Robertson, S. S. D., and Steyn, K. (1971). "Proteolytic and related enzymes in the venom of African snakes," in A. De Vries and E. Kochava, Eds., *Toxins of Animal and Plant Origins,* Vol. I., Gordon and Breach, New York, pp. 483–489.

Deutsch, H. F. and Diniz, C. R. (1955). Some proteolytic activities of snake venoms, *J. Biol. Chem.,* **216,** 17.

Deutsch, H. F. and Diniz, C. R. (1956). Some proteolytic activities of snake venoms, *Venoms,* American Association for the Advancement of Science, Washington, D.C., p. 199.

Dimitrov, G. D. (1971). Purification and partial characterization of two proteolytic enzymes from the venom of *Vipera russellii, Toxicon,* **9,** 33.

Drujan, B. D., Segal, J., and Tovar, E. (1963). Estudios bioquimicos de venenos de tres serpientes Venezolanas: *Crotalus terrificus, Bothrops atrox,* y *Bothrops venezuelae, Acta Cient. Venez.,* Supl. **92,** 1.

Ensnouf, M. P. and Tunnah, G. W. (1967). The isolation and properties of the thrombin-like activity from *Ancistrodon rhodostoma* venom, *Br. J. Haematol.,* **13,** 581.

Friederich, C. and Tu, A. T. (1971). Role of metals in snake venoms for hemorrhagic, esterase and proteolytic activities, *Biochem. Pharmacol.,* **20,** 1549.

Ghosh, B. N. and Chowdhury, D. K. (1941). Enzymes in snake venoms, *Ann. Biochem. Exp. Med.,* **1,** 31.

Ghosh, B. N., Dutt, P. K., and Chowdhury, D. K. (1939). Detection of dipeptidase, polypeptidase, carboxypeptidase and esterase in different snake venoms, *J. Indian Chem. Soc.,* **16,** 75.

Hadidian, Z. (1956). "Proteolytic activity and physiologic and pharmacologic actions of *Agkistrodon piscivorus* venom," in E. E. Buckley and N. Porges, Eds., *Venoms,* American Association for the Advancement of Science, Washington, D.C., pp. 205–215.

Hagihara, S. (1974). Isolation and properties of proteinase from habu snake (*Trimeresurus flavoviridis*) venom, *Acta Med. Univ. Kagoshima.,* **16,** 127.

Henriques, O. B. and Evseeva, L. (1969). Proteolytic, esterase, and kinin-releasing activities of some Soviet snake venoms, *Toxicon,* **6,** 205.

Henriques, O. B., Lavras, A. A. C., Fichman, M., Mandelbaum, F. R., and Henriques, S. B. (1958). The proteolytic activity of the venom of *Bothrops jararaca, Biochem. J.,* **68,** 597.

Houssay, B. A. (1930). Classification des actions des venine de serpents sur l'organisme animal, *C. R. Soc. Biol.,* **105,** 308.

Iwanaga, S., Sato, T., Mizushima, Y., and Suzuki, T. (1965). Studies on snake venoms. XVII. Properties of bradykinin releasing enzyme in the venom of *Agkistrodon halys blomhoffii, J. Biochem. (Tokyo),* **58,** 123.

Jentsch, J. (1969a). Weitere Untersuchungen zur Aminosäuresequenz des Melittins. II. Bevorzugte Spaltung der Valin-, Leucin- und Isoleucinbrindungen durch α-Protease aus *Crotalus atrox* Gift, *Z. Naturforsch.,* **24b,** 415.

Jentsch, J. (1969b). Enzymatische Spaltung nativer cystinhaltiger Polypeptide durch Thermolysin (E.C. 3.4.4). II. Vergleich von Thermolysin mit α-protease aus *Crotalus* Gift und Subtilisin, *Z. Naturforsch.*, **24**, 1290.

Kaiser, E. and Raab, E. (1967). Collagenolytic activity of snake and spider venoms, *Toxicon*, **4**, 251.

Kocholaty, W. F., Ledford, E. B., Daly, J. G., and Billings, T. A. (1971). Toxicity and some enzymatic properties and activities in the venoms of Crotalidae, Elapidae and Viperidae, *Toxicon*, **9**, 131.

Lacerda, J. B., de (1884). *Leçons sur le venin des serpents du Brésil et sur la methode de traitement des morsures venimeuses par le permaganate de potasse*, Lombaerts, Rio de Janeiro.

Lodi, W. R. (1969). Studies on pancreatic trypsin from *Crotalus durissus terrificus:* Comparison with proteases of venoms, *Toxicon*, **7**, 267.

Maeno, H. and Mitsuhashi, S. (1961). Studies on habu snake venom. IV. Fractionation of habu snake venom by chromatography on CM-cellulose, *J. Biochem. (Tokyo)*, **50**, 434.

Mandelbaum, F. R., Carrillo, M., and Henriques, S. B. (1967). Proteolytic activity of *Bothrops* protease A on the B chain of oxidized insulin, *Biochim. Biophys. Acta*, **132**, 508.

Markland, F. S. and Damus, P. S. (1971). Purification and properties of a thrombin-like enzyme from the venom of *Crotalus adamananteus* (eastern diamondback rattlesnake), *J. Biol. Chem.*, **246**, 6460.

Mebs, D. (1968a). Vergleichende Enzymuntersuchungen an Schlangengiften unter besonderer Berücksichtigung ihrer Casein-spaltenden Proteasen, *Hoppe-Seylers Z. Physiol. Chem.*, **349**, 1115.

Mebs, D. (1968b). Analysis of *Leptodeira annulata* venom, *Herpetologica*, **24**, 338.

Murata, Y., Satake, M., and Suzuki, T. (1963). Studies on snake venom. XII. Distribution of proteinase activities among Japanese and Formosan snake venoms, *J. Biochem.*, **53**, 431.

Ohsaka, A. (1960). Proteolytic activities of Habu snake venom and their separation from lethal toxicity, *Jap. J. Med. Sci. Biol.*, **13**, 33.

Omori-Satoh, T. and Ohsaka, A. (1970). Purification and some properties of hemorrhagic principle I in the venom of *Trimeresurus flavoviridis, Biochim. Biophys. Acta*, **207**, 432.

Oshima, G., Iwanaga, S., and Suzuki, T. (1968a). Studies on snake venoms. XVIII. An improved method for purification of the proteinase b from the venom of *Agkistrodon halys blomhoffii* and its physicochemical properties, *J. Biochem.*, **64**, 215.

Oshima, G., Matsuo, Y., Iwanaga, S., and Suzuki, T. (1968b). Studies on snake venoms. XIX. Purification and some physicochemical properties of proteinase a and c from the venom of *Agkistrodon halys blomhoffii, J. Biochem.*, **64**, 227.

Oshima, G., Sato-Ohmori, T., and Suzuki, T. (1969). Proteinase, arginine ester hydrolase and a kinin releasing enzyme in snake venoms, *Toxicon*, **7**, 229.

Oshima, G., Iwanaga, S., and Suzuki, T. (1971). Some properties of proteinase b in the venom of *Agkistrodon halys blomhoffii, Biochim. Biophys. Acta*, **250**, 416.

Pfleiderer, G. and Krauss, A. (1965). Wiikungsspezifität von Schlangengift-Proteasen (*Crotalus atrox*), *Biochem. Z.*, **342**, 85.

Pfleiderer, G. and Sumyk, G. (1961). Investigation of snake venom enzymes. I. Separation of rattlesnake venom proteinase by cellulose ion-exchange chromatography, *Biochim. Biophys. Acta*, **51**, 482.

Robertson, S. S. D., Steyn, K., and Delpierre, G. R. (1969). Studies on African snake venoms. III. The caseinase activity of some African Elapidae venoms, *Toxicon*, **6**, 243.

Satake, M., Murata, Y., and Suzuki, T. (1963a). Studies on snake venom. XIII. Chromatographic separation and properties of three proteinases from *Agkistrodon halys blomhoffii* venom, *J. Biochem.*, **53**, 438.

Satake, M., Omori, T., Iwanaga, S., and Suzuki, T. (1963b). Studies on snake venoms. XIV. Hydrolyses of insulin B chain and glucagon by proteinase c from *Agkistrodon halys blomhoffii* venom, *J. Biochem.*, **54**, 8.

Shipolini, R., Ivanov, Ch., and Dimitrov, G. (1964). Venom of the Bulgarian viper. II. The proteolytic, hemolytic and toxic activity of fractions separated by paper electrophoresis, *God. Khin-Tekhnol. Inst.*, **11**, 87.

Simpson, J. W. (1971). Collagenolytic activity of snake venom: The absence of collagenolytic activity in the trypsin-like enzyme from *Crotalus atrox* venom, *Comp. Biochem. Physiol.*, **40B**, 633.

Simpson, J. W. (1972). Collagenolytic activity of snake venom: The effects of enzyme inhibitors on the collagenolytic and trypsin-like enzymes derived from *Crotalus atrox* venom, *Int. J. Biochem.*, **3**, 243.

Simpson, J. W. (1975). Distribution of collagenolytic enzyme activity among snake venoms, *Comp. Biochem. Physiol.*, **51B**, 425.

Simpson, J. W. and Rider, L. J. (1971). Collagenolytic activity from venom of the rattlesnake *Crotalus atrox*, *Proc. Soc. Exp. Biol. Med.*, **137**, 893.

Simpson, J. W. and Taylor, A. C. (1973). Elastrolytic activity from the venom of the rattlesnake *Crotalus atrox*, *Proc. Soc. Exp. Biol. Med.*, **144**, 380.

Simpson, J. W., Taylor, A. C., and Levy, B. M. (1971). Collagenolytic activity in some snake venoms, *Comp. Biochem. Physiol.*, **39B**, 963.

Soh, K. S. and Chan, K. E. (1974). Caseinolytic and esteratic activities of Malayan pit viper venom and its proteolytic and thrombin-like fractions, *Toxicon*, **12**, 151.

Spiekerman, A. M., Fredericks, K. K., Wagner, F. W., and Prescott, J. M. (1973). *Leucostoma* peptidase A: A metalloprotease from snake venom, *Biochim. Biophys. Acta*, **293**, 464.

Takahashi, T. and Ohsaka, A. (1970). Purification and characterization of a proteinase in the venom of *Trimeresurus flavoviridis*, *Biochim. Biophys. Acta*, **198**, 293.

Toom, P. M., Squire, P. G., and Tu, A. T. (1969). Characterization of the enzymatic and biological activities of snake venoms by isoelectric focusing, *Biochim. Biophys. Acta*, **181**, 339.

Toom, P. M., Solie, T. N., and Tu, A. T. (1970). Characterization of a nonproteolytic arginine ester hydrolyzing enzyme from snake venom, *J. Biol. Chem.*, **245**, 2549.

Tu, A. T. and Toom, P. M. (1967). Hydrolysis of peptides by snake venoms of Australia and New Guinea, *Aust. J. Exp. Biol. Med. Sci.*, **45**, 561.

Tu, A. T. and Toom, P. M. (1968). Hydrolysis of peptides by Crotalidae and Viperidae venoms, *Toxicon*, **5**, 201.

Tu, A. T. and Toom, P. M. (1971). Isolation and characterization of the toxic component of *Enhydrina schistosa* (common sea snake) venom, *J. Biol. Chem.*, **246**, 1012.

Tu, A. T., Chua, A., and James, G. P. (1965a). Peptidase activities of snake venoms, *Comp. Biochem. Physiol.*, **15**, 517.

Tu, A. T., James, G. P., and Chua, A. (1965b). Some biochemical evidence in support of the classification of venomous snakes, *Toxicon*, **3**, 5.

Tu, A. T., Chua, A., and James, G. P. (1966a). Proteolytic enzyme activities in a variety of snake venoms, *Toxical. Appl. Pharmacol.*, **8**, 218.

Tu, A. T., Passey, R. B., and Tu, T. (1966b). Proteolytic enzyme activities of snake venoms, *Toxicon*, **4**, 59.

Tu, A. T., Toom, P. M., and Murdock, D. S. (1967). "Chemical differences in the venoms of genetically different snakes," in F. E. Russell and P. R. Saunders, *Animal Toxins*, Pergamon, New York, pp. 351–362.

Tu, A. T., Homma, M., and Hong, B. (1969). Hemorrhagic, myonecrotic, thrombotic, and proteolytic activities of viper venoms, *Toxicon*, **6**, 175.

Uriel, J., Della Santina, M., and De Rizzo, E. (1968). Characterization of enzymes in *Bothrops jararaca* venom by immunodiffusion methods, *Bull. Soc. Chim. Biol.*, **50**, 938.

Van der Walt, S. J. (1972). Studies on puff adder *(Bitis arietans)* venom. IV. Association of protease A, *Hoppe-Seylers Z. Physiol. Chem.*, **353**, 1217.

Van der Walt, S. J. and Joubert, F. J. (1971). Studies on puff adder (*Bitis arietans*) venom. I. Purification and properties of protease A, *Toxicon*, **9**, 153.

Van der Walt, S. J. and Joubert, F. J. (1972a). Studies on puff adder (*Bitis arietans*) venom. II. Specificity of protease A, *Toxicon*, **10**, 341.

Van der Walt, S. J. and Joubert, J. F. (1972b). Studies on puff adder (*Bitis arietans*) venom. III. Ultracentrifuge and ORD studies on protease A, *Toxicon*, **10**, 351.

Wagner, F. W. and Prescott, J. M. (1966). A comparative study of proteolytic activities in the venoms of some North American snakes, *Comp. Biochem. Physiol.*, **17**, 191.

Wagner, F. W., Spiekerman, A. M., and Prescott, J. M. (1968). *Leucostoma* peptidase A: Isolation and physical properties, *J. Biol. Chem.,* **243,** 4486.

Zwilling, V. R. and Pfleiderer, G. (1967). Eigenschaften der α-Protease aus dem Gift von *Crotalus atrox, Hoppe-Seylers Z. Physiol. Chem.,* **348,** 519.

8 Arginine Ester Hydrolase and Other Esterases

1 ARGININE ESTER HYDROLASE   127
    1.1 Differentiation from Proteolytic Enzymes, 128
    1.2 Isolation, 128
    1.3 Biological Activities, 129

2 OTHER ESTERASES   129

    References   131

Snake venoms contain a variety of noncholinesterases. The most extensively studied of these is the arginine ester-hydrolyzing enzyme. When synthetic substrates of arginine esters and tryosine esters began to be used in convenient assay methods for trypsin and chymotrypsin, a number of investigators employed these same substrates for snake venom proteolytic enzyme assays. Among them was the author, who erroneously reported indication of proteolytic activity in venoms (Tu et al., 1965; 1966a, b). At that time we recognized that snake venoms, while hydrolyzing arginine esters, do not hydrolyze tyrosine esters. This behavior suggested the presence of trypsin on a trypsinlike protein. However, we noticed one drastic difference between the venom enzymes and trypsin – their behavior in the presence of trypsin inhibitors. No venom enzyme activity was inhibited by soybean trypsin inhibitor or ovomucoid. Thus we concluded that venom proteolytic enzyme is different from typsin (Tu et al., 1966a). Later, a number of investigators, including us, recognized the presence of an arginine esterase activity distinct from a venom proteolytic activity, and two types of enzymes were eventually separated.

1 ARGININE ESTER HYDROLASES

Not all snake venoms contain such enzymes. Usually arginine ester hydrolases are present in the venoms of Crotalidae and Viperidae (Tu et al., 1965, 1966a; Oshima et al., 1969;

Kocholaty et al., 1971), but are not found in the venoms of Elapidae and Hydrophiidae (sea snakes). A notable exception is *Ophiophagus hannah* venom, which exhibits a weak activity toward arginine esters (Tu et al., 1966b; Kocholaty et al., 1971). In general, however, the differences suggest that the ability to hydrolyze arginine esters has taxonomic significance at the family level.

1.1 Differentiation from Proteolytic Enzymes

Since snake venoms do not hydrolyze *N*-acetyl-L-tyrosine ethyl ester and *N*-benzoyl-L-tyrosine ethyl ester, the presence of chymotrypsin is eliminated. As stated earlier, some snake venoms hydrolyze *p*-toluenesulfonyl-L-arginine methyl ester (TAME) and *N*-benzoyl-L-arginine methyl ester (BAME), but the activity is not inhibited by trypsin inhibitors. Thus snake proteases were considered at one time to be similar to trypsin but not identical.

Deutsch and Diniz (1955) observed that venom protease using hemoglobin as a substrate was inhibited by EDTA, but the hydrolysis of BAEE (*N*-benzoyl-L-arginine ethyl ester) was not affected. This suggested that two enzymes were present. In the fractionation of *Agkistrodon halys blomhoffii* venom, casein activity was separated from activity toward BAEE or TAME (Satake et al., 1963). Delpierre (1968) observed that TAME activity and casein activity were clearly separated in the fractionation patterns of venoms of *Bitis arietans, B. gabonica*, and *Causus rhombeatus*. Henriques and Evseeva (1969) assayed the protease, arginine ester hydrolase, and kinin-releasing activities of a number of venoms, including those of *Vipera lebetina, V. ursini, Echis carinatus, Agkistrodon halys halys, A. halys, A. halys blomhoffii*, and *Naja naja oxiana*, and found that all the venoms showed these activities with the exception of *N. naja oxiana*. They also observed that the kinin-releasing activity was not inhibited by protease inhibitors. Since the bradykinin-releasing activity is normally associated with arginine ester hydrolase activity, their findings suggested that venom arginine ester hydrolase is different from nonspecific proteases.

Using isoelectric focusing techniques, Toom et al. (1969) showed that there were four distinct-BAEE hydrolyzing activities and three caseinolytic activities in the venom of *Agkistrodon rhodostoma*. Two of these fractions had activities toward both casein and BAEE. Soh and Chan (1974) observed that a fraction obtained from the venom of *A. rhodostoma* had a strong caseinolytic activity with a weak activity toward BAEE or TAME. On the other hand, a fraction with arginine ester-hydrolyzing activity did not have caseinolytic activity. These findings suggest that the two activities are due to two separate enzymes, a view that was eventually proved correct by the evidence of separate purified enzymes from snake venoms.

A proteolytic enzyme, *leucostoma* peptidase A, was isolated in pure form from the venom of *Agkistrodon piscivorus leucostoma*; it was free of activity toward BAEE and L-leucyl-β-naphthylamide (Wagner et al., 1968). Thus it was unequivocally proved that arginine ester hydrolase is different from proteolytic enzyme. Isolation by Toom et al. (1970) of an arginine ester hydrolase that showed no protease activity (Fig. 1 of Chapter 7) provided additional evidence that the two activities are due to two different enzymes.

1.2 Isolation

In the first step of purification, Toom et al. (1970) observed that the venom of *Agkistrodon contortrix lacticinctus* contained two arginine ester hydrolases. The most active fraction was further purified, and eventually they isolated an arginine ester hydrolase in homogeneous form. The purified enzyme has a sedimentation coefficient of 2.7.S, a diffusion coefficient of 8.3×10^{-7} cm^2 sec^{-1} and a molecular weight of 30,000. The isoelectric point is 9.1, indicating that it is a basic protein. Enzymatic assays showed the enzyme to be specific for arginine esters, and there was no activity toward N-benzoyl-L-alanine ethyl ester, benzyloxycarbonyl lysine methyl ester, benzyloxycarbonyl lysine benzyl ester, L-lysine ethyl ester, L-lysine-p-nitrophenyl ester, L-lysine-p-nitroanilide, N-benzoyl-L-arginine-p-nitroanilide, N-benzoyl-L-arginine-β-naphthylamide, N-benzoyl-L-arginine amide, N-benzoyl-L-tryosine ethyl ester, acetyl-L-tyrosine ethyl ester, indophenyl acetate, fibrin, casein, hemoglobin, congocoll, or azocoll. The K_m values determined with N-benzoyl-L-arginine ethyl ester and p-toluenesulfonyl-L-arginine methyl ester are 1.17×10^{-4} M and 1.49×10^{-3} M, respectively. The enzyme was inhibited by diisopropyl fluorophosphate, phenylmethylsulfonyl fluoride, and N-bromosuccinimide. Optimum rates of hydrolysis were obtained at pH 7.5 to 8.5. Unfortunately, the biological activity of the purified enzyme was not tested.

1.3 Biological Activities

It is rather important to know what kind of biological activities arginine ester hydrolases possess. After all, the arginine esters used for the enzyme assay are synthetic compounds that normally do not occur in the natural system. As early as 1957, Hamburg and Rocha e Silva (1957a) noted that the bradykinin-releasing activity of the venom of *Bothrops jararaca* is not related to its proteolytic activity on casein. Later they found that the release of bradykinin activity is associated with esterase activity on BAEE (Hamberg and Rocha e Silva, 1957b). Sato et al. (1965) subsequently separated three arginine ester hydrolases with three distinct biological activities. The first enzyme fraction had bradykinin-releasing enzyme activity, the second induced an increase in capillary permeability but did not possess bradykinin-releasing or clotting activity, and the third showed clotting activity.

Arvin, a special blood coagulation enzyme isolated from the venom of *Agkistrodon rhodostoma*, has arginine ester hydrolase activity (Collins et al., 1971; Exner and Koppel, 1972). Soh and Chan (1974) also observed that a fraction with thrombinlike activity (*A. rhodostoma* venom) had marked activity toward TAME and BAEE but no caseinolytic activity.

It is well known that clotting activity and bradykinin-releasing activity are highly specific proteases. Thus arginine ester hydrolase activities are associating with proteases having very high specificity. The arginine ester hydrolase activities of bradykinin and Arvin are discussed in detail in Chapters 24 and 21, respectively.

2 OTHER ESTERASES

Snake venoms are rich in esterase. For instance, they contain phosphomono- and diesterases, arginine ester hydrolase, and acetylcholinesterase. Those enzymes have specific biological functions and have been extensively investigated. In addition to these well-known enzymes, snake venoms also contain miscellaneous esterases whose biological functions are not yet defined. Many substrates used, such as indoxyl acetate and α-naphthylamide (Tu and Toom, 1967), are not naturally occurring compounds. It is still not clear whether the hydrolysis is due to a side reaction of some well-known enzyme or to a specific enzyme. Isolation of each enzyme and extensive investigation of substrate specificities should clarify this question in the future.

The rest of this section provides a brief summary of esterases not covered previously. For example, snake venoms contain enzymes that hydrolyze indoxyl acetate and α-naphthyl acetate (McLean et al., 1971). It is not known whether one enzyme is responsible for these hydrolyses or whether two separate enzymes catalyze the reactions. The venoms capable of hydrolyzing these substrates are indicated by a plus (+) sign in Table 1.

Table 1 Ester Hydrolysis by Snake Venoms

Venoms	Substrates		
	Indoxylacetate	α-Naphthylacetate	α-Naphthylbutyrate
Naja naja	+	+	+
N. naja atra	+	+	+
N. nigricollis	+	+	+
N. melanoleuca	+	+	-
N. nivea	+	+	+
N. haje	+	+	+
Ophiophagus hannah	+	+	+
Bungarus fasciatus	+	+	+
B. caeruleus	+	+	+
Dendroaspis jamesoni	+	+	-
D. polyepsis	+	+	-
D. angusticeps	+	+	-
Demansia textilis	-	-	+

+, hydrolysis; -, no hydrolysis. The table is constructed based on the data by McLean et al, 1971.

Crotalus adamanteus venom can hydrolyze *N*-methylindoxyl acetate, which has been used as a substrate for serum cholinesterase. However, cholinesterase has not been demonstrated in the venom and therefore is not responsible for the hydrolysis of this substrate. The hydrolysis is also not due to lipase or acetylphenylalanine-3-naphthyl esterase (Elliott and Panagides, 1972).

Munjal and Elliott (1972) used immunological and histochemical reactions to find the identity of certain antigenic components and esterases in several snake venoms. The types of esterases they investigated were those that hydrolyzed β-naphthyl acetate, indoxyl acetate, and β-carbonaphthoxycholine. They found that several of the esterases present in *Naja* spp. are immunologically identical, whereas esterases from *Dendroaspis, Micrurus, Bitis, Agkistrodon*, and bee venom are not immunologically identical to those in the *Naja* spp. venoms.

REFERENCES

Collins, J. P., Basford, J. M., and Jones, J. G. (1971). Some catalytic properties of IRC–50 Arvin, *Biochem. J.*, **125**, 71.

Delpierre, G. R. (1968). African snake venoms. II. Differentiation between proteinase and amino-acid esterase activities of some African viperidae venoms, *Toxicon*, **6**, 103.

Deutsch, H. F. and Diniz, C. R. (1955). Some proteolytic activities of snake venoms, *J. Biol. Chem.*, **216**, 17.

Elliott, W. B. and Panagides, K. (1972). Hydrolysis of *N*-methylindoxyl acetate by a non-choline esterase type of esterase, *Anal. Biochem.*, **45**, 345.

Exner, T. and Koppel, J. L. (1972). Observations concerning the substrate specificity of Arvin, *Biochim. Biophys. Acta*, **258**, 825.

Hamberg, U. and Rocha e Silva, M. (1957a). On the release of bradykinin by trypsin and snake venoms, *Arch. Int. Pharmacodyn.*, **110**, 222.

Hamberg, U. and Rocha e Silva, M. (1957b). Release of bradykinin as related to the esterase activity of trypsin and of the venom of *Bothrops jararaca*, *Experientia*, **13**, 489.

Henriques, O. B. and Evseeva, L. (1969). Proteolytic, esterase, and kinin-releasing activities of some Soviet snake venoms, *Toxicon*, **6**, 205.

Kocholaty, W. F., Ledford, E. B., Daly, J. G., and Billings, T. A. (1971). Toxicity and some enzymatic properties and activities in the venoms of Crotalidae, Elapidae and Viperidae, *Toxicon*, **9**, 131.

McLean, R. L., Massaro, E. J., and Elliott, W. B. (1971). A comparative study of the homology of certain enzymes in elapid venoms, *Comp. Biochem. Physiol.*, **39B**, 1023.

Munjal, D. and Elliott, W. B. (1972). Immunological and histochemical identity of esterases and other antigens in elapid venoms, *Toxicon*, **10**, 47.

Oshima, G., Sato-Ohmori, T., and Suzuki, T. (1969). Proteinase, arginine ester hydrolase and a kinin releasing enzyme in snake venoms, *Toxicon*, **7**, 229.

Satake, M., Murata, Y., and Suzuki, T. (1963). Studies on snake venom. XIII. Chromatographic separation and properties of three proteinases from *Agkistrodon halys blomhoffii* venom, *J. Biochem.*, **53**, 438.

Sato, T., Iwanaga, S., Mizushima, Y., and Suzuki, T. (1965). Studies on snake venoms. XV. Separation of arginine ester hydrolase of *Agkistrodon halys blomhoffii* venom into three enzymatic entities: "bradykinin releasing", "clotting" and "permeability increasing," *J. Biochem. (Tokyo)*, **57**, 380.

Soh, K. S. and Chan, K. E. (1974). Caseinolytic and esteratic activities of Malayan pit viper venom and its proteolytic and thrombin-like fractions, *Toxicon*, **12**, 151.

Toom, P. M., Squire, P. G., and Tu, A. T. (1969). Characterization of the enzymatic and biological activities of snake venoms by isoelectric focusing, *Biochim. Biophys. Acta*, **181**, 339.

Toom, P. M., Solie, T. N., and Tu, A. T. (1970). Characterization of a nonproteolytic arginine ester hydrolyzing enzyme from snake venom, *J. Biol. Chem.*, **245**, 2549.

Tu, A. T. and Toom, P. M. (1967). The presence of a L-leucyl-β-napthylamide hydrolyzing enzyme in snake venoms, *Experientia*, **23**, 439.

Tu, A. T., James, G. P., and Chua, A. (1965). Some biochemical evidence in support of the classification of venomous snakes, *Toxicon*, **3**, 5.

Tu, A. T., Chua, A., and James, G. P. (1966a). Proteolytic enzyme activities in a variety of snake venoms, *Toxical. Appl. Pharmacol.*, **8**, 218.

Tu, A. T., Passey, R. B., and Tu, T. (1966b). Proteolytic enzyme activities of snake venoms, *Toxicon*, **4**, 59.

Wagner, F. W., Spiekerman, A. M., and Prescott, J. M. (1968). *Leucostoma* peptidase A: Isolation and physical properties, *J. Biol. Chem.*, **243**, 4486.

9 Other Enzymes

1 HYDROLYTIC ENZYMES 132
    1.1 Hyaluronidase, 132
    1.2 Amylase, 134
    1.3 NAD Nucleosidase, 134

2 NONHYDROLYTIC ENZYMES 135
    2.1 Transaminase, 136
    2.2 Lactate Dehydrogenase, 136

References 136

The enzymes we have discussed thus far are the better known ones. Snake venoms, however, contain many other enzymes, some not fully characterized and others not yet studied at all. In this chapter, the enzymes not mentioned previously are discussed.

1 HYDROLYTIC ENZYMES

1.1 Hyaluronidase (Hyaluronate Glycanohydrolase   EC 3.2.1.d)

Hyaluronidase is present in many venoms, including snake venoms. Zeller (1948) summarized the effects of venom hyaluronidase in a review article, and the role of hyaluronidase in animal parasites was discussed in an article by Favilli (1956). Hyaluronic acid, a mucopolysaccharide present in skin, connective tissues, and bone joint serves to promote intercellular adhesion. The enzyme is also referred to as the "spreading factor" because hydrolysis of hyaluronic acid facilitates toxin diffusion into the tissues of a victim (Duran-Reynals, 1936, 1939).

    Aqueous extracts of mammalian testicle contain a factor that increases the permeability of the tissues to injected fluids. Similar factors were later observed in snake and spider venoms (Duran-Reynals, 1939; Favilli, 1940a). Favilli (1940b) noted that venoms possess a marked mucolytic effect which can be inhibited by antivenin. McLean and Hale (1941) observed that N-acetylhexosamine was released from potassium hyaluronate by the action of *Vipera russellii* venom and that the liberation of

1 Hydrolytic Enzymes

N-acetylhexasmine was correlated to a reduction in viscosity. Thus the spreading factor nature of hyaluronidase was inferred.

Chiu (1960) investigated the hyaluronidase activities of six venoms. Partial fractionation of venom of *Naja naja atra* indicated that the toxic fraction was different from the fractions containing hyaluronidase activity. In *N. oxiana* venom, the toxic fraction was separated from the hyaluronidase fraction (Turakulov et al., 1969).

The hyaluronidase activities of several Egyptian snake venoms have been investigated (Mohamed and El-Karimi, 1969; Mohamed et al., 1973). The enzyme in *N. haje* venom is heat labile, partially inhibited by sodium gentisate, but not affected by EDTA. In regard to these properties, it resembles testicular hyaluronidase. Although venom hyaluronidase is not toxic per se, it enhances the effect of the toxins by increasing their spread.

Venoms of the following snakes contain the enzyme:

Elapidae

Acanthophis antarcticus	Duran-Reynals, 1939
Bungarus multicinctus	Chiu, 1960; Yang et al., 1960
Denisonia superba	Duran-Reynals, 1939; Chain and Duthie, 1940
Naja haje	Tarabini-Castellani, 1938; Duran-Reynals, 1939; Mohamed and El-Karimi, 1969
N. naja atra	Chiu, 1960; Brisbois et al., 1968a, b, c; Simon et al., 1969
N. nigricollis	Tarabini-Castellani, 1940; Jaques, 1956; Boquet et al., 1967; Mohamed and El-Karimi, 1969
N. oxiana	Sakhibov et al., 1968; Turakulov et al., 1969; Sorokin et al., 1972
N. tripudians	Jaques, 1956
Notechis scutatus	Duran-Reynals, 1939; Chain and Duthie, 1940
Walterinnesia aegytea	Mohamed and El-Karimi, 1969

Viperidae

Cerastes cornutus	Mohamed and El-Karimi, 1969
C. vipera	Tarabini-Castellani, 1940; Mohamed and El-Karimi, 1969
Echis carinatus	Tarabini-Castellani, 1940; Favilli, 1940a, b; Jaques, 1956; Mohamed and El-Karimi, 1969
Vipera ammodytes	Tarabini-Castellani, 1940; Favilli, 1940a, b
V. aspis	Tarabini-Castellani, 1940; Favilli, 1940a, b; McClean and Hale, 1940, 1941; Jaques, 1956
V. lebetina	Berdyeva, 1960
V. russellii	McClean and Hale, 1940, 1941; Jaques, 1956
V. russellii formosensis	Chiu, 1960

Crotalidae

Agkistrodon acutus	Chiu, 1960
A. piscivorus	Duran-Reynals, 1939
Bothrops alternatus	Tarabini-Castellani, 1938; Favilli, 1940a, b
B. jararaca	Tarabini-Castellani, 1938; Favilli, 1940a, b

Crotalidae

Crotalus adamanteus — Duran-Reynals, 1939
C. atrox — Duran-Reynals, 1939; Madinaveitia, 1941
C. terrificus — Tarabini-Castellani, 1938; Duran-Reynals, 1939; Favilli, 1941; Madinaveitia, 1941; Slotta and Ballester, 1954
Trimeresurus gramineus — Chiu, 1960
T. mucrosquamatus — Chiu, 1960; Ouyang and Shiau, 1970

1.2 Amylase (α-1,4-Glucan 4-glucanohydrolase   EC 3.2.1.1)
The presence of amylase was investigated by Mohamed et al. (1969) in the venoms of Egyptian snakes. No amylase could be found in venoms of *Naja haje, N. nigricollis, Walterinessia aegyptia, Echis carinatus, E. coloratus, Cerastes cornutus,* and *C. vipera.*

1.3 NAD Nucleosidase (NAD Glycohydrolase   EC 3.2.2.5), NADase
Nicotinamide adenine dinucleotide nucleosidase is an enzyme that hydrolyzes the nicotinamide N-ribosidic linkage of NAD (long arrow). (See Scheme 18.) The products of this reaction are nicotinamide and adenosine diphosphate ribose

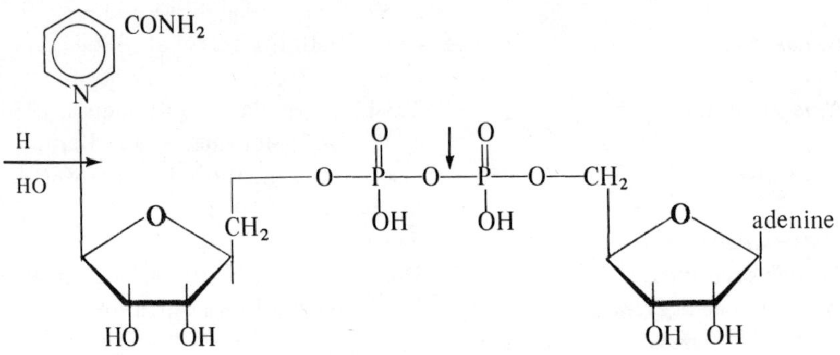

Scheme 18

The enzyme NADase should not be confused with nucleotide pyrophosphatase, whose action is indicated by the small arrow in Scheme 18. The products resulting from the latter reaction are nicotinamide mononucleotide and 5'-AMP; in snake venoms, the enzyme responsible for this reaction is exonuclease (phosphodiesterase).

NADase was first reported in the venom of *Bungarus fasciatus* (Bhattacharya, 1953). Later Suzuki et al. (1960) found it in the venoms of *B. multicinctus* and *Trimeresurus gramineus.*

Brisbois et al. (1968a, b) also found that NADase is present in *Naja naja atra* venom More recently, Tatsuki et al. (1975) made a systematic study of 37 snake venoms and found the enzyme in the following:

Agkistrodon halys blomhoffi
A. piscivorus piscivorus
A. contortrix
A. contortrix mokasen
A. acutus

A. contortrix laticinctus
Causus rhombeatus
Bungarus multicinctus
B. fasciatus

The venoms of the following species do not contain this enzyme:

Crotalus adamanteus
C. atrox
C. durissus terrificus
C. viridis viridis
C. basiliscus
Trimeresurus okinavensis
T. flavoviridis
T. mucrosquamatus
Bothrops atrox
Bitis gabonica
B. arietans
Vipera russellii
V. palestinae
V. ammodytes
Echis carinatus
Naja nivea
N. haje
N. naja atra
N. melanoleuca
N. naja samarensis
N. nigricollis
Ophiophagus hannah
Dendroaspis angusticeps
D. polyepis
Hemachatus haemachatus

The enzyme was purified from the venom of *Trimeresurus gramineus*, increasing the specific activity 5 times (Suzuki et al., 1960). The purified enzyme is separated from venom exonuclease (phosphodiesterase). The enzyme as purified by Tatsuki et al. (1975) has a pH optimum at 6.5 to 8.5. The NAD analogs, such as β-NAD$^+$ and NADP$^+$, can be hydrolyzed at equal rates, and 3-acetyl-PyAD is hydrolyzed at about one-third the rate. Reduced forms of the substrate (NADH, NADPH), α-NAD$^+$, and NMN$^+$ are not hydrolyzed. The enzyme is heat labile and loses activity at 60°C. The fact that iodoacetic acid and *p*-chloromercuric benzoic acid do not inhibit the enzyme activity suggests that a free sulfhydryl group is not essential for NADase action.

2 NONHYDROLYTIC ENZYMES

The best known nonhydrolytic enzyme found in snake venoms is L-amino acid oxidase, which was discussed in Chapter 5. Other nonhydrolytic enzymes reported in snake venom are certain types of transaminases.

2.1 Glutamic-Pyruvic Transaminase (L-Alanine: 2-Oxoglutarate Aminotransferase EC 2.6.1.2)

Glutamic-pyruvic transaminase activity was found in all venoms tested except that of *Naja nigricollis*. The enzymatic activity of *Walterinessia aegyptia* venom was destroyed by boiling but not by incubation of the venom for 1 hr at $56°$ (Mohamed et al., 1969).

No glutamic-oxalacetic transaminase activity was found in any venoms of *Naja haje, N. nigrocollis, Walterinessia aegyptia, Echis carinatus, E. coloratus, Cerastes cornutus*, and *C. vipera*, although they contain glutamic-pyruvic transaminase activity.

2.2 Lactate Dehydrogenase (L-Lactate: NAD Oxidoreductase EC 1.1.1.27)

This enzyme reversibly catalyzes the conversion of lactate to pyruvic acid (see Scheme 19).

$$CH_2-\underset{OH}{CH}-COOH \rightleftharpoons CH_3COCOOH$$

Scheme 19

The presence of the enzyme in the venoms of the following species was confirmed by McLean et al. (1971):

Naja naja
N. naja atra
N. nigricollis
N. melanoleuca
N. nivea
N. haje
Dendroaspis jamesoni
D. polylepsis
Demansia textilis

The enzyme was not found in the venoms of *Ophiophagus hannah, Bungarus caeruleus*, and *B. fasciatus*.

REFERENCES

Berdyeva, A. T. (1960). Comparative characteristics of hyaluronidase activities of *Vipera lebetina* and cobra venoms, *Tr. Turkmensk. Med. Inst.,* **10,** 129.

Bhattacharya, K. L. (1953). Effect of snake venoms on coenzyme. I. *J. Indian Chem. Soc.,* **30,** 685.

Boquet, P., Izard, Y., Meaume, J., Jouannet, M., Ronsseray, A. M., Dumarey, C., Ozenne, S., and Casseault, S. (1967). Biochemical and immunological studies on snake venom. II. Study of enzymic and toxic properties of fractions separated by filtration of *Naja nigricollis* venom on Sephadex P, *Ann. Inst. Pasteur,* **112,** 213.

Brisbois, L., Delori, P., and Gillo, L. (1968a). Venins de serpents: Fractionnement d'un venin de cobra (*Naja naja atra*) par chromatographies sur gels de dextrane, *L'Ing. Chim.,* **50,** 45.

Brisbois, L., Rabinovitch-Mahler, N., Delori, P., and Gillo, L. (1968b). Étude des fractions obteneus par chromatographie du venin de *Naja naja* atra sur sulphoéthyl-Sephadex, *J. Chromatogr.,* **37,** 463.

Brisbois, L., Rabinovitch-Mahler, N., Delori, P., and Gillo, L. (1968c). Études des fractions obtenues par chromatographie du venin de *Naja naja atra* sur sulphoéthyl-Sephadex, *J. Chromatogr.,* **37,** 463.

Chain, E. and Duthie, E. (1940). Identity of hyaluronidase and spreading factor, *Br. J. Exp. Pathol.,* **21,** 325.

References

Chiu, W. C. (1960). Studies on the snake venom enzyme. VII. On the hyaluronidase activity of Formosan snake venoms, *J. Yamaguchi Med. Assoc.,* **9,** 1355.

Duran-Reynals, F. (1936). The invasion of the body by animal poisons, *Science,* **83,** 286.

Duran-Reynals, F. (1939). A spreading factor in certain snake venoms and its relation to their mode of action, *J. Exp. Med.,* **69,** 69.

Elliot, W. B. and Panagides, K. (1972). Hydrolysis of *N*-methylindoxyl acetate by a non-choline esterase type of esterase, *Anal. Biochem.,* **45,** 345.

Favilli, G. (1940a). Proprieta mucinolitiche dei fattori diffusori, *Boll. Ist. Sieroter. Milan,* **19,** 481.

Favilli, G. (1940b). Mucolytic effect of several diffusing agents and of a diazotized compound, *Nature,* **145,** 866.

Favilli, G. (1956). "Occurrence of spreading factors and some properties of hyaluronidases in animal parasites and venoms," in E. E. Buckley and N. Porges, Eds., *Venoms,* American Association for the Advancement of Science, Washington, D.C., pp. 281–289.

Jaques, R. (1956). "The hyaluronidase content of animal venoms," in E. E. Buckley and N. Proges, Eds., *Venoms,* American Association for the Advancement of Science, Washington D.C., pp. 291–293.

Madinaveitia, J. (1941). Diffusion factors. VII. Concentration of mucinase from testicular extracts and from *Crotalus atrox* venom, *Biochem. J.,* **35,** 447.

McClean, D. and Hale, C. W. (1940). Mucinase and tissue permeability, *Nature,* **145,** 867.

McClean, D. and Hale, C. W. (1941). Studies on diffusing factors, the hyaluronidase activity of testicular extracts, bacterial culture filtrates and other agents that increase tissue permeability, *Biochem. J.,* **35,** 159.

McLean, R. L., Massaro, E. J., and Elliot, W. B. (1971). A comparative study of the homology of certain enzymes in elapid venoms, *Comp. Biochem. Physiol.,* **39B,** 1023.

Mohamed, A. H. and El-Karimi, M. M. A. (1969). Biological and viscoimetric measurement of hyaluronidase content of some Egyptian snake venoms, *J. Trop. Med. Hyg.,* **72,** 79.

Mohamed, A. H., El-Serougi, M., and Khaled, L. Z. (1969). Effect of *Cerastes cerastes* venom on blood coagulation mechanism, *Toxicon,* **7,** 181.

Mohamed, A. H., Kamel, A., and Ayobe, M. H. (1973). Hyaluronidase activity of Egyptian snake and scorpion venoms, *Ain Shams Med. J.,* **24,** 445.

Ouyang, C. and Shiau, S. Y. (1970). Relationship between pharmacological actions and enzymatic activities of the venom of *trimeresurus gramineus, Toxicon,* **8,** 183.

Sakhibov, D. N., Sorokin, V. M., and Yukel'son, L. Y. (1968). Purification and extraction of biologically active substances from cobra venom, *Uzb. Biol. Zh.,* **12,** 65.

Simon, J., Brisbois, L., and Gillo, L. (1969). Fractionation of cobra venom by electrofocusing, *J. Chromatogr.,* **44,** 209.

Slotta, K. and Ballester, A. (1954). Determinačao colorimetrica da hialuronidase das venenos ofidicos, *Mem. Inst. Butantan,* **26,** 311.

Sorokin, V. M., Nigmatov, Z., and Yukel'son, L. Ya. (1972). Fractionation of middle-Asian cobra venom on sulfoethyl-Sephadex and investigation of the biological activity of fractions obtained, *Biokhimiya,* **37,** 112.

Suzuki, T., Iizuka, K., and Murata, Y. (1960). Studies on snake venom. IX. On the studies of diphosphopyridine nucleotidase in snake venom. *Yakugaku Zasshi,* **80,** 868.

Tarabini-Castellani, G. (1938). Presence of diffusion factors in some animal venoms, *Arch. Ital. Med. Sper.,* **2,** 969.

Tarabini-Castellani, G. (1940). Recent investigation on diffusion factors in snake venoms: Specificity of the neutralization with immune serum, *Boll. Ist. Sieroter. Milan,* **19,** 332.

Tatsuki, T., Iwanaga, S., Oshima, G., and Suzuki, T. (1975). Snake venom NAD nucleosidase: Its occurrence in the venoms from the genus *Agkistrodon* and purification and properties of the enzyme from the venom of *A. halys blomhoffii, Toxicon,* **13,** 211.

Turakulov, Ya. Kh., Sakhibov, D. N., Sorokin, V. M., and Yukel'son, L. Ya. (1969). Separation of central-Asian cobra venom by means of gel filtration through Sephadex and determination of biological activity of the resulting fractions, *Biokhimiya,* **34,** 1119.

Yang, C. C., Kao, K. C., and Chiu, W. C. (1960). Biochemical studies on the snake venoms. VIII. Electrophoretic studies on banded krait (*Bungarus multicinctus*) venom and the relation of toxicity with enzyme activities, *J. Biochem.* (*Tokyo*), **48**, 714.

Zeller, E. A. (1948). Enzymes of snake venoms and their biological significance, *Adv. Enzymol.*, **8**, 459.

10 Enzyme Inhibitors

1 PHOSPHOLIPASE A$_2$ INHIBITOR	139
2 ACETYLCHOLINESTERASE INHIBITOR	140
3 ANGIOTENSINASE INHIBITORS	140
4 PROTEINASE INHIBITORS	141
5 OTHERS	145
References	145

It was well known many years ago that snake venoms inhibit fermentation and glycolysis reactions (Chain, 1936; Chain and Goldsworthy, 1938), dehydrogenase activity (Chain, 1939; Fleckenstein and Gayer, 1950), and the oxidase system (Ghosh and Chatterjee, 1948). Braganca and Quastel (1953) extensively investigated the inhibitory effects of various snake venoms on many enzyme systems and found that pyruvate dehydrogenase of brain, succinic dehydrogenase of brain and heart, cytochrome oxidase of brain, and choline oxidase of liver were all extensively inhibited. Of course all such inhibitory effects of snake venoms on various enzyme systems may well be due, not to the specific inhibitors present in snake venoms, but to the actions of nonspecific proteolytic enzymes, phospholipase A$_2$, and other hydrolytic enzymes. Only recently have true venom enzyme inhibitors actually been isolated and their mode of inhibitory actions studied. These specific inhibitors are reviewed in this chapter.

1 PHOSPHOLIPASE A$_2$ INHIBITOR

A specific inhibitor of phospholipase A$_2$ has been isolated and characterized from the venom of *Naja naja*. The molecular weight of the inhibitor is 5000, and it has a p*I* of 8.6. The inhibitor action arises from stoichiometric combination to phospholipase A$_2$. The fact that the complex disintegrates into its respective components during starch gel electrophoresis suggests that it is formed through weak electrostatic interaction of the two components. Some basic proteins such as salmine, polylysine, and cobra cytotoxin do not inhibit phospholipase A_2, and the phospholipase A_2 inhibitor does not combine with the basic proteins isolated from cobra venom (Braganca et al., 1970).

An inhibitor of phospholipase A_2 is also present in the venoms of *Bothrops* (Vidal and Stoppani, 1970). This is the reason why the phospholipase A_2 activity of venoms of *B. neuwiedii, B. jararaca, B. jararacussu*, and *B. atrox* is low compared to the values for other snake venoms (Marinetti, 1965). Inhibitory action is lost upon treatment with trypsin, suggesting that the inhibitor is a polypeptide (Vidal and Stoppani, 1970). The inhibitor has a p*I* of 6.8 with no tyrosine or tryptophan residues. The enzyme–inhibitor complex can be isolated from a chromatographic column at pH 4.5 but not at 7.5. The inhibitor contains a free sulfhydryl group, whereas phospholipase A_2 does not. When Hg(II) is added, the inhibitor combines with it and the activity is lost. Phospholipase A_2 activity is not affected in the presence of Hg(II). Thus, when Hg(II) is added to a mixture of the inhibitor and phospholipase A_2, an apparent enzyme activation occurs. The presence of the inhibitor is the cause of the "lag period" in phospholipase A_2 action; the lag period does not occur in the absence of inhibitor or in the presence of Hg(II).

There are two phospholipases A_2 in the venom of *Naja oxiana*. One has a molecular weight of 11,000 daltons; the other, of 18,000 daltons. Since the activity of the latter one is low, it is believed to be attached to the enzyme inhibitor (Sakhibov et al., 1974). Phospholipase A_2 inhibitor is also present in the serum of *Trimeresurus flavoviridis* (Kihara, 1976).

2 ACETYLCHOLINESTERASE INHIBITOR

It has been known that the acetylcholinesterase activity of cobra venom in aqueous solution is not very stable (Mounter, 1951; Augustinsson and Grahn, 1952). The acetylcholinesterase activity of *Naja naja atra* venom is very low relative to that of *Bungarus multicinctus* venom. The low activity of *N. naja atra* venom is attributed to the combination of the enzyme with a specific inhibitor (Chang and Lee, 1955). The inactivation of cholinesterase by the inhibitor is irreversible, indicating that the protein–protein interaction is strong. The inhibitor has its maximum action in the pH range of 6.0 to 8.5 and is inactive below pH 5 and above pH 10 (Lee et al., 1956). The inhibitor in *N. naja atra* venom was also reported by Brisbois et al. (1968b, c). Tazieff-Depierre and Pierre (1966) obtained three fractions from *N. nigricollis* venom with antiacetylcholinesterase activity.

3 ANGIOTENSINASE INHIBITORS

The kidney is involved in homeostatic control of arterial blood pressure by releasing the hypertensive peptide angiotensin. Angiotensin I is converted into angiotensin II by the enzyme angiotensinase (or angiotensin I-converting enzyme). (See Scheme 20.)

Venom of *Bothrops jararaca* contains an inhibitor of angiotensinase (Ondetti et al., 1971). The structures of angiotensinase inhibitors are shown in Table 1. Since these inhibitors act on angiotensinase, angiotensin II formation from angiotensin I is inhibited, resulting in the inactivation of central pressor activities (Solomon et al., 1974).

$$\text{Asp−Arg−Val−Tyr−Ile−His−Pro−Phe−His−Leu} \xrightarrow{\text{Angiotensinase}} \text{Asp−Arg−Val−Tyr−Ile−His−Pro−Phe + His−Leu}$$

Scheme 20

Table 1 Amino Acid Sequences of Angiotensinase Inhibitors

Name	Sequence[a]
V-9	pGlu-Gly-Gly-Trp-Pro-Arg-Pro-Gly-Pro-Glu-Ile-Pro-Pro
V-8	pGlu-Ser- - -Trp-Pro- - - - -Gly-Pro-Asn-Ile-Pro-Pro
V-7	pGlu-Asn- - -Trp-Pro-His- - - - -Pro-Gln-Ile-Pro-Pro
V-6-II	pGlu-Asn- - -Trp-Pro-Arg- - - - -Pro-Gln-Ile-Pro-Pro
V-6-I	pGlu- - - - -Trp-Pro-Arg- - - - -Pro-Gln-Ile-Pro-Pro
V-2	pGlu- - - - -Trp-Pro-Arg-Pro-Thr-Pro-Gln-Ile-Pro-Pro

The relationship between angiotensinase inhibitors and the bradykinin-potentiating factor, as well as the mechanism of inhibition, will be discussed in more detail in Chapter 24, "Autopharmacological Action." Apparently, they are identical substances called by different names.

Angiotensinase can be inactivated at inhibitor concentration ranges of 5×10^{-9} to $1 \times 10^{-5} M$ (Bakhle, 1972).

4 PROTEINASE INHIBITORS

Some snake venoms contain proteinase inhibitors that inactivate pancreatic trypsin, chymotrypsin, plasma kallikrein, and plasmin. Such inhibitors have been found in the venoms of *Hemachatus haemachatus, Dendroaspis anguisticepts, D. polylepis, Naja nivea, N. haje*, and *Vipera russellii* (Takahashi et al., 1974a). Toxins such as α-bungarotoxin, cytotoxins I and II, and cardiotoxin (*Naja naja atra* venom) do not possess proteinase inhibitor activity.

Creighton (1975) compared the amino acid sequences of proteinase inhibitors and found remarkable similarities among them (Table 2). The inhibitors he compared were the bovine pancreatic inhibitor (Kassell and Laskowski, 1965), bovine colostrum trypsin inhibitor (Čechová et al., 1969), turtle egg white inhibitor (Laskowski et al., 1974), snail inhibitor K (Dietl and Tschesche, 1974), proteinase inhibitor II from Russell's viper venom (Takahashi et al., 1974b, d), and inhibitors K and I from black mamba venom (Strydom, 1973). As can be seen in Table 2, the positions of sulfur atoms are identical, suggesting the identical positions of disulfide bridges within the inhibitor molecules.

Table 2 Amino Acid Sequences of Proteinase Inhibitors from Snake Venoms and Other Sources

Bovine pancreatic inhibitor	Arg-Pro-Asp-Phe-Cys-Leu-Glu-Pro-Pro-Tyr-Thr-Gly-Pro-
Bovine colostrum inhibitor	Phe-Gln-Thr-Pro-Pro-Asp-Leu-Cys-Gln-Leu-Pro-Gln-Ala-Arg-Gly-Pro-
Turtle egg white inhibitor	(Lys,Glx,Asx,Gly,Arg)Asp-Ile-Cys-Arg-Leu-Pro-Pro-Glu-Gln-Gly-Pro-
Snail inhibitor K	Glu-Gly-Arg-Pro-Ser-Phe-Cys-Asn-Leu-Pro-Ala-Glu-Thr-Gly-Pro-
Vipera russelli inhibitor	His-Asp-Arg-Pro-Thr-Phe-Cys-Asn-Leu-Ala-Pro-Glu-Ser-Gly-Arg-
Dendroaspis polylepis polylepis toxin I	Ala-Ala-Lys-Tyr-Cys-Lys-Leu-Pro-Leu-Arg-Ile-Gly-Pro-
Dendroaspis polylepis polylepis toxin K	(Gln,Pro)Leu-Arg-Lys-Leu-Cys-Ile-Leu-His-Arg-Asn-Pro-Gly-Arg-

Bovine pancreatic inhibitor	Cys-Lys-Ala-Arg-Ile-Ile-Arg-Tyr-Phe-Tyr-Asn-Ala-Lys-Ala-Gly-Leu-Cys-Gln-Thr-Phe-Val-Tyr-Gly-Gly-
Bovine colostrum inhibitor	Cys-Lys-Ala-Ala-Leu-Leu-Arg-Tyr-Phe-Tyr-Asx-Ser-Thr-Ser-Asn-Ala-Glu-Pro-Phe-Thr-Gly-Gly-
Turtle egg white inhibitor	Cys-Lys-Gly-Arg-Ile-Pro-Arg-Tyr-Phe-Tyr-Asn-Pro-Ala-Ser-Arg-Met-Cys-Glu-Ser-Phe-Ile-Tyr-Gly-Gly-
Snail inhibitor K	Cys-Lys-Ala-Ser-Phe-Arg-Gln-Tyr-Tyr-Tyr-Asn-Ser-Lys-Ser-Gly-Gly-Cys-Gln-Phe-Ile-Tyr-Gly-Gly-
Vipera russelli inhibitor	Cys-Arg-Gly-His-Leu-Arg-Arg-Ile-Tyr-Tyr-Asn-Leu-Glu-Ser-Asn-Lys-Cys-Lys-Val-Phe-Phe-Tyr-Gly-Gly-
Dendroaspis polylepis polylepis toxin I	Cys-Lys-Arg-Lys-Ile-Pro-Ser-Phe-Tyr-Tyr-Lys-Trp-Lys-Ala-Lys-Gln-Cys-Leu-Pro-Phe-Asp-Tyr-Ser-Gly-
Dendroaspis polylepis polylepis toxin K	Cys-Tyr-Gln-Lys-Ile-Pro-Ala-Phe-Tyr-Tyr-Asn-Gln-Lys-Lys-Lys-Gln(Glx,Gly)Phe-Thr-Trp-Ser-Gly-

		References
Bovine pancreatic inhibitor	Cys-Arg-Ala-Lys-Arg-Asn-Asn-Phe-Lys-Ser-Ala-Glu-Asp-Cys-Met-Arg-Thr-	Kassell and Laskowski, 1965
Bovine colostrum inhibitor	Cys-Gln-Gly-Asn-Asn-Asx-Asn-Phe-Glu-Thr-Thr-Glu-Met-Cys-Leu-Arg-Ile-	Cechova et al, 1969
Turtle egg white inhibitor	Cys-Lys-Gly-Asn-Lys-Asn-Asn-Phe-Lys-Thr-Lys-Ala-Glu-Cys-Val-Arg-Thr-	Laskowski, Jr. et al, 1974
Snail inhibitor K	Cys-Arg-Gly-Asn-Gln-Asn-Arg-Phe-Asp-Thr-Thr-Gln-Gln-Cys-Gln-Gly-Val-	Dietle and Tschesche, 1974
Vipera russelli inhibitor	Cys-Gly-Gly-Asn-Ala-Asn-Asn-Phe-Glu-Thr-Arg-Glu-Asp-Cys-Arg-Gln-Thr-	Takahashi et al, 1974e
Dendroaspis polylepis polylepis toxin I	Cys-Gly-Gly-Asn-Ala-Asn-Arg-Phe-Lys-Thr-Ile-Glu-Glu-Cys-Arg-Arg-Thr-	Strydom, 1973
Dendroaspis polylepis polylepis toxin K	Cys-Gly-Gly-Asn-Ser-Asn-Arg-Phe-Lys-Thr-Ile-Glu-Glu-Cys-Arg-Arg-Thr-	Strydom, 1973

Bovine pancreatic inhibitor	Cys-Gly-Gly-Ala
Bovine colostrum inhibitor	Cys-Glu-Pro-Pro-Gln-Gln-Thr-Asp-Lys-Ser
Turtle egg white inhibitor	Cys-Gly-Pro-Gly-Ile-Cys-Leu....60 more residues
Snail inhibitor K	Cys-Val
Vipera russelli inhibitor	Cys-Gly-Gly-Lys
Dendroaspis polylepis polylepis toxin I	Cys-Val-Gly
Dendroaspis polylepis polylepis toxin K	Cys-Ile-Arg-Lys

Inhibitor II from *Vipera russellii* venom combines with trypsin in a unimolar stoichiometric ratio (Takahashi et al., 1972). The K_i value with trypsin is $7.6 \times 10^{-10} M$, with chymotrypsin is $1.4 \times 10^{-10} M$, with bovine plasmin is $1.0 \times 10^{-9} M$, and with bovine plasma kallikrein is $2.9 \times 10^{-10} M$ (Takahashi et al., 1974e). It is rather interesting to note that snake venom proteinase inhibitor is ineffective against proteinases of snake venom origins, such as bradykinin-releasing enzyme and reptilase (*Bothrops atrox* origin). Inhibitor II does not inhibit the action of human and bovine thrombin, bromelanin, papain, ficin, nagarse, thermolysin, or carboxypeptidases A and B (Takahashi et al.,

Table 3 Amino Acid Compositions of Proteinase Inhibitors Isolated from Snake Venoms

Genus Species Subspecies	Dendroaspis polylepis polylepis		Hemachatus hemachatus	Vipera russelli	
Origin	S. Africa		Africa	Asia	
Name	I.	K.	II.	I	II
Amino Acid					
Lysine	8	7	2	2	3
Histidine	0	1	1	2	2
Arginine	5	7	4	5	7
Aspartic acid	3	4	6	6	8
Threonine	2	3	3	2	3
Serine	2	2	2	2	2
Glutamic acid	3	7	6	4-5	5
Proline	4	3	2	4-5	2
Glycine	5	5	6	6-7	8
Alanine	4	1	5	2-3	2
Valine	1	0	1	1	1
Methionine	0	0	0	0	0
Isoleucine	3	4	3	2	1
Leucine	3	3	4	1	3
Tyrosine	4	3	2	1	3
Phenylalanine	3	3	4	3	4
Half-cystine	6	6	6	4	6
Tryptophan	1	1	0	0	0
Total residue	57	60	57	47-51	60
References	Strydom, 1973		Takahashi et al., 1974e		

1974c). The venom also contains another proteinase inhibitor, inhibitor I (Takahashi et al., 1974c). Inhibitor I from *V. russellii* venom is smaller than inhibitor II and consists of 47 to 51 amino acid residues, as compared to 60 residues for inhibitor II (Takahashi et al., 1974e). The actions of the two inhibitors are very similar.

Inhibitor II isolated from *Hemachatus haemachatus* venom has 57 amino acid residues (Table 3). All three inhibitors have similar inhibition spectra.

Inhibitors I and K were isolated from the venom of *Dendroaspis polylepis polylepis* by Strydom (1973). The toxicity of these inhibitors is very weak, 70 times less than that of the original venom. The LD_{50} values of crude venom and of inhibitors I and K are 0.55, 30, and 36 $\mu g\,g^{-1}$, respectively, in mice. The concentrations of I and K in the venoms are 8 and 2.7%, respectively.

5 OTHERS

Naja flava venom contains a potent inhibitor of succinic dehydrogenase (Radomski and Deichman, 1958). The enzyme inhibitor is not important for the toxic action of cobra venom. Venom of *N. naja atra* contains inhibitors of glyceraldehyde-3-phosphodehydrogenase, cytochrome oxidase, and all respiration and oxidative phosphorylation enzymes (Gillo, 1967; Brisbois et al., 1968a, b). Some of these inhibitors are probably not true enzyme inhibitors, as they may not interact with enzymes. Instead, they probably act on the organization of cell membranes where the respective enzymes are located in specific arrangement. Disruption of this arrangement inactivates such enzyme systems. It is also known that *N. naja atra* venom inhibits anerobic glycolysis, but the site of inhibition is not known (Gillo, 1967).

A number of snake venoms (*N. naja, Bothrops neuwiedi, B. alternatus, Crotalus terrificus*) inhibit reduced NAD oxidase and succinic acid oxidase activities (Bodano and Stoppani, 1962). The inhibition is not at the enzyme or substrate level, but rather at the electron transfer level, and is closely related to venom phospholipase A_2 type effects.

Serotonin can be released from platelets by the action of thrombin, and *Echis colorata* venom can inhibit the release of serotonin (Biran et al., 1973). The mechanism of this inhibition is not known and may or may not be due to enzyme inactivation.

REFERENCES

Augustinsson, K. B. and Grahn, M. (1952). Stability of cobra venom acetylcholinesterase and the stabilizing effects of various metallic ions, *Ark. Kömi,* **4,** 277.

Bakhle, Y. S. (1972). "Inhibition of converting enzyme by venom peptides," in J. Genest, Ed., *Hypertension,* Springer, Berlin, pp. 541–547.

Biran, H., Dvilansky, A., Nathan, I., and Livne, A. (1973). Impairment of human platelet aggregation and serotonin release caused in *vitro* by *Echis colorata* venom, *Thromb. Diath. Haemorrh.,* **30,** 191.

Bodano, B. N. and Stoppani, O. A. M. (1962). Action of snake venoms on enzymic activities of the nicotinamide adenine dinucleotide (NAD) oxidase and succinic oxidase of the myocardium, *Rev. Soc. Arg. Biol.,* **38,** 342.

Braganca, B. M. and Quastel, J. H. (1953). Enzyme inhibitions by snake venoms, *Biochem. J.,* **53,** 88.

Braganca, B. M., Sambray, Y. M., and Sambray, R. Y. (1970). Isolation of polypeptide inhibitor of phospholipase A from cobra venom, *Eur. J. Biochem.,* **13,** 410.

Brisbois, L., Delori, P., and Gillo, L. (1968a). Venins de serpents: Fractionnement d'un venin de cobra (*Naja naja atra*) par chromatographies sur gels de detrane, *L'Ing. Chim.,* **50,** 45.

Brisbois, L., Rabinovitch-Mahler, N., Delori, P., and Gillo, L. (1968b). Étude des fractions obtenues par chromatographie du venin de *Naja naja atra* sur sulphoéthyl-Sephadex, *J. Chromatogr., 37*, 463.

Brisbois, L., Rabinovitch-Mahler, N., and Gillo, L. (1968c). Isolation from cobra venom of a factor inhibiting glycolysis in Ehrlich ascites carcinoma cells, *Experientia, 24,* 673.

Cechová, D., Svestková, V., Keil, B., and Sorm, F. (1969). Similarities in primary structures of cow colostrum trypsin inhibitor and bovine basic pancreatic trypsin inhibitor, *FEBS Lett., 4,* 155.

Chain, E. (1936). Effect of snake venoms on glycolysis and fermentation in cell-free extracts, *Quart. J. Exp. Physiol., 26,* 299.

Chain, E. (1939). Inhibition of dehydrogenase by snake venom, *Biochem. J., 33,* 407.

Chain, E. and Goldsworthy, L. (1938). Studies on the chemical nature of the antifermenting principle in black tiger snake venom, *Quart. J. Exp. Physiol., 27,* 375.

Chang, C. C. and Lee, C. Y. (1955). Cholinesterase and anticholinesterase activities in snake venoms, *J. Formosan Med. Assoc., 54,* 103.

Creighton, T. E. (1975). Homology of protein structures, proteinase inhibitors. *Nature, 255,* 743.

Dietl, T. and Tschesche, H. (1974). In H. Fritz, H. Tschesche, L. J. Greene, and E. Truscheit, Eds., *Proteinase Inhibitors,* Springer, Berlin, pp. 254–264.

Fleckenstein, A. and Gayer, J. (1950). Dehydrasenhemmung durch Schlangengifte, *Klin. Wohnschr., 28,* 789.

Ghosh, B. N. and Chatterjee, A. K. (1948). Effect of snake venoms on the oxidation of glucose and its metabalities in cell suspensions, *J. Indian Chem. Soc., 25,* 359.

Gillo, L. (1967). Analyse et separation des constituants d'un complexe proteique: Le venin de cobra, *J. Int. Chim. Anal., 15,* 70.

Kassell, B. and Laskowski, M., Sr. (1965). The basic trypsin inhibitor of bovine pancreas. V. The disulfide linkages, *Biochem. Biophys. Res. Commun., 20,* 463.

Kihara, H. (1976). Studies on phospholipase A in *Trimeresurus flavoviridis* venom. III. Purification and some properties of phospholipase A inhibitor in Habu serum, *J. Biochem., 80,* 341.

Laskowski, M., Jr., Kato, I., Leary, T. R., Schrode, J., and Sealock, R. W. (1974). In H. Fritz, H. Tschesche, L. J. Greene, and E. Truscheit, Eds., *Proteinase Inhibitors,* Springer, Berlin, pp. 597–611.

Lee, C. Y., Chang, C. C., and Kamijo, K. (1956). Cholinesterase inactivation by snake venoms, *Biochem. J., 62,* 582.

Marinetti, G. V. (1965). The action of phospholipase A on lipoproteins, *Biochim. Biophys. Acta, 98,* 554.

Mounter, L. A. (1951). The specificity of cobra venom cholinesterase, *Biochem. J., 50,* 122.

Ondetti, M. A., Williams, N. J., Sabo, E. F., Pluscec, J., Weaver, E. R., and Kocy, O. (1971). Angiotensin-converting enzyme inhibitors from the venom of *Bothrops jararaca:* Isolation, elucidation of structure, and synthesis, *Biochemistry, 22,* 4033.

Radomski, J. L. and Deichmann, W. B. (1958). The relationship of certain enzymes in cobra and rattlesnake venoms to the mechanism of action of these venoms, *Biochem. J., 70,* 293.

Sakhibov, D. N., Yukel'son, L. Ya., and Salikhov, R. (1974). Phospholipase A from central Asia cobra venom, *Khim. Prir. Soedin., 3,* 387.

Solomon, T. A., Cavero, I., and Buckley, J. P. (1974). Inhibition of central pressor effects of angiotensin I and II, *J. Pharm. Sci., 63,* 511.

Strydom, A. J. C. (1973). Snake venom toxins: The amino acid sequence of two toxins from *Dendroaspis jamesoni kaimosae, Biochim. Biophys. Acta, 328,* 491.

Takahashi, H., Iwanaga, J., and Suzuki, T. (1972). Isolation of a novel inhibitor of kallikrein, plasmin and trypsin from the venom of Russell's viper (*Vipera russellii*), *FEBS Lett., 27,* 207.

Takahashi, H., Iwanaga, S., and Suzuki, T. (1974a). Distribution of proteinase inhibitors in snake venoms, *Toxicon, 12,* 193.

Takahashi, H., Iwanaga, S., Hokama, Y., Suzuki, T., and Kitagawa, T. (1974b). Primary structure of proteinase inhibitor. II. Isolated from the venom of Russell's viper (*Vipera russellii*), *FEBS Lett., 38,* 217.

References

Takahashi, H., Iwanaga, S., and Suzuki, T. (1974c). Snake venom proteinase inhibitors. I. Isolation and properties of two inhibitors of kallikrein, trypsin, plasmin and α-chymotrypsin from the venom of Russell's viper (*Vipera russellii*), *J. Biochem. (Tokyo)*, **76,** 709.

Takahashi, H., Iwanaga, S., Kitagawa, T., Hokama, Y., and Suzuki, T. (1974d). Snake venom proteinase inhibitors. II. Chemical structure of inhibitor. II. Isolated from the venom of Russell's viper (*Vipera russellii*), *J. Biochem.,* **76,** 721.

Takahashi, H., Iwanaga, S., Kitagawa, T., Hokama, Y., and Suzuki, T. (1974e). Novel proteinase inhibitors in snake venoms: Distribution, isolation, and amino acid sequence. *Bayer-Symp.*, (Proteinase inhibitors, Proc. Int. Res. Conf. 2nd. 1973), pp. 265–276.

Tazieff-Depierre, F. and Pierre, J. (1966). Action curarisante de la toxine α de *Naja nigricollis, C. R. Acad. Sci. Paris,* **263,** 1785.

Vidal, J. C. and Stoppani, A. O. M. (1970). Inhibition of phospholipase A by a naturally occurring peptide in *Bothrops* venoms, *Experientia,* **26,** 831.

III SNAKE VENOMS: Properties and Actions

11 Venoms of Hydrophiidae (Sea Snakes)

1 SEA SNAKE POISONING	152
2 TOXICOLOGY AND PHARMACOLOGY	152
2.1 Yield of Venoms, 152	
2.2 Toxicity, 152	
2.3 Immunology, 157	
2.4 Pharmacology, 160	
Laticauda semifasciata, 160	
Laticauda laticaudata, 161	
Enhydrina schistosa, 161	
Hydrophis cyanocinctus and Lapemis hardwickii, 163	
Pelamis platurus, 163	
3 PATHOLOGY	163
4 BIOCHEMISTRY	164
4.1 Composition, 164	
4.2 Isolation of Lethal Toxins, 167	
4.3 Structure of Neurotoxins, 170	
References	174

The venoms of sea snakes (Hydrophiidae) contain potent neurotoxins that are among the most toxic substances in the world. They are more toxic than venoms of terrestrial snakes, including rattlesnakes, copperheads, kraits, and cobras (Pickwell and Evans, 1972). Sea snake venom neurotoxins act on the neuromuscular junctions and paralyze the victims.

Sea snakes are more numerous than terrestrial venomous snakes and are common in tropical and subtropical regions bordering the Indian and Pacific oceans (Smith, 1926; Taylor, 1965; Tu and Tu, 1970; Tu, 1974a, b, 1976).

1 SEA SNAKE POISONING

There have been many conflicting reports concerning the aggressiveness and danger of sea snakes. Apparently, some sea snakes are more aggressive than others. A possible explanation is that aggressiveness depends on many other factors, such as the season and the breeding cycle. Sea snakebites are quite common and death is not unusual among fishermen in Asia. Of 115 patients with sea snakebites treated personally by Reid from 1957 to 1964 in Penanag, Malaya, 8 died (Reid, 1973).

Most reported sea snakebite cases are from Malaya because of Reid's effort to document these cases in scientific journals. Barme (1963, 1968) reported that sea snakebites are frequent among Vietnamese fishermen and are often fatal. Not many sea snake deaths have been recorded in other regions because sea snakebites usually occur in remote places and frequently are not reported.

As human activity expands into the oceans, the frequency of human contact with sea snakes will inevitably increase. Because of the highly toxic nature of the venoms, sea snake poisoning is a potential public health hazard and should not be regarded lightly (Halstead, 1970).

Sea snake envenomation in man is characterized by generalized myalgia, usually starting 30 min to 1 hr after the bite. Aches, pains, and stiffness occur when arm, thigh, neck, or trunk muscles are moved. Myoglobinuria becomes evident 3 to 6 hr after the bite. The clinical features of the sea snakebite poisoning in human victims differ markedly from those recorded in animal experiments. In human beings sea snake venom appears to be primarily myotoxic, whereas in animals it is neurotoxic (Reid, 1961, 1963). Apparently there are different symptoms, depending on the host. However, in a case report of death due to sea snakebite in Japanese seamen operating in southwestern Pacific waters, there was no indication of myotoxic effect (Okonogi, 1973). Therefore the myotoxic effect of snake poisoning in human beings may depend on the species of sea snake. In a reported case of sea snakebite in Sri Lanka (Karunaratne and Panabokke, 1972), the postmortem examination revealed that the main abnormalities were in the renal tubules. In certain tubular degeneration and necrosis were observed. There were centrilobular fatty changes in the liver, and interstitial edema of the heart had occurred.

2 TOXICOLOGY AND PHARMACOLOGY

2.1 Yield of Venoms

In contrast to land snakes, sea snakes and their venoms have not been investigated extensively. The fact that many of these snakes spend their entire lives in the ocean makes capturing enough specimens for scientific study of their venoms and neurotoxins very difficult. Moreover, the amount of crude venom that can be obtained from sea snakes is very small, ranging from 0.2 to 19 mg per snake, depending on the size of the snake (Tu and Tu, 1970). A comparison of the yields of venom from sea snakes and land snakes is shown in Table 1.

2.2 Toxicity

As may be seen from the summarized data in Table 2, crude sea snake venoms are extremely toxic. After the nontoxic components are removed, the pure toxins are even more toxic than the crude venoms.

Table 1   Yield of Sea Snake Venoms

Snakes	Yield mg/snake	Reference
Sea snakes		
Acalyptophis peronii	1.5	Tu, 1974b
Aipysurus eydouxii	0.6	Tu, 1974a,b
A. laevis	6.6	Tamiya and Puffer, 1974
Enhydrina schistosa	6.9-14.0	Tu, 1974a,b
	4.6	Reid, 1956
Hydrophis belcheri	2.4	Tamiya and Puffer, 1974
H. cyanocinctus	18.0	Tu, 1974a,b
	4.3	Reid, 1956
H. elegans	5.4	Tamiya and Puffer, 1974
H. ornatus	19.0	Tu, 1974
	8.8	Tamiya and Puffer, 1974
H. torquatus diadema	3.9	Tu, 1974a
Lapemis hardwickii	2.4-5.2	Tu, 1974a,b
Pelamis platurus	2.0	Tu and Tu, 1970
	2.8	Pickwell et al, 1974
	2.6	Tu, 1976
Praescutata viperina	0.2	Tu, 1974a,b
Land snakes		
Naja naja kaouthia	263	Ganthavorn, 1969
Ophiophagus hannah	420	Ganthavorn, 1969
Bungarus fasciatus	114	Ganthavorn, 1969
Agkistrodon rhodostoma	59	Ganthavorn, 1969
Vipera russellii siamensis	133	Ganthavorn, 1969

Table 2 Toxicities of Venoms and Toxins of Sea Snakes and Snakes of Terrestrial Origin

	Route of administration	LD_{50} (µg/g) in mice	Reference
Sea Snakes (Hydrophiidae)			
Venoms			
Astrotia stokesii	im	3.5	Barber et al, 1974
Aipysurus eydouxii	iv	>4.0	Tu, 1974a
A. laevis	im	0.5	Barber et al, 1974
Enhydrina schistosa	ip	0.11	Carey and Wright, 1960a
	iv	0.09	Tu and Ganthavorn, 1969
	iv	0.14	Tu, 1974a,b
	iv	0.21	Tu, 1974a,b
	iv	0.35	Cheymol et al, 1967
Hydrophis belcheri	im	0.07	Barber et al, 1974
H. cyanocinctus	ip	0.24	Carey and Wright, 1960a
	iv	0.67	Cheymol et al, 1967
	iv	0.35	Tu and Ganthavorn, 1969
H. elegans	im	0.30	Barber et al, 1974
H. klossi	ip	0.2-0.53	Carey and Wright, 1960a
H. melanosoma	ip	0.40	Carey and Wright, 1960a
H. spiralis	ip	0.25-0.38	Carey and Wright, 1960a
Kerilia jerdoni	ip	0.53	Carey and Wright, 1960a
Lapemis hardwickii	ip	0.26	Carey and Wright, 1960a
	iv	0.71	Tu and Ganthavorn, 1969
	iv	0.70	Tu and Hong, 1971
	iv	1.37	Tu, 1974a
	iv	1.40	Tu, 1974a
	iv	0.44	Cheymol et al, 1967a
Laticauda colubrina	sc	0.42	Tu, T. et al, 1963
	sc	0.45	Levey, 1969
	iv	0.4	Sato et al, 1969
L. laticaudata	iv	0.17	Sato et al, 1969
	iv	0.16	Tu and Salafranca, 1974
	iv	0.16	Vick et al, 1975

Table 2   *Continued*

	Route of administration	LD$_{50}$ (μg/g) in mice	Reference
Laticauda semifasciata	sc	0.34	Tu, T., 1961
	iv	0.28	Tu et al, 1971
	iv	0.39	Tu and Salafranca, 1974
	iv	0.21	Tu, T., 1961
	iv	0.30	Vick et al, 1975
Pelamis platurus	iv	0.18	Tu and Ganthavorn, 1969
	iv	0.09	Pickwell et al, 1974
	iv	0.11	Pickwell et al, 1974
	iv	0.44	Tu and Salafranca, 1974
Praescutata viperina	iv	4.5	Tu and Salafranca, 1974
Pure Toxins			
Enhydrina schistosa	iv	0.04	Tu and Toom, 1971
Lapemis hardwickii	iv	0.06	Tu and Hong, 1971
Pelamis platurus	iv	0.044	Tu and Salafranca, 1974
Laticauda semifasciata	iv	0.07	Tu et al, 1971
		0.05	Tu et al, 1971
Land Snakes			
Venoms			
Elapidae (Elapids)			
Naja naja	iv	0.13	Friederich and Tu, 1971
Naja naja atra	iv	0.29	Friederich and Tu, 1971
Viperidae (Vipers)			
Bitis arietans	im	2.0	Tu et al, 1969
Bitis gabonica	im	5.2	Tu et al, 1969
Bitis nasicornis	im	8.6	Tu et al, 1969
Vipera aspis	im	4.1	Tu et al, 1969
Vipera russellii siamensis	im	2.1	Tu et al, 1969

Table 2 *Continued*

	Route of administration	LD$_{50}$ (μg/g) in mice	Reference
Crotalidae			
Agkistrodon acutus	iv	0.38	Friederich and Tu, 1971
Bothrops atrox	iv	1.4	Tu and Homma, 1970
Bothrops nasuta	iv	4.6	Tu and Homma, 1970
Bothrops nummifer	iv	2.4	Tu and Homma, 1970
Bothrops picadoi	iv	1.6	Tu and Homma, 1970
Bothrops schlegelii	iv	1.6	Tu and Homma, 1970
Crotalus atrox	iv	3.6	Friederich and Tu, 1971
Crotalus adamanteus	iv	2.4	Friederich and Tu, 1971
Crotalus horridus	iv	2.6	Friederich and Tu, 1971

No attempt is made to cover the toxicity of all land snake venoms.

Some authors have expressed toxicity in terms of minimum lethal doses (LD$_{100}$) rather than LD$_{50}$ values. The LD$_{100}$ doses for several snakes are as follows (Baxter and Gallichio, 1974):

Species	LD$_{100}$ (μg g^{-1}) in Mice
Aipysurus laevis	0.39
Astrotia stokesii	0.30
Enhydrina schistosa	0.22
Hydrophis cyanocinctus	0.65
H. elegans	0.43
H. major	0.22
H. spiralis	0.65
Lapemis hardwickii	0.43
Laticauda semifasciata	0.35

The susceptibility of the host to venom depends on the species of animal used. For instance, the LD$_{50}$ values for i.v. injection of *Laticauda semifasciata* venom are 0.211 μg g^{-1} for mice, 0.0631 for guinea pigs, and 0.0486 for rabbits (Tu, 1959). The toxicity also depends on the route of injection. Thus the LD$_{50}$ values for subcutaneous injection with the same venom are 0.338 μg g^{-1} for mice, 0.0897 for guinea pigs, and 0.211 for rabbits.

2 Toxicology and Pharmacology

Table 3 Stability of Toxins and pH Values

pH	Pelamis platurus[a] major toxin	Laticauda semifasciata[b] a	b
1	—	0.06	0.04
2	0.083	0.07	0.04
4	0.080	0.07	0.04
6	0.078	0.07	0.04
8	0.063	0.07	0.05
10	0.080	0.05	0.04
11	—	0.05	0.04

The figures were the LD_{50} values expressed as µg/g in mice. All injections were made intravenously.

The data were obtained from: [a], Tu et al, 1975; [b], Tu et al, 1971.

The thermostability of *L. semifasciata* venom is well recognized. Homma et al. (1964) demonstrated that the venom retained its toxicity even after heating at 100°C for 5 min. The venom is also stable in acidic as well as basic pH.

Pure toxins are also stable under heat treatment. Two toxins isolated from *L. semifasciata* retained their toxicity at 100°C (Tu et al., 1971), while the major toxin from *Pelamis platurus* was somewhat stable at 50°C for 30 min (Tu et al., 1975). The above toxins were also stable in both acidic and basic solutions, as can be seen from the constant LD_{50} values (Table 3).

2.3 Immunology

Venoms of various species of sea snakes seem to contain a common antigen. By using the double-diffusion method, Carey and Wright (1960a) showed that venoms of *Enhydrina schistosa, Kerilia jerdonii, Microcephalus gracilis, Lapemis hardwickii, Hydrophis cyanocinctus,* and *H. spiralis* formed precipitin lines with antisera of *Enhydrina schistosa* venom. Tu and Ganthavorn (1969) demonstrated that venoms of *Hydrophis cyanocinctus, Pelamis platurus,* and *Lapemis hardwickii* cross-reacted immunologically with anti-*Enhydrina schistosa* sera (Figs. 1A, B, C and Fig. 2).

Similarly, venoms of *Praescutata viperina* show a precipitation line in immunoelectrophoresis with anti-*Enhydrina schistosa* sera, but *Laticauda semifasciata* and *L. laticaudata* venoms do not form precipitins with antisera (Tu and Salafranca, 1974). Tu and Ganthavorn (1969) also showed that antivenin for *Enhydrina schistosa* is effective for neutralizing the venoms of *Hydrophis cyanocinctus, Pelamis platurus,* and *Lapemis hardwickii.* Antivenin for *Laticauda semifasciata* is effective for the neutralization of *Hydrophis cyanocinctus* venom (Okonogi et al., 1967). Barme (1963) reported that

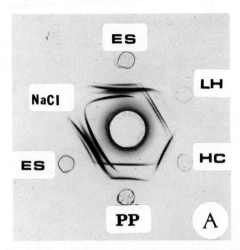
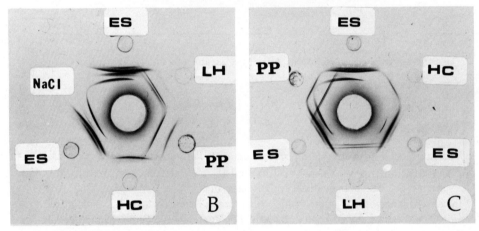

Figure 11.1 Immunological relationship of different sea snake venoms shown in Ouchterlony reaction. Antibody used was anti-*Enhydrina schistosa* venom. ES: *Enhydrina schistosa* venom, LM: *Lapemis hardwickii* venom, PP: *Pelamis platurus* venom, HC: *Hydrophis cyanocinctus* venom, reproduced from Tu and Ganthavarn (1969) by permission of the copyright owner, the American Society of Tropical Medicine and Hygiene.

antivenin for *Lapemis hardwickii* manufactured in Vietnam effectively neutralized the venoms of *Enhydrina schistosa* and *Hydrophis cyanocinctus*. More recent work (Tu and Salafranca, 1974) indicates that the commercial antivenin for *Enhydrina schistosa* manufactured at the Commonwealth Serum Laboratory, Melbourne, Australia, is effective for neutralizing the venoms of sea snakes *Pelamis platurus* from Central America, *Praescutata viperina* from the Gulf of Thailand, *Laticauda semifasciata* from the Philippines, and *L. laticaudata* from Japan. Similarity in the toxins of cobras and sea snakes is also reflected in their immunological properties, since different cobra venoms are neutralized by sea snake antivenin for *Enhydrina schistosa* (Minton, 1967).

Baxter and Gallicho (1974) observed that the antivenins for tiger snake (*Notechis scutatus*) and sea snake (*Enhydrina schistosa*) neutralized the venoms of *Aipysurus laevis*, *Astrotia stokesii*, *Enhydrina schistosa*, *Hydrophis cyanocinctus*, *H. elegans*, *H. major*, *H.*

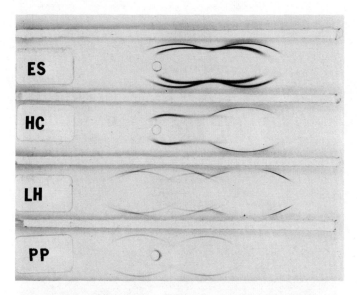

Figure 11.2 Immunological relationship of different sea snake venoms shown in immunoelectrophoresis. Antibody used was anti-*Enhydrina schistosa* venom. Reproduced from Tu and Ganthavorn (1969) by permission of the copyright owner the American Society of Tropical Medicine and Hygiene.

spiralis, Lapemis hardwickii, and *Laticauda semifasciata.* The polyvalent horse antivenin for *Lapemis hardwickii, Hydrophis cyanocinctus,* and *Laticauda semifasciata* venoms can be maufactured from formaline-denatured venom antigens. The antivenin is effective *in vitro* as well as *in vivo* (Okonogi et al., 1972).

Antivenin for sea snake venom can also be made in animals other than the horse. For instance, antivenin for the venom of *Laticauda semifasciata* was manufactured in the rabbit (Okonogi et al., 1967) and the goat (Okonogi, 1970).

Only commercially available sea snake antivenin is manufactured by the Commonwealth Serum Laboratory of Australia and used in Malaysia clinically for treatment of sea snakebites (Reid, 1975a, b). The antivenin, made from *Enhydrina schistosa* venom, not only is effective against homologous venom but also neutralizes a wide variety of heterologous venoms (Tu and Ganthavorn, 1969; Baxter and Gallichio, 1974; Vick et al., 1975). The antivenin is effective even for *Pelamis platurus* venom from Central America (Tu and Salafranca, 1974).

Several studies using pure sea snake toxins have been made. When the tryptophan residue of two toxins isolated from *Laticauda semifasciata* (Philippine origin) venom was chemically modified with N-bromosuccinimide, the toxicity was completely lost (Hong and Tu, 1970; Tu and Hong, 1971). However, detoxified neutrotoxins formed a precipitation line with antibody made to original venom, as can be seen in Fig. 3 (Tu et al., 1971). Thus the tryptophan residue is important for toxic action but not necessarily for immunological reaction. Although erabutoxins a and b of *L. semifasciata* (Japan origin) differ slightly, they formed a fused precipitation line with antierabutoxin b (Sato and Tamiya, 1970). Moreover, erabutoxin b with iodinated histidine also binds to antierabutoxin b. When erabutoxin a was modified chemically, the toxicity was lost. However, modified erabutoxin a formed a precipitation line with antierabutoxin a (Seto et al., 1970).

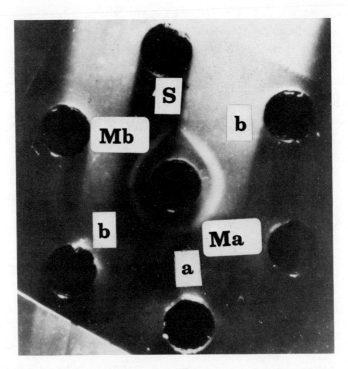

Figure 11.3 Immunodiffusion patterns of native and tryptophan-modified toxins a and b isolated from the venom of *Laticauda semifasciata*. Rabbit antiserum was applied in the center well. The samples in the side wells are as follows: a, toxin a; b, toxin b; Ma, modified toxin a; Mb, modified toxin b; S, saline. The concentration used was 20 mg toxin per milliliter. (Reproduced from Tu et al., 1971, by permission of the copyright owner, the American Chemical Society.)

2.4 Pharmacology

Laticauda semifasciata. When a venom of *Laticauda semifasciata* was injected into frogs, it produced general depression without convulsions (Tu, 1959). The venom causes death in mice, guinea pigs, and rabbits with symptoms of severe respiratory disturbance accompanied by convulsions.

Tu (1957, 1961) reported that *L. semifasciata* venom produced a marked inhibition of respiration and caused an initial rise, followed by a fall, in blood pressure. The venom also stimulated isolated frog heart and slightly stimulated rabbit heart. The coronary outflow of the rabbit heart sometimes showed a slight increase. The venom has a curarelike action, as well as a direct paralytic effect, on isolated frog gastrocnemius muscle (Tu, 1961).

Contraction of isolated sciatic nerve–sartorius muscle preparation induced by electrical stimulation through the nerve ceased in 20 to 25 min. However, even when isolated *L. semifasciata* toxin was added to the medium, contraction by direct electrical stimulation of the muscle was not affected (Tamiya et al., 1967). Contraction of the rectus abdominis muscle of a frog induced by acetylcholine was inhibited by the isolated toxin, and the inhibition was not removed by washing the muscle with Ringer's solution. In contrast, contraction of the muscle by potassium chloride was not affected by the toxin.

Crude venom of *L. semifasciata* stimulates the peristaltic movement of isolated rabbit small intestine, whereas pure toxins do not (Uwatoko-Setoguchi, 1970). Crude venom, as well as erabutoxins a and b, block the neuromuscular junction of isolated sciatic nerve–gastrocnemius muscle or sciatic nerve–satorius muscle preparation of the frog (Uwatoko-Setoguchi, 1970). Autoradiographic studies of radioactive erabutoxin b indicate that the toxin specifically binds to the end plates of diaphragm (Sato et al., 1970). Both erabutoxins a and b block the end-plate receptors without affecting the muscle fibers or acetylcholine output at the nerve ending (Cheymol et al., 1972).

The toxin obtained from the venom of *L. semifasciata* reversibly binds to a postsynaptic site of the neuromuscular junction (Maeda et al., 1974). Venom of *L. semifasciata* showed hemolytic activity, caused constriction of the visceral blood vessels of the frog, and induced vasoconstriction of the vena abdominalis in the frog (Tu, 1963). The venom also possesses the ability to increase capillary permeability in the rabbit (Tu et al., 1963).

Laticauda laticaudata. The pharmacological properties of *Laticauda laticaudata* venom are somewhat similar to those of *L. semifasciata* (Tu, 1963, 1967). Tu (1963, 1967) reported that the venom inhibited contraction of isolated rectus abdominis muscle of the frog induced by acetylcholine. The venom had no effect on isolated gastrocnemius muscle of the frog, but showed an inhibitory effect at the neuromuscular junction and produced a marked inhibition of respiration.

Enhydrina schistosa. The paralyzing action of *Enhydrina schistosa* venom on isolated rat phrenic nerve diaphragm preparation clearly demonstrates a peripheral action of the venom on the neuromuscular junction (Carey and Wright, 1961).

Chan and Geh (1967) observed that small doses of *E. schistosa* venom abolished twitches caused by intraarterial acetylcholine without interfering with muscle contraction in cat tibialis anterior muscle preparation. However, at high doses of the venom (>250 μg kg^{-1}) neuromuscular block was seen. Geh and Chan (1973) suggested that the venom produces this phenomenon by blocking acetylcholine action at the prejunctional site of the motor endplate. The presynaptic blocking nature of *E. schistosa* venom may be a rather unique property.

The *in vitro* neuromuscular blocking properties of *E. schistosa* venom were investigated by Walker and Yeoh (1974) in a variety of skeletal muscle preparations. The species variation observed in the neurotoxin response indicated that there may be different forms of receptors. The blocking properties of the venom, presumably due to the presence of neurotoxins, were markedly different from those of d-tubocurine and diallylbisnortoxiferine, which were very similar to each other.

The effects of venom from a common sea snake, *E. schistosa*, on the actions of acetylcholine in the superior cervical ganglion, in the adrenal medulla, and in the atropinized cat were investigated by Yeoh and Walker (1974). The crude venom reduced nictitating membrane responses to preganglionic nerve stimulation and to intraarterial injection of acetylcholine, without lowering the responses to postganglionic nerve stimulation or to intraarterial adrenaline. In the eviscerated cat, the venom also antagonized rises in heart rate and blood pressure induced by injections into the celiac artery of acetylcholine, but not of histamine. Pressor effects due to splanchnic nerve stimulation and carotid artery occlusion were blocked. In the atropinized cat, the venom depressed heart rate and pressor responses to intravenous acetylcholine, but not to adrenaline. Yeoh and Walker concluded that the venom antagonized the actions of acetylcholine at autonomic ganglia and the adrenal medulla.

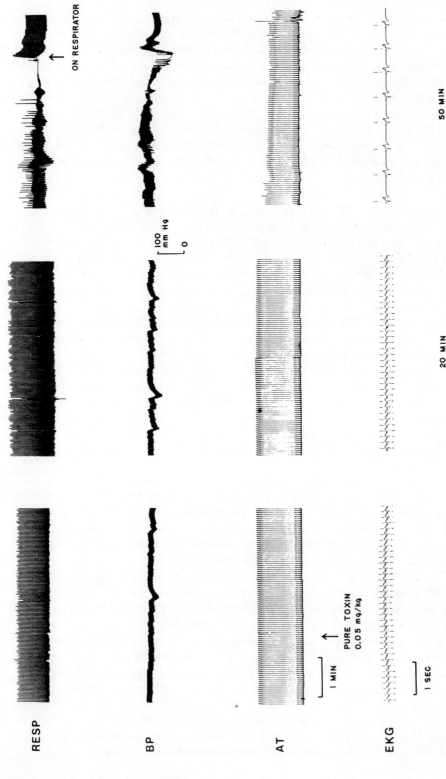

Figure 11.4 Pharmacologic effect of a major toxin isolated from the venom of *Pelamis platurus* (from Costa Rica). The dose was 0.05 mg per kilogram of rabbit. RESP: respiration, BP: blood pressure, AT: twitch response of tibialis anterior muscle to stimulation of sciatic nerve; EKG: electrocardiogram.

Hydrophis cyanocinctus and *Lapemis hardwickii*. Cheymol et al. (1967) studied the neuromuscular blocking action of venoms of *Enhydrina schistosa*, *Hydrophis cyanocinctus*, and *Lapemis hardwickii* from Vietnam. They concluded that all three venoms behaved similarly and that specific receptors of the postsynaptic membrane were blocked almost irreversibly. On the other hand, muscle fibers and nerve fibers were not affected directly.

Pelamis platurus. The major symptoms arising in dogs from *Pelamis platurus* venom were observed by Pickwell et al. (1974). Arterial blood pressure, heart rate, and electrocardiogram remained unchanged for periods of minutes. When death was near, arterial pressure rose, simultaneously accompanied by a decrease in inspiratory volume, while the heart rate gradually declined.

A major toxin was isolated from the venom of the sea snake *P. platurus* from Central America (Tu et al., 1975). The effect of this toxin, as well as of toxic and nontoxic fractions, on respiration, blood pressure, heart, and skeletal muscle of rabbits were studied by Tu et al. (1976). It was shown that a pure toxin and the crude venom had identical modes of action and produced respiratory paralysis. As can be seen from Fig. 4, the toxin caused initial respiratory stimulant effects, followed by respiratory paralysis. In most cases, respiratory paralysis occurred before a profound fall in arterial pressure. Depression of the twitch response (AT in Fig. 4) to nerve stimulation was observed in the tibialis anterior muscle. On the other hand, three nontoxic fractions induced transient respiratory stimulant effects. When artificial respiration was administered immediately after the cessation of respiration, bradycardia could be reversed, and declining blood pressure rose to nearly normal levels, and the animal survived an additional 60 to 120 min. After death, the rabbit chest was opened. Electrical stimulation of the phrenic nerves did not cause any response of the diaphragm, while stimulation of the diaphragm muscle produced contraction. Indirect stimulation of the sciatic nerve and direct stimulation of the tibialis anterior muscle induced responses. It appears that nerve impulse transmission at the neuromuscular junction is interrupted by *P. platurus* toxin.

3 PATHOLOGY

It is well known that sea snake venoms contain potent neurotoxins that act on neuromuscular junctions. However, from clinical observation of myoglobinurea in human victims, Reid (1973) concluded that sea snake venom is myotoxic.

Acute tubular necrosis in human sea snakebite cases was reported by Sitprija et al. (1973). They confirmed the observations of Marsden and Reid (1961) and Karunaratne and Panabokke (1972) that myoglobinuria and acute renal failure occurred in the victim of sea snake poisoning.

Chan and Geh (1967) observed that the injection of *Enhydrina schistosa* venom into guinea pigs produced muscle weakness and respiratory paralysis, but there was no histological evidence of muscle necrosis.

To confirm the occurrence of tubular necrosis, Schmidt et al. (1976) injected *Laticauda semifasciata* venom intramuscularly into mice. Unlike the venom of *Crotalus*, sea snake venom showed a very weak effect on the kidney. Sea snake envenomation resulted in focal organellar swelling and focal intracellular edema of the visceral

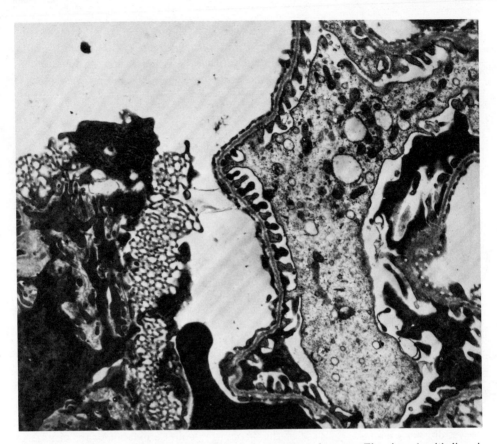

Figure 11.5 Section of glomerulus of a sea snake-enenomated mouse. The visceral epithelium is swollen. The mitochondria and endoplasmic reticulum are dilated. The plasma membranes of some of the visceral epithelial cells appear to be ruptured. (Electronmicrograph by Mr. M. C. Schmidt of the author's laboratory.)

epithelium (Fig. 5). The venom did not affect the cells of the proximal convoluted tubule, as observed in the electron microscope (Schmidt et al., 1976).

To confirm the observation of myonecrosis in human cases, we injected *Laticauda semifasciata* venom into mice intramuscularly (unpublished data). As can be seen from Fig. 6, the muscle showed a normal pattern and even the capillaries were not damaged. Thus the myoglobinuria and myotoxic effect observed by Reid may be restricted to human victims. In our experiments in mice, all sea snake venoms tested proved nonhemorrhagic. Since only one sea snake venom (*L. semifasciata*) was tested for myotoxic activity, it is not possible to generate a conclusion.

4 BIOCHEMISTRY

4.1 Composition

That sea snakes are probably a relatively primitive form of venomous snake is manifested in their venoms, which are simpler in composition than those of other families of snakes (Toom et al. 1969). This is illustrated in Fig. 7. In Fig. 7A, the protein distribution in the

Figure 11.6 Injection of sea snake (*Laticauda semifasciata*) venom intramuscularly into mice does not cause myonecrosis. Muscle tissue examined under electron microscope appears normal. (Electronmicrograph by Mr. Michael Stringer of the author's laboratory.)

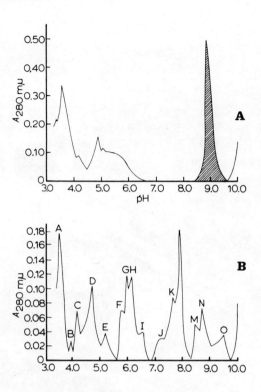

Figure 11.7 Comparison of venoms of sea snake and land snake. (A) Isoelectric focusing pattern of sea snake (*Enhydrina schistosa*) venom. (B) Isoelectric focusing pattern of land snake (*Agkistrodon rhodostoma*) venom. (Reproduced from Toom et al., 1969, by permission of the copyright owner, the Elsevier Publishing Company.)

venom of *Enhydrina schistosa* is shown as an example of sea snake venom. *Agkistrodon rhodostoma* venom is used as an example of the Crotalidae family in Fig. 7B. Isoelectric point distribution indicates that sea snake venom is rich in basic protein with a pI of 9.0 (Fig. 7A). There are also considerable amounts of acidic proteins, but the venom is low in neutral proteins. On the other hand, Crotalidae venom shows a complex distribution of proteins with all ranges of pI (Fig. 7B).

Banerjee et al. (1973) compared the electrophoretic patterns of the venoms of various snakes. The numbers of bands that appear are as follows:

Hydrophiidae (Sea Snakes)

Laticauda semifasciata
    Japan   3
    Philippines   3

Elapidae

Bungarus fasciatus   7
Micrurus fulvius   6
Naja naja   7

4 Biochemistry

Crotalidae

Agkistrodon piscivorus	10
Bothrops atrox	13
B. jararaca	13
Crotalus adamanteus	14
C. basiliscus	8
C. horridus	12
C. durissus terrificus	9

From these results, it can be seen that the venoms of sea snakes are indeed simpler in protein composition than those of terrestrial snakes. Moreover, sea snake venoms not only are simpler in overall protein composition (Fig. 7), but also contain fewer enzymes than the venoms of land snakes.

The simplicity of protein composition in sea snake venom as compared to land snake venom was also demonstrated by Shipman and Pickwell (1973). They showed that venom of *Pelamis platurus* (yellow-bellied sea snake) contained proteins with molecular weights of 6000, 8700, 9600, 11,700, 24,500, and 27,500. Venom of *Naja naja* (common cobra) was more complex, with proteins ranging from 1000 to 70,000 daltons.

The venom of *Enhydrina schistosa* lacks proteolytic and arginine ester-hydrolyzing activity (Toom et al., 1969). Venoms of *Laticauda colubrina* and *L. semifasciata* do not hydrolyze *p*-toluenesulfonyl-L-arginine methyl ester and *N*-benzoyl-L-arginine ethyl ester, whereas these substrates are easily hydrolyzed by venoms of some terrestrial snakes (Tu and Toom, 1971).

Protease and acetylcholinesterase are not found in the venom of *L. semifasciata* (Uwatoko et al., 1966a, b).

The presence of phospholipase A_2 in *Enhydrina schistosa* venom was recognized by Carey and Wright (1960b) and Ibrahim (1970). The enzyme was isolated from the venom of *Laticauda semifasciata* (Tu et al., 1970; Tu and Passey, 1971; Uwatoko-Setoguchi et al., 1968; Uwatoko-Setoguchi and Obo, 1969). The venom of *L. semifasciata* also contains phosphomonoesterase and phosphodiesterase (Uwatoko-Setoguchi, 1970).

Enhydrina schistosa venom shows the following enzyme activities: filbrinogen clotting activity, hyaluronidase, alkaline phosphatase, phosphodiesterase, deoxyribonuclease acetylcholinesterase, and leucine aminopeptidase (Tu and Toom, 1971). Leucine aminopeptidase activity can also be detected in the venoms of *Laticauda laticaudata* and *Hydrophis cyanocinctus* (Tu and Toom, 1967).

Copley et al. (1973) investigated thrombinlike activity (conversion of fibrinogen to fibrin) in 43 venoms (1 Hydrophiidae, 15 Elapidae, 7 Viperidae, 20 Crotalidae). They found that none of the Hydrophiidae and Elapidae venoms and only one Viperidae venom (that of *Vipera ammodytes*) showed such activity, but all Crotalidae venoms contained thrombinlike activity.

4.2 Isolation of Lethal Toxins

Lethal toxins have been isolated from a variety of sea snake venoms by a number of investigators. Toxins isolated from the venom of *Laticauda semifasciata*, laticatoxins III and IV (Uwatoko et al., 1966a, b), seem identical to erabutoxin a and b (Arai et al., 1964; Tamiya and Arai, 1966) judging from the fractionation pattern. Many of the toxins were isolated in crystalline form, and some of them are illustrated in Figs. 8 and 9.

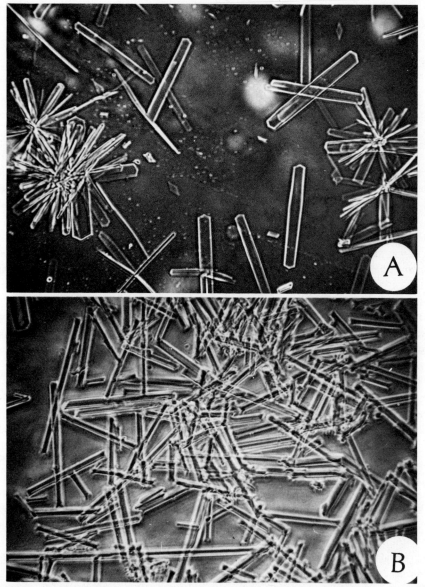

Figure 11.8 Crystals of neurotoxins obtained from *Laticauda semifasciata*. (A) Crystal laticatoxin III isolated from the venom of *L. semifasciata* captured on Okinawa. (B) Crystal laticatoxin IV isolated from the venom of *L. semifasciata* captured on Okinawa. (C) Toxin a isolated from the venom of *L. semifasciata* captured on Gato Island, the Philippines. (D). Toxin b isolated from the venom of *L. semifasciata*. Photographs of laticatoxins III and IV kindly supplied by Dr. Y. Uwatoko-Setoguchi and originally published in Uwatoko et al., 1966b. Photographs of toxin a and b taken in our laboratory and originally published in Tu et al., 1971. Reproduction by permission of the copyright owner, the American Chemical Society.

Sea snake venoms usually contain more than one lethal fraction. However, usually the most toxic component is more abundant (Chang and Tang, 1972).

The amino acid compositions of all sea snake venom toxins isolated in pure form are summarized in Table 4. Most sea snake toxins are Type I neurotoxins (see p. 000), with the exception of toxin III. Toxins from the subfamily Hydrophiinae contain no

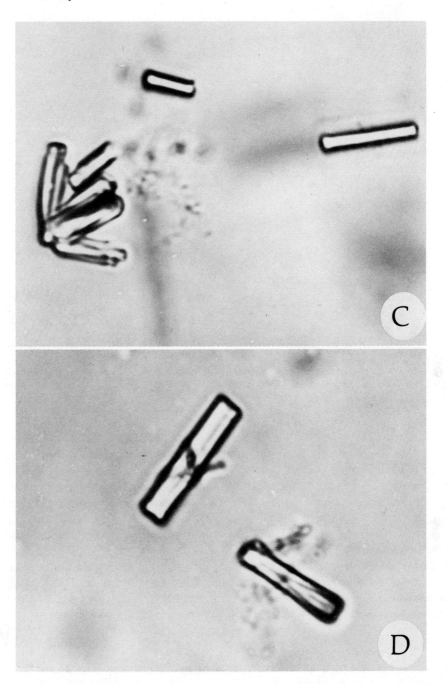

phenylalanine and only 1 mole of tryptophan, tyrosine, leucine, valine, alanine, and methionine. In Laticaudinae toxins, the amino acid compositions differ more markedly than those of Hydrophiinae toxins. One interesting difference is that no Laticaudinae toxin contains any methionine.

As mentioned before, the protein compositions of sea snake venoms are much simpler than those of terrestrial snakes. Thus the purification of neurotoxin from the venoms of

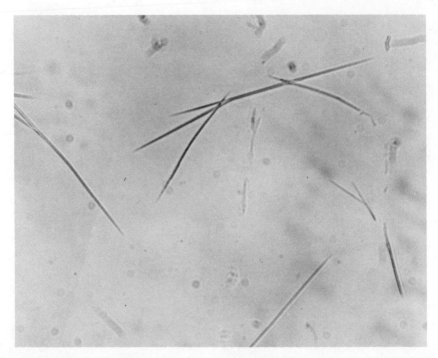

Figure 11.9 Crystal of the purified toxin isolated from *Lapemis Hardwickii* venom. (Reproduced from the paper of Tu and Hong, 1971, by permission of the copyright owner, *The Journal of Biological Chemistry*.)

sea snakes is not very difficult. In our laboratory, we found that a combination of Sephadex and carboxymethyl cellulose column chromatography is very satisfactory. Usually we use Sephadex 50 for a first-step purification. As can be seen from Fig. 10, the neurotoxic fraction appears in the middle of the overall fractionation pattern. This means that we eliminate 88% of the proteins with molecular weights larger than that of neurotoxin (fractions I, II, III), as well as the proteins with smaller molecular weights (fractions V and VI).

In the second-step purification using CM cellulose, the neurotoxin can be separated from other proteins on the basis of charge. Since neurotoxins are highly basic, any proteins that elute out earlier than neurotoxins usually have pI's lower than 9. Also some basic proteins (VIII, IX, X) with low toxicity are present in sea snake venoms.

4.3 Structure of Neurotoxins

It has been shown by many investigators that neurotoxins from Elapidae and sea snake venoms consist of either 60 to 62 or 70 to 74 amino acid residues. Thus they can be conveniently divided into two groups on the basis of number of amino acids. The first group is designated as Type I neurotoxins, and the second as Type II (Figs. 4, 5, and 6 of Chapter 18). Detailed descriptions of these neurotoxins are given in review articles (Tu, 1973, 1974b).

However, the readers should be aware that some toxins fall somewhere between Type I and Type II. Other toxins do not even fit into this classification. Toxins obtained from

Table 4 Amino Acid Compositions of Sea Snake Neurotoxins

Subfamily Genus Species	Hydrophiinae Pelamis platurus		Lapemis hardwickii	Enhydrina schistosa			
Origin	Costa Rica	Formosa	Thailand	Malaysia			
Amino Acid	Pelamis toxin a		venom from 1969 collection	venom from 1972 collection	venom from 1967 collection	venoms from 1966, 67, and 69 collection	Toxin 4	Toxin 5
Lysine	5	5	5	5	5	5	5	5
Histidine	2	2	2	2	2	2	2	2
Arginine	3	3	3	3	3	3	3	3
S-Carboxy-methylcysteine	-	-	-	-	-	-	-	-
Aspartic acid	6	6	6	6	6	6	6	6
Threonine	7	7	8	8	8	7	7	7
Serine	5	6	6	6	6	5	5	6
Glutamic acid	7	8	8	8	8	8	8	8
Proline	1	2	3	3	3	3	3	2
Glycine	3	4	4	4	5	4	4	4
Alanine	1	1	1	1	1	1	1	1
Half-cystine	8	9	8	8	9	9	9	9
Valine	1	1	1	1	1	1	1	1
Methionine	1	1	1	1	0	1	1	1
Isoleucine	2	2	2	2	2	2	2	2
Leucine	1	1	1	1	1	1	1	1
Tyrosine	1	1	1	1	1	1	1	1
Phenylalanine	0	0	0	0	0	0	0	0
Tryptophan	1	1	1	1	1	1	1	1
Total residues	55	60	61	61	62	60	60	60
References	Tu et al, 1975	Liu et al, 1975	Tu and Hong, 1971	Yu et al, 1975	Tu and Toom, 1971	Yu et al, 1975	Karlsson et al, 1972b	

Table 4 *Continued*

Subfamily Genus Species	Hydrophiinae Hydrophis cyanocinctus		Laticaudinae Laticauda semifasciata						Laticauda taticaudata	Laticauda colubrina	Aipysarus laevis		
Origin	Formosa		Philippines			Okinawa			Okinawa	Okinawa	Australia		
Amino Acid	Hydrophitoxin		Toxin	Toxin		Erabutoxin			Laticotoxin	Laticotoxin			
	a	b	a	b		a	b	c	Toxin III		a	b	c
Lysine	6	5	4	5		4	4	3	4	4	6	5	7
Histidine	2	2	1	1		1	2	2	1	2	1	1	1
Arginine	3	3	3	2		3	3	3	2	5	3	4	3
S-Carboxy-methylcysteine	-	8	-	-		-	-	-	10	-	-	-	-
Aspartic acid	6	6	5	4		5	4	5	6	9	8	8	8
Threonine	7	7	6	5		5	5	5	6	4	7	7	7
Serine	5	6	7	6		8	8	8	6	6	3	3	3
Glutamic acid	8	8	8	8		8	8	8	6	7	8	7	7
Proline	2	2	4	4		4	4	4	4	5	3	3	3
Glycine	4	4	5	6		5	5	5	4	5	4	4	4
Alanine	1	1	0	0		0	0	0	4	0	1	1	1
Half-cystine	8	-	8	8		8	8	8	-	8	9	9	9
Valine	1	1	2	3		2	2	2	3	1	1	1	1
Methionine	1	1	0	0		0	0	0	0	0	0	1	0
Isoleucine	2	2	4	4		4	4	4	2	2	2	2	2
Leucine	1	1	1	1		1	1	1	2	1	2	2	2
Tyrosine	1	1	1	1		1	1	1	3	1	1	1	1
Phenylalanine	0	0	2	2		2	2	2	1	1	0	0	0
Tryptophan	1	1	1	1		1	1	1	2	1	1	1	1
Total residues	59	60	62	61		62	62	62	66	62	60	60	60
References	Liu et al, 1973	Liu and Blackwell, 1974	Tu et al, 1971			Tamiya and Avai, 1966	Tamiya and Abe, 1972		Maeda and Tamiya, 1974	Sato et al, 1969	Maeda and Tamiya, 1976		

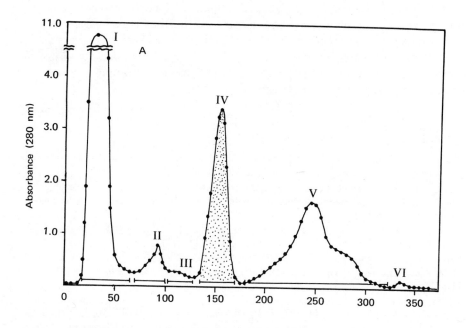

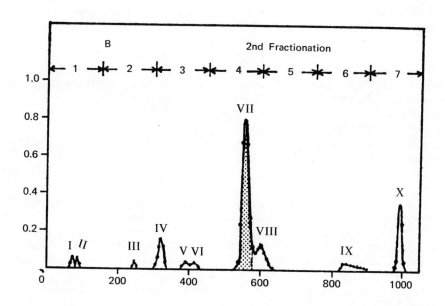

Figure 11.10 An example of sea snake neurotoxin purification. Most sea snake venom neurotoxin can be purified by a combination of Sephadex G-50 (A) and CM-cellulose column (B) chromatography. The example used here is isolation of *Pelamis* toxin a from *Pelamis platurus* venom. (For details, see Tu et al., 1975.)

Australian snakes are much bigger in size and quite different from neurotoxins obtained from Asian and African venoms (Karlsson et al., 1972b; Halpert and Eaker, 1975).

Nearly all neurotoxins isolated from sea snake venoms are Type I.

Since neurotoxins obtained from sea snake and Elapidae venoms are very similar in chemical structure, it is best that they be discussed together. Therefore a separate chapter (Chapter 18) is devoted to the chemistry of neurotoxins.

REFERENCES

Arai, H., Tamiya, N., Toshioka, S., Shinonaga, S., and Kano, R. (1964). Studies on sea snake venoms. I. Protein nature of the neurotoxic component, *J. Biochem.*, **56**, 568.

Banerjee, S., Devi, A., and Copley, A. L. (1973). Studies of actions of snake venoms on blood coagulation. II. Electrophoretic analysis of venoms of Viperidae, Crotalidae, Elapidae and Hydrophidae, *Thromb. Res.*, **3**, 451.

Barber, D. W., Puffer, H. W., Tamiya, N., and Shynkar, T. P. (1974). Aspects of the neuromuscular activity of sea snake venom, *Proc. West. Pharmacol. Soc.*, **17**, 235.

Barme, M. (1963). "Venomous sea snakes of Viet Nam and Their Venoms," in H. L. Keegan and W. V. MacFarlane, Eds., *Venomous and Poisonous Animals and Noxious Plants of the Pacific Region*, MacMillan, New York, pp. 373–378.

Barme, M. (1968). "Venomous sea snakes (Hydrophiidae)," in W. Bücherl, E. E. Buckley, and V. Deulofew, Eds., *Venomous Animals and Their Venoms*, Academic, New York, pp. 286–308.

Baxter, E. H. and Gallichio, H. A. (1974). Cross-neutralization by tiger snake (*Notechis scutatus*) antivenine and sea snake (*Enhydrina schistosa*) antivenine against several sea snake venoms, *Toxicon*, **12**, 273.

Carey, J. E. and Wright, E. A. (1960a). The toxicity and immunological properties of some sea-snake venoms with particular reference to that of *Enhydrina schistosa*, *Trans. Roy. Soc. Trop. Med. Hyg.*, **54**, 50.

Carey, J. E. and Wright, E. A. (1960b). Isolation of the neurotoxic component of the venom of the sea snake, *Enhydrina schistosa*, *Nature*, **185**, 103.

Carey, J. E. and Wright, E. A. (1961). The site of action of the venom of the sea snake, *Enhydrina schistosa*, *Trans. Roy. Soc. Trop. Med. Hyg.*, **55**, 153.

Chan, K. E. and Geh, S. L. (1967). Antagonism of intraarterial acetylcholine induced contraction of skeletal muscle by sea snake venom, *Nature*, **213**, 5081.

Chang, T. W. and Tang, N. (1972). Selection pressures on homologous proteins of varied activities, *Nature New Biol.*, **239**, 207.

Cheymol, J., Barme, M., Bourillet, F., and Roch-Arveiller, M. (1967). Action neuromusculaire de trois venins d'hydrophides, *Toxicon*, **5**, 111.

Cheymol, J., Tamiya, N., Bourillet, F., and Roch-Aveiller, M. (1972). Action neuromusculaire du venin de serpent marin "Erabu" (*Laticauda semifasciata*) et des erabutoxines a et b, *Toxicon*, **10**, 125.

Copley, A. L., Banerjee, S., and Devi, A. (1973). Studies of snake venoms on blood coagulation. I. The thromboserpentin (thrombinlike) enzyme in the venoms, *Thromb. Res.*, **2**, 487.

Friederich, C. and Tu, A.T. (1971). Role of metals in snake venoms for hemorrhagic, esterase and proteolytic activities, *Biochem. Pharmacol.*, **20**, 1549.

Ganthavorn, S. (1969). Toxicities of Thailand snake venoms and neutralization capacity of antivenin, *Toxicon*, **7**, 239.

Geh, S. L. and Chan, K. E. (1973). The prejunctional site of action of *Enhydrina schistosa* venom at the neuromuscular junction, *Eur. J. Pharmacol.*, **21**, 115.

Halpert, J. and Eaker, D. (1975). Amino acid sequence of a presynaptic neurotoxin from the venom of *Notechis scutatus scutatus* (Australian tiger snake), *J. Biol. Chem.*, **250**, 6990.

Halstead, B. W. (1970). *Poisonous and Venomous Marine Animals of the World*, U.S. Government Printing Office, Washington, D.C.

References

Homma, M., Okonogi, T., and Mishima, S. (1964). Studies on sea snake venom. 1. Biological toxicities of venoms possessed by three species of sea snakes captured in coastal waters of Amami Oshima, *Gunma J. Med. Sci.,* **13,** 283.

Hong, B. and Tu, A. T. (1970). Importance of tryptophan residue for toxicity in sea snake venom toxins, *Fed. Proc.,* **29,** 883.

Ibrahim, S. A. (1970). A study on sea snake venom phospholipase A, *Toxicon,* **8,** 221.

Karlsson, E., Eaker, D., Fryklund, L., and Kadin, S. (1972a). Chromatographic separation of *Enhydrina schistosa* (common sea snake) venom and the characterization of two principal neurotoxins, *Biochemistry,* **11,** 4628.

Karlsson, E., Eaker, D., and Ryden, L. (1972b). Purification of a presynaptic neurotoxin from the venom of the Australian tiger snake *Notechis scutatus scutatus, Toxicon,* **10,** 405.

Karunaratne, K. E. and Panabokke, R. G. (1972). Sea snake poisoning – case report, *J. Trop. Med. Hyg.,* **75,** 91.

Levey, H. A. (1969). Toxicity of the venom of the sea-snake, *Laticauda colubrina,* with observations on a Malay "folk cure," *Toxicon,* **6,** 269.

Liu, C. S. and Blackwell, R. Q. (1974). Hydrophitoxin b from *Hydrophis cyanocinctus* venom, *Toxicon,* **12,** 543.

Liu, C. S., Huber, G. S., Lin, C. S., and Blackwell, R. Q. (1973). Fractionation of toxins from *Hydrophis cyanocinctus* venom and determination of amino acid composition and end groups of hydrophitoxin a, *Toxicon,* **11,** 73.

Liu, C. S., Wang, C., and Blackwell, R. Q. (1975). Isolation and partial characterization of pelamitoxin a from *Pelamis platurus* venom, *Toxicon,* **13,** 31.

Maeda, N. and Tamiya, N. (1974). The primary structure of the toxin *Laticauda semifasciata* III, a weak and reversibly acting neurotoxin from the venom of a sea snake, *Laticauda semifasciata, Biochem. J.,* **141,** 389.

Maeda, N. and Tamiya, N. (1976). Isolation, properties and amino acid sequences of three neurotoxins from the venom of a sea snake, *Aipysurus laevis, Biochem. J.,* **153,** 79.

Maeda, N., Takagi, K., Tamiya, N., Shen, Y., and Lee, C. (1974). The isolation of an easily reversible post-synaptic toxin from the venom of a sea snake, *Laticauda semifasciata, Biochem. J.,* **141,** 383.

Marsden, A. T. H. and Reid, H. A. (1961). Pathology of sea-snake poisoning, *Br. Med. J.,* **1,** 1290.

Minton, S. A., Jr. (1967). Paraspecific protection of elapid and sea snake antivenins, *Toxicon,* **5,** 74.

Okonogi, T. (1970). Neutralization of sea snake venom with goat antivenin, *Snake,* **2,** 18.

Okonogi, T. (1973). Venomous sea snake bite, *Snake,* **5,** 156.

Okonogi, T., Hattori, Z., and Isoarashi, I. (1967). Experimental study of immunology of sea snake venoms, *Jap. J. Bacteriol.,* **22,** 173.

Okonogi, T., Hattori, Z., Amagai, E., Sawai, Y., and Kawamura, Y. (1972). Studies of immunity against the venom of *Lapemis hardwickii* with a special reference to a pilot production of therapeutic antivenin horse serum, *Snake,* **4,** 84.

Pickwell, G. V. and Evans, W. E. (1972). *Handbook of Dangerous Animals for Field Personnel,* U.S. Navy.

Pickwell, G. V., Vick, J. A., Shipman, W. H. and Grenan, M. M. (1974). "Production, toxicity and preliminary pharmacology of venom from the sea snake, *Pelamis platurus,*" in L. R. Worthen, Ed., *Food and Drugs from the Sea, Proceedings in 1972,* Marine Technology Society, Washington, D.C., pp. 247–265.

Reid, H. A. (1956). Sea-snake bite research, *Trans. Roy. Soc. Trop. Med. Hyg.,* **50,** 517.

Reid, H. A. (1961). Myoglobinuria and sea snake poisoning, *Br. Med. J.,* **1,** 1284.

Reid, H. A. (1963). "Snakebite in Malaya," in H. L. Keegan and W. V. MacFarlane, Eds., *Venomous and Poisonous Animals and Noxious Plants of the Pacific Region,* Macmillan, New York, pp. 355–362.

Reid, H. A. (1973). "Clinical aspects of animal toxins," in A. De Vries and E. Kochva, Eds., *Toxins of Animal and Plant Origin,* Vol. III, Gordon and Breach, New York, pp. 957–984.

Reid, H. A. (1975a). Antivenom in sea-snake bite poisoning, *Lancet,* **1,** 622.

Reid, H. A. (1975b). Epidemiology of sea snake bites, *J. Trop. Med. Hyg.,* **78,** 106.

Sato, S. and Tamiya, N. (1970). Iodination of erabutoxin b: diiodohistidine formation, *J. Biochem.*, **68**, 867.

Sato, S., Yoshida, H., Abe, H., and Tamiya, N. (1969). Properties and biosynthesis of a neurotoxic protein of the venoms of sea snakes *Laticauda laticaudata* and *Laticauda colubrina, Biochem. J.*, **115**, 85.

Sato, S., Abe, T., and Tamiya, N. (1970). Binding of iodinated erabutoxin b, a sea snake toxin, to the endplates of the mouse diaphragm, *Toxicon*, **8**, 313.

Schmidt, M. E. Abdelbaki, Y. Z., and Tu, A. T. (1976). Nephrotoxic action of rattlesnake and sea snake venoms: An electron-microscopic study, *J. Pathol.*, **118**, 75.

Seto, A., Sato, S., and Tamiya, N. (1970). The properties and modification of tryptophan in a sea snake toxin, erabutoxin a, *Biochem. Biophys. Acta*, **214**, 483.

Shipman, W. H. and Pickwell, G. V. (1973). Venom of the yellow-bellied sea snake (*Pelamis platurus*): Some physical and chemical properties, *Toxicon*, **11**, 375.

Sitprija, V., Sribhibhadh, R., Benyajati, C., and Tangchai, P. (1973). "Acute renal failure in snakebite," in A. De Vries and E. Kochva, Eds., *Toxins of Animal and Plant Origin*, Vol. III, Gordon and Breach, New York, pp. 1013–1028.

Smith, M. (1926) *Monograph of the Sea Snakes (Hydrophiidae)*, The British Museum, London.

Tamiya, N. and Abe, H. (1972). The isolation, properties, and amino acid sequence of erabutoxin, c, a minor neurotoxic component of the venom of a sea snake *Laticauda semifasciata, Biochem. J.*, **130**, 547.

Tamiya, N. and Arai, H. (1966). Crystallization of erabutoxins a and b from *Laticauda semifasciata* venom, *Biochem. J.*, **99**, 624.

Tamiya, N. and Puffer, H. (1974). Lethality of sea snake venoms, *Toxicon*, **12**, 85.

Tamiya, N., Arai, H., and Sato, J. (1967). "Studies on sea snake venoms: Crystallization of erabutoxins a and b from *Laticauda semifasciata* venom, and of laticotoxin a from *Laticauda laticaudata* venom," in F. E. Russel and P. R. Saunders, Eds., *Animal Toxins*, Pergamon, Oxford, pp. 249–258.

Taylor, E. H. (1965). The serpents of Thailand and adjacent waters, *Univ. Kansas Bull.* No. XIV.

Toom, P. M., Squire, P. G., and Tu, A. T. (1969). Characterization of the enzymatic and biological activities of snake venoms by isoelectric focusing, *Biochim. Biophys. Acta*, **181**, 339.

Toom, P. M., Solie, T. N., and Tu, A. T. (1970). Characterization of a nonproteolytic arginine ester hydrolyzing enzyme from snake venom, *J. Biol. Chem.*, **245**, 2549.

Tu, A. T. (1973). Neurotoxins of animal venoms: Snakes, *Ann. Rev. Biochem.*, **42**, 235.

Tu, A. T. (1974a). Sea snake investigation in the Gulf of Thailand, *J. Herpetol.*, **8**, 201.

Tu, A. T. (1974b). Sea snake venoms and neurotoxins, *J. Agr. Food Chem.*, **22**, 36.

Tu, A. T. (1976). Investigation of the sea snake, *Pelamis platurus* (Reptilia, Serpentes, Hydraphiidae), on the Pacific coast of Costa Rica, Central America, *J. Herpetol.*, **10**, 13–20.

Tu. A. T. and Ganthavorn, S. (1969). Immunological properties and neutralization of sea snake venoms from Southeast Asia, *Am. J. Trop. Med. Hyg.*, **18**, 151.

Tu, A. T. and Homma, M. (1970). Toxicological study of snake venoms from Costa Rica, *Toxicol. Appl. Pharmacol.*, **16**, 73.

Tu, A. T. and Hong, B. (1971). Purification and chemical studies of a toxin from the venom of *Lapemis hardwickii* (Hardwick's sea snake), *J. Biol. Chem.*, **246**, 2772.

Tu, A. T. and Passey, R. B. (1971). "Phospholipase A from sea snake venom and its biological properties," in A. De Vries and E. Kochva, Eds., Gordon and Breach, New York, pp. 419–436.

Tu, A. T. and Salafranca, E. S. (1974). Immunological properties and neutralization of sea snake venoms. II. *Am. J. Trop. Med. Hyg.*, **23**, 135.

Tu, A. T. and Toom, P. M. (1967). The presence of a L-leucyl-β-naphthylamide hydrolyzing enzyme in snake venoms, *Experientia*, **23**, 439.

Tu, A. T. and Toom, P. M. (1971). Isolation and characterization of the toxic component of *Enhydrina schistosa* (common sea snake) venom, *J. Biol. Chem.*, **246**, 1012.

Tu, A. T. and Tu, T. (1970). "Sea snakes from Southeast Asia and Far East and their venoms," in B. W. Halstead, Ed., *Poisonous and Marine Animals of the World*, Vol. 3, U.S. Government Printing Office, Washington, D.C., pp. 885–903.

References

Tu, A. T., Homma, M., and Hong, B. (1969). Hemorrhagic, myonecrotic, thrombotic, and proteolytic activities of viper venoms, *Toxicon,* **6,** 175.

Tu, A. T., Homma, M., Hong, B. S., and Terrill, J. B. (1970). Neutralization of rattlesnake venom toxicities by various compounds, *J. Clin. Pharmacol.,* **10,** 323.

Tu, A. T., Hong, B. S., and Solie, T. N. (1971). Characterization and chemical modifications of toxins isolated from the venoms of the sea snake, *Laticauda semifasciata,* from Philippines, *Biochemistry,* **10,** 1295.

Tu, A. T., Lin, T. S., and Bieber, A. L. (1975). Purification and chemical characterization of the major neurotoxin from the venom of *Pelamis platurus, Biochemistry,* **14,** 3408.

Tu, T. (1957). Toxicological studies on the venom of a sea snake, *Laticauda semifasciata* (Reinwardt), in Formosan waters: First report. *J. Formosan Med. Assoc.,* **56,** 609.

Tu, T. (1959). Toxicological studies on the venom of a sea snake, *Laticauda semifasciata* (Reinwardt), in Formosan waters, *J. Formosan Med. Assoc.,* **58,** 182.

Tu, T. (1961). Toxicological studies on the venom of a sea snake, *Laticauda semifasciata, Biochem. Pharmacol.,* Abs. No. 244.

Tu, T. (1963). Toxicological studies on the venom of a sea snake, *Laticauda semifasciata* (Reinwardt), in Formosan waters: Fifth Report, *J. Formosan Med. Assoc.,* **62,** 100.

Tu, T. (1967). "Toxicological studies on the venom of the sea snake *Laticauda laticaudata affinis,*" in F. E. Russel and P. R. Saunders, Eds., *Animal Toxins,* Pergamon, Oxford, pp. 245–248.

Tu, T., Lin, M. J., Yang, H. M., Lin, H. J., and Chen, C. N. (1963). Toxicological studies on the venom of a sea snake, *Laticauda colubrina, J. Formosan Med. Assoc.,* **62,** 122.

Tu, T., Tu, A. T., and Lin, T. S. (1976). Some pharmacological properties of the venom, venom fractions and pure toxin of yellow-bellied sea snake, *Pelamis platurus, J. Pharm. Pharmacol.,* **28,** 139.

Uwatoko, Y., Nomura, Y., Kojima, K., and Obo, F. (1966a). Investigation of sea snake venom. I. Fractionation of sea snake (*Laticauda semifasciata*) venom, *Acta Med. Univ. Kagoshima.,* **8,** 141.

Uwatoko, Y., Nomura, Y., Kojima, K., and Obo, F. (1966b). Investigation of sea snake venom. II. Crystallization of toxic compounds in sea snake (*Laticauda semifasciata*) venom, *Acta Med. Univ. Kagoshima.,* **8,** 151.

Uwatoko-Setoguchi, Y. (1970). Studies on sea snake venom. VI. Pharmacological properties of *Laticauda semifasciata* venom and purification of toxic components, acid phosphomonoesterase and phospholipase A in the venom, *Acta Med. Univ. Kagoshima.,* **12,** 73.

Uwatoko-Setoguchi, Y. and Obo, F. (1969). Studies on sea snake venom. V. Some properties of phospholipase A in *Laticauda semifasciata* venom, *Acta Med. Univ. Kagoshima.,* **11,** 139.

Uwatoko-Setoguchi, Y., Minamishima, Y., and Obo, F. (1968). Studies on sea snake venom. IV. Purification of phospholipase A in *Laticauda semifasciata* venom, *Acta Med. Univ. Kagoshima.,* **10,** 219.

Vick, J. A., Von Bredow, J., Grenan, M. M., and Pickwell, G. V. (1975). "Sea snake antivenin and experimental envenomation therapy," in W. A. Dunson, Ed., *The Biology of Sea Snakes,* University Park Press, Baltimore.

Walker, M. J. A. and Yeoh, P. N. (1974). The *in vitro* neuromuscular blocking properties of sea snake (*Enhydrina schistosa*) venom, *Eur. J. Pharmacol.,* **28,** 199.

Yeoh, P. N. and Walker, M. J. A. (1974). Effect of sea snake (*Enhydrina schistosa*) venom on the ganglionic nicotinic actions of acetylcholine, *J. Pharm. Pharmacol.,* **20,** 441.

Yu, N., Lin, T., and Tu, A. T. (1975). Laser Raman scattering of neurotoxins isolated from the venoms of sea snakes *Lapemis hardwickii* and *Enhydrina schistosa, J. Biol. Chem.,* **250,** 1782.

12 Elapidae Venoms

1 COBRA VENOMS … 179
   1.1 Effect on Motor End Plate, 179
   1.2 Effect on Ganglions, 180
   1.3 Effect on Respiration and Circulation, 181
   1.4 Effect on Muscles, 181
   1.5 Effect on Central Nervous System, 182
   1.6 Immunology, 182
   1.7 Other Effects, 183

2 KRAIT VENOMS … 184
   2.1 Toxicology, 184
   2.2 Pharmacology, 184
      Early History, 184
      α-Bungarotoxin, 185
      β-Bungarotoxin, 185
   2.3 Other Actions, 190

3 CORAL SNAKE VENOMS … 190
   3.1 Biological Action, 191
   3.2 Biochemistry, 192

4 OTHER ELAPIDAE VENOMS … 192
   4.1 Australian Venoms, 192
   4.2 African Venoms, 193
      Walterinnesra aegyptia, 193
      Dendraspis Spp. (Mambas), 194

References … 195

Elapidae venoms contain potent neurotoxins and also, like many other snake venoms, several varieties of proteins, both enzymes and nonenzymes. But because of the potent toxicity of neurotoxins, they are chiefly responsible for fatal poisoning by whole Elapidae venoms.

Often a venom contains several neurotoxins that have similar actions. For instance, two toxins isolated from *Naja oxiana* venom both have postsynaptic effects on frog rectus abdominus muscle (Zakhayou and Spiridonov, 1974).

In this chapter the emphasis is placed on the biological properties of venoms and their toxic components. The chemical properties of neurotoxins (Chapter 18), cardiotoxins and other basic proteins (Chapter 19), and the binding of neurotoxins to acetylcholine receptors (Chapter 17) are discussed in the chapters indicated.

1 COBRA VENOMS

Cobras have a wide distribution and can be found in Africa, the Middle East, central Asia, south Asia, and southeast Asia. Cobra venom is a mixture of many different proteins. Although the most toxic components are neurotoxins, lethal cardiotoxins are also present in venoms. Cobra venom contains a variety of enzymes, but fewer than the venoms of Viperidae and Crotalidae. Unlike krait (*Bungarus* spp.) venom, which contains both pre- and postsynaptic toxins, the neurotoxins of cobra venoms are postsynaptic acting.

1.1 Effect on Motor End Plate

The curarelike action of cobra venoms and their toxins is well known (Vick et al., 1965; Meldrum, 1965a, b; Tazieff-Depierre and Pierre, 1966; Cheymol et al., 1971).

Although snake neurotoxins are known to block nerve transmission at the site of the neuromuscular junction, they do not affect transmission within the axon (Su, 1960; Schmidt et al., 1964; Lester, 1966; Chang and Lee, 1966; Eaker et al., 1971; Osman et al., 1974; Cheymol et al., 1974). The axon is quite resistant to the action of venom. When frog spinal nerve stems were incubated with *Naja oxiana* venom, there was little penetration of the venom. Only the epineurium was damaged, and there was a swelling of the myelin sheath (Orlov et al., 1971).

The nondepolarizing effect of cobra venoms and toxins is similar to that of curare. The action of cobra neurotoxin differs from that of curare, however, and, strictly speaking, should not be referred to as curarization because its mechanism is not one of pure competition (Cheymol et al., 1971).

Differences in the actions of curare and cobra neurotoxins are diagrammatically illustrated in Fig. 1. Because cobra neurotoxins bind to receptors in the motor end plate, they do not affect the electrical properties of the muscle fibers (Chang and Lee, 1966; Eaker et al., 1971; Loots et al., 1973; Osman et al., 1974). Binding of neurotoxins to the extrajunctional cholinergic receptors is very strong and essentially irreversible.

Cobra venoms, however, contain other components that can depolarize the muscle membrane. Earl and Excell (1972) isolated two fractions of high toxicity from *N. nivea* venom, which produced a nondepolarizing postsynaptic block of neuromuscular transmission. In addition, they isolated two fractions of relatively low toxicity, which depolarized the muscle membrane. The depolarizing action of the venom was not attributed to phospholipase A_2 or protease, but rather to heat-stable, basic proteins. Earl and Excell proposed that the depolarizing fractions produced a nonspecific increase in membrane permeability by displacing membrane calcium. These fractions are probably identical to cardiotoxins.

Neurotoxins are the principal lethal toxins in cobra venoms, and the crude venoms therefore produce the same neurotoxic effect as pure neurotoxins. The crude venoms are a mixture of various proteins and contain other lethal toxins in addition to neurotoxins. Cardiotoxins are the next most toxic components, and hence the crude venoms provoke cardiotoxic as well as neurotoxic effects. Cheymol et al. (1971) observed that the crude

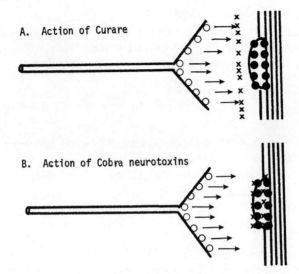

Figure 12.1 Diagrams showing the actions of curare (A) and snake neurotoxins (B) of postsynaptic type. Note the layer of curare in front of the receptor (A). The snake neurotoxins are actually attached to the receptor at the postsynaptic site (B).

venom of *N. nigricollis* produced cardiovascular failure or myocardial contracture in an isolated preparation. The pure α-naja toxin isolated from the same venom was devoid of this effect and produced only a blocking action on impulse transmission in the heart.

Neurotoxins and cardiotoxins of cobra venoms are structurally similar (see Chapters 18 and 19), but they have different pharmacologic effects. A neurotoxin obtained from *N. naja* venom blocked the response of lamprey lingual retractor fibers to acetylcholine. However, the response of lamprey isolated heart to acetylcholine, propionylcholine, succinylcholine, or suberylcholine was not affected by the neurotoxins, even at 10 μg ml^{-1} (Lukomskaya and Magazanik, 1974).

The molecular pharmacologic aspects of binding neurotoxins to receptors in motor end plates is discussed in Chapter 17.

1.2 Effect on Ganglions

The nondepolarizing blocking nature of neuromuscular transmission by cobra neurotoxin is somewhat similar to that of curare. Since curare also blocks ganglia, it is of interest to determine whether snake neurotoxins have a similar action. This was investigated by Chou and Lee (1969), who studied the effect of cobra venom (*Naja naja atra*) and its components on ganglionic transmission *in vivo* (superior cervical ganglion of the cat) and *in vitro* (hypogastric nerve–vas deferens preparation of the guinea pig). Cobra venom applied directly to the ganglion produced complete ganglionic blockade after a transient phase of stimulation. This effect was caused by the venom cardiotoxin, but not by cobra neurotoxin or phospholipase A_2. Since cardiotoxin causes irreversible depolarization of the cell membrane, its ganglionic stimulating and blocking actions are believed to be due to nonspecific, persistent depolarization of the ganglion cells. *In vitro* experiments revealed that cardiotoxin in higher concentrations affected, first, ganglionic transmission, then possibly the postganglionic neuroeffector junction, and, finally, the smooth muscle itself.

1 Cobra Venoms

1.3 Effect on Respiration and Circulation

Envenomation of cobras has serious effects on the respiratory systems of the victims. It is important to determine whether the respiratory paralysis is central or peripheral in origin.

Peng (1952a) concluded from an experiment on rabbits that the fall of blood pressure due to *Naja naja atra* envenomation is of peripheral origin and is mainly due to vasodilation in the splanchnic area and also sometimes in the skin and muscles. The rise in blood pressure after administration of the venom and just before death is due to stimulation of the vasomotor center by asphyxia. Lee and Peng (1961) further investigated the cause of respiratory failure in dogs and cats by recording the respiratory discharge down the phrenic nerve and the action potentials in the intercostal and diaphragmatic muscles. The phrenic nerve discharge was found even after complete paralysis of the respiratory muscles, if the animals were kept alive by artificial ventilation. The discharge was intensified and prolonged if asphyxia developed. The continued activity of the central respiratory mechanism was indicated by the presence of the Hering–Breuer reflex. Lee and Peng concluded that the respiratory failure caused by *N. naja atra* venom is indeed peripheral in origin.

Vick et al. (1965) studied the effect of *N. naja* venom on the peripheral blocking of respiratory effort in the dog and concluded that the effect is due to interference with nerve impulse transmission at the neuromuscular junction. Although modified, respiratory center activity, as judged by phrenic discharges, remained functional after envenomation. The continued ability of the nerve to conduct was also established by observing action potentials at all points tested on the nerve. The muscle of the diaphragm is not affected by venom, since direct stimulation produced a contraction. This response cannot be reproduced by stimulating the peripheral end of the cut phrenic nerve. Thus it can be concluded that venom interferes with transmission at the neuromuscular junction. Use of an artificial ventilator can increase the survival time, but animals receiving toxic levels of venom usually die of cardiovascular failure.

1.4 Effect on Muscles

The most pronounced effects of cobra venom are, first, its neurotoxic action and, secondarily, its cardiotoxic action. However, the effects of cobra venoms are not restricted to these two. The venoms also cause local myonecrosis (Reid, 1964), although they do not produce local hemorrhage (Tu et al., 1967). Myonecrosis induced by *Naja naja kaouthia* (*N. naja siamensis*) venom in mice was studied by Stringer et al. (1971), using electron microscopy. Degeneration of the entire muscle fiber and its constituents was observed (Figs. 15 and 16 of Chapter 23). The myofilaments coalesced to form an amorphous mass, and the sarcotubular system disappeared. The mitochondria swelled into vacuoles containing fragmented cristae; the vacuoles ruptured and were lysed. The degenerated fiber became a less electron-dense mass containing a few cellular remnants enclosed by a vestige of the plasma membrane. These results clearly demonstrate that envenomation with cobra venom causes necrosis of skeletal muscle. In human victims this may easily be overlooked, however, because the initial local reaction may be slight, and the patient may be discharged before myonecrosis is evident (Reid, 1968).

Naja nivea venom contains a fraction that causes depolarization of isolated frog skeletal muscle in calcium-deficient Ringer's solution (Earl and Excell, 1971). This is probably due to the removal of membrane calcium with a resultant increase in membrane permeability.

1.5 Effect on Central Nervous System

While the inhibitory action of cobra venoms on the peripheral nerve is well documented, there are also reports of blocking actions of venoms and their components on the central nervous system.

Tseng et al. (1968) investigated the distribution of *Naja naja atra* venom and purified neurotoxin and cardiotoxin and found only small amounts of radioactivity in the cerebrospinal fluid of rabbits after i.v. injection of venom or components. They concluded that the penetration of venom to the brain is too small to account for the rapid respiratory paralysis. In view of this result it seems rather odd that some cobra venoms produce a central effect. It may be that different cobra venoms have different effects on the central nervous system. The reported data are presented here without the present author's personal comment.

Vick et al. (1964) observed that three major physiological changes follow the i.v. injection of *N. naja* venom. The initial change is a complete and irreversible loss of cortical electrical activity, which occurs within 30 to 60 sec. The second change, a sharp rise in portal venous pressure and a precipitous fall in systemic blood pressure, occurs in 3 to 5 min (Morales et al., 1961; Bhanganada and Perry, 1963). The final change is respiratory paralysis due to blockade at the neuromuscular junction of the diaphragm (Sarkar et al., 1942; Rosenberg and Ehrenpreis, 1961).

Vick et al. (1966) fractionated *N. naja* venom into 12 components, of which 3 are physiologically active. One fraction produced a loss of cortical electrical activity, as seen in the electroencephalogram. The second fraction caused respiratory paralysis, and the third affected the cardiovascular system, ultimately producing irreversible hypotension.

Naja oxiana venom initially inhibited induced electric potentials and later blocked desynchronization in the cerebral cortex of rabbits, suggesting that the venom blocks nonspecific functional systems of the brain, especially the ascending activating effect in the reticular formation. Cobra venom also depressed the central action of both nicotine and strychnine in anesthetized cats, suggesting a blocking effect of the venom on central N-cholinoreactive systems (Orlov and Gelashvili, 1972).

Neurotoxins isolated from the venom of *N. oxiana* also inhibited Mg(II)-ATPase and Na(I), K(I)-ATPase of microsomes from the gray matter of cattle brain *in vitro* (Mirsalikhova et al., 1974).

1.6 Immunology

Cobras have a wide distribution ranging from Asia and the Middle East to Africa. From a clinical viewpoint, it is important to know how the specific antivenin is effective against different venoms. In an extensive study Minton (1967) investigated 17 Elapidae antivenins and 1 sea snake antivenin against 15 Elapidae venoms and 1 sea snake venom. He found that cobra antivenins are largely genus specific and show considerable variation in neutralization capacity. Antivenin for the Thailand cobra, *Naja naja siamensis*, can also neutralize the venom of *N. naja atra* from Formosa (Tu and Ganthavorn, 1968).

Polyvalent antivenin, including antivenin for *N. naja* manufactured at the Haffkine Institute in India, neutralized the venoms of *N. naja kaouthia*, *N. naja oxiana*, and *Ophiophagus hannah*. There were also a number of cross-reacting antigens in the *Naja* species (Kankonkar et al., 1972).

Normally, antivenin is most effective on its own venom and potency decreases against heterologous venoms, depending on closeness of phylogenecity (Ganthavorn, 1969; Mohamed et al., 1973a, b). However, Mohamed et al. (1973a) pointed out that one

1 Cobra Venoms

cannot exactly correlate the *in vivo* neutralization results with the Ouchterlony diffusion pattern.

Antibody to cobrotoxin from *Naja naja atra* was purified by Chang and Yang (1969). The anticobrotoxin can be digested with papain, and the univalent fragments combined with cobrotoxin to form a soluble antigen–antibody complex. Estimation of the molecular weight of the soluble complex formed from the univalent fragments and cobrotoxin provides evidence that cobrotoxin consists of three antibody-combining sites per molecule. Antibodies to α_1-toxin of *N. nigricollis* venom can be isolated by fixing α_1-toxin on Sepharose 4B (Detrait and Boquet, 1972). The antigen–antibody complex can be dissociated with $4M$ $MgCl_2$, and a soluble specific antibody to the α_1-toxin is thereby obtained. The antibody is an IgT.

Boquet et al. (1972, 1973) and Bergeot-Poilleux and Boquet (1975) investigated a number of neurotoxins from Elapidae and Hydrophiidae venoms and found that neurotoxins can be grouped into three types immunologically:

Group I consists of neurotoxins belonging to Type I, which have four disulfide bonds.
Group II also consists of Type I neurotoxins. Only γ-toxin of *N. nigricollis* and toxin F_7 of *N. naja philippinensis* belong to Group II.
Group III consists of Type II neurotoxins, which have five disulfide bonds.

Naja naja siamensis toxin does not cross-react immunologically with α-bungarotoxin. A chemical modification of *N. naja siamensis* toxin by polyalanylation resulted in a partial decrease in the antigenic activity of the toxin (Aharonov et al., 1974).

Treatment of *N. nigricollis* toxin with formaldehyde allows retention of its antigenicity. It can then be used as a toxoid (Dumarey and Boquet, 1972).

To produce high titers of antibody against neurotoxin is difficult. First, the toxicity of neurotoxins is very high, and animals frequently die during the process of immunization. Second, the neurotoxins are small proteins and therefore are usually rather poor antigens (Moroz et al., 1967). When *N. naja* venom is cross-linked with gluteraldehyde, the toxicity is reduced while the capacity for antibody formation is increased (Brade and Vogt, 1971).

The immunological properties of snake venoms are also used to study the phylogenetic relationships of snakes. Venoms of *N. naja siamensis* and *N. naja atra* gave identical immunodiffusion patterns but showed slight differences in immunoelectrophoretic pattern. This indicates that cobras from Thailand and Formosa are very similar but not identical (Tu and Ganthavorn, 1968).

1.7 Other Effects

A century ago, Brunton and Fayrer (1873, 1874) demonstrated that the death of a dog after cobra envenomation is often preceded by convulsion. They concluded that these were secondary effects to the anoxia produced by respiratory paralysis. To investigate whether cobra venom has any direct convulsant action, Bhargava et al. (1970) injected *Naja naja* venom or neurotoxin directly into the hippocampus of the rabbit. The principal action of crude venom and of the neurotoxin fraction on the somatosensory-evoked potential was to produce an abnormal negative wave with latency to a peak of about 26 msec. This effect is closely similar to that produced by cortical application of curare or strychnine, although the venom had a slower onset of action. The effect of strychnine or curare is readily reversed by washing, but that of the venom persists for at least 8 hr after washing. This sustained effect is similar to the action of neurotoxin at the neuromuscular junction, where the block in transmission is also irreversible. This may suggest the possibility that blockade of a cortical cholinergic synapse is responsible for the effect observed.

Many snake venoms, including *N. naja* venom, cause swelling of rat liver mitochondria (Taub and Elliott, 1964). Mitochondrial respiration with succinate was strongly inhibited by a 15-min incubation with the venom of *N. naja* and other snakes. Venom of the sea snake *Enhydrina schistosa* caused swelling of mitochondria in a similar manner. Sea snakes (Hydrophiidae) are closely related to Elapidae, and their neurotoxins are also similar to those of elapids. The effects on mitochondria of the venoms of these closely related snakes are very similar. Conversely, venom of Viperidae (*Bitis lachesis*) caused less swelling, and Crotalidae (*Crotalus horridus*) venom produced no swelling. Decreased oxygen consumption in rabbit liver tissues was observed after *Naja naja atra* envenomation (Ukino, 1939a); the decrease was venom concentration dependent within the sublethal dose. The venom also caused a decrease in rabbit blood pH (Matsuda, 1944a).

Naja naja venom contains a factor that releases amines from rat peritoneal mast cells (Morrison et al., 1975). The active component, called the CVA protein, has a molecular weight of 18,500. The initiation of mast cell degranulation by the CVA protein required both cell energy and Ca(II).

The lethal action of *N. mossambica mossambica* venom toward anthropods is due to protein factors that differ from the mammalian toxins (Zlotkin et al., 1975). Neurotoxins to mammals are nontoxic to larvae and isopods. The insect toxic component demonstrates a high phospholipase A_2 activity.

2 KRAIT VENOMS

Kraits (*Bungarus* spp.) can be found in India, Burma, southern China, Malaysia, and Formosa and are more common in populated areas than are cobras (Ditmars, 1966). There are several species within the genus *Bungarus*: *B. bungaroides*, *B. caeruleus*, *B. candidus*, *B. ceylonicus*, *B. fasciatus*, *B. flaviceps*, *B. javanicus*, *B. lividus*, *B. magnimaculatus*, *B. multicinctus*, *B. niger*, and *B. walli*. *Bungarus multicinctus* venom is well known as the source of α- and β-bungarotoxins. Although *B. fasciatus* and *B. multicinctus* venoms have been relatively well investigated, there has been little study of the venoms of other species.

2.1 Toxicology

Bungarus multicinctus venom in the crude state is fairly toxic with an LD_{50} value of 0.16 (0.12 to 0.22) $\mu g\ g^{-1}$ in mice by subcutaneous injection. The LD_{50} values for the isolated toxins are 0.3, 0.089, and 0.12 $\mu g\ g^{-1}$ for α-, β-, and γ-bungarotoxin, respectively (Table 1). Chang and Lee (1963) and Eterovic et al. (1975) agreed that β-type (presynaptic) toxins are more toxic than α-type (postsynaptic) toxins.

Like *B. multicinctus* venom, many fractions of *B. fasciatus* venom are toxic. The LD_{50} values of pure toxins confirm that the neurotoxin is more toxic than the cardiotoxin.

2.2 Pharmacology

Early History. The curare-like properties of *Bungarus multicinctus* venom were recognized by To and Tin (1944a, b). Lee and Peng (1961) concluded that the respiratory failure caused by the venom was primarily peripheral in origin. The term "curarelike action" should not be confused with "curare action" (Fig. 1). "Curarelike action" refers only to the fact that the venom causes neuromuscular block, which does not involve the loss of excitability of the motor nerve and the muscle. The site of action was further studied by Chang (1960a), who found that the blocking was confined to the neuromuscular junction, since the conductivity of the phrenic nerve and the twitch response to direct stimulation were unaffected. Acetylcholine release from the motor nerve after complete neuromuscular blockade was not depressed.

For further study, Chang (1960b) used frog rectus abdominis muscle. The venom inhibited the acetylcholine response of the muscle, but the potassium chloride contracture was not affected appreciably. The effect was progressive and could not be reversed by washing out the venom. Chang concluded that the curarelike action of the venom resulted from progressive and irreversible occupation of the specific receptor for acetylcholine in the motor end plate. Further progress was made in studying the mode of action by using isolated α-, β-, and γ-bungarotoxins (Chang and Lee, 1963). Chang and Lee reached the important and definitive conclusion that α-bungarotoxin blocks neuromuscular transmission by an irreversible combination with the motor end-plate acetylcholine receptor, whereas β- and γ-bungarotoxins exhibit their actions at presynaptic sites. β-Bungarotoxin destroys synaptic vesicles and thus inhibits the release of acetylcholine from presynaptic sites (Chen and Lee, 1970).

α-Bungarotoxin. The effects of α-bungarotoxin on neuromuscular transmission are very similar to those of cobra neurotoxin. The difference is that the paralytic effect of the bungarotoxin is much more irreversible and is not reversed by neostigmine (Lee and Chang, 1966).

Stalc and Zupancic (1972) reported that α-bungarotoxin became attached to membrane-bound acetylcholinesterase and that this binding was irreversible. They concluded that acetylcholine receptor sites and acetylcholine-hydrolyzing sites are located on the same macromolecule.

Magazanik and Vyskočil (1973) observed that α-bungarotoxin not only blocked the receptors, but also modified the function of other links of the cholinergic transmembrane system that are involved in electrogenesis at the postjunctional membrane.

From *Bungarus multicinctus* venom Dryden et al. (1973, 1974) obtained four fractions with postjunctional blockade activity and three fractions with presynaptic blocking activity. Apparently venoms from different sources have different neurotoxic fractions.

The binding of α-bungarotoxin to acetylcholine receptor is discussed in detail in Chapter 17.

β-Bungarotoxin. It is now well established that β-bungarotoxin causes neuromuscular blockade at the presynaptic site. Initially it increased spontaneous acetylcholine release. This in turn increased the miniature end-plate potential frequency. At a later stage inhibition of the nerve impulse and the release of acetylcholine occurred (Lee and Chang, 1966).

Although β-bungarotoxin inhibits the release of acetylcholine, however, it does not affect the receptors. Because low concentrations of Ca(II) can delay the inhibitory effect of β-toxin (Chang and Lee, 1963; Chang et al., 1973a), it is considered important to investigate the relation between Ca(II) release and the action of β-bungarotoxin. Wagner

Table 1 Toxicities of Elapidae Venoms and Their Components in Mice

Venoms or Components	LD$_{50}$ (μg/g)	Route of Injection	Reference
Acanthophis antarcticus	0.25	i.v.	Minton, 1974
	0.50	s.c.	Minton, 1974
Bungarus multicinctus	0.16 (0.12-0.22)	s.c.	Chang and Lee, 1963
α-bungarotoxin	0.30 (0.28-0.32)	s.c.	Chang and Lee, 1963
β-bungarotoxin	0.089 (0.084-0.094)	s.c.	Chang and Lee, 1963
γ-bungarotoxin	0.12 (0.096-0.14)	s.c.	Chang and Lee, 1963
fraction 1 (α-type)	>15	i.p.	Lee et al, 1972
" 2 (α-type)	0.15-0.3	"	"
" 3 (β-type)	0.4 -0.5	"	"
" 4 (β-type)	0.5 -0.6	"	"
" 5 (β-type)	0.02-0.03	"	"
" 6 (β-type)	0.04-0.06	"	"
" 7 (β-type)	0.02-0.04	"	"
" 8 (β-type)	0.04-0.05	"	"
" 2	5	s.c.	Eterovic et al, 1975
" 3	0.21	"	"
α-bungarotoxin	0.14	"	"
fraction 4	1.93	"	"
" 5	2.65	"	"
" 6	1.85	"	"
" 7	0.17	"	"
" 8	0.69	"	"
" 9	0.50	"	"
β-bungarotoxin	0.06	"	"
" 11	0.16	"	"
" 12	0.09	"	"
" 13	0.04	"	"
" 14	0.15	"	"
α-bungarotoxin	0.11	i.p.	"
fraction 4 (α-type)	1.0	"	"
" 6 (α-type)	1.2	"	"
" 7	0.15	"	"
" 8	0.62	"	"
" 9	0.4	"	"
" 10	0.01	"	"
" 11 (β-type)	0.022	"	"
" 12 (β-type)	0.014	"	"
" 13 (β-type)	0.007	"	"
" 14 (β-type)	0.045	"	"
B. caeruleus	0.09	i.v.	Minton, 1974
	0.09	i.p.	Minton, 1974
	0.45	s.c.	Minton, 1974

Table 1 *Continued*

	Venoms or Components	LD$_{50}$ (μg/g)	Route of Injection	Reference
B. fasciatus				
	fraction 3	>2.5	i.p.	Lu and Lo, 1974
	" 4	>2.5	"	"
	" 5 (cardiotoxin)	0.24	"	"
	" 6 (cardiotoxin)	0.18	"	"
	" 7	0.70	"	"
	" 8	0.06	"	"
	" 9	>2.5	"	"
	" 10	>2.5	"	"
	" 11	>2.5	"	"
	Cardiotoxin fraction VI-I-1	0.16	"	"
	neurotoxin VIII-3	0.04	"	"
Dendroaspis angusticeps		1.50	i.v.	Minton, 1974
		3.05	s.c.	Minton, 1974
D. jamesoni		0.3-0.4	i.p.	Excell and Patel, 1972
D. polylepsis		0.25	i.v.	Minton, 1974
		0.32	s.c.	Minton, 1974
	toxin α	0.09	s.c.	Strydom, 1972
	toxin γ	0.12	s.c.	Strydom, 1972
D. viridis				
	toxin I	0.045	—	Bechis et al, 1976
	toxin V	0.080	—	Bechis et al, 1976
	toxin	0.08	i.p.	Shipolini et al, 1973
	fraction 4-1	5.8	"	"
	" 4-2	>10	"	"
	" 4-3	>10	"	"
	" 4-4	2.5	"	"
	" 4-5	5.0	"	"
	" 4-6	1.3	"	"
	" 4-7	0.25	"	"
	" 4-9	0.60	"	"
	" 4-10	0.40	"	"
	" 4-11	0.10	"	"
	" 4-12	1.0	"	"
	" 4-13	8.3	"	"
	" 4-14	5.0	"	"
	" 4-15	11.7	"	"
	" 4-16	2.7	"	"
Denisonia superba		0.30	i.v.	Minton, 1974
		1.12	s.c.	"
Hemachatus hemachatus		1.10	i.v.	"
		2.65	s.c.	"

Table 1   *Continued*

Venoms or Components	LD$_{50}$ (µg/g)	Route of Injection	Reference
Micrurus carinicauda dumerilii	0.96	i.p.	Cohen et al, 1971
M. frontalis frontalis	0.63	"	"
M. fulvius fulvius	0.90	"	"
M. mipartitus hertwigii	0.55	"	"
M. nigrocinctus	0.58	"	"
M. spixii obscurus	1.61	"	"
Naja flava	0.40	"	Mohamed et al, 1973a
N. haje (from Egypt)	0.12	"	"
(from Ethiopia)	0.25	"	"
neurotoxin I	0.06	i.v.	Chicheportiche et al, 1972
toxin α	0.069	—	Botes et al, 1971
N. melanoleuca			
F$_5$	0.15	i.v.	Bergeot-Poilleux and Boquet, 1975
F$_6$	0.09	"	"
F$_7$	0.10	"	"
N. naja	0.44	i.v.	Kankonkar et al, 1972
	0.17	i.p.	Mohamed et al, 1973a
	0.13	i.v.	Friederich and Tu, 1971
Toxin A	0.15	i.v.	Nakai et al, 1970
N. naja atra	0.67	s.c.	Lee et al, 1962
	0.29	i.v.	Friederich and Tu, 1971
cobrotoxin	0.065	i.v.	Yang, 1965
N. naja kaothia	0.40	i.v.	Kankonkar et al, 1972
N. naja oxiana	0.96	"	"
Toxin I	0.56	"	Turakulov et al, 1971
Toxin II	0.13	"	Nishankhodzhaeva et al, 1972
N. naja philippeninsis			
major toxin	0.05-0.06	"	Hauert et al, 1974
N. nigricollis	0.44	i.m.	Mohamed and Nawar, 1975
venom	0.40	i.p.	Mohamed et al, 1973a
venom	0.630	i.v.	Kopeyan et al, 1973
Toxin I	0.036	"	"
Toxin II	0.037	"	"
N. nigricollis mossambica			
venom	2.00	—	Rochat et al, 1974
Toxin I	0.04	—	"
Toxin II	0.04	—	"
Toxin III	0.05	—	"

Table 1 *Continued*

Venoms or Components	LD$_{50}$ (µg/g)	Route of Injection	Reference
Naja nivea	0.42	—	Botes et al, 1971
toxin α	0.069	—	"
toxin β	0.072	—	"
toxin δ	0.77	—	"
Ophiophagus hannah	0.90	i.v.	Kankonkar et al, 1972
Oxyuranus scutellatus	0.02	i.v.	Minton, 1974
	0.12	s.c.	"
Pseudechis australis	0.30	i.v.	"
	1.5	s.c.	"
P. porphyriacus	0.50	i.v.	"
	2.0	s.c.	"
Pseudonaja textilis	0.01	i.v.	"
	0.25	s.c.	"
Walterinnesia aegyptia	0.25	i.p.	Mohamed et al, 1973a

i.v.: intravenous injection
s.c.: subcutaneous injection
i.p.: intraperitoneal injection

et al. (1974) found that β-bungarotoxin inhibited Ca(II) accumulation into subcellular fractions from rat brain at very low concentrations (2 to 8 pmole toxin per milligram protein). Since the inhibited calcium uptake had the characteristics of mitochondrial calcium accumulation, they concluded that the toxin affects mitochondria. Lau et al. (1974) investigated the effect of β-bungarotoxin on calcium uptake by sarcoplasmic reticulum from rabbit skeletal muscle. They suggested that β-bungarotoxin induces Ca(II) leakage in sarcoplasmic reticulum membranes.

Unlike α-bungarotoxin, β-bungarotoxin is an inhibitor at the presynaptic site and has no postsynaptic action on the membrane potential, action potential, or the sensitivity to acetylcholine at the motor end plate (Chang et al., 1973a). The presynaptic action of β-bungarotoxin was compared with that of botulinum toxin (Chang and Huang, 1974). In mouse and rat diaphragms, botulinum toxin was about 10 to 100 times more potent than β-bungarotoxin. On chick biventer cervicis muscle, β-bungarotoxin was 3 to 10 times more potent. The paralytic actions of β-bungarotoxin and botulinum toxin appear to take place in two processes. First, they bind with the respective target sites, and, second, they inhibit changes in the target macromolecule of the nerve terminals, leading to failure of transmitter release. The two toxins show a mutual antagonism, especially when β-bungarotoxin is added before or simultaneously with botulinum toxin (Chang et al., 1973a, b; Chang and Huang, 1974).

Howard (1975) reported that mitochondrial respiration can be inhibited by β-bungarotoxin just as by phospholipase A_2. The important question is whether this inhibition is due to phospholipase A_2 contaminated in the β-bungarotoxin preparation or to β-bungarotoxin itself which has phospholipase A_2 activity. A presynaptic toxin isolated from *Notechis scutatus scutatus* has neurotoxic as well as phospholipase A_2 activity (Halpert et al., 1976). The relationship between presynaptic neurotoxin and phospholipase A_2 is discussed in more detail in Chapter 18.

2.3 Other Actions

Bungarus fasciatus venom uncoupled the energy transfer in oxidative phosphorylation of rat liver mitochondria (Ziegler et al., 1965). The uncoupling activity is associated with oxidation of reduced diphosphopyridine nucleotide, other carriers being little affected. After treatment of mitochondria with 5 to 10 µg of venom per milligram of protein, ADP inhibits respiration (reverse acceptor control). As this inhibition occurs, cytochrome b and flavoprotein become reduced while pyridine nucleotide is oxidized. This reverse acceptor control is produced only by freshly reconstituted or freshly collected venom. The sites and mechanism of inhibition are summarized in Scheme 21 by Ziegler et al. (1965). Activity I is responsible for uncoupling the phosphorylation reaction, and activity II blocks electron transport at the cytochrome c level.

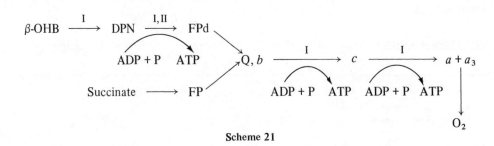

Scheme 21

Unlike the venoms of *Trimeresurus mucrosquamatus, T. gramineus, Agkistrodon acutus,* and *Naja naja atra*, the venom of *Bungarus multicinctus* does not have an excitory effect on respiration. Instead, the venom paralyzes the animal with respiratory arrest immediately on envenomation (Rai, 1937). Like *N. naja atra* venom, *B. multicinctus* venom also inhibits the oxygen consumption of liver tissues after envenomation (Ukino, 1939b), as well as causing a slight decrease in the pH of rabbit blood (Matsuda, 1944b). Nonneurotoxic venoms such as those of *Agkistrodon acutus, Trimeresurus elegans*, and *T. mucrosquamatus* increase the tissue respiration rate as determined by the method of Warburg.

3 CORAL SNAKE VENOMS

Coral snakes (*Micrurus* spp. and *Micruroides* spp.) are distributed from the southeastern United States to the region of the Mexican border. *Micrurus* and *Micruroides* are the only Elapidae snakes found in the United States. The best known coral snake in the United States is *Micrurus fulvius. Micruroides euryxanthus* inhabits southern Arizona, New Mexico, and Sonora (Minton, 1971). Relatively few studies have been made of the venom of coral snakes as compared to other Elapidae venoms. One reason is that the venom of coral snakes is very expensive because these snakes are relatively small, making it difficult to extract venom in large quantities. From 39,664 venom extracts of *Micrurus carinicauda dumerilii* (the eastern coral snake), Cohen et al. (1968) obtained 60.71 g of venom, an average of 1.5 mg per extraction. For *M. fulvius* venom, they obtained 27.42 g from 9143 milkings, an average yield of 3.0 mg per extraction.

3.1 Biological Action

The symptoms of *Micrurus* poisoning are visual perturbation, eyelid ptosis, muscular hypotonicity, increased salivary secretion, and muscle paralysis of the pharynx, larynx, and locomotive system (Brazil, 1965). Breathing stops before the heart does. The use of artificial respiration at the advent of respiratory muscle failure usually only prolongs the survival time of the victim without preventing death.

Micrurus fulvius venom is different in action from typical cobra venoms, according to the study of Weis and McIsaac (1971). They observed that the contractility of cardiac and skeletal muscle decreased, primarily because of a direct effect of venom on the muscle membrane. Histologically the envenomated muscle showed a swelling of the individual fibers with hyaline degeneration. The resting membrane potential of the surface muscle fibers decreased rapidly to extremely low values. Immediate hypotension was also observed. The action of the venom was usually irreversible. Although the venom has its principal action directly on muscle fibers, Weis and McIsaac proposed that an effect on neuromuscular transmission cannot be entirely ruled out. Respiratory paralysis occurred before cardiac failure in cats, and paralysis of the twitch response to nerve stimulation paralleled respiratory depression. Death occurred later through cardiac depression despite artificial ventilation. Unlike sea snake poisoning, coral snake envenomation caused no release of myoglobin in rats.

Ramsey et al. (1972b) observed very similar results. They found that intravenous injection of *M. fulvius* venom produced a precipitous fall in aortic pressure concurrently with a marked reduction in cardiac output and an elevation of hepatic and portal vein and pulmonary artery pressures. The initial profound decrease in cardiac output and aortic pressure occurred secondarily to sequestration of venous return in the hepatosplanchnic bed. These alterations were followed by interference with myocardial contraction, which resulted in death despite controlled ventilation. The pressure changes that occurred within 30 sec after venom infusion were maximal by 5 to 8 min, followed by increases in pressure and cardiac output toward control values.

Snyder et al. (1973) isolated the fraction that possessed neuromuscular blocking action similar to the actions of Elapidae and sea snake neurotoxins. The action of the crude venom on chick muscle was typically characterized by an augmented contraction with progressive failure of relaxation. On the other hand, the neurotoxic fraction produced neither an augmented contraction nor a depolarizing failure. After treatment with neurotoxin, the muscle remained fully responsive to electrical stimulation. The acetylcholine dose-responsive curve was progressively shifted to the right with increasing neurotoxin concentration. Like the response to indirect stimulation, the acetylcholine response was not recovered by repeated washing with Tyrode solution. It was therefore concluded that the active fraction of coral snake venom is a neurotoxin similar to Elapidae and Hydrophiidae neutrotoxins. Pellegrini and Brazil (1976) also concluded that the mechanism of respiratory paralysis induced by *Micrurus frontalis* venom is exclusively peripheral in origin, like that of other Elapidae venoms. This conclusion differs from the results of Parnas and Russell (1967), who failed to find any neuromuscular blockade when coral snake venom was applied to a crustacean muscle preparation. These differences may be attributed to the muscle preparation used (Lee, 1970).

The effect of coral snake venom on the coagulation system is not clear, since there have been very few reports. Eagle (1937) observed that *Micrurus* spp. coagulated citrated horse plasma. Coral snake venom did not destroy fibrinogen, indicating that it probably lacks fibrinogenolytic activity. It may be that the venom contains thrombinlike activity.

3.2 Biochemistry

The major component of coral snake venom is protein (Stevan and Seligmann, 1970), and venom of *Micrurus fulvius fulvius* contains both acidic and basic toxins (Ramsey et al., 1972a). None of the fractions or the crude venom is proteolytic. This is rather typical for Elapidae venoms, which are usually nonproteolytic. The coral snake venom contains phospholipase A_2, anticoagulant, and hemolytic activities. The venom of *M. fulvius* causes uncoupling and reverse acceptor control to develop in rat liver mitochondrial preparations. Both effects appear to be the result of the release of fatty acids from the lipids of the mitochondria (Ziegler et al., 1967).

In the presence of methylene blue and visible light, *M. fulvius* venom can be photooxidized, resulting in detoxification. Rabbits inoculated with the photooxidized venom produced an immunoglobulin that protected mice against a manyfold lethal dose. The effectual and harmless toxoid obtained can be used for active immunization, as well as for antisera for passive immunization (Kocholaty et al., 1967). Antivenin produced against the venom of *M. mipartitus hertwigi* can also neutralize the venoms of *M. m. anomalus, M. carinicaudus dumerilii*, and *M. alleni* (Bolanos et al., 1975).

The neutralization ability of antisera from five different coral snake venoms against homologous and heterologous venoms were investigated by Cohen et al. (1971). Since *M. frontalis* venom produced the most cross-reactive neutralizing antibody, neutralizing five of six venoms, they concluded that *M. frontalis* venom would be most suitable for use in the production of a polyvalent coral snake antivenin.

4 OTHER ELAPIDAE VENOMS

4.1 Australian Venoms

Many plants and animals in Australia and New Guinea are different from those in Asia. Normally, plants and animals in the Oriental regions are separated from those of Australian regions at the Wallace line. The continental zoogeographic areas are divided into several regions, such as the Palaearctic, Nearctic, Neotropical, Ethropran, Oriental, and Australian (Stirton, 1963). This is also true for the distribution of poisonous snakes. Snakes in New Guinea and Australian are similar, but they are different from the species in Asia. For instance, there are many varieties of Viperidae and Crotalidae snakes in southeast Asia, but none in Australia and New Guinea. Therefore it is not surprising that the venoms of Australian snakes are different from those of Asian snakes.

Tu and Toom (1967) investigated the substrate specificities of venom peptidases of Australian snakes and found that Australian snake venoms hydrolyzed most of the dipeptides tested and some tripeptides. They also found that the substrate specificities of *Oxyuranus scutellatus scutellatus* and *O. scutellatus canni* venoms are identical. The former snake is found in southern New Guinea, while the latter is distributed throughout northern Australia. Even though geographically separated, these two venoms showed identical peptidase substrate specificities.

Of the 140 species of Australian terrestrial snakes only about 10 are deadly, and antivenins for all of these are readily available (Challen, 1975).

The toxicological properties of New Guinean *Acanthophis antarcticus* venom was investigated by Peng (1952b). It had curarelike action, especially on the respiratory muscle, and a vasoconstrictor effect. The venom stimulated the intestine and uterus at low doses. It had a feeble action on the heart. The venom did not produce any local irritation or change in blood sugar level.

The neurotoxic fraction was isolated from the Australian tiger snake, *Notechis scutatus scutatus* (Karlsson et al., 1972). The main neurotoxic component, notexin, constituted 6% of the crude venom. Unlike cobra neurotoxins, notexin is a presynaptic toxin and inhibits the release of acetylcholine from motor nerves. The chemical properties of notexin, together with those of other neurotoxins, are discussed in Chapter 18. The whole venom of *N. scutatus scutatus* acted at both the presynaptic nerve terminals and the postsynaptic acetylcholine receptors of sciatic–sartorius neuromuscular junctions of toads. It seems that the venom contains not only notexin but also the typical postsynaptic neurotoxin, like many other Elapidae and sea snake venoms (Datyner and Gage, 1973a b).

The effect of *N. scutatus scutatus* venom on the morphology of motor nerve terminals was studied by Lane and Gage (1973), using sciatic–sartorius nerve–muscle preparations of the toad. They reported that the venom caused damage of the presynaptic vesicle similar to that observed by Chen and Lee (1970), using β-bungarotoxin. The degree of degenerative change depended on the concentration of venom used. Electrophysiological recordings showed an initial increase, followed by a decline in the miniature end-plate potential frequency and ultimately in the potential. This suggests that the venom acts directly on the presynaptic terminal. Although notexin inhibits the transmitter release, it does not affect muscle fiber membrane. In this respect, it is similar in action to β-bungarotoxin and botulinum toxin (Harris et al., 1973).

Notexin is a myonecrotic agent. The subcutaneous injection of notexin causes edema and necrosis of rat skeletal muscle. Myofibrillar ATPase activity decreases or is absent in severely necrotic fibres (Harris et al., 1974). Notexin also exhibits phospholipase A_2 activity (Halpert et al., 1967); like phospholipase A_2, notexin binds to Ca(II) in a 1 : 1 molar ratio. The chemical aspects of notexin are discussed further in Chapter 18.

The venom of *Tropidechis carinatus* was briefly investigated by Trinca et al. (1971). It contains phospholipase A_2 activity and possesses a powerful coagulant action on human plasma. The toxicity of this venom can be neutralized by antivenin made for *Notechis scutatus* venom.

4.2 African Venoms

In this section, African Elapidae venoms other than cobra venom will be discussed. Africa is rich in a variety of Elapidae, including such genera as *Boulengerina* (water cobras), *Naja* (cobras), *Dendraspis* (mambas), *Elapsoidea* (African garter snakes), *Walterinnesra* (desert black snake), *Aspidelaps* (shield-nose snake), *Hemachatus* (ringhals), *Paranaja* (burrowing cobra), and *Pseudohaje* (tree cobras).

Walterinnesra aegyptia. *Walterinnesra aegyptia* envenomation produces muscular paralysis (Gitter et al., 1962; Mohamed and Zaki, 1958). Using toad gastrocnemius–sciatic preparation, Mohamed and Zaki (1958) reported that the motor end plates remained sensitive to acetylcholine after envenomation. Thus they concluded that

W. aegyptia venom produces a different type of neurotoxic effect than does curare or other cobra venoms. This aspect was reexamined by Lee et al. (1971), who observed, contrary to the earlier finding, that the response to acetylcholine was completely abolished by the venom in toad gastrocnemius–sciatic and chick biventer cervicis muscle. Therefore Lee et al. (1971) concluded that the mode of neurotoxic action of *W. aegyptia* venom is the same type as that of cobra venoms.

In addition to the blocking action at the postsynaptic site, the effect of *W. aegyptia* venom on the presynaptic site was investigated by Lee and Tsai (1972), who found that the venom did not prevent the release of acetylcholine from the motor nerve endings. This shows that *W. aegyptia* venom is a postsynaptic blocking type and has no presynaptic poisoning activity.

Dendraspis Spp. (Mambas). *Dendraspis* snakes are usually slender and can reach 7 ft or longer. The venom of *D. polyepis* (black mamba) has diverse reactions. It is neurotoxic, myonecrotic, and anticoagulant (Zaki et al., 1970). The venom causes edema, as well as vacuolar and hyaline degeneration of the striated muscle fibers. The kidney shows congestion and hemorrhage after the injection of venom, and hemorrhage is also produced in the liver, myocardium, lung, and brain.

The LD_{50} of *D. jamesoni* venom by i.p. injection in mice is 0.3 to 0.6 $\mu g\, g^{-1}$ (Excell and Patel, 1972). The LD_{50} of *Dendroaspis* venom is unaffected by heat (Christensen and Anderson, 1967), and both the neurotoxic and cardiotoxic activities are heat stable, as with *Naja* venoms (Patel and Excell, 1974). Fractions 9, 10, and 12 rapidly blocked miniature end-plate potentials recorded at neuromuscular junctions in both frog sartorius muscle and rat diaphragm; the resting membrane potentials were not affected (Excell and Patel, 1972). The results indicate that the basic, heat-stable toxic components of *D. jamesoni* venom resemble those found in *Naja* (cobra) venoms.

There are at least three postsynaptic, curare-like neurotoxins in the venom of *D. jamesoni* (Excell and Patel, 1972). This venom also contains a component that abolishes the direct excitability of both frog sartorius and rat diaphragm muscle and produces depolarization of these fibers independently of the end-plate region (Patel and Excell, 1974). These actions are similar to the effect of cardiotoxin in *Naja* venoms (Meldrum, 1965a, b; Earl and Excell, 1972). However, in comparison to *Naja* venoms, there was less depolarizing activity; this may be due to the virtual absence of phospholipase A_2 and its accelerating action. Rat diaphragm fibers were more resistant to the cardiotoxic action than were frog sartorius fibers. High calcium ion content completely prevented depolarization, as previously shown for the cardiotoxin from *Naja nivea* venom.

Among the 14 fractions separated from the venom of *D. jamesoni*, the presence of at least four neurotoxins was ascertained (Patel and Excell, 1975). They reduced the sensitivity of the end plate to acetylcholine with no detectable impairment of quantal release. The depolarizing fraction rapidly arrested isolated, perfused rat heart, confirming the presence of cardiotoxin.

The LD_{50} of *D. viridis* venom is 0.33 $\mu g\, g^{-1}$ in mice by i.p. injection. The purified toxin is more toxic and has a LD_{50} value of 0.08 $\mu g\, g^{-1}$ (Shipolini et al., 1973). Three neutrotoxins with postsynaptic action were isolated by Banks et al. (1974).

The clinical features of envenomation by *D. angusticeps* include unconsciousness, sweating, vomiting, respiratory arrest, and hypotension (Chapman, 1968). The hypotension persists for more than 5 days after envenomation. Venom of *D. angusticeps* caused hypotension in the cat, dog, rat, and guinea pig (Osman et al., 1973). The hypotensive effect of the crude venom was due to acetylcholine, which was blocked by atropine. The dialysed and the desalted venoms, which contain little or no acetylcholine, produced only slight hypotension. Venom in large doses produced a direct depression of the isolated heart, which was not mediated through cholinergic receptors and was apparently due to cardiotoxin present in the venom.

REFERENCES

Aharonov, A., Grurari, D., and Fuchs, S. (1974). Immunochemical characterization of *Naja naja siamensis* toxin and of a chemically modified toxin, *Eur. J. Biochem.,* **45,** 297.

Banks, B. E. C., Mildei, R., and Shipolini, R. A. (1974). The primary sequences and neuromuscular effects of three neurotoxic polypeptides from the venoms of *Dendroaspis viridis, Eur. J. Biochem.,* **45,** 457.

Bechis, G., Van Rietschoten, J., Granier, C., Jover, E., Rochat, H., and Miranda, F. (1976). On the characterization of two long toxins from *Dendroaspis viridis, Bull. Inst. Pasteur,* **74,** 35.

Bergeot-Poilleux, G. and Boquet, P. (1975). Remarques à propos de trois neurotoxines du venin de *Naja melanoleuca, C. R. Acad. Sci. Paris,* Ser. D, **280,** 1757.

Bhanganada, K. and Perry, J. F., Jr. (1963). Cardiovascular effects of cobra venom, *J. Am. Med. Assoc.,* **183,** 257.

Bhargava, V. K., Horton, R. W., and Meldrum, B. S. (1970). Long-lasting convulsant effect on the cerebral cortex of *Naja naja* venom, *Br. J. Pharmacol.,* **39,** 455.

Bolaños, R., Cerdas, L., and Taylor, R. (1975). The production and characteristics of a coral snake (*Micrurus mipartitus hertwigi*) antivenin, *Toxicon,* **13,** 139.

Boquet, P., Izard, Y., and Ronsseray, A. (1972). An attempt to classify by serological techniques the toxic proteins of low molecular weight extracted from Elapidae and Hydrophiidae venoms, *J. Formosan Med. Assoc.,* **71,** 307.

Boquet, P., Poilleux, G., Dumarey, C., Izard, Y., and Ronsseray, A. M. (1973). An attempt to classify the toxic proteins of Elapidae and Hydrophiidae venoms, *Toxicon,* **11,** 333.

Botes, D. P., Strydom, D. J., Anderson, C. G., and Christensen, P. A. (1971). Snake venom toxins: Purification and properties of three toxins from *Naja nivea* (Linnaeus) (Cape cobra) venom and the amino acid sequence of toxin d, *J. Biol. Chem.,* **246,** 3132.

Brade, V. and Vogt, W. (1971). Immunization against cobra venom, *Experientia,* **27,** 1338.

Brazil, O. V. (1965). Ação neuromuscular da peçonha de *Micrurus, O Hospital,* **68,** 183.

Brunton, T. L. and Fayrer, J. (1873). On the nature and physiological action of the poison of *Naja tripudians* and other Indian venomous snakes, I, *Proc. Roy. Soc.,* **21,** 358.

Brunton, T. L. and Fayrer, J. (1874). On the nature and physiological action of the poison of *Naja tripudians* and other Indian venomous snakes, II, *Proc. Roy. Soc.,* **22,** 68.

Challen, R. G. (1975). Venomous Australian snakes and antivenenes, *Aust. J. Hosp. Pharm.,* **5,** 4.

Chang, C. C. (1960a). Studies on the mechanism of curare-like action of *Bungarus multicinctus* venom, *J. Formosan Med. Assoc.,* **59,** 315.

Chang, C. C. (1960b). Studies on the mechanism of curare-like action of *Bungarus multicinctus* venom. II. Effect on response of rectus abdominis muscle of the frog to acetylcholine, *J. Formosan Med. Assoc.,* **59,** 416.

Chang, C. C. and Huang, M. C. (1974). Comparison of the presynaptic actions of botulinum toxin and β-bungarotoxin on neuromuscular transmission, *Naunyn-Schmiedebergs Arch. Exp. Pathol. Pharmakol.,* **282,** 129.

Chang, C. C. and Lee, C. Y. (1963). Isolation of neurotoxins from the venom of *Bungarus multicinctus* and their modes of neuromuscular blocking action, *Arch. Int. Pharmacodyn.,* **144,** 144.

Chang, C. C. and Lee, C. Y. (1966). Electrophysiological study of neuromuscular blocking action of cobra neurotoxin, *Br. J. Pharmacol. Chemother.,* **28,** 172.

Chang, C. C. and Yang, C. C. (1969). Immunochemical studies on cobrotoxin, *J. Immunol.,* **102,** 1437.

Chang, C. C., Chen, T. F., and Lee, C. Y. (1973a). Studies of the presynaptic effect of β-bungarotoxin on neuromuscular transmission, *J. Pharmacol. Exp. Ther.,* **184,** 339.

Chang, C. C., Huang, M. C., and Lee, C. Y. (1973b). Mutual antagonism between botulinum toxin and β-bungarotoxin, *Nature,* **243,** 166.

Chapman, D. S. (1968). "The symptomatology, pathology and treatment of the bites of venomous snakes of Central and Southern Africa, in W. Bürcherl, E. Buckley, and V. Deulofeu, Eds., *Venomous Animals and Their Venoms,* Academic, New York, pp. 463–527.

Chen, I. L. and Lee, C. Y. (1970). Ultrastructural changes in the motor nerve terminals caused by β-bungarotoxin, *Virchows Arch. Pathol., Anat., Physiol.,* **6,** 318.

Cheymol, J., Gonçalves, J. M., Bourillet, F., and Roch-Arveiller, M. (1971). Action neuromusculaire comparee de la crotamine et du venin de *Crotalus durissus terrificus* var. *Crotaminicus.* I. Sur preparations neuromusculaires *in situ, Toxicon,* **9,** 279.

Cheymol, J., Karlsson, E., Bourillet, F., and Roch-Arveiller, M. (1974). Biological activity of various fractions isolated from *Hemachatus haemachates* venom, *Arch. Int. Pharmacodyn. Ther.,* **208,** 81.

Chicheportiche, R., Rochat, C., Sampieri, F., and Lazdunski, M. (1972). Structure–function relationships of neurotoxins isolated from *Naja haje* venom: Physicochemical properties and identification of the active site, *Biochemistry,* **11,** 1681.

Chou, T. C. and Lee, C. Y. (1969). Effect of whole and fractionated cobra venom on sympathetic ganglionic transmission, *Eur. J. Pharmacol.,* **8,** 326.

Christensen, P. A. and Anderson, C. G. (1967). "Observation on *Dendroaspis* venom," in. F. E. Russell and P. R. Saunders, Eds., *Animal Toxins,* Pergamon, Oxford, pp. 223–234.

Cohen, M. and Sumyk, G. B. (1966). Cardiovascular and respiratory effects of cobra venom and a venom fraction, *Toxicon,* **3,** 291.

Cohen, P., Dawson, G. H., and Seligmann, E. B., Jr. (1968). Cross-neutralization of *Micrurus fulvius fulvius* (coral snake) venom by anti-*Micrurus carinicauda dumerilii* serum, *Am. J. Trop. Med. Hyg.,* **17,** 308.

Cohen, P., Berkeley, W. H., and Seligmann, E. B., Jr. (1971) Coral snake venoms: *In vitro* relation of neutralizing and precipitating antibodies, *Am. J. Trop. Med. Hyg.,* **20,** 646.

Datyner, M. E. and Gage, P. W. (1973a). Australian tiger snake venom: Inhibitor of transmitter release, *Nature New Biol.,* **241,** 246.

Datyner, M. E. and Gage, P. W. (1973b). Presynaptic and postsynaptic effects of the venom of Australian tiger snake at the neuromuscular junction, *Br. J. Pharmacol.,* **49,** 340.

Detrait, J. and Boquet, P. (1972). Isolement des anti-corps antitoxine α_1 du venin de *Naja nigricollis* au moyen du Sépharose, *C. R. Acad. Sci. Paris,* Ser. D, **274,** 1765.

Ditmars, R. L. (1966). *Snakes of the World,* Macmillan, New York.

Dryden, W. F., Marshall, I. G., and Harvey, A. L. (1973). The neuromuscular blockage caused by fractions obtained by chromatography of commercial *B. multicinctus* venom, *J. Pharm. Pharmacol.,* **25,** 125.

Dryden, W. F., Harvey, A. L., and Marshall, I. G. (1974). Pharmacological studies on the bungarotoxins: Separation of the fractions and their neuromuscular activity, *Eur. J. Pharmacol.,* **26,** 256.

Dumarey, C. and Boquet, P. (1972). Pouvoir immunogène de la toxine α du venin de *Naja nigricollis* polymérisée par l'aldéhyde formique, *C. R. Acad. Sci. Paris,* Ser. D, **275,** 3053.

Eagle, H. (1937). Coagulation of the blood by snake venoms and its physiologic significance, *J. Exp. Med.,* **65,** 613.

Eaker, D., Harris, J. B., and Thesleff, S. (1971). Action of a cobra neurotoxin on denervated rat skeletal muscle, *Eur. J. Pharmacol.,* **15,** 154.

Earl, J. E. and Excell, B. J. (1971). The action of a depolarizing fraction from *Naja nivea* venom on frog skeletal muscle, *J. Physiol.,* **214,** 27p.

Earl, J. E. and Excell, B. J. (1972). The effects of toxic components of *Naja nivea* (cape cobra) venom on neuromuscular transmission and muscle membrane permeability, *Comp. Biochem. Physiol.,* **41A,** 597.

Eterovic, V. A., Hebert, M. S., Hanley, M. R., and Bennett, E. L. (1975). The lethality and spectroscopic properties of toxins from *Bungarus multicinctus* (Blyth) venom, *Toxicon,* **13,** 37.

References

Excell, B. and Patel, R. (1972). Characterization of toxic fractions from *Dendroaspis jamesoni* venom, *J. Physiol. (London)*, **225**, 29.

Friederich, C. and Tu, A. T. (1971). Role of metals in snake venoms for hemorrhagic esterase and proteolytic activities, *Biochem. Pharmacol.*, **20**, 1549.

Ganthavorn, S. (1969). Toxicities of Thailand snake venoms and neutralization capacity of antivenin, *Toxicon*, **7**, 239.

Gitter, S., Moroz-Perlmutter, C., Boas, J. H., Livni, E., Rechnic, J., Goldblum, N., and De Vries, A. (1962). Studies on the snake venoms of the Near East: *Walterinnesia aegyptia* and *Pseudecerastes fieldii*, *Am. J. Trop. Med. Hyg.*, **11**, 861.

Halpert, J., Eaker, D., and Karlsson, E. (1976). The role of phospholipase activity in the action of a presynaptic neurotoxin from the venom of *Notechis scutatus scutatus* (Australian tiger snake), *FEBS Lett.*, **61**, 72.

Harris, J. B., Karlsson, E., and Thosleff, S. (1973). Effects of an isolated toxin from Australian tiger snake (*Notechis scutatus scutatus*) venom at the mammalian neuromuscular junction, *Br. J. Pharmacol.*, **47**, 141.

Harris, J. B., Johnson, M. A., and Karlsson, E. (1974). Histological and histochemical aspects of the effect of notexin on rat skeletal muscle, *Br. J. Pharmacol.*, **52**, 152p.

Hauert, J., Maire, M., Sussman, A., and Bargetzi, J. P. (1974). The major lethal neurotoxin of the venom of *Naja naja philippinensis:* Purification, physical and chemical properties, partial amino acid sequence, *Int. J. Peptide Proteins Res.*, **6**, 201.

Howard, N. L. (1975). Phospholipase A_2 from puff adder (*Bitis arietans*) venom, *Toxicon*, **13**, 21.

Kamel, A. (1974). Fractionation of Egyptian cobra venom, *Toxicon*, **12**, 495.

Kankonkar, R. C., Rao, S. S., Vad, N. E., and Sant, M. V. (1972). Efficacy of Haffkine Institute polyvalent antivenin against Indian snake venoms, *Indian J. Med. Res.*, **60**, 512.

Karlsson, E., Heibbronn, E., and Widlund, L. (1972). Isolation of the nicotinic acetylcholine receptor by biospecific chromatography on insolubilized *Naja naja* neurotoxin, *FEBS Lett.*, **28**, 107.

Kocholaty, W. F., Ashley, B. D., and Billings, T. A. (1967). An immune serum against the North American coral snake (*Micrurus fulvius fulvius*) venom obtained by photooxidative detoxification, *Toxicon*, **5**, 43.

Kopeyan, C., Van Rietschoten, J., Martinez, G., Rochat, H., and Miranda, F. (1973). Characterization of five neurotoxins isolated from the venoms of two Elapidae snakes, *Naja haje* and *N. nigricollis*, *Eur. J. Biochem.*, **35**, 244.

Lane, N. J. and Gage, P. W. (1973). Effect of tiger snake venom on the ultrastructure of motor nerve terminals, *Nature New Biol.*, **244**, 94.

Lau, Y. H., Chiu, T. H., Caswell, A. H., and Potter, L. T. (1974). Effects of β-bungarotoxin on calcium uptake by sarcoplasmic reticulum from rabbit skeletal muscle, *Biochem. Biophys. Res. Commun.*, **61**, 510.

Lee, C. Y. (1970). Elapid neurotoxins and their mode of action, *Clin. Toxicol.*, **3**, 457.

Lee, C. Y. and Chang, C. C. (1966). Modes of actions of purified toxins from elapid venoms on neuromuscular transmission, *Mem. Inst. Butantan*, **33**, 555.

Lee, C. Y. and Peng, M. T. (1961). An analysis of the respiratory failure produced by the Formosan elapid venoms, *Arch. Int. Pharmacodyn.*, **133**, 180.

Lee, C. Y. and Tsai, M. C. (1972). Does the desert black snake venom inhibit release of acetylcholine from motor nerve endings? *Toxicon*, **10**, 659.

Lee, C. Y., Chang, C. C., Su, C., and Chen, Y. W. (1962). The toxicity and thermostability of Formosan snake venoms, *J. Formosan Med. Assoc.*, **61**, 239.

Lee, C. Y., Huang, P. P., and Tsai, M. C. (1971). Mode of neuromuscular blocking action of the desert black snake venom, *Toxicon*, **9**, 429.

Lee, C. Y., Lin, J. S., and Lin, Shiau, S. Y. (1972). A study of carcinolytic factor of Formosan cobra venom, *Proc. Natl. Sci. Counc. Taiwan*, **5**, 9.

Lester, H. A. (1966). Postsynaptic action of cobra toxin at the myoneural junction, *Nature*, **227**, 727.

Loots, J. M., Meij, H. S., and Meyer, B. J. (1973). Effects of *Naja nivea* venom on nerve, cardiac and skeletal muscle activity of the frog, *Br. J. Pharmacol.*, **47**, 576.

Lu, M. S. and Lo, T. B. (1974). Chromatographic separation of *Bungarus fasciatus* venom and preliminary characterization of its components, *J. Chin. Biochem. Soc.,* **3,** 57.

Lukomskaya, N. Y. and Magazanik, L. G. (1974). Blocking effect of snake venom polypeptides on cholinoreceptive membranes the lamprey *Lampetra fluvitilia, Zh. Evol. Biokhim. Fiziol.,* **10,** 524.

Magazanik, L. C. and Vyskoyil, F. (1973). Some characteristics of end-plate potentials after partial blockade by α-bungarotoxin in *Rana temporaria, Experientia,* **29,** 157.

Matsuda, S. (1944a). The effect of Formosan snake venom on rabbit blood acid-base equilibrium. 4. Formosan cobra, *Nihon Daigaku Med. J.,* **7,** 47.

Matsuda, S. (1944b). The effect of Formosan snake venom on rabbit blood acid-base equilibrium. 5. *Bungarus multicinctus* venom, *Nihon Daigaku Med. J.,* **7,** 61.

Meldrum, B. S. (1965a). Action of whole and fractionated Indian cobra (*Naja naja*) venom on skeletal muscle, *Br. J. Pharmol. Chemother.,* **25,** 197.

Meldrum, B. S. (1965b). The actions of snake venoms on nerve and muscle: The pharmacology of phospholipase A and of polypeptide toxins, *Pharmacol. Rev.,* **17,** 393.

Minton, S. A., Jr. (1967). Paraspecific protection by elapid and sea snake antivenins, *Toxicon,* **5,** 47.

Minton, S. A. (1971). *Snake Venoms and Envenomation,* Marcel Dekker, New York.

Minton, S. A., Jr. (1974). *Venom Diseases,* Charles C Thomas, Springfield, SU.

Mirsalikhova, N. M., Ziyamukhamedov, R., and Yukel'son, L. Y. (1974). Effect of some components of Central Asian cobra venom on the level of magnesium-dependent and (magnesium–sodium–potassium)-dependent ATPase of brain cortex cells, *Dokl. Akad. Nauk Uzb. SSR,* **31,** 62.

Mohamed, A. H. and Nawar, N. N. Y. (1975). Dysmelia in mice after maternal *Naja nigricollis* envenomation: A case report, *Toxicon,* **13,** 475.

Mohamed, A. H. and Zaki, O. (1958). Effect of the black snake toxin on the gastrocnemius–sciatic preparation, *J. Exp. Biol.,* **35,** 20.

Mohamed, A. H., Darwish, M. A., and Hani-Ayobe, M. (1973a). Immunological studies on Egyptian cobra antivenin, *Toxicon,* **11,** 31.

Mohamed, A. H., Darwish, M. A., and Hani-Ayobe, M. (1973b). Immunological studies on *Naja nigricollis* antivenin, *Toxicon,* **11,** 35.

Mohamed, A. H., Saleh, A. M., Ahmed, S., and Beshir, S. R. (1975). Histopathological and histochemical effects of *Naja haje* venom on kidney tissue of mice, *Toxicon,* **13,** 409.

Morales, F., Root, H. D., and Perry, J. F., Jr. (1961). Protection against *Naja naja* venom in dogs by hydrocortisone, *Proc. Soc. Exp. Biol. Med.,* **108,** 522.

Moroz, C., Grotto, L., Goldblum, N., and De Vries, A. (1967). "Enhancement of immunogenicity of snake venom neurotoxins," in F. E. Russell and P. R. Saunders, Eds., *Animal Toxins,* Pergamon, Oxford, pp. 299–302.

Morrison, D. C., Roser, J. F., Henson, P. M., and Cochran, C. G. (1975). Isolation and characterization of a noncytotoxic mast-cell activator from cobra venom, *Inflammation,* **1,** 103.

Nakai, K., Nakai, C., Sasaki, T., Kakiuchi, K., and Hayashi, K. (1970). Purification and some properties of toxin A from the venom of the Indian cobra, *Naturwissenschaften,* **57,** 387.

Nishankhodzhaeva, S. A., Sorokin, V. M., and Yukel'son, L. Ya. (1972). Extraction and characterization of toxin II of Central Asian cobra venom, *Med. Zh. Uzb.,* **7,** 44.

Orlov, B. N. and Gelashvili, D. B. (1972). Central mechanisms of the neurotropic action of cobra venom, *Uch. Zap. Gor'k. Gos. Univ.,* **20.**

Orlov, B. N., Cherepnov, V. L., and Pishchik, A. (1971). Histological changes in nerve tissue during the action of animal venoms, *Uch. Zap. Gor'k. Gos. Univ.,* **132.**

Osman, O. H., Ismail, M., and Hamadein, H. A. (1974). Neuromuscular blocking activity of snake (*Naja melanoleuca* Hallowell) venom, *Toxicon,* **12,** 501.

Parnas, I. and Russell, F. E. (1967). "Effects of venoms on nerve, muscle and neuromuscular junction," in F. E. Russell and P. R. Saunders, Eds., *Animal Toxins,* Pergamon, Oxford, pp. 401–415.

Patel, R. and Excell, B. J. (1974). The modes of action of whole *Dendroaspis jamesoni* venom on skeletal nerve–muscle preparations, *Toxicon,* **12,** 577.

Patel, R. and Excell, B. J. (1975). The effects of lethal components of *Dendroaspis jamesoni* snake venom on neuromuscular transmission and on muscular membrane permeability, *Toxicon,* **13,** 295.

Pellegrini, A. and Brazil, O. V. (1976). Origem da Paralisia Respiratória Determinada Pela Peçonha de *Micrurus frontalis, Cienc. Cult.,* **28,** 199.

Peng, M. (1952a). Action of the venom of *Naja naja atra* on respiration and circulation, *Mem. Fac. Med. Nat. Taiwan Univ.,* **2,** 170.

Peng, M. (1952b). The toxicological study on New Guinean *Acanthophis antarcticus* venom, *Mem. Fac. Med. Nat. Taiwan Univ.,* **2,** 1952.

Rai, K. (1937a). The influence of the poisons of Siamese snakes on the blood-sugar of the rabbit, *Jap. J. Med. Sci.,* **10,** 163.

Rai, K. (1937b). Über den Einfluss der wichtigeren formosanischen Schlangengifte auf die Atmung des Kaninchens. I. Mitteilung. Die nach einer Einzeldosis der verschiedenen Schlangengifte eintretenden Veränderungen der Atmung bei Abwesenheit therapentischer Masnahmen, *Folia Pharm. Jap.,* **24,** 123.

Ramsey, H. W., Snyder, G. K., Kitchen, H., and Taylor, W. J. (1972a). Fractionation of coral snake venom: Preliminary studies on the separation and characterization of the protein fractions, *Toxicon,* **10,** 67.

Ramsey, H. W., Taylor, W. J., Boruchow, I. B., and Snyder, G. K. (1972b). Mechanism of shock produced by an elapid snake (*Micrurus f. fluvius*) venom in dogs, *Am. J. Physiol.,* **222,** 782.

Reid, H. A. (1964). Cobra bites, *Br. Med. J.,* **2,** 540.

Reid, H. A. (1968). "Symptomatology, pathology, and treatment of land snake bite in India and Southeast Asia," in W. Bücherl, E. Buckley, and V. Deulofeu, Eds., *Venomous Animals and Their Venoms,* Academic, New York, pp. 611–641.

Rochat, H., Gregoire, J., Martin-Moutot, N., Menashe, M., Kopeyan, C., and Miranda, F. (1974). Purification of animal neurotoxins: Isolation and characterization of three neurotoxins from the venom of *Naja nigricollis mossambica* Peters, *FEBS Lett.,* **42,** 335.

Rosenberg, P. and Ehrenpreis, S. (1961). Reversible block of axonal conduction by curare after treatment with cobra venom, *Biochem. Pharmacol.,* **8,** 192.

Sarkar, B. B., Maitra, S. R., and Ghosh, B. N. (1942). The effect of neurotoxin, haemolysin and cholinesterase isolated from cobra venom on heart, blood pressure and respiration, *Indian J. Med. Res.,* **30,** 453.

Schmidt, J. L., Goodsell, E. B., Brondyk, H. D., Kueter, K. E., and Richards, R. K. (1964). Prolonged effect of cobra venom on the sensitivity to α-tubocurarine, *Arch. Int. Pharmacodyn.,* **147,** 569.

Shipolini, R. A., Bailey, G. S., Edwardson, J. A., and Banks, B. E. C. (1973). Separation and characterization of polypeptides from the venom of *Dendroaspis viridis, Eur. J. Biochem.,* **40,** 337.

Snyder, G. K., Ramsey, H. W., Taylor, W. J., and Chiou, C. Y. (1973). Neuromuscular blockade of chick biventer cervicis nerve–muscle preparations by a fraction from coral snake venom, *Toxicon,* **11,** 505.

Stalc, A. and Zupanic, A. O. (1972). Effect of α-bungarotoxin on acetylcholinesterase bound to mouse diaphragm endplates, *Nature,* **239,** 91.

Steven, L. J. and Seligmann, E. B., Jr. (1970). Agar-gel and acrylamide-disc electrophoresis of coral snake venoms, *Toxicon,* **8,** 11.

Stirton, R. A. (1963). *Time, Life, and Man, The Fossil Record,* Wiley, New York.

Stringer, J. M., Kainer, R. A., and Tu, A. T. (1971). Ultrastructural studies of myonecrosis induced by cobra venom in mice, *Toxicol. Appl. Pharmacol.,* **18,** 442.

Strydom, D. J. (1972). Snake venom toxins: The amino acid sequences of two toxins from *Dendroaspis polylepis polylepis* (black mamba) venom, *J. Biol. Chem.,* **247,** 4029.

Su, C. (1960). Mode of curare-like action of cobra venom, *J. Formosan Med. Assoc.,* **59,** 1083.

Taub, A. M. and Elliott, W. B. (1964). Some effects of snake venoms on mitochondria, *Toxicon,* **2,** 87.

Tazieff-Depierre, F. and Pierre, J. (1966). Action curarisante de la toxine α de *Naja nigricollis, C. R. Acad. Sci. Paris,* **263,** 1785.

To, S. and Tin, S. (1944a). Toxikologische Untersuchungen betreffs des Giftes von *Bungarus multicinctus.* I. Teil: Allgemeine Intoxikatioserscheinungen infolge experimenteller Injektion von Bungarusgift und desren Wirkung auf Atmung und Blutdruck, *J. Formosan Med. Assoc.,* **42,** 1.

To, S. and Tin, S. (1944b). Toxikologische Untersuchungen betreffs des Giftes von *Bungarus multicinctus.* II. Teil: Wirkung des Bungarusgiftes auf ausgeschnittene Organe, *J. Formosan Med. Assoc.,* **42,** 17.

Trinca, J. C., Graydon, J. J., Covacevich, J., and Limpus, C. (1971). The rough-scaled snake (*Tropidechis carinatus*), a dangerously venomous Australian snake, *Med. J. Aust.*, **2**, 801.

Tseng, L. F., Chiu, T. H., and Lee, C. Y. (1968). Absorption and distribution of ^{131}I-labeled cobra venom and its purified toxins, *Toxicol. Appl. Pharmacol.*, **13**, 526.

Tu, A. T. and Ganthavorn, S. (1968). Comparison of *Naja naja siamensis* and *Naja naja atra* venoms, *Toxicon*, **5**, 207.

Tu, A. T. and Toom, P. M. (1967). Hydrolysis of peptides by snake venoms of Australia and New Guinea, *Aust. J. Exp. Biol. Med. Sci.*, **45**, 561.

Tu, A. T., Toom, P. M., and Ganthavorn, S. (1967). Hemorrhagic and proteolytic activities of Thailand snake venoms, *Biochem. Pharmacol.*, **16**, 2125.

Turakulov, Ya. Kh., Sorokin, V. M., Nishankhodzhaeva, S. A., and Yukel'son, L. Ya. (1971). Toxins of middle Asian cobra venom, *Biokhimiya*, **36**, 1282.

Ukino, T. (1939a). Über den Einfluss der wichtigen formosanischen Schlangengifte auf die Gewebsatmung der Kaninchenleber. 4. Mitteilung: Wirkung des Giftes von *Naja naja atra* Cantor, *J. Formosan Med. Assoc.*, **38**, 832.

Ukino, T. (1939b). Über den Einfluss der wichtigen formosanischen Schlangengifte auf die Gewebsatmung der Kaninchenleber. 5. Mitteilung: Wirkung des Giftes von *Bungarus multicinctus* Blyth, *J. Formosan Med. Assoc.*, **38**, 841.

Vick, J. A., Ciuchta, H. P., and Polley, E. H. (1964). Effect of snake venom and endotoxin on cortical electrical activity, *Nature*, **203**, 1387.

Vick, J. A., Ciuchta, H. P., and Polley, E. H. (1965). The effect of cobra venom on the respiratory mechanism of the dog, *Arch. Int. Pharmacodyn. Thér.*, **153**, 424.

Vick, J. A., Ciuchta, H. P., Broomfield, C., and Currie, B. T. (1966). Isolation and identification of toxic fractions of cobra venom, *Toxicon*, **3**, 237.

Wagner, G. M., Mart, P. E., and Kelly, R. B. (1974). β-bungarotoxin inhibition of calcium accumulation by rat brain mitochondria, *Biochem. Biophys. Res. Commun.*, **58**, 475.

Weis, R. and McIsaac, R. J. (1971). Cardiovascular and muscular effects of venom from coral snake, *Micrurus fulvius*, *Toxicon*, **9**, 219.

Yang, C. C. (1965). Crystallization and properties of cobrotoxin from Formosan cobra venom, *J. Biol. Chem.*, **240**, 1616.

Zakharov, M. D. and Spiridonov, V. K. (1974). Isolation and anticholinergic properties of Middle Asiatic cobra (*Naja oxiana*) toxins, *Izv. Sib. Otd. Akad. Nauk SSSR, Ser. Biol. Nauk*, **1**, 77.

Zaki, O. A., Khogali, A., and Zak, F. (1970). Black mamba venom and its fractions, *Arch. Pathol.*, **89**, 30.

Ziegler, F. D., Vázquez-Colón, L., Elliott, W. B., Taub, A., and Gans, C. (1965). Alteration of mitochondrial function by *Bungarus fasciatus* venom, *Biochemistry*, **4**, 555.

Ziegler, F. D., Vázquez-Colón, L., Elliott, W. B., Gans, C., and Taub, A. (1967). "Production by snake venoms of uncoupling activity and reverse acceptor control in rat liver mitochondrial preparation. II. Studies on energy metabolism following treatment of mitochondria with several snake venoms," in F. E. Russell and P. R. Saunders, Eds., *Animal Toxins,* Pergamon, Oxford, pp. 236–243.

Zlotkin, E., Menashé, M., Rochat, H., Miranda, F., and Lissitzky, S. (1975). Proteins toxic to arthropods in the venom of elapid snakes, *J. Insect Physiol.*, **21**, 1605.

13 Viperidae Venoms

1   CHEMISTRY   201

    1.1   *Vipera palestinae* Venom, 201
    1.2   *Vipera ammodytes* Venom, 203
    1.3   *Vipera russellii* Venom, 203

2   BIOLOGICAL EFFECTS   206

    2.1   Toxicology, 206
    2.2   Pharmacological Effects, 206
    2.3   Cytotoxicity, 207

    References   208

Viperidae (viperids, true vipers) are found in Asia, Europe, and Africa but not in the Americas or Australia. Snakes of Viperidae and Crotalidae (pit vipers) have taxonomic similarities, and therefore some biologists classify them in the same family. The pharmacological properties of Viperidae venoms have some similarity to those of Crotalidae venoms and are also somewhat similar to those of Elapidae venoms. In other words, Viperidae venoms have properties between those of Crotalidae and Elapidae. Like Crotalidae venoms, the lethal toxins of Viperidae have not been studied extensively. In this chapter, we review the chemical aspects of Viperidae venoms and their biological effects.

1   CHEMISTRY

1.1   *Vipera palestinae* Venom

Snake venoms, no matter what their species origin, are mixtures of different proteins, toxic and nontoxic or enzymatic and nonenzymatic. However, venoms of Viperidae and Crotalidae have higher molecular weight proteins than do Elapidae venoms. Mebs (1969) investigated the dialyzabilities of 10 Elapidae, 2 Viperidae, and 3 Crotalidae venoms. He found that 34 to 75% of Elapidae venom proteins, but only 5 to 13% of Viperidae and 10 to 17% of Crotalidae venom proteins are dialyzable.

Venoms of *Vipera palestinae* contain neurotoxins, hemorrhagic toxins, spreading factor (hyaluronidase), anticoagulant, procoagulant, protease, phospholipase A_2, and L-amino acid oxidase (Kochwa et al., 1960). The venom also contains many other proteins, but these are the fractions that were tested for biological activities. The venom contains two neurotoxic and three hemorrhagic fractions. Venom protease can be isolated

Table 1   Neurotoxins from the Venoms of Vipers (Family: Viperidae)

Genus Species Subspecies	Vipera palestinae	Vipera ammodytes
Origin	Israel	Bulgaria
Name	Viperotoxin	Toxin
Amino Acid		
Lysine	9	5
Histidine	3	2
Arginine	6	6
Aspartic acid	10	23
Threonine	4	9
Serine	6	6
Glutamic acid	10	18
Proline	12	1
Glycine	10	14
Alanine	6	12
Valine	4	5
Methionine	1	1
Isoleucine	4	5
Leucine	4	6
Tyrosine	4	11
Phenylalanine	5	7
Half-cystine	6	14
Tryptophan	4	no data
Total residue	108	145
Reference	Moroz et al. 1966	Aleksiev and Shipolini, 1971

free from neurotoxic activity; however, it can enhance the lethal activity of neurotoxin (Moroz-Perlmutter et al., 1965).

One of the neurotoxins, viperotoxin, was isolated and characterized (Moroz et al., 1966). It contains 108 amino acid residues and has a molecular weight of 11,600 daltons (Table 1). It is composed of one polypeptide chain cross-linked intramolecularly by three disulfide bridges. Lysine is in the amino-terminal position, and proline in the carboxy-terminal position. Venom proteins are synthesized in the venom gland. When the biosynthesis of viperotoxin (neurotoxin) was at its maximum in the venom gland, the rate of ^{32}P incorporation into ribosomal and soluble RNA was also at its peak (Shaham and Kochva, 1969; Bamberg et al., 1967). This explains the fact that, when the lumina are filled with venom, the secretory epithelium becomes quite inactive, and the rate of venom production slows down considerably (Kochva, 1960).

1.2 Vipera ammodytes Venom

Vipera ammodytes venom can be separated into several biologically active components such as lethal, proteolytic, phosphodiesterase (exonuclease), L-amino acid oxidase, and phospholipase A_2 fractions (Shipolini et al., 1968). Partial purification of lethal toxin resulted in a toxicity of 0.3 $\mu g\ g^{-1}$ in mice (Shipolini et al., 1970a). Purified toxin has a molecular weight of 31,500 to 35,500 daltons, but this value is reduced to 11,130 after carboxymethylation. The toxin apparently consists of three subunits with an approximate molecular weight of 11,000 (Shipolini et al., 1970b). Of the 145 total amino acid residues, 28% are acidic. The amino acid composition, which is shown in Table 1, has no resemblance to the composition of other known toxins of Elapidae, Hydrophiidae, and Crotalidae (Aleksiev and Shipolini, 1971).

Gubešek et al. (1974) separated the venom into four fractions: lethal, hemorrhagic proteolytic, and arterial blood pressure depressing.

The snake possesses autoantibodies in the serum against its own poison (Yomtov and Nedjalkov, 1962). It also contains common antigens between its serum and its venom.

1.3 Vipera russellii Venom

Vipera russellii venom contains a number of enzymes such as exonuclease (phosphodiesterase), phosphomonoesterase, 5′-nucleotidase, protease, L-amino acid oxidase, and phospholipase A_2 (Dimitrov and Kakonkar, 1968).

The toxicity of the venom is sensitive to heat treatment and loses its potency when the venom is incubated at 95°C (Bandyopadhyay et al., 1970).

The venom contains four toxins (Bolar and Master, 1976), one of which is a glycoprotein with two subunits. Their molecular weights and other characteristics are as follows:

Toxin	Molecular Weight	Content	Number of Subunits
T_1	55,000	Glycoprotein	2
T_2	15,000	Basic protein	4
T_3	<4,000	Basic protein	0
T_4	13,000	Basic protein	3

Table 2. Toxicities of Viperidae Venoms in Mice

Venoms	LD_{50} (µg/g)	Route of Injection	References
Bitis arietan (B. lachesis)	2.00	i.v.	Friederich and Tu, 1971
	0.50	i.p.	Mohamed et al, 1973
	7.75	s.c.	Minton, 1974
B gabonica	4.95	i.v.	Friederich and Tu, 1971
	1.0	i.p.	Mohamed et al, 1973
	12.5	s.c.	Minton, 1974
Causus rhombeatus	9.25	i.v.	Minton, 1974
	15.0	s.c.	Minton, 1974
Cerastes cerastes	0.30	i.p.	Mohamed et al, 1973
	0.45	i.v.	Hassan and El-Hawary, 1975
	15.0	s.c.	Minton, 1974
C. cornutus	0.48	i.v.	Boquet, 1968
C. vipera	0.25	i.p.	Mohamed et al, 1973
	0.64	i.v.	Hassan and El-Hawary, 1975
Echis carinatus	1.9	i.v.	Kankonkar et al, 1972
	4.8	i.v.	Latifi et al, 1973
	0.21	i.p.	Mohamed et al, 1973
	1.3-6.5	—	Kornalick and Taborska, 1973
	1.2	i.v.	Christiensen, 1955
	1.2	i.v.	Boquet, 1968
(From Iran)	0.44	i.v.	Hassan and El-Hawary, 1975
(From Egypt)	0.65	i.v.	Hassan and El-Hawary, 1975
	6.55	s.c.	Minton, 1974
E. Coloratus	0.38	i.p.	Mohamed et al, 1973

Table 2 *Continued*

Venoms	LD$_{50}$ (μg/g)	Route of Injection	References
Vipera ammodytes	6.59	s.c.	Minton, 1974
	0.19-0.64	i.p.	Novak et al, 1973
V. aspis	1.0	i.v.	Boquet, 1968
V. berus	0.55	i.v.	Minton, 1974
	0.80	i.p.	Minton, 1974
	6.45	s.c.	Minton, 1974
V. Latifi	4.8	i.v.	Latifi et al, 1973
V. Lebetina	7.7	i.v.	Latifi et al, 1973
	0.64	i.v.	Hassan and El-Hawary, 1975
	16.0	s.c.	Minton, 1974
hemorrhagic fraction	6.0	s.c.	Krylova, 1967
thromboplastic fraction	0.3	i.v.	Krylova, 1967
V. palestinae	1.9	i.p.	Krupnick et al, 1968
V. persica	0.83	i.v.	Hassan and El-Hawary, 1975
V. russellii	0.31	i.v.	Kankonkar et al, 1972
	0.08	i.v.	Minton, 1974
	0.40	i.p.	Minton, 1974
	4.75	s.c.	Minton, 1974
V. russellii formosensis	1.37	s.c.	Lee, 1948
	34.8 (frogs)	s.c.	Lee, 1948
V. russellii siamensis	2.11	i.v.	Friederich and Tu, 1971
V. xanthina palaestinae	0.18	i.v.	Minton, 1974
	9.40	s.c.	Minton, 1974

2 BIOLOGICAL EFFECTS

2.1 Toxicology

Viperidae venoms are less toxic than Elapidae and Hydrophiidae venoms, but are comparable to or stronger than Crotalidae venoms in this respect. The LD_{50} values of Viperidae venoms in mice are listed in Table 2.

Bivalent antivenin made for *Cerastes cerastes* and *C. vipera* effectively neutralized the toxicities of homologous venoms. The neutralization capacity of these antivenins for other Viperidae venoms (*Vipera persica, V. lebetina, Echis carinatus*) is very low, and they offer practically no protection toward Elapidae venoms (Hassan and El-Hawary, 1975). It is of interest that there is no definite correlation between the capacity to neutralize toxicity and the Ouchterlony reaction. For instance, the *Echis* venoms give the poorest immunodiffusion precipitation lines but show high *in vivo* neutralization capacity.

2.2 Pharmacological Effects

Venom of *Vipera russellii formosensis* produced diverse effects. For instance, it produced a paralytic effect in the central nervous system, circulatory failure, intravascular clot, hemorrhage, and direct paralytic action on isolated gastrocnemius muscle of the frog (Lee, 1948).

Many people are sensitive to antivenin and may suffer anaphylactic shock. The use of antibody that consists of small fragments, rather than relatively large γ-globular protein, may reduce the risk of serum reaction in snakebite treatment. Sein et al. (1972), therefore, digested antibody to *V. russellii* venom with pepsin and found the fragments of antibody to be as potent as the original antivenin in neutralization capacity.

The i.v. injection of *Vipera palestinae* venom (3 mg kg^{-1}) caused the death of rabbits in approximately 2 hr. The normal alpha rhythm of the electrocorticogram was replaced shortly after injection by a delta rhythm that lasted for a few minutes. The pattern was followed by a gradual decrease in the amplitude and frequency of the cortical potentials, resulting, in several cases, in complete disappearance of detectable cortical activity (Krupnick et al., 1968).

Although some venoms are neurotoxic on peripheral nerve, the venom of *Vipera ammodytes* is not (Sket, 1971; Sket et al., 1973a). The crude venom of *V. ammodytes* and two basic fractions induced a rapid decrease in arterial blood pressure and respiratory failure in rats but produced no neuromuscular blockade of rat diaphragm–phrenicus preparation after prolonged stimulation. They decreased muscular tension upon direct stimulus. Thus the respiratory depression may be due to inhibition of the respiratory center and not to paralysis of the respiratory muscles.

As far back as 1936, Schaumann (1936) reported that the venom of *V. ammodytes* temporarily weakened the contractions of striated muscle *in vitro*. Absence of inhibitory action on synaptic transmission in a neuromuscular preparation of the crayfish was reported by Parnas and Russell (1967). Injection of the venom and the basic protein caused coma in rats. On the assumption that coma is due to lower oxygen consumption by the rat brain, Sket et al. (1973b) investigated the effect of venom and the basic protein on the brain oxygen activity. They found no differences in oxygen consumption activity between normal brain and the brain of intoxicated comatose rats. Therefore coma is not due to lower oxygen activity in the brain.

Injection of all snake venoms caused a very rapid fall in arterial blood pressure, with a return to normal in a few minutes. By fractionation of *V. ammodytes* venom, Novak et al. (1973) obtained three fractions responsible for the depression of arterial blood pressure. They attributed this action to the effect of phospholipase A_2.

Similarly, venoms of *Vipera aspis, V. berus, Bitis lachesis,* and *B. gabonica* did not affect the amplitude of the contractions of neuromuscular preparation (Cheymol et al., 1972). The lack of neurotoxic effect of *B. gabonica* venom on isolated rat phrenic nerve–diaphragm preparation was also shown by Whaler (1972). He suggested that the factors contributing to death are direct cardiac damage and failure of O_2 uptake.

Echis carinatus (from Ethiopia) venom has a paralyzing effect at the site of the postsynaptic receptors as studied in isolated or *in situ* neuromuscular preparations (Cheymol et al., 1971). The neuromuscular blocking action of this Ethiopian Viperidae venom seems to be due to the presence of toxin similar to Elapidae α-toxin. Venoms of *V. russellii, Cerastes cerastes*, and *Echis carinatus* induced contractions, followed by slow paralysis, in isolated rat diaphragms (Cheymol et al., 1972).

Viperidae venoms are often very similar in action to Crotalidae venoms. They have pronounced effects on blood coagulation and hemolytic activity, as discussed in detail in Chapters 21 and 20, respectively.

2.3 Cytotoxicity

Venom of *Bitis lachesis (B. arietans)* has strong cytoxic activity toward bovine spleen cells and mouse embryo cells (Tu and Passey, 1969). In strength of cytopathic action *B. lachesis* venom is somewhere between Elapidae and Crotalidae venoms. Similarly *B. nasicornis* venom has been shown to cause the death of tissue culture cells of human fibroblasts of the MRC 5 strain (March and Glatston, 1974). Venoms of *B. gabonica rhinoceros, B. arietans*, and *Vipera russelli* on KB cells have been found to be cytotoxic (Tu and Giltner, 1974).

The effect of *Vipera ammodytes ammodytes* venom on isolated heart was studied (Ivančević et al., 1963). The venom produced interstitial edema, particularly about the capillaries and veins and between the endocardium and myocardium. Normal striations were obliterated. The cytoplasm became markedly swollen and vacuolated. It was concluded that the action of the venom is quite different from that produced by histamine and histamine liberator, 48/80.

The effect of *Vipera palestinae* venom on the kidney, thymus, and lymph nodes was investigated (Ickowicz et al., 1966). The venom produced edema within the tubules, swelling of the epithelial cells, partial destruction of many cells, and desquamation within the lumen of the tubules. A slight congestion of the blood vessels between the tubules was noticeable after any dose of venom, becoming more pronounced with larger doses. There was no difference in the response regardless of the injection route used. Normally the stress reaction to a toxic agent is manifested by pyknosis and karyorrhexis in the thymus and lymphatic glands. The fact that there were no changes in the tissues of the thymus and lymphatic glands after venom poisoning suggests that stress does not play a major role in snake venom poisoning.

The venom of *Cerastes cerastes* blocked muscular contraction to indirect stimulation (Mohamed and Khaled, 1966). It was considered that this action is due to a blockage of the conductance of the nerve. The blockage of the muscular response when the venom was added to the muscle chamber could be a result of the inactivation of the fine nerve branches. The muscle itself, however, was not affected by the venom. Partially curarized muscle can be recovered when potassium chloride is added (Wilson and Wright, 1936). With the venom, however, no such recovery was observed; thus KCl is not antagonistic to the venom. Eserine and physostigmine did not reverse the blocking effect of the venom, since no transmitter was released. When acetylcholine per se was added, there was partial recovery, indicating that the postsynaptic membrane was still sensitive to the transmitter substance.

Viperidae venoms also affect mitochondria. For instance, *Vipera ursini* venom caused swelling of rat liver mitochondria and stimulated the Mg-activated mitochondrial ATPase (Sattyev et al., 1974). The factor responsible for this action is phospholipase. The effect of phospholipase A_2 is discussed in detail in Chapter 2.

REFERENCES

Aleksiev, B. and Shipolini, R. (1971). Weitere Untersuchunger zur Fraktionierung and Reinigung der toxischen Proteine aus dem Gift der bulgarischen Viper (*Vipera ammodytes ammodytes*), *Hoppe-Seylers Z. Physiol. Chem.,* **352,** 1183.

Bamberg, E., Rotenberg, D., Sharf, F., and Kochva, E. (1967). Ribonucleic acid metabolism in venom glands of the snake *Vipera palaestinae, Israel, J. Chem.,* **5,** 116.

Bandyopadhyay, R., Roy, S. N., Koley, B. N., and Maitra, S. R. (1970). Effect of different grades of temperature on the toxicity of cobra (*Naja naja*) and viper (*Vipera russellii*) venoms, *Indian J. Physiol. Allied Sci.,* **24,** 117.

Bolar, H. V. and Master, R. W. P. (1976). The complex toxic components of the Russell's viper venom, *Biochem. Biophys. Res. Commun.,* **70,** 573.

Boquet, P. (1968). "Pharmacology and toxicology of snake venoms of Europe and the Mediterranean regions," in W. Bücherl, E. E. Buckley, and V. Deulofeu, Eds., *Venomous Animals and Their Venoms,* Vol. 1, Academic, New York, pp. 339–357.

Cheymol, J., Bourillet, F., and Roch-Arveiller, M. (1971). Neuromuscular effect of a venom from *Echis carinatus* (Viperidae) from Ethiopia, *Therapie,* **26,** 1007.

Cheymol, J., Bourillet, F., and Roch-Arveiller, M. (1972). Neuromuscular activity in the venom of some Viperidae, *C. R. Soc. Biol.,* **166,** 1283.

Christensen, P. A. (1955). South African snake venoms and antivenoms, *S. Afr. Inst. Med. Res.* 142 pages.

Dimitrov, G. D. and Kankonkar, R. C. (1968). Fractionation of *Vipera russellii* venom by gel filtration. I. *Toxicon,* **5,** 213.

Friederich, C. and Tu, A. T. (1971). Role of metals in snake venoms for hemorrhagic, esterase and proteolytic activities, *Biochem. Pharmacol.,* **20,** 1549.

Gubensek, F., Sket, D., Turk, V., and Lebez, D. (1974). Fractionation of *Vipera ammodytes* venom and seasonal variation of its composition, *Toxicon,* **12,** 167.

Hassan, F. and El-Hawary, M. F. S. (1975). Immunological properties of antivenins. I. Bivalent *Cerastes cerastes vipera* antivenin, *Am. J. Trop. Med. Hyg.,* **24,** 1031.

Ickowicz, M., Shulov, A., and Naor, D. (1966). The effect of *Vipera palestina* venom of the thymus, lymph nodes, and kidneys, *Toxicon,* **3,** 305.

Ivancević, I., Marian, N., and Knezević, M. (1963). Effects of *Vipera ammodytes ammodytes* venom on isolated heart, *Toxicon,* **1,** 65.

Kankonkar, R. C., Rao, S. S., Vad, N. E., and Sant, M. V. (1972). Efficacy of Haffkine Institute polyvalent antivenin against Indian snake venoms, *Indian J. Med. Res.,* **60,** 512.

Kochva, E. A. (1960). A quantitative study of venom secretion by *Vipera palestinae, Am. J. Trop. Med. Hyg.,* **9,** 381.

Kochwa, S., Perlmutter, C., Gitter, S., Rechnic, J., and De Vries, A. (1960). Studies on *Vipera palestinae* venom: Fractionation by ion exchange chromatography, *Am. J. Trop. Med.,* **9,** 374.

Kornalik, F. and Taborska, E. (1973). "Individual interspecies variability in the composition of some viperidae venoms," in E. Kaiser, Ed., *Animal and Plant Toxins,* Wilhelm Goldman, Munich, pp. 99–103.

References

Krupnick, J., Bicher, H. I., and Gitter, S. (1968). Central neurotoxic effects of the venoms of *Naja naja* and *Vipera palestinae, Toxicon,* **6,** 11.

Krylova, E. S. (1967). Hemorrhagic and thromboplastic activities of electrophoretic fractions of the venom of *Vipera lebetina, Tr. Konf. Biokhim. Respub. Srednei Azii Kaz., 1st, Alma-Ata 1966,* 294.

Latifi, M., Farzanpay, R., and Tabatabai, M. (1973). "Comparative studies of Iranian snake venoms by gel diffusion and neutralization tests," in E. Kaiser, Ed., *Animal and Plant Toxins,* Wilhelm Goldmann, Munich, pp. 201–206.

Lee, C. Y. (1948). Toxicological studies on the venom of *Vipera russellii formonsensis* Maki, *J. Formosan Med. Assoc.,* **47,** 65.

Marsh, N. and Glatston, A. (1974). Some observations on the venom of the rhinoceros horned viper, *Bitis nasicornis* Shaw, *Toxicon,* **12,** 621.

Mebs, D. (1969). Preliminary studies on small molecular toxic components of elapid venoms, *Toxicon,* **6,** 247.

Minton, S. A. (1974). *Venom Diseases,* Charles C Thomas, Springfield, Ill.

Mohamed, A. H. and Khaled, L. Z. (1966). Effect of the venom of *Cerastes cerastes* on nerve tissue and skeletal muscle, *Toxicon,* **3,** 223.

Mohamed, A. H., Darwish, M. A., and Havi-Ayobe, M. (1973). Immunological studies on Egyptian cobra antivenin, *Toxicon,* **11,** 31.

Moroz, C., De Vries, A., and Sela, M. (1966). Isolation and characterization of a neurotoxin from *Vipera palestinae* venom, *Biochim. Biophys. Acta,* **124,** 136.

Moroz-Perlmutter, C., Goldblum, N., and De Vries, A. (1965). Biochemical and antigenic properties of a purified neurotoxin of *Vipera palestinae* venom, *J. Immunol.,* **94,** 164.

Novak, V., Sket, D., and Gubensek, F. (1973). "Arterial blood pressure depressing proteins in *Vipera ammodytes* venom," in E. Kaiser, Ed., *Animal and Plant Toxins,* Wilhelm Goldmann, Munich, pp. 159–162.

Parnas, I. and Russell, F. E. (1967). "Effects of venoms on nerve, muscle and neuromuscular junction," in F. E. Russell and P. R. Saunders, Eds., *Animal Toxins,* Pergamon, Oxford, pp. 401–415.

Sattyev, R., Yukel'son, L. Y., Sakhibov, D. N., and Gagel'gans, A. I. (1974). Effect of venoms of Central Asian snakes (viper and cobra) on the structural state of rat liver mitochondria, *Uzb. Biol. Zh.,* **18,** 64.

Schaumann, O. (1936). Pharmakologische Versuche mit Schlangengiften und Schlangensera, *Behringwerk Mitt.,* **7,** 33.

Sein, H., Moroz, C., and De Vries, A. (1972). Isolation of immunologobulin and fragment from donkey anti-Russell's viper serum, *Ann. Inst. Pasteur,* **122,** 213.

Shaham, N. and Kochva, E. (1969). Localization of venom antigens in the venom gland of *Vipera palaestinae* using a fluorescent-antibody technique, *Toxicon,* **6,** 263.

Shipolini, R. A., Dimitrov, G. D., and Ivanov, Ch. P. (1968). Fractionation of the venom of *Vipera ammodytes* by gel filtration and starch-gel electrophoresis, *C. R. Acad. Bulg. Sci.,* **21,** 573.

Shipolini, R., Ivano, C., Aleksiev, B. V., and Ganchev, K. (1970a). Isolation and physiological action of fractions from the venom of *Vipera ammodytes ammodytes, Dokl. Bolg. Akad. Nauk,* **23,** 279.

Shipolini, R. A., Vernon, Ch. A., Ivanov, Ch. P., Aleksiev, B. V., and Ganchev, K. (1970b). Microstructural study of the main toxic component of the venom from the Bulgarian viper, *God. Vissh. Khimikotekhnol. Inst. Sofia,* **15,** 47.

Sket, D. (1971). Mechanism of action of partially purified basic protein components from *Vipera ammodytes* venom, *Inst. Jozef. Stefan IJS Porocilo,* p. 267.

Sket, D., Gubenšek, F., Adamič, S., and Lebez, D. (1973a). Action of a partially purified basic protein fraction from *Vipera ammodytes* venom, *Toxicon,* **11,** 47.

Sket, D., Gubenšek, F., Pavlin, R., and Lebez, D. (1973b). Oxygen consumption of rat brain homogenates after *in vitro* and *in vivo* addition of the basic protein from *Vipera ammodytes* venom, *Toxicon,* **11,** 193.

Tu, A. T. and Giltner, J. B. (1974). Cytotoxic effects of snake venoms of KB and Yoshida sarcoma cells, *Res. Commun. Chem. Pathol. Pharmacol.,* **9,** 783.

Tu, A. T. and Passey, R. B. (1969). Effects of snake venoms om mammarian cells in tissue culture, *Toxicon*, **6**, 277.

Whaler, B. C. (1972). Gaboon viper venom and its effects, *J. Physiol.*, **222**, 61.

Wilson, A. T. and Wright, S. (1936). Anticurare action of KCl, *Quart J. Exp. Physiol.*, **26**, 127.

Yomtov, M. and Nedjalkov, S. (1962). Untersuchengen über das Gift der bulgarischen Sandotter. III. Vorhandensein von Autoantikärpern gegen das Eigengift bei der *Viper ammodytes* und von gemeinsamen Antigenen zwischen ihrem Serum und ihrem Gift, *Z. Immunitästsforsch. Exp. Ther.*, **124**, 212.

14 Venoms of Crotalidae (Crotalids, Pit Vipers)

1	**CHEMISTRY OF CROTOXINS AND RELATED COMPOUNDS**	213
	1.1 Crotoxin, 213	
	1.2 Crotamine, 216	
	1.3 Convulxin, 217	
2	**PHARMACOLOGY OF CROTOXINS AND RELATED COMPOUNDS**	217
	2.1 Crotoxin, 217	
	2.2 Crotamine, 218	
3	**OTHER RATTLESNAKE TOXINS**	219
	3.1 Chemistry, 219	
	3.2 Biological Effect, 222	
4	**OTHER CROTALIDAE VENOMS**	227
	References	229

Snake venoms, especially Crotalidae venoms, contain a rather large number of pharmacologically and biochemically active proteins. It has been shown that the isoelectric point distribution pattern of Crotalidae (*Agkistrodon rhodostoma*) venom is much more complicated than that of sea snake (*Enhydrina schistosa*) venom (Toom et al., 1969). Also, the electrophoretic patterns of various Crotalidae venoms are much more complicated than those of Elapidae venoms (Bertke et al., 1966).

Crotoline venoms not only are more complicated than the venoms of other families of snakes, but also contain more high molecular weight proteins. Yang (1963) compared the fractionation patterns obtained by separation of eight venoms from the families Crotalidae, Viperidae, and Elapidae on Sephadex G-75 and G-50. All Crotalidae venoms investigated (*Agkistrodon acutus, Crotalus adamanteus, C. terrificus terrificus, Trimeresurus gramineus formosensis, T. mucrosquamatus*) had elution patterns located closer to the void volume than did the venoms of other families. This suggests that Crotalidae venoms contain higher molecular weight fractions than do the others.

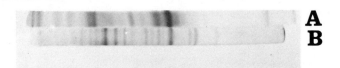

Figure 14.1 An example of the complexicity of Crotalidae venoms, using *Crotalus scutalatus* venom. (A) Acrylamide gel electrophoretic pattern; 16 bands are visible. (B) Isoelectric focusing gel pattern of same venom; 22 bands are visible. (Reproduced from Bieber et al., 1975, by permission of the copyright owner. Elsevier Scientific Publishing Company.)

As an example of the complexity, venom of *Crotalus scutulatus* as examined by acrylamide disk gel electrophoresis and by polyacrylamide gel isoelectric focusing is shown in Fig. 1.

Among Crotalidae snakes, rattlesnakes (genera *Crotalus* and *Sistrurus*) are of particular interest to Americans, since these snakes are abundant in North, Central, and South America. The presence of rattlesnakes was recognized by our ancestors in ancient times, as witnessed by a stone sculpture of a Mexican rattlesnake, *Crotalus* sp., made by the Aztec people several thousand years ago (Fig. 2).

Unlike the genus *Agkistrodon*, rattlesnakes are found only in the New World, and North America is especially rich in its variety of these snakes. Since rattlesnake and other Crotalidae venoms contain a greater array of biologically active components than

Figure 14.2 Stone sculpture of a rattlesnake, a typical poisonous snake in America. This stone carving was made by the Aztec people many thousands of years ago. (Photograph taken in the National Anthropology Museum in Mexico City by the author.)

1 Chemistry of Crotoxins and Related Compounds

Elapidae and Hydrophiidae (sea snake) venoms, they have a more complex spectrum of pharmacological actions.

Because of the great complexity of Crotalidae venom composition, fewer studies have been made of these venoms because purification of each component is difficult. The venom from Brazilian rattlesnakes (*Crotalus durissus terrificus*), however, has undergone extensive chemical study from as far back as 1938. This investigation, due mainly to the remarkable efforts and achievements of Dr. K. Slotta and his co-workers, especially Dr. H. Fraenkel-Conrat, really opened the era of modern biochemical research on snake venoms. It is logical, therefore, to start this chapter with the chemistry of crotoxin.

1 CHEMISTRY OF CROTOXINS AND RELATED COMPOUNDS

The fact that many similar names are associated with the components of Brazilian rattlesnake venom sometimes leads to a rather confusing situation. For clarity, these terms may be briefly defined as follows:

CROTACTIN: The toxic component of crotoxin after removing a basic phospholipase A_2.
CROTOXIN ACIDIC PROTEIN; CROTAPOTIN: The acidic protein component of crotoxin.
CROTOXIN BASIC PROTEIN: The basic protein component of crotoxin; actually, a basic phospholipase A_2.
BASIC *CROTALUS* PHOSPHOLIPASE A: Same as crotoxin basic protein.
CROTAMINE: A basic protein of low toxicity found in the venom of only one variety of Brazilian rattlesnake; a different toxic protein from crotoxin.
CONVULXIN: A new neurotoxic protein in the venom of Brazilian rattlesnake, different from crotoxin and crotamine.

1.1 Crotoxin

The first isolation of a neurotoxin from snake venom was achieved by Slotta and his co-worker from the venom of *Crotalus durissus terrificus* (Slotta, 1938; Slotta and Fraenkel-Conrat, 1938a, b, 1939). The crystalline toxin was called crotoxin. From studies by ultracentrifugation and by Tiselius electrophoresis (Gralen and Svedberg, 1938), crotoxin appeared to be homogeneous (Li and Fraenkel-Conrat, 1942). The molecular weight of crotoxin is 30,000, and the pI is 4.7. A molecular weight value of 30,900 was obtained by Paradies and Breithaupt (1975).

Crotoxin exhibits not only neurotoxicity, but also indirect hemolytic, smooth muscle-stimulating, phospholipase A_2, and hyaluronidase activities. Neumann and Habermann (1955) and Habermann (1957a, b) attributed the neurotoxic activity and the phospholipase A_2 activity to two separate components. Definite evidence of molecular heterogeneity was presented by Fraenkel-Conrat and Singer (1956), who separated two biologically inactive 2,4-dinitrophenylated proteins. The presence of phospholipase A_2 in crotoxin was well recognized. For example, Feeney et al. (1954) observed that the lipoprotein fractions of yolk, lipovitellin, and lipovitellenin were attacked by crotoxin, obviously as a result of the action of phospholipase A_2. Nygaard and Sumner (1953) isolated phospholipase A_2 from crotoxin.

Since crotoxin is heterogeneous, although it appeared to be homogeneous in some physicochemical studies, the toxic component was called crotactin by Habermann and Neumann (1956). Crotactin had higher toxicity than crotoxin and its original venom, as the following values indicate:

Substance	$LD_{50}(\mu g\ g^{-1})$ in mice (i.p. injection)
Venom	0.15
Crotoxin	0.1
Crotactin	0.06–0.07

The heterogeneity of crotoxin was further proved by fractionation of this toxin into six components (Habermann and Rübsamen, 1971). The toxicity of crotoxin is attributable to more than one of these components. The two phospholipases possess some genuine toxicity, which is greatly enhanced by nontoxic, strongly acidic constituents. A toxic fraction exhibiting little enzyme activity may be identical with crotactin. Habermann and Rübsamen conclusively proved that crotoxin is a mixture of acidic and basic constituents.

Independently, Hendon and Fraenkel-Conrat (1971) and Horst et al. (1972) also showed that crotoxin is composed of acidic and basic proteins and demonstrated that the neurotoxicity of the toxin requires the synergistic action of both. The acidic protein lacked the hemolytic and neurotoxic activity of crotoxin, and the basic protein showed only the high, indirect hemolytic activity. A mixture of the two components, however, generated high toxicity. The p*I* values of the two components are 3.7 and 8.6, respectively.

Breithaupt et al. (1974) isolated the major acidic and basic proteins from crotoxin. The acidic protein is called crotapotin and has a pI of 3.4. The basic protein is phospholipase A_2 and has a pI of 9.7. Apparently, the "crotapotin" isolated by Breithaupt et al. (1974) is identical to the "crotoxin acidic component" isolated by Hendon and Fraenkel-Conrat (1971) and Horst et al. (1972). The molecular weights of these acidic and basic proteins as determined by the two groups are as follows:

Protein	Molecular Weight	Reference
Crotoxin acidic protein	9,000	Horst et al., 1972
(Crotapotin)	8,900	Breithaupt et al., 1974
Crotoxin basic protein	12,000	Horst et al., 1972
(Basic *Crotalus* phospholipase A)	12,500	Breithaupt et al., 1974

The amino acid compositions of crotoxin and related toxins are shown in Table 1.

Crotoxin can be reconstituted from its components with regeneration of its full toxicity (Rübsamen et al., 1971).

Actually there are three forms of phospholipase A_2 in crotoxin. Two of them are basic proteins, and one is acidic. Crotapotin (the acidic protein of crotoxin) forms a complex with the basic phospholipase A_2 components and inhibits the enzyme activity. Crotapotin, however, does not interact with the acidic phospholipase A_2 component (Breithaupt et al., 1975).

Table 1 Amino Acid Compositions of Crotoxin and Related Toxins Isolated from the Venom of *Crotalus durissus terrificus*

Name Amino Acid	Crotoxin	Crotoxin	Crotoxin Acidic Protein	Crotapotin	Crotoxin Basic Protein	Crotoxin phospho- lipase A	Crotamin	Crotamin
Lysine	12	11	8	2	9	11	11	9
Histidine	4	3	1	1	2	2	3	2
Arginine	14	10	2	2	8	12	2	2
Aspartic acid	26	19	10	12	9	11	3	2
Threonine	12	10	4	4	6	8	0	0
Serine	14	11	5	6	6	7	3	3
Glutamic acid	27	20	12	14	8	10	2	2
Proline	13	10	4	5	5	5	4	3
Glycine	25	19	9	10	10	13	5	5
Alanine	15	11	5	6	6	7	0	0
Valine	4	3	1	1	2	2	0	0
Methionine	4	2	1	1	1	2	1	1
Isoleucine	7	7	3	2	4	5	1	1
Leucine	8	7	1	1	6	7	1	1
Tyrosine	15	11	2	3	9	12	1	1
Phenylalanine	17	8	2	3	6	7	2	2
Half-cystine	30	21	10-11	14	10-11	16	4	6
Tryptophan	6	3	1	1	2	3	3	2
Total residues	253	186	75-76	88	109-110	140	46	42
References	Fischer and Dörfel, 1954	Hendon and Fraenkel-Conrat, 1971	Hendon and Fraenkel-Conrat, 1971	Breithaupt et al, 1974	Hendon and Fraenkel-Conrat, 1971	Breithaupt et al, 1974	Goncalves and Giglio, 1964	Laure, 1975

1.2 Crotamine

Crotamine is a basic polypeptide toxin that, unlike crotoxin, is present in the venoms of only one variety of Brazilian rattlesnake, called *Crotalus durissus terrificus* var. *crotaminicus* (Brazil, 1972). It produces skeletal muscle spasms in mammals, leading to spastic paralysis of peripheral origin (Bourillet, 1970; Cheymol et al., 1971b). The molecular weight is 4760 to 4802 (Giglio, 1975).

When the venom solution is mixed with acetone or alcohol, cortoxin is precipitated first and then the crotamine fraction is precipitated from the supernatant liquid. By electrophoresis, partially purified crotamine can be separated into different fractions, crotamine, toxin III, and enzymes (Habermann, 1957b). The amino acid composition of crotamine is listed in Table 1. The amino acid sequence was identified by Laure (1975), who showed that the number of amino acid residues is 42 rather than 46; crotamine is a rather small toxic protein. The sequence is as follows:

$$\text{Tyr-Lys-Gln-Cys-His-Lys-Lys-Gly-Gly-}\overset{10}{\text{His}}\text{-Cys-Phe-Pro-Lys-}$$

$$\text{Glu-Lys-Ile-Cys-Leu-}\overset{20}{\text{Pro}}\text{-Pro-Ser-Ser-Asp-Phe-Gly-Lys-Met-}$$

$$\text{Asp-}\overset{30}{\text{Cys}}\text{-Arg-Trp-Arg-Trp-Lys-Cys-Cys-Lys-Lys-}$$

$$\overset{40}{\text{Gly}}\text{-Ser-Gly}$$

Crotalidae are quite different from the snakes of Elapidae and Hydrophiidae, and the venoms of Crotalidae also differ from those of other families. The fact that the sequence of one Crotalidae venom component, crotamine, is now available will permit direct comparisons with the sequences of Type I neurotoxins of Elapidae and sea snake venoms. There are some similarities, which are shown in Fig. 3. There are two fewer sulfur atoms in crotamine. The corresponding sulfur atoms in crotamine and Type I neurotoxins are connected through the dotted line in the figure. Careful examination of the sequences of crotamine and Type I neurotoxins of Elapidae and Hydrophiidae shows that there is little homology in each loop. It is very likely that crotamine is a much more primitive form of molecule than typical Type I neurotoxins of Elapidae and Hydrophiidae. Since Crotalidae are considered more advance forms of snakes from the viewpoint of evolution, crotamine is probably a primitive molecule left over from venoms of ancestrial snakes common also to Elapidae and Hydrophiidae.

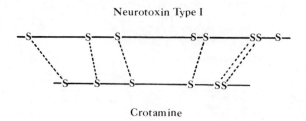

Figure 14.3 Comparison of Type I neurotoxin of Hydrophiidae and Elapida with crotamine, a toxin isolated from a Crotalidae venom. Note some similarity in the relative positions of the half-cystine residues.

1.3 Convulxin

Prado-Franceschi and Brazil (1969) isolated a new neurotoxic protein from *Crotalus durissus terrificus* venom. Convuluxin is different from crotoxin and crotamine. Mice injected with convulxin show tachypnea within 20 sec. followed by apnea of short duration. Large doses cause convulsions that end, as a general rule, in the death of the animal. Neither crotoxin nor crotamine produces such convulsions (Brazil et al., 1969). The chemical nature of convulxin has not yet been identified.

2 PHARMACOLOGY OF CROTOXIN AND RELATED COMPOUNDS

The venom of Brazilian rattlesnakes contains complex mixtures of pharmacologically active proteins and therefore shows complex biological activities. The venom is neurotoxic and also has blood clotting activity and hemolytic action, and causes an immediate and transitory fall in blood pressure (Brazil et al., 1967). Although crude venom causes respiratory disturbance, crotoxin does not produce respiratory failure (Brazil et al., 1966b). The neurotoxin that causes respiratory disturbance has been isolated and is different from crotoxin and crotamine.

2.1 Crotoxin

There is a great difference in sensitivity to crotoxin, depending on the species of host animal. Pigeons are extremely sensitive, whereas rats are very resistant to its lethal action. The LD_{50} is 0.0022 $\mu g\ g^{-1}$ by i.v. injection in pigeons, and 0.756 $\mu g\ g^{-1}$ by i.p. injection in rats (Brazil et al., 1966a).

The neurotoxic action of crotoxin is of peripheral origin and is very similar to the effect elicited by curare. Crotoxin evokes paralysis in pigeons, guinea pigs, mice, rabbits, and rats. In nearly all of them paralysis and flaccidity of the skeletal muscles are typical symptoms. Artificial respiration maintains the life of the animals after spontaneous respiration has ceased. Consciousness and sensitivity seem to be preserved in animals presenting crotoxin paralysis. In addition to paralysis, defecation, vomiting, salivation, albuminuria, hemoglobinuira, oliguria, and in some instances opacity and exfoliation of the cornea were observed in dogs. When administered by intraventricular route in cats, crotoxin elicited convulsions instead of paralysis (Brazil et al., 1966a). Crotoxin interrupts neuromuscular transmission by producing a nondepolarization type of blockade. A decrease in the sensitivity of the end plate to the depolarizing action of acetylcholine is the unique or the main cause of crotoxin neuromuscular blockade. No ganglionic blockade is caused by crotoxin.

In human envenomation by Brazilian rattlesnakes, renal lesions could be observed in autopsy and biopsy tissues (Amorin and Mello, 1952, 1954). Similar lesions could be produced experimentally by injecting the crude venom (Amorim et al., 1960). Hadler and Brazil (1966) investigated the effect of crotoxin on dog kidneys and observed early lesions within the first 4 days. In these early lesions, the renal tubules were less attacked than the glomeruli, which showed capillary congestion, thickness of the basement membrane, deposition of PAS positive material between the capillary loops, and nuclear pycnosis of some glomerular cells. The degenerative lesions of the proximal convolutions reached only some epithelial cells in very restricted areas. In late lesions, tubular damage predominated, the segment of the tubules most affected being the proximal one. The tubules showed intense microvacuolar degeneration, as well as nuclear pycnosis and necrosis of many epithelial cells.

Thus it has been amply demonstrated that crotoxin has a wide spectrum of pharmacologic action. Since crotoxin is a complex compound consisting of many components, the fact that it has a variety of biological activities is not surprising.

Brazil et al. (1973) investigated the action of crotoxin and its components phospholipase A_2 and crotapotin on the neuromuscular junction of the frog. Crotapotin itself had no effect on miniature end-plate potential (m.e.p.p.) frequency or amplitude. End-plate potentials (e.p.p.) recorded in curare—Ringer's solution were also unaffected by crotapotin. Quantal release of transmitter, calculated from m.e.p.p. and e.p.p. recordings, showed a reduction to block after the addition of phospholipase A_2 and crotapotin, confirming the presynaptic action. Basic phospholipase A_2 is the essential component for the action of crotoxin.

Both crotoxin and crotactin completely blocked contraction elicited by nervous stimulation, though spontaneous contractions usually preceded complete blockade in sartorius muscle—sciatic nerve preparation from *Rana temporaria* (Brazil and Excell, 1970). Neither crotoxin nor crotactin altered the muscle membrane potential.

At one time crotactin was thought to be a pure protein and free from phospholipase A_2. It has been demonstrated, however, that crotactin is a mixture of crotapotin and small amounts of phospholipase A_2 (Rübsamen et al., 1971).

For the study of *in vivo* distribution of crotoxin, [^{131}I] crotoxin was prepared (Lomba et al., 1966). The iodinated crotoxin does not induce change in the biochemical and biological properties of the toxin.

The neurotoxicity of the basic component of crotoxin can be enhanced by the addition of an acidic protein, volvatoxin, from the mushroom. Other acidic proteins have no synergistic effect (Jeng and Fraenkel—Conrat, 1976).

2.2 Crotamine

Crotamine has a much lower toxicity than does the crude venom, crotoxin, or convulxin. By intravenous injection, the LD_{50} values in mice for crude venom, crotoxin, crotamine, and convulxin are 0.169, 0.082, 1.5, and 0.052 $\mu g\ g^{-1}$, respectively (Brazil, 1972).

There is a difference in the pharmacological properties of the venoms of two Brazilian rattlesnakes; one contains crotamine, whereas the other does not (Cheymol et al., 1966, 1969). The main action of the venom that contains crotamine is a peripheral contracture, affecting muscular fibers by direct depolarization. This effect is a tachyphylactic one. The venom without crotamine shows a neuromuscular blocking effect with a peripheral, slowly induced, irreversible action. This effect is principally due to a direct action on muscular fibers and also, at large doses, to a blockage of the specific receptors on the postsynaptic membrane. Since snakes with different venom compositions have identical morphologies, it is necessary to use polyvalent antivenins.

There is a difference between the action of crotamin and that of the crude venom. In the rat, crotamine action is on the muscle fiber membrane: a change in Ca or Na ion permeability produced contracture, which induced a K ion efflux, responsible for the later phases of the crotamine action. A long-lasting interaction with Ca ion sites explains the appearance of spontaneous contractures and tachyphylaxis (Cheymol et al., 1971a). In phrenic nerve—diaphragms, crotamine and crude venom show identical effects. Immediate contracture, followed by other spontaneous and irregular contractures and tachyphylaxis, occurs on the diaphragm of the rat (Cheymol et al., 1971b). This effect is inhibited by tetrodotoxin, Ca(II), Mg(II), and K(I). Small doses of crotamine sensitized the rectus abdominis muscle of the frog to contracture by K(I), and large doses produced irregular contractions due to instability of the muscle membrane. This action was inhibited by Ca(II) and Mg(II).

3 OTHER RATTLESNAKE TOXINS

Snakes of the Crotalidae family, especially the rattlesnake (*Crotalus* spp. and *Sistrurus* spp.), copperhead, cottonmouth, and water moccasin (*Agkistrodon* spp.), are important snakes commonly found in North America. It has been estimated that, of the 6000 individuals bitten each year in the United States, 16 cases are fatal (Parrish et al., 1965). Unfortunately, no reliable snakebite statistics are available for the country as a whole because no one has accurately assembled all of the cases. However, a good record was kept by the Venomous Snakebite Committee of the Florida State Board of Health, which reported 382 verified and complete records of snakebites (Andrew et al., 1968). Of this number, 168 were inflicted by pigmy rattlesnakes (*Sistrurus miliarius*) and 71 by larger rattlesnakes. Cottonmouth moccasins (*Agkistrodon piscivorus*) accounted for 69 bites, copperheads (*A. contortrix*) for 4, coral snakes (*Micrurus* sp.) for 14, and unidentified venomous snakes for 58. Of the 18 species of rattlesnakes in the United States, envenomation by *Crotalus atrox* (western diamondback rattlesnake) is responsible for the majority of deaths. The main reason there are fewer deaths in the United States than in Asia is the widespread use of antivenin in this country.

The chemistry of North American rattlesnakes will be reviewed first.

3.1 Chemistry

Unlike the venoms of Elapidae and Hydrophiidae, Crotalidae venoms are rich in neutral and acidic proteins. They also contain basic toxic proteins, but the content is rather low.

Highly basic toxic proteins of low molecular weight have been isolated and purified from the venoms of several rattlesnake species (Bonilla, 1969; Bonilla and Fiero, 1971). A basic neurotoxin isolated from the venom of *Crotalus adamanteus* has only 32 amino acid residues (Bonilla et al., 1971). Its amino acid composition is shown in Table 2, together with the constituents of other purified rattlesnake toxins. Chromatographic separation of the components in crude venom from *C. viridus helleri* produced two toxic fractions, one of which was cationic at pH 5.6 whereas the other was anionic (Dubnoff and Russell, 1971).

A major lethal toxin with cardiotoxic properties was isolated from the venom of *C. scutulatus* and designated as Mojave toxin (Bieber et al., 1975). Unlike basic neurotoxins or cytotoxins isolated from venoms of cobras, kraits, and sea snakes, Mojave toxin is an acidic protein with an isoelectric point of 4.7. Mojave toxin is also different from crotoxin and rattlesnake basic toxin, isolated by Bonilla and his co-workers. The molecular weight is 22,000 daltons, and the toxin can be dissociated into a single protein with a molecular weight of 12,000 in SDS gel electrophoresis of reduced Mojave toxin. Separation in isoelectric focusing acrylamide gel in the presence of $8M$ urea also showed a single protein band, suggesting that the toxin is composed of subunits. The content of Mojave toxin in the whole venom is high and is estimated at 10%. The high content of Mojave toxin is probably responsible for the high toxicity (LD_{50} 0.18 $\mu g\ g^{-1}$ in mice by i.v. injection) of *C. scutulatus* venom as compared to other rattlesnake venoms. Bieber et

Table 2 Amino Acid Compositions of Lethal Toxins Isolated from the Venom of North American Rattlesnakes

Genus Species	Crotalus adamanteus	Crotalus scutulatus
Origin	U.S.A.	U.S.A.
Name	Basic protein	Mojave toxin
Amino Acid		
Lysine	5	4
Histidine	2	1
Arginine	2	3
Aspartic acid	2	7
Threonine	1	3
Serine	2	3
Glutamic acid	1	6
Proline	2	2
Glycine	4	6
Alanine	1	3
Valine	1	1
Methionine	1	1
Isoleucine	1	3
Leucine	1	3
Tyrosine	1	5
Phenylalanine	2	3
Half-cystine	3	8
Tryptophan	0	
Total residue	32	
References	Bonilla et al, 1971	Bieber et al, 1975

al. (1975) suggested that all rattlesnakes of North and Central America contain Mojave toxin as the common toxin. If this is proved to be true, Mojave toxin should be called rattler toxin.

Because of the unique properties of Mojave toxin as compared to other known snake toxins, Tu et al. (1976b) undertook a study of its conformation to determine whether Mojave toxin also differs structurally from the neurotoxins isolated from sea snake venoms. As the conformations of Elapidae toxins have been well studied by fluorescence, circular dichroic, and Raman spectroscopy, a direct comparison can be made.

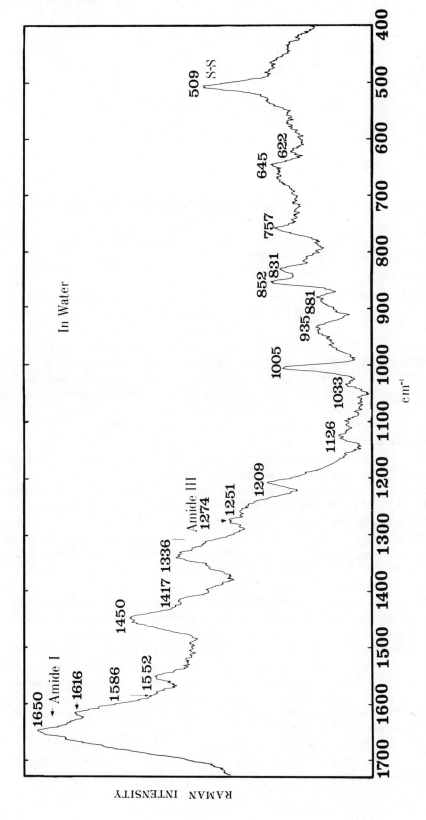

Figure 14.4 Raman spectrum of Mojave toxin isolated from *Crotalus scutulatus* venom in water. (Reproduced from Tu et al., 1976b, by permission of the copyright owner, Academic Press.)

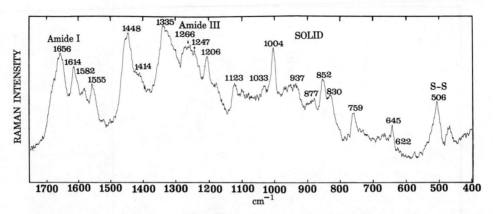

Figure 14.5 Raman spectrum of Mojave toxin in solid state. (Reproduced from Tu et al., 1976, by permission of the copyright owner, Academic Press.)

Raman spectra of the Mojave toxin in aqueous solution and in the solid state are shown in Figs. 4 and 5. The Raman spectra reveal that the Mojave toxin contains a predominantly α-helical secondary structure and that the tyrosyl residues, on the basis of the Raman frequencies and intensities, are exposed to the solvent. These features of the Mojave toxin distinguish it structurally from the neurotoxins of sea snake venoms, which have an antiparallel β configuration (Yu et al., 1975; Tu et al., 1976a). However, like the sea snake venom toxins, Mojave toxin contains four disulfide bridges and is not greatly altered in structure by removal of the aqueous solvent.

3.2 Biological Effect

Rattlesnake venoms produce complicated symptoms since they contain a variety of proteins. Although *Crotalus* venom is less toxic than cobra, krait, coral snake, or sea snake venom, a rattlesnake bite can be fatal because of the injection of a relatively large dose to the victim. Examples of LD_{50} values for rattlesnake and other venoms are listed in Table 3. No attempt has been made to include all venoms in the table.

To simulate actual poisoning, dogs were deliberately forced to be bitten by rattlesnakes and the pharmacologic effects were investigated by Vick (1971). Death occurred within 8 to 24 hr. There was a significant increase in both heart rate and blood pressure immediately after the bites. The symptoms were quite different from those observed when venoms were injected intravenously. In the latter case, a dramatic decrease in heart rate and blood pressure occurred. Many animals showed edema, necrosis, and generalized tissue breakdown. Tissue damage is discussed in detail in Chapter 23.

Vick concluded that death was produced primarily by respiratory failure. He also observed that lyophilized venom produced very similar effects to those resulting from actual biting. The drying process does not alter the toxic effect of rattlesnake venoms.

In contrast to the venoms of Elapidae and Hydrophiidae, not all Crotalidae venoms are neurotoxic. The notable exception is *Crotalus durissus terrificus* venom, which was discussed in Section 1 of this chapter.

When *C. adamanteus* venom was injected directly into the brain of a cat (Russell and Bohr, 1962), it failed to produce the typical cardiovascular responses observed after i.v. injection, indicating that the cerebrospinal fluid–blood–brain barrier relationship was not altered to any great extent by this venom. The *Crotalus* venoms induce behavior changes, however, as well as motor and parasympathetic dysfunction.

Table 3 Toxicities of Crotalidae Venoms and Their Components in Mice

Venoms or Components	LD50 (µg/g)	Route of Injection	Reference
Agkistrodon acutus	9.2	s.c.	Lee et al., 1962
	0.38	i.v.	Friederich and Tu, 1971
A. bilineatus	2.4	i.v.	a
A. contortrix	10.9	i.v.	a
	10.5	i.p.	a
	25.6	s.c.	b
A. contortrix laticinctus	20.0	i.m.	Schmidt et al., 1972
A. halys blomhoffii	1.2	—	Omori et al., 1964
	20.0	s.c.	b
A. piscivorus	5.1	i.p.	a
	25.8	s.c.	b
	4.0	i.v.	a
A. rhodostoma	6.2	i.v.	a
	4.7	i.v.	b
	5.0	i.p.	b
	23.4	s.c.	b
Bothrops alternatus	1.96	i.v.	b
	15.8	s.c.	b
B. atrox	4.27	i.v.	a
	3.80	i.p.	a
	1.4	i.v.	Tu and Homma, 1970
	22.0	s.c.	b
B. jararaca	1.1	i.v.	b
	7.0	s.c.	b
B. jararacussu	0.46	i.v.	b
	13.0	s.c.	b
B. nasuta	4.6	i.v.	Tu and Homma, 1970
B. neuwiedi	2.3	i.v.	b
	14.0	s.c.	b
B. nummifer	2.4	i.v.	Tu and Homma, 1970
B. picadoi	1.6	i.v.	Tu and Homma, 1970
B. schlegelii	1.6	i.v.	Tu and Homma, 1970
	33.2	s.c.	b
Crotalus adamanteus	2.4	i.v.	Friederich and Tu, 1971
	14.6	s.c.	b
	1.95-2.06	i.v.	Russell and Eventov, 1964
	1.89	i.p.	a
	1.68	i.v.	a
	2.2	i.v.	Bonilla et al., 1971
basic protein	5.2	i.v.	Bonilla et al., 1971

Table 3 *Continued*

Venoms or Components		LD50 (μg/g)	Route of Injection	Reference
C. atrox		3.6	i.v.	Friederich and Tu, 1971
		3.7	i.p.	a
		4.2	i.v.	a
C. basiliscus		2.8	s.c.	b
C. cerastes		4.0	i.p.	a
C. durissus terrificus		0.169	i.v.	Brazil, 1972
		0.35	i.v.	Friederich and Tu, 1971
		0.3	i.p.	a
	crotoxin	0.108	i.v.	Rübsamen et al., 1971
		0.082	i.v.	Brazil, 1972
		0.5	s.c.	Rübsamen et al., 1971
	crotapotin (acidic protein)	>50	i.v.	Rübsamen et al., 1971
	crotactin	0.06-0.07	—	Habermann and Neumann, 1956
	crotamin	1.5	i.v.	Brazil, 1972
	convulxin	0.52	i.v.	Brazil, 1972
	basic phospholipase A_2	0.54	i.v.	Rübsamen et al., 1971
		>100	s.c.	Rübsamen et al., 1971
C. horridus horridus		2.6	i.v.	Friederich and Tu, 1971
		2.6	i.v.	a
		2.9	i.p.	a
		3.1	s.c.	Tu et al., 1970
C. ruber ruber		3.7	i.v.	a
		6.7	i.p.	a
C. scutulatus		0.18	i.v.	Bieber et al., 1975
		0.21	i.v.	a
		0.23	i.p.	a
	Mohave toxin	0.039	i.v.	Bieber et al., 1975
		0.26	i.p.	Hendon, 1975
C. viridis		1.61	i.v.	a
		2.25	i.p.	a
		1.01	i.v.	Friederich and Tu, 1971
C. viridis helleri		1.29	i.v.	a
		1.60	i.p.	a
		1.27	i.v.	Schaeffer et al., 1973
		1.00 (cats)	i.v.	Schaeffer et al., 1973
		0.05 (dogs)	i.v.	Schaeffer et al., 1973
C. viridis lutosus		2.20	i.p.	a
Lachesis mutus		5.93	i.p.	a
		4.51	i.v.	b
		6.41	i.p.	b
		36.90	s.c.	b

3 Other Rattlesnake Toxins

Table 3 *Continued*

Venoms or Components	LD50 (µg/g)	Route of Injection	Reference
Sistrurus milarius barbouri	12.6	i.v.	Friederich and Tu, 1971
	2.8	i.v.	b
	6.8	i.p.	b
	24.3	s.c.	b
Trimeresurus albolabris	0.37	i.v.	Minton, 1968
	12.75	s.c.	Minton, 1968
T. flavoviridis	30.8-149.8	i.v.	Tadokoro et al., 1964
	4.25	i.v.	b
	5.07	i.p.	b
	27.3	i.p.	b
T. gramineus	4.0	s.c.	Lee et al., 1962
T. mucrosquamatus	8.6	s.c.	Lee et al., 1962
T. okinavensis	3.8	i.v.	Minton, 1968
	15.0	i.p.	Minton, 1968
T. wagleri	0.75	i.v.	Minton, 1968
	3.58	i.p.	Minton, 1968
	4.63	s.c.	Minton, 1968

a. Data obtained from Poisonous Snakes of the World, A Manual for Use by U.S. Amphibious Forces, published by U.S. Government Printing Office.
b. Data obtained from the book, Venom Diseases, by S.A. Minton, 1974.

The effect of venom on consciousness in monkeys was investigated by Vick and Lipp (1970). The electroencephalographic pattern was quite similar after the administration of venoms of *Crotalus adamanteus, C. atrox*, and *Naja naja*. There was an initial dyssynchronization, followed by the appearance of high-voltage, slow waves and eventual flattening of the record before death. This type of activity is indicative of cerebral depression. There was a decrease in the level of consciousness of the monkeys, along with cardiac arrhythmias and progressive respiratory depression, which appeared simultaneously with the high-voltage, slow-wave pattern. Vick and Lipp believed that respiratory depression was the cause of the electroencephalographic change.

The effect on consciousness has also been observed for the venoms of other Crotalidae, such as *Trimeresurus mucrosquamatus, T. gramineus*, and *Agkistrodon acutus* (Oh, 1936).

Crotalus viridis helleri and other rattlesnake venoms induce swelling, erythema, local hemorrhage, and immediate hypotensive crisis (Russell et al., 1962; Schaeffer et al., 1973; Whigham et al., 1973; Pattabhiraman et al., 1974). They also cause reduction in total blood volume index, plasma volume index, and red cell mass index (Carlson et al., 1975). It is believed that perfusion failure following rattlesnake envenomation is associated with hypovolemia due to increases in vascular permeability and hemorrhage.

Unlike the toxins of cobras and sea snakes, rattlesnake toxin is not stable. After incubation for 30 min at 45°C, Mojave toxin isolated from *C. scutulatus* venom lost some

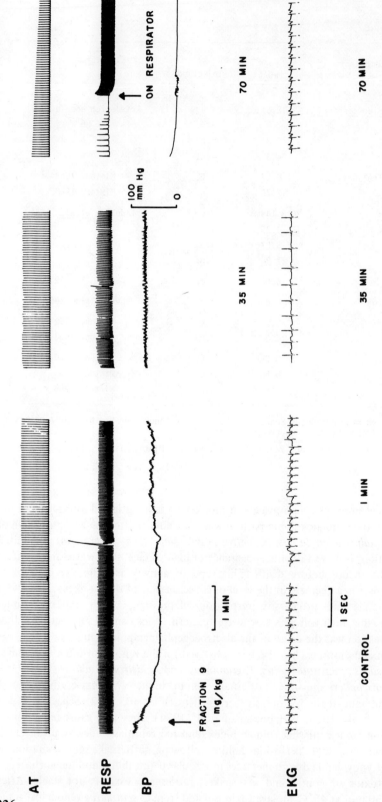

Figure 14.6 Pharmacological effect of Mojave toxin in rabbits AT: twitch response of tibialis anterior muscle to stimulation of sciatic nerve, RESP: change of respiration, BP: pressor response elicited on systemic arterial pressure, EKG: change of electrocardiogram. (Reproduced from Bieber et al., 1975, by permission of the copyright owner, Elsevier Scientific Publishing Company.)

toxicity; it lost all its toxicity at 60°C (Bieber et al., 1975). The toxin is completely stable between pH 5 and 9 but is quite unstable in the acid pH range.

Since Mojave toxin has been well characterized chemically and some of its pharmacologic properties are known (Bieber et al., 1975), I used it as an example of rattlesnake venom in this book. Intravenous injection of Mojave toxin caused a precipitous fall in arterial pressure, which began immediately and reached its lowest point within 2 min (Fig. 6). The electrocardiograms showed an occasionally increased T-wave voltage, superposition of P and T waves, and an increase in amplitude of the S wave during the initial period of arterial pressure decline 30 to 40 min after injection. The changes in the electrocardiogram were a decrease in amplitude of the P wave, an increase in the R/S ratio, an increase in the voltage of the QRS complex, and bradycardia. Between 70 and 120 min after injection, the blood pressure gradually declined and respiration was markedly impaired; the electrocardiograms showed an increase in amplitude of the S wave, a notched T wave with increased voltage, and bradycardia. Finally, cardiovascular collapse and respiratory failure occurred simultaneously. When artificial respiration was carried out immediately after the cessation of breathing, the blood pressure did not rise again. Before marked respiratory depression was observed, the contraction of the neurally excited tibialis anterior muscle was not affected by Mojave toxin. The twitch response of the muscle, however, gradually decreased after respiration was severely depressed. Both the phrenic nerve and the diaphragm muscle responded well to electrical stimulation immediately after death occurred.

The effects of crude venom of *C. scutulatus* on the circulatory system and skeletal muscle were different from those of Mojave toxin. The i.v. administration of crude venom induced a fall in the blood pressure, followed by restoration of the preinjection blood pressure value within 60 min. An increase in the pulse pressure was observed for a few minutes after injection. Apenea of 5- to 10-sec duration appeared immediately after injection. The fairly constant findings in the electrocardiogram were increased T-wave voltage, followed by a decrease of the voltage of the QRS complex and very tall, slender peaked T waves. No change in the twitch response of the tibialis anterior muscle was observed. From this study of the mode of action, it is evident that Mojave toxin has remarkable cardiotoxic effects.

Mojave toxin decreases the uptake of Ca(II) as well as increases the efflux of Ca(II) from the sarcoplasmic reticulum (Cate and Bieber, 1976).

Since rattlesnake venoms are composed mainly of proteins, they are antigenic. Antibodies to *Crotalus atrox* venom disappear rapidly unless frequent immunization is made (Glenn et al., 1970).

4 OTHER CROTALIDAE VENOMS

So far in this chapter, the chemistry and the biological effects of purified rattlesnake toxins and of crude venoms have been discussed. Many varieties of pit vipers (Crotalidae) are related to rattlesnakes. Unfortunately, however, no lethal toxins have been isolated from these other Crotalidae, and thus we know little about the chemical properties and pharmacologic effects of the pure toxins. Most of the work in this area has been done using crude venoms.

Identification of the main lethal toxin is difficult in some cases. For instance, *Agkistrodon piscivorus piscivorus* venom yielded 16 fractions by column chromatography

(Clark and Higginbotham, 1971). Several fractions had lethal action, but none of them showed higher toxicity than the original venom. It is possible than some venoms do not have major lethal toxins. Instead, the toxicity of the whole venom may be due to the synergistic action of several components. This important question has not yet been answered completely.

Agkistrodon rhodostoma from Malaya was separated into 15 components in an isoelectric focusing column. On a weight basis, the lethal fraction accounted for less than 5% of the total, but the toxicity increased only three times as compared to the original venom. If all the lethal action resided in the toxic fraction, the toxicity of the fractionated components should have increased 20 times. A lower toxicity of the lethal fraction than the theoretical level suggested that other components also contributed to the toxic effect in a synergistic manner (Toom et al., 1969).

Venoms of Crotalidae are much more heat labile than Elapidae venoms. Practically all toxicities were lost when venoms of *Trimeresus gramineus, T. mucrosquamatus*, and *Agkistrodon acutus* were incubated at 70°C for 30 min. Under the same conditions *Naja naja atra* venom retained full toxicity, and *Bungarus multicinctus* venom retained 77% potency (Lee et al., 1962).

Crotalidae envenomation produces strong local tissue damage, which is discussed in detail in Chapter 23. In severe poisoning, the victim dies of complicated systemic effects. The exact cause of death due to Crotalidae poisoning is not known.

Like many other snake venoms, venoms of *Trimeresurus mucrosquamatus, T. gramineus*, and *Agkistrodon acutus* produce immediate hypotensive crisis (Kyu, 1933a, b, c).

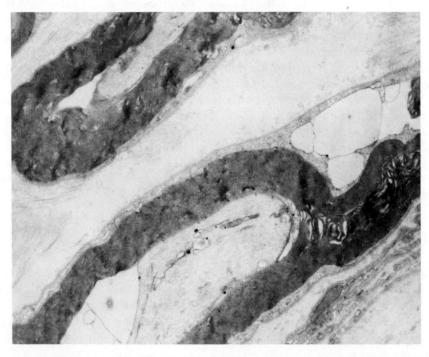

Figure 14.7 Effect of *Agkistrodon contortrix laticinctus* venom on the axon. There is a loss of collagen in the endoneurial space. (Reproduced from Schmidt et al., 1972, by permission of the copyright owner, Springer-Verlag.)

References

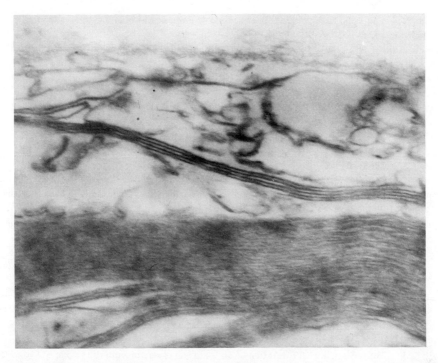

Figure 14.8 Effect of *Agkistrodon contortrix laticinctus* venom on the axon. Note severe vacuolation of a Schwann cell and myelin splitting, which is considered a "demyelinative" effect of the venom. (Reproduced from Schmidt et al., 1972, by permission of the copyright owner, Springer-Verlag.)

Okonogi et al. (1960) studied the effect of *Trimeresurus flavoviridis* venom on the central nervous system and found degeneration of the nerve cells of the cerebral cortex. There was vacuolation of nerve cell nuclei.

Venom of *Agkistrodon piscivorus* caused vacuolation of the cytoplasm of Schwann cells of giant squid axons (Martin and Rosenberg, 1968). Since phospholipase A_2 produced similar vacuolation, probably phospholipase A_2 was responsible for this action. In our study, sciatic nerves of Swiss white mice were used (Schmidt et al., 1972). We found that the sciatic nerve was quite resistant to venom action. There was no change in the ultrastructure of the axon from i.m. injection of *A. contortrix laticinctus* venom. However, when the nerve was incubated with venom solution, various neuropathies were observed. Schwann cells were markedly vacuolated, and this effect became more pronounced with increased concentration and time of incubation. Minor separation and splitting of the myelin occurred, and occasionally demyelinated change was noted (Figs. 7 and 8). The venom did not affect the basal lamina.

There is a common antigen among the venoms of *Trimeresurus flavoviridis, T. okinavensis, T. gramineus*, and *Agkistrodon acutus* (Iizuka et al., 1960), but not between these four venoms and *Naja naja atra* venom.

REFERENCES

Amorim, M. F. and Mello, R. F. (1952). Nefrose de nefron intermediá rio no envenenamento crotalice humano: Estudo anatomopatologico, *Mem. Inst. Butantan*, **23**, 281.

Amorim, M. F. and Mello, R. F. (1954). Intermediate nephron nephrosis from snake poisoning in man: Histopathologic study, *Am. J. Pathol.,* **30,** 479.

Amorim, M. F., Mello, R. F., and Saliba, F. (1960). Intermediate nephron nephrosis experimentally induced with *Crotalus d. terrificus* venom in the dog, *Rev. Braz. Biol.,* **20,** 359.

Andrew, C. E., Dees, J. E., and Edwards, R. O. (1968). Venomous snake-bite in Florida, *J. Fla. Med. Assoc.,* **55,** 308.

Bertke, E. M., Watt, D. D., and Tu, T. (1966). Electrophoretic patterns of venoms from species of Crotalidae and Elapidae snakes, *Toxicon,* **4,** 73.

Bieber, A. L., Tu, T., and Tu, A. T. (1975). Studies of an acidic cardiotoxin isolated from the venom of Mojave rattlesnake (*Crotalus scutulatus*), *Biochim. Biophys. Acta,* **400,** 178.

Bonilla, C. A. (1969). Rapid isolation of basic proteins and polypeptides from salivary gland secretions by adsorption chromatography on polyacrylamide gel, *Anal. Biochem.,* **32,** 522.

Bonilla, C. A. and Fiero, M. K. (1971). Comparative biochemistry and pharmacology of salivary gland secretions, *J. Chromatogr.,* **56,** 253.

Bonilla, C. A., Fiero, M. K., and Frank, L. P. (1971). "Isolation of a basic protein neurotoxin from *Crotalus adamanteus* venom," in A. De Vries and E. Kochva, Eds., *Toxins of Animal and Plant Origin,* Vol. 1, Gordon and Breach, New York, pp. 343–360.

Bourillet, F. (1970). Action neuromusculaires comparées de la crotamine et de la vératrine, *Ann. Pharm. Fr.,* **28,** 535.

Brazil, O. V. (1966). Pharmacology of crystalline crotoxin. II. Neuromuscular blocking action, *Mem. Inst. Butantan,* **33,** 981.

Brazil, O. V. (1972). Neurotoxins from the South American rattlesnake venom, *J. Formosan Med. Assoc.,* **71,** 394.

Brazil, O. V. and Excell, B. J. (1970). Action of crotoxin and crotactin from the venom of *Crotalus durissus terrificus* (South American rattlesnake) on the frog neuromuscular junction, *J. Physiol.,* **212,** 34P.

Brazil, O. V., Franceschi, J. P., and Waisbich, E. (1966a). Pharmacology of crystalline crotoxin. I. Toxicity, *Mem. Inst. Butantan,* **33,** 973.

Brazil, O. V., Fariña, R., Yoshida, L., and De Oliveira, V. A. (1966b). Pharmacology of crystalline crotoxin. III. Cardiovascular and respiratory effects of crotoxin and *Crotalus durissus terrificus* venom, *Mem. Inst. Butantan,* **33,** 993.

Brazil, O. V., Franceschi, J. P. and Waisbich, E. (1967). Neurotoxic factor in *Crotalus durissus terrificus* venom different from crotoxin and crotamine, *Cienc. Cult.,* **19,** 658.

Brazil, O. V., Laszlo, G. M., and Eugenio, O. A. G. B. (1969). Origem da paralisia respiratoria dausada pela crotoxina, *Congr. Lation Am. Cienc. Fisiol.,* Abstract IX, Brazil.

Brazil, O. V., Excell, B. J., and Sanatana de Sa, S. (1973). The importance of phospholipase A in the action of the crotoxin complex at the frog neuromuscular junction, *J. Physiol.,* **234,** 63.

Breithaupt, H., Rübsamen, K., and Habermann, E. (1974). Biochemistry and pharmacology of the crotoxin complex: Biochemical analysis of crotapotin and the basic *Crotalus* phospholipase A, *Eur. J. Biochem.,* **49,** 333.

Breithaupt, H., Omori-Satoh, T., and Lang, J. (1975). Isolation and characterization of three phospholipases A from the crotoxin complex, *Biochim. Biophys. Acta,* **403,** 355.

Brown, J. H. (1973). *Toxiocology and Pharmacology of Venoms from Poisonous Snakes,* Charles C Thomas, Springfield, Ill.

Carlson, R. W., Schaeffer, R. C., Jr., Whigham, H., Michaels, S., Russell, F. E., and Weil, M. H. (1975). Rattlesnake venom shock in the rat: Development of a method, *Am. J. Physiol.,* **229,** 1668.

Cate, R. L. and Bieber, A. L. (1976) Effects of Mojave toxin on rat skeletal muscle sarcoplasmic reticulum, *Biochem. Biophys. Res. Comm.,* 72, 295.

Cheymol, J., Bourillet, F., and Roch, M. (1966). Action neuromusculaire des venins de quelques Crotalidae, Elapidae, et Hydrophiidae, *Mem. Inst. Butantan,* **33,** 541.

Cheymol, J., Bourillet, F., Roch-Arveiller, M., and Toan, T. (1969). Effects neuromusculaires des venins des deux variétés de *Crotalus durissus terrificus, Arch. Int. Pharmacodyn.,* **179,** 40.

Cheymol, J., Bourillet, F., and Roch-Arveiller, M. (1971a). Action neuromusculaire comparée de venin

References

de *Naja nigricollis* et de la neurotoxine (α-najatoxine) qui en est extraite, *Arch. Int. Pharmacodyn.,* **192,** 26.

Cheymol, J., Gonçalves, J. M., Bourillet, F., and Roch-Arveiller, M. (1971b). Action neuromusculaire comparée de la crotamine et du venin de *Crotalus durissus terrificus* var. *Crotaminicus.* V. Sur preparations isolees, *Toxicon,* **9,** 287.

Clark, J. M. and Higginbotham, R. D. (1971). Cottonmouth moccasin venom: Fractionation of toxic and allergenic components and interaction with tissue mast cells, *Tex. Rept. Biol. Med.,* **29,** 181.

Dubnoff, J. W. and Russell, F. E. (1971). "Separation and purification of *Crotalus* venom fractions," in A. De Vries and E. Kochva, Eds., *Toxins of Animal and Plant Origin,* Vol. 1, Gordon and Breach, New York, pp. 361–368.

Feeney, R. E., MacDonnel, L. R., and Fraenkel-Conrat, H. (1954). Effects of crotoxin (lecithinase A) on egg yolk and yolk constituents, *Arch. Biochem. Biophys.,* **48,** 130.

Fischer, F. G. and Dörfel, H. (1954). Die Aminosäuren-Zusammensetzung von Crotoxin, *Z. Physiol. Chem.,* **297,** 278.

Fraenkel-Conrat, H. and Singer, B. (1956). Fractionation and composition of crotoxin, *Arch. Biochem. Biophys.,* **60,** 64.

Friederich, C. and Tu, A. T. (1971). Role of metals in snake venoms for hemorrhagic, esterase and proteolytic activities, *Biochem. Pharmacol.,* **20,** 1549.

Giglio, J. R. (1975). Analytical studies on crotamine hydrochloride, *Anal. Biochem.,* **69,** 207.

Glenn, W. G., Becker, R. E., and Buysere, M. D. (1970). Goat antibody response to rattlesnake venom during and after immunization, *Am. J. Vet. Res.,* **31,** 1237.

Gonçalves, J. M. and Giglio, J. R. (1964). Amino acid composition and terminal group analysis of crotamine, *Int. Congr. Biochem. New York,* Abstract II-134, p. 170.

Gralen, N. and Svedberg, F. (1938). The molecular weight of crotoxin, *Biochem. J.,* **32,** 1375.

Habermann, E. (1957a). Pharmacology of the venom of the Brazilian rattlesnake, *Naunyn-Schmiedebergs Arch. Exp. Pathol. Pharmakol.,* **232,** 244.

Habermann, E. (1957b). Gewinnung und Eigenschaften von Crotactin, phospholipase A, Crotamin und "Toxin III" aus dem Gift der brasilianischen Klapperschlange, *Biochem. Z.,* **329,** 405.

Habermann, E. and Neumann, W. P. (1956). Crotactin, ein neues pharmakologisches Wirkprinzip aus dem Gift von *Crotalus terrificus, Naunyn-Schmiedebergs Arch. Exp. Pathol. Pharmakol.,* **228,** 217.

Habermann, E. and Rübsamen, K. (1971). "Biochemical and pharmacological analysis of the so-called crotoxin," in A. De Vries and E. Kochva, Eds., *Toxins of Animal and Plant Origin,* Vol. 1, Gordon and Breach, New York, pp. 333–342.

Halder, W. A. and Brazil, O. V. (1966). Pharmacology of crotoxin. IV. Nephrotoxicity, *Mem. Inst. Butantan,* **33,** 1001.

Hendon, R. A. (1975). Preliminary studies on the neurotoxin in the venom of *Crotalus scutulatus* (Mojave rattlesnake), *Toxicon,* **13,** 477.

Hendon, R. A. and Fraenkel-Conrat, H. (1971). Biological roles of the two components of crotoxin, *Proc. Natl. Acad. Sci. USA,* **68,** 1560.

Horst, J., Hendon, R. A., and Fraenkel-Conrat, H. (1972). The active components of crotoxin, *Biochem. Biophys. Res. Commun.,* **46,** 1042.

Iizuka, K., Murata, Y., and Satake, M. (1960). Studies on snake venom. X. On the antigen–antibody reaction of Formosan and Japanese snake venoms with commercial antiserum, *Yakugaku Zasshi,* **80,** 1035.

Jeng, T-W. and Fraenkel-Conrat, H. (1976). Activation of crotoxin B by volvatoxin A2, *Biochem. Biophys. Res. Comm.,* **70,** 1324.

Kyu, K. (1933a). Toxikologische untersuchugen über die Gifte der Crotalinae Formosa's. I. Mitteilung. Studien über das Gift von *Trimeresurus mucrosquamatus* Cantor, *J. Formosan Med. Assoc.,* **32,** 69.

Kyu, K. (1933b). Toxikologische untersuchungen über die Gifte der Crotalinae Formosa's II. Mitteilung. Studien über das Gift von *Trimeresurus gramineus* Shaw, *J. Formosan Med. Assoc.,* **32,** 123.

Kyu, K. (1933c). Toxikologische untersuchungen uber die Gifte der Crotalinae Formosa's. III. Mitteilung. Studien über das Gift von *Agkistrodon acutus* Günther, *J. Formosan Med. Assoc.*, **32**, 22.

Laure, C. J. (1975). Die Primärstruktur des Crotamins, *Hoppe-Seylers Z. Physiol. Chem.*, **356**, 213.

Lee, C. Y., Chang, C. C., Su, C., and Chen, Y. W. (1962). The toxicity and thermostability of Formosan snake venoms, *J. Formosan Med. Assoc.*, **61**, 239.

Li, C. H. and Fraenkel-Conrat, H. (1942). Electrophoresis of crotoxin, *J. Am. Chem. Soc.*, **64**, 1586.

Lomba, M. G., Kieffer, J., Warsbich, E., and Brazil, O. V. (1966). Preparation and properties of I^{131}-labeled crotoxin, *Mem. Inst. Butantan*, **33**, 921.

Martin, R. and Rosenberg, P. (1968). Fine structural alterations associated with venom action on squid nerve fibers, *J. Cell. Biol.*, **36**, 341.

Minton, S. A., Jr. (1968). Antigenic relationships of the venom of *Atractaspis microlepidota* to that of other snakes, *Toxicon*, **6**, 59.

Minton, S. A., Jr. (1974). *Venom Diseases,* Charles C Thomas, Springfield, Ill.

Neumann, W. P. and Habermann, E. (1955). Über Crotactin, das Haupttoxin des Giftes der brasilianischen Klapperschlange (*Crotalus terrificus terrificus*), *Biochem. Z.*, **327**, 170.

Nygaard, A. P. and Sumner, J. B. (1953). The effect of lecithinase A on the succinoxidase system, *J. Biol. Chem.*, **200**, 723.

Oh, J. (1936). Über die Wirkung des Giftes gewisser formosanischer Schlangen auf die motorischen und sensiblen peripheren Nerven, *J. Formosan Med. Assoc.*, **35**, 2092.

Okonogi, T., Hoshi, S., Homma, M., Mitsuhashi, S., Maeno, H., and Sawai, Y. (1960). Experimental studies on habu snake venom. III. Experimental histopathological studies on the central nerve system of guinea pigs, *Jap. J. Microbiol.*, **4**, 297.

Omori, T., Iwanaga, S., and Suzuki, T. (1964). The relationship between the hemorrhagic and lethal activities of Japanese mamushi (*Agkistrodon halys blomhoffii*) venom, *Toxicon*, **2**, 1.

Paradies, H. H. and Breithaupt, H. (1975). Subunit structure of crotoxin: Hydrodynamic and shape properties of crotoxin, phospholipase A, and crotapotin, *Biochem. Biophys. Res. Commun.*, **66**, 496.

Parrish, H. M., Silberg, S. L., and Groldner, J. C. (1965). Snakebite: A pediatric problem, *Clin. Pediatr.*, **4**, 237.

Pattabhiraman, T. R., Buffkin, D. C., and Russell, F. E. (1974). Some chemical and pharmacological properties of toxic fractions fron the venom of the southern Pacific rattlesnake, *Crotalus viridis helleri*. II, *Proc. West. Pharmacol. Soc.*, **17**, 227.

Prado-Franceschi, J. and Brazil. O. V. (1969). Convulxina, uma nova neurotoxina da peconha da *Crotalus durissus terrificus, Congr. Latamer. Cienc. Fisiolo.*, Abstract IX, Brazil.

Rübsaman, K., Breithau, H., and Habermann, E. (1971). Biochemistry and pharmacology of crotoxin complex. I. Subfractionation and recombination of crotoxin complex, *Naunyn-Schmiedebergs Arch. Exp. Pathol. Pharmacol.*, **270**, 274.

Russell, F. E. and Bohr, V. C. (1962). Intraventricular injection of venom, *Toxicol. Appl. Pharmacol.*, **4**, 165.

Russell, F. E. and Eventov, R. (1964). Lethality of crude and lyophilized *Crotalus* venom, *Toxicon*, **2**, 81.

Russell, F. E., Buess, F. W., and Strassberg, J. (1962). Cardiovascular response to *Crotalus* venom, *Toxicon*, **1**, 5.

Schaeffer, R. C., Jr., Carlson, R. W., Whigman, H., Russell, F. E., and Weil, M. H. (1973). Some hemodynamic effects of rattlesnake (*Crotalus viridis helleri*) venom, *Proc. West. Pharmacol. Soc.*, **16**, 58.

Schmidt, M. E., Abdelbaki, Y. Z., and Tu, A. T. (1972). Fine structural changes of myelinated nerve associated with copperhead envenomation, *Acta Neuropathol.*, **21**, 68.

Slotta, K. H. (1938). A crotaxina, primeira substância pura dos venenos ofidicos, *Ann. Acad. Brazil. Sci. Rio*, **10**, 195.

Slotta, K. H. and Fraenkel-Conrat, H. (1938a). Schlangengifte. III. Mitteil. Reinigung und Krystallisation des Klapperschlangen-Giftes, *Ber. Deut. Chem. Ges.*, **71**, 1076.

Slotta, K. H. and Fraenkel-Conrat H. L. (1938b). Two active proteins from rattlesnake venoms, *Nature*, **142**, 213.

References

Slotta, C. H. and Fraenkel-Conrat, H. L. (1939). Crotoxin, *Nature,* **144,** 290.

Tadokoro, S., Kurihara, N., Ogawa, H., and Horikawa, K. (1964). Comparison of toxicity between habu-snake venoms collected at different times, *Gunma J. Med. Sci.,* **13,** 301.

Toom, P. M., Squire, P. G., and Tu, A. T. (1969). Characterization of the enzymatic and biological activities of snake venoms by isoelectric focusing, *Biochim. Biophys. Acta,* **181,** 339.

Tu, A. T. and Homma, M. (1970). Toxicologic study of snake venoms from Costa Rica, *Toxicol. Appl. Pharmacol.,* **16,** 73.

Tu, A. T., Homma, M., Hong, B. S., and Terrill, J. B. (1970). Neutralization of rattlesnake venom toxicities by various compounds, *J. Clin. Pharmacol.,* **10,** 323.

Tu, A. T., Jo, B. H., and Yu, N. T. (1976a). Laser Raman spectroscopy of snake venom neurotoxins, *Int. J. Peptide Protein Res.* 8,337.

Tu, A. T., Prescott, B., Chou, C. H., and Thamas, G. J., Jr. (1976b). Structural properties of Mojave toxin of *Crotalus scutulatus* (Mojave rattlesnake) determined by laser Raman spectroscopy, *Biochem. Biophys. Res. Commun.* 68, 1139.

Vick, J. A. (1971). "Symptomatology of experimental and clinical crotalid envenomation," in L. L. Simpson, Ed., *Neuropoisons: Their Pathophysiological Actions,* Vol. 1: *Poisons of Animal Origin,* Plenum, New York. pp. 71–86.

Vick, J. A. and Lipp, J. (1970). Effect of cobra and rattlesnake venoms on the central nervous system of the primate, *Toxicon,* **8,** 33.

Whigham, H., Russell, F. E., and Weil, M. H. (1973). Circulatory and metabolic alterations in rats following intravenous infusion of rattlesnake (*Crotalus viridis*) venom, *Proc. West. Pharmacol. Soc.,* **16,** 223.

Yang, C. C. (1963). Fractionation of snake venom on Sephadex, *J. Formosan Med. Assoc.,* **62,** 611.

Yu, N., Lin, T., and Tu, A. T. (1975). Laser Raman scattering of neurotoxins isolated from the venoms of sea snakes *Lapemis hardwickii* and *Enhydrina schistosa, J. Biol. Chem.,* **250,** 1782.

15 Colubridae Venoms

Colubridae constitute by far the largest family of snakes. It includes about 250 genera with over 1000 species (Ditmars, 1966). Not all of the snakes that belong to this family are poisonous, however; many species of Colubridae possess venom glands with posterior maxillary fangs. Because of the awkward position of the rear fangs, they cause relatively little clinically significant envenomation (Minton, 1974).

Nevertheless, a case report of death due to *Dispholidus typus* was reported (Spies et al., 1962). The patient had nonclotting blood, attributed to afibrinogenemia, and died of hemorrhage 8 days after the bite. Two other cases of bites from *Dispholidus typus* were reported by Jenkins and Russell (1973).

Dispholidus typus yields a very small amount of venom, only 1.6 to 1.8 mg (Robertson and Delpierre, 1969), on milking. No activity toward *N*-acetyltyrosine ethyl ester (a synthetic substrate for chymotrypsin) could be found. Casein activity can be separated from arginine ester-hydrolyzing activity, indicating that two separate enzymes are responsible for these activities.

Very little is known about the venoms of snakes in this family. The LD_{50} values in mice by i.v. injection for two species are as follows: *Dispholidus typus*, 0.071 $\mu g\ g^{-1}$; *Thelotornis kirtlandi*, 1.24 $\mu g\ g^{-1}$. Both venoms showed phospholipase A_2 activity (Christensen, 1968), and proteolytic activity was also detected in the venom of *D. typus* (Delpierre et al., 1971).

The enzymatic activity of venom of another opisthoglyphous snake, *Leptodeira annulata*, was studied by Mebs (1968). This venom showed very strong proteolytic activity with casein as a substrate. The enzyme has a pH optimum of 8.0 to 9.0 and is activated by Ca(II) and Mg(II) but inhibited by EDTA. Soybean trypsin inhibitor and kallikrein trypsin inhibitor do not affect the proteolytic activity. Weak phospholipase A_2 and phosphodiesterase activities were also found in *L. annulata* venom. The enzyme activities that were not detected were 5′-nucleotidase, ATPase, cholinesterase, L-amino acid oxidase, fibrinogen coagulation enzyme, and arginine ester-hydrolyzing enzyme. Meb (1968) concluded that *L. annulata* venom has a simpler composition than other snake venoms. An absence of clotting enzyme in the venoms of Colubridae was also reported by Eagle (1937).

Normally the African snake *Atractaspis microlepidota* is considered a member of the Viperidae family, but because of different biological characteristics the removal of this species from the Viperidae has been suggested (Bourgeois, 1961; Kochva et al., 1967; Underwood, 1966). Minton (1968), however, considered *A. microlepidota* to be a typical Colubridae, and therefore the venom of this snake is discussed in this chapter. There are a considerable number of immunological cross reactions of *A. microlepidota* venom with antisera of other families of snakes. For instance, *A. microlepidota* venom reacted with 10 of 16 Elapidae antisera, giving one to three bands of precipitate, with 1 of 6 Viperidae antisera, and with 2 of 7 Crotalidae antisera. *Atractapsis* venom gave no reaction, however, with antiserum of *Dispholidus typus* (boomslang) or *Enhydrina schistosa* (common sea snake).

Boquet and St. Giron (1973) investigated immunological cross reactions between the antibodies of Elapidae and Viperidae venoms and glands of four Colubridae (*Enhydris bocourti, Homalopsis buccata, Boiga cyanea*, and *Dyspholidus* typus). Since they found precipitation lines, they concluded that there are common antigens between Colubridae salivary glands and venoms from *Naja nigricollis, N. haje, N. melanoleuca, N. naja, Dendroaspis viridis, Echis carinatus, Vipera berus*, and *V. aspis*.

REFERENCES

Boquet, P. and Saint Girons, H. (1972). Etude immunologique des glandes salivaires du vestibule buccal de quelques colubridae opistoglyphes, *Toxicon*, **10**, 635.

Bourgeois, M. (1961). *Atractaspis*—a misfit among the Viperdae? *News Bull. S. Afr. Zool. Soc.*, **3**, 29.

Christensen, P. A. (1968). "The venoms of Central and South African snakes," in W. Bürcherl, E. E. Buckley, and V. Deulofeu, Eds., *Venomous Animals and Their Venoms*, Vol. 1, Academic, New York, pp. 437–461.

Delpierre, G. R., Robertson, S. S. D., and Steyn, K. (1971). "Proteolytic and related enzymes in the venom of African snakes," in A. De Vries and E. Kochava, Eds., *Toxins of Animal and Plant Origins*, Vol. 1, Gordon and Bleach, New York, pp. 483–489.

Ditmars, R. L. (1966). *Snakes of the World*, Macmillan, New York.

Eagle, H. (1937). Coagulation of the blood by snake venoms and its physiologic significance, *J. Exp. Med.*, **65**, 613.

Grasset, E. and Schaafsma, A. W. (1940). Recherches sur les venins des colubridés opisthoglyphes Africans. I. *Dispholidus typus*. I. Envenimation expérimentale: Propriétés toxiques et antigéniques. *Bull. Soc. Pathol. Exot.*, **50**, 33.

Jenkins, M. S. and Russell, F. E. (1973). "Physical therapy for injuries produced by rattlesnakes," in E. Kaiser, Ed., *Tier- und Pflanzengifte*, Wilhelm Goldmann, Verlag, Munich, pp. 195–199.

Kochva, E., Shayer-Wollberg, M., and Sobol, R. (1967). The special pattern of the venom gland in *Atractaspis* and its bearing on the toxonomic status of the genus, *Copeia*, **4**, 763.

Mebs, D. (1968). Analysis of *Leptodeira annulata* venom, *Herpetologica*, **24**, 338.

Minton, S. A., Jr. (1968). Antigenic relationships of the venom of *Atractaspis microlepidota* to that of other snakes, *Toxicon*, **6**, 59.

Minton, S. A., Jr. (1974). *Venom Diseases*, Charles C Thomas, Springfield, Ill.

Robertson, S. S. D. and Delpierre, G. R. (1969). Studies on African snake venoms. IV. Some enzymatic activities in the venom of the boomslang *Dispholidus typus, Toxicon*, **7**, 189.

Spies, S. K., Malherbe, L. F., and Pepler, W. J. (1962). Boomslangbyt met Afibrinogenemie. Beskrywing van'n geval met Nekropsie-Bevindings, *S. Afr. Med. J.*, **36**, 834.

Underwood, G. (1966). *A Contribution to the Classification of Snakes* Trustees of the British Museum (Natural History), London.

16 Distribution of Venoms in Envenomated Animals

1 RADIOACTIVE TRACER STUDY 236
   1.1 *In Vitro* Labeling, 236
   1.2 *In Vivo* Labeling, 238
   1.3 Amount of Venom Injected in Natural Bites, 238

2 NONRADIOACTIVE METHOD 238

   References 239

The question of the quantity and the distribution of venom in the body of a snakebite victim is very important. There are several ways to study this problem. The radioactive tracer method is frequently used. One technique is to label venoms or toxins with a radioactive element *in vitro*. The other method is to label venom with a radioactive element biosynthetically; for instance, a ^{35}S-containing compound is injected into snakes, which eventually synthesize radioactive venom *in vivo* (Lomba et al., 1969). There are also nonradioactive methods for studying venom distribution. One is to identify venoms in different organs, either chemically or immunologically. The other nonradioactive method is to study tissue damage in different organs at different times after envenomation. However, tracer studies, especially the labeled biosynthetic method, probably produce the best results.

1 RADIOACTIVE TRACER STUDY

1.1 *In Vitro* Labeling

Using ^{131}I-labeled *Bungarus multicinctus* venom, Lee and Tseng (1966) observed that the highest radioactivity was found in the following organs, listed in decreasing order of magnitude: kidneys, lungs, spleen, heart, stomach, intestine, diaphragm, skeletal muscle, and brain. Pure toxins and α- and β-bungarotoxins also showed similar results. α-Bungarotoxin was found to localize on the end-plate zone of the mouse diaphragm, whereas β-bungarotoxin did not show such localization.

Similarly *Naja naja atra* and its purified toxins were labeled with ^{131}I (Tseng et al., 1968). Whereas only 30% of the injected cardiotoxin was absorbed within 4 hr, absorption of neurotoxin was about 60% complete within 2 hr. After i.v. injection into rabbits, the plasma level of cardiotoxin declined much more rapidly than that of neurotoxin. Cardiotoxin was taken up by various organs, especially by the kidneys, liver, spleen, and lungs; whereas neurotoxin did not accumulate in any particular organ except the kidney, where a high dose was found. Mice gave a distribution pattern similar to that of rabbits. Radioautography showed that neurotoxin localized in the motor end-plate zone of the mouse diaphragm, whereas cardiotoxin spread widely over the entire diaphragm. Tseng et al. concluded that the venom concentration in the cerebrospinal fluid was too low to account for the rapid respiratory paralysis caused by cobra venom or its neurotoxin.

Neurotoxin can be eliminated from the body rapidly. When Shü et al. (1968) tagged ^{131}I to cobrotoxin of *N. naja atra* venom, about 70% of the radioactivity was excreted in urine within 5 hr after i.v. injection of a sublethal dose. Most of the radioactivity excreted in the urine 20 min postinjection was still in the intact cobrotoxin, as analyzed by gel filtration, paper electrophoresis, and the toxicity test. Four hours after injection, half of the radioactivity had spread into the iodine. Apparently, labeled iodine can be detached rapidly from cobrotoxin, a finding which indicates that radioactivity location should not be directly correlated with cobrotoxin distribution.

Tritium can also be used for labeling venoms. Huang et al. (1972) injected (i.p.) ^{3}H-labeled venoms of *Crotalus atrox, C. adamanteus*, and *Agkistrodon piscivorus* into albino rats to study venom distribution and excretion patterns. Fecal excretion, not urine, was the major route of elimination of venoms. The average excretion of snake venoms during the 80-hr period postenvenomation was 42 to 48%. Tissue distribution indicated that the greater part of the administered venoms was picked up by the liver and excreted in the intestines via bile secretion. Huang et al. also observed that the exchange of tritium between venoms and neuroamines was negligible, indicating that tritium labeling is quite firm and does not detach readily *in vivo*.

After i.m. injection of ^{125}I-labeled venom of *Bitis arietans*, venom localized in the kidneys, diaphragm, brain, liver, and uterus (Gumaa et al., 1974). Some of the radioactivity incorporated into the tissues is due to specific binding. The rest is due to nonspecific binding and is in equilibrium with the blood in particular tissues. For instance, the spleen, lung, and heart do not specifically bind venom, and the radioactivity is due to venom in the blood in these tissues. The highest radioactive incorporation was in the thyroid gland; this was due to specific binding of the venom. The significant binding of ^{125}I-labeled venom to rat uterus is in keeping with its known stimulatory action on uterine tissue.

The distribution of *Naja naja* venom in mice was studied by Sumyk et al. (1963), using ^{125}I. One or two minutes after injection, the mice were sacrificed. High concentrations of radioactivity were found in the blood, lungs, liver, kidneys, and heart. No or little radioactivity was present in the spleen, brain, stomach, or intestines. At 5 min after injection, the highest radioactivity level was found in the kidneys, and there was a marked increase in the intestines and spleen. Animals surviving for 10 min showed most of the radioactivity concentrated in the cortex of the kidneys.

1.2 *In Vivo* Labeling

When $Na_2{}^{35}SO_4$ was administered (i.p.) to snakes of *Crotalus durissus terrificus*, ^{35}S was incorporated into the molecular structure of venom proteins, particularly crotamine and crotoxin (Lomba et al., 1969). It would be interesting to know how this biosynthetically labeled venom is distributed into envenomated animals.

In other studies ^{65}Zn and ^{75}Se were used for *in vivo* labeling of *Vipera ammodytes* venom (Lebez et al., 1968a, b). The distribution of ^{65}Zn-labeled venom in the guinea pig followed this order: spleen, liver, kidneys, lungs, and heart. No ^{65}Zn was found in the central nervous system, cerebellum, and nervous ishiaticus. From this experiment it is clear that the venom does not penetrate the blood–brain barrier, and thus it is unlikely that the venom is neurotoxic to the central nervous system.

1.3 Amount of Venom Injected in Natural Bites

How much venom is actually injected into a snakebite victim is of great concern to everybody. This important question was successfully answered by the use of isotopically labeled venom. Allon and Kochva (1974) labeled the venom of *Vipera palestinae* with $[^{14}C]$-amino acids. When mice and rats were bitten by snakes having radioactive venoms, the amounts of venom injected into either prey were found to be variable, ranging from 0 to more than 200 mg. In most cases, 50 mg was injected, approximately 8% of the total venom available in the glands. For three snakes the average amount injected into rats was 2 to 3 times higher than that injected into mice; for two snakes the mice received 3 to 10 times more than the rats; and for two snakes there was no difference between the amounts injected into the two preys. Allon and Kochva concluded that the size of the prey does not necessarily influence the amount of venom injected by the snake. This finding is very interesting and parallels the statistical results for human victims. In about 40% of all human cases no adverse symptoms arose because no venom was injected. In some cases there were toxic symptoms, but these were not fatal. Allon and Kochva's finding that in most bites only 8% of the total venom in the glands is injected into the prey is alarming because it indicates that any snakebite can be potentially fatal if the snake injects a greater amount of venom.

2 NONRADIOACTIVE METHOD

Eaker et al. (1969) analyzed *Naja naja* venom in different organs of a person who committed suicide using cobra venom. There was a slight swelling on the back of the right hand, wrist, and forearm. There was also a large amount of fluid in the lungs, with isolated hemorrhages, superficial ruptures in the gastric mucosa, and moderate cerebral edema. Microscopic examination of the heart, liver, kidneys, pancreas, and spleen showed no pathological changes. In the cerebral cortex there were moderate nuclear changes with chromatolysis in the pyramid cells. Eaker et al. were able to fractionate and identify the toxic fraction of *N. naja* venom from the edema fluid of the right forearm.

Systemic toxic effects were investigated in mice by i.v. injection of venoms of *Bothrops atrox, B. schlegelii, B. nummifer, B. picadoi*, and *B. nasuta* (Tu and Homma, 1970). Most of the snake venoms caused hemorrhage in the heart, lungs, mesentery, and small intestine. Some venoms produced hemorrhage in the stomach and muscle after i.v. injection.

The immunofluorescence technique is valuable in the visual demonstration of sites of venom localization. This method was successfully utilized by Tiru-Chelvam (1972), who traced the venoms of *Bothrops atrox* and *Crotalus adamanteus* by the use of fluorescein isothiocyanate-labeled antivenin and fluorescence microscopy. The two venoms showed similar patterns of localization in the medulla and cervical cord. In addition, the *Crotalus* venom showed strong localization to renal tubules and blood vessel walls.

Coulter et al. (1973) and Sutherland et al. (1975) used solid-phase radioimmunoassay to detect 210 ng ml^{-1} postmortem in the serum of a child.

REFERENCES

Allon, N. and Kochva, E. (1974). The quantities of venom injected into prey of different size by *Vipera palaestinae* in a single bite, *J. Exp. Zool.,* **188,** 71.

Coulter, A. R., Sutherland, S. K., and Broad, A. J. (1973). Assay of snake venoms in tissue fluids, *J. Immunol. Methods,* **4,** 297.

Eaker, D. K., Karlsson, E., Lic, F., Rammer, L., and Saldeen, T. (1969). Isolation of neurotoxin in a case of fatal cobra bite, *J. Forensic Med.,* **16,** 96.

Gumaa, K. A., Osman, O. H., and Kertsz, G. (1974). Distribution of I^{125}-labelled *Bitis arietans* venom in the rat, *Toxicon,* **12,** 565.

Huang, C. L., Mir, G. N., Liu, S. J., Hemnani, K. L., and Yau, E. T. (1972). Distribution and excretion of ^{3}H-serum of *Lampropeltis getulus* in rats, *J. Pharm. Sci.,* **61,** 119.

Lebez, D., Maretic, Z., Gubensek, F., and Kristan, J. (1968a). Studies on labelled animal poisons. IV. Incorporation of selenium-75 and phosphorus-32 in spider venoms, *Biol. Vestn.,* **16,** 11.

Lebez, D., Gubensek, F., and Maretić, Z. (1968b). Studies on labelled animal poisons. III. *In vivo* with radioactive isotopes, *Toxicon,* **5,** 263.

Lee, C. Y. and Tseng, L. F. (1966). Distribution of *Bungarus multicinctus* venom following envenomation, *Toxicon,* **3,** 281.

Lomba, M., Brazil, O. V., Keifer, J., and Barberio, J. C. (1969). Biosynthetic method of labelling snake venoms, *Publ. Int. Energ. At.,* **189,** 49.

Shü, I. C., Ling, K. H., and Yang, C. C. (1968). Study on I^{131}-labeled cobrotoxin, *Toxicon,* **5,** 295.

Sumyk, G., Lal, H., and Hawrylewicz, E. J. (1963). Whole-animal autoradiographic localization of radioiodine labeled cobra venom in mice, *Fred. Proc.,* **22,** 668, Abstract No. 3035.

Sutherland, S. K., Coulter, A. R., and Broad, A. J. (1975). Human snake bite victims: The successful detection of circulating snake venom by radioimmunoassay, *Med. J. Aust.,* **1,** 27.

Tiru-Chelvam, R. (1972). Demonstration of sites of snake-venom localization by immunofluorescence techniques, *J. Pathol.,* **107,** 303.

Tseng, L. F., Chiu, T. H., and Lee, C. Y. (1968). Absorption and distribution of ^{131}I-labelled cobra venom and its purified toxins, *Toxicol. Appl. Pharmacol.,* **12,** 526.

Tu, A. T. and Homma, M. (1970). Toxicologic study of snake venoms from Costa Rica, *Toxicol. Appl. Pharmacol.,* **16,** 73.

17 Binding of Neurotoxins to Acetylcholine Receptors

1 ACETYLCHOLINE RECEPTOR	240
2 BINDING WITH CHOLINERGIC RECEPTORS	243
2.1 Electric Eels and Rays, 243	
2.2 Skeletal Muscle, 247	
2.3 Diaphragm, 248	
2.4 Cultured Cells, 249	
2.5 Brain, 251	
3 LACK OF SNAKE NEUROTOXIN INTERACTION WITH ACETYLCHOLINESTERASE	251
4 LOCALIZATION OF RECEPTORS	252
References	253

One of the most fascinating applications of snake venom is the utilization of its neurotoxins to isolate the various postsynaptic cholinergic receptor components. The first neurotoxin used for this purpose was α-bungarotoxin obtained from the venom of *Bungarus multicinctus*, a snake found in Formosa (Chang and Lee, 1963). The toxin was isolated by Lee and his colleagues at the National Taiwan University. They showed that the toxin acted as an irreversible antagonist of cholinergic receptors at the site of the vertebrate nerve–muscle junctions. Their finding that this effect could be prevented by d-tubocurarine, a specific but reversible antagonist of neuromuscular cholinergic receptors, indicated that α-bungarotoxin interacted with the cholinergic receptors.

To fully understand the action of neurotoxins on cholinergic receptors, a brief review of these receptors will be helpful.

1 ACETYLCHOLINE RECEPTOR

The receptor is functionally distinct from acetylcholinesterase (AChE), another protein present in the postsynaptic membrane which interacts with acetylcholine (ACh) by hydrolyzing it after depolarization of the membrane has occurred. The acetylcholine

1 Acetylcholine Receptor

receptor (AChR) is one of several receptors for specific neurotransmitters. Among the other receptors are the serotonin receptor, a protein (or site) that recognizes serotonin, and the alpha and beta adrenergic receptors, which recognize the catecholamines, epinephrine, and norepinephrine. These receptors are found in such organs as the brain and heart and autonomically controlled muscles (e.g., those of the stomach and the uterus).

The reduction of the disulfide bonds in the AChR has a deleterious effect on receptor activity, which involves the postsynaptic membrane function of nerves. The receptor of *Torpedo californica* has a molecular weight of 270,000 ± 30,000 daltons (Martinez-Carrion et al., 1975). However, Raftery (1973) considers subunits with molecular weight of 40,000 daltons to be present in the receptor of the torpedo fish. The molecular weight of the receptor of *Electrophorus electricus* is 42,000 (Reiter et al., 1972). Electroplax cells contain three to four times as much protein as do synaptic vesicles or synaptosome membranes. The amino acid compositions of receptor protein and acetylcholinesterase are listed in Table 1.

Several authors have proposed that the active site of AChE and the site of the AChR are the same (Ehrenpreis, 1967; Belleau et al., 1970). Indeed, they have common properties: they are present in the same tissues, they have similar molecular weights, and both are affected by certain chemical agents. For example, diisopropylfluorophosphate (DFP) and dibenamine inhibit the hydrolysis of ACh by AChE and produce or inhibit the binding of cholinergic ligands to the AChR.

However, a great deal of evidence suggests that the active site of AChE and the site of the AChR are different. The biochemical reactivities of the AChR and of AChE toward several series of compounds appear to be quite different. For example, ACh, acetylthiocholine, and acetylselenocholine are hydrolyzed by AChE at similar rates, but have different depolarizing potencies (Mautner et al., 1966; O'Brien et al., 1972). Also, analogs of benzoquinonium and ambenonium derivatives have effects on carbamylcholine depolarization of excitable membranes that differ greatly from their effects on AChE *in vitro* (Webb, 1965; O'Brien et al., 1972).

The functions of the AChR and of AChE can also be differentiated chemically. For instance, carbamylcholine (1×10^{-5} to $1 \times 10^{-4} M$) causes depolarization of excitable membranes but is not hydrolyzed by AChE. However, α-acetyl-β-methylcholine ($1 \times 10^{-2} M$) is a specific substrate of AChE (not hydrolyzed by other cholinesterases) that does not depolarize excitable membranes (Webb, 1965; Bartels, 1968). In another example, *trans*-3,3'-bis[α-(trimethylammonium)methyl]azobenzene dibromide (*trans*-Bis-Q) depolarizes electroplax membranes at concentrations less than $1 \times 10^{-7} M$, whereas *cis*-Bis-Q shows no activity even at much higher concentrations. Both isomers are equally active as inhibitors of AChE (50% inhibition at $1 \times 10^{-5} M$) (Bartels et al., 1971).

Treatment of the AChR and of AChE with certain chemical agents also shows different effects on the two macromolecules. Treatment with *p*-chloromercuribenzoate (PCMB) or 1,4-dithiothreitol (DDT) had no effect on the K_m or V_{max} value of either soluble or membrane-bound AChE, but 0.5 mM PCMB or 1.0 mM DDT inhibited the depolarization of membranes induced by carbamylcholine. ACh, or trimethylbutylammonium ion. The apparent dissociation constant of the receptor–carbamylcholine complex was increased three to four times by treatment with such sulfhydryl reagents (Karlin, 1967). Finally, ACh agonists and antagonists, drugs such as muscarone, nicotine, decamethonium ion, curare, and atropine, have much higher affinities for the AChR than for AChE.

Table 1 Comparison of Amino Acid Compositions of Acetylcholine Receptor and Acetylcholinesterase

Amino Acid	AcCH Receptor	Acetylcholinesterase		
Lysine	6.1	4.3	4.8	4.6
Histidine	2.1	2.3	2.1	2.3
Arginine	3.5	5.4	5.0	5.2
Aspartic acid	11.8	10.8	12.6	13.1
Threonine	6.3	4.3	4.1	4.5
Serine	7.1	6.9	6.8	6.8
Glutamic acid	10.7	9.4	11.1	10.4
Proline	6.2	8.1	7	5.9
Glycine	6.4	7.7	8.8	8.7
Alanine	6.0	5.5	7.4	6.2
Half cystine	2.0	1.1	0.9	1.6
Valine	5.5	7.0	6.9	7.1
Methionine	1.7	3.0	1.3	2.7
Isoleucine	5.2	3.7	4.0	3.8
Leucine	9.3	9.0	8.2	8.6
Tyrosine	3.6	3.8	2.9	3.6
Phenylalanine	4.4	5.3	5.1	5.3
Tryptophan	2.1	2.0	—	2.0
Hexosamine	0	1.6	1.3	—
Reference	Eldefrawi and Eldefrawi, 1973	Leuzinger and Baker, 1967	Dudai et al, 1972	Rosenberry et al, 1972

The allosteric ("anionic") site of AChE has also been proposed by several authors as the possible site of the AChR (Zupancic, 1967, 1970; Podleski et al., 1969; Changeux et al., 1969). However, it now appears that the active site of AChE and the site of the AChR are on different polypeptide chains. First of all, selective binding of the receptor to the α-toxin of *Naja nigricollis* has been observed through the covalent linkage of this complex to Sepharose granules (Meunier et al., 1971a; Changeux et al., 1971). In addition, after the binding of α-bungarotoxin or α-toxin of *N. nigricollis* venom to the AChR, the latter can be separated from AChE (Miledi et al., 1971; Meunier et al., 1972).

Fractionation of membrane particles from *Torpedo californica* electroplax indicates that the AChR fraction can be separated from the AChE fraction (Duguid and Raftery, 1973). This clearly indicates that the receptor proteins are localized in membrane structures different from those containing the AChE molecule.

All of these experiments indicate that the acetylcholine receptor is a protein distinct from acetylcholinesterase, although the two are located in the same tissue.

2   BINDING WITH CHOLINERGIC RECEPTORS

There have been many studies on the binding of snake neurotoxins to acetylcholine receptors of various tissues. Receptors obtained from the electroplax of electric eels and rays are especially plentiful, because of their high concentration in the tissues of these organisms. Other sources such as skeletal muscle, diaphragm, and brain are also used to obtain acetylcholine receptors. In this section, the emphasis will be on the interactions of neurotoxins with these receptors rather than on the properties of the receptors themselves.

2.1   Electric Eels and Rays

An attempt to isolate the acetylcholine receptor from purified membrane fragments of eel electroplax by the use of α-bungarotoxin has been reported. The toxin binds irreversibly to the AChR and blocks the excitation of ACh, but carbamylcholine d-tubocurarine protects the AChR from α-bungarotoxin binding. The binding of decamethonium to AChR also irreversibly blocks α-bungarotoxin binding to AChR. The toxin does not bind to the catalytic site of AChE. Extracts of electroplax prepared from membrane treated with sodium deoxycholate also bind the toxin (Changeux et al., 1970a, b).

Selective adsorption of the AChR to the α-toxin of *Naja nigricollis* venom was achieved by coupling the complex to Sepharose granules. The preparation used was a homogenate of eel electric tissue. Meunier et al. (1971a) estimated that 75 to 100% of the receptor activity was present on the granules and that 85 to 100% of the AChE activity remained in the supernatant, as measured by the binding of decamethonium ($1 \times 10^{-5} M$) and by the displacement by Flaxedil ($1 \times 10^{-5} M$).

Raftery et al (1971) attempted to isolate the AChR by labeling electroplax membrane fragments with [^{125}I]α-bungarotoxin and then extracting the membranes with 1% Triton X-100. The extract was then chromatographed on Sepharose 6-B, desalted on Sephadex G-75, and put through an isoelectrofocusing column in a pH 3 to 10 gradient. The free unlabeled toxin focused at pH 9.5, and the suspected receptor peak, bound to the radioactive toxin, focused at pH 5.2. The yield was about 3.5×10^{-11} mole of AChR per gram of electroplax tissue.

The acetylcholine receptor was characterized by reacting [^{3}H]α-toxin of *Naja nigricollis* with membrane fragments from homogenized eel electroplax. The membranes were then extracted with 1% sodium deoxycholate, and the extract centrifuged at $100,000 \times g$ for 60 min in a $0.7M$ sucrose gradient containing 1% deoxycholate. The peak containing the nonradioactive AChE was definitely separated from the radioactive AChR–toxin complex peak, as the peak of AChE activity was near the bottom of the tube and that of the AChR–toxin complex was near the top. Gel filtration on Sepharose 6-B showed that the receptor–toxin complex has about the same molecular weight as β-galactosidase in the presence of 1% deoxycholate (Meunier et al., 1971b).

Iodination of bungarotoxin is quite common and gives a 60% yield (Eldefrawi and Fertuck, 1974). α-Bungarotoxin labeled with ^{131}I is irreversibly bound to the AChR in isolated postsynaptic membranes from *Torpedo* electroplax. d-Tubocurarine and carbamylcholine slowed the binding of the toxin to the receptor. Several means of solubilizing the membranes were tried, but only with Triton X-100 did the toxin bind to the soluble proteins and not dissociate from the receptor. The toxin-bound protein, fractionated on Sephadex G-200, was eluted in the void volume before the AChE peak. Ultracentrifugation of the complex in density gradients containing 1% Triton X-100 showed that the toxin-bound protein sedimented primarily as a single peak, which sedimented faster than the AChE marker peak. Pretreatment of the complex with SDS showed that the toxin-labeled protein was an aggregated form. The molecular weight of the subunits was estimated as 80,000 daltons. The primary aggregate, in the absence of Triton X-100 or SDS, seemed to be a tetramer having four binding sites for the toxin. With time, these tended to form larger aggregates in the range of 500,000 to 2,000,000 daltons (Miledi et al., 1971).

Utilizing the fact that the receptor and snake neurotoxin bind to each other, Karlsson et al. (1972) isolated the receptor, using affinity chromatography, by immobilizing *Naja naja siamensis* neurotoxin.

Lubrol WX was used to solubilize membranes from *Torpedo* electroplax. The solubilized receptor, in addition to binding ACh, bound various nicotinic drugs, α-bungarotoxin, and cobra toxin (α-toxin of *Naja naja*). The soluble receptor showed the drug profile of the "nicotinic" receptor. The ACh binding was reversible; the two binding sites had dissociation constants of $1.0 \times 10^{-9} M$ and $2.2 \times 10^{-5} M$. Partial separation of the AChR and AChE was achieved by ultrafiltration. Only a low degree of separation was accomplished, as neither protein was pure (Eldefrawi et al., 1972).

$[^3H]$α-toxin of *Naja nigricollis* was used to label the AChR in membrane fragments of eel electroplax. Several reversible cholinergic agents — carbamylcholine, decamethonium, d-tubocurarine, and gallamine — and two affinity labeling agents — dinaphthyldecamethonium mustard ($DNC_{10}M$) and TDF — protected against toxin binding. The receptor–toxin complex was separated easily from the free toxin by ammonium sulfate precipitation in the presence of 1% Triton X-100. Initially, the membranes were extracted with 1% sodium deoxycholate or 1% Triton X-100, and the extract was then reacted with the toxin. The toxin remains bound to the receptor, and the complex migrates as a single band during electrophoresis on acrylamide gel containing 1% sodium deoxycholate. Ultracentrifugation in sucrose gradients of the toxin–receptor complex or of the dissociated complex (in the presence of either sodium doxycholate or Triton X-100) showed a large sedimentation coefficient (about 9.5S). Acrylamide gel electrophoresis in SDS gave a molecular weight of 55,000 daltons for the receptor subunit (Meunier et al., 1972).

Reconstituted membrane can still bind to α-toxin of *Naja nigricollis*. The reconstituted membrane had 12.8 nmole of receptor sites binding 1 g of the toxin, as compared to 4.65 nmole of receptor in native membranes binding the same amount of toxin (Changeux et al., 1972). Michaelson and Raftery (1974) also reconstituted the membranes; these not only bound to α-bungarotoxin but also were excitable by acetylcholine and carbamylcholine.

Clark et al. (1972) investigated the binding of $[^{125}I]$α-bungarotoxin to eel and ray electroplax membrane fragments and 1% Triton X-100 extracts of membranes, by treatment of the membrane components with enzymes and chemical agents. Phospho-

2 Binding with Cholinergic Receptors

Table 2 Apparent Dissociation Constants for the Reaction of the Receptor and α-Toxin in the Presence of Cholinergic Agents

Cholinergic Agents	By ^{3}H method (in vitro)	By ^{22}Na$^+$ method (in vitro)	By depolarization of the electroplaque (in vivo)
Agonists	M	M	M
decamethonium	0.3×10^{-6}	1.2×10^{-6}	1.2×10^{-6}
nicotine	1.8×10^{-5}	---	2.3×10^{-5}
Antagonists			
d-tubocuraine	1.7×10^{-7}	1.5×10^{-7}	1.6×10^{-7}
flaxedil	4.4×10^{-7}	3.3×10^{-7}	3.0×10^{-7}
hexamenthonium	6.1×10^{-5}	6.2×10^{-5}	3.0×10^{-5}

Data were obtained from the paper of Weber et al, 1972

lipase C and trypsin lowered the ability of the membrane fragments to bind the toxin. In the detergent extracts, the binding of the toxin was more rapidly destroyed by trypsin, while detergent seemed to decrease the sensitivity of binding to the action of phospholipase C. This was interpreted to mean that the receptor is a protein associated with lipid in the membrane and that the lipid is displaced when the receptor is extracted from the membrane with detergent. The inactivation effect of PCMB was also increased by solubilization of the membrane, indicating that the process of extraction exposes the receptor fully to the solvent. Clark et al. concluded that the receptor is a membrane-bound phospholipoprotein.

α-Bungarotoxin bound to proteolipid of molecular weight 37,000 daltons in cholinergic receptor of the electroplax (Fiszer De Plazas and De Robertis, 1972).

The constant rate for the reaction of the receptor and *Naja nigricollis* α-toxin is $(1.7 \pm 0.5) \times 10^7 M/\text{min}^{-1}$ (Weber et al., 1972). The apparent dissociation constants in the presence of several cholinergic agents are summarized in Table 2, where it can be seen that the constants are in the order of magnitude of 10^{-6} to 10^{-7}. The local anesthetics tetracaine and procaine:

$$\text{BuNH}-\underset{}{\bigcirc}-\text{COOCH}_2-\text{CH}_2-\text{N}(\text{CH}_3)_2$$

block α-toxin binding in a different manner than do cholinergic agents (Weber et al., 1972).

The production of antibodies for the acetylcholine receptor has been demonstrated. The AChR was isolated from detergent-treated membranes of eel electroplax by affinity chromatography with α-toxin of *Naja naja* coupled to Bio-Gel A-50M. The absorbed receptor was eluted by cholinergic ligands and then purified by sucrose-gradient centrifugation. The purified preparation (0.32 to 0.43 mg of the preparation emulsified in 1.4 ml of complete Freund's adjuvant) was injected into rabbits subcutaneously near the spine. About 14 days later, the rabbits were given a second injection. Four out of seven rabbits developed extreme paralysis of both peripheral and respiratory muscles within 3 weeks of the second injection. Apparently the rabbits had produced antibodies against the eel neuromuscular AChR, which cross-reacted with their own neuromuscular AChR. By an immunodiffusion technique, it was shown that serum obtained from the paralyzed rabbits contained precipitating antibodies to eel electroplax AChR but not to extracts from which the AChR had been removed by DEAE-cellulose chromotography (Patrick and Lindstrom, 1973).

Acetylcholine receptor was isolated from *Torpedo* electroplax by affinity adsorption, using cobra toxin of *Naja naja siamensis* venom coupled to Sepharose 4-B (Eldefrawi and Eldefrawi, 1973). The subunit had a molecular weight of 83,000 to 112,000. The amino acid composition of the receptor protein is very similar to that of acetylcholinesterase, as shown in Table 1, but differs somewhat from that of AChR from other sources.

For binding studies, neurotoxin is quite frequently tagged with radioactive elements. There is some question as to whether the radioactively labeled toxin behaves the same as the native toxin. A study using tritiated α-toxin of *Naja nigricollis* indicated that it behaves exactly as does the native, unlabeled toxin (Weber and Changeux, 1974a). The number of [^{3}H]α-toxin binding sites on membrane fragments is about 10 nmole g^{-1} of membrane protein for *Electrophorus* and 1000 nmole g^{-1} for *Torpedo*. The kinetics of association of [^{3}H]α-toxin with the membrane is compatible with a bimolecular mechanism of bind to a homogeneous class of sites. The second-order rate constant of association is $2.5 \times 10^{-7} M \min^{-1}$. The half-time for the dissociation of the [^{3}H]α-toxin–membrane complex in the presence of an excess of unlabeled toxin is about 60 hrs.

The initial rate of [^{3}H]α-toxin binding to membrane fragments decreases when the concentration of decamethonium increases. It is believed that the cholinergic effectors and [^{3}H]α-toxin bind to a common membrane site in a mutually exclusive manner (Weber and Changeux, 1974b).

Snake neurotoxin has also been used for the study of local anesthetic action. Weber and Changeux (1974c) observed that local anesthetics, such as dimethisoquin and dibucaine, decreased the initial rate of binding of tritiated α-neurotoxin from *Naja nigricollis* to membrane fragments purified from *Electrophorus* and *Torpedo* electric tissues. It appears that local anesthetics bind to the cholinergic receptor site with a low affinity, but that they inhibit the depolarization of the electroplax by binding at different sites situated on or near the cholinergic receptor protein.

It is well known that the receptor molecule is composed of subunits. Thus the question arises as to whether the receptor molecule may consist of functionally different types of polypeptide chains. Radioactively labeled *Naja naja siamensis* toxin binds covalently to only one of the subunits. Therefore it may be concluded that there are different kinds of subunits in the receptor molecules to *Torpedo* electric tissue, and that these have different functions (Gordon et al., 1974).

Acetylcholine receptor I from *Electrophorus electricus* was isolated and studied in detail by Chang (1974). The major subunit has a molecular weight of 41,500 and readily undergoes polymerization by forming disulfide linkages. Modification of free –SH groups by *p*-chloromercuribenzenesulfonate has no effect on the binding capacity of the receptor to α-bungarotoxin.

By employing *Naja naja siamensis* toxin, Heilbronn and Mattson (1974) isolated nicotinic cholinergic receptor from *Torpedo marmorate* electric organs. The neurotoxin-binding capacity was lost when rabbit antiserum was added to the receptor.

2 Binding with Cholinergic Receptors

The orientation of the receptor protein is such that all neurotoxin-binding sites are on the outer surface (Michaelson and Raftery, 1974).

The binding of α-bungarotoxin is indeed to the receptor protein and not to acetylcholinesterase. Robaire and Kato (1974) found that the toxin had no effect on the kinetic parameters of either membrane-bound or solubilized electroplax or electric eel acetylcholinesterase.

Acetylcholinesterase-rich membrane particles from *Torpedo californica* can be separated from acetylcholine receptor-rich particles (Duguid and Raftery, 1973; Reed et al., 1975). This is additional evidence that the AChR and AChE are two different molecules.

Falpius et al. (1975) isolated the AChR with a high degree of purity from the electric organ of the *Electrophorus electricus* eel by combining the receptor with a neurotoxin obtained from the venom of *Naja naja siamensis*.

2.2 Skeletal Muscle

A toxin obtained from *Naja naja siamensis* blocks transmission at the frog myoneural junction. The toxin causes an irreversible exponential decline in the amplitude of the end-plate potential as a result of a decrease in the sensitivity of the postsynaptic receptors. The toxin does not affect acetylcholine release, acetylcholinesterase activity, or the passive electrical properties of the muscle fiber membrane. The rate constant for the toxin binding to the receptor is $1.5 \times 10^5 M \sec^{-1}$ (Lester, 1972a).

Inactivation of frog myoneural acetylcholine receptors by cobra toxin was studied in the presence of reversible antagonists (d-tubocurarine and dihydro-β-erythroidine) and agonists (carbachol and nicotine) by Lester (1972b). During treatment with agonists, most receptors were desensitized rather than activated. Neither agonists nor antagonists reversed the blockade produced by the toxin. At 26 μM d-tubocurarine, receptors were partially protected from the toxin. At 5 μM or lower concentrations of d-tubocurarine, curare-blocked and free receptors were equally vulnerable to the toxin. Receptors were prevented from binding to the toxin at 7 to 140 μM carbachol, a concentration range that desensitizes 60 to 100% of the receptors.

Some substances diminish acetylcholine potentials at concentrations many times lower than the doses that noticeably influence the natural end-plate potentials. This difference in effective concentration is due to the marked ability of these substances to increase the desensitization. The acetylcholine potential was measured by iontophoretic microapplication of acetylcholine from a micropipette near the muscle end-plate zone. Vyskočil and Magzanik (1972) found that α-bungarotoxin and N3 toxin from *Naja naja siamensis* venom markedly increased the desensitization rate, in parallel with their irreversible cholinolytic activity.

The binding of α-bungarotoxin to neuromuscular junction occurs not at only one site but at two sites (Albuquerque et al., 1973). One site, which is blocked competitively by α-bungarotoxin and by curare, is presumably the acetylcholine receptor. The binding of bungarotoxin at this site is responsible for an irreversible blockade of neuromuscular transmission. The second site, which is blocked competitively by bungarotoxin and perhydorhistrionicotoxin, is proposed to be part of the cholinergic ion conductance modulator. The binding of bungarotoxin to this site does not result in an irreversible blockade. This explains why the overall effect of bugarotoxin is partially reversible.

In 1973 Chiu et al., using [^{3}H]α-bungarotoxin, attempted to isolate the AChR from rat skeletal muscle and mouse diaphragm. The muscles were removed, treated with the toxin, and then extracted with 1.5% Triton X-100. Both innervated and denervated muscles were used. The denervated muscle bound up to 23 times more toxin than the normal muscle. The extracts were then chromatographed on Sepharose 6-B columns with $0.05M$ Tris buffer, pH 8, containing 1.5% Triton X-100. The mouse diaphragms gave two radioactive peaks in addition to the free toxin peak. The peaks in the mouse extracts had apparent molecular weights of 550,000 daltons and higher. The extracts from both the normal and the denervated rat muscle contained radioactive receptor–toxin peaks in the same place; these peaks had an apparent molecular weight of 200,000 daltons. No AChE activity was detected in any of the extracts.

The density of the extrajunctional AChRs was calculated by Famborough (1974a) from the number of $[^{125}I]\alpha$-bugarotoxin binding sites. The receptor density increased from <6 sites per square micron in innervated muscle to 635 ± 29 sites 14 days after denervation.

Acetylcholine receptor protein was isolated from denervated muscle of cat, using α-bungarotoxin binding as a marker. The receptor bound cholinergic agonists and antagonists with high affinities (Dolly and Barnard, 1974).

α-Bungarotoxin is also helpful in studying the localization of the AChR. Fertuck and Salpeter (1974) used electron microscope autoradiography to study the binding of $[^{125}I]\alpha$-bungarotoxin at mouse motor end plates and found that the label was localized at the top of the functional folds, that is, at the postjunctional membrane nearest the axon. Since the α-bungarotoxin had fully eliminated the response of the muscle, these results indicated that the active AChR occupies a limited area of the junctional folds and is not distributed uniformly.

Utilizing the fact that α-bungarotoxin can bind to the AChRs, Famborough et al. (1973) determined the number of receptors in the neuromuscular junctions of patients with myasthenia gravis. Fewer junctional AChRs were present in the myasthenic muscles than in the muscles of normal subjects. It was concluded that this reduction in receptors may account for the defect in neuromuscular transmission in myasthenia gravis.

Also in patients with myasthenia gravis, a serum globulin, which inhibits the binding of α-bungarotoxin to rat muscle receptors, has been detected (Almon et al., 1974). The significance of the presence of the binding inhibitor is speculated to be similar to that of the antibodies against nuclei, muscle structural proteins, and other tissues found in a proportion of patients with myasthenia gravis.

2.3 Diaphragm

An attempt was made to isolate the acetylcholine receptor from rat diaphragms by treating the diaphragms with α-bungarotoxin, which binds to the end plates. The binding was inhibited by d-tubocurarine and carbamylcholine, but not by atropine. The diaphragms were homogenized, and the membranes solubilized with 1% Triton X-100. The toxin–receptor complex was removed by zone sedimentation in 5 to 20% sucrose in $0.02M$ Tris buffer, pH 7.4, containing $0.15M$ NaCl, $0.4mM$ EGTA, and 1% Triton X-100. The toxin–receptor complex was recovered as a single band which differed from the bands of toxin and of AChE. Treatment of the complex with 2-mercaptoethanol and SDS resulted in the recovery of free toxin, but the AChR was denatured by this treatment (Berg et al., 1972).

Using $[^{125}I]\alpha$-bugarotoxin, Fambrough and Hartzell (1972) found $(4.1 \pm 0.2) \times 10^7$ AChRs per motor end-plate in the rat diaphragm. With the same approach, Hartzell and Fambrough (1972) studied the time course of the changes in AChR distribution and density occurring in rat diaphragm after denervation. The extrajunctional receptor density begins to increase between 2 and 3 days after denervation and increases approximately linearly to 1695 receptors per square micron at 45 days. Using α-bungarotoxin labeled with tritium, Chang et al. (1973) confirmed that the toxin is localized chiefly in the motor end-plate region of the rat diaphragm.

In normal muscles, most of the toxin–receptor complexes that formed appeared to be stable over a 24-hr period. In contrast, the radioactivity associated with the extrajunctional toxin–receptor complex of denervated muscle was lost rapidly from the tissue. Only 19% of the radioactivity originally associated with regions devoid of end plates from denervated hemidiaphragms remained after 25 hr in culture (Berg and Hall, 1974).

The chemical properties of the AChR isolated from normal rat diaphragm muscle (junctional receptors) and that from extrajunctional regions of denervated diaphragm are very similar. Both receptors are believed to be glycoproteins. The receptor–toxin (from *Naja naja siamensis* venom) complexes from the two receptor sources were indistinguishable by gel filtration and by zone sedimentation in sucrose gradients, and they showed identical precipitation curves with rabbit antiserum to eel AChR. Both toxin-receptor complexes bound concanavalin A. The only difference observed was that the isoelectric point of the junctional complex was 0.15 pH unit lower than that of the extrajunctional complex (Brockes and Hall, 1975a). The binding of receptor protein to α-bungarotoxin is very strong, as the dissociation constants for the junctional and extrajunctional receptors with α-bungarotoxin are $3.7 \times 10^{-10} M$ and $1.7 \times 10^{-10} M$, respectively (Brockes and Hall, 1975b). The association reaction between toxin and receptor did not obey simple second-order kinetics. This may reflect the fact that there are two classes of binding sites, corresponding to two rates of dissociation.

2.4 Cultured Cells

Instead of whole muscle, Patrick et al. (1972) used a muscle cell line (A clone, L6) in tissue culture. *Naja naja* neurotoxin bound to acetylcholine receptors during differentiation of a clonal myoblast cell line. Bound toxin was released from the cells with a half-life of about 7 hrs. This release was not associated with a decrease in the total number of toxin-binding sites. The appearance of toxin-binding sites paralleled that of fused fibers during differentiation of the muscle cells in tissue culture.

The pharmacologic properties of the receptor of cultured embryonic chick muscle resembled those of the nicotinic AChR of adult vertebrate muscle. Vogel et al. (1972) used monoiodo and diiodo derivatives of α-bungarotoxin to study the binding of the toxin to the AChR of cultured embryonic chick and rat muscle cells. Autoradiography of muscle cells labeled with toxin showed that the AChRs were distributed over the entire cell surface. In addition, discrete areas with a high receptor concentration were found.

The α_1-isotoxin of *Naja nigricollis* venom binds to the cholinergic receptor in the muscle of chick embryos. By utilizing this property, Giacobini et al. (1973) studied the development of the cholinergic receptor in chick muscle in embryos. They found that the cholinergic receptor concentration reached a maximum after the twelfth day of incubation.

α-Bungarotoxin binds to the AChRs of cultured chick embryo sympathetic neurons (Greene et al., 1973) and thus can be used to probe these receptors in intact neurons. The binding of α-bungarotoxin to cultured mouse muscle cells was investigated extensively by Fambrough and his co-workers (Powell and Fambrough, 1973; Hartzell and Fambrough,

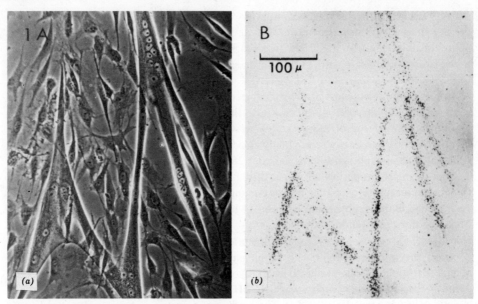

Figure 17.1 An autoradiograph of tissue culture rat muscle labeled with [^{125}I]-α-bungarotoxin. (A) Phase-contrast photomicrograph of a living myotube immediately after this treatment. (B) Bright-field photomicrograph of the grains produced by the bound [^{125}I]-α-bungarotoxin. (Photographs kindly supplied by Dr. Fambrough. Data reproduced from Hartzell and Fambrough, 1973.)

1973; Fambrough et al., 1974; Ritchie and Fambrough, 1975; De Vreotes and Fambrough, 1975).

An autoradiographic study by Powell and Fambrough (1973) showed that myotubes containing two to four nuclei bound α-bungarotoxin, whereas fibroblasts and some mononucleated myoblasts did not. Similarly the long myotubes in 13-day and 20-day cultures had numerous α-bungarotoxin-binding sites.

The α-bungarotoxin–receptor complexes were found to turn over slowly (Hartzell and Fambrough, 1973). Less than 2% of the bound [^{125}I]α-bungarotoxin was released into the medium during a 90-min washout, and less than 5% in a 13-hr wash. The slow turnover number indicates the irreversible nature of α-bungarotoxin binding to receptors.

The distribution of [^{125}I]α-bungarotoxin binding sites in myotubes is not uniform (Fig. 1). The toxin did not bind to cultured heart cells, perhaps because of the difference in the types of receptors in heart muscle and in skeletal muscle. In heart muscle, the receptors are muscarinic, whereas those in skeletal muscle are nicotinic (Hartzell and Fambrough, 1973).

The interaction of α-bungarotoxin with the receptor of embryonic skeletal muscle was almost irreversible (Fambrough, 1974b). However, heating the complex at 70°C caused the release of undegraded α-bungarotoxin, suggesting that the toxin was not bound covalently to the receptor.

α-Bungarotoxin is a very useful reagent for studying the appearance of new receptor sites on myotubes. Fambrough et al. (1974) blocked all of the available receptor sites with unlabeled α-bungarotoxin and then measured the appearance of new binding sites with [^{125}I]α-bungarotoxin. The rate of appearance of new receptors during vigorous myotube growth was rapid. A net increase of about 35 receptors per square micron of surface membrane occurred during a 60-min period.

De Vreotes and Fambrough (1975) used [^{125}I] α-bungarotoxin as a specific marker to measure the half-life of AChR metabolism in the surface membranes of chick and rat myotubes developing in cell culture. The half-life was found to be 22 to 24 hrs.

2.5 Brain

Although brain contains acetylcholine receptors, the concentration is very low; thus the isolation of brain AChR is difficult. The AChR isolated from guinea pig cerebral cortex by Bosmann (1972) had a molecular weight of 86,000 daltons for the receptor subunit, as measured by acrylamide gel electrophoresis. Two kinds of binding sites for [^{3}H] acetyl-α-bungarotoxin were found, with dissociation constants of $6.6 \times 10^{-8} M$ and $2.4 \times 10^{-7} M$ (Bosmann, 1972).

[^{125}I] toxin of *Bungarus fasciatus* venom reacted with protein fractions extracted from eel electoplax and hog cerebral cortex. The toxin, with a molecular weight of 7600 daltons, bound irreversibly to the AChR and blocked neuromuscular transmission. The extracts were fractionated on Sephadex G-100 superfine columns with $0.05M$ Tris (pH 8) buffer containing $0.1M$ NaCl. In the case of the eel extract, the radioactive toxin–receptor complex eluted somewhat ahead of the AChE peak, but the separation was not complete. There appeared to be very little AChE activity in the hog brain extract, and the radioactive toxin–receptor complex appeared to have a molecular weight between 40,000 and 80,000 daltons, as measured by gel filtration. When a control run was done with guinea pig liver tissue that had been treated in the same way as the brain tissue, no significant peaks of toxin binding were seen. The authors estimated that the concentration of AChR in hog brain (gray matter) was only 1.5% of that in electroplax (Moore and Loy, 1972).

Using [^{125}I] α-bungarotoxin, Salvaterra and Moore (1973) found that the nicotinic AChR concentration was 3.4 pmole g^{-1} for rat brain particulate and 2.1 pmole g^{-1} for guinea pig brain particulate tissue.

It is well known that α-bungarotoxin is highly specific for the AChR in diaphragm tissue. However, Enterovic and Bennett (1974) found it to be less specific toward the receptors in brain. They fractionated rat brains into myelin, nerve endings, and mitochondria and observed that α-bungarotoxin bound mostly to the nerve ending fraction.

α-Bungarotoxin does not pass the brain–blood barrier and therefore does not enter the central nervous system (Lee and Tseng, 1966). Thus injection of the toxin into the ventricle of the brain caused the binding of toxin to only the nicotinic receptors of the brain (Polz-Tejera et al., 1975).

3 LACK OF SNAKE NEUROTOXIN INTERACTION WITH ACETYLCHOLINESTERASE

As discussed many times in this chapter, α-bungarotoxin binds specifically and irreversibly to acetylcholine receptors of the motor end plate and to other nicotinic cholinergic receptors. It has no inhibitory effect on acetylcholinesterase in the neuromuscular junction (Chang and Lee, 1963; Changeux et al., 1970a). Using end-plate membrane acetylcholinesterase of rat diaphragm, Chang and Su (1974) showed that α-bungarotoxin does not alter the rate of hydrolysis of acetylcholine. Robaire and Kato (1974) further studied this possible interaction by examining kinetic parameters and found that none of these parameters was affected by the toxin. Furthermore, the toxin did not bind to solubilized acetylcholinesterase, whereas it did bind to electroplax membranes. Thus the AChR is not identical with the anionic site of AChE.

4 LOCALIZATION OF RECEPTORS

Since snake neurotoxins bind specifically to acetylcholine receptors, this reaction can be used to observe the localization of receptors in diaphragm tissue by means of radiosotope techniques (Lee and Tseng, 1966; Sato et al., 1970).

By utilizing the property of specific binding, Bourgeois et al. (1971) observed the localization of the receptor in the electroplax, using the immuno fluorescence technique.

In the electoplax of the electric eel (Menez et al., 1971), the amount of *Naja nigricollis* toxin binding to the receptor was found to be 8.4 nmole g^{-1} protein. This means that there are approximately 2.7 times more catalytic sites of acetylcholinesterase than toxin binding sites in this tissue.

Using [^{3}H]α-bungarotoxin, Barnard et al. (197,) were able to demonstrate that the AChR sites are localized in mouse skeletal muscle end plates.

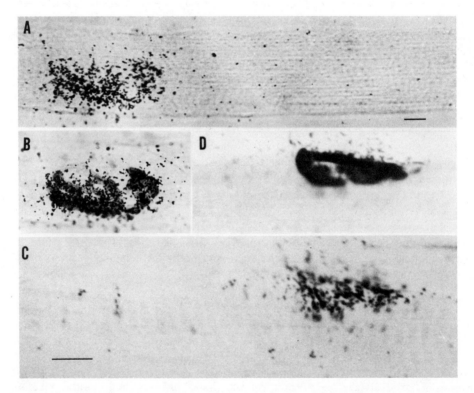

Figure 17.2 Autoradiographs of the end-plate regions of isolated muscle fibers previously incubated with [^{125}I]-α-bungarotoxin. An end-plate is shown before (A) and after (B) the preparation was stained for acetylcholinesterase; a second end-plate is also seen oriented toward one side of the fiber, before (C) and after (D) staining for acetylcholinesterase. Magnification bars in (A) and (C) represent 10 μ. (Photographs kindly supplied by Dr. D. M. Fambrough. Data reproduced from Fambrough and Hartzell, 1972. Copyright 1972 by the American Association for the Advancement of Science.)

On the other hand, Fambrough and Hartzel (1972) labeled α-bungarotoxin with ^{125}I instead of tritium. They also found that virtually all of the receptors are localized in the end plate (Fig. 2).

REFERENCES

Albuquerque, E. X., Barnard, E. A., Chiu, T. H., Lapa, A. J., Dolly, J. O., Jansson, S. E., Daly, J., and Witkop, B. (1973). Acetylcholine receptor and ion conductance modulator sites at the murine neuromuscular junction: Evidence from specific toxin reactions, *Proc. Natl. Acad. Sci. U.S.A.*, **70**, 949.

Almon, R. R., Andres, C. G., and Appel, S. H. (1974). Serum globulin in myasthenia gravis: Inhibition of α-bungarotoxin binding to acetylcholine receptors, *Science*, **186**, 55.

Barnard, E. A., Wieckowski, J., and Chiu, T. H. (1971). Cholinergic receptor molecules and cholinesterase molecules at mouse skeletal muscle junctions, *Nature*, **234**, 207.

Bartels, E. (1968). Reactions of acetylcholine receptor and esterase studied on the electroplax, *Biochem. Pharmacol.*, **17**, 945.

Bartels, E., Wassermann, N. H., and Erlanger, B. R. (1971). Photochromic activators of the acetylcholine receptor, *Proc. Natl. Acad. Sci. U.S.A.*, **68**, 1820.

Belleau, B., Di Tullio, V., and Tsai, Y. H. (1970). Kinetic effects of leptocurares and pachycurares on the methanesulfonylation of acetylcholinesterase, *Mol. Pharmacol.*, **6**, 41.

Berg, D. K. and Hall, Z. W. (1974). Fate of α-bungarotoxin bound to acetylcholine receptors of normal and denervated muscle, *Science*, **184**, 473.

Berg, D. K., Kelly, R. B., Sargent, P. B., Williamson, P., and Hall, Z. W. (1972). Binding of α-bungarotoxin to acetylcholine receptors in mammalian muscle, *Proc. Natl. Acad. Sci. U.S.A.*, **69**, 147.

Bosmann, H. B. (1972). Acetylcholine receptor. I. Identification and biochemical characteristics of a cholinergic receptor of guinea pig cerebral cortex, *J. Biol. Chem.*, **247**, 130.

Bourgeois, J. P., Tsuji, S., Boquet, P., Pillot, J., Ryter, A., and Changeux, J. P. (1971). Localization of the cholinergic receptor protein by immunofluorescene in eel electroplax, *FEBS Lett.*, **16**, 92.

Brockes, J. P. and Hall, Z. W. (1975a). Acetylcholine receptors in normal and denervated rat diaphragm muscle. I. Purification and interaction with [^{125}I] α-bungarotoxin, *Biochemistry*, **14**, 2092.

Brockes, J. P. and Hall, Z. W. (1975b). Acetylcholine receptors in normal and denervated rat diaphragm muscle. II. Comparison of junctional and extrajunctional receptors, *Biochemistry*, **14**, 2100.

Chang, C. C. and Lee, C. Y. (1963). Isolation of neurotoxins from the venoms of *Bungarus multicinctus* and their modes of neuromuscular blocking action, *Arch. Int. Pharmacodyn.*, **144**, 144.

Chang, C. C. and Su, M. J. (1974). Does α-bungarotoxin inhibit motor endplate acetylcholinesterase? *Nature*, **247**, 480.

Chang, C. C., Chen, T. F., and Chuang, S. T. (1973). N, O-Di- and N, N,O-tri [^{3}H]acetyl-α-bungarotoxins as specific labelling agents of cholinergic receptors, *Br. J. Pharmacol.*, **47**, 147.

Chang, H. W. (1974). Purification and characterization of acetylcholine receptors. I. From *Electrophorus electricus*, *Proc. Natl. Acad. Sci. U.S.A.*, **71**, 2113.

Changeux, J. P., Podleski, T., and Meunier, J. C. (1969). On some structural analogies between acetylcholinesterase and the macromolecular receptor of acetylcholine, *J. Gen. Physiol.*, **54**, 225.

Changeux, J. P., M. Kasai, M. Huchet, and Meunier, J. C. (1970a). Extraction à partir du tissu électrique de gymnote d'une protéine présentant plusieurs propriétés carachtéristiques du récepteur physiologique de l'acétylcholine, *C. R. Acad. Sci. Paris*, Ser. D., **270**, 2864.

Changeux, J. P., Kasai, M., and Lee, C. Y. (1970b). Use of snake venom toxin to characterize the cholinergic receptor protein, *Proc. Natl. Acad. Sci. U.S.A.*, **67**, 1241.

Changeux, J. P., Meunier, J. C., and Huchet, M. (1971). Studies on the cholinergic receptor protein of *Electrophorus electricus*. I. An assay *in vitro* for the cholinergic receptor site and solubilization of the receptor protein from electric tissue, *Mol. Pharmacol.*, **7**, 538.

Changeux, J. P., Huchet, M., and Cartaud, J. (1972). Reconstitution partielle d'une membrane excitable apres dissolution par le deoxycholate de sodium, *C. R. Acad. Sci. Paris*, **274**, 122.

Chiu, T. H., Dolly, J. O., and Barnard, E. A. (1973). Solubilization from skeletal muscle of two components that specifically bind α-bungarotoxin, *Biochem. Biophys. Res. Commun.*, **51**, 205.

Clark, D. G., Wolcott, R. B., and Raftery, M. A. (1972). Partial characterization of an α-bungarotoxin binding component of electroplax membranes, *Biochem. Biophys. Res. Commun.,* **48,** 1061.

De Vreotes, P. N. and Fambrough, D. M. (1975). Acetylcholine receptor turnover in membranes of developing muscle fibers, *J. Cell Biol.,* **65,** 335.

Dolly, J. O. and Barnard, E. A. (1974). Affinity of cholinergic ligands for the partially purified acetylcholine receptor from manomalian skeletal muscle, *FEBS Lett.,* **46,** 145.

Dudai, Y., Silman, I., Kalderon, N., and Blumberg, S. (1972). Purification by affinity chromatography of acetylcholinesterase from electric organ tissue of the electric eel subsequent to tryptic treatment, *Biochim. Biophys. Acta,* **268,** 138.

Duguid, J. R. and Raftery, M. A. (1973). Fractionation and partial characterization of membrane particles from *Torpedo californica electroplax, Biochemistry,* **12,** 3593.

Ehrenpreis, S. (1967). Possible nature of the cholinergic receptor, *N. Y. Acad. Sci.,* **144,** 720.

Eldefrawi, M. E. and Eldefrawi, A. T. (1973). Purification and molecular properties of the acetylcholine receptor from *Torpedo electroplax, Arch. Biochem. Biophys.,* **159,** 362.

Eldefrawi, M. E. and Fertuck, H. C. (1974). Rapid method for the preparation of iodine-125-labeled α-bungarotoxin, *Anal. Biochem.,* **58,** 63.

Eldefrawi, M. E., Eldefrawi, A. T., Seifert, S., and O'Brien, R. D. (1972). Properties of lubrol-solubilized acetylcholine receptor from *Torpedo electroplax, Arch. Biochem. Biophys.,* **150,** 210.

Enterović, V. A. and Bennett, E. L. (1974). Nicotinic cholinergic receptor in brain detected by binding of [^{3}H]α-bungarotoxin, *Biochim. Biophys. Acta,* **363,** 346.

Fambrough, D. M. (1974a). Acetylcholine receptors: Revised estimates of extrajunctional receptor density in denervated rat diaphragm, *J. Gen. Physiol.,* **64,** 468.

Fambrough, D. M. (1974b). "Cellular and developmental biology of acetylcholine receptors in skeletal muscle," in P. E. De Robertis and J. Schacht, Eds., *Neurochemistry of Cholinergic Receptors,* An American Society for Neurochemistry Monograph, Raven, New York, pp. 85–113.

Fambrough, D. M. and Hartzell, H. C. (1972). Acetylcholine receptors: Number and distribution at neuromuscular junctions in rat diaphragm, *Science,* **176,** 189.

Fambrough, D. M., Drachman, D. B., and Satyamurti, S. (1973). Neuromuscular junction in myasthenia gravis: Decreased acetylcholine receptors, *Science,* **182,** 293.

Fambrough, D. M., Hartzell, H. C., Rash, J. E., and Ritchie, A. K. (1974). Receptor properties of developing muscle, *Ann. N.Y. Acad. Sci.,* **228,** 47.

Fertuck, H. C. and Salpeter, M. M. (1974). Localization of acetylcholine receptor by ^{125}I-labeled α-bungarotoxin binding at mouse motor endplates, *Proc. Natl. Acad. Sci. U.S.A.,* **71,** 1376.

Fiszer De Plazas, S. and De Robertis, E. (1972). Binding of α-bungarotoxin to the cholinergic receptor proteolipid from *Electophorus electroplax, Biochim. Biophys. Acta,* **274,** 258.

Fulpius, B. W., Maelicke, A., Klett, R., and Rerch, E. (1975). "The acetylcholine receptor: Studies of the interaction with α-neurotoxin from *Naja naja siamensis,"* in P. G. Waser, Ed., *Cholinergic Mechanisms,* Raven, New York, pp. 375–380.

Giacobini, G., Filogamo, G., Weber, M., Boquet, P., and Changeux, J. P. (1973). Effects of a snake α-neurotoxin on the development of innervated skeletal muscles in chick embryo, *Proc. Natl. Acad. Sci. U.S.A.,* **70,** 1708.

Gordon, A., Bandini, G., and Hucho, R. (1974). Investigation of the *Naja naja siamensis* toxin binding site of the cholinergic receptor protein from *Torpedo* electric tissue, *FEBS Lett.,* **47,** 204.

Greene, L. A., Sytkowski, A. J., Vogel, Z., and Nirenberg, M. W. (1973). α-Bungarotoxin used as a probe for acetylcholine receptors of cultured neurones, *Science,* **243,** 163.

Hartzell, H. C. and Fambrough, D. M. (1972). Acetylcholine receptors: Distribution and extrajunctional density in rat diaphragm after denervation correlated with acetylcholine sensitivity, *J. Gen. Physiol.,* **60,** 248.

Hartzell, H. C. and Fambrough, D. M. (1973). Acetylcholine receptor production and incorporation into membranes of developing muscle fibers, *Develop. Biol.,* **30,** 153.

Heilbronn, E. and Mattson, C. (1974). Nicotinic cholinergic receptor protein: Improved purification method, preliminary amino acid composition, and observed autoimmuno response, *J. Neurochem.,* **22,** 315.

References

Karlin, A. (1967). Chemical distinctions between acetylcholinesterase and the acetylcholine receptor, *Biochim. Biophys. Acta,* **139**, 358.

Karlsson, E., Eaker, D., and Ponterius, G. (1972). Modification of amino groups in *Naja naja* neurotoxins and the preparation of radioactive derivatives, *Biochim. Biophys. Acta,* **257**, 235.

Lee, C. Y. and Tseng, L. F. (1966). Distribution of *Bungarus multicinctus* venom following envenomation, *Toxicon,* **3**, 281.

Lester, H. A. (1972a). Blockade of acetylcholine receptors by cobra toxin: Electrophysiological studies, *Mol. Pharmacol.,* **6**, 623.

Lester, H. A. (1972b). Vulnerability of desensitized or curare-treated acetylcholine receptors to irreversible blockade by cobra toxin, *Mol. Pharmacol.,* **8**, 632.

Leuzinger, W. and Baker, A. L. (1967). Acetylcholinesterase, I. Large-scale purification, homogeneity, and amino acid analysis, *Proc. Nat. Acad. Sci., USA,* **7**, 446.

Martinez-Carrion, M., Sator, V., and Raftery, M. A. (1975). The molecular weight of an acetylcholine receptor isolated from *Torpedo californica, Biochem. Biophys. Res. Commun.,* **65**, 129.

Mautner, H. G., Bartels, E., and Webb, G. D. (1966). Sulfur and selenium isotopes related to acetylcholine and choline. IV. Activity in the electroplax preparation, *Biochem. Pharmacol.,* **15**, 187.

Menez, A., Morgat, J., Fromageot, P., Ronseray, A., Boquet, P., and Changeux, J. P. (1971). Tritium labeling of the α-neurotoxin of *Naja nigricollis, FEBS Lett.,* **17**, 333.

Meunier, J. A., Huchet, M., Boquet, P., and Changeux, J. P. (1971a). Séparation de la protéine réceptrice de l'acétylcholine et de l'acétylcholinéstérase, *C. R. Acad. Sci. Paris,* Ser. D, **272**, 117.

Meunier, J. C., Olsen, R. W., Menez, A., Morgat, J. L., Fromageot, P., Ronseray, A. M., Boquet, P., and Changeux, J. P. (1971b). Quelques propriétès physiques de la protéine receptrice de l'acétylcholine étudiées à l'acide d'une neurotoxine radioactive, *C. R. Acad. Sci. Paris,* Ser. D, **273**, 595.

Meunier, J. C., Olsen, R. W., Menez, A., Fromageot, P., Boquet, P., and Changeux, J. P. (1972). Some physical properties of the cholinergic receptor protein from *Electrophorus electrious* revealed by a tritiated α-toxin from *Naja nigricollis* venom, *Biochemistry,* **11**, 1200.

Michaelson, D. M. and Raftery, M. A. (1974). Purified acetylcholine receptor: Reconstitution to a chemically excitable membrane, *Proc. Natl. Acad. Sci. Paris,* **71**, 4768.

Miledi, R., Molinoff, P., and Potter, L. T. (1971). Isolation of cholinergic receptor protein of *Torpedo* electric tissue, *Nature,* **229**, 554.

Moore, W. J. and Loy, N. J. (1972). Irreversible binding of a krait neurotoxin to membranes from eel electroplax and hog brain, *Biochem. Biophys. Res. Commun.,* **46**, 2093.

O'Brien, R. D., Eldefrawi, M. E., and Eldefrawi, A. T. (1972). Isolation of acetylcholine receptors, *Ann Rev. Pharmacol.,* **12**, 19.

Patrick, J. and Lindstrom, J. (1973). Autoimmune response to acetylcholine receptor, *Science,* **180**, 871.

Patrick, J., Heineman, S. F., Lindstrom, J., Schubert, D., and Steinbach, J. H. (1972). Appearance of acetylcholine receptors during differentiation of a myogenic cell line, *Proc. Natl. Acad. Sci. U.S.A.,* **69**, 2762.

Podleski, T. R., Meunier, J. C., and Changeux, J. P. (1969). Compared effects of dithiothreithol on the interaction of an affinity-labeling reagent with acetylcholinesterase and the excitable membrane of the electroplax, *Proc. Natl. Acad. Sci. U.S.A.,* **63**, 1239.

Polz-Tejera, G., Schmidt, J., and Karten, H. J. (1975). Autoradiographic localization of α-bungarotoxin-binding sites in the central nervous system, *Nature,* **258**, 349.

Powell, J. A., and Fambrough, D. M. (1973). Electrical properties of normal and dysgenic mouse skeletal muscle in culture, *J. Cell Physiol.,* **82**, 21.

Raftery, M. A. (1973). Isolation of acetylcholine receptor α-bungarotoxin complexes from *Torpedo californica* electroplax, *Arch. Biochem. Biophys.,* **154**, 270.

Raftery, M. A., Schmidt, J., Clark, D. G., and Wolcott, R. G. (1971). Demonstration of a specific α-bungarotoxin binding component in *Electrophorus electricus* electroplax membranes, *Biochem. Biophys. Res. Commun.,* **45**, 1622.

Reed, K., Vandlen, R., Bode, J., Duquid, J., and Raftery, M. A. (1975). Characterization of acetylcholine receptor-rich and acetylcholinesterase-rich membrane particles from *Torpedo californica* electroplax, *Arch. Biochem. Biophys.*, **167**, 138.

Reiter, M. J., Cowburn, D. A., Prives, J. M., and Karlin, A. (1972). Affinity labeling of the acetylcholine receptor in the electroplax: electrophoretic separation in sodium dodecyl sulfate, *Proc. Natl. Acad. Sci. U.S.A.*, **69**, 1168.

Ritchie, A. K. and Fambrough, D. M. (1975). Ionic properties of the acetylcholine receptor in cultured rat myotubes, *J. Gen. Physiol.*, **65**, 751.

Robaire, B. and Kato, G. (1974). Acetylcholinesterase and α-bungarotoxin: A study of their possible interaction, *FEBS Lett.*, **46**, 218,

Rosenberry, T. L., Chang, H. W., and Chen, Y. T. (1972). Purification of acetylcholinesterase by affinity chromatography and determination of active site stoichiometry, *J. Biol. Chem.* **247**, 1555.

Salvaterra, P. M. and Moore, W. J. (1973). Binding of [^{125}I] α-bungarotoxin to particulate fractions of rat and guinea pig brain, *Biochem. Biophys. Res. Commun.*, **55**, 1311.

Sato, S., Abe, T., and Tamiya, N. (1970). Binding of iodinated erabutoxin b, a sea snake toxin, to the endplates of the mouse diaphragm, *Toxicon*, **8**, 313.

Vogel, Z., Sytkowski, A. J., and Nirenberg, M. W. (1972). Acetylcholine receptors of muscle grown *in vitro*, *Proc. Natl. Acad. Sci. U.S.A.*, **69**, 3180.

Vyskoycil, F. and Magazanik, L. G. (1972). The desensitization of postjunctional muscle membrane after intracellular application of membrane stabilizers and snake venom polypeptides, *Brain Res.*, **48**, 417.

Webb, G. D. (1965). Affinity of benzoquinonium and ambenonium derivatives for the acetylcholine receptor, tested on the electroplax, and for acetylcholinesterase in solution, *Biochim. Biophys. Acta*, **102**, 172.

Weber, M. and Changeux, J. P. (1974a). Binding of *Naja nigricollis* [^{3}H] α-toxin to membrane fragments from *Electrophorus* and *Torpedo* electric organs, *Mol. Pharmacol.*, **10**, 1.

Weber, M. and Changeux, J. P. (1974b). Binding of *Naja nigricollis* [^{3}H] α-toxin to membrane fragments from *Electrophorus* and *Torpedo* electric organs. II. Effect of cholinergic agonists and antagonists on the binding of the tritiated α-neurotoxin, *Mol. Pharmacol.*, **10**, 15.

Weber, M. and Changeux, J. P. (1974c). Binding of *Naja nigricollis* [^{3}H] α-toxin to membrane fragments from *Electrophorus* and *Torpedo* electric organs. III. Effects of local angethetics on the binding of the tritiated α-neurotoxin, *Mol. Pharmacol.*, **10**, 35.

Weber, M., Menez, A., Fromageot, P., Boquet, P., and Changeux, J. P. (1972). Effet des agents cholinergigues et des anesthésiques locaux sur la cinétique de liaison de la toxins α tritiée de *Naja nigricollis* au récepteur cholinergique, *C. R. Acad. Sci. Paris*, **274**, 1575.

Zupancic, A. O. (1967). Evidence for the identity of anionic centers of cholinesterases with cholinoreceptors, *Ann. N.Y. Acad. Sci.*, **144**, 689.

Zupancic, A. O. (1970). Kinetic response of a membrane-bound acetylcholinesterase to cholinergic activating and blocking agents, *FEBS Lett.*, **11**, 277.

18 Neurotoxins: Chemistry and Structural Aspects

1 GENERAL TYPES OF NEUROTOXINS   258
   1.1 Axonic Transmission Inhibition, 258
   1.2 Synaptic Transmission Inhibition, 258
        Presynaptic Site, 258
        Postsynaptic Site, 262

2 ISOLATION   262

3 SEQUENCE   266
   3.1 Type I Neurotoxins, 266
   3.2 Type II Neurotoxins, 266

4 STRUCTURE–TOXIC ACTION RELATIONSHIP   272
   4.1 Tryptophan Residue, 272
   4.2 Tyrosine Residue, 275
   4.3 Arginine Residue, 276
   4.4 Lysine and N-Terminal Amino Acids, 276
   4.5 Disulfide Bond, 277
   4.6 Carboxyl Group, 278
   4.7 Histidine Residue, 278
   4.8 Size and Toxicity, 279

5 CONFORMATION AND STABILITY   279
   5.1 Importance of Disulfide Linkages in Neurotoxins, 279
   5.2 Stability, 281
   5.3 X-Ray Diffraction, 281
   5.4 Fluorescence Spectroscopy, 282
   5.5 Optical Rotatory Dispersion and Circular Dichroism, 282
   5.6 Laser Raman Spectroscopy, 283
        Peptide Backbones, 284
        Disulfide Linkages (S–S) and C–S, 288
        Side-Chain Residues, 289
        Conformation in Solid and in Solution, 290
        Conformation and Heat Treatment, 291

6 STRUCTURAL SIMILARITY OF MANY BASIC PROTEINS AND NEUROTOXINS   291

7 PRESYNAPTIC TOXINS AND THEIR RELATION TO PHOSPHOLIPASE A_2 ACTIVITY   293

   References   294

The term "neurotoxin" has been in common use by many people, yet there are different types of neurotoxins having different modes of neurotoxic action. Some toxins are neurotoxic because they inhibit axonic transmission. Others are neurotoxic because they inhibit synaptic transmission. Moreover, there are many different mechanisms of inhibition, even for synaptic transmission. It is not the objective of this author to describe neurotoxic mechanism in depth, but rather to review it briefly in order to facilitate understanding of snake neurotoxic actions. Details can be found in advanced pharmacology books and some reviews such as the book *Neuropoisons* by Simpson (1971) and the article by O'Brien (1969), "Poisons As Tools in Studying The Nervous System".

1 GENERAL TYPES OF NEUROTOXINS

For convenience, the different types of neurotoxic actions are classified into two groups: axonic transmission inhibition and synaptic transmission inhibition.

1.1 Axonic Transmission Inhibition

When a nerve is stimulated, depolarization of the axon takes place and the signal passes along the axon. This involves continuous changes of the Na(I) and K(I) concentrations in the nerve fibers (Fig. 1). Any substance that affects the normal movement of Na(I) or K(I) can be a neuropoison. Tetrodotoxin, a poison obtained from puffer fish and newt eggs, is known to affect Na(I) efflux while K(I) flow is unchanged (Narahashi et al., 1964). The site of the action is on the Na(I) gate itself rather than on Na(I). Thus internally perfused tetrodotoxin does not exert a neurotoxic effect (Narahasi and Haas, 1967). On the other hand, DDT is neurotoxic because it affects the Na pump as well as the K pump (Narahashi and Haas, 1967). Scorpion venoms and toxins also affect the conductivity of the axon because they delay the inactivation of Na(I) permeability and prolong the duration of action potential (see Chapter 28).

Snake venom neurotoxins and other basic proteins do not change the action potential of axonic transmission. However, venom phospholipase A_2 facilitates the penetration of basic proteins such as direct lytic factor into axons by splitting axonal phospholipids, thus causing disarrangement of the membrane matrix (see Chapter 2, "Phospholipase A_2").

1.2 Synaptic Transmission Inhibition

The neuromuscular junction or motor end plate is one place where a nerve fiber meets a skeletal muscle fiber and is a special form of synapasis (Fig. 2). When the nerve impulse arrives at the presynaptic site of the neuromuscular junction, Ca(II) moves into the membranes of the sole feet and triggers the releases of the transmitter substance, acetylcholine. Acetylcholine (about 10 Å in diameter) passes across the synaptic (about 200 Å) and combines with acetylcholine receptors at the postsynaptic site of neuromuscular junction. This causes depolarization of the muscle fiber and eventually muscle contraction. The acetylcholine molecules used for transmission are immediately hydrolyzed by the enzyme acetylcholinesterase. Thus any substances that interfere with normal processes of synaptic transmissions can be neurotoxic.

Presynaptic Site. The best example of this type of neurotoxin is β-bugarotoxin (Chang et al., 1973a), which exhibits an interesting property. Botulinum toxin and β-bungarotoxin are mutually antagonistic. In the presence of botulinum toxin, the inhibitory effect

1 General Types of Neurotoxins

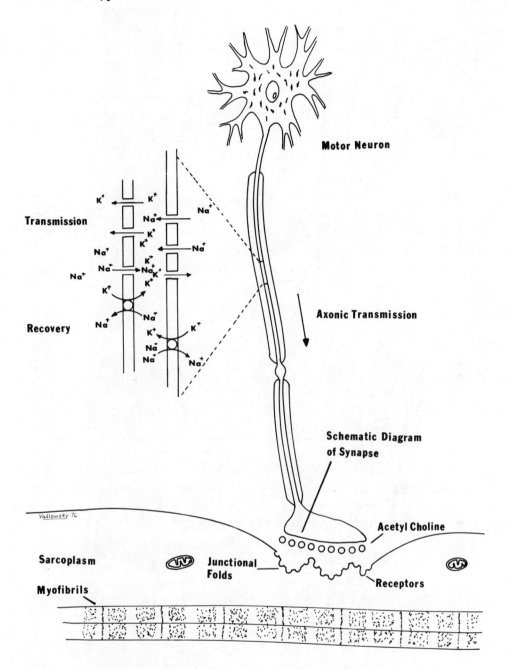

Figure 18.1 Diagram showing axonic transmission and synaptic transmission.

of β-bungarotoxin on transmitter release is antagonized. The converse is also true; the inhibitory effect of botulinum toxin is decreased in the presence of β-bungarotoxin (Chang et al., 1973b). β-Bungarotoxin has a molecular weight of 21,800 and consists of two subunits of molecular weights 8800 and 12,400, which are held together by disulfide bonds (Kelly and Brown, 1974). Biological activity is extremely stable and resistant to

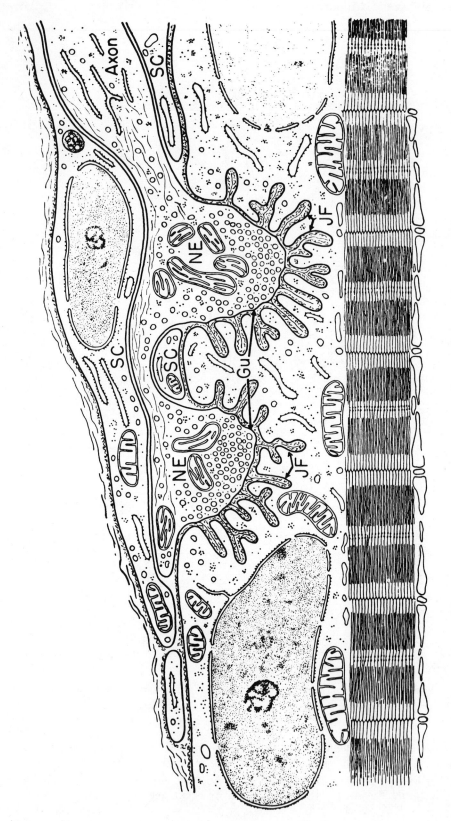

Figure 18.2 Diagram showing detailed view of the neuromuscular junction. SC: terminal Schwann cells, GU: synaptic gutters, JF: subneural folds of the postsynaptic membrane, NE: cross section of the terminal nerve branches at the motor end plate. (Photograph supplied by Dr. Keith R. Porter.)

1 General Types of Neurotoxins

heat. In a rat phrenic nerve–diaphragm preparation, the toxin produced failure of neuromuscular transmission, after which subthreshold end plate potentials could be detected.

Scorpion venom toxins are also presynaptic neurotoxins. They accelerate release of acetylcholine, thereby continuously depolarizing the muscle or postjunctional membrane. Under this condition, normal nerve impulses cannot be transmitted to the muscle. The neurotoxic action of scorpion venom is discussed in more detail in Chapter 28.

There are several modes of action whereby certain compounds can be designated as neurotoxic:

1. Inhibition of acetylcholinesterase biosynthesis.
2. Destruction of acetylcholinesterase before it is used for transmission.

Any compound that combines or competes with acetylcholinesterase can be neurotoxic. Such compounds are usually called antiacetylcholinesterases.

One antiacetylcholinesterase neurotoxic substance is (diisopropylflourophosphate) (DFP), which is commonly used as an inhibitor for enzymes involving serine as an active site.

Snake venom contain an inhibitor, antiacetylcholinesterase (see Chapter 10 "Enzyme Inhibitors"), but this does not seem to behave as a neurotoxic agent at the concentration present in the venom.

3. Destruction of synaptic vesicles. It is in the vesicles that acetylcholine is stored.

β-Bungarotoxin obtained from *Bungarus multicinctus* venom is a nondepolarizing neurotoxin that acts on the presynaptic site (Chang and Lee, 1963; Lee and Chang, 1966;

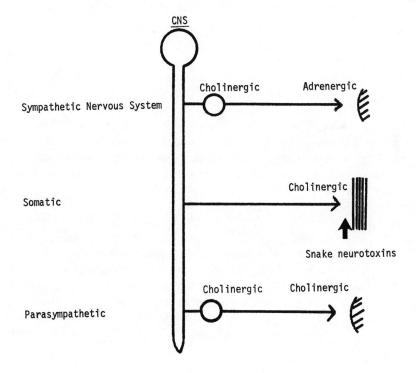

Figure 18.3 Diagram of the efferent nervous system. CNS: Central nervous system. The diagram indicates that there are different types of synaptic transmissions.

Lee, 1972; Chang et al., 1973a, b). Ultrastructural investigation indicates that the vesicles are completely eliminated in the axonal terminal (Chen and Lee, 1970).

4. Changes in sodium ion permeability. An example of this type of natural neurotoxin is batrochotoxin, obtained from the skin of a Colombian frog of the genus *Phyllobates*. The toxin increases Na(I) permeability, resulting in depolarization of the presynaptic terminal and a concomitant Ca(II)-dependent increase in acetylcholine release. The subsequent block in transmitter release appears due to complete depolarization of the nerve terminal. Thus the effects of batrochotoxin can be prevented by tetrodotoxin, which blocks passive diffusion of Na ions through excitable membranes (Daly, 1970; Daly and Witkop, 1971).

Postsynaptic Site. For the somatic cells to be depolarized for eventual muscle contraction, the acetylcholine receptors in the postsynaptic membrane must receive the transmitter substance, acetylcholine. Figure 2 shows the ultrastructure of the motor end plate, drawn schematically for better understanding of the neuromuscular junction. Most snake venom toxins, such as sea snake and cobra neurotoxins and α-bungarotoxin, strongly bind to the acetylcholine receptor; thus they are neurotoxic or, more specifically, myoneural toxicants. (For further detail, see Chapter 17, "Binding of Neurotoxins to Acetylcholine Receptors.")

In the peripheral nervous systems, somatic nerve transmission is only one of three types (Fig. 3). Whether snake neurotoxins have any effects on other synaptic transmissions, such as cholinergic junctions in the parasympathetic and sympathetic ganglia and noradrenalic junctions in the sympathetic nervous system, is a question of great interest.

2 ISOLATION

A large number of neurotoxins have been isolated from the venoms of Elapidae. The amino acid compositions of these neurotoxins from Asia and Africa are listed in Tables 1 and 2, respectively. Neurotoxins isolated from Asian and African Elapidae venoms consist of either about 60 to 62 or more than 70 amino acid residues. Thus they can be conveniently classified accordingly, the first group being designated as Type I neurotoxins and the second as Type II (Fig. 4). Sometimes, Type I neurotoxins are referred to as short-chain toxins; Type II neurotoxins, as long-chain toxins. The two types have similar biological actions, and they both block acetylcholine receptors in skeletal muscle (Szczepaniak, 1974).

A neurotoxin has been isolated from the venom of the Australian tiger snake, *Notechis suctatus scutatus* (Karlsson et al., 1972). Notexin is totally different from Types I and II neurotoxins of Asian and African snakes, as can be seen from its amino acid composition (Table 3). As pointed out in Chapter 12, plants and animals in Australia and New Guinea are normally quite distinct from Asian species; the well-known dividing line is the Wallace line. In view of the zoogeographic distributions of animals and plants in Australia, it is to be expected that a toxin from Australia would be different from toxins from Asian and African snakes. The sequence of notexin, which was identified by Halpert and Eaker (1975), is as follows:

10
Asn–Leu–Val–Gln–Phe–Ser–Tyr–Leu–Ile–Gln–Cys–Ala–Asn–His–Gly–Lys–Arg–

```
                    20                                30
Pro–Thr–Trp–His–Tyr–Met–Asp–Tyr–Gly–Cys–Tyr–Cys–Gly–Ala–Gly–Gly–Ser–
              40                              50
Gly–Thr–Pro–Val–Asp–Glu–Leu–Asp–Arg–Cys–Cys–Lys–Ile–His–Asp–Asp–Cys–
                        60
Tyr–Asp–Glu–Ala–Gly–Lys–Lys–Gly–Cys–Phe–Pro–Lys–Met–Ser–Ala–Tyr–Asp–
      70                              80
Tyr–Tyr–Cys–Gly–Glu–Asn–Gly–Pro–Tyr–Cys–Arg–Asn–Ile–Lys–Lys–Lys–
                  90                                      100
Cys–Leu–Arg–Phe–Val–Cys–Asp–Cys–Asp–Val–Glu–Ala–Ala–Phe–Cys–Phe–Ala–
                          110
Lys–Ala–Pro–Tyr–Asn–Asn–Ala–Asn–Trp–Asn–Ile–Asp–Thr–Lys–Lys–Arg–Cys–
Gln.
```

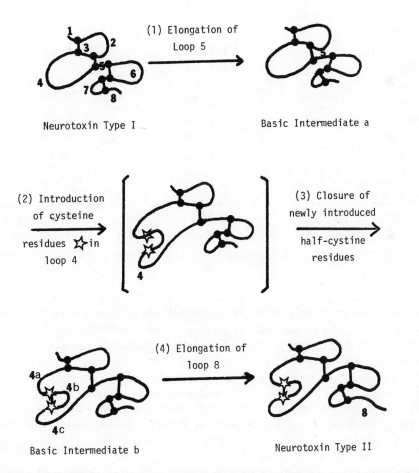

Figure 18.4 The evolutionary conversion of Type I neurotoxin to nonneurotoxic basic proteins and Type II neurotoxin.

Table 1 Neurotoxins from the Venoms of Krait and Asian Cobra (Family: Elapidae)

Genus Species Subspecies	Bungarus Multicinctus			Naja naja					Naja naja atra		Naja naja atra	Naja naja kaouthia
Origin	Formosa			India					Pakistan		Formosa	India
Name	α	α	β	A	B	C	D	II	3	4	cobrotoxin	toxin
Amino Acid												
Lysine	6	6	13	4	4	5	4	5	4	4	3	5
Histidine	2	2	5	1	1	1	1	2	1	1	2	1
Arginine	3	3	14	6	6	5	5	4	6	6	6	5
Aspartic acid	4	4	22	9	9	9	9	7	9	9	8	9
Threonine	7	7	11-12	9	9	9	8	6	9	9	8	9
Serine	6	6	6	3	4	3	3	4	3	4	4	3
Glutamic acid	5	5	12	1	1	1	2	6	1	1	7	1
Proline	8	8	8	6	6	6	6	4	6	6	2	6
Glycine	4	4	16	5	5	4	5	6	5	5	7	4
Alanine	5	5	11	2	2	3	2	0	2	2	0	3
Valine	5	4	4	4	4	4	4	2	4	4	1	4
Methionine	1	1	2	0	0	0	0	0	0	0	0	0
Isoleucine	2	2	8	5	4	5	5	2	5	4	2	5
Leucine	2	2	7	1	1	1	1	2	1	1	1	1
Tyrosine	2	2	13	1	1	1	1	1	1	1	2	1
Phenylalanine	1	1	6	3	3	3	3	0	3	3	0	3
Half-cystine	10	10	19-20	10	10	10	11	7	10	10	8	10
Tryptophan	1	2	3	1	1	1	2	1	1	1	1	1
Total residue	74	74	180-182	71	71	71	72	59	71	71	62	71
References	Mebs et al, 1971	Clark et al, 1972	Lee et al, 1972	Nakai et al, 1971				Hayashi and Ohta, 1975	Karlsson et al, 1971		Yang et al, 1969a	Karlsson and Eaker, 1972

It is more similar to venom and pancreatic phospholipase A_2 than to Type II neurotoxins.

Type I neurotoxins contain eight half-cystine or four disulfide bonds. All the sea snake (Hydrophiidae) neurotoxins belong to Type I. One remarkable finding is that all of the Type I neurotoxins obtained from Elapidae and Hydrophiidae are very similar in their chemical structures, regardless of species or geographical origin.

Sometimes, there are small amino acid composition differences in the venoms, depending on the geographical origin of the species (see examples in Tables 1 and 2). Currently, classification of snakes depends entirely on morphology. The chemical differences observed in venoms and snakes serum proteins should aid in the taxonomy of poisonous snakes, especially at the species and subspecies level.

Both Type I and Type II neurotoxins contain only small numbers of residues to histidine (usually 1 or 2), alanine (usually 0 to 4), methionine (0 to 2), leucine (1 or 2),

2 Isolation

Table 1 *Continued*

	Naja naja philippinensis			Naja naja oxiana			Naja oxiana		Naja naja siamensis	Ophiophagus hannah				
	Philippine			C. Asia U.S.S.R.		Iran	C. Asia U.S.S.R.		Thailand	Thailand				
										Minor Neurotoxins				
	F-2	F-1	I	I	α	toxin	I	II	Principal Neurotoxin	3c	5	7c	a	b
	3	3	6	5	5	5	6	6	5	3	5	2	6	8
	2	2	2	2	2	2	1	2	1	2	2	2	0	1
	6	5	4	4	4	4	2	4	5	6	4	7	4	3
	8	8	7	8	8	8	5	8	9	7	8	8	7	9
	8	8	6	6	6	6	7	6	9	10	6	8	9	9
	4	4	4	4	4	4	3	4	3	3	4	4	3	4
	7	7	6	6	6	6	6	7	1	8	6	7	4	2
	2	2	3	4	4	4	6	4	6	2	3	2	7	6
	7	7	5	5	5	5	4	5	4	7	5	7	4	5
	0	0	1	0	0	0	4	0	3	0	0	0	4	3
	1	1	2	2	2	2	1	2	4	1	2	1	5	3
	0	0	0	0	0	0	0	0	0	0	0	0	0	0
	2	2	2	2	2	2	4	2	5	0	2	2	3	3
	1	1	2	2	2	2	2	2	1	2	2	1	1	1
	2	2	1	1	1	1	2	1	1	2	1	2	2	2
	0	0	0	0	0	0	0	0	3	0	0	0	1	2
	8	8	8	8	8	8	8	8	10	8	8	8	10	10
	1	1	2	2	2	2	1	1	1	1	2	1	3	2
	62	61	61	61	61	61	62	62	71	62	61	62	73	73
	Nakai et al, 1970		Hauert et al, 1974	Grishin et al, 1973	Arnberg et al, 1974	Karlsson and Eaker, 1972	Turakulov et al, 1972		Karlsson et al, 1971	Joubert, 1973				

tryosine (1 or 2), phenylalanine (0 to 3), and tryptophan (1 or 2). It is of interest to note that these amino acids are hydrophobic. Neurotoxins are highly basic proteins with high pI values; yet they contain large numbers of acidic amino acids. This indicates that many of them are amidated, a conclusion verified by sequence study. It is also interesting that neurotoxins are rich in nonionizable hydroxyl groups containing amino acids, threonine, and serine. This suggests that there must be extensive networks of hydrogen bonds among charged amino acids (lysine, histidine, arginine, aspartic acid, glutamic acid) and noncharged polar amino acids (threonine and serine). These hydrogen bonds add extra stability to neurotoxins in addition to that provided by the four covalently linked-disulfide bonds.

A neurotoxin with the sequence of cobrotoxin was synthesized by a solid-phase method (Aoyagi et al., 1972). The synthesized toxin demonstrated 20% of the toxicity of native cobrotoxin. A synthetic approach should further clarify the molecular mechanism of toxic action in the future.

The toxicities of purified neurotoxins are usually much higher than those of the original venoms. This is reasonable; as nontoxic materials in venoms are removed, the toxicity of the pure toxins should increase. Another reason for the high toxicity of neurotoxins is that they are very stable and do not lose potency during the purification process.

3 SEQUENCE

Information regarding the sequence of various snake toxins is very important as it provides insight on structure–function relationships, the molecular evolution of various proteins, and molecular mechanisms of toxic actions. For convenience, two separate tables have been constructed, one for Type I neurotoxins and a second for Type II neurotoxins. Structurally related, relatively low-toxic basic proteins are discussed in Chapter 19.

3.1 Type I Neurotoxins

One of the most remarkable findings from the sequence studies of Type I neurotoxins is that the relative positions of cysteine residues are identical regardless of species and geographical origin (Table 4). For convenience, each portion of the toxin is identified by a loop number from the N-terminal to the next cysteine residue. Since the sixth and seventh cysteines are next to each other, no loop number is assigned. Notable conclusions can be reached by comparing the different toxins of Type I.

LOOP
1. Contains only 2 residues but is a variable region as far as the specific amino acid residue is concerned.
2. Contains 13 amino acid residues; most of the sequences are remarkably similar to each other.
3. Varies in the number of amino acid residues, which ranges from 4 to 5.
4. Contains 16 residues and has a high degree of homology in amino acid sequences. The first residue is always tyrosine, next to the disulfide bond cysteine. The last two residues are always arginine and glycine.
5. Contains only 1 residue, always glycine.
6. Contains 10 residues but is a fairly variable region with the exception of the first residue which is always proline.
7. Contains 4 residues with one exception, which has 5.
8. Contains 2 residues, and the first is always asparagine.

The positions of the disulfide bonds of several Type I neurotoxins have been identified (Yang et al., 1970; Endo et al., 1971). The relative positions are the same, and it can be assumed that all neurotoxins of Type I have identical disulfide bonds at the same positions.

3.2 Type II Neurotoxins

Whereas Type I neurotoxins contain 8 cysteines or 4 disulfide bonds, those of Type II possess 10 cysteines or 5 disulfide bonds (Table 5). When the two types are compared,

Table 2 Neurotoxins from the Venoms of African Cobra and Mamba (Family: Elapidae)

Genus Species Subspecies	Dendroaspis angusticeps			D. jamesoni kaimosae		D. polylepis polylepis		D. viridis			Hemachates hemachates				Naja haje					N. haje annulifera				Naja melanoleuca				N. nigricollis		N. nigricollis mossambica			N. nivea		
Origin	S. Africa			S. Africa		S. Africa		Guinea			Africa				Africa					S. Africa			Africa	S. Africa			Guinea	S. Africa		Africa		Africa	S. Africa		Africa
Name	F_{VII}	Ta_1	Ta_2	Vn^I_I	Vn^{III}_I	α	γ	4-7-3	4-9-3	4-11-3	I	II	III	IV	α	I^a	II^a	I^b	II^b	III^b	CM-10	CM-12	CM-14	III	CM-8	CM-11	CM-13a	F_5	F_7	b	d	α	I	II	I	II	III	α	β	δ
Lysine	4	4	7	6	8	6	9	7	7	6	6	6	4	6	6	6	6	6	6	4	7	7	7	4	9	10	11	7	6	4	6	6	6	6	4	4	5	6	7	6
Histidine	2	2	2	4	0	3	0	1	1	4	1	1	2	2	2	2	2	2	2	0	1	1	2	0	1	0	1	1	2	1	2	2	2	2	2	3	3	1	2	2
Arginine	5	5	3	3	5	5	4	4	4	4	4	4	5	4	4	4	4	4	4	5	3	4	5	5	1	1	1	2	3	5	3	3	3	3	7	6	6	6	6	4
Aspartic acid	7	7	2	6	6	5	7	6	6	6	6	6	9	5	7	7	7	7	5	10	4	4	5	10	6	7	4	5	7	10	7	7	7	7	7	8	8	9	5	7
Threonine	6	6	4	6	8	5	6	7	7	6	7	7	7	9	7	7	6	7	5	7	5	5	5	8	4	4	6	7	6	8	7	8	8	8	8	7	7	5	5	7
Serine	5	5	5	5	4	4	4	3	3	5	3	3	3	4	4	4	4	3	3	3	3	3	3	3	2	2	3	3	4	3	4	2	2	2	3	4	3	3	3	4
Glutamic acid	2	2	5	5	4	5	6	6	6	4	5	5	5	8	7	7	6	7	7	3	7	6	5	3	1	0	2	4	8	1	7	6	6	5	7	6	5	1	5	7
Proline	5	5	4	2	5	2	4	7	7	2	7	7	5	4	4	4	4	4	4	6	3	3	4	6	5	5	5	7	4	6	4	5	5	5	4	4	4	6	3	4
Glycine	5	5	3	5	5	5	5	5	5	5	6	5	5	5	5	5	5	5	6	5	6	6	6	5	2	2	2	3	5	4	5	5	5	5	6	5	4	5	6	5
Alanine	1	1	4	1	2	1	4	4	4	2	4	4	0	0	0	0	0	0	0	3	0	0	0	3	1	1	3	2	0	4	0	0	0	0	0	0	1	3	0	0
Valine	2	2	4	1	4	2	3	4	4	2	5	5	1	1	1	1	1	1	2	4	2	2	2	4	7	6	1	2	2	5	2	2	2	2	1	1	1	5	2	1
Methionine	2	2	2	0	0	0	0	0	0	0	0	0	0	0	0	0	1	0	1	1	1	1	1	1	2	2	0	1	1	0	1	0	0	1	0	0	1	2	1	0
Isoleucine	1	1	3	4	2	4	3	2	2	3	2	2	3	1	3	3	5	3	5	4	6	6	5	1	1	2	3	1	3	4	3	3	3	4	1	1	1	3	5	3
Leucine	3	3	2	1	1	0	1	1	1	1	1	1	2	2	1	1	0	1	0	1	1	1	0	1	5	5	8	2	0	1	0	2	2	1	2	2	2	1	0	1
Tyrosine	3	3	2	2	4	4	1	2	2	2	2	2	1	1	1	1	1	1	2	1	2	2	2	1	3	3	1	2	1	1	1	1	1	1	2	2	2	1	2	1
Phenylalanine	0	0	0	0	2	0	3	1	1	0	1	1	0	0	0	0	0	0	0	3	1	1	0	3	1	1	2	0	0	3	0	0	0	0	0	0	0	3	0	0
Half-cystine	8	8	8	8	10	8	10	10	10	8	10	10	8	8	8	8	8	8	8	10	8	8	8	10	8	8	8	10	8	10	8	8	8	8	8	8	8	10	8	8
Tryptophan	0	0	0	1	2	1	2	2	2	1	3	3	1	1	1	1	1	1	1	1	1	1	1	1	1	1	0	2	1	1	1	1	1	1	1	1	2	1	1	1
Total residue	61	61	60	60	72	60	72	72	71	60	73	72	61	61	61	61	61	61	61	71	61	61	61	72	60	60	61	61	61	71	61	61	61	61	63	63	63	71	61	61
References	Viljoen and Botes, 1973	Viljoen and Botes, 1973	Viljoen and Botes, 1974	Strydom, 1973	Strydom, 1973	Strydom, 1972	Strydom, 1972	Shipolini et al, 1973	Shipolini et al, 1973	Shipolini et al, 1973	Bechis et al, 1976	Bechis et al, 1976	Strydom and Botes, 1971	Botes and Strydom, 1969	Miranda et al, 1970	Miranda et al, 1970	Miranda et al, 1970	Miranda et al, 1970	Miranda et al, 1970	Miranda et al, 1970	Joubert, 1975	Joubert, 1975	Joubert, 1975	Kopeyan et al, 1975	Joubert, 1976	Joubert, 1976	Joubert, 1976	Poilleux and Boquet, 1972	Poilleux and Boquet, 1972	Botes, 1972	Botes, 1972	Karlsson et al, 1966	Kopeyan et al, 1973	Kopeyan et al, 1973	Rochat et al, 1974	Rochat et al, 1974	Rochat et al, 1974	Botes et al, 1971	Botes et al, 1971	Botes et al, 1971

Table 4 Amino Acid Sequences of Type I Neurotoxins

Snake (Reference)	Toxin[a]	Loop 1 – Loop 8 Sequence
Elapidae		
Dendroaspis angusticeps (Viljoen and Botes, 1973)	F$_{VII}$ (Ta1)	Thr-Met-Cys-Tyr-Ser-His-Thr-Thr-Thr-Ser-Arg-Ala-Ile-Leu-Thr-Asn-Cys- - - -Gly-Glu-Asn-Ser-Cys-Tyr-Arg-Lys-Ser-Arg-Arg-His-Pro-Pro-Lys-Met-Val-Leu-Gly-Arg-Gly-Cys-Gly-Cys-Pro-Pro-Gly-Asp-Asp-Asn-Leu-Glu-Val-Lys-Cys-Cys-Thr-Ser-Pro-Asp-Lys-Cys-Asn-Tyr
(Viljoen and Botes, 1974)	Ta2	Met-Ile-Cys-Tyr-Ser-His-Lys-Thr-Pro-Gln-Pro-Ser-Ala-Thr-Ile-Thr-Cys- - - -Glu-Asp-Lys-Thr-Cys-Tyr-Lys-Lys-Ser-Val-Arg-Lys-Leu-Pro-Ala-Val-Val-Ala-Gly-Arg-Gly-Cys-Gly-Cys-Pro-Ser-Lys-Glu-Met-Leu-Val-Ala-Ile-His-Cys-Cys-Arg-Ser- - -Asp-Lys-Cys-Asn-Glu
Dendroaspis jamesoni kaimosae (Strydom, 1973)	V$_n$^{I}I	Arg-Ile-Cys-Tyr-Asn-His-Gln-Ser-Thr-Thr-Pro-Ala-Thr-Thr-Lys-Ser-Cys- - - -Gly-Glu-Asn-Ser-Cys-Tyr-Lys-Lys-Thr-Trp-Ser-Asp-His-Arg-Gly-Thr-Ile-Ile-Glu-Arg-Gly-Cys-Gly-Cys-Pro-Lys-Val-Lys-Gln-Gly-Ile-His-Leu-His-Cys-Cys-Gln-Ser- - -Asp-Lys-Cys-Asn-Asn
Dendroaspis polylepis polylepis (Strydom, 1972)	α	Arg-Ile-Cys-Tyr-Asn-His-Gln-Ser-Thr-Thr-Arg-Ala-Thr-Thr-Lys-Ser-Cys- - - -Glu-Asn-Ser-Cys-Tyr-Lys-Lys-Tyr-Trp-Arg-Asp-His-Arg-Gly-Thr-Ile-Ile-Glu-Arg-Gly-Cys-Gly-Cys-Pro-Lys-Val-Lys-Pro-Gly-Val-Gly-Ile-His-Cys-Cys-Gln-Ser- - -Asp-Lys-Cys-Asn-Tyr
Dendroaspis viridis (Banks et al, 1974)	4.11.3	Arg-Ile-Cys-Tyr-Asn-His-Gln-Ser-Thr-Thr-Pro-Ala-Tyr-Thr-Lys-Ser-Cys- - - -Gly-Glu-Asn-Ser-Cys-Tyr-Lys-Lys-Thr-Trp-Ser-Asp-His-Arg-Gly-Thr-Ile-Ile-Glu-Arg-Gly-Cys-Gly-Cys-Pro-Lys-Val-Lys-Arg-Gly-Val-His-Leu-His-Cys-Cys-Gln-Ser- - -Asp-Lys-Cys-Asn-Asn
Hemachatus haemachatus (Strydom and Botes, 1971)	II	Leu-Glu-Cys-His-Asn-Gln-Gln-Ser-Thr-Gln-Pro-Thr-Thr-Lys-Ser-Cys-Pro- - -Gly-Asp-Thr-Asn-Cys-Tyr-Asn-Lys-Arg-Trp-Arg-Asp-His-Arg-Gly-Thr-Ile-Ile-Glu-Arg-Gly-Cys-Gly-Cys-Pro-Thr-Val-Lys-Pro-Gly-Ile-Asn-Leu-Lys-Cys-Cys-Thr-Thr- - -Asp-Arg-Cys-Asn-Asn
(Strydom and Botes, 1971)	IV	Leu-Glu-Cys-His-Asn-Gln-Gln-Ser-Ser-Gln-Thr-Pro-Thr-Thr-Gln-Thr-Cys-Pro- - -Gly-Asp-Thr-Asn-Cys-Tyr-Lys-Lys-Gln-Trp-Arg-Asp-His-Arg-Gly-Ser-Arg-Thr-Glu-Arg-Gly-Cys-Gly-Cys-Pro-Thr-Val-Lys-Pro-Gly-Ile-Lys-Leu-Lys-Cys-Cys-Thr-Thr- - -Asp-Arg-Cys-Asn-Lys
Naja haje annulifera (Joubert, 1975a)	CM-10	Met-Ile-Cys-Tyr-Lys-Gln-Gln-Ser-Leu-Gln-Phe-Pro-Ile-Thr-Thr-Val-Cys-Pro- - -Gly-Glu-Lys-Asn-Cys-Tyr-Lys-Lys-Gln-Trp-Ser-Gly-His-Arg-Gly-Thr-Ile-Ile-Glu-Arg-Gly-Cys-Gly-Cys-Pro-Ser-Lys-Lys-Gly-Ile-Glu-Ile-Asn-Cys-Cys-Thr-Thr- - -Asp-Lys-Cys-Asn-Arg
(Joubert, 1975a)	CM-12	Met-Ile-Cys-Tyr-Lys-Gln-Arg-Ser-Leu-Gln-Phe-Pro-Ile-Thr-Thr-Val-Cys-Pro- - -Gly-Glu-Lys-Asn-Cys-Tyr-Lys-Lys-Gln-Trp-Ser-Asp-His-Arg-Gly-Thr-Ile-Ile-Glu-Arg-Gly-Cys-Gly-Cys-Pro-Ser-Lys-Lys-Gly-Ile-Glu-Ile-Asn-Cys-Cys-Thr-Thr- - -Asp-Lys-Cys-Asn-Arg
(Joubert, 1975a)	CM-14	Met-Ile-Cys-His-Gln-Gln-Ser-Ser-Gln-Pro-Pro-Thr-Ile-Lys-Thr-Cys-Pro- - -Gly-Glu-Thr-Asn-Cys-Tyr-Lys-Lys-Arg-Trp-Asp-Asp-Arg-Gly-Thr-Ile-Ile-Glu-Arg-Gly-Cys-Gly-Cys-Pro-Lys-Lys-Gly-Val-Gly-Ile-Tyr-Cys-Cys-Lys-Thr- - -Asn-Lys-Cys-Asn-Arg
Naja haje haje (Botes and Strydom, 1969)	α	Leu-Gln-Cys-His-Asn-Gln-Gln-Ser-Ser-Glu-Pro-Pro-Thr-Thr-Lys-Thr-Cys-Pro- - -Gly-Glu-Thr-Asn-Cys-Tyr-Lys-Lys-Arg-Trp-Arg-Asp-His-Arg-Gly-Ser-Ile-Thr-Glu-Arg-Gly-Cys-Gly-Cys-Pro-Ser-Val-Lys-Lys-Gly-Ile-Glu-Ile-Asn-Cys-Cys-Thr-Thr- - -Asp-Lys-Cys-Asn-Asn
Naja melanoleuca (Botes, 1972)	d	Met-Glu-Cys-His-Asn-Gln-Gln-Ser-Ser-Gln-Thr-Pro-Thr-Thr-Lys-Thr-Cys-Pro- - -Gly-Glu-Thr-Asn-Cys-Tyr-Lys-Lys-Gln-Trp-Asp-Asp-His-Arg-Gly-Thr-Ile-Ile-Glu-Arg-Gly-Cys-Gly-Cys-Pro-Ser-Val-Lys-Lys-Gly-Val-Lys-Ile-Asn-Cys-Cys-Thr-Thr- - -Asp-Arg-Cys-Asn-Asn
Naja nigricollis (Eaker and Porath, 1967) (Botes and Strydom, 1969)	α	Leu-Glu-Cys-His-Asn-Glx-Glx-Ser-Ser-Glx-Pro-Pro-Thr-Thr-Lys-Thr-Cys-Pro- - -Gly-Glx-Thr-Asx-Cys-Tyr-Lys-Lys-Val-Trp-Arg-Asp-His-Arg-Gly-Thr-Ile-Ile-Glu-Arg-Gly-Cys-Gly-Cys-Pro-Thr-Val-Lys-Pro-Gly-Ile-Lys-Leu-Asn-Cys-Cys-Thr-Thr- - -Asx-Lys-Cys-Asn-Asn
Naja naja atra (Yang et al, 1969a)	Cobrotoxin	Leu-Glu-Cys-His-Asn-Gln-Gln-Ser-Ser-Gln-Thr-Pro-Thr-Thr-Thr-Gly-Cys-Ser-Gly-Gly-Thr-Asn-Cys-Tyr-Lys-Lys-Arg-Arg-Asp-His-Arg-Gly-Tyr-Arg-Thr-Glu-Arg-Gly-Cys-Gly-Cys-Pro-Ser-Val-Lys-Asn-Gly-Ile-Glu-Ile-Asn-Cys-Cys-Thr-Thr- - -Asp-Arg-Cys-Asn-Asn
Naja naja oxiana (Grishin et al, 1973) (Arnberg et al, 1974)	II / Oxiana α	Leu-Glu-Cys-His-Asn-Gln-Gln-Ser-Ser-Gln-Thr-Pro-Thr-Thr-Lys-Thr-Cys-Ser- - -Gly-Glu-Thr-Asn-Cys-Tyr-Lys-Lys-Trp-Trp-Ser-Asp-His-Arg-Gly-Thr-Ile-Ile-Glu-Arg-Gly-Cys-Gly-Cys-Pro-Lys-Val-Lys-Pro-Gly-Val-Asn-Leu-Asn-Cys-Cys-Arg-Thr- - -Asp-Arg-Cys-Asn-Arg
Naja nivea (Botes, 1971)	β	Met-Ile-Cys-His-Asn-Gln-Gln-Ser-Ser-Gln-Arg-Pro-Thr-Ile-Lys-Thr-Cys-Pro- - -Gly-Glu-Thr-Asn-Cys-Tyr-Lys-Lys-Arg-Trp-Arg-Asp-His-Arg-Gly-Thr-Ile-Ile-Glu-Arg-Gly-Cys-Gly-Cys-Pro-Ser-Val-Lys-Lys-Gly-Val-Gly-Ile-Tyr-Cys-Cys-Lys-Thr- - -Asp-Lys-Cys-Asn-Arg
(Botes et al, 1971) (Same as *Naja haje haje* - α)	δ	Leu-Gln-Cys-His-Asn-Gln-Gln-Ser-Ser-Gln-Pro-Pro-Thr-Thr-Lys-Thr-Cys-Pro- - -Gly-Glu-Thr-Asn-Cys-Tyr-Lys-Lys-Arg-Trp-Arg-Asp-His-Arg-Gly-Ser-Ile-Thr-Glu-Arg-Gly-Cys-Gly-Cys-Pro-Ser-Val-Lys-Lys-Gly-Ile-Glu-Ile-Asn-Cys-Cys-Thr-Thr- - -Asp-Lys-Cys-Asn-Asn
Sea snake (Hydrophiidae)		
Enhydrina schistosa (Fryklund et al, 1972)	schistosa 4	Met-Thr-Cys-Cys-Asn-Gln-Gln-Ser-Ser-Gln-Pro-Lys-Thr-Thr-Thr-Asn-Cys- - - -Ala-Glu-Ser-Ser-Cys-Tyr-Lys-Lys-Thr-Trp-Ser-Asp-His-Arg-Gly-Thr-Arg-Ile-Glu-Arg-Gly-Cys-Gly-Cys-Pro-Gln-Val-Lys-Pro-Gly-Ile-Lys-Leu-Glu-Cys-Cys-His-Thr- - -Asn-Glu-Cys-Asn-Asn
(Fryklund et al, 1972)	schistosa 5	Met-Thr-Cys-Cys-Asn-Gln-Gln-Ser-Ser-Gln-Pro-Lys-Thr-Thr-Thr-Asn-Cys- - - -Ala-Glu-Ser-Ser-Cys-Tyr-Lys-Lys-Thr-Trp-Arg-Asp-His-Arg-Gly-Thr-Arg-Ile-Glu-Arg-Gly-Cys-Gly-Cys-Pro-Gln-Val-Lys-Ser-Gly-Ile-Lys-Leu-Glu-Cys-Cys-His-Thr- - -Asn-Glu-Cys-Asn-Asn
Laticauda semifasciata (Sato and Tamiya, 1971)	Erabutoxin α	Arg-Ile-Cys-Phe-Asn-Gln-His-Ser-Ser-Gln-Pro-Gln-Thr-Thr-Lys-Thr-Cys-Pro-Ser-Gly-Ser-Glu-Ser-Cys-Tyr-Asn-Lys-Gln-Trp-Ser-Asp-Phe-Arg-Gly-Thr-Ile-Ile-Glu-Arg-Gly-Cys-Gly-Cys-Pro-Thr-Val-Lys-Pro-Gly-Ile-Lys-Leu-Ser-Cys-Cys-Glu-Ser- - -Glu-Val-Cys-Asn-Asn
(Sato and Tamiya, 1971)	Erabutoxin β	Arg-Ile-Cys-Phe-Asn-Gln-His-Ser-Ser-Gln-Pro-Gln-Thr-Thr-Lys-Thr-Cys-Pro-Ser-Gly-Ser-Glu-Ser-Cys-Tyr-His-Lys-Gln-Trp-Ser-Asp-Phe-Arg-Gly-Thr-Ile-Ile-Glu-Arg-Gly-Cys-Gly-Cys-Pro-Thr-Val-Lys-Pro-Gly-Ile-Lys-Leu-Ser-Cys-Cys-Glu-Ser- - -Glu-Val-Cys-Asn-Asn
(Tamiya and Abe, 1972)	Erabutoxin c	Arg-Ile-Cys-Phe-Asn-Gln-His-Ser-Ser-Gln-Pro-Gln-Thr-Thr-Lys-Thr-Cys-Pro-Ser-Gly-Ser-Glu-Ser-Cys-Tyr-His-Lys-Gln-Trp-Ser-Asp-Phe-Arg-Gly-Thr-Ile-Ile-Glu-Arg-Gly-Cys-Gly-Cys-Pro-Thr-Val-Lys-Pro-Gly-Ile-Asn-Leu-Ser-Cys-Cys-Glu-Ser- - -Glu-Val-Cys-Asn-Asn

[a] For the sequences of toxins a, b, and c isolated from the venom of *Aipysarus laevis*, see the paper by Maeda and Tamiya, 1976. The paper appeared after the table was constructed and could not be included.

3 Sequence

Table 3 Neurotoxin from the Venom of Australian Snake

Genus Species Subspecies	Notechis scutatus scutatus
Origin	Australia
Amino Acid	Toxin
Lysine	11
Histidine	3
Arginine	5
Aspartic acid	18
Threonine	3
Serine	3
Glutamic acid	7
Proline	5
Glycine	10
Alanine	9
Valine	4
Methionine	2
Isoleucine	4
Leucine	4
Tyrosine	10
Phenylalanine	5
Half-cystine	14
Tryptophan	2
Total residue	119
References	Karlsson et al, 1972

certain similarities appear; these are shown diagrammatically in Fig. 5. If we consider the fourth and fifth cysteine residues of Type II as insertions to loop 4 of Type I, the relative positions of the rest of the residues correspond well for the two types of toxins. Therefore assignment of loops 4a, 4b, and 4c is made for the portion from the third to the sixth cysteine residues in Type II neurotoxin.

Table 5 Amino Acid Sequences of Type II Neurotoxins

Snake	Toxin	Loop 1	Loop 2	Loop 3
Bungarus multicinctus	α-bungarotoxin	Ile-Val-Cys-His-Thr-Thr-Ala-Thr-Ile-Pro-Ser-Ser-Ala-Val-Thr-Cys-Pro-Pro-Gly-Glu-Asn-Leu-		
Dendroaspis jamesoni kaimosae	V_NIII_1	Arg-Thr-Cys-Tyr-Lys- - - -Thr-Tyr-Ser-Asp-Lys-Ser-Lys-Thr-Cys-Pro-Pro-Gly-Glu-Asp-Ile-		
Dendroaspis polylepis polylepis	γ	Arg-Thr-Cys-Asn-Lys- - - - -Thr-Phe-Ser-Asp-Gln-Ser-Lys-Ile-Cys-Pro-Pro-Gly-Glu-Asn-Ile-		
Dendroaspis polylepis polylepis	δ	Arg-Thr-Cys-Tyr-Lys- - - - -Thr-Pro-Ser-Asp-Gln-Ser-Lys-Ile-Cys-Pro-Pro-Gly-Glu-Asn-Ile-		
Dendroaspis viridis^a	4.7.3	Arg-Thr-Cys-Tyr-Lys- - - - -Thr-Pro-Ser-Val-Lys-Pro-Glu-Ile-Cys-Pro-Pro-Gly-His-Gly-Asn-Ile-		
	4.9.3 (has oxidized Trp residue — circled)			
Naja haie	III	Ile-Arg-Cys-Phe-Ile- - - - -Thr-Pro-Asp-Val-Thr-Ser-Gln-Ala-Thr-Cys-Pro-Asp-Gly-Gln-Asn-Ile-		
Naja melanoleuca	b	Ile-Arg-Cys-Phe-Ile- - - - -Thr-Pro-Asp-Val-Thr-Ser-Gln-Ile-Cys-Ala-Asp-Gly- - -His-Val-		
Naja naja	3.9.4	Lys-Arg-Cys-Tyr-Arg- - - - -Thr-Pro-Asp-Leu-Lys-Ser-Gln-Thr-Cys-Pro-Pro-Gly-Glu- - -His-Val-		
Naja naja oxiana	A	Ile-Arg-Cys-Phe-Ile- - - - -Thr-Pro-Asp-Ile-Thr-Ser-Lys-Asp-Cys-Pro-Asn-Gly- - -His-Val-		
Naja nivea	I	Ile-Arg-Cys-Phe-Ile- - - - -Pro-Ile-Pro-Ile-Thr-Ser-Glu-Thr-Cys-Ala-Pro-Gly-Gln-Asn-Leu-		
Ophiophagus hannah	α	Thr-Arg-Cys-Phe-Ile- - - - -Thr-Pro-Asp-Val-Thr-Ser-Ala-Thr-Cys-Pro-Ala-Gly-Gln-Asp-Ile-		
	a	Thr-Lys-Cys-Tyr-Val- - - - -Thr-Pro-Asp-Ala-Thr-Ser-Gln-Thr-Cys-Pro-Asp-Gly- - -His-Val-		
	b	Thr-Lys-Cys-Tyr-Val- - - - -Thr-Pro-Asp-Ala-Thr-Ser-Gln-Thr-Cys-Pro-Asp-Gly-Gln-Asp-Ile-		

Snake	Toxin	Loop 4a	Loop 4b	Loop 4c
Bungarus multicinctus	α-bungarotoxin	Cys-Tyr-Arg-Lys-Met-Trp-Cys-Asp-Ala-Phe-Cys-Ser-Ser-Arg-Gly-Lys-Val-Val-Glu-Leu-Gly-		
Dendroaspis jamesoni kaimosae	V_NIII_1	Cys-Tyr-Thr-Lys-Thr-Trp-Cys-Asp-Gly-Phe-Cys-Ser-Gln-Arg-Gly-Lys-Arg-Val-Glu-Leu-Gly-		
Dendroaspis polylepis polylepis	γ	Cys-Tyr-Thr-Lys-Thr-Trp-Cys-Asp-Ala-Trp-Cys-Ser-Gln-Arg-Gly-Lys-Arg-Val-Glu-Leu-Gly-		
Dendroaspis polylepis polylepis	δ	Cys-Tyr-Thr-Lys-Thr-Trp-Cys-Asp-Ala-Trp-Cys-Ser-Gln-Arg-Gly-Lys-Ile-Val-Glu-Leu-Gly-		
Dendroaspis viridis^a	4.7.3	Cys-Tyr-Thr-Glu-Thr-Trp-Cys-Asp-Ala-Trp-Cys-Ser-Gln-Arg-Gly-Lys-Glu-Val-Glu-Leu-Gly-		
	4.9.3 (has oxidized Trp residue)			
Naja haie	III	Cys-Tyr-Thr-Lys-Thr-Trp-Cys-Asp-Asn-Phe-Cys-Gly-Met-Arg-Gly-Lys-Arg-Val-Asp-Leu-Gly-		
Naja melanoleuca	b	Cys-Tyr-Thr-Lys-Thr-Trp-Cys-Ala-Ser-Arg-Gly-Lys-Val-Ile-Glu-Leu-Gly-		
Naja naja	3.9.4	Cys-Tyr-Thr-Lys-Thr-Trp-Cys-Ala-Asp-Pro-Cys-Thr-Ser-Arg-Gly-Lys-Val-Ile-Glu-Leu-Gly-		
Naja naja oxiana	A	Cys-Tyr-Thr-Lys-Thr-Trp-Cys-Asp-Gly-Phe-Cys-Ser-Ile-Arg-Gly-Lys-Arg-Val-Asp-Leu-Gly-		
Naja nivea	I	Cys-Tyr-Thr-Lys-Met-Trp-Cys-Asp-Gly-Phe-Cys-Gly-Met-Arg-Gly-Lys-Val-Asp-Leu-Gly-		
Ophiophagus hannah	α	Cys-Tyr-Thr-Lys-Thr-Trp-Cys-Asp-Ala-Phe-Cys-Ser-Ser-Arg-Gly-Lys-Val-Val-Asp-Leu-Gly-		
	a	Cys-Tyr-Thr-Lys-Thr-Trp-Cys-Asp-Gly-Phe-Cys-Ser-Ser-Arg-Gly-Lys-Arg-Val-Asp-Leu-Gly-		
	b	Cys-Tyr-Thr-Lys-Thr-Trp-Cys-Asp-Gly-Phe-Cys-Ser-Arg-Gly-Ile-Arg-Asp-Leu-Gly-		

Snake	Toxin	Loop 5	Loop 6	Loop 7
Bungarus multicinctus	α-bungarotoxin	Cys-Ala-Ala-Thr-Cys-Pro-Ser-Lys-Lys-Pro-Tyr-Glu-Glu-Val-Thr-Cys	Cys-Ser-Thr-Asp-Lys-
Dendroaspis jamesoni kaimosae	$V_N III_1$	Cys-Ala-Ala-Thr-Cys-Pro-Lys-Val-Lys-Thr-Gly-Val-Glu-Ile-Lys-Cys	Cys-Ser-Thr-Asp-Tyr-
Dendroaspis polylepis polylepis	γ	Cys-Ala-Ala-Thr-Cys-Pro-Lys-Val-Lys-Ala-Gly-Val-Glu-Ile-Lys-Cys	Cys-Ser-Thr-Asp-Asp-
Dendroaspis polylepis polylepis	δ	Cys-Ala-Ala-Thr-Cys-Pro-Lys-Val-Lys-Ala-Gly-Val-Gly-Ile-Lys-Cys	Cys-Ser-Thr-Asp-Asn-
Dendroaspis viridis[a]	4.7.3	Cys-Ala-Ala-Thr-Cys (has oxidized Trp residue — circled)	Cys-Ser-Thr-Asp-Asp-
	4.9.3			
Naja haje	III	Cys-Ala-Ala-Thr-Cys-Pro-Thr-Val-Lys-Pro-Gly-Val-Asp-Ile-Lys-Cys	Cys-Ser-Thr-Asp-Asn-
Naja melanoleuca	b	Cys-Ala-Ala-Thr-Cys-Pro-Thr-Val-Lys-Pro-Gly-Val-Asn-Ile-Lys-Cys	Cys-Ser-Thr-Asp-Asn-
Naja naja	3.9.4	Cys-Val-Ala-Thr-Cys-Pro-Lys-Val-Lys-Pro-Tyr-Glu-Gln-Ile-Thr-Cys	Cys-Ser-Thr-Asp-Asp-
Naja naja oxiana	A	Cys-Ala-Ala-Thr-Cys-Pro-Thr-Val-Arg-Thr-Gly-Val-Asp-Ile-Lys-Cys	Cys-Ser-Thr-Asp-Asp-
Naja naja oxiana	I	Cys-Ala-Ala-Thr-Cys-Pro-Lys-Val-Lys-Pro-Gly-Val-Gln-Ser-Tyr-Cys	Cys-Ser-Thr-Asp-Asp-
Naja nivea	α	Cys-Ala-Ala-Thr-Cys-Pro-Lys-Val-Lys-Pro-Gly-Val-Asn-Ile-Lys-Cys	Cys-Ser-Arg-Asp-Asn-
Ophiophagus hannah	a	Cys-Ala-Ala-Thr-Cys-Pro-Ile-Val-Lys-Pro-Gly-Val-Glu-Ile-Lys-Cys	Cys-Ser-Thr-Asp-Asn-
	b	Cys-Ala-Ala-Thr-Cys-Pro-Lys-Val-Lys-Pro-Gly-Val-Asp-Ile-Lys-Cys	Cys-Ser-Thr-Asp-Asn-

Snake	Toxin	Loop 8	Reference
Bungarus multicinctus	α-bungarotoxin	Cys-Asn-His-Pro-Pro-Lys – -Arg-Gln-Pro-Gly	Mebs et al, 1971
Dendroaspis jamesoni kaimosae	$V_N III_1$	Cys-Asn-Pro-Phe-Pro-Val-Trp- - Asn-Pro-Arg	Strydom, A. J. C., 1973
Dendroaspis polylepis polylepis	γ	Cys-Asp-Lys-Phe-Gln-Phe- - Gly-Lys-Pro-Arg	Strydom, D. J., 1972
Dendroaspis polylepis polylepis	δ	Cys-Asn-Lys-Phe-Lys-Phe- - Gly-Lys-Pro-Arg	Strydom, D. J., 1973
Dendroaspis viridis[a]	4.7.3	Cys-Asp-Pro-Phe-Pro-Val- - Lys-Asn-Pro-Arg	Banks et al, 1974
	4.9.3	(has oxidized Trp residue — circled)	
Naja haje	III	Cys-Asn-Pro-Phe-Pro-Thr- - Arg-Glu-Arg-Ser	Kopeyan et al, 1975
Naja melanoleuca	b	Cys-Asn-Pro-Phe-Pro-Thr- - Arg-Asn-Arg-Pro	Botes, 1972
Naja naja	3.9.4	Cys-Asn-Pro-His-Pro-Lys- - Met-Lys-Pro	Shipolini et al, 1974
Naja naja oxiana	A	Cys-Asp-Pro-Phe-Pro-Thr- - Arg-Lys-Arg-Pro	Nakai et al, 1971
Naja naja oxiana	I	Cys-Asn-Pro-His-Pro-Lys- - Gln-Lys-Arg-Pro	Grishin et al, 1974
Naja nivea	α	Cys-Asn-Pro-Phe-Pro-Thr- - Arg-Lys-Arg-Ser	Botes, 1971
Ophiophagus hannah	a	Cys-Asn-Pro-Phe-Pro-Thr-Trp-Arg-Lys-Arg-Pro	Joubert, 1973
	b	Cys-Asn-Pro-Phe-Pro-Thr-Trp-Arg-Lys-His	

[a] For the sequences of toxins I and V of *Dendroaspis viridis*, see the paper by Bechis et al, 1976. The paper appeared after the table was constructed and could not be included.

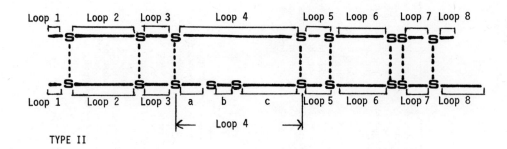

Characteristic points within Type II neurotoxins are as follows:

LOOP

1. As in Type I neurotoxins, contains only 2 residues and is a variable region.
2. Contains 10 to 12 residues and is more heterologous in amino acid sequences than loop 2 of Type 1.
3. Contains 5 or 6 residues and is fairly constant in amino acid residue distribution.
4. Contains 18 residues in all toxins so far investigated.
4a. Contains 5 residues; as in Type I, the first residue is always tyrosine. However, that is the only similarity between Type I and Type II.
4b. Contains 3 residues and shows variations.
4c. Always contains 10 residues and shows considerable variations in amino acid residues, with the exception of the third (Arg), fourth (Gly), fifth (Lys), ninth (Leu), and tenth (Gly).
5. Contains 3 residues is the most constant region as far as the amino acid residues are concerned. The sequence is always Ala-Ala-Thr, regardless of species, genus, and geographical origin. Is also different from loop 5 of Type I in that the latter contains only 1 residue.
6. Contains 10 residues; the first is always proline.
7. Contains only 4 residues, and the sequences are quite similar.
8. Contains 9 or 10 residues; the first is either Aspartic acid or aparagine.

The position of disulfide bonds in Type II has also been identified (Botes, 1971; Ohta and Hayashi, 1973), and it can be assumed that all toxins of Type II have the same configuration. The two-dimensional structures of Types I and II are shown diagrammatically in Fig. 6, together with weakly toxic basic proteins.

Even though Types I and II neurotoxins differ in the total number of amino acid and cysteine residues, there are remarkable similarities between them. Especially when the two extra cystein residues in Type II are considered as part of loop 4, the resemblance between the two types is striking.

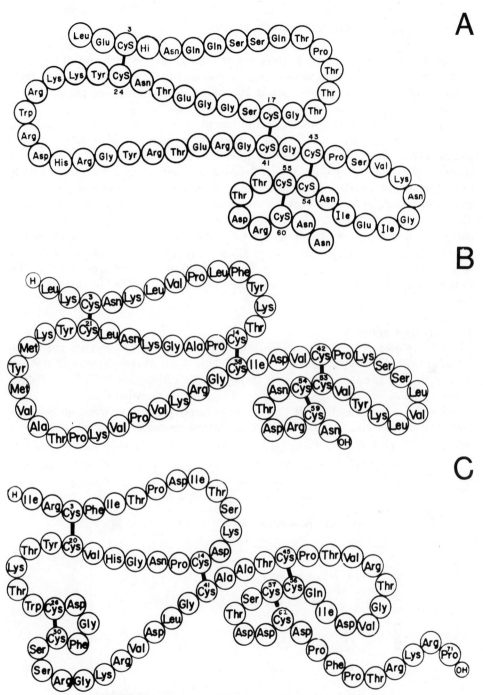

Figure 18.6 Actual examples of different types of neurotoxins and snake venom basic proteins. (A) Type I neurotoxin. The example is cobrotoxin from *Naja naja atra* venom. (Reproduced from Yang et al., 1970, by permission of the copyright owner, Elsevier Scientific Publishing Company.) (B) Nonneurotoxic basic proteins (basic intermediate a). The example is cytotoxin II of *Naja naja* venom. (Reproduced from Takechi and Hayashi, 1972, by permission the copyright owner, Academic Press.) (C) Type II neurotoxin. The example is toxin B of *Naja naja* venom. (Reproduced from Ohta and Hayashi, 1973, by permission of the copyright owner, Academic Press.)

4 STRUCTURE–TOXIC ACTION RELATIONSHIP

Natural poisons such as venoms are very potent toxic substances, and purified toxins are even more toxic than crude venoms. It is important for us to understand how these poisons produce such remarkable effects. Human beings have long been puzzled by the fact that a small amount of venom can kill a person so quickly.

Since many toxins are proteins consisting of only common amino acids, their extremely toxic action must arise from a particular arrangement of certain functional groups within the molecule. Chemical modification techniques used to investigate structure–function relationships of enzymes have been applied successfully to the study of toxicity–structure relationships of neurotoxins. However, strictly speaking, a chemical modification technique alone does not reveal the role of conformation in toxic action. The loss of toxicity after chemical modification may be due to the conformational change of the peptide backbone in the toxin molecule or to change in the spatial orientation of a particular side chain. Since study of the secondary and tertiary structures of toxins is not easy, relatively few investigations have been made. In this section, modification of the primary structure of toxins and its effect on toxicity will be discussed.

4.1 Tryptophan Residue

The indole chromophore of tryptophan, absorbing strongly at 280 nm, is converted on oxidation with N-bromosuccinimide to oxinodole, a much weaker chromophore at this wavelength. The spectral changes after chemical modification with NBS, 2-nitrophenylsulfenyl chloride, and 2-hydroxy-5-nitrohenzyl bromide are shown in Fig. 7.

Normally, neurotoxins I and II contain one or two tryptophan residues, one being the predominant case. The fact that the tryptophan residue can be modified readily by a variety of specific reagents indicates that this residue is relatively exposed to the outside.

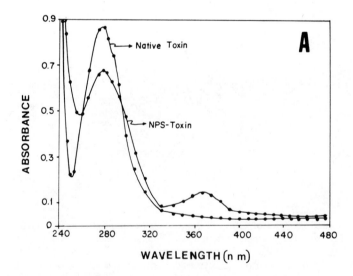

Figure 18.7 Absorption spectra of native and modified toxin obtained from *Lapemis hardwickii* venom. (A) Modified with 2-nitrophenylsulfenyl chloride.

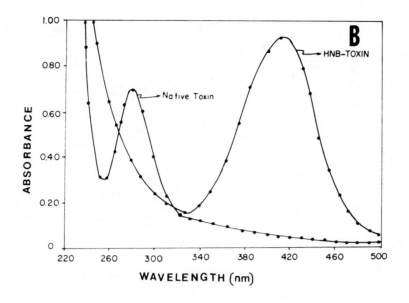

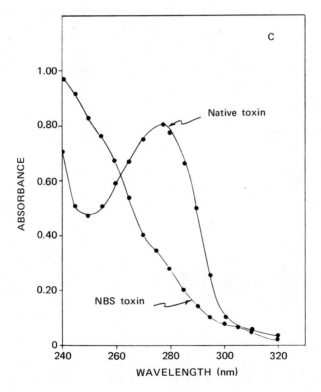

(B) Modified with 2-hydroxy-5-nitrobenzyl bromide. (C) Modified with N-bromosuccinimide. (Reproduced from Tu and Hong, 1971, by permission of the copyright owner, *The Journal of Biological Chemistry*.)

Figure 18.7   *Continued*

Table 6 Summary of Tryptophan Modification in Snake Type I Neurotoxins

Toxin	Venom	Reagent	Toxicity	Reference
Cobrotoxin	Naja naja atra	2-hydroxy-5-nitrobenzyl bromide	None	Chang and Hayashi, 1969
a, b	Laticauda semifasciata	NBS	None	Hong and Tu, 1970; Tu et al, 1971
erabutoxin a		NBS, 2-hydroxy-5-nitrobenzyl bromide	None	Seto et al, 1970
major toxin	Enhydrina schistosa	NBS	None	Tu and Toom, 1971
major toxin	Lapemis hardwickii	NBS; 2-hydroxy-5-nitrobenzyl bromide; 2-nitrophenylsulfenyl chloride	None	Tu and Hong, 1971
toxin	Naja haje	HCOOH	50%	Chicheportiche et al, 1972
3, 7c	Naja naja	2-hydroxy-5-nitrobenzyl bromide	20-50%	Karlsson et al, 1973
		HCOOH + O_3	8%	Karlsson et al, 1973
cobrotoxin	Naja naja atra	HCOOH + O_3; 2-hydroxy-5-nitrobenzyl bromide; 2-nitrophenylsulfenyl chloride	None	Chang and Yang, 1973

4 Structure–Toxic Action Relationship

This view agrees with the results of studies of fluorescence laser Raman spectroscopy and X-ray crystallography.

Chemical modifications of the tryptophan residue in Type I toxins carried out by many investigators are summarized in Table 6. Such modification eliminated toxicity. Since there is only one residue of tryptophan in the Type I neurotoxins investigated, clearly it is this residue that is essential for toxic action. Chicheportiche et al. (1972) considered that Try-28 takes part in the noncovalent interactions that stabilize the association with the receptor. Unlike the situation with Type I neurotoxins, chemical modification of the tryptophan residue in Type II neurotoxin does not alter the toxicity (Ohta and Hayashi, 1974a). In this respect, Type I and Type II neurotoxins show a difference. Although modification of the tryptophan residue in Type I neurotoxins eliminates their toxicity, it does not alter the capacity to bind to the antibody (Tu et al., 1971; Tamiya et al., 1971).

Ramachandran and Witkop (1959) reported that NBS modification may cause cleavage of proteins at the peptide bond containing tryptophan. This possibility was excluded for the case of snake venom neurotoxin (Tu et al., 1971).

4.2 Tyrosine Residue

Snake neurotoxins normally contain one or two residues of tyrosine. One of the tyrosine residues is always present next to the disulfide linkage half-cystine residue in loop 4. Cobrotoxin from *Naja naja atra* contains 2 moles of tyrosine, one of which is important for lethal action (Chang et al., 1971a; Huang et al., 1973). When Tyr-35 is nitrated with tetranitromethane, there is no change in biological activity. However, in the presence of guanidine hydrochloride, the originally buried Tyr-25 is exposed and modified with a loss of toxicity. It is apparent that Tyr-35 is more exposed and can be modified more readily than Tyr-25. This relationship is shown diagramatically in Fig. 8.

Chicheportiche et al. (1972) also found from studies of nitration and acetylation that Tyr-24 of neurotoxin I of *Naje haje* is masked in the native conformation. The Tyr-24 of *N. haje* toxin is comparable to Tyr-25 of *N. naja atra* cobrotoxin and is essential for the stabilization of an active structure. Its nitration abolishes toxicity. The masked nature of tyrosine is also demonstrated in the neurotoxin of *Lapemis hardwickii* (Raymond and Tu, 1972). Upon iodination, 84% of the tyrosine residue can be modified, whereas nitration results in only 50% of the tyrosine incurring loss of toxicity. The nitration product is 3-nitrotyrosine rather than 3,5-dinitrotyrosine. The "masked" or "buried" nature of the

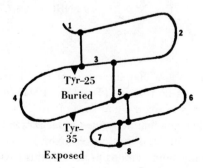

Figure 18.8 Position of two tyrosine residues in cobrotoxin of *Naja naja atra* venom. Some neurotoxins contain only one residue. In such cases, the position of the tyrosine corresponds to Tyr-25, next to the first disulfide bridge.

single tyrosine residue in snake neurotoxin is further confirmed by the low fluorescence yield of the tyrosine residue and by Raman spectroscopic data (Bukolova-Orlova et al., 1974; Hauert et al., 1974; Yu et al., 1975; Tu et al., 1976a).

In Type II neurotoxin, Ohta and Hayashi (1974b) found that the tyrosine residue in toxin B of *Naja naja* is not essential for toxic actions.

4.3 Arginine Residue

Toxins a and b from *Laticauda semifasciata* contain 3 and 2 moles of arginine, respectively. On modification with 1,2-cyclohexanedione, one arginine residue is modified in both toxins, without change in toxicities. This merely indicates that one residue is not essential for toxic action. The rest of the arginine residues cannot be modified under the experimental conditions employed, and thus it cannot be determined whether they are essential for toxicity (Tu et al., 1971). In cobrotoxin from *Naja naja atra* venom, toxicity is retained when only Arg-28 is modified. With the additional modification of Arg-33, the lethality drops dramatically. Upon further modification of Arg-30 the cobrotoxin loses its toxicity completely. Thus Arg-30 and Arg-33 are functionally important (Yang et al., 1974).

4.4 Lysine and N-Terminal Amino Acids

The origin of neurotoxicity is found in the fact that basic neurotoxins combine strongly with acidic receptor proteins. Since all known neurotoxins are basic proteins with very high isoelectric points (higher than 9), it is logical to assume the importance of arginine and lysine residues. The results of actual experiments, however are not clear for the following reasons:

1. Some arginine and lysine residues are masked and therefore cannot be modified completely.

2. *O*-Methylisourea is used frequently for the modification of lysine. This converts lysine to homoarginine residue, which is is very much like arginine:

$$\begin{array}{c} NH_2(\epsilon) \\ | \\ | \\ -Lys- \end{array} + \begin{array}{c} NH \\ \| \\ C-NH \\ | \\ OCH_3 \end{array} \longrightarrow \begin{array}{c} \overset{+}{N}H_2 \\ \| \\ C-NH_2 \\ | \\ NH(\epsilon) \\ | \\ -Lys- \end{array} + CH_3OH$$

Actual experimental results show that three out of four lysines can be modified with *O*-methylisourea without loss of toxicity (Tu et al., 1971). This merely indicates that three of four lysine residues in toxin a and four of five in toxin b are not essential for toxic action. It does not indicate that no lysine residue is essential. Karlegon et al. (1972) used the same reagent for the modification of toxin 3 from the venom of *Naja naja siamensis* concluded that the amino groups present in this toxin are not essential for its toxic action. In the case of cobrotoxin from *N. naja atra* venom, all of three lysine residues are modified, but the N-terminal amino acid of leucine is not modified. The modified toxin shows full toxicity (Chang et al., 1971b).

4 Structure–Toxic Action Relationship

Cooper and Reich (1972) modified lysine and N-terminal amino residues of *N naja siamensis* toxin with pyridoxal phosphate, followed by reduction of the Schiff's base with tritiated sodium borohydride. The pyridoxal phosphate-coupled toxin retained the activity and blocked the action of acetylcholine at the motor end plate.

Modification of free amino groups in cobrotoxin by fluorescein thiocarbamylation decreases the toxicity without affecting the antigenicity (Yang et al., 1967a). This suggests that the antigenic sites of cobrotoxin are different from the active sites of toxicity.

Stepwise, modification of lysine residues in cobrotoxin can be accomplished by the use of trinitrobenzenesulfonate. When Lys-27 is selectively modified, no loss of toxicity is observed. However, when both Lys-27 and Lys-47 are modified, complete loss of activity ensues. This suggests that the ϵ-amino group of Lys-47 is essential for toxic action (Chang et al., 1971b). Thus two specific modifying reagents produce different effects.

Chemical modification of Lys-27 and Lys-53 in *Naja haje* neurotoxins by acetylation or dansylation significantly decreased but did not suppress binding to the toxin receptor in *Torpedo* membranes (Chicheportiche et al., 1975). After modification the affinity of the lysine residues was decreased by a factor of about 200. These authors concluded that Lys-53 (and possibly also Lys-27) is an important residue for the recognition of the toxin by its receptor, but is not essential for toxic action.

4.5 Disulfide Bond

We have already discussed the fact that the disulfide bonds are an important factor for stabilizing neurotoxin conformation. The disulfide bonds are also essential for the lethal

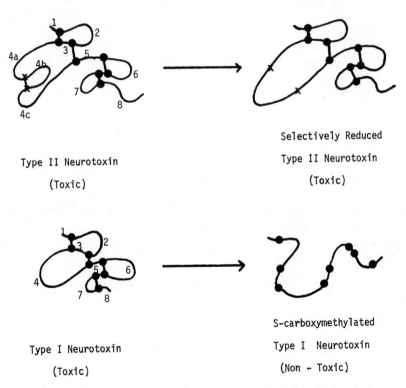

Figure 18.9 Reduction of disulfide bonds in Type I and Type II neurotoxins.

action of toxins. Actually, all four disulfide bonds in Type I neurotoxins, but only four out of five in Type II neurotoxins, are essential for toxic action.

When all the disulfide bond in cobrotoxin from *Naja naja atra* venom is reduced with β-mercaptoethanol (Yang, 1967) or oxidized with performic acid (Yang, 1965), toxicity is lost. Reduction of the disulfide bonds of cobra neurotoxin by dithiothreitol results in decreased activity on electroplax preparation (Bartels and Rosenberry, 1971). Activity can be restored completely, however, by reoxidation with 3,3'-dithiobis(6-nitrobenzoic acid). Reduction of the disulfide bonds in the vicinity of the receptor does not decrease the effect of the cobrotoxin.

When all four disulfide bonds of toxin a of *Pelamis platurus* venom were reduced and alkylated to convert all cysteine, no toxicity was observed, even upon injection of 230 times the LD_{50} dose of modified toxin (Tu et al., 1975). This relationship is shown diagrammatically in Fig. 9.

As can be seen from Fig. 9, Type II neurotoxins consist of five S—S bonds. When the disulfide bond within loop 4 is selectively modified, there is no change in toxicity. This was demonstrated for α-toxin of *Naja nivea* (Botes, 1974). Similarly, when the disulfide bond in the loop was selectively modified for toxin III of *N. haje*, the curarizing efficacy of the toxin was not affected (Chicheportiche et al., 1975).

4.6 Carboxyl Group

Only one of seven free carboxyl groups is essential for the toxic action of cobrotoxin of *Naja naja atra* venom. Without unfolding the molecule, six out of seven free carboxyl groups were modified with glycine methyl ester after activation with carbodiimide; there was no loss of toxicity (Chang et al., 1971c). Apparently, the masked carboxyl group Glu-21 is essential for toxicity (Chang et al., 1971c).

The *C*-terminal asparagine in cobrotoxin can be tritiated without loss of its lethality or antigenicity, or change in its lethality, antigenicity, or electrophoretic mobility (Huang and Ling, 1974).

4.7 Histidine Residue

The His-26 of erabutoxin b from *Laticauda semifasciata* was iodinated, resulting in the diiodo derivative. The modified toxin showed the same toxicity and capacity to bind with the antibody as the native toxin (Sato and Tamiya, 1970).

Histidine residues in cobrotoxin of *Naja naja atra* can be photooxidized in the presence of methylene blue (Huang et al., 1972). The reaction can proceed rapidly, and His-32 is more accessible to photooxidation than is His-4. This can be readily understood from the positions of the respective histidine residues in two-dimensional structure (Fig. 10). The His-4 at loop 2 neighbors with disulfide bond cysteine, which restricts the exposure of His-4. On the other hand, His-32 is located in the middle part of loop 4, which is one of

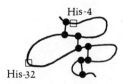

Figure 18.10 Positions of two histidine residues in Type I neurotoxins.

the longest loops in a neurotoxin and is, therefore, relatively more flexible. This makes His-32 readily accessible for modification.

4.8 Size and Toxicity

Noncontrolled proteolytic digestion of neurotoxins causes loss of toxicity. The lethality of cobrotoxin of *Naja naja atra* was lost upon 4-hr digestion by trypsin. Toxicity was also lost when the toxin was treated with chymotrypsin, nagarse, and papain (Yang, 1965). Thus the smaller-size fragments of the toxin do not retain toxicity.

Under more controlled conditions, removal of very small peptide fragments does not cause complete loss of toxicity. For instance, Karlsson et al. (1972) used *Arthrobacter* proteinase to remove the dipeptide Arg–Pro from the C-terminal end of neurotoxin 3 (71 amino acid residues) from *Naja naja siamensis* venom. The remaining toxin with 69 amino acid residues retained 70% of the original toxicity. After removal of the tetrapeptide Arg–Lys–Arg–Pro, however, the toxin retained only half of the original toxicity.

In the toxin of central Asian cobra venom, the removal of C-terminal asparagine caused no loss of toxicity, whereas the removal of N-terminal valine or leucine from toxin I or II resulted in loss of toxicity. It can be seen from the two-dimensional view of Type II neurotoxin in Figs. 4 and 6 that there are only two residues in loop 1 from the N-terminal end. Thus it is readily understandable why the removal of N-terminal amino acid is more critical than the loss of C-terminal amino acid.

5 CONFORMATION AND STABILITY

The conformation of snake neurotoxins plays an important role in their stability. Normally proteins with small molecular weights contain relatively high numbers of disulfide bonds, as compared to proteins of larger molecular weights. In large-size proteins, the specific conformation is stabilized and maintained by many secondary noncovalent bonds such as hydrogen bonds, ionic bonds, hydrophobic interactions, and Van der Waals forces. However, in smaller proteins there are not enough secondary-type bonds to stabilize the specific configuration. Therefore strong covalent bonds of disulfide linkages become important for maintaining the specific three-dimensional structures of small proteins.

5.1 Importance of Disulfide Linkages in Neurotoxins

As can be seen from Tables 1 and 2, all neurotoxins, cytotoxins, and cardiotoxins (see Chapter 19) contain either four or five disulfide bonds and have molecular weights of about 7000 or 9000. This is a rather high content of disulfide bonds for the size of the proteins. Other toxic proteins of small molecular weight also contain a rather high number of disulfide linkages. Some investigators consider that all small-size neurotoxins from plant and marine organisms are similar to snake neurotoxins because they contain a similar number of disulfide bonds. This is rather an oversimplification, since the similarity usually ends at the number of disulfide bonds. Careful examination of amino acid composition usually reveals dissimilarity.

As can be seen from Table 7, the smaller the protein, the higher the relative disulfide bond content. For instance, sea anemone toxin II uses 25% of its total residues to maintain its own specific configuration through disulfide bonds. Snake toxins use 10 to 15% of total residues to form disulfide bridges. On the other hand, in larger protein

Table 7 Percentage of Total Half-Cystine Residues Forming Disulfide Bridges in Various Proteins

Proteins	Total Residues	No. of Disulfide Bonds	Half-Cystine Used For Disulfide Bridges		References
			mole	mole %	
Anemonia sulcata					
Toxin III	24	3	6	25.0	
Toxin II	44	3	6	13.6	Béress et al, 1975
Toxin I	45	3	6	13.3	
Viscotoxin 1-Ps	46	3	6	13.0	Samuelsson and Jayawardene, 1974
Viscotoxin A-2	46	3	6	13.0	Olson and Samuelsson, 1974
Phoratoxin	46	3	6	13.0	Mellstrand and Samuelsson, 1974
Viscotoxin A_3	46	3	6	13.0	Samuelsson et al, 1968
Viscotoxin B	46	3	6	13.0	Samuelsson, 1973
Pelamis toxin a	55	4	8	14.5	Tu et al, 1975
Cobrotoxin	62	4	8	12.9	Yang et al, 1969a
Proinsulin	84	3	6	7.1	Chance et al, 1968
Bovine pancreatic RNase	124	4	8	6.5	Stein and Moore, 1963
Lysozyme	129	4	8	6.2	Dickerson and Geis, 1969
Chymotrypsinogen A	246	4	8	3.3	Hartley, 1964

molecules such as chymotrypsinogen A, eight half-cysteines, or only 3.3% of the total residues, are used to make four disulfide bridges.

5.2 Stability

In the preceding section we mentioned that disulfide bridges play an important role in maintaining the specific conformations of neurotoxins. In Type I neurotoxins, it will be recalled, there are always four disulfide linkages, and in type II, five. These are rather numerous for proteins having molecular weights of 7000 to 8000 daltons. The high content of disulfide bonds in these small proteins makes neurotoxins very stable because of the highly compact structure of the toxin molecules.

The compact conformation of neurotoxins also possess the lowest energy (Gabel et. al., 1976).

Homma et al. (1964) investigated the thermostability and pH stability of sea snake venoms from *Laticauda semifasciata, L. laticaudata,* and *Hydrophis cyanocinctus* and found that these venoms are very stable. Cobrotoxin of *Naja naja atra* venom is fully active even in $8M$ urea (Yang et al., 1969a). Moreover, cobrotoxin does not lose its toxicity when heated at 80°C for 30 min (Yang, 1965). Toxins a and b from *L. semifasciata* venom are stable at 100°C for 30 min and over a pH range of 1 to 11 (Tu et al., 1971). The fact that Types I and II neurotoxins of *Naja haje* are also stable is attributed to the high content of disulfide bonds for small-size proteins. Full toxicity is retained even after exposure to anhydrous formic acid or to $1N$ HCl for 100 min (Chicheportiche et al., 1972).

Toxin B of *Naja naja* venom is also stable at 100°C for 30 min and over a wide pH range without loss of toxicity (Hayashi and Ohta, 1975).

5.3 X-Ray Diffraction

Of all the physical methods available for the determination of tertiary structures of proteins, X-ray crystallography is a very superior technique as it can deduce the absolute configurations of protein molecules. In snake neurotoxins, complete spatial assignment of atoms has been made (Low *et al,* 1976; Tsernoglou and Petsko, 1976).

For cobrotoxin from the venom of *Naja naja atra*, the unit cell dimensions are $a = b = 40.40$ Å and $c = 71.16$ Å. The space group is either $P4_1 2_1$ or $P4_3 2_1$. There are eight molecules per unit cell (Wong et al., 1972). Erabutoxin a and b were also studied and found to have dimensions of $a = 50.0$ Å, $b = 46.5$ Å, $c = 21.2$ Å and $a = 49.76$ Å, $b = 46.77$ Å, $c = 21.58$ Å, respectively. There is no change in unit cell dimension when His-26 of erabutoxin b is modified by iodination. Both toxins have space group $P2_1 2_1 2_1$ (Low et al., 1971). Toxins a and b of *Laticauda semifasciata* have space groups similar to the one for erabutoxins a and b and have one molecule per asymmetric unit. The unit cell dimensions are $a = 49.5$, $b = 46.6$, and $c = 21.1$ Å for toxin a; the corresponding values for toxin b are 49.9, 46.6, and 21.3 Å. Pseudosymmetry of the $hk0$ plane suggests that the molecule may have two similar halves (Tsernoglou et al., 1972; Tu et al., 1973).

Two orthorhombic forms of crystals were made for erabutoxin c obtained from *L. semifasciata* venom. The space group is $P2_1 2_1 2_1$, and the cell dimensions are $a = 55.8$, $b = 53.9$, $c = 41.4$ Å, and $a = 56.2$, $b = 55.4$, $c = 38.5$ Å. There are two molecules in both asymmetric units. A new form of erabutoxin b crystallizes in $P2_1 2_1 2_1$ and appears to be isomorphous with one of the erabutoxin c forms. The cell dimensions are $a = 55.7$, $b = 53.8$, and $c = 41.5$ Å (Preston et al., 1975).

5.4 Fluorescence Spectroscopy

Simple proteins have fluorescence due to tryptophan and tyrosine residues. By studying fluorescent spectra, one can deduce the environment of tryptophan and tyrosine in proteins. Bukolova-Orlova et al. (1974, 1976) observed that fluoresence maxima for toxins I and II obtained from *Naja oxiana* venom were at 347 and 345 nm, respectively. This suggests that tryptophan is the chromophore and there is no contribution from a tyrosine residue, a clear indication that the single tryptophan residue in both neurotoxin molecules is located on the surface of the macromolecules and is accessible for freely relaxing water. The tyrosine residues of both toxins have a very low fluorescence yield that is intrinsic for buried phenolic chromophores. Normally tyrosine fluoresces at 290 to 310 nm. The fact that the emission spectra showed no peaks or shoulders due to tyrosine residue is further evidence that this residue is buried.

The tyrosine residue is normally positioned next to the disulfide bond at loop 4:

```
         Loop 3                    Loop 4
      ⎧‾‾‾‾‾‾‾‾⎫              ⎧‾‾‾‾‾‾‾‾⎫
——————————————Cys——Tyr——————————————
```

This location of the tyrosine residue may restrict the free rotation of the aromatic ring and provides limited access to solvent molecules.

Exactly the same results were obtained by Hauert et al. (1974) for a neurotoxin from *N. naja philippinensis* venom, and the same conclusions were drawn by these authors.

5.5 Optical Rotatory Dispersion and Circular Dichroism

Optical rotatory dispersion (ORD) is a spectrum of optical rotation as a function of wavelength. Circular dichroism (CD) is the absorptivity difference between left and right

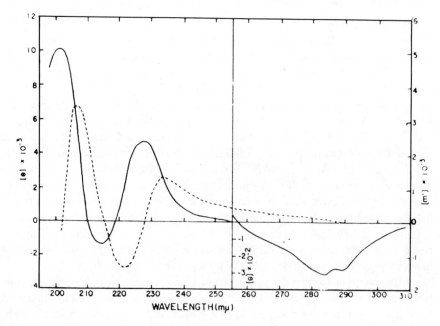

Figure 18.11 Optical rotatory dispersion and circular dichroism spectra of cobrotoxin obtained from *Naja naja atra* venom. (Based on Yang et al, 1968, by permission of the copyright owner, Elsevier Scientific Publishing Company.)

5 Conformation and Stability

circularly polarized light due to the optically active compounds. Proteins are optically active compounds because of many factors. The primary structure of a protein is inherently asymmetric. The secondary structure of many proteins is helical, resulting in additional optical activity. The tertiary structure of a protein may be such that an inherently symmetric group is thrust into an asymmetric environment (Van Holde, 1971).

Several investigators applied ORD and CD techniques to snake neurotoxin study. The ORD of cobrotoxin of *Naja naja atra* shows a maximum at 233 nm and a shoulder at 270 nm (Yang et al., 1967b). Normally CD gives higher resolution and more useful information than ORD. As can be seen in Fig. 11, the CD spectrum of cobrotoxin gives negative maxima at 285 and 215 nm and positive maxima at 228 and 201 nm. From a comparison of polypeptides of known structures and their CD and ORD spectra, Yang et al. (1968) concluded that cobrotoxin consists of β-structure. The spectra, however, are quite different from those of most globular proteins at nearby the 220 nm band. This may well be due to the effect of aromatic side chains. The same result was obtained by Hauert et al. (1974) for *N. naja philippinensis*, and they concluded that the neurotoxin consists predominantly of β-sheet structure. Denaturation with 6M guanidine–HCl completely converts the toxin to random structure.

The ORD measurement of neurotoxin I of *N. haje* indicates that it has two conformational changes at acidic pH (Chicheportiche et al., 1972). Definite conclusions regarding the conformation of neurotoxins based solely on CD spectra may be erroneous because of the contribution of aromatic side chains.

5.6 Laser Raman Spectroscopy

When light is directed at a molecule, the incident light can interact with the molecule and give two types of scattering: elastic or Rayleigh scattering, and inelastic or Raman

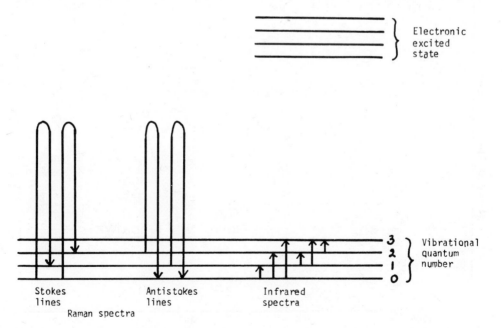

Figure 18.12 Origin of Raman spectra. Note the symmetrical relationship of the Stokes and anti-Stokes lines.

scattering. When a molecule that is vibrating with its own unique frequency interacts with incident light, the vibrational frequency will be changed either by adding the characteristic frequency ($+\Delta\nu$) or by subtracting it ($-\Delta\nu$). The Raman frequency ($\Delta\nu$) is independent of the frequency of incident light as long as the virtual state does not correspond to an electronic absorption state of the molecule (Fig. 12). The frequency line $h(\nu_0 + \Delta\nu)$ is called the anti-Stokes line (h is Planck's constant, ν_0 is zero the frequency of incident light); the line $h(\nu_0 - \Delta\nu)$, the Stokes line. A modern Raman spectrometer automatically measures the Raman frequency ($\Delta\nu$).

Peptide Backbones. In recent years, laser Raman spectroscopy has been applied successfully to different proteins, especially in the elucidation of their conformations. The first Raman study applied to snake venom component was on cobramine B obtained from *Naja naja* venom (Yu et al., 1973). Cobramine B is a basic protein structurally similar to neurotoxins but lacking neurotoxicity (Tu, 1973).

Subsequently, laser Raman spectroscopy was applied to neurotoxins (Yu et al., 1975; Tu et al., 1976a) and cardiotoxin (Tu et al., 1976b). Laser Raman spectra for neurotoxins of *Lapemis hardwickii*, *Enhydrina schistosa*, and *Pelamis platurus* (Figs. 13 and 14, Table 8) are very similar. These snakes all belong to the subfamily Hydrophiinae. A distinct amide III band was observed at 1240 to 1245 cm^{-1} for these three toxins (Table 8). The amide I band appeared at 1672 cm^{-1} for all three toxins. This clearly indicates that the neurotoxins do not contain α-helix and that the backbone

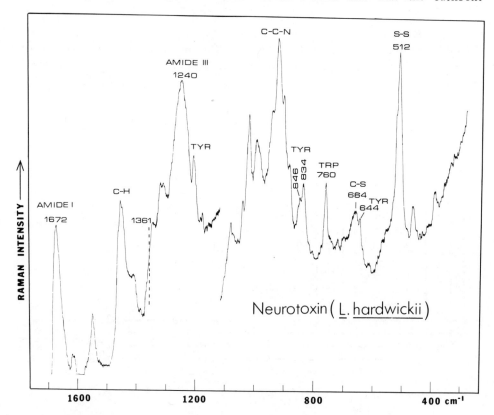

Figure 18.13 Raman spectrum of purified neurotoxin isolated from the venom of *Lapemis hardwickii*. (Reproduced from Yu et al., 1975).

5 Conformation and Stability

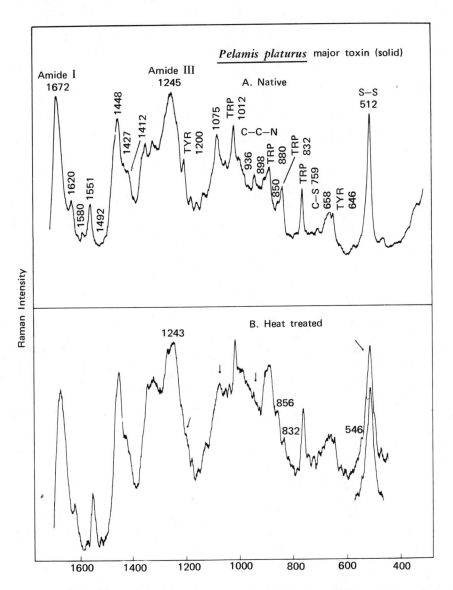

Figure 18.14 Raman spectra of purified neurotoxin isolated from the venom of *Pelamis platurus*, before (A) and after (B) heat treatment at 100°C for 30 min. (Reproduced from Tu et al., 1976a).

conformation is the antiparallel β type. These results are in good agreement with the data obtained from CD by Yang et al. (1967b) and Hauert et al. (1974); these investigators also concluded that neurotoxins predominantly consist of β-sheet structure.

The amino acid sequences of both Type I and Type II neurotoxins have been identified by many investigators (see Section 3 of this chapter). Type I neurotoxins consist of eight loops and four disulfide bridges. As shown diagrammatically in Fig. 15, the direction of peptide backbones in these loops of Type I neurotoxins is antiparallel. Therefore, the antiparallel-β structure observed in the laser Raman spectra must originate

Table 8 Some Characteristic Raman Lines of Snake Neurotoxins

Family	Hydrophiidae				Elapidae
Subfamily	Hydrophiinae			Laticaudinae	
Genus	Pelamis	Lapemis	Enhydrina	Laticauda	Naja
Species	platurus	hardwickii	schistosa	semifasciata	naja
Origin	Costa Rica	Thailand	Malaya	Philippines	India
Name \ Bands	Pelamis Toxin a	Toxin	Toxin	Toxin b	Cobramine B
	(cm^{-1})	(cm^{-1})	(cm^{-1})	(cm^{-1})	(cm^{-1})
Amide I	1672	1672	1672	1672	1672
Heating	1664	—	—	—	—
Amide III	1245	1240	1242	1248	1235
Heating	1243 (broader)	—	—	—	—
-S-S	512	512	512	512	510
Heating	512 (major) 546 (minor)	—	—	—	—
Reference	Tu et al, 1976	Yu et al, 1975	Yu et al, 1975	Tu et al, 1976	Yu et al, 1973

in these regions of Type I neurotoxin molecule (Fig. 15). The presence of the β-turn in neurotoxins was first predicted by Chen et al. (1975) using the method of Chou and Fasman. Eventually, the presence of anti-parallel β and β-turn structures was confirmed by X-ray diffraction study (Tsernoglou and Petsko, 1976; Low et al., 1976).

The amide I band arises from the coupled C=O stretching vibrations of peptide linkages, and the amide III band from "in-plane" vibration of the peptide bond, the main contribution from the in-plane bending motion of an —NH (Fig. 16). Therefore the exchange of hydrogen with deuterium should shift the band to a shorter wave number by a factor of $\sqrt{2}$. Indeed, this was the case as the original amide III bands at 1240 cm^{-1} shifted to 980 cm^{-1} for the toxins of E. schistosa and L. hardwickii (Fig. 17), while the amide I band remained the same.

Toxin b obtained from *Laticauda semifasciata*, which belongs to the subfamily Laticaudinae, also has antiparallel-β structure (Tu et al., 1976a).

5 Conformation and Stability 287

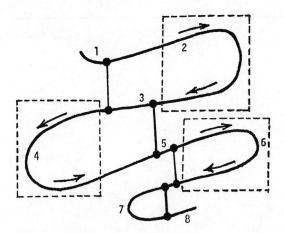

Figure 18.15 Two-dimensional structure of Type I neurotoxin, showing the antiparallel direction of the peptide backbone (represented by dotted retangular boxes). The numbers represent loop numbers from the NH_2 terminal to the next cysteine residue. The turns in the peptide backbone are β-turns.

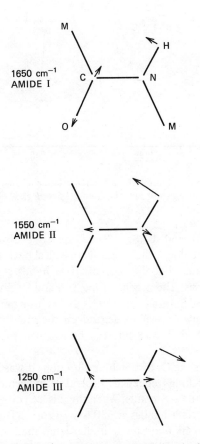

Figure 18.16 Different in-plane vibrational modes of peptide bonds. (Reproduced from Miyazawa, 1967 by permission of the copyright owner, Marcel Dekker.)

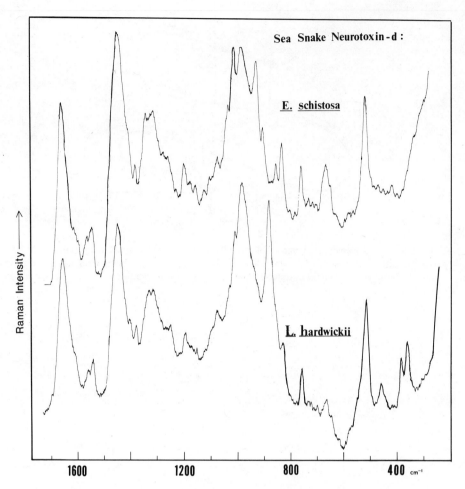

Figure 18.17 Raman spectra of neurotoxins after deuterium exchange. (Reproduced from Yu et al., 1975, by permission of the copyright owner, *The Journal of Biological Chemistry*.)

Disulfide Linkages (S–S) and C–S. There are four disulfide linkages in Type I neurotoxins. The fact that a very sharp and symmetrical peak at 512 cm^{-1} was obtained for all neurotoxins indicates that the four disulfide bonds present in the native toxins have very similar geometries. The origins of the S–S and C–S stretching vibrations have been extensively studied by Sugeta et al., (1972, 1973), using model compounds. The S–S stretching frequency is now known to depend on the internal rotation about the C–S bonds. The observed 512-cm^{-1} S–S line can be assigned to the gauche-gauche-gauche form (Figs. 18 and 19).

The line at 658 cm^{-1} is the C–S stretching vibration. It was also shown by Sugeta et al. (1972) that the C–S stretching frequencies depend on molecular conformation about the C–C bonds adjacent to the C–S bonds. When there is a hydrogen atom trans to the C–S bond, the S–S stretching vibration lies at 630 to 670 cm^{-1}. Thus the stretching vibration at 658 cm^{-1} for neurotoxins indicates that the conformation about the C–C bond is mostly in the ν(H, H) form, which was originally assigned by Sugeta et al. (1972). The 701-cm^{-1} peak for neurotoxins other than *Laticauda semifasciata* is the C–S stretching

5 Conformation and Stability

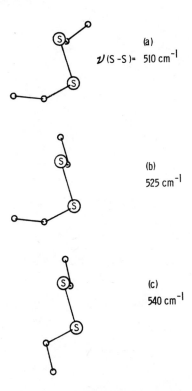

Figure 18.18 Conformations of disulfide bonds in snake neurotoxins. Top: Gauche-gauche-gauche conformation. Middle: Trans-gauche-gauche conformation. Lower: Trans-gauche-trans conformation.

of the methionyl internal rotation about the CH_2-CH_2 and CH_2-S bonds in $-CH_2-CH_2-S-CH_3$, which is in trans-gauche form. Toxin b of *L. semifasciata* does not contain methionine, and there is no peak at 700 cm^{-1}.

Side-Chain Residue. Studies of glycyltyrosine (Yu et al., 1973) and lysozyme (Yu and Jo, 1973) have shown that the relative intensities of the Raman lines at 846 and 836 cm^{-1} are related to the environment of the tyrosine side chain. From the spectra of neurotoxins, it is apparent that the single tyrosine residue present in sea snake neurotoxins, with the exception of *Laticauda semifasciata* toxin b, is not readily accessible to water molecules and perhaps is involved in unusual binding with other side-chain residues. The laser Raman spectra are in full agreement with the fact that only 50% of the tyrosine molecule is nitrated with tetranitromethan (Raymond and Tu, 1972). The tyrosine cannot be iodinated in *Naja nigricollis* toxin (Menez et al., 1971), a finding also in full agreement with the result of fluorescent study (Bukolova-Orlova et al., 1974; Hauert et al., 1974).

Most of snake neurotoxins, both of Type I and Type II, contain one or two residues of tryptophan, one residue being predominant (Tu, 1973). The single tryptophan residue of neurotoxin can be readily modified by different selective reagents. The 1361-cm^{-1} band of tryptophan in laser Raman spectra is very sensitive to whether the tryptophan is "buried" or "exposed." When the indole ring is buried and involved in certain interactions within a protein molecule, the 1361-cm^{-1} band shows a sharp peak. As the indole ring becomes accessible to water molecule, the 1361-cm^{-1} line diminishes. The

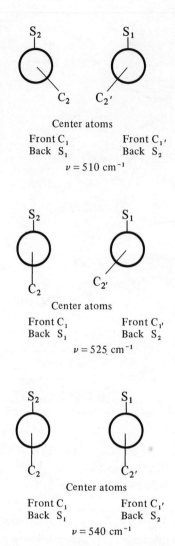

Figure 18.19 Conformations of disulfide bonds in snake neurotoxins. Top: Gauche-gauche-gauche conformation. Middle: Trans-gauche-gauche conformation. Lower: Trans-gauche-trans conformation.

lack of a distinct peak at 1361 cm^{-1} in Raman spectra of neurotoxins indicates that the single residue is exposed. The result of Raman study not only agrees with that of chemical study, but also is in full accord with fluorescence results (Bukolova-Orlova et al, 1974; Hauert et al., 1974).

Conformation in Solid and in Solution. There is an important question as to whether the conformations of neurotoxins in aqueous solution are identical to those of solid neurotoxins. There are no changes in the amide I and III regions of the spectra, indicating that the backbone conformations of the neurotoxins are the same in the solid state and in solution (Yu et al., 1975). In view of the compact and stable molecule of neurotoxins, this is a rather reasonable finding.

6 Structural Similarity of Many Basic Proteins and Neurotoxins

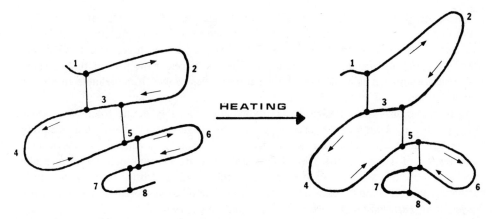

Figure 18.20 Diagram representing the change in conformation of neurotoxins after heat treatment. There is no basic change in the geometry of disulfide bonds. Partial denaturation occurs within each loop. (Reproduced from Tu et al., 1976a.)

Conformation and Heat Treatment. Raman spectroscopy of *Pelamis* toxin a indicated partial formation of a random coil in the peptide backbone conformation after heat denaturation (Fig. 20). The amide I band shifted to 1664 from 1672 cm^{-1}. The amide III peak remained at 1243 cm^{-1} but broadened somewhat. This is reasonable since the four disulfide bonds hold the molecule rather tightly (Fig. 4) in a folded conformation. The slight changes observed in the S—S Raman band (512 cm^{-1}) and in the C—C—N, C—C, and C—N bands (900 to 1100 and 1300 to 1400 cm^{-1}) suggest that some loops of the toxin molecule were modified somewhat by heat treatment. However, the integrity of the four disulfide bonds maintained the tightly folded structure of the peptide backbone because the integrity and intensity of 512-cm^{-1} band remained unchanged. The heat-treated toxin thus retained its original backbone structure, but with a slight twist. This finding, which is shown diagramatically in Fig. 20, is consistent with the fact that the neurotoxin is relatively stable to heat treatment (Tu et al., 1976a).

The heat treatment experiment indicated that the relative intensities of the 850- and 832-cm^{-1} lines were reversed from those observed in the native toxin and, therefore, that the tyrosine is more exposed after heat treatment. After vigorous heat treatment, the major peak of disulfide bond stretching vibration at 512 cm^{-1} remained, but a new shoulder appeared at 546 cm^{-1} (Fig. 14). This suggests that a new type of S—S stretching vibration was produced as a result of heat treatment, although most of the S—S vibration remained in the same gauche-gauche-gauche conformation. The newly formed S—S stretching vibration must be the trans-gauche-trans type similar to the one observed by Sugeta et al. (1972, 1973) on the model compounds (Figs. 18 and 19).

6 STRUCTURAL SIMILARITY OF MANY BASIC PROTEINS AND NEUROTOXINS

There are many different types of basic proteins in the venoms of Elapidae and Hydrophiidae. The most important basic proteins, some of them well characterized, are neurotoxins. Some cardiotoxins and cytotoxins are also well characterized. But not all basic proteins have been studied for their structures and biological activities.

When the structures of various basic proteins are examined very carefully, striking similarities are observed. This similarity in structure of basic proteins of different biological function is more than accidental. It is very likely that these proteins are interrelated and are probably molecules at different stages in the process of evolution. The basic assumption is made that the toxin molecules evolve from a simple to a more complex form.

Most cytotoxins, cardiotoxins, weak toxins, and reversibly acting toxins are "Basic Intermediate a" compounds. Like Type I neurotoxins, they have four disulfide bonds, but there are three amino acid residues in loop 5 instead of the one residue characteristic of Type I neurotoxins. "Basic Intermediate b" type compounds are like Type II neurotoxins which consist of five disulfide bonds and three residues in loop 5, but have a shorter chain in loop 8. "Basic Intermediate b" type protein was isolated from the venom of *Laticauda semifasciata* (Maeda and Tamiya, 1974).

Usually Elapidae venoms contain both Type I and Type II neurotoxins and "Basic Intermediate a" compounds. Hydrophiidae (sea snakes) venoms, however, usually contain only Type I neurotoxin. There are many other basic proteins of smaller molecular weight than Type I neurotoxin. If we can map the sequences of all the basic proteins in the venoms of Hydrophiidae and Elapidae, we certainly should find many precursor molecules Type I to neurotoxin.

It is rather fascinating that, whereas both Type I and Type II neurotoxins are highly toxic, the intermediate molecules are relatively weak toxins. Carlsson (1975) isolated a very interesting basic protein, designated as $S_4 C_{11}$ from the venom of *Naja melanoleuca*. Although $S_4 C_{11}$ has a low toxicity, it is structurally related to neurotoxin or other venom basic proteins. It has a total of 65 amino acid residues with five disulfide bridges. The most striking difference from Type II neurotoxin is that the fifth disulfide bond is located in loop 2. Apparently, $S_4 C_{11}$ protein is derived from Type I, rather than Type II, neurotoxin, from the viewpoint of molecular evolution. This relationship is shown in Fig. 21. The $S_4 C_{11}$ basic protein is not an intermediate molecule in the evolution of Type I to Type II. Instead, it branched out from Type I independently and therefore is designated as a type of "Basic Side-Product a." The neurotoxin CM-13 B, isolated from *Naja haje annulifera* venom (Joubert, 1975b), is structurally very similar to the $S_4 C_{11}$ basic protein of *N. melanoleuca* venom. They both have 65 amino acid residues and five disulfide linkages. However, unlike $S_4 C_{11}$, which has very low toxicity, CM-13 B is highly neurotoxic. This illustrates the fact that non-curare-type basic proteins have structures very similar to those of neurotoxins, although they differ in toxicity.

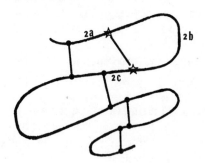

Figure 18.21 Special type of nonneurotoxin basic protein isolated from Elapidae venom. Note the fifth disulfide bond in loop 2.

7 Presynaptic Toxins and their Relation to Phospholipase A_2 Activity

Toxin with nine half-cysteine residues was isolated from the venom of *Enhydrina schisotsa* (Tu and Toom, 1971). According to the study of Fryklund et al. (1972), the extra cysteine residue is located in loop 2 (Fig. 21). This basic toxin may be an intermediate molecule toward "Basic Side-Product a."

7 PRESYNAPTIC TOXINS AND THEIR RELATION TO PHOSPHOLIPASE A_2 ACTIVITY

In addition to β-bungarotoxin, notexin from *Notechis scutatus scutatus* venom is also a presynaptic toxin. Notexin also exhibits phospholipase A_2 activity. When the sequence of notexin is compared to that of porcine pancreatic (Fig. 22) and that of *Naj melanoleuca* venom (Fig. 23) phospholipase A_2, one is fascinated to find a high degree of homology. Notexin exhibited a 99.8% loss of both phospholipase A_2 activity and lethal neurotoxicity upon treatment with *p*-bromophenacyl bromide (Halpert et al., 1976). The modified residue is His-48, which corresponds to the histidine residue shown to be part of the catalytic site of pancreatic phospholipase A_2.

The respiratory rate of brain mitochondria failed to increase upon addition of ATP in the presence of β-bungarotoxin (Howard, 1975). Pure phospholipase A_2 induces the same effect.

It may well be that presynaptic toxin and phospholipase A_2 originated from the same ancestral molecule. Thus some presynaptic toxins still retain phospholipase A_2 activity.

As discussed in Section 2.4 of Chapter 2, most phospholipase A_2 is relatively nontoxic. The homology of sequences for presynaptic toxins and phospholipase A_2 eliminates the

```
       1             5                  10                 15
 Ala-Leu-Trp-Gln-Phe-Arg-Ser-Met-Ile-Lys-CYS-Ala-Ile-Pro-Gly-Ser-
 Asn-    -Val-   -  -Ser-Tyr-Leu-    -Gln-    -   -Asn-His-    -Lys-
                 20                    25                 30
 His-Pro-Leu-Met-Asp-Phe-Asn-Asn-Tyr-Gly-CYS-Tyr-CYS-Gly-Leu-Gly-
 Arg-    -Thr-Trp-His-Tyr-Met-Asp-   -   -   -   -   -   -Ala-   -
           35                  40                 45
 Gly-Ser-Gly-Thr-Pro-Val-Asn-Glu-Leu-Asn-Arg-CYS-----Glu-His-Thr-
    -   -   -   -   -Asp-   -   -Asp-   -   -   -CYS-Lys-Ile-His-
       50                  55                 60
 Asp-Asn-CYS-Tyr-Arg-Asp-Ala-Lys-Asn-Leu-Asn-Asp-Ser-CYS-Lys-Phe-
    -Asp-   -   -Asp-Glu-   -   -Ala-Gly-Lys-Lys-Gly-   ---------
    65             70                  75                 80
 Leu-Val-Asp-Asn-Pro-Tyr-Thr-Glu-Ser-Tyr-Ser-----Tyr-CYS-Ser-Ser-
 -----------Phe-   -Lys-Met-Ser-Ala-   -Asp-Tyr-   -   -Gly-Glu-
              85                  90                 95
 Asn-Thr-Glu-Ile-Thr-CYS-Asn-Ser-Lys-Asn-Asn-Ala-CYS-Glu-Ala-Phe-
    -----Gly-Pro-Tyr-   -Arg-Asn-Ile-Lys-Lys-Lys-   -   -Leu-Arg-   -
              100                 105                110
 Ile-CYS-Asn-----Asp-Arg-Asn-Ala-Ala-Ile-CYS-Phe-Ser-Lys-Ala-Pro-
 Val-    -Asp-CYS-   -Val-Glu-   -   -Phe-   -   -Ala-   -   -   -
              115                 120                125
 Tyr-Asn-Lys-Glu-His-Lys-Asn-Leu-Asn-Thr-Lys-Lys-Tyr-CYS----
    -    -Asn-Ala-Asn-Trp-   -Ile-Asp-   -   -Arg-   -Gln
```

Figure 18.22 Alignment showing homology between notexin (lower sequence) and porcine pancreatic phospholipase A_2. (Reproduced from Halpert and Eaker, 1975.)

```
             1              5                  10                 15
          Asn-Leu-Val-Gln-Phe-Ser-Tyr-Leu-Ile-Gln-CYS-Ala-Asn-His-Gly-
            -   -Tyr-   -   -Lys-Asn-Met-  -His-   -Thr-Val-Pro-----
            16             20                 25                 30
          Lys-Arg-Pro-Thr-Trp-His-Tyr-Met-Asp-Tyr-Gly-CYS-Tyr-CYS-Gly-
          Asn-   -Ser-Trp-   -   -Phe-Ala-Asn-   -   -   -   -   -
            31             35                 40                 45
          Ala-Gly-Gly-Ser-Gly-Thr-Pro-Val-Asp-Glu-Leu-Asp-Arg-CYS-CYS-
          Arg-  -   -   -   -   -   -   -Asp-   -   -   -   -   -
            46             50                 55                 60
          Lys-Ile-His-Asp-Asp-CYS-Tyr-Asp-Glu-Ala-Gly-Lys-Lys-----Gly-
          Gln-   -   -   -Asn-   -   -Gly-   -   -Glu-   -Ile-Ser-  -
            61             65                 70                 75
          CYS-Phe-Pro-Lys-Met-Ser-Ala-Tyr-Asp-Tyr-Tyr-----CYS-Gly-Glu-
            -   -Trp-   -Tyr-Ile-Lys-Thr-   -Thr-   -Asp-Ser-  -Gln-Gly-
            76             80                 85                 90
          Asn-Gly-Pro-Tyr-CYS-Arg-Asn-Ile-Lys-Lys-Lys-CYS-Leu-Arg-Phe-
          Thr-Leu-Thr-Ser-   -Gly-Ala-Ala-Asn-Asn-----   -Ala-Ala-Ser-
            91             95                100                105
          Val-CYS-Asp-CYS-Asp-Val-Glu-Ala-Ala-Phe-CYS-Phe-Ala-Lys-Ala-
            -   -   -   -   -Arg-Val-   -Asn-   -   -   -Arg-  -
           106            110                115                120
          Pro-Tyr-Asn-Asn-Ala-Asn-Trp-Asn-Ile-Asp-Thr-Lys-Lys-Arg-CYS-Gln
            -   -Ile-Asp-Lys-  -Tyr-   -   -   -Phe-Asn-Ala-   -   -
```

Figure 18.23. Alignment showing homology between notexin (upper sequence) and a phospholipase A_2 from *Naja melanoleuca* venom. (Reproduced from Halpert and Eaker, 1975.)

possibility that the enzyme action of presynaptic toxin is due to contaminated phospholipase A_2. On the basis of the available information, it is most logical to state that phospholipase A_2 is nonneurotoxic, but some presynaptic toxins possess phospholipase A_2 activity in addition to neurotoxicity.

REFERENCES

Aoyagi, H., Yonezawa, H., Takahashi, N., Kato, T., Izumiya, N., and Yang, C. C. (1972). Synthesis of a peptide with cobrotoxin activity, *Biochim. Biophys. Acta,* **263,** 823.

Arnberg, H., Eaker, D., Frykland, L., and Karlsson, E. (1974). Amino acid sequence of oxiana *a*, the main neurotoxin of the venom of *Naja naja oxiana, Biochim. Biophys. Acta,* **359,** 222.

Banks, B. E. C., Miledi, R., and Shipolini, R. A. (1974). The primary sequences and neuromuscular effects of three neurotoxic polypeptides from the venoms of *Dendroaspis viridis, Eur. J. Biochem.,* **45,** 457.

Bartels, E. and Rosenberry, T. L. (1971). Sanke neurotoxins: Effect of disulfide reduction on interaction with electroplax, *Science,* **174,** 1236.

Bechis, G., Granier, C., Van Rietschoten, J., Jover, E., Rochat, H., and Miranda, F. (1976). Purification of six neurotoxins from the venom of *Dendroaspis viridis.* Primary structure of two long toxins *Eur. J. Biochem.,* **68,** 445.

Béress, L., Béress, R., and Wunderer, G. (1975). Isolation and characterization of three polypeptides with neurotoxic activity from *Anemonia sulcata, FEBS Lett.,* **50,** 311.

Botes, D. P. (1971). Snake venom toxins: The amino acid sequences of toxins α and β from *Naja nivea* venom and the disulfide bonds of toxin α, *J. Biol. Chem.,* **246,** 7383.

References

Botes, D. P. (1972). Snake venom toxins: The amino acid sequences of toxin b and d from *Naja melanoleuca* venom, *J. Biol. Chem.*, **247**, 2866.

Botes, D. P. (1974). Snake venom toxins: The reactivity of the disulfide bonds of *Naja nivea* toxin α, *Biochim. Biophys. Acta*, **359**, 242.

Botes, D. P. and Strydom, D. J. (1969). A neurotoxin, toxin α from Egyptian cobra (*Naja haje haje*) venom, *J. Biol. Chem.*, **244**, 4147.

Botes, D. P., Strydom, D. J., Anderson, C. G., and Christensen, P. A. (1971). Snake venom toxins: Purification and properties of three toxins from *Naja nivea* (Linnaeus) (Cape cobra) venom and the amino acid sequence of toxin d, *J. Biol. Chem.*, **246**, 3132.

Bukolova-Orlova, T. G., Burstein, E. A., and Yukel'son, L. Ya. (1974). Fluorescence of neurotoxins from middle-Asian cobra venom, *Biochim. Biophys. Acta*, **342**, 275.

Bukolova-Orlova, T. G., Permykov, E. A., Burstein, E. A., and Yukel'son, L. Ya (1976). Reinterpretation of luminescence properties of neurotoxin from the venom of Middle-Asian cobra *Naja oxiana* Eichw., *Biochim. Biophys. Acta*, **439**, 426.

Carlsson, F. H. H. (1975). Snake venom toxins: The primary structure of protein S_4C_{11}, a neurotoxin homologue from the venom of forest cobra (*Naja melanoleuca*), *Biochim. Biophys. Acta*, **400**, 310.

Chance, R. E., Ellis, R. M., and Bromer, W. W. (1968). Porcine insulin: Characterization and amino acid sequence, *Science*, **161**, 165.

Chang, C. C. and Hayashi, K. (1969). Chemical modification of the tryptophan residue in cobrotoxin, *Biochem. Biophys. Res. Commun.*, **37**, 841.

Chang, C. C. and Lee, C. Y. (1963). Isolation of neurotoxins from the venom of *Bungarus multicinctus* and their modes of neuromuscular blocking action, *Arch. Int. Pharmacodyn.*, **144**, 241.

Chang, C. C. and Lee, C. Y. (1966). Electrophysiological study of neuromuscular blocking action of cobra neurotoxin, *Br. J. Pharmacol. Chemother.*, **28**, 172.

Chang, C. C. and Yang, C. C. (1973). Immunochemical studies on the tryptophan-modified cobrotoxin, *Biochim. Biophys. Acta*, **295**, 595.

Chang, C. C., Yang, C. C., Hamaguchi, K., Nakai, K., and Hayashi, K. (1971a). Studies on the status of tyrosyl residues in cobrotoxin, *Biochim. Biophys. Acta*, **236**, 164.

Chang, C. C., Yang, C. C., Nakai, K., and Hayashi, K. (1971b). Studies on the status of free amino and carboxyl groups in cobrotoxin, *Biochim. Biophys. Acta*, **251**, 334.

Chang, C. C., Yang, C. C., Kurobe, M., Nakai, K., and Hayashi, K. (1971c). The identification of the special glutamic acid residue essential for activity of cobrotoxin, *Biochem. Biophys. Res. Commun.*, **43**, 429.

Chang, C. C., Chen, T. F., and Lee, C. Y. (1973a). Studies of the presynaptic effect of β-bungarotoxin on neuromuscular transmission, *J. Pharmacol. Exp. Ther.*, **184**, 339.

Chang, C. C., Huang, M. C., and Lee, C. Y. (1973b). Mutual antagonism between botulinum toxin and β-bungarotoxin, *Nature*, **243**, 166.

Chen, I. L. and Lee, C. Y. (1970). Ultrastructural changes in the motor nerve terminals caused by β-bungarotoxin, *Virchows Arch. Pathol. Anat. Physiol.*, **6**, 318.

Chen, Y. H., Lu, H. S., and Lo, T. B. (1975). Conformation of snake neurotoxins: prediction and comparison, *J. Chin. Biochem. Soc.*, **4**, 69.

Chicheportiche, R., Rochat, C., Sampieri, F., and Lazdunski, M. (1972). Structure–function relationships of neurotoxins isolated from *Naja haje* venom: Physicochemical properties and identification of the active site, *Biochemistry*, **11**, 1681.

Chicheportiche, R., Vincent, J. P., Kopeyan, C., Schweitz, H., and Lazdunski, M. (1975). Structure–function relationship in the binding of snake neurotoxins to the *Torpedo* membrane receptor, *Biochemistry*, **14**, 2081.

Clark, D. G., Machurchie, D. D., Elliot, E., Wolcott, R. G., Landel, A. M., and Raftery, M. A. (1972). Elapid neurotoxins: Purification, characterization, and immunochemical studies of α-bungarotoxin, *Biochemistry*, **11**, 1663.

Cooper, D. and Reich, E. (1972). Neurotoxin from venom of cobra, *Naja naja siamensis*, *J. Biol. Chem.*, **247**, 3008.

Daly, J. (1970). Batrachotoxin, a novel steroidal alkaloid with selective effects of biomembrane permeability, *Aldrichim. Acta,* **3,** 3.

Daly, J. and Witkop, B. (1971). Batrachotoxin, an extremely active cardio- and neurotoxin from the Columbian arrow poison frog, *Clin. Toxicol.,* **4,** 331.

Dickerson, R. E. and Geis, I. (1969). *The Structure and Action of Proteins,* Benjamin, Menlo Park, Calif.

Eaker, D. L. and Porath, J. (1967). The amino acid sequence of a neurotoxin from *Naja nigricollis* venom, *Jap. J. Microbiol.,* **11,** 353.

Endo, Y., Sato, S., Ishii, S., and Tamiya, N. (1971). The disulfide bonds of erabutoxin a, a neurotoxic protein of a sea-snake (*Laticauda semifasciata*) venom, *Biochem. J.,* **122,** 463.

Fryklund, L., Eaker, D. and Karlsson, E. (1972). Amino acid sequences of two principal neurotoxins of *Enhydrina schistosa* venom, *Biochemistry,* **11,** 4633.

Gabel, D., Rasse, D., and Scheraga, A. (1976). Search for low-energy conformations of a neurotoxic protein by means of predictive rules, tests for hard-sphere overlaps, and energy minimization, *Int. J. Peptide Protein Res.,* **8,** 237.

Grishin, E. V., Sukhikh, A. P., Lukyanchuk, N. N., Slobodyan, L. N., Lipkin, V. M., and Ovchinnikov, Yu. A. (1973). Amino acid sequence of neurotoxin II from *Naja naja oxiana* venom, *FEBS Lett.,* **36,** 77.

Grishin, E. V., Sukhikh, A. P., Slobodyan, L. N., and Ovchinnikov, Yu. A. (1974). Amino acid sequence of neurotoxin I from *Naja naja oxiana* venom, *FEBS Lett.,* **45,** 118.

Halpert, J. and Eaker, D. (1975). Amino acid sequence of a presynaptic neurotoxin from the venom of *Notechis scutatus scutatus* (Australian tiger snake), *J. Biol. Chem.,* **250,** 6990.

Halpert, J., Eaker, D., and Karlsson, E. (1976). The role of phospholipase activity in the action of a presynaptic neurotoxin from the venom of *Notechis scutatus scutatus* (Australian tiger snake), *FEBS Lett.,* **61,** 72.

Hartley, B. S. (1964). "The structure and activity of chymotrypsin," in T. W. Goodwin, J. I. Harris, and B. S. Hartley, Eds., *Structure and Activity of Enzymes,* Academic, New York.

Hauert, J., Maire, M., Sussman, A., and Bargetzi, J. P. (1974). The major lethal neurotoxin of the venom of *Naja naja philippinensis:* Purification, physical and chemical properties, partial amino acid sequence, *Int. J. Peptide ProteinsRes.,* **6,** 201.

Hayashi, K. and Ohta, M. (1975). Neurotoxic proteins in snake venoms, *Tanpakushitsu Kakusan Koso,* **20,** 53.

Homma, M., Okonogi, T., and Mishima, S. (1964). Studies on sea snake venom: Biological toxicities of venoms possessed by three species of sea snakes captured in coastal waters of Amami Oshima, *Gunma J. Med. Sci.,* **13,** 283.

Hong, B. and Tu, A. T. (1970). Importance of tryptophan residue for toxicity in sea snake venom toxins, *Fed. Proc.,* **29,** 888.

Howard, B. D. (1975). Effects of β-bungarotoxin on mitochondrial respiration are caused by associated phospholipase A activity, *Biochem. Biophys. Res. Commun.,* **67,** 58.

Huang, J. S. and Ling, K. H. (1974). Tritiation of cobrotoxin, *Toxicon,* **12,** 435.

Huang, J. S., Liu, S. S., Ling, K. H., Chang, C. C., and Yang, C. C. (1972). Photooxidation of cobrotoxin, *J. Formosan Med. Assoc.,* **71,** 383.

Huang, J. S., Liu, S. S., Liang, K. H., Chang, C. C., and Yang, C. C. (1973). Iodination of cobrotoxin, *Toxicon,* **11,** 39.

Joubert, F. J. (1973). Snake venom toxins: The amino acid sequences of two toxins from *Ophiophagus hannah* (king cobra) venom, *Biochim. Biophys. Acta,* **317,** 85.

Joubert, F. J. (1975a). The amino acid sequences of three toxins (CM–10, CM–12, and CM–14) from *Naja haje annulifera* (Egyptian cobra) venom, *Hoppe-Seylers Z. Physiol. Chem.,* **356,** 53.

Joubert, F. J. (1975b). The purification and amino acid sequence of toxin CM–13b from *Naja haje annulifera* (Egyptian cobra) venom, *Hoppe-Seylers Z. Physiol. Chem.,* **356,** 1901.

Joubert, F. J. (1976). Snake venom toxins: The amino acid sequences of three toxins (CM–8, CM–11, and CM–13a) from *Naja haje annulifera* (Egyptian cobra) venom, *Eur. J. Biochem.,* **64,** 219.

References

Karlsson, E. and Eaker, D. (1972). Isolation of the principal neurotoxins of *Naja naja* subspecies from the Asian mainland, *Toxicon,* **10,** 217.

Karlsson, E., Arnberg, H., and Eaker, D. (1971). Isolation of the principal neurotoxins of two *Naja naja* subspecies, *Eur. J. Biochem.,* **21,** 1.

Karlsson, E., Eaker, D., and Ryden, L. (1972). Purification of a presynaptic neurotoxin from the venom of the Australian tiger snake, *Notechis scutatus scutatus, Toxicon,* **10,** 405.

Karlsson, E., Eaker, D., and Drevin, H. (1973). Modification of the invariant tryptophan residue of two *Naja naja* neurotoxins, *Biochim. Biophys. Acta,* **328,** 510.

Kelly, R. B. and Brown, F. R., III (1974). Biochemical and physiological properties of a purified snake venom neurotoxin which acts presynaptically, *J. Neurobiol.,* **5,** 135.

Kopeyan, C., Van Rietschoten, J., Martinez, G., Rochat, H., and Miranda, F. (1973). Characterization of five neurotoxins isolated from the venoms of two Elapidae snakes, *Naja haje* and *N. nigricollis, Eur. J. Biochem.,* **35,** 244.

Kopeyan, C., Miranda, F., and Rochat, H. (1975). Amino-acid sequence of toxin III of *Naja haje, Eur. J. Biochem.,* **58,** 117.

Lee, C. Y. (1972). Chemistry and pharmacology of polypeptide toxins in snake venoms, *Ann. Rev. Pharmacol.,* **12,** 265.

Lee, C. Y. and Chang, C. C. (1966). Modes of action of purified toxins from elapid venoms on neuruomuscular transmission, *Mem. Inst. Butantan,* **33,** 555.

Lee, C. Y., Chang, S. L., Kau, S. T., and Luh, S. H. (1972). Chromatographic separation of the venom of *Bungarus multicinctus* and characterization of its components, *J. Chromatogr.,* **72,** 71.

Low, B. W., Potter, R., Jackson, R. B., Tamiya, N., and Sato, S. (1971). X-ray crystallographic study of the erabutoxins and of diiodo derivatives, *J. Biol. Chem.,* **246,** 4366.

Low, B. W., Preston, H. S., Sato, A., Rosen, L., Searl, J. E., Rudko, D. D. and Richardson, J. S. (1976). Three dimentional structure of erabutoxin b neurotoxic protein: Inhibitor of acetylcholine receptor, *Proc. Nat. Acad. Sci.,* U.S.A., **73,** 2991.

Maeda, N. and Tamiya, N. (1974). The primary structure of the toxin *Laticauda semifasciata.* III. A weak and reversibly acting neurotoxin from the venom of a sea snake, *Laticauda semifasciata, Biochem. J.,* **141,** 389.

Mebs, D., Narita, K., Iwanaga, S., Samejima, Y., and Lee, C. Y. (1971). Amino acid sequence of α-bungarotoxin from the venom of *Bungarus multicinctus, Biochem. Biophys. Res. Commun.,* **44,** 711.

Mellstrand, S. T. and Samuelsson, G. (1974). Phoratoxin, a toxic protein from the mistletoe *Phoradendron tomentosum* subsp. *macrophyllum* (*Lorantaceta*), *Acta Pharm. Suec.,* **11,** 347.

Menez, A., Morgat, J., Fromageot, P., Ronseray, A., Boquet, P., and Changeux, J. P. (1971). Tritium labeling of the α-neurotoxin of *Naja nigricollis, FEBS Lett.,* **17,** 333.

Miranda, F., Kupeyan, C., Rochat, H., Rochat, C., and Lissitzky, S. (1970). Purification of animal neurotoxins: Isolation and characterization of four neurotoxins from two different sources of *Naja haje* venom, *Eur. J. Biochem.,* **17,** 477.

Miyazawa, T. (1967). "Infrared spectra and helical conformations," in G. D. Fasman, Ed., *Poly-α-Amino Acids,* Marcel Dekker, New York, pp. 69–103.

Nakai, K., Nakai, C., Sasaki, T., Kakiuchi, K., and Hayashi, K. (1970). Purification and some properties of toxin A from the venom of the Indian cobra, *Naturwissenschaften,* **57,** 387.

Nakai, K., Sasaki, T., and Hayashi, K. (1971). Amino acid sequence of toxin A from the venom of the Indian cobra (*Naja naja*), *Biochem. Biophys. Res. Commun.,* **44,** 893.

Narahashi, T., and Haas, H. G. (1967). DDT: Interaction with nerve membrane conductance changes, *Science,* **157,** 1438.

Narahashi, T., Moore, J. W., and Scott, W. R. (1964). Tetrodoxin blockage of sodium conductance increase in lobster giant axon, *J. Gen. Physiol.,* **50,** 1413.

Narahashi, T., Anderson, N. C., and Moore, J. W. (1967). Comparison of tetrodotoxin and procaine in internally perfused squid giant axon, *J. Gen. Physiol.,* **50,** 1413.

O'Brien, R. D. (1969). "Poisons as tools in studying the nervous system," in F. R. Blood, Ed., *Essays in Toxicology,* Vol. 1, Academic, New York, pp. 1–59.

Ohta, M. and Hayashi, K. (1973). Localization of the five disulfide bridges in toxin B from the venom of the Indian cobra (*Naja naja*), *Biochem. Biophys. Res. Commun.,* **55,** 431.

Ohta, M. and Hayashi, K. (1974a). Chemical modification of the tryptophan residue in toxin B from the venom of the Indian cobra, *Biochem. Biophys. Res. Commun.,* **57,** 973.

Ohta, M. and Hayashi, K. (1974b). Chemical modification of tyrosine residue in toxin B from the venom of the Indian cobra, *Naja naja, Biochem. Biophys. Res. Commun.,* **56,** 981.

Olson, T. and Samuelsson, G. (1974). The disulfide bonds of viscotoxin A2 from the European mistletoe (*Viscum album* L. *Loranthaceae*), *Acta Pharm. Suec.,* **11,** 381.

Poilleux, G. and Boquet, P. (1972). Proprietes de trois isolees du venin d'un Elapidae: *Naja melanoleuca, C. R. Acad. Sci. Paris,* Ser. D, **274,** 1953.

Preston, H. S., Kay, J., Sato, A., and Low, B. W. (1975). Crystalline erabutoxin, C, *Toxicon,* **13,** 273.

Ramachandran, L. K. and Witkop, B. (1959). Selective cleavage of C-tryptophyl peptide bonds in proteins and peptides, *J. Am. Chem. Soc.,* **81,** 4028.

Raymond, M. L. and Tu, A. T. (1972). Role of tyrosine in sea snake neurotoxin, *Biochim. Biophys. Acta,* **285,** 498.

Rochat, H., Gregoire, J., Martin-Moutot, N., Menashe, M., Kopeyan, C., and Miranda, F. (1974). Purification of animal neurotoxins: Isolation and characterization of three neurotoxins from the venom of *Naja nigricollis mossambica* Peters, *FEBS Lett.,* **42,** 335.

Samuelsson, G. (1973). Mistletoe toxins, *Syst. Zool.,* **22,** 566.

Samuelsson, G. and Jayawardene, A. L. (1974). Isolation and characterization of viscotoxin 1-Ps from *Viscum album* L. ssp. *austriacum* (Wiesb.) Vollmann, growing on *Pinus silvestris, Acta Pharm. Suec.,* **11,** 175.

Samuelsson, G., Seger, L., and Olson, T. (1968). The amino acid sequence of oxidized viscotoxin A3 from the European mistletoe (*Viscum album* L. *Loranthaceae*), *Acta Chem. Scand.,* **22,** 2624.

Sato, S. and Tamiya, N. (1970). Iodination of erabutoxin b: Diiodohistidine formation, *J. Biochem.,* **68,** 867.

Sato, S. and Tamiya, N. (1971). The amino acid sequences of erabutoxins, neurotoxic proteins of sea snake (*Laticauda semifasciata*) venom, *Biochem. J.,* **122,** 453.

Seto, A., Sato, S., and Tamiya, N. (1970). The properties and modification of tryptophan in a sea snake toxin, erabutoxin a, *Biochim. Biophys. Acta,* **214,** 483.

Shipolini, R. A., Bailey, G. S., Edwardson, J. A., and Banks, B. E. C. (1973). Separation and characterization of polypeptides from the venom of *Dendroaspis viridis, Eur. J. Biochem.,* **40,** 337.

Shipolini, R. A., Bailey, G. S., and Banks, B. E. C. (1974). The separation of a neurotoxin from the venom of *Naja melanoleuca* and primary sequence determination, *Eur. J. Biochem.,* **42,** 203.

Simpson, L. L. (1971). *Neuropoisons: Their Pathophysiological Actions,* Vol. 1, Plenum, New York, pp. 1–361.

Strydom, D. J. (1973). Snake venom toxins: The evolution of some of the toxins found in snake venoms, *System. Zool.,* **22,** 596.

Strydom, A. J. C. (1973). Snake venom toxins: The amino acid sequence of two toxins from *Dendroaspis jamesoni kaimosae, Biochim. Biophys. Acta,* **328,** 491.

Strydom, A. J. C. and Botes, D. P. (1971). Snake venom toxins: Purification, properties, and complete amino acid sequence of two toxins from ringhals (*Hemachatus haemachatus*) venom, *J. Biol. Chem.,* **246,** 1341.

Strydom, D. J. (1972). Snake venom toxins: The amino acid sequences of two toxins from *Dendroaspis polylepis polylepis* (black mamba) venom, *J. Biol. Chem.,* **247,** 4029.

Sugeta, H., Go, A., and Miyazawa, T. (1972). S–S and C–S stretching vibrations and molecular conformations of dialkyl disulfides and cysteine, *Chem. Lett.,* 83–86.

Sugeta, H., Go. A., and Miyazawa, T. (1973). Vibrational spectra and molecular conformation of dialkyl disulfides, *Bull. Chem. Soc.,* **46,** 3407.

Szczepaniak, A. C. (1974). Effect of α-bungarotoxin and dendroaspis neurotoxins on acetylcholine responses of snail neurones, *J. Physiol.* (*London*), **24,** 55.

Takechi, M. and Hayashi, K. (1972). Localization of the four disulfide bridges in cytotoxin II from the venom of the Indian cobra (*Naja naja*), *Biochem. Biophys. Res. Commun.,* **49,** 584.

Tamiya, N. and Abe, H. (1972). The isolation, properties, and amino acid sequence of erabutoxin c, a minor neurotoxic component of the venom of a sea snake, *Laticauda semifasciata, Biochem. J.,* **130,** 547.

Tamiya, N., Sato, S., and Seto, A. (1971). "Structure and action of the sea snake neurotoxins," in A. De Vries and E. Kochva, Eds., *Toxins of Animal and Plant Origin,* Vol. I, Gordon and Breach, New York, pp. 237–250.

Tsernoglou, D. and Petsko, G. A. (1976). The crystal structure of a post-synaptic neurotoxin from sea snake at 2.2 Å resolution. *FEBS Letters,* **68,** 1.

Tsernoglou, D., Raymond, M., and Tu, A. T. (1972). *Am. Chem. Soc. Rocky Mt. Reg. Mett.,* Abstract, p. 12.

Tu, A. T. (1973). Neurotoxins of animal venoms: Snakes, *Ann. Rev. Biochem.,* **42,** 235.

Tu, A. T. and Hong, B. (1971). Purification and chemical studies of a toxin from the venom of *Lapemis hardwickii* (Hardwick's sea snake), *J. Biol. Chem.,* **246,** p. 2772.

Tu, A. T. and Toom, P. M. (1971). Isolation and characterization of the toxic component of *Enhydrina schistosa* (common sea snake) venom, *J. Biol. Chem.,* **246,** 1012.

Tu, A. T., Hong, B. S., and Solie, T. N. (1971). Characterization and chemical modifications of toxins isolated from the venoms of sea snake, *Laticauda semifasciata* from the Philippines, *Biochemistry,* **10,** 1295.

Tu, A. T., Hong, B., Toum, P. M., and Tsernoglou, D. (1973). "Chemical study of sea snake venom toxins from three species in Southeast Asia," in E. Kaiser, Ed., *Animal and Plant Toxins,* Wilhelm Goldman, Munich, pp. 45–50.

Tu, A. T., Lin, T. S., and Bieber, A. L. (1975). Purification and chemical characterization of the major neurotoxin from the venom of *Pelamis platurus, Biochemistry,* **14,** 3408.

Tu, A. T., Jo, B. H., and Yu, N. T. (1976a). Laser Raman spectroscopy of snake venon neurotoxins, *Int. J. Peptide Protein Res.,* **8,** 337.

Tu, A. T., Prescott, B., Chou, C. H., and Thomas, G. J., Jr. (1976b). Structural properties of Mojave toxin of *Crotalus scutulatus* (Mojave rattlesnake) determined by laser Raman spectroscopy, *Biochem. Biophys. Res. Commun.,* **68,** 1139.

Turakulov, Ya. Kh., Sorokin, V. M., and Nishankhodzhaeva, S. A. (1972). Amino acid composition of middle Asian cobra venom, *Biokhimiya,* **37,** 124.

Van Holde, K. E. (1971). *Physical Biochemistry,* Prentice-Hall, Englewood Cliffs., N. J.

Viljoen, C. C. and Botes, D. P. (1973). Snake venom toxins: The purification and amino acid sequence of toxin Fv_{11} from *Dendroaspis angusticeps* venom, *J. Biol. Chem.,* **248,** 4915.

Viljoen, C. C. and Botes, D. P. (1974). Snake venom toxins: The purification and amino acid sequence of toxin TA2 from *Dendroaspis angusticeps* venom, *J. Biol. Chem.,* **249,** 366.

Wong, C. H., Chang, T. W., Lea, T. J., and Yang, C. C. (1972). X-ray crystallographic study of cobrotoxin, *J. Biol. Chem.,* **247,** 608.

Yang, C. C. (1965). Enzymic hydrolysis and chemical modification of cobrotoxin, *Toxicon,* **3,** 19.

Yang, C. C. (1967). The disulfide bonds of cobrotoxin and their relationship to lethality, *Biochim. Biophys. Acta,* **133,** 356.

Yang, C. C., Chang, C. C., and Wei, H. C. (1967a). Studies on fluorescent cobrotoxin, *Biochim., Biophys. Acta,* **147,** 600.

Yang, C. C., Chang, C. C., Hamaguchi, K., Ikeda, K., Hayashi, K., and Suzuki, T. (1967b). Optical rotatory dispersion of cobrotoxin, *J. Biochem. (Japan),* **61,** 272.

Yang, C. C., Chang, C. C., Hayashi, K., Suzuki, T., Ikeda, K., and Hamaguchi, K. (1968). Optical rotatory dispersion and circular dichroism of cobrotoxin, *Biochim. Biophys. Acta,* **168,** 373.

Yang, C. C., Yang, H. J., and Huang, J. S. (1969a). The amino acid sequence of cobrotoxin, *Biochim. Biophys. Acta,* **188,** 65.

Yang, C. C., Chang, C. C., Hayashi, K., and Suzuki, T. (1969b). Amino acid composition and end group analysis of cobrotoxin, *Toxicon,* **7,** 43.

Yang, C. C., Yang, H. J., and Chiu, R. H. C. (1970). The position of disulfide bonds in cobrotoxin, *Biochim. Biophys. Acta,* **214,** 355.

Yang, C. C., Chang, C. C., and Liou, I. F. (1974). Studies on the status of arginine residues in cobrotoxin, *Biochim. Biophys. Acta,* **365,** 1.

Yu, N. T. and Jo, B. H. (1973). Comparison of protein structure in crystals and in solution by laser Raman scattering, *Arch. Biochem. Biophys.*, **156**, 469.

Yu, N. T., Jo, B. H., and O'Shea, D. C. (1973). Laser Raman scattering of cobramine B, a basic protein from cobra venom, *Arch. Biochem. Biophys.*, **156**, 71.

Yu, N. T., Lin, T. S., and Tu, A. T. (1975). Laser Raman scattering of neurotoxins isolated from the venoms of sea snakes *Lapemis hardwickii* and *Enhydrina schistosa, J. Biol. Chem.*, **250**, 1782.

19 Nonneurotoxic Basic Proteins (Cardiotoxins, Cytotoxins, and Others)

1 CARDIOTOXINS	302
1.1 Pharmacological Activity, 302	
1.2 Biochemistry, 303	
2 CYTOTOXINS	306
2.1 Isolation and Chemical Characterization, 306	
2.2 Biological Activity, 308	
3 OTHERS	316
References	317

Some snake venoms, especially the Elapidae venoms, contain a variety of basic proteins other than potent neurotoxins. They all have high isoelectric points comparable to those of neurotoxins and small molecular weights comparable to or slightly smaller than those of Type I and Type II neurotoxins. The chemical structures of these basic proteins are very similar to those of potent neurotoxins (Tu, 1973), but their toxicities are usually much lower. These basic proteins exhibit different types of pharmacologic actions, including cardiotoxic, cell lysis, cytotoxic, and iodine accumulation activities. There are many different sizes of basic proteins within a venom. Most of the investigations directed toward the basic proteins other than neurotoxins are fairly recent.

In this chapter, we will review all of the basic proteins. For convenience, cardiotoxins will be discussed first, and then cytotoxins, although there is some opinion that these two types of proteins are chemically identical. Probably, some cardiotoxins may be identical to so-called cytotoxins, but this author does not believe this to be true for all cardiotoxins. Although some basic proteins may show both cardiotoxic and cytotoxic activities, differences can exist.

1 CARDIOTOXINS

The cardiotoxic action of cobra (*Naja naja*) venom was first recognized by Sarkar (1947), who observed that a substance from *N. naja* venom obtained by salt precipitation stopped perfused toad heart and arrested the heart-beat of cats. Cardiotoxin is a distinctly different protein from neurotoxin, which does not cause systolic arrest of heart (Devi and Sarkar, 1966). The protein responsible for this cardiotoxic action was thought to be of high molecular weight, but it was found that one active component remained in a dialysis bag while the other active one dialyzed out (Raudonat and Holler, 1958). The presence of cardiotoxins in cobra venom adequately explains the physiological effects of the venom. It is known that animals dying of cobra poisoning experience both peripheral respiratory paralysis, which is caused by neurotoxins, and cardiac arrest.

1.1 Pharmacological Activity

The toxicity of cardiotoxin is much weaker than that of neurotoxin. Partially purified cardiotoxin obtained from the venom of *Naja naja atra* has an LD_{50} (i.p.) value of 1.48 $\mu g\ g^{-1}$, while the corresponding value of partially purified neurotoxin is 0.074 $\mu g\ g^{-1}$ in mice (Lee et al., 1968). The cardiotoxic fraction causes contraction, followed by paralysis, of chick biventer cervicis muscle, frog sartorius muscle, and rat diaphragm. These symptoms result as a consequence of irreversible depolarization of the muscle cell membrane. Cardiotoxin causes systolic arrest of isolated frog heart and rat atrium. It also produced a contraction of guinea-pig ileum, which was largely antagonized by atropine but not by hexamethonium or pyribenzamine. In cats, cardiotoxin causes a reduction in the systolic pressure, accompanied by various electrocardiogram changes, such as P–R interval prolongation, decreased amplitude of QRS, S–T and T changes, ventricular premature beats, complete A–V block, and idioventricular rhythm. Thus cardiotoxin has a broad spectra of pharmacologic effects. Lee et al. (1968) and his co-workers (Ho et al., 1975) concluded that cardiotoxin causes irreversible depolarization of the cell membrane, consequently impairing its functions.

It is now believed that the cardiotoxin affects a membrane calcium-binding site and induces contracture by releasing the membrane calcium, rather than by increasing the Na(I) permeability of the muscle membrane (Shiau et al., 1976).

Cardiotoxin isolated from *N. naja* venom had cytolytic action toward Ehrlich ascites tumor cells (Leung et al., 1976). The rate of cardiotoxin-induced cytolysis is dose dependent and is not affected by cell concentration. Calcium ion inhibits the cytolysis reversibly. Addition of Ca ion stops the cytolytic action, whereas removal of Ca ion by EDTA abolishes the inhibitory effect. The ability of the Ca ion to inhibit cardiotoxin-induced cytolysis is probably due to interference with the binding of cardiotoxin molecules to the cell membrane.

The most basic protein obtained from *N. naja* venom has a low direct lytic activity but a strong cardiotoxic activity. Therefore Slotta and Vick (1969) proposed that the basic protein be called cardiotoxin rather than direct lytic factor. Cardiotoxin and phospholipase A_2 display a synergistic action *in vitro* and *in vivo*. The lethality of cardiotoxin increases markedly when additional phospholipase A_2 is injected either before or after the injection of cardiotoxin. Melittin behaves like cardiotoxin, and the property of this direct lytic factor is discussed in the next chapter, "Hemolysis." The amino acid sequence of cardiotoxin from *N. naja atra* has been determined (Narita and Lee, 1970) and is shown in Table 1.

Both the contracture-inducing and direct hemolytic activities of cardiotoxin can be potentiated by phospholipase A_2 (Lee et al., 1972). Since cardiotoxin is a highly basic

1 Cardiotoxins

protein, it will combine with acidic compounds such as gangliosides, RNA, or heparin. The binding of such polyanionic compounds can neutralize the actions of cardiotoxin.

Because of the similarity in amino acid composition of cardiotoxin and cobramine B, Lee et al. (1971) concluded that they are identical substances (Table 1). Cobramine A and B have been isolated from *N. naja* venom (Larsen and Wolff, 1968a) and been shown to inhibit iodide accumulation by thyroid slices (Wolff et al., 1968). Cobramine B also possesses a weak hemolytic activity but strong cardiotoxicity (Larsen and Wolff, 1967, 1968b).

Chang et al. (1972a, b) found that cardiotoxin (from *N. naja atra* venom) is the only agent that blocks nerve conduction and membrane depolarization. Phospholipase A_2 and lysolecithin have no effect. Therefore the biological synergistic effect of phospholipase A_2 reported by many investigators is believed to be due to easier penetration of cardiotoxin after the membrane matrix is disorganized by the action of phospholipase A_2. *Naja naja* cardiotoxins (CM 11 and 12), isolated by the method of Takechi et al. (1971), gave results identical to those obtained with *N. naja atra* cardiotoxin (Chang et al., 1972b). Toxin A obtained from the venom of *N. naja* (Nakai et al., 1970) has no effect on nerve conduction and diaphragm membrane potential (Chang et al., 1972b). The fact that cardiotoxin of *N. naja atra* also causes myonecrosis (Lai et al., 1972) substantiates the observation of myonecrosis due to cobra envenomation by Reid (1964) and by Stringer et al. (1971). With regard to cancer cells, cardiotoxin is cytotoxic to HeLa, KB, Yoshida sarcoma, and Ehrlich ascites tumor cells *in vitro* but has no antitumor activity *in vivo* (Lee et al., 1972).

The cardiotoxin obtained from *N. naja siamensis* venom irreversibly prevents spontaneous contraction of embryonic heart cells both in culture and *in vivo* (Arms and McPheeters, 1975). It was also noted that early embryonic heart is less susceptible than mature heart to the inhibitory effects of cardiotoxin.

Most of the cardiotoxins or nonneurotoxic basic proteins have been isolated from the venoms of cobra (*Naja spp*). A cardiotoxinlike substance was isolated from *Bungarus fasciatus* venom (Lin et al., 1972). In this species, two homogeneous cardiotoxic principles (VI-A, VI-B) possess contracture-producing activity on chick biventer cervicis muscle preparations, produce local irritation of rabbit eye conjunctiva, and show a depressing action on isolated frog heart and rat atrium. Both cardiotoxins lack direct hemolytic action. The contracture-inducing activity of cardiotoxin VI-B can be abolished by EDTA or high Ca(II) concentrations (12 mM). However, it is accelerated by low Ca(II) concentrations. Slight differences in the pharmacologic properties of VI-B and *N. naja atra* cardiotoxin indicate that the two toxins may not be identical (Lin et al., 1975). The fact that the amino acid compositions of these two cardiotoxic principles are quite different from those of the known cardiotoxins suggest that cardiotoxin is not limited to one type of protein; rather, many different proteins can have cardiotoxic activity.

Three cardiotoxins were isolated from *Bungarus fasciatus* venom by Lu and Lo (1974). The LD_{50} values of these cardiotoxins are 4.8 μg g^{-1} for V-2, 38 μg g^{-1} for V-3, and 3.1 μg g^{-1} for V1-I-1 fractions. The neurotoxin has a much lower LD_{50} value of 0.91 μg g^{-1}, indicating that it is about four to five times more toxic than these cardiotoxins.

1.2 Biochemistry

A cardiotoxin (toxin III) isolated from the venom of a *Naja naja* specimen from Hong Kong is very similar to that of *N. naja atra* from Formosa (Keung et al., 1975a). It has a molecular weight of 7250 daltons with a pI of 11.1. *Naja naja* cardiotoxin contains

Table 1 Amino Acid Compositions of Cardiotoxins and Other Nonneurotoxic Basic Proteins from Snake Venoms

Genus Species Subspecies	Bungarus fasciatus		Haemachatus hemachatus	Naja naja			N. naja atra	N. nigricollis
Origin	S.E. Asia		Africa	Cambodia	Hong Kong	India	Formosa	Ethiopia
Name	Cardiotoxin VI	VB	Direct lytic factor	Cardiotoxin	Cardiotoxin III	Cobramine B	Cardiotoxin	Cardiotoxin
M.W.	12,700	13,652	6,334		7,250	5,840	6,777	
Lysine	5	9	10	8	8-9	8	9	9
Histidine	1	3	1	0	0	0	0	0
Arginine	3	4	1	2	2	2	2	2
Aspartic acid	11	16	6	8	7	5	6	6
Threonine	6	10	3	3	2	3	3	3
Serine	1	1	3	3	2	2	2	2
Glutamic acid	5	6	1	0	0	0	0	1

Proline	3	5	5	4	5	4	5	6
Glycine	8	11	2	2	2	2	2	2
Alanine	8	11	1	2	2	2	2	2
Valine	2	3	4	4	6	6	7	3
Methionine	1	1	2	3	2	2	2	4
Isoleucine	3	5	2	4	1	1	1	3
Leucine	3	6	6	6	5-6	5	6	5
Tyrosine	5	9	1	2	3	3	3	2
Phenylalanine	3	4	1	1	3	1	2	1
Half-cystine	8	18	8	8	8	6	8	8
Tryptophan	1	2	0	0	0	0	0	1
Total residue	77	124	57	60	58-60	52	60	60
References	Lin _et al_, 1975		Aloof-Hirsch _et al_, 1968	Frykland and Eaker, 1975a	Keung _et al_, 1975a	Larsen and Wolff, 1968a	Lee _et al_, 1971	Frykland and Eaker, 1975b

3 moles of tyrosine residues, two of which can be nitrated with a concomitant loss of cytolytic activity. However, even when all tyrosine residues are nitrated, there is no change in immunological specificity (Keung et al., 1975b).

The amino acid compositions of cardiotoxins isolated from different venoms are shown in Table 1. Some have 58 to 60 amino acid residues and are probably similar to Type I neurotoxin in primary structure. Others have 77 or 124 residues, indicating that not all cardiotoxins are identical chemically. The amino acid sequences of some cardiotoxins are essentially identical to those of cytotoxins. Therefore the primary structures of cardiotoxins will be discussed in Section 2, together with those of cytotoxins.

Cardiotoxin stimulates ACTH-induced lipolysis and steroidogenesis. At high doses (>3 μg ml^{-1}), however, these reactions are inhibited. The inhibitory and stimulating effects of cardiotoxin are accompanied by a decrease and an increase, respectively, in the total cyclic AMP level. Cardiotoxin inhibition cannot be reversed by increasing the ACTH concentration in isolated adrenal or isolated fat cells.

Practically all of the cardiotoxins isolated have been obtained from the venoms of Elapidae, but it should not be assumed that cardiotoxin is restricted to the venoms from snakes of this family. Bieber et al. (1975) have isolated cardiotoxin from the venom of *Crotalus scutulatus*, which belongs to the Crotalidae family. Surprisingly, this cardiotoxin has a high molecular weight of 22,000 daltons and a low isoelectric point of 4.7, properties quite different from those of cardiotoxic basic proteins from Elapidae venoms. This cardiotoxin was discussed in Chapter 14.

2 CYTOTOXINS

Histolysis is a typical sign of snake envenomation, expecially by Crotalidae and Viperidae, and cytotoxic actions of snake venom on many different cells are quite common (Gaertner et al., 1962; Yoshikura et al. 1966; Tu and Passey, 1969). Because different types of cells differ in their susceptibilities to lysis, snake cytotoxins are useful tools in studying the molecular structures of cell membranes (Braganca, 1974). A number of cytotoxins have been isolated, and their chemical and biological activities studied. In this section we will discuss first the chemistry of cytotoxins, and then the biological effects.

2.1 Isolation and Chemical Characterization

The first isolation of a cytotoxin in the homogeneous state was accomplished by Braganca et al. (1967) from the venom of *Naja naja*. Takechi et al. (1971) also isolated two cytotoxins from the same venom which had high cytotoxicity toward Yoshida sarcoma and ascites hepatoma cells. The amino acid compositions of these and other toxins are shown in Table 2.

The cytopathic effects of cytotoxins can be visualized by observing cellular morphological changes by microscopy or by release of enzymes from the lysed cells. Dimari et al. (1975) found that the concentration of lactate dehydrogenase in SV101 cells was linearly proportional to the number of cells in a given culture. By studying the amount of enzyme released, they could measure the cytolytic activity of cytotoxins. They postulated that cell lysis arises from the interaction of the toxin with membrane receptors, leading to alteration of cell membrane structure. Heparin can reduce the cytotoxic action, presumably because of its ability to bind the cytotoxin in an acidic compound—basic protein interaction.

2 Cytotoxins

Cytotoxic protein P_6 from *N. naja* venom has been shown to inhibit cellular Na(I) and K(I)–ATPase activity, which has a linear correlation to its lytic activity (Zaheer et al., 1975). These authors theorized that the lysis of the cell is due to an imbalance of K(I) and Na(I) in the cell, which leads to swelling and disintegration of the membrane structure Direct lytic factor isolated from the same venom by Lankishch et al. (1972) also inhibited Na(I) and K(I)-ATPase activity. This may indicate that the so-called DLF is indeed identical to cytotoxin.

Louw (1974c) investigated the degree of similarity between different cytotoxins from the same venom and those from other venoms. He found that there is a marked similarity among cytotoxins; some are even identical immunologically. However, he observed that cytotoxins are immunologically distinct from neurotoxins. As can be seen in Fig. 4 of Chapter 18, cytotoxins are basically similar to Type I neurotoxins, which also contain eight half-cystines. The locations of disulfide bridges had previously been determined and reported to be identical to those in neurotoxins (Takechi and Hayashi, 1972). Moreover, cytotoxin is similar to Type II neurotoxins at loop 5, which consists of three amino acids residues (Fig. 4 of Chapter 18).

Although the similarity in disulfide linkage backbone is remarkable, there is considerable difference in amino acid composition between cytotoxins and neurotoxins. The primary differences are as follows:

1. Cytotoxins have high lysine, methionine, and tyrosine contents relative to those of neurotoxins. For instance, there are 7 to 10 lysine residues in most cytotoxins, as compared to 2 to 7 residues in neurotoxins.

2. Cytotoxins are low in arginine and glutamic acid contents, as compared to neurotoxins.

As can be seen from Table 3, which shows the amino acid sequences of cytotoxins and cardiotoxins, they are nearly identical. On the basis of chemical similarity, it is very likely that some cytotoxins and cardiotoxins are identical substances. Some investigators have assigned their isolated fractions as cytotoxins and showed them to contain cytotoxic activity. Many others, however, did not investigate cardiotoxic activity. Because the progress of chemical investigation is faster than that of pharmacological research, not all the fractions isolated have been properly tested for pharmacological activity. Therefore it is highly probable that many cytotoxins would show cardiotoxic activity if they were tested; on the other hand, there is a possibility that some cytotoxins may not possess such activity. For this reason this author did not combine these two groups of chemically similar compounds. In this chapter the designation of a compound as a cytotoxin or cardiotoxin is based entirely on the name originally assigned by the investigators.

The similarity in the primary structures of cardiotoxins and cytotoxins is remarkable (Table 3). There are also no chemical differences in primary protein structure between specimens of different geographical origins. The characteristic structural properties are as follows:

LOOP
1. Always has 2 residues, usually leucine and lysine.
2. Has 10 or 11 amino acids (usually 10), with a high degree of homology.
3. Always consists of 6 amino acid residues and has a high degree of homology.
4. Always consists of 16 residues, but has more variability in amino acid residues than other loops.

Table 2 Amino Acid Compositions of Snake Venom Cytotoxins

Genus Species Subspecies	Naja haje annulifera		N. melanoleuca			N. mossambica mossambica				N. naja	
Origin	Egypt		S. Africa			S. Africa				India	
Name	V^{II}_1	V^{II}_2	V^{II}_1	V^{II}_2	V^{II}_3	V^{II}_1	V^{II}_2	V^{II}_3	V^{II}_4	(CM-XI) cytotoxin I	(CM-XII) cytotoxin II
Amino Acid											
Lysine	9	9	7	9	10	9	10	9	10	8-9	9
Histidine	1	1	1	1	1	0	0	0	0	0	0
Arginine	1	1	2	1	1	2	2	3	2	2	2
Aspartic acid	5	6	6	6	5	6	6	6	6	7	7-8
Threonine	4	3	3	7	7	3	3	3	2	3	3
Serine	4	2	4	2	2	2	3	2	3	2	2
Glutamic acid	1	1	2	2	2	1	1	1	1	1	0
Proline	5	6	4	4	4	6	5	6	4	5	5
Glycine	2	2	2	3	3	2	3	2	2	2	2
Alanine	1	2	2	2	2	2	1	1	3	2	2
Valine	8	6	6	2	2	3	3	3	5	6	7
Methionine	2	4	2	0	0	4	3	3	3	2	2
Isoleucine	1	1	3	2	2	3	3	3	3	2	1
Leucine	4	5	5	6	6	5	5	6	5	6	6
Tyrosine	2	3	3	3	3	2	2	2	3	5	4
Phenylalanine	1	1	0	3	3	1	1	1	0	0	1
Half-cystine	8	8	8	8	8	8	8	8	8	8	8
Tryptophan	1	Nd	0	0	0	1	1	1	0	0	0
Total residue..	60	61	60	61	61	60	60	60	60	61-62	61-62
References	Weise et al 1973		Carlsson and Joubert, 1974			Louw, 1974a		Louw, 1974b		Takechi et al, 1971	

5 Consists of 3 residues, characteristically isolucine, aspartic acid, and valine. Since Type I neurotoxins have only 1 residue in this loop and Type II toxins have 3, cardiotoxins and cytotoxins are identical to Type II neurotoxin in this aspect.

6 Always has 10 residues with a high degree of homology.

7 Always has 4 residues, and the sequences are highly similar to one another.

8 Consists of only 1 residue, without exception asparagine.

2.2 Biological Activity

As previously mentioned, snake venoms are quite commonly cytotoxic to many types of cells. The venoms of *Vipera palestinae* and *Echis colorata* are cytopathic to mouse

2 Cytotoxins

Table 2 *Continued*

	N. naja naja			N. naja ceylonicus		N. naja oxiana		N. naja samarensis	N. naja siamensis
	India		Pakistan	Ceylon		Central Asia	Kirghiz, U.S.S.R.	Philippine	Thailand
cytotoxin I	NNI-A	NNI-B	NNP-A	NNC-A	NNC-B	NNO-C	cytotoxin	NNSe-C	NNS-B
9	9	9	9	9	9-10	10	10	10	9
0	0	0	0	0	0	1	1	1	0
2	2	2	2	2	2	1	1	1	2
8	8-9	7	8-9	8-9	7	5	5	5	8
3	3	3	3	3	3	2	2	2	3
2	2	2	2	2	2	3	3	3	2
1	1	0	1	1	0	0	0	0	0
4	4-5	5	4	4	7	5	5	5-6	5
2	2	2	2	2	2	2	2	2	2
2	2	2	2	2	2	3	3	2-3	3
5	5	5	4-5	4	6	6	7	5-6	4
2	2	2	2	2	2	2	2	2	2
2	2	1	2	2	1	1	1	1	1
6	6	6	6	6	6	6	6	6	6
4	4	4	4	4	4	2	2	2	3
0	0	1	0	0	1	2	2	2	1
8	8	8	8	8	8	8	8	8	8
0	0	0	0	0	0	0	0	0	0
60	60-62	59	59-61	59-60	62-63	59	60	57-60	59
Hayashi et al., 1971	Dimari et al, 1975					Grishin et al, 1974		Dimari et al, 1975	

embryo and chick embryo cells in tissue culture (Gaertner et al., 1962), and *Trimeresurus flavoviridis* venom is cytotoxic to mouse embryo cells, HeLa cells, and human embryonic fibroblasts in tissue culture (Sato et al., 1964; 1970). Penso and Balducci (1963) observed that *Vipera aspis* caused degeneration and cytolysis of Chang cells in tissue culture. Tu and Passey (1969) also found that venoms of Crotalidae and Viperidae showed very strong cytolytic activity toward bovine spleen cells and mouse embryo cells in tissue cultures. The cytolytic actions of Hydrophiidae and Elapidae, however, were very weak. The cytotoxic actions of snake venoms on bovine spleen cells are illustrated in Fig. 1. In all cases the cytolytic actions of snake venoms could be neutralized by antivenins.

The cytopathic effects of snake venoms on Yoshida sarcoma cells are shown in Fig. 2.

Table 3 Amino Acid Sequences of Cardiotoxins and Cytotoxins Isolated from Snake Venoms

	Snake	Name	Loop 1	Loop 2	Loop 3
Cardiotoxins					
1.	Naja naja (Cambodian cobra)	—	Leu-Lys-Cys-	-Asn-Lys-Leu-Ile-Pro-Ile-Ala-Ser-Lys-Thr-Cys-	-Pro-Ala-Gly-Lys-Asn-Leu-
2.	N. naja atra	—	Leu-Lys-Cys-	-Asn-Lys-Leu-Val-Pro-Leu-Phe-Tyr-Lys-Thr-Cys-	-Pro-Ala-Gly-Lys-Asn-Leu-
3.		Analogue I	Leu-Lys-Cys-	-Asn-Lys-Leu-Ile-Pro-Ile-Ala-Ser-Lys-Thr-Cys-	-Pro-Ala-Gly-Lys-Asn-Leu-
4.	N. nigricollis	—	Leu-Lys-Cys-	-Asn-Gly-Leu-Ile-Pro-Pro-Phe-Thr-Lys-Thr-Cys-	-Pro-Lys-Gly-Lys-Asn-Leu-
Cytotoxins					
1.	Naja naja	Cytotoxin I	Leu-Lys-Cys-	-Asn-Lys-Leu-Ile-Pro-Leu-Ala-Tyr-Lys-Thr-Cys-	-Pro-Ala-Gly-Lys-Asn-Leu-
2.		Cytotoxin II	Leu-Lys-Cys-	-Asn-Lys-Leu-Val-Pro-Leu-Phe-Tyr-Lys-Thr-Cys-	-Pro-Ala-Gly-Lys-Asn-Leu-
3.	N. naja oxiana	—	Leu-Lys-Cys-	-Lys-Lys-Leu-Val-Pro-Leu-Phe-Ser-Lys-Thr-Cys-	-Pro-Ala-Gly-Lys-Asn-Leu-
4.	N. haje annulifera	VII₁	Leu-Lys-Cys-His-	-Lys-Leu-Val-Pro-Val-Trp-Lys-Thr-Cys-	-Pro-Glu-Gly-Lys-Asn-Leu-
5.		VII₂	Leu-Lys-Cys-His-	-Lys-Leu-Val-Pro-Pro-Phe-Trp-Lys-Thr-Cys-	-Pro-Glu-Gly-Lys-Asn-Leu-
6.		VII₂A	Leu-Lys-Cys-His-	-Lys-Leu-Val-Pro-Pro-Phe-Trp-Lys-Thr-Cys-	-Pro-Glu-Gly-Lys-Asn-Leu-
7.	N. melanoleuca	VII₁	Leu or Glu-Ile	-Asn-Lys-Leu-Val-Pro-Ile-Ala-His-Lys-Thr-Cys-	-Pro-Ala-Gly-Lys-Asn-Leu-
8.		VII₂	Ile-Lys-Cys-His-	-Asn-Thr-Leu-Leu-Pro-Phe-Ile-Tyr-Lys-Thr-Cys-	-Pro-Glu-Gly-Gln-Asn-Leu-
9.		VII₃	Ile-Lys-Cys-His-	-Asn-Thr-Leu-Leu-Pro-Phe-Ile-Tyr-Lys-Thr-Cys-	-Pro-Glu-Gly-Gln-Asn-Leu-
10.		3·20	Ile-Lys-Cys-His-Asn-Thr-Leu-	-Pro-Phe-Ile-Tyr-Lys-Thr-Cys-	-Pro-Glu-Gly-Asn-Asn-Leu-
11.	N. mossambica mossambica	VII₁	Leu-Lys-Cys-	-Asn-Gln-Leu-Ile-Pro-Phe-Trp-Lys-Thr-Cys-	-Pro-Lys-Gly-Lys-Asn-Leu-
12.		VII₂	Leu-Lys-Cys-	-Asn-Gln-Leu-Ile-Pro-Pro-Phe-Trp-Lys-Thr-Cys-	-Pro-Glu-Gly-Lys-Asn-Leu-
13.		VII₃	Leu-Lys-Cys-	-Asn-Arg-Leu-Ile-Pro-Pro-Phe-Trp-Lys-Thr-Cys-	-Pro-Glu-Gly-Lys-Asn-Leu-
14.		VII₄	Leu-Lys-Cys-	-Asn-Lys-Leu-Ile-Pro-Ile-Ala-Tyr-Lys-Thr-Cys-	-Pro-Glu-Gly-Lys-Asn-Leu-

			Loop 4	Loop 5
Cardiotoxins				
1.	Naja naja (Cambodian cobra)		Cys-Thr-Lys-Met-Phe-Met-Met-Ser-Asp-Leu-Thr-Ile-Pro-Val-Lys-Lys-Arg-Gly-Cys-	Ile-Asp-Val-
2.	N. naja atra		Cys-Tyr-Lys-Met-Phe-Met-Val-Ala-Thr-Pro-Lys-Val-Pro-Val-Lys-Arg-Gly-Cys-	Ile-Asp-Val-
3.		Analogue I	Cys-Tyr-Lys-Met-Phe-Met-Ser-Asp-Leu-Thr-Ile-Pro-Val-Lys-Arg-Gly-Cys-	Ile-Asp-Val-
4.	N. nigricollis		Cys-Tyr-Lys-Met-Thr-Met-Arg-Ala-Ala-Pro-Met-Val-Pro-Val-Lys-Arg-Gly-Cys-	Ile-Asp-Val-
Cytotoxins				
1.	Naja naja	Cytotoxin I	Cys-Tyr-Lys-Met-Tyr-Met-Val-Ser-Asn-Lys-Thr-Val-Pro-Val-Lys-Arg-Gly-Cys-	Ile-Asp-Val-
2.		Cytotoxin II	Cys-Tyr-Lys-Met-Tyr-Met-Val-Ala-Thr-Pro-Lys-Val-Pro-Val-Lys-Arg-Gly-Cys-	Ile-Asp-Val-
3.	N. naja oxiana		Cys-Tyr-Lys-Met-Phe-Met-Val-Ala-Ala-Pro-His-Val-Pro-Val-Lys-Arg-Gly-Cys-	Ile-Asp-Val-
4.	N. haje annulifera	VII 1	Cys-Tyr-Lys-Met-Phe-Met-Val-Ser-Thr-Val-Pro-Val-Lys-Arg-Gly-Cys-	Ile-Asp-Val-
5.		VII 2	Cys-Tyr-Lys-Met-Tyr-Met-Val-Ala-Thr-Pro-Met-Leu-Pro-Val-Lys-Arg-Gly-Cys-	Ile-Asp-Val-
6.		VII 2A	Cys-Tyr-Lys-Met-Tyr-Met-Val-Ala-Thr-Pro-Met-Leu-Pro-Val-Lys-Arg-Gly-Cys-	Ile-Asp-Val-
7.	N. melanoleuca	VII 1	Cys-Tyr-Gln-Met-Tyr-Met-Val-Ser-Lys-Ser-Thr-Ile-Pro-Val-Lys-Arg-Gly-Cys-	Ile-Asp-Val-
8.		VII 2	Cys-Phe-Lys-Gly-Thr-Leu-Lys-Phe-Pro-Lys-Lys-Thr-Thr-Tyr-Asn-Arg-Gly-Cys-	Ala-Ala-Thr-
9.		VII 3	Cys-Phe-Lys-Gly-Thr-Leu-Lys-Phe-Pro-Lys-Lys-Thr-Thr-Tyr-Arg-Gly-Cys-	Ala-Ala-Thr-
10.		3.20	Cys-Phe-Lys-Gly-Thr-Leu-Lys-Phe-Pro-Lys-Lys-Ile-Thr-Tyr-Lys-Arg-Gly-Cys-	Ala-Asp-Ala-
11.	N. mossambica mossambica	VII 1	Cys-Tyr-Lys-Met-Thr-Met-Arg-Ala-Ala-Pro-Met-Val-Pro-Val-Lys-Arg-Gly-Cys-	Ile-Asp-Val-
12.		VII 2	Cys-Tyr-Lys-Met-Thr-Met-Arg-Gly-Ala-Ser-Lys-Val-Pro-Val-Lys-Arg-Gly-Cys-	Ile-Asp-Val-
13.		VII 3	Cys-Tyr-Lys-Met-Thr-Met-Arg-Leu-Ala-Pro-Val-Lys-Arg-Gly-Cys-	Ile-Asp-Val-
14.		VII 4	Cys-Tyr-Lys-Met-Leu-Ala-Ser-Lys-Met-Val-Pro-Val-Lys-Arg-Gly-Cys-	Ile-Asn-Val-

Table 3 *Continued*

		Loop 6	Loop 7	Loop 8
Cardiotoxins				
1. Naja naja (Cambodian cobra)	—	Cys-Pro-Lys-Asn-Ser-Leu-Leu-Val-Lys-Tyr-Val-Cys-Cys-Asn-Thr-Asp-Arg-Cys-Asn		
2. N. naja atra	—	Cys-Pro-Lys-Ser-Ser-Leu-Leu-Val-Lys-Tyr-Val-Cys-Cys-Asn-Thr-Asp-Arg-Cys-Asn		
3.	Analogue I	Cys-Pro-Lys-Ser-Asn-Leu-Val-Lys-Tyr-Val-Cys-Cys-Asn-Thr-Asp-Arg-Cys-Asn		
4. N. nigricollis	—	Cys-Pro-Lys-Ser-Ser-Leu-Ile-Lys-Tyr-Met-Cys-Cys-Asn-Thr-Asp-Lys-Cys-Asn		
Cytotoxins				
1. Naja naja	Cytotoxin I	Cys-Pro-Lys-Asn-Ser-Leu-Val-Lys-Tyr-Glu-Cys-Cys-Asn-Thr-Asp-Arg-Cys-Asn		
2.	Cytotoxin II	Cys-Pro-Lys-Ser-Ser-Leu-Val-Lys-Tyr-Val-Cys-Cys-Asn-Thr-Asp-Arg-Cys-Asn		
3. N. naja oxiana	VII1	Cys-Pro-Lys-Ser-Ser-Leu-Val-Lys-Tyr-Val-Cys-Cys-Asn-Thr-Asp-Lys-Cys-Asn		
4. N. haje annulifera	VII2	Cys-Pro-Lys-Asn-Ser-Ala-Leu-Val-Lys-Tyr-Val-Cys-Ser-Thr-Asp-Lys-Cys-Asn		
5.	VII2A	Cys-Pro-Lys-Asp-Ser-Ala-Leu-Val-Lys-Tyr-Met-Cys-Asn-Thr-Asp-Lys-Cys-Asn		
6.	VII1	Cys-Pro-Lys-Asp-Ser-Ala-Leu-Val-Lys-Tyr-Val-Cys-Ser-Thr-Asp-Lys-Cys-Asn		
7. N. melanoleuca	VII1	Cys-Pro-Lys-Ser-Ser-Leu-Val-Lys-Tyr-Val-Cys-Cys-Asn-Thr-Asp-Arg-Cys-Asn		
8.	VII2	Cys-Pro-Lys-Ser-Ser-Leu-Val-Lys-Tyr-Val-Cys-Cys-Asn-Thr-Asp-Lys-Cys-Asn		
9.	VII3	Cys-Pro-Lys-Ser-Ser-Leu-Val-Lys-Tyr-Val-Cys-Cys-Asn-Thr-Asp-Lys-Cys-Asn		
10.	3·20	Cys-Pro-Lys-Thr-Ser-Ala-Leu-Val-Lys-Tyr-Val-Cys-Cys-Asn-Thr-Asp-Lys-Cys-Asn		
11. N. mossambica mossambica	VII1	Cys-Pro-Lys-Ser-Ser-Leu-Ile-Lys-Tyr-Met-Cys-Cys-Asn-Thr-Asn-Lys-Cys-Asn		
12.	VII2	Cys-Pro-Lys-Ser-Ser-Leu-Ile-Lys-Tyr-Met-Cys-Cys-Asn-Thr-Asp-Lys-Cys-Asn		
13.	VII3	Cys-Pro-Lys-Ser-Ser-Leu-Ile-Lys-Tyr-Met-Cys-Cys-Asn-Thr-Asn-Lys-Cys-Asn		
14.	VII4	Cys-Pro-Lys-Asn-Ser-Ala-Leu-Val-Lys-Tyr-Val-Cys-Ser-Thr-Asp-Arg-Cys-Asn		

References. Cardiotoxins: 1. Fryklund and Eaker (1975a); 2. Narita and Lee (1970); 3. Hayashi et al. (1975); 4. Fryklund and Eaker (1975b). Cytotoxins: 1. Hayashi et al. (1971); 2. Takechi and Hayashi (1972); 3. Grishin et al. (1974); 4. Weise et al. (1973); 5, 6. Joubert (1975); 7. Carlsson and Joubert (1974); 8, 9. Carlsson (1974); 10. Shipolini et al. (1975); 11 to 13. Louw (1974a); 14. Louw (1974b)

2 Cytotoxins

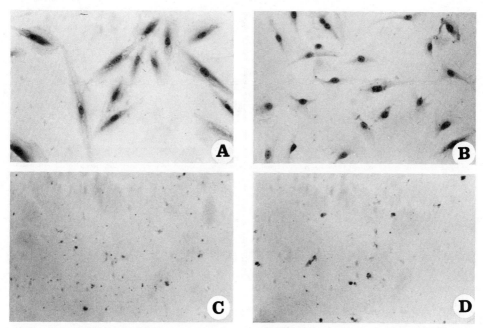

Figure 19.1 Effect of snake venoms on bovine spleen cells. (A) Control. (B) Incubated with 500 μg ml^{-1} *Naja naja atra* venom for 1 hr. (C) Incubated with 500 μg ml^{-1} *Crotalus viridis viridis* 1 hr. Note strong lytic action of this venom as compared to *N. naja atra* venom. (D) Incubated with 500 μg ml^{-1} *Bitis lachesis* venom for 1 hr. Note strong lytic action. (For details see Tu and Passey, 1969.)

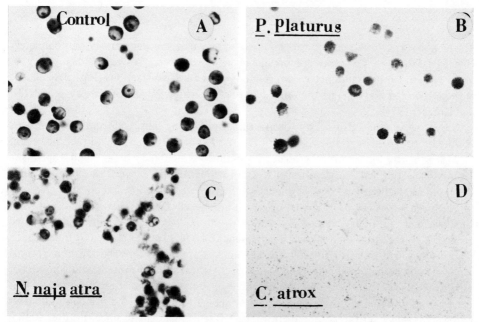

Figure 19.2. Cytotoxic effect of various snake venoms on Yoshida sarcoma cells. (A) Control: Yoshida sarcoma cells without addition of venom. (B) With *Pelamis platurus* venom, 1 mg ml^{-1}: no significant effect exhibited. (C) With *Naja naja* venom, 0.25 mg ml^{-1}: plasma membrane lysis and subsequent clumping exhibited. (D) With *Crotalus atrox* venom, 1 mg ml^{-1}: complete lysis of cells exhibited.

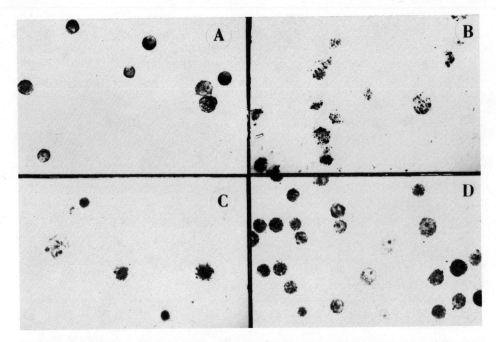

Figure 19.3 Cytotoxic effects of fractions obtained from *Crotalus atrox* venom on Yoshida sarcoma cells. (A) Fraction I, 0.25 mg ml^{-1}: no significant effect observed. (B) Fraction II, 0.031 mg ml^{-1}: both plasma and nuclear membranes lysed. Chromatin granules can still be seen. (C) Fraction III, 0.062 mg ml^{-1}: effect similar to fraction II at a lower concentration. (D) Fraction IV, 0.02 mg ml^{-1}: no significant effect observed.

The action of *Crotalus atrox* venom is so strong that the cells are lysed completely (Fig. 2D). However, not every component in snake venom is cytotoxic. For instance, *C. atrox* venom can be fractionated into four components. Two fractions (Figs. 3A and 3D) are nontoxic to cells, whereas the other two fractions (Figs. 3B and 3C) are cytotoxic (Tu and Giltner, 1974).

Regarding venom cytotoxicity toward tumor-producing cells, Gillo and his co-workers (Gillo, 1966; Wirtheimer and Gillo, 1966; Brisbois et al., 1968a, b) found that the venom of *Naja naja atra* inhibited the growth of spontaneous mammary tumors and Koprowska tumors. Isolated Ehrlich cancerous cells, after treatment with *N. naja atra* venom failed to produced ascites when injected into mice, while the lethal fraction and phospholipase A$_2$ of the venom showed *no* activity on the Ehrlich cell (Wirtheimer and Gillo, 1966). Brisbois et al. (1968b) isolated a fraction from *N. naja atra* venom that inhibited anaerobic glycolysis of ascitic Ehrlich carcinoma cells. Braganca et al. (1967) found that a fraction they had isolated from *N. naja* venom was preferentially destructive to Yoshida sarcoma cells over normal cells. The active fraction was a basic protein with a molecular weight of 10,000, free of lethality and various enzyme activities. They also showed that the fraction dealys the growth of solid fibrosarcoma tumors in mice from 34 to 48 days but has no effect on the development of L1210 leukemia cells in mice. Further studies (Patel et al., 1969) on the mechanism of action have indicated initial binding of the cobra venom fraction to cell membrane constituents (probably phospholipids) of susceptible cells.

Table 4 Cytotoxic Activity of Snake Venoms (or Fractions) on KB, Yoshida Sarcoma and Normal Peritoneal Cells

	KB Cells*								Yoshida Sarcoma Cells†								Normal Peritoneal Cells†							
	\multicolumn{24}{c}{Number of Serial Dilutions}																							
	1	2	3	4	5	6	7	8	1	2	3	4	5	6	7	8	1	2	3	4	5	6	7	8
CROTALIDAE (pit vipers)																								
Crotalus atrox	+	+	+	+	±	−	−	−	+	+	+	+	+	+	±	−	+	+	+	+	+	±	−	−
Fraction I	+	+	±	±	−	−	−	−	+	+	±	+	−	−	−	−	+	±	−	−	−	−	−	−
Fraction II	+	−	−	−	−	−	−	−	+	+	+	+	+	+	+	±	+	+	+	+	+	+	+	+
Fraction III	+	−	−	−	−	−	−	−	+	+	+	+	+	±	±	−	+	+	+	+	+	+	+	+
Fraction IV	+	−	−	−	−	−	−	−	+	±	−	−	−	−	−	−	+	±	−	−	−	−	−	−
C. horridus atricaudatus	+	+	±	±	−	−	−	−																
C. terrificus terrificus	+	+	±	−	−	−	−	−																
C. viridis viridis	+	+	+	±	−	−	−	−																
Agkistrodon piscivorus	+	+	+	±	−	−	−	−																
A. contortrix laticinctus	+	+	+	±	−	−	−	−																
VIPERIDAE (vipers)																								
Bitis gabonica rhinoceros	+	+	+	±	−	−	−	−																
B. arietans	+	+	+	±	±	−	−	−																
Vipera russellii	+	+	+	±	±	−	−	−	+	+	+	+	±	−	−	−	+	+	+	±	−	−	−	−
ELAPIDAE (elapids)																								
Ophiophagus hannah	+	+	±	−	−	−	−	−																
Naja naja	+	+	−	−	−	−	−	−																
N. naja atra	+	+	+	±	−	−	−	−	+	+	±	−	−	−	−	−	+	+	+	+	+	+	±	−
N. melanoleuca	+	+	+	+	±	−	−	−																
N. naja philippinensis	+	+	+	+	±	−	−	−																
HYDROPHIIDAE (sea snake)																								

Venoms of *Laticauda semifasciata* (from Okinawa and Philippines), *Enhydrina schistosa* and its fractions I, IV, V, VI, VII, VIII (isolated by the method reported in reference 4), *Lapemis hardwickii* and its fractions I, II, III (isolated by the method reported in reference 5), and *Pelamis platurus* (from Costa Rica) showed negative results on KB, Yoshida sarcoma, and normal peritoneal cells at all dilutions.

*For KB cells, complete inhibition of cell growth designated as +, partial inhibition as ±, and no inhibition as −.
†For Yoshida sarcoma and normal peritoneal cells, plasma membrane lysis in 50 to 100% designated as +, in 20 to 50% as ±, and in less than 20% as −.

Of the various species of snake venoms tested, *Naja naja* venom is the most active in destroying tumor cells (Braganca et al., 1967). Cotte et al. (1972) has reported that venoms of *Bothrops* and *Crotalus* caused cytopathogenic effects on a wide variety of cell cultures such as HeLa cells, human breast carcinoma, and Hep-2. The γ toxin isolated from the venom of *Naja nigricollis* is cytotoxic to KB cells *in vitro* (Boquet, 1970), and cytotoxin CM-XI (cytotoxin I) and cytotoxin-XII (cytotoxin II) showed high cytotoxicity to Yoshida sarcoma and ascites hepatoma cells (Takechi et al., 1971). Lee et al. (1972), testing cardiotoxin from *N. naja atra* venom on different cancer cells, found it to be toxic to HeLa, KB, and Yoshida sarcoma cells *in vitro*, but observed no anticancer activity when cardiotoxin was used *in vivo* on Yoshida rat sarcomas or Ehrlich ascites tumors in mice. Tu and Giltner (1974) investigated the venoms of six Crotalidae, three Viperidae, five Elapidae, and four Hydrophiidae, as well as some of the venom fractions. Venoms of Crotalidae and Viperidae have very strong cytotoxic actions on KB and Yoshida sarcoma cells and actually lyse these cells. Venoms of Elapidae are also weakly cytotoxic, while Hydrophiidae (sea snake) venoms do not lyse the cell even if fractionated components are used (Table 4).

In order for snake venom or a component to be considered an antitumor agent, selective cytotoxicity toward tumor cells must be established. When normal peritoneal cells are tested, venoms and their fractions are toxic. Thus snake venoms and venom fractions can destroy tumor cells but can also be equally destructive to normal peritoneal cells. Braganca et al. (1967) reported the isolation of a cytotoxic fraction specific toward Yoshida sarcoma cells. The nontumor cells used by these investigators were rat and human erythrocytes, human leucocytes, and rat bone marrow cells; no normal peritoneal cells were used. Tu and Giltner (1974) have concluded that snake venoms and venom fractions are not useful antitumor agents because of the lack of selective cytotoxicity toward tumor cells.

3 OTHERS

Many other basic proteins with low toxicity have been isolated from snake venoms and called by various names, such as cobramines A and B and direct lytic factor (DLF). There is considerable debate about the identity of these basic proteins. It may well be that these basic proteins are indeed identical to cardiotoxin and cytotoxin Some day probably in the near future, this nomenclature will be clarified and simplified as the structural identities of these basic proteins are elucidated. There is already some evidence that DLF is probably identical to cytotoxin (or cardiotoxin). For instance, DLF isolated from *Naja naja* venom inhibited Na(I) and K(I)-ATPase activity (Lankishch et al. 1972), as does the cytotoxin isolated from the same venom (Zaheer et al., 1975) Furthermore, *Haemachatus haemachatus* DLF is very similar in amino acid composition to cytotoxins (Table 1).

Cobramines A and B were originally isolated from the venom of *Naja naja*. They are heat stable, have inhibitory action on iodide accumulation by thyroid slices, and are not identical to phospholipase A_2 (Larsen and Wolff, 1968a). Cobramine B inhibits amino acid and 3-*O*-methyl-D-glucose transport by cells of the small intestine and the uptake of *p*-aminohippurate and amino acids by kidney cortex slices (Larsen and Wolff, 1967). These proteins appear to be present only in Elapidae venoms, and not in venoms of snakes of the Crotalidae and Viperidae families. The amino acid composition of cobramine

B is shown in Table 1. Wolff et al. (1968) postulated that cobramine B interacts with the negative surface charges of thyroid cells. It increases iodine efflux from thyroid slices, leads to K^+ loss from the cells, and increases the rate of equilibration of sucrose or inulin spaces.

Shipolini and Banks (1974) isolated a basic polypeptide from the venom of *Dendroaspis viridis* and called it 4.9.6. Peptide 4.9.6 does not block neuromuscular transmission in frogs and therefore is not a neurotoxin. It is also not a lytic factor, and consequently not a cytotoxin or hemolytic toxin. The toxicity of this basic protein is relatively weak; it has an LD_{50} value of 2 $\mu g\ g^{-1}$ in mice, as compared to 0.1 $\mu g\ g^{-1}$ for the main neurotoxin. In chemical structure, this relatively weak toxin, 4.9.6, is very similar to relatively nontoxic components Ta 1 and Ta 2 of the venom of *D. angusticeps*, isolated by Viljoen and Botes (1974). Since the specific pharmacological action of 4.9.6 is still unknown, it is not possible to define the structure–function relationship. However, this example serves to illustrate the fact that Elapidae venoms contain many basic proteins that are very similar in structure to neurotoxins but are not neurotoxins.

The presence of nonneurotoxic basic proteins is not restricted to the venoms of Elapidae. A nontoxic basic protein was isolated in the homogeneous state from *Crotalus adamanteus* venom (Sulkowski et al., 1975). In comparison to cobramine, cardiotoxin, or cytotoxins, the *C. adamanteus* basic protein has a much higher molecular weight of 24,000, as compared to 6000 to 7000 daltons for proteins of Elapidae origin. Moreover, this *C. adamanteus* basic protein is nonneurotoxic and noncardiotoxic, shows no stimulation or inhibition activity on human fibroblasts or Burkitt lymphoma cells, and has no proteinase-inhibiting activity. Although there is nerve growth factor (NGF) activity, the specific activity did not increase as the protein was purified. This excludes the possibility that *C. adamanteus* basic protein is identical to NGF. Therefore the exact nature of this protein is still not clear.

A basic protein isolated from *Vipera ammodytes* venom (Sket et al., 1973a, b) proved to be highly toxic and therefore is similar to Elapidae basic neurotoxin rather than to Elapidae nonneurotoxic basic proteins. Likewise, basic proteins of low molecular weight have been isolated from the venoms of several rattlesnake species (Bonilla and Fiero, 1971; Bonilla, 1969). Again, isolated Crotalidae basic proteins are highly toxic; therefore their activities are different from those of Elapidae nonneurotoxic basic proteins.

REFERENCES

Aloof-Hirsch, S., De Vries, A., and Berger, A. (1968). The direct lytic factor of cobra venom; purification and chemical characterization, *Biochim. Biophys. Acta,* **154,** 53.

Arms, K. and McPheeters, D. (1975). Sensitivity of cultured embryonic heart cells to cardiotoxin obtained from *Naja naja siamensis* venom, *Toxicon,* **13,** 333.

Bieber, A. L., Tu, T., and Tu, A. T. (1975). Studies of an acidic cardiotoxin isolated from the venom of Mojave rattlesnake (*Crotalus scutulatus*), *Biochim. Biophys. Acta,* **400,** 178.

Bonilla, C. A. (1969). Rapid isolation of basic proteins and polypeptides from salivary gland secretions by absorption chromatography of polyacrylamide gel, *Anal. Biochem.,* **32,** 522.

Bonilla, C. A. and Fiero, M. K. (1971). Comparative biochemistry and pharmacology of salivery gland secretions, *J. Chromatogr.,* **56,** 253.

Boquet, P. (1970). Venins. Action de la toxine γ du venin de *Naja nigricollis* sur les cellules KB cultivees *in vitro, C. R. Acad. Sci. Paris,* Ser. D, **271,** 2422.

Braganca, B. M. (1974). Cytotoxic proteins from cobra venom as probes for the study of membrane structure, *Biomembranes: Archit. Biog. Bioenerg., Differ. Proc. Int. Symp.,* p. 201.

Braganca, B. M., Patel, N. T., and Badrinath, P. G. (1967). Isolation and properties of a cobra venom factor selectively cytotoxic to Yoshida sarcoma cells, *Biochim. Biophys. Acta,* **136,** 508.

Brisbois, L., Rabinovitch-Mahler, N., and Gillo, L. (1968a). Isolation from cobra venom of a factor inhibiting glycolysis in Ehrlich ascites carcinoma cells, *Experientia,* **24,** 673.

Brisbois, L., Rabinovitch-Mahler, N., Delori, P., and Gillo, L. (1968b). Etudes des fractions obtenues par chromatographie du venin de *Naja naja atra* sur sulphoethyl-Sephadex, *J. Chromatogr.,* **37,** 463.

Carlsson, F. H. H. (1974). Snake venom toxins: Primary structures of two novel cytotoxin homologs from the venom of forest cobra (*Naja melanoleuca*), *Biochem. Biophys. Res. Commun.,* **59,** 269.

Carlsson, F. H. H. and Joubert, F. J. (1974). Snake venom toxins: Isolation and purification of three cytotoxin homologs from the venom of the forest cobra (*Naja melanoleuca*) and the complete amino acid sequence of toxin $V^{11}1$, *Biochim. Biophys. Acta,* **336,** 453.

Chang, C. C., Wei, J. W., Chuang, S. T., and Lee, C. Y. (1972a). Are the blockade of nerve conduction and depolarization of skeletal muscle induced by cobra venom due to phospholipase A, neurotoxin or cardiotoxin? *J. Formosan Med. Assoc.,* **71,** 323.

Chang, C. C., Chuang, S. T., Lee, C. Y., and Wei, J. W. (1972b). Role of cardiotoxin and phospholipase A in the blockade of nerve conduction and depolarization of skeletal muscle induced by cobra venom, *Br. J. Pharmacol.,* **44,** 752.

Cotte, C. A., Essenfeld-Yahr, E., and Lairet, A. C. (1972). Effects of *Crotalus* and *Bothrops* venom on normal and malignant cells cultivated *in vitro, Toxicon,* **10,** 157.

Devi, A. and Sarkar, N. K. (1966). Cardiotoxic and cardiostimulating factors in cobra venom, *Mem. Inst. Butantan,* **33,** 573.

Dimari, S. J., Lembach, K. J., and Chatman, V. B. (1975). The cytotoxins of cobra venoms: Isolation and partial characterization, *Biochim. Biophys. Acta,* **393,** 320.

Frykland, L. and Eaker, D. (1975a). The complete amino acid sequence of a cardiotoxin from the venom of *Naja naja* (Cambodian cobra), *Biochemistry,* **14,** 2860.

Frykland, L. and Eaker, D. (1975b). The complete covalent structure of a cardiotoxin from the venom of *Naja nigricollis* (African black-necked spitting cobra), *Biochemistry,* **14,** 2865.

Gaertner, C., Goldblum, N., Gitter, S., and De Vries, A. (1962). The action of various snake venoms and their chromatographic fractions on animal cells in culture, *J. Immunol.,* **88,** 526.

Gillo, L. (1966). Les venins de serpents, source d'enzyme anticancereux. I. Aspects biochimiques fondamentaux du probleme, *Mem. Inst. Butantan,* **33,** 933.

Grishin, E. V., Sukhikh, A. P., Adamovich, T. B., and Ovchinnikov, Yu. A. (1974). The isolation and sequence determination of a cytotoxin from the venom of the middle-Asian cobra *Naja naja oxiana, FEBS Lett.,* **48,** 179.

Hayashi, K., Takechi, M., and Sasaki, T. (1971). Amino acid sequence of cytotoxin I from the venom of the Indian cobra (*Naja naja*), *Biochem. Biophys. Res. Commun.,* **45,** 1357.

Hayashi, K., Takechi, M., Sasaki, T., and Lee, C. Y. (1975). Amino acid sequence of cardiotoxin-analogue I from the venom of *Naja naja atra, Biochem. Biophys. Res. Commun.,* **64,** 360.

Ho, C. L., Lee, C. Y., and Lu, H. H. (1975). Electrophysiological effects of cobra cardiotoxin on rabbit heart cells, *Toxicon,* **13,** 437.

Joubert, F. J. (1975). The amino acid sequence of toxin $V^{11}2$, a cytotoxin homologue from banded Egyptian cobra (*Naja haje annulifera*) venom, *Hoppe-Seylers Z. Physiol. Chem.,* **356,** 1893.

Keung, W. M., Yip, T. T., and Kong, Y. C. (1975a). The chemistry and biological effects of cardiotoxin from the Chinese cobra (*N. naja* Linn) on hormonal responses in isolated cell systems, *Toxicon,* **13,** 239.

Keung, W. M., Leung, W. W., and Kong, Y. C. (1975b). Studies on the status of disulfide linkages and tyrosine residues in cardiotoxin, *Biochem. Biophys. Res. Commun.,* **66,** 383.

Lai, M. K., Wen, C. Y., and Lee, C. Y. (1972). Local lesions caused by cardiotoxin isolated from Formosan cobra venom, *J. Formosan Med. Assoc.,* **71,** 328.

Lankisch, P. G., Schoner, K., Shoner, W., Kunze, H., Bohn, E., and Vogt, W. (1972). Inhibition of Na^+- and K^+-activated ATPase by the direct lytic factor of cobra venom (*Naja naja*), *Biochim. Biophys. Acta,* **266,** 133.

Larsen, P. R. and Wolff, J. (1967). Inhibition of accumulative transport by a protein from cobra venom, *Biochem. Pharmacol.,* **16,** 2003.

Larsen, P. R. and Wolff, J. (1968a). The basic proteins of cobra venom. I. Isolation and characterization of cobramines A and B, *J. Biol. Chem.,* **243,** 1283.

Larsen, P. R. and Wolff, J. (1968b). The toxic proteins of cobra venom, *Biochem. Pharmacol.,* **17,** 503.

Lee, C. Y., Chang, C. C., Chiu, T. H., Chiu, P. J. S., Tseng, T. C., and Lee, S. Y. (1968). Pharmacological properties of cardiotoxin isolated from Formosan cobra venom, *Naunyn-Schmiedebergs Arch. Exp. Pathol. Pharmokol.,* **259,** 360.

Lee, C. Y., Lin, J. S., and Wei, J. M. (1971). "Identification of cardiotoxin with cobramine B, DLF, toxin γ and cobra venom cytotoxin," in A. De Vries and E. Kochva, Eds., *Toxins of Animal and Plant Origin,* Vol. 1, Gordon and Breach, New York, pp. 307–317.

Lee, C. Y., Lin, J. S., and Lin, Shiau, S. Y. (1972). A Study of carcinolytic factor of Formosan cobra venom, *Proc. Natl. Sci. Counc. Taiwan,* **5,** 9.

Leung, W. W., Keung, W. M., and Kong, Y. C. (1976). The cytolytic effect of cobra cardiotoxin on Ehrlich ascites tumor cells and its inhibition by Ca^{2+}, *Naunyn-Schmiedebergs Arch. Exp. Pathol. Pharmakol.,* **292,** 193.

Lin, Shiau S. Y., Huang, M. C., and Lee, C. Y. (1972). Isolation of cardiotoxic and neurotoxic principles from the venom of *Bungarus fasciatus, J. Formosan Med. Assoc.,* **71,** 350.

Lin, Shiau S. Y., Huang, M. C., and Lee, C. Y. (1975). A study of cardiotoxic principles from the venom of *Bungarus fasciatus* (Schneider), *Toxicon,* **13,** 189.

Louw, A. I. (1974a). Snake venom toxins: Amino acid sequences of three cytotoxin homologs from *Naja mossambica mossambica* venom, *Biochim. Biophys. Acta,* **336,** 481.

Louw, A. I. (1974b). Snake venom toxins. Complete amino acid sequence of cytotoxin $V^{11}4$ from the venom of *Naja mossambica mossambica, Biochem. Biophys. Res. Commun.,* **58,** 1022.

Louw, A. I. (1974c). The purification and properties of five non-neurotoxic polypeptides from *Naja mossambica mossambica* venom, *Biochim. Biophys. Acta,* **336,** 470.

Lu, M. S. and Lo, T. B. (1974). Chromatographic separation of *Bungarus fasciatus* venom and preliminary characterization of its components, *J. Chin. Biochem. Soc.,* **3,** 57.

Nakai, K., Nakai, C., Sasaki, T., Kakiuchi, K., and Hayashi, K. (1970). Purification and some properties of toxin A from the venom of the Indian cobra, *Naturwissenschaften,* **57,** 387.

Narita, K. and Lee, C. Y. (1970). The amino acid sequence of cardiotoxin from Formosan cobra (*Naja naja atra*) venom, *Biochem. Biophys. Res. Commun.,* **41,** 339.

Patel, T. N., Braganca, B. M., and Bellare, R. A. (1969). Changes produced by cobra venom cytotoxin on the morphology of Yoshida sarcoma cells, *Exp. Cell. Res.,* **57,** 289.

Penso, G. and Balducci, D. (1963). *Tissue Culture in Biological Research,* Elsevier, Amsterdam.

Raudonat, H. H. and Holler, B. (1958). Heart-active component of cobra venom (cardiotoxin), *Naunyn-Schmiedebergs Arch. Exp. Pathol. Pharmokol.,* **233,** 431.

Reid, H. A. (1964). Cobra bites, *Br. Med. J.,* **2,** 540.

Sarkar, N. K. (1947). Existence of a cardiotoxic principle in cobra venom, *Ann. Biochem. Exp. Med.,* **8,** 11.

Sato, I., Ryan, K. W., and Mitsuhashi, S. (1964). Studies on habu snake venom. VI. Cytotoxic effect of habu (*Trimeresurus flavovirides* Hallowell) and cobra (*Naja naja*) venoms on the cells *in vitro, Jap. J. Exp. Med.,* **34,** 119.

Sato, H., Yogi, M., Hirakawa, Y., Kazama, M., and Takaki, S. (1970). Studies on immunoelectrophoretic separation of habu venom and its detoxification *in vitro, Acta Med. Univ. Kagoshima.,* **12,** 129.

Shiau, S. Y. Lin, Huang, M. C., and Lee, C. Y. (1976). Mechanism of action of cobra cardiotoxin in the skeletal muscle, *J. Pharmacol. Exp. Therap.,* **196,** 758.

Shipolini, R. A. and Banks, B. E. C. (1974). The amino acid sequence of a polypeptide from the venom of *Dendroaspis viridis, Eur. J. Biochem.,* **49,** 399.

Shipolini, R. A., Kissonerghis, M., and Banks, B. E. C. (1975). The primary structure of a major polypeptide component from the venom of *Naja melanoleuca, Eur. J. Biochem.,* **56,** 449.

Sket, D., Gubensek, F., Adamic, S., and Lebez, D. (1973a). Action of a partially purified basic protein fraction from *Vipera ammodytes* venom, *Toxicon,* **11,** 47.

Sket, D., Gubensek, F., Pavlin, R., and Lebez, D. (1973b). Oxygen consumption of rat brain homogenates after *in vitro* and *in vivo* addition of the basic protein from *Vipera ammodytes* venom, *Toxicon,* **11,** 193.

Slotta, K. H. and Vick, J. A. (1969). Identification of the direct lytic factor from cobra venom as cardiotoxin, *Toxicon,* **6,** 167.

Stringer, J. M., Kainer, R. A., and Tu, A. T. (1971). Ultrastructural studies of myonecrosis induced by cobra venom in mice, *Toxicol. Appl. Pharmacol.,* **18,** 442.

Sulkowski, E., Kress, L. F., and Laskowski, M., Sr. (1975). Crystalline basic protein from venom of *Crotalus adamanteus, Toxicon,* **13,** 149.

Takechi, M. and Hayashi, K. (1972). Localization of the four disulfide bridges in cytotoxin II from the venom of the Indian cobra (*Naja Naja*), *Biochem. Biophys. Res. Commun.,* **49,** 584.

Takechi, M., Sasaki, T., and Hayashi, K. (1971). The N-terminal amino acid sequences of two basic cytotoxic proteins from the venom of the Indian cobra, *Naturwissenshaften,* **58,** 323.

Takechi, M., Hayashi, K., and Sasaki, T. (1972). The amino acid sequence of cytotoxin II from the venom of the Indian cobra (*Naja naja*), *Mol. Pharmacol.,* **8,** 446.

Tu, A. T. (1973). Neurotoxins of animal venoms: snakes, *Ann. Rev. Biochem.,* **42,** 235.

Tu, A. T. and Giltner, J. B. (1974). Cytotoxic effects of snake venoms on KB and Yoshida sarcoma cells, *Res. Commun. Chem. Pathol. Pharmacol.,* **9,** 783.

Tu, A. T. and Passey, R. B. (1969). Effects of snake venoms on mammarian cells in tissue culture, *Toxicon,* **6,** 277.

Viljoen, C. C. and Botes, D. P. (1974). Snake venom toxins: The purification and amino acid sequence of toxin TA2 from *Dendroaspis angusticeps* venom, *J. Biol. Chem.,* **249,** 366.

Weise, H. K., Carlsson, F. H. H., Joubert, F. J., and Strydom, D. J. (1973). Snake venom toxins: The purification of toxins $V^{11}1$ and $V^{11}2$, two cytotoxin homologues from banded Egyptian cobra (*Naja haje annulifera*) venom, and the complete amino acid sequence of toxin $V^{11}1$, *Hoppe-Seylers Z. Physiol. Chem.,* **354,** 1317.

Wirtheimer, C. and Gillo, L. (1966). Les venins de serpents, source d'enzymes anticancereux. II. Etude experimentale, *Mem. Inst. Butantan,* **33,** 937.

Wolff, R., Salabe, H., Ambrose, M., and Larsen, P. R. (1968). The basic proteins of cobra venom. II. Mechanism of action of cobramine B on thyroid tissue, *J. Biol. Chem.,* **243,** 1290.

Yoshikura, H., Ogawa, H., Ohsaka, A., and Omori-Sato, T. (1966). Action of *Trimeresurus flavoviridis* venom and the partially purified hemorrhagic principles on animal cells cultivated *in vitro, Toxicon,* **4,** 183.

Zaheer, A., Noronha, S. H., Hospattankar, A. V., and Braganca, B. M. (1975). Inactivation of (Na^+ + K^+)-stimulated ATPase by a cytotoxic protein from cobra venom in relation to its lytic effects on cells, *Biochim. Biophys. Acta,* **394,** 293.

20 Hemolysis

1	BIOLOGICAL ASPECT	322
	1.1 Venoms of Different Species, 322	
	1.2 Erythrocytes of Different Species, 322	
	1.3 Other Factors, 322	
	1.4 Relation to Lethality, 323	
2	CHEMISTRY OF HEMOLYTIC FACTOR	323
	2.1 Property, 323	
	2.2 Structure, 323	
3	ROLE OF PHOSPHOLIPASE A_2 IN HEMOLYSIS	324
	3.1 Phospholipase A_2 Alone, 324	
	3.2 Synergistic Action with Direct Hemolytic Factor, 325	
	References	326

Sometimes there is slight confusion about the two phenomena of hemolysis and hemorrhage. Hemolysis is the rupture of the erythrocyte membrane so that the hemoglobin contained inside diffuses to surrounding media. Hemorrhage is the rupturing of capillary tubes so that the blood penetrates to surrounding tissues.

Snake venoms contain direct and indirect lytic factors. Direct hemolytic factor (DHF), sometimes called direct lytic factor (DLF), itself can hemolyze the red cell. Indirect lytic factor lyses the red cell very slowly; however, its lytic action can be greatly accelerated by the addition of phosphatidylcholine (Roy, 1955). Indirect hemolytic factor is identified as phospholipase A_2.

Hemolytic action is frequently expressed as HU_{50}, the concentration required to hemolyze 50% of the erythrocytes (De Hurtado and Layrisse, 1964; Tu et al., 1970; Tu and Passey, 1971). Thus we can measure hemolytic action quantitatively. The terms "direct lytic action" and "indirect lytic action" seem convenient and divide the hemolytic action of snake venoms into two groups. However, there is no distinct separation between them. The proportion of these two factors in each venom is different, so that the rate of hemolysis varies, depending on the venom used. Venom-induced hemolysis is also dependent on the species of animal from which the erythrocytes originated and on many other factors.

1 BIOLOGICAL ASPECT

1.1 Venoms of Different Species

Venoms of Elapidae usually contain both direct and indirect factors (De Vries et al., 1962). Hemoglubinuria is one of the clinical symptoms of Australian Elapidae poisoning (Kellaway, 1929). This is a manifestation of *in vivo* hemolysis due to Elapidae venom DLF action.

Sakhibov and Davlyatov (1969) observed that cobra venom has strong hemolytic activity whereas the venom of the pit viper, *Agkistrodon halys*, is weak in this respect. This is readily understandable as the venoms of Elapidae contain DLF whereas those of Crotalidae have little of this factor. Moreover, DLF and phospholipase A_2 enhance hemolytic action synergistically. Since most snake venoms contain phospholipase A_2, venoms of Elapidae cause direct hemolysis because of the actions of both DLF and phospholipase A_2. The hemolytic activity of Crotalidae venoms can be accelerated by the addition of serum or phospholipids. Because of the difference in hemolytic behavior of heated venoms and nonheated crotaline venoms, Byrd and Johnson (1970) concluded that there must be a heat-labile component that is required for indirect hemolysis by crotaline venom in addition to phospholipase A_2.

1.2 Erythrocytes of Different Species

Erythrocytes of different species show different sensitivities to hemolysis (Rosenfeld et al., 1968; Mitchell and Reichert, 1886; Pestana, 1908; Brazil and Pestana, 1910). Chopra and Roy (1936) observed that Russell's viper venom hemolyzed human and guinea pig red cells but not sheep cells. Dass et al. (1970) investigated the hemolytic activity of Russell's viper venom, using erythrocytes of 12 different animals. The erythrocytes of each species showed a different susceptivity to hemolysis.

1.3 Other Factors

There are many phosphatidylcholines in plasma and in other tissues. Therefore phospholipase A_2 can liberate lysolecithin from tissues other than erythrocytes, causing lysis of the erythrocytes. If this is true, all venoms containing phospholipase A_2 should be hemolytic. But this is not the case. For instance, *Echis coloratus* venom is nonhemolytic in dogs, although it contains phospholipase A_2 (Klibansky et al., 1966). There are many reasons for nonhemolysis. A possible reason is the absorption of lysolecithin by the red cell surface or the interaction of lysolecithin with other compounds. For instance, serum albumin binds to lysolecithin and inhibits its hemolytic action (Luzzio, 1967). In connection with the role of albumin, Philpot and Higginbotham (1972) made an interesting observation. Normal human serum does not activate snake venom hemolysis. However, serum from patients with various diseases in which the level of serum albumin was decreased activated water moccasin venom-induced hemolysis. Therefore it seems that albumin plays an important role in *in vivo* hemolysis.

The addition of guinea pig serum enhances the hemolytic action of *Naja naja* venom (Vincent et al., 1972). Curiously, however, the serum does not activate the hemolytic action of DLF or phospholipase, and there is still no satisfactory explanation for the role of guinea pig serum in hemolysis.

Schroeter et al. (1972). reported that the glutathione reductase level in erythrocytes is related to the hemolytic activity of DLF. They found that cells with high reductase activity were also highly sensitive to DLF.

1.4 Relation to Lethality

Hemolysis is, no doubt, one of the toxic effects that snake venoms produce. However, it is not a main lethal factor (Condrea et al., 1969). Purified neurotoxin of *Naja nigricollis* venom is nonhemolytic (Izard et al., 1969). The hemolytic factor isolated from the venom of *N. oxiana* has an LD_{50} value of 2.4 μg g^{-1} in mice by i.p. injection (Yukel'son et al., 1973, 1974a), a toxicity level certainly much lower than that of the original venoms and even lower than that of the purified toxin. The value of 2.4 μg g^{-1} indicates, however that the hemolytic factor is not entirely nontoxic.

Potency of toxicity and degree of hemolysis were investigated in the venoms of *Vipera ammodytes, Dendroaspis angusticeps, Naja melanoleuca, N. nigricollis, Bitis arietans*, and *B. gabonica* (Nedyalkov and Marchev, 1971). The conclusion from these studies was that no correlation existed between the toxicity and hemolytic activities.

2 CHEMISTRY OF HEMOLYTIC FACTORS

2.1 Property

Because the hemolytic factor lyses the red cell membrane, the effect on membrane components and permeability has been extensively investigated.

Direct hemolytic factor of *Naja naja* venom binds strongly to the erythrocyte ghost membrane (Schroeter et al., 1973; Damerau et al., 1974). Hemolysis induced by cobramine B, which is believed to be identical to DHF of cobra venom, is not dependent on an ATPase-inhibiting effect (Lankishch et al., 1972). Treatment of the red cell with hemolytic factor of *N. naja* venom increases the activities of glyceraldehyde-3-phosphate dehydrogenase, adenylate kinase, 3-phosphoglycerate kinase, and aldolase in the red cell membrane, but catalase activity is not affected (Fajnholc et al., 1972). When the concentration of hemolytic factor in the red cell suspensions was increased, catalase and hemoglobin were leaked from red cell membrane. *Naja oxiana* hemolytic factor increases the membrane permeability to K(I) and also enhances the membrane conductivity of human erythrocytes (Yukel'son et al., 1974b). Since fatty acids can inhibit hemolytic action, venom hemolytic factor must affect the phospholipid moiety of the red cell membrane (Mirsalikhova et al., 1975).

2.2 Structure

Direct lytic factor isolated from *Haemachates haemachatus* is a basic peptide with 57 amino acid residues (Aloof-Hirsch et al., 1968). It has a molecular weight of 7000 and contains four disulfide bridges. Hemolytic factor was also isolated from the same venom and called hemolytic protein 12B (Porath, 1966). The sequence of 12B was identified by Fryklund and Eaker (1973) and is shown in Fig. 1.

From the position of Sulfur atoms, it is evident that hemolytic factors belong to the class of nonneurotoxic basic protein (see Chapter 19) and are also structurally related to neurotoxins (see Fig. 4 of Chapter 18).

```
Leu-Lys-Cys-His-Asn-Lys-Leu-Val-Pro-Phe-Leu-Ser-Lys-Thr-Cys-Pro-

Glu-Gly-Lys-Asn-Leu-Cys-Tyr-Lys-Met-Thr-Met-Leu-Lys-Met-Pro-Lys-

Ile-Pro-Ile-Lys-Arg-Gly-Cys-Thr-Asp-Ala-Cys-Pro-Lys-Ser-Ser-Leu-

Leu-Val-Lys-Val-Val-Cys-Cys-Asn-Lys-Asp-Lys-Cys-Asn-OH
```

Figure 20.1 Amino acid sequence of hemolytic factor isolated from the venom of *Haemachatus haemachatus*. (Data obtained from Fryklund and Eaker, 1973.)

Heparin inhibits the hemolytic action of DHF purified from *Naja oxiana* venom (Yukel'son et al., 1975). This effect is believed to be due to the combination of heparin and DHF. SInce heparin is an acidic compound and DHF is basic, this appears to be a typical combination of an acidic and a basic compound.

3 ROLE OF PHOSPHOLIPASE A_2 IN HEMOLYSIS

3.1 Phospholipase A_2 Alone

Lysolecithin, produced by the action of phospholipase A_2 on phosphatidyl choline, has been established as one of the causative agents producing hemolysis. Pure phospholipase A_2 itself causes no hemolysis or a very weak hemolytic activity. It has been shown that phospholipase A_2 from the venoms of *Naja naja* and *Haemachatus haemachatus* is devoid of hemolytic activity and causes no significant breakdown of phospholipids in the intact erythrocytes (Condrea et al., 1964). When pure phospholipase A_2 isolated from the venom of *Laticauda semifasciata* is used, the HU_{50} (50% hemolysis point) is 16 μg g^{-1}. In the presence of 0.1 mg of beef phosphatidylcholine, however, only 0.4 μg of pure phospholipase A_2 is required to cause 50% hemolysis. This means that pure phospholipase A_2 is only 2.5% as hemolytic as a mixture of phospholipase A_2 and phospholipids (Tu et al., 1970; Tu and Passey, 1971). Apparently, the phospholipase A_2 reaction produces lysophosphatide (lysolecithin), which is the true hemolytic agent rather than phospholipase A_2 per se.

As stated before, there is really no clear-cut distinction between direct and indirect hemolytic action. Normally, we consider phospholipase A_2 to be an indirect hemolytic agent. However, under special conditions, phospholipase A_2 can be directly hemolytic. In a hypotonic condition when the erythrocytes are swollen, phospholipase A_2 can cause direct hemolysis. In the hereditary disease sphaerocytosis, in which the patient has swollen cells, the erythrocytes are much more susceptible to phospholipase A_2 than are normal cells. They are lysed even in isotonic medium (Lankisch and Vogt, 1972). It is believed that, when the red cells are swollen, there is a rearrangement of membrane structure so as to be more susceptible to phospholipase A_2. Similarly, the ability for hemolysis by *Crotalus atros* venom is also dependent on hypotonicity of the solution used (Bethell and Bleyl, 1942).

Hemolysis by phospholipase A_2 also depends on the pressure applied to the erythrocyte membrane. Phospholipase A_2 of *Naja naja* and *Crotalus adamanteus* venoms cannot hydrolyze the intact red cell membrane phospholipids. Hydrolysis takes place, however, when a pressure >3 dynes cm^{-1} for *Naja naja* enzyme and >31 dynes cm^{-1} for *C. adamanteus* enzyme is applied (Demel et al., 1975).

In the special cases when venom phospholipase A_2 itself is hemolytic because of its high basicity. There is a wide range of isoelectric points for phospholipase A_2 (see Chapter 2). Some phospholipase A_2 enzymes are acidic proteins; others, basic proteins. For example, phospholipase A_2 isolated from the venom of *Agkistrodon halys blomhoffii* is basic and can hydrolyze erythrocyte membrane phospholipids as well as cause hemolysis without the presence of DHF (Martin et al., 1975).

3.2 Synergistic Action with Direct Hemolytic Factor

Cobra venoms contain a basic protein called the direct hemolytic factor, which itself is hemolytic but has no phospholipase A_2 activity. When phospholipase A_2 and DHF are combined, the mixture becomes strongly hemolytic (Condrea et al., 1964, 1971; Oldigs et al., 1971; Yukel'son et al., 1975). Venoms of *Vipera palestinae* and *V. russellii* contain phospholipase A_2 but not DHF, and both are nonhemolytic. When other surface-active agents are used instead of DHF, phospholipase A_2 causes hemolysis. It is thus clear that the function of DHF is to disrupt the membrane surface, thereby allowing phospholipase A_2 to penetrate into the hydrophobic region of the membrane. Eventually, phospholipase reacts with cell membrane phospholipids to produce lysophosphatides, which cause even more disruption of the cell membranes.

Gul and Smith (1972) observed that DHF can be replaced by albumin in phospholipase A_2-induced hemolysis. They first observed that *Naja naja* phospholipase A_2 hydrolyzed human erythrocyte phospholipids without causing significant hemolysis in an isotonic medium. When albumin was added to the incubation medium either before or after the enzyme action, it caused hemolysis without any further increase in phospholipid splitting. This experiment indicates that, although DHF, together with phospholipase A_2, is a naturally occurring, potent inducer of hemolysis, DHF is not an absolutely indispensable substance in phospholipase A_2-induced hemolysis.

A marked synergism on erythrocyte membrane with phospholipase A_2 and melitin and polymyxin has been demonstrated (Vogt et al., 1970; Mollay and Kreil, 1974). Actually melitin forms a complex with lecithins (Mollay and Kreil, 1973), suggesting that such a complex may be an intermediate step in hemolysis.

The hemolytic action of phospholipase A_2 of the venoms of *Crotalus terrificus terrificus* and *C. adamanteus* requires deoxycholate (Hölzl and Wagner, 1968). Apparently, deoxycholate can also replace DHF.

There is a good correlation between the degrees of phospholipid hydrolysis in the red cell membrane and of hemolysis (Gul et al., 1974).

Dextran and sucrose, inhibitors of osmotic hemolysis, have no effect on hemolysis induced by the hemolytic factor of cobra venom. Therefore Lanisch et al. (1974) conclude that a nonosmotic mechanism is involved in hemolysis induced by a mixture of phospholipase A_2 and hemolytic factor. Yukel'son and Krasil'nikov (1974) found that phospholipase A_2 did not affect the membrane conductivity, whereas venom hemolytic factor increased it. It is also known that the former hydrolyzes phospholipids, but the latter does not. Therefore it is clear that phospholipase A_2 and hemolytic factor are complementary in action.

Although we still do not know the exact mechanism of hemolysis, it is logical to assume that the main function of hemolytic factors or other basic compounds is to disrupt the matrix of cell membrane organization in such a way that venom phospholipase A_2 can attack membrane phosphatydylcholine more effectively. Lysophosphatydylcholine produced as a result of phospholipase A_2 hydrolysis further enhances the hemolytic action.

REFERENCES

Aloof-Hirsch, S., De Vries, A., and Berger, A. (1968). The direct lytic factor of cobra venom: Purification and chemical characterization, *Biochim. Biophys. Acta*, **154**, 53.

Bethell, F. H. and Bleyl, K. (1942). The production of microspherocytosis of red cells and hemolytic anemia by the injection of rattlesnake (*Crotalus atrox*) venom, *J. Clin. Invest.*, **21**, 641.

Brazil, V. and Pestana, B. R. (1910). Nova contribuicão ao estudo do envenenamento ophidico: VIII. Hemolise, *Rev. Med. S. Pualo*, **13**, 61.

Byrd, F. E. and Johnson, B. D. (1970). Some effects of heat-labile venom components on indirect hemolysis by crotalid venoms, *Am. J. Trop. Med. Hyg.*, **19**, 724.

Chopra, R. N. and Roy, A. C. (1936). The hemolysis caused by snake venoms: A preliminary report, *Indian Med. Gaz.*, **71**, 21.

Condrea, E., De Vries, A., and Mager, J. (1964). Hemolysis and splitting of human erythrocyte phospholipids by snake venoms, *Biochim. Biophys. Acta*, **84**, 60.

Condrea, E., Barzilay, M., and De Vries, A. (1969). Study of hemolysis in the lethal effect of *Naja naja* venom in the mouse and guinea pig, *Toxicon*, **7**, 95.

Condrea, E., Barzilay, M., and De Vries, A. (1971). Action of cobra venom lytic factor on sialic acid – depleted erythrocytes and ghosts, *Naunyn-Schmiedebergs Arch. Exp. Pathol. Pharmakol.*, **268**, 458.

Damerau, B., Schroeter, R., and Vogt, W. (1974). Binding of the direct lytic factor (DLF) of cobra venom (*Naja naja*) to red cell membranes, *Protides Biol. Fluids, Proc. Colloq.*, **21**, 267.

Dass, B., Chatterjee, S. C., and Devi, P. (1970). Haemolytic activity of Russell's viper venom, *Indian J. Med. Res.*, **58**, 399.

De Hurtado, I. and Layrisse, M. (1964). A quantitative method for the assay of snake venom hemolytic activity, *Toxicon*, **2**, 43.

Demel, R. A., Geurts van Kessel, W. S. M., Zwaal, R. F. A., Roelofsen, B., and Van Deenen, L. L. M. (1975). Relation between various phospholipase actions on human red cell membranes and the interfacial phospholipid pressure in monolayers, *Biochim. Biophys. Acta*, **406**, 97.

De Vries, A., Kirschmann, C., Klibansky, C., Condrea, E., and Gitter, S. (1962). Hemolytic action of indirect lytic snake venom *in vivo*, *Toxicon*, **1**, 19.

Fajnholc, N. E., Condrea, E., and De Vries, A. (1972). Activation of enzymes in red blood cell membranes by a basic protein isolated from cobra venom, *Biochim. Biophys. Acta*, **255**, 850.

Fryklund, L. and Eaker, D. (1973). Complete amino acid sequence of a nonneurotoxic hemolytic protein from the venom of *Haemachatus haemachates* (African ringhals cobra), *Biochemistry*, **12**, 661.

Gul, S. and Smith, A. D. (1972). Haemolysis of washed human red cells by the combined action of *Naja naja* phospholipase A_2 and albumin, *Biochim. Biophys. Acta*, **288**, 237.

Gul, S., Khara, J. S., and Smith, A. D. (1974). Hemolysis of washed human red cells by various snake venoms in the presence of albumin and Ca^{2+}, *Toxicon*, **12**, 311.

Hölzl, J. and Wagner, H. (1968). Über die Hämolyseaktivität von Lysolecithin, das durch Lecithinspaltung mit Phospholipase A in wässrigem Milieu erhalten Wird, *Z. Naturforsch.*, B, **23**, 449.

Izard, Y., Boquet, P., Golémi, E., and Goupil, D. (1969). La toxine y n'est pas le facteur lytique direct du venin de *Naja nigricollis*, *C. R. Acad. Sci. Paris*, **269**, 666.

Kellaway, C. H. (1929). A preliminary note on the venom of *Pseudechis guttattus*, *Med. J. Aust.*, **1**, 372.

Klibansky, C., Ozcan, E., Joshua, D., Djaldetti, M., Bessler, H., and De Vries, A. (1966). Intravascular hemolysis in dogs induced by *Echis coloratus* venom, *Toxicon*, **3**, 213.

Lankisch, P. G. and Vogt, W. (1972). Direct hemolytic activity of phospholipase A, *Biochim. Biophys. Acta*, **270**, 241.

References

Lankisch, P. G., Schoner, K., Schoner, W., Kunze, H., Bohn, E., and Vogt, W. (1972). Inhibition of Na$^+$- and K$^+$ activated ATPase by the direct lytic factor of cobra venom (*Naja naja*), *Biochim. Biophys. Acta,* **266,** 133.

Lankisch, P. G., Damerau, B., and Vogt, W. (1974). Osmotic and non-osmotic components of the haemolysis induced by the direct lytic factor (DLF) of cobra venom, *Naunyn-Schmiedebergs Arch. Exp. Parthol. Pharmakol.,* **282,** 255.

Luzzio, A. J. (1967). Inhibitory properties of serum proteins on the enzymatic sequence leading to lysis of red blood cells by snake venom, *Toxicon,* **5,** 97.

Martin, J. K., Luthra, M. A., Wells, M. A., Watts, R. P., and Hanahan, D. J. (1975). Phospholipase A$_2$ as a probe of phospholipid distribution in erythrocyte membranes: Factors influencing the apparent specificity of the reaction, *Biochemistry,* **14,** 5400.

Mirsalikhova, N. M., Yukel'son, L. Y., and Ziyamukhamedov, R. (1975). Role of calcium ions and fatty acids in inhibition of (magnesium ion)- and (sodium, potassium ion)-dependent ATPase by direct hemolysin and phospholipase A of cobra venom, *Ukr. Biokhim. Zh.,* **47,** 61.

Mitchell, S. W. and Reichert, E. T. (1886). *Researches upon the Venoms of poisonous Serpents,* Smithsonian Institution, Washington, D. C.

Mollay, C. and Kreil, G. (1973). Fluorometric measurements on the interaction of melittin with lecithin, *Biochim. Biophys. Acta,* **316,** 196.

Mollay, C. and Kreil, G. (1974). Enhancement of bee venom phospholipase A$_2$ activity by melitin, direct lytic factor from cobra venom and polymyxin B, *FEBS Lett.,* **46,** 141.

Nedyalkov, S. and Marchev, N. (1971). Comparative biologic and immunochemical studies on the viper (*Vipera ammodytes ammodytes*) venom and the venoms of African poisonous snakes, *Kongr. Mikrobiol. Mater. Kongr. Mikrobiol. Bulg., 2nd 1969,* **1,** 297.

Oldigs, H. D., Lege, L., and Lankisch, P. G. (1971). Vergleichende Hämolyseversuche an Meerschweinchenund Rattenerythrocyten *in vitro* mit Phospholipase A und Direkt-Lytischem-Faktor uas Cobragift (*Naja naja*), *Naunyn-Schmiedebergs Arch. Exp. Pathol. Pharmakol.,* **268,** 27.

Pestana, B. R. (1908). Notas sôbre a aoão hemolitica dos venenos de diversas especies de covras brasileiras, *Rev. Méd. S. Paulo,* **11,** 436.

Philpot, V. B. and Higginbotham, K. (1972). Activation of water mocassin venom hemolysis of human erythrocytes by pathological and artificial altered human serum, *Trans. N.Y. Acad. Sci.,* **34,** 344.

Porath, J. (1966). Some separation methods based on molecular size and charge and their application to purification of polypeptides and proteins in snake venoms, *Mem. Inst. Butantan,* **33,** 379.

Rosenfeld, G., Nahas, L., and Kelen, E. M. (1968). "Coagulant, proteolytic, and hemolytic properties of some snake venoms," in W. Bücherl, E. E. Buckley, and V. Deulofeu, Eds., *Venomous Animals and Their Venoms,* Vol. 1, Academic, New York, pp. 229–274.

Roy, A. C. (1955). Lecithin and venom hemolysis, *Nature,* **155,** 696.

Sakhibov, D. N. and Davlyatov, Y. (1969). Comparative study of the direct hemolytic activity of the venom from Central Asian snakes, *Ref. Zh. Farmakol. Khimioter. Siedstva Toksikol.,* Abstract No. 254882.

Schroeter, R., Lankisch, P. G., Lege, L., and Vogt, W. (1972). Possible implication of glutathione reductase in hemolysis by the direct lytic factor of cobra venom (*Naja naja*), *Naunyn-Schmiedebergs Arch. Exp. Pathol. Pharmakol.,* **275,** 203.

Schroeter, R., Damerau, B., and Vogt, W. (1973). Differences in binding of the direct lytic factor (DLF) of cobra venom (*Naja naja*) in intact red cells and ghosts, *Naunyn-Schmiedebergs Arch. Exp. Pathol. Pharmakol.,* **280,** 201.

Tu, A. T. and Passey, R. B. (1971). "Phospholipase A from sea snake venom and its biological properties," in A. De Vries and E. Kochva, Eds., *Toxins of Animal and Plant Origin,* Gordon and Breach, New York, pp. 419–436.

Tu, A. T., Passey, R. B., and Toom, P. M. (1970). Isolation and characterization of phospholipase A from sea snake, *Laticauda semifasciata,* venom, *Arch. Biochem. Biophys.,* **140,** 96.

Vincent, J. E., Bonta, I. L., and Noordhoek, J. (1972). Some effects of guinea pig serum and heparin on hemolysis induced by *Naja naja, Agkistrodon piscivorus* and *Apis mellifera* venom, *Toxicon,* **10,** 415.

Vogt, W., Patzer, P., Lege, L., Oldigs, H. D., and Wille, G. (1970). Synergism between phospholipase A and various peptides and SH-reagents in causing haemolysis, *Naunyn-Schmiedebergs Arch. Exp. Pathol. Pharmokol.*, **265**, 442.

Yukel'son, L. Y. and Krasil'nikov, O. V. (1974). Effect of phospholipase and direct hemolysin of cobra venom on artificial bimolecular membranes, *Khim. Prir. Soedin*, **5**, 687.

Yukel'son, L. Y., Sadykov, E. S., and Sorokin, V. M. (1973). Direct hemolytic factor of the venom of *Naja oxiana, Uzb. Biol. Zh.*, **17**, 12.

Yukel'son, L. Y., Sadykov, E. I., and Sorokin, V. M. (1974a). Isolation and characterization of a direct hemolytic factor of Central Asian cobra venom, *Biokhimiya*, **39**, 816.

Yukel'son, L. Y., Krasil'nikov, O. V., Isaev, P. I., and Tashmukhamedov, B. A. (1974b). Effect of the direct hemolytic factor of cobra venom on the conductivity of bimolecular phospholipid membranes, *Khim. Prir. Soedin.*, **5**, 688.

Yukel'son, L. Y., Sadykov, É., Sakhibov, D. N., and Sorokin, V. M. (1975). Effect of "direct" hemolytic factor and phospholipase A_2 Central-Asian cobra venom on erythrocytes, *Biokhymiya*, **40**, 589.

21 Blood Coagulation

1	**COAGULATION AND ANTICOAGULATION**	**330**
	1.1 General Statement, 330	
	1.2 Isolation of Anticoagulant, 331	
2	**EFFECT ON FIBRINOGEN**	**333**
	2.1 Arvin (Ancrod), 333	
	Chemistry, 334	
	Microclot and Effect on Coagulation System, 337	
	Physiological Effect, 338	
	Clinical Trials on Human Beings, 340	
	2.2 Reptilase, 341	
	2.3 Comparison of Arvin and Reptilase, 343	
	2.4 Rattlesnake Venoms (Crotalus spp.), 343	
	2.5 Other Pit Viper (Crotalidae) Venoms, 344	
	2.6 Viperidae Venoms, 344	
	2.7 Fibrinogenolytic Action, 345	
3	**EFFECT ON FIBRIN**	**345**
4	**EFFECT ON FACTOR X (STEWART FACTOR)**	**346**
5	**EFFECT ON FACTOR V (Ac-GLOBULIN)**	**348**
6	**EFFECT ON FACTOR III (THROMBOPLASTIN)**	**349**
7	**EFFECT ON FACTOR II (PROTHROMBIN)**	**350**
	7.1 Activation, 350	
	7.2 Inhibition, 350	
8	**EFFECT ON FACTOR VIII (ANTIHEMOPHILIC FACTOR)**	**350**
9	**EFFECT ON PLATELETS**	**351**
	References	353

Many venoms exert profound effects on the blood coagulation system. Some accelerate the coagulation process, and others retard it. Thus snake venoms are often divided into two types, coagulant (procoagulant) and anticoagulant. However, this is an oversimplification. Some venoms contain both coagulant and anticoagulant factors simultaneously, and sometimes a venom becomes coagulant or anticoagulant, depending on the concentration used. There are many different mechanisms that cause coagulant or anticoagulant action.

In this chapter we will discuss first the procoagulant or anticoagulant property of venom. Then we will consider the action of venom or venom components on each blood coagulation factor.

1 COAGULATION AND ANTICOAGULATION

1.1 General Statement

Purified enzymes are extremely useful in the study of the mechanism of venom effects on blood coagulation. With crude venoms, the situation is extremely complicated because of one venom's having many different components that affect different stages of the blood coagulation scheme. The terms "procoagulation" and "anticoagulation" seem to describe the action of snake venom on blood coagulation adquately, but actually, as mentioned above, the situation is much more complicated in that some venoms contain both procoagulant and anticoagulant factors. When fibrinolytic activity is stronger than coagulant activity, a fibrin clot is not observed since fibrin is hydrolyzed as soon as it is formed. When, on the other hand, coagulant action is stronger than fibrinolytic action, a fibrin clot is first formed and then slowly disappears (Tu et al., 1966). Moreover, since the relative proportions of fibrinolytic and coagulant fractions in each venom are different, a venom sometimes becomes coagulating or anticoagulating, depending on the concentration in which it is present. Meaume et al. (1966) reported that, depending on the amount of *Naja nigricollis* venom added, the venom either decreases or increases the coagulation time of human plasma.

There are many examples of venoms that contain both coagulant and anticoagulant components. For instance, in venoms of *Agkistrodon acutus, Trimeresurus gramineus* (Ouyang, 1957ab; Shiau and Ouyang, 1965), *Echis colorata* (Moav et al., 1963), *E. carinatus* (Schieck et al. 1972a, b), *Vipera lebetina* (Roitman, 1966), and *V. aspis* (Boffa and Boffa, 1971, 1973; Boffa et al., 1972a, b) there are both coagulant and anticoagulant factors that can be separated by various chemical methods.

Although the effect of crude venoms on blood coagulation is extremely complicated, it is important to know the combined effects of whole venoms as this information becomes important in cases of snake poisoning.

Various mechanisms can induce procoagulation or anticoagulation. A venom may act as a coagulant for the following reasons:

1. It has a thromboplastinlike activity.
2. It contains a factor X activator.
3. It can activate factor V.
4. It can activate prothrombin.
5. It has a thrombinlike activity.

Similarly, a venom may serve an anticoagulant function for these reasons:

1. It contains fibrinogenolytic activity.
2. It has fibrinolytic activity.
3. It can activate plasminogen to release plasmin.

1 Coagulation and Anticoagulation

4. It has inhibitory or destructive action toward any of the blood coagulation factors preceding thrombin. For instance, venoms that can inhibit factors II, III, or V or thrombin are anticoagulant.

5. It has an antithrombic action.

As early as 1938, Haunt reported that *Naja tripudians* venom was able to inactivate thrombin *in vitro*. There have been very few reports on the antithrombic action of snake venoms since then, and it is not clear whether venoms really contain an antithrombic principle.

Normally, cobra venoms are considered to be primarily neurotoxic. Actually, however, cobra venoms have diverse actions, although neurotoxicity is the most apparent effect. MacKay et al. (1968, 1969) investigated the venoms of three mambas (*Dendroaspis angusticeps, D. polylepis* and *D. jamesoni*) and three cobras (*Naja melanoleuca, N. nigricollis,* and *Ophiophagus hannah*) and found that all of the venoms were anticoagulant. The action appears to be due to an effect on both the extrinsic and the blood thromboplastin mechanisms. *Naja flava* venom also acts as an anticoagulant by interfering with thromboplastin and thrombin generation (Mohamed and Hanna, 1973).

Let us use some specific venoms as examples of the complex actions of snake venoms on the blood coagulation system. Since the venom of *Agkistrodon rhodostoma* is one of the more extensively studied venoms, it will be used first to demonstrate this complexity. It is well known that *A. rhodostoma* is a coagulant venom, as it coagulates fibrinogen to fibrin. However, the clot formed is different from an ordinary fibrin clot and really consists of microclots, which are rapidly lysed by the subsequent extensive fibrinolytic reaction of tissues (Regoeczi et al., 1966). The result is a prolonged defibrinogenated state and prevention of coagulation. In this respect, the venom appears to be anticoagulant. This is one of the very special examples; we cannot simply say that *A. rhodostoma* venom is coagulant or anticoagulant. Both statements are correct, yet neither describes the true actions of the venom accurately.

Bothrops jararaca venom is coagulant, but two different mechanisms work to make it coagulant. One factor activates factor X, and the other one is a thrombinlike activity (Denson and Rousseau, 1970).

Bitis nasicornis venom has anticoagulant activity (MacKay et al., 1970). The anticoagulant action appears to be due to an effect on both the extrinsic and the intrinsic blood thromboplastin mechanisms. The venom has proteolytic and caseinolytic activities and is capable of activating plasminogen, enhances the activation of plasminogen by streptokinase, and potentiates the action of plasmin. In the euglobulin clot lysis system, a high concentration of venom produces inhibition.

The venom of *Bitis arietans* is also anticoagulant (Brink and Steytler, 1974). The action is due to fibrinolysis as well as to fibrinogenolysis.

1.2 Isolation of Anticoagulant

The venom of *Agkistrodon acutus* has a coagulant action at high concentration and an anticoagulant action at low concentration (Ouyang, 1957b). The coagulant action is due to a thrombinlike activity, while the anticoagulant action results chiefly from the inactivation of prothrombin, thromboplastin, Ac-globulin, and fibrinolysis. When the coagulant and anticoagulant fractions were separated from each other (Cheng and Ouyang, 1967), three anticoagulant principles were found; two fractions were associated with caseinolytic activity, but none of the three fractions had any amino acid esterase activity. The main anticoagulant principle was isolated from the venom of *A. acutus* (Ouyang and Teng, 1972; Ouyang et al., 1972); it has a molecular weight of 20,650 with a sedimentation velocity of 2.00S. The pI is 4.7, indicating an acidic protein. The principle is a glycoprotein and is thermolabile. The purified fraction has no caseinolytic, tosyl-L-arginine methyl ester esterase, phospholipase A_2, phosphodiesterase, alkaline phosphomonoesterase, or fibrinolytic activity. The amino acid composition is shown in Table 1. The fact that 25% of the total amino acid residues are acidic amino acids

Table 1   Amino Acid Composition of Anticoagulant Principles from Snake Venoms

Genus Species Subspecies	Agkistrodon acutus	Trimeresurus gramineus
Origin	Formosa	Formosa
Amino Acid		
Lysine	14	17
Histidine	5	3
Arginine	5	3
Aspartic acid	17	18
Threonine	8	7
Serine	16	6
Glutamic acid	23	12
Proline	0	7
Glycine	10	11
Alanine	10	7
Valine	8	8
Methionine	3	3
Isoleucine	6	7
Leucine	9	8
Tyrosine	7	7
Phenylalanine	10	4
Half-cystine	9	15
Tryptophan	0	4
Total residue	160	150
References	Ouyang and Teng, 1972	Ouyang and Yang, 1975

accounts for the low isoelectric point. The content of carbohydrate is rather low, only 2%. The toxicity is low; the LD_{50} value exceeds $30\ \mu g\ g^{-1}$ in mice by i.p. injection.

The mechanism of anticoagulation of this protein is due to an interaction with prothrombin but not to the destruction of prothrombin or of prothrombin activation factors (Ouyang and Teng, 1973). The anticoagulant principle does not digest fibrinogen or fibrin or inactive thrombin.

An anticoagulant principle isolated from the venom of *Trimeresurus gramineus* (Ouyang and Yang, 1975) has a molecular weight of 19,500, a sedimentation velocity of 1.70S, and a p*I* of 4.5. The amino acid composition of this principle is considerably different from that of *Agkistrodon acutus*, as can be seen from Table 1. Like the *A. acutus* principle, the anticoagulant from *T. gramineus* does not have caseinolytic, tosyl-L-arginine methyl ester esterase, phospholipase A_2, phophodiesterase, alkaline phosphomonoesterase, fibrinolytic, hemorrhagic, or local irritating activities. The purified anticoagulant principle does not destroy fibrinogen, induce fibrinolysis, inactive thrombin (antithrombin), interfere with the interaction between thrombin and fibrinogen. The inhibition of prothrombin activation is due, not to the destruction of prothrombin or its activation factors, but to an interference in the interaction between prothrombin and its activation factors because of the reversible binding of these factors with the anticoagulant principle of the venom.

2 EFFECT ON FIBRINOGEN

We will discuss the effect of venom and its components on fibrinogen in detail. This is logical, as enormous efforts have been made to study the venom enzymes similar to thrombin that convert fibrinogen to fibrin. Many such enzymes have been isolated from snake venoms, and their properties extensively studied. Among them, Arvin (Ancrod) is a well-known example. In this section we will discuss the action and properties of Arvin in detail.

Since the mechanism of blood coagulation itself is very complicated, it is helpful to understand the coagulation process before considering the effects of venom enzymes. For this reason the mechanism of coagulation is summarized in Scheme 22. There are two clotting processes, the intrinsic and the extrinsic process. In the extrinsic system, factor X is activated by the action of factors III and VII, which are found in tissues. In the intrinsic system of blood coagulation, factor X is activated by the action of factor VIII, which itself has been activated through factors IX, XI, and XII in platelets. Factor X activates the other blood coagulation factors shown in the lower portion of Scheme 22, and eventually fibrinogen is converted to fibrin by the active form of specific protease, thrombin.

Blood clot, fibin itself, can be lysed by the action of plasmin which is formed from the zymogen, plasminogen.

2.1 Arvin (Ancrod)

In studies of the effect of snake venoms on blood coagulation, Arvin, a very specific coagulant isolated from the venom of *Agkistrodon rhodostoma*, has been investigated more extensively than any other venom component. Arvin is a very specific protease whose action has some similarity to that of thrombin and occurs in the terminal sequence of a complex blood coagulation mechanism. Because of its limited proteolysis of

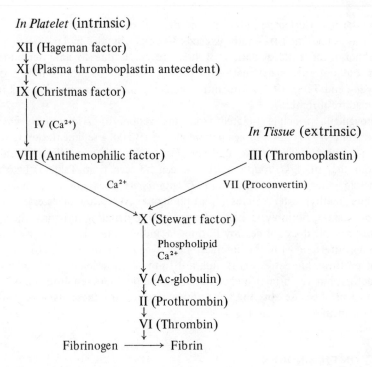

Summary of blood coagulation mechanism. Many factors have both inactive and active forms; these have been combined as single factors in this figure. For instance, inactive and active forms of factor X are simply denoted as X.

Scheme 22

fibrinogen, Arvin yields a readily dispersible fibrin clot (Ewart et al., 1970). Moreover, Arvin has no effect on other clotting factors (Bell et al. 1968a, b; Sharp et al., 1968). For these reasons it is used clinically for patients requiring anticoagulant therapy.

Chemistry. The final stage of blood coagulation is the conversion of fibrogen to a fibrin clot catalyzed by thrombin. The fibrogen molecule is composed of three peptide chains, α (A), β (B), and γ.

Thrombin is actually a very specific proteolytic enzyme releasing two major peptides and other peptides from fibrinogen (see Scheme 23). Fibrinopeptides arise from the

$$\text{Fibrinogen} \xrightarrow{\text{Thrombin}} \text{Fibrin + Fibrinopeptide A + Fibrinopeptide B} \\ \text{+ Fibrinopeptide AP + Fibrinopeptide AY}$$

Scheme 23

N-terminal ends of the α (A) and β (B) chains because of proteolytic action of thrombin. The result is a polymerized product of fibrin. The molecular weights of fibrinopeptides A and B are 1900 and 2400 daltons, respectively (Gladner et al, 1959), and the sequences of bovine fibrinopeptides are as follows (Folk et al, 1959a, b; Folk, 1960):

Fibrinopeptide A: Glu–Arg–Gly–Ser–Asp–Pro–Pro–Ser–Gly–Asp–Phe–Leu–Thr–Glu–Gly–Gly–Gly–Val–Arg

2 Effect on Fibrinogen

Fibrinopeptide B: AcThr–Glu–Phe–Pro–Asp–Tyr(OSO_3^-)–Asp–Glu–Gly–Glu–Asp–Asp–Arg–Pro–Lys–Val–Gly–Leu–Gly–Ala–Arg

The conversion of fibrinogen to fibrin by Arvin results in the release of one major and two minor peptides. The notable difference from thrombin is that Arvin does not give peptide B as a product (Blömback, 1958). These Arvin fibrinpeptides from human fibrinogen were characterized by Holleman and Coen (1970) and Ewart et al. (1970). Peptide AP contains phosphate, while peptide A and AY do not contain phosphorus. The sequences of Arvin fibrinopeptides are as follows:

Fibrinopeptide A: Ala–Asp–Ser–Gly–Glu–Gly–Asp–Phe–Leu–Ala–Glu–Gly–Gly–Gly–Val–Arg

Fibrinopeptide AP: Ala–Asp–SerP–Gly–Glu–Gly–Asp–Phe–Leu–Ala–Glu–Gly–Gly–Gly–Val–Arg

Fibrinopeptide AY: Asp–Ser–Gly–Glu–Gly–Asp–Phe–Leu–Ala–Glu–Gly–Gly–Gly–Val–Arg

Arvin interacts with at least two serum proteins. One of these is α_2-macroglobulin, and the other may be antithrombin III. Thus the coagulant activity of Arvin is neutralized by incubation with normal human serum, as reported by Pitney and Regoeczi (1970). They also observed that a considerable amount of Arvin is trapped in the fibrin mesh when fibrinogen is clotted with Arvin.

Loss of fibrinopeptide A leads to the production of a polymer of the type $[(\alpha\beta(B)\gamma)_2]_n$ rather than the $[(\alpha\beta\gamma)_2]_n$ type fibrin polymer resulting from the action of thrombin (Ewart et al., 1970).

The weight average molecular weight (M_W) of Arvin is 45,000, and the z average molecular weight (M_z) is 51,000 daltons (Esnouf and Tunnah, 1967).

Hessel and Blombäck (1971) used disulfide knot (DSK) as a substrate; DSK is the fragment remaining after treatment of fibrinogen with CNBr, which retains disulfide linkages to hold the three peptide chains of fibrinogen together at the vicinity of the N-terminal ends. They found that both thrombin and reptilase released fibrinopeptides A, AP, and Gly–Pro–Arg. However, Arvin did not release Gly–Pro–Arg. The tripeptide is a new fibrinopeptide found by Hessel and Blombäck (1971).

Like many other serine proteinases, Arvin is also inactivated by diisopropyl phosphorofluoride and phenylmethanesulfonyl fluoride (Collins et al., 1971; Collins and Jones, 1972). Arvin can also be inactivated by nitrous acid (Collins, 1972). The loss of esterase activity is first order with respect to enzyme concentration. The loss of coagulant activity is biphasic, an initial fast inactivation being followed by a slower loss of coagulant activity occurring at the same rate as the loss of esterase activity. Collins (1972) also found that the loss of enzyme active sites proceeds in parallel with the loss of the N-terminal valine residue. Thus the N-terminal valine appears essential for catalytic activity.

Arvin has marked activity on TAME and BAEE but no caseinolytic activity (Beaumont et al., 1972; Soh and Chan, 1974). A difference in the specificity of Arvin from that of trypsin was pointed out by Collins et al. (1971). Although Arvin hydrolyzes methyl. ethyl, and cyclohexyl ester of α-N-benzoyl-L-arginine at equal rates, it fails to catalyze the hydrolysis of simple amides and the trypsin-susceptible bonds in casein, glucagon, and

Table 2 Amino Acid Composition of Arvin Isolated from the Venom of *Agkistrodon rhodostoma*

Amino Acid	Content (mole/10^5 g protein)	
Lysine	34	31
Histidine	22	23
Arginine	60	57
Aspartic acid	99	89
Threonine	23	25
Serine	28	43
Glutamic acid	47	38
Proline	41	42
Glycine	53	53
Alanine	35	33
Valine	26	40
Methionine	17	16
Isoleucine	56	47
Leucine	45	43
Tyrosine	19	16
Phenylalanine	21	23
Half-cystine	43	14
Tryptophan	—	9
Total residue		642
References	Esnouf and Tunnah 1967	Hatton 1973

insulin. Both the catalytic activity and the rate of inactivation of Arvin are dependent on the ionization of a group with an apparent pK_a of 7.0, which probably reflects the involvement of a histidine residue at the active site (Collins and Jones, 1972). It is also believed that the esterase and coagulant activities of the enzyme are associated with the same active site. Titration of the active site of the enzyme with *p*-nitrophenol-*p'*-guanidinobenzoate indicates that Arvin has one active site per mole, based on a molecular weight of 55,000.

The Amino acid composition of Arvin was determined by Esnouf and Tunnah (1967) and by Hatton (1973) and is shown in Table 2. There is considerable discrepancy in the

2 Effect on Fibrinogen

composition reported by the two groups. The difference may be due to contamination by other proteins or to the polymorphic nature of Arvin, as pointed out by Hatton (1973), who observed five forms of the enzyme in polyacrylamide gel electrophoresis. It is not clear whether all five bands show coagulant activity. Arvin is a glycoproein, and 29% of its weight is carbohydrate (Hatton, 1973). It contains about 86 moles of nonnitrogenous sugars, 60 moles of hexosamines, and 18 moles of sialic acid. A large number of half-cystine residues, 43 moles, was reported by Esnouf and Tunnah (1967), as compared to 14 moles by Hatton (1973). Hatton pointed out that the large number reported by Esnouf and Tunnah may be due to merged peaks of half-cystine and glucosamine.

Arvin is quite specific immunologically as the antibody to it does not cross-react with many other venoms (Lewis et al., 1971).

Microclot and Effect on Coagulation System. When Arvin is injected into mice, fibrin microclots are formed which initially localise in the lungs and to a lesser extent in the kidneys, liver, and spleen. This can be studied by means of immunofluorescence (Silverman et al., 1971). These microclots are rapidly dispersed, as indicated by the lack of severe symptoms of respiratory distress. After 30 min, the microclots are more diffused as fluorescence materials are more evenly distributed throughout the septa. All fluorescence materials disappear within 12 hrs indicating that fibrinolytic activation occurs in the tissues. This rapid lysis of microclots has significance in clinical defibrination. The fall in plasma fibrinogen is very fast, occurring within 5 min after the injection of Arvin into mice (Fig. 1). Plasma fibrinogen slowly appears after 12 hr, as also can be seen in Fig. 1 (Silverman et al., 1973).

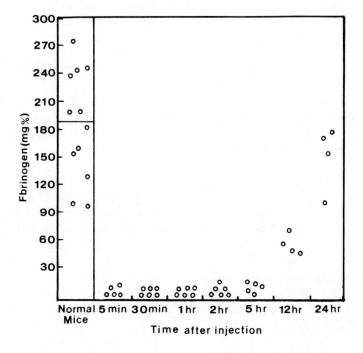

Figure 21.1 Defibrination effect of Arvin in mice. (Reproduced from Silverman et al., 1973) by permission of Dr. E. V. Potter and of the copyright owner, British Journal of Haematology.

Arvin has no effect on the levels of other blood coagulation factors, such as factors II, V, VII, VIII, IX, X, and XIII, and on platelet count. In these aspects Arvin is different from thrombin, which lowers the concentrations of these factors and the platelet count (Bell et al., 1968a, b; Kwann and Barlow, 1971).

Arvin does not activate fibrin-stabilizing factor (factor XIII). This results in the formation of a non-cross-linked clot that is easily dissolved. In the presence of activated fibrin-stabilizing factor, Arvin gives cross-linked fibrin clots (Barlow et al., 1970).

Because of microclot formation, which is quite different from ordinary fibrin clotting, the administration of Arvin does not produce thrombi. The rapid clearance of microclots from the circulation is believed to be due to reticuloendothelial phagocytosis (Ashford et al., 1968). Formation of microclots by Arvin does not affect the plasma levels of free fatty acids, triglycerides, or total cholesterol, suggesting that Arvin has little effect on the intermediate metabolism (Barboriak, 1971).

Although there is some similarity of Arvin to thrombin, there are also some distinct differences. For instance, Arvin clots platelet fibrinogen but, unlike thrombin, does not release potassium, serotonin, ADP, or ATP from platelets (Brown et al., 1972).

Physiological Effect. Defibrination effects by Arvin *in vivo* were observed by many investigators (Moberg et al., 1971; Bowell et al., 1970). As early as 1963, Reid et al. (1963) observed that in patients envenomated by *Agkistrodon rhodostoma* blood coagulation was retarded and fibrinogen content decreased. Chan et al. (1965) observed that the venom can convert fibrinogen to fibrin directly in the absence of the other clotting factors. They also found that plasma can be defibrinated without loss of factor VIII by adjusting the venom concentration. The venom also accelerates the formation of prothrombin activator in the thromboplastin generation test and in the Hicks-Pitney screening test. In 1966 Marsten et al. found that the venom is highly effective in preventing experimental thrombosis. They attributed this antithrombotic effect to the induced hypofibrinogenemia and accelerated fibrinolysis. With the isolation of Arvin from the venom (Esnouf and Tunnah, 1967), extensive investigation began, using better defined components rather than the whole venom.

Because of the important therapeutic use of Arvin in patients with thromboembolic problems, Klein et al. (1969) investigated the effect of Arvin on cardiac function of dogs. They found that it did not cause any significant changes in left ventricular function, both tension-generating and velocity-shortening attributes remaining unchanged.

The effects of Arvin on wound healing and collagen formation were also investigated. Defibrinogenation produces impaired wound healing and defective connective tissue formation in the rabbit (Holt et al., 1969). Olsen and Pitney (1969) observed that Arvin therapy produced alterations in the histological structure of experimental pulmonary emboli in rabbits. Surgical wounds in dogs treated with Arvin do not heal as readily as wounds in nontreated animals. Necrosis of emboli can be the result of a direct lytic effect of Arvin on the fibrin in the embolus. However, Arvin produces clearance of the pulmonary vessels faster than occurs in control animals, and there is no permanent alteration in the histology of the vessels after the obstruction has disappeared.

Ashford and Bunn (1970) studied the effect of Arvin on reticuloendothelial activity, using the clearance rate of colloidal carbon as the criterion of effect. They found that a single i.v. or i.m. dose of Arvin in rabbit caused a profound increase in the rate of carbon clearance, the effect lasting for 48 to 72 hr. The reticuloendothelial stimulant effect of Arvin appears to be unrelated to its coagulant action on fibrinogen. The mechanism may

2 Effect on Fibrinogen

involve a direct action on phagocytic cells and an effect on colloidal particles via serum opsonins.

Chan (1969) obtained partially purified fraction and called it thrombinlike fraction, but it is now known as Arvin. From the effects of this fraction in rats, Chan concluded that his partially purified fraction had a marked antithrombotic effect. Heparin does not have this action unless given in high doses that result in excessive hemorrhage.

It has been known that fibrin produces symptoms similar to those of chronic arthritis. This is believed to be due to an autoimmune response to the antigen fibrin. Ford et al. (1970) thought that a defibrinogenic agent such as Arvin might have some use in an experimentally induced, chronic immune arthritis but found no curative effect, although defibrination in rabbits was nearly total.

It is also known that fibrinogen is a cause of intraabdominal adhesions as a manifestation of a fibrotic diathesis. Ashby et al. (1970) suggested that Arvin might be a useful drug for preventing postoperative adhesions in the peritoneum and elsewhere. They found no evidence of impaired healing either of the abdominal wall or of the excised 1-cm-square peritoneal defects.

Van der Ziel et al. (1970) reported that dogs remained healthy despite a total absence of fibrinogen for 6 weeks. Unlike the situation with heparin, no spontaneous bleeding was observed. All other clotting factors, including platelet counts, and liver and renal functions remained normal. In another experiment, a silastic, femoral–jugular, venovenous shunt was inserted subcutaneously in dogs 12 hr after Arvin injection. The animals retained shunt patency. The disappearance rate of Arvin from the blood is not affected by nephrectomy in rabbits or by a series of injections of unlabeled Arvin. Labeled Arvin dimer disappears from the plasms faster than momomer. Most of the Arvin is degraded *in vivo*, probably in the reticuloendothelial system; a small fraction is eliminated unchanged by glomerular filtration. ^{131}I[Arvin] conjugated both *in vitro* and *in vivo* with a specific antivenin by forming small and completely soluble complexes.

Although there is no evidence that Arvin promotes lysis of thrombi *in vivo* (Tuppie et al., 1971), Arvin may be a useful drug for thromboembolism therapy by prevention of the propagation of existing thrombi or rethrombosis by removing the precursor of the fibrin substrate. In addition, the reduction in plasma fibrinogen reduces blood viscosity and may improve blood flow through vessels partially obstructed by thrombi. Pitney et al. (1970) injected Arvin into patients with embolic episodes. The results were still not clear but they found that Arvin is a safe anticoagulant.

Arvin is antigenic; thus repeated injection of it may produce antibodies in the recipient, who eventually shows resistance to Arvin treatment. When Pitney and Regoeczi (1970) examined serum samples from patients clinically resistant to Arvin, they found that Arvin had formed a complex with γ-globulin fraction, confirming the presence of anti-Arvin antibody in the patients.

Rahimtoola et al. (1970) showed that Arvin is very effective in preventing initial thrombosis when given prophylactically after damaging the vascular endothelium of the femoral vein with phenol, but it has no thrombolytic action. Pitney et al. (1971) treated venous thrombosis in dogs with streptokinase and Arvin. Arvin treatment prevented rethrombosis in only 5 of the 14 veins treated. This relative ineffectiveness in preventing rethrombosis, may be accounted for, however, by failure to clear the veins completely with streptokinase and failure to maintain a satisfactory degree of hypofibrinogenemia during Arvin therapy because of the problem of bleeding.

The toxicity of Arvin was tested in pregnant rabbits (Penn et al., 1971). Arvin did not delay or prevent implantation but caused a high rate of embryonic death and resorption during early organogenesis. Administration during days 11 to 15 resulted in hemorrhage and subsequent abortion in 3 of 11 rabbits, together with a high death rate of embryos in 5 of the remainder. Subcutaneous injection of Arvin into pregnant mice resulted in the death of many of the embryos up to about day 11, the percentage loss being proportional to the dose. Two mice in the higher dose group showed signs of vaginal hemorrhage at about day 11.

In surgery, hypofibrinogenemia might be useful in protecting the surfaces of foreign bodies implanted in the blood stream against platelet fibrin deposition. Turina et al. (1972) investigated the possibility of substituting Arvin for more conventional anticoagulant agents and found that it could not be used for this purpose. Platelet fibrin thrombi formed even at extremely low levels of plasma fibrinogen, and operative stress caused a marked increase in fibrinogen production, necessitating increasing amounts of the venom extract to reduce the fibrinogen level to an acceptable therapeutic range. At these dosages the complication of intraoperative and postoperative hemorrhage becomes significant and renders this overall approach to the problem an unacceptable one.

A radioactive tracer study indicates that most of the Arvin is degraded *in vivo* (Regoeczi and Bell, 1969). Arvin can also combine with antibody, forming small, completely soluble complexes.

The defibrination effect of Arvin is useful in the treatment of experimentally induced nephritis in rabbits, by protecting the structure and function of the glomeruli (Naish et al., 1972).

Clinical Trials on Human Beings. Arvin has been extensively investigated for possible therapeutic use against many types of diseases, especially those involving fibrinogen and fibrin. Regoeczi and Bell (1969) and Bell and Regoeczi (1970) studied the effect of Arvin on the rate of clearance of plasma fibrinogen by injecting radioactive ^{131}I into human subjects. Disappearance of the clottable radioactively labeled fibrinogen from the circulation proceeded exponentially with an average half-life of 0.85 hr. Thus the mean clearance rate is 12% of the intravascular pool per hour.

An essential condition for successful hemodialysis is maintenance of an adequate blood flow through the dialyser. To prevent blood coagulation, heparin is used. Although heparin is an effective drug for this purpose, it may cause some complications such as hemorrhage, hypersensitivity, and alopecia. Hall et al. (1970) investigated whether Arvin can replace heparin and found Arvin to be an effective and safe anticoagulant that reduced the deposition of fibrin and leucocytes on the cellophane membrane.

Arvin has an effect on erythrocyte flexibility. The lowering of the plasma fibrinogen content after Arvin treatment in human beings produces a slower rate of packing, which is indicative of an increased rigidity of the erythrocytes (Myers et al., 1973).

Intravascular sludging of sickled red cells, with subsequent fibrin deposition, gives rise to the painful crises that characterize sicle-cell anemia. The prevention of further thrombus formation might, in theory, restrict tissue infarction and hasten clinical recovery. On this assumption Mann et al. (1972) tried Arvin on human sickle-cell anemia patients, but observed no significant clinical improvement from the treatment.

Arvin also has an effect on platelet electrophoretic mobility. It is known that a change in platelet electrophoretic mobility is induced by ADP and noradrenaline in patients with acute illness and in those with chronic vascular disease. In patients receiving defibrinogen treatment with Arvin, platelet electrophoretic mobility returned to the normal pattern (Tuppie et al., 1972).

2 Effect on Fibrinogen

2.2 Reptilase

The procoagulant nature of *Bothrops jararaca* has been known for some time (Henriques et al., 1959). Another extensively studied thrombinlike enzyme somewhat similar to Arvin is the reptilase isolated from the venom of *B. jararaca*. Type I of the enzyme was isolated from the venom of *B. atrox moojeni*, and Type II from the venom of a different subspecies, *B. atrox marajoensis* (Holleman and Weiss, 1976). The molecular weights of Types I and II are 29,000 and 31,400, respectively. Type I reptilase contains 27% carbohydrate, composed mainly of hexose, glucosamine, and sialic acid. The two Types have identical p*I*s and amino acid composition (Table 3).

Like Arvin, reptilase causes defibrination of plasma (Egberg and Nordström, 1969, 1970; Funk et al., 1971; Rådegran et al., 1972). Reptilase releases three fibrinopeptides

Table 3 Amino Acid Compositions of Thrombinlike Enzymes from Snake Venoms

Genus Species Subspecies	Agkistrodon rhodostoma		Bothrops atrox	Crotalus adamanteus	C. horidus	A. acutus	Trimeresurus gramineus	Non-snake origin	
Origin	Malaya	Malaya	S. America	U.S.A.	U.S.A.	Formosa	Formosa	—	
Name	Arvin	Arvin	Type I Type II					Thrombin	
M.W.	33,300	33,300	29,000 31,400	32,700	39,000	33,500	29,500	33,700	
Amino Acid									
Lysine	11	10	11	11	14	15	7	18	22
Histidine	7	8	6	9	8	10	6	5	6
Arginine	20	19	8	12	8	10	7	17	20
Aspartic acid	33	30	22	31	44	24	27	28	24
Threonine	8	8	9	14	14	9	11	13	12
Serine	9	14	10	16	24	7	11	14	14
Glutamic acid	16	13	12	23	30	18	14	29	29
Proline	14	14	14	22	10	10	10	14	14
Glycine	18	18	17	20	20	10	17	21	22
Alanine	12	11	13	11	16	7	12	12	12
Valine	9	13	11	17	18	7	10	16	16
Methionine	6	5	4	2	2	5	3	4	4
Isoleucine	19	16	14	18	5	8	11	11	11
Leucine	15	14	12	21	16	11	12	24	24
Tyrosine	6	5	8	7	12	14	7	10	10
Phenylalanine	7	8	8	13	12	7	6	10	10
Half-cystine	14	5	8	14	10	15	16	6	3
Tryptophan	—	3	6	6	10	—	2	6	6
Total residue		214	193	267	278	187	189	258	259
References	Esnouf and Tunnah 1967	Hatton 1973	Holleman, and Weiss, 1976	Markland and Damus 1971	Bonilla 1975	Ouyang et al. 1971	Ouyang and Yang 1974	Seegers et al. 1968	Harmison et al. 1961

when fibrinogen is converted to fibrin (Clegg and Bailey, 1962). Using S-sulfonated fibrinogen as a substrate, it was found that reptilase removes only the A chain, whereas thrombin cleaves both the A and the B chains in fibrinogen.

Chemical differences in the actions of reptilase and Arvin were observed by Hessel and Blombäck (1971). With CNBr-cleaved fibrinogen as a substrate, thrombin, reptilase, and Arvin all released fibrinopeptides A and AP. Thrombin and reptilase also released Gly—Pro—Arg, but Arvin did not. These results are summarized in Scheme 24, where T = thrombin, R = reptilase, and Ar = Arvin.

$$\begin{array}{l}
\overset{\displaystyle\begin{array}{c}T\\R\\Ar\\\downarrow\end{array}}{}\\
\underset{1}{\text{Ala}}-\text{Asp}-\text{Ser}-\underset{5}{\text{Gly}}-\text{Glu}-\text{Gly}-\text{Asp}-\text{Phe}-\text{Leu}-\underset{10}{\text{Ala}}-\text{Glu}-\text{Gly}-\text{Gly}-\text{Gly}-\underset{15}{\text{Val}}-\text{Arg}-\text{Gly}-\\
\overset{\displaystyle\begin{array}{c}T\\R\\\downarrow\end{array}}{\text{Pro}}-\underset{20}{\text{Arg}}-\text{Val}-\text{Val}-\text{Glu}-\overset{\displaystyle\begin{array}{c}Ar\\\downarrow\end{array}}{\text{Arg}}-\underset{25}{\text{His}}-\text{Gln}-\text{Ser}-\text{Ala}-\text{Cys}-\underset{30}{\text{Lys}}-\text{Asp}-\text{Ser}-\text{Asp}-\text{Trp}-\text{Pro}-\\
\underset{35}{\text{Phe}}-\text{Cys}-\text{Ser}-\text{Asp}-\text{Glu}-\underset{40}{\text{Asp}}-\text{Trp}-\text{Asn}-\text{Tyr}-\text{Lys}-\underset{45}{\text{Cys}}-\text{Pro}-\text{Ser}-\text{Gly}-\text{Cys}-\underset{50}{\text{Arg}}-\text{Met}
\end{array}$$

Scheme 24

Reptilase can also be isolated from the venom of *Bothrops atrox* (Follana et al., 1972). When reptilase is injected into dogs, the level of fibrinogen drops within 6 to 24 hr but returns to normal by the fourth to fifth day. Plasminogen is rapidly converted to plasmin by reptilase, and the total plasma antiplasmin activity increases 35 to 73% above the initial level. Like Arvin, reptilase can be used for controlled lowering of the plasma fibrinogen level. However, reptilase shows more species-dependent variation in activity. For instance, the rabbit is resistant to reptilase action. For some reason reptilase has a low activity toward rabbit fibrinogen (Wik et al., 1972).

Arvin and reptilase are not identical immunologically. Immunodiffusion of purified Arvin and reptilase with their respective antibodies shows no cross-reaction between the two enzymes. There is a cross-reaction between antireptilase and the *Agkistrodon rhodostoma* venom, however, showing that a component other than Arvin is immunologically similar to reptilase. Venoms from *Bothrops jararaca* and *B. jararacussu* also show cross-reaction with antireptilase (Barlow et al., 1973). This finding can be significant in Arvin therapy. As stated in the Section 2.1, Arvin is antigenic; thus patients eventually form an antibody to it and thereby become resistant to Arvin treatment. Since reptilase is immunologically different from Arvin, reptilase can be used in defibrination therapy for patients who develop resistance to Arvin.

In general, venoms of *Bothrops* snakes contain thrombinlike enzymes and, often, more than two enzymes (Stocker et al., 1974). The symptoms of fibrinopenia in human cases of *Bothrops* envenomation are quite common (Rosenfeld, 1966). Although some thrombinlike enzymes have been isolated from the venoms of *Bothrops* spp. (Fichman and Henriques, 1962; Donner and Houskova, 1967), it is not clear how they differ from reptilase because of lack of a well-defined comparative study.

The condition of fibrinogenopenia cannot be prevented by the use of heparins, ϵ-aminocaproic acid, or kallikrein inactivator. The only effective agent is anti-*Bothrops* serum (Ruiz Reyes et al., 1967).

2 Effect on Fibrinogen

Fibrin gels produced from fibrinogen with *Bothrops atrox* venom are different from those made with thrombin. Gelation time experiments indicate that the fibrin gel formed by the venom collapses readily (Scheinthal et al., 1970). The fibrin fibers formed by thrombin, however, are joined together by strong bonds that are not easily dissociable.

2.3 Comparison of Arvin and Reptilase

The actions of Arvin and reptilase are very similar. Both release fibrinopeptide A and its derivatives only and do not release fibrinopeptide B (Blömback, 1958). However, Arvin does not activate fibrin-stabilizing factor; thus it is different from reptilase and thrombin. Both reptilase and thrombin activate fibrin-stabilizing factor and thereby promote cross-link formation (Kopeć et al., 1969). Pizzo et al. (1972) found that Arvin progressively and totally digested the α-chains of fibrin monomers at sites different from plasmin. No such digestion was observed, however, when fibrin monomer was treated with reptilase or thrombin. They also observed that Arvin-induced fibrin is markedly susceptible to digestion by plasmin. There is also a difference in the subunit structures of human fibrins formed by the action of Arvin, reptilase, and thrombin (Mattock and Esnouf, 1971).

2.4 Rattlesnake Venoms (*Crotalus* spp)

Thrombinlike enzyme is quite common in snake venoms and can also be found in the venom of rattlesnakes. The enzyme has been isolated from the venom of *Crotalus adamanteus*. The purified enzyme has a significant esterase activity on p-tosyl-L-arginine methyl ester and N-benzoyl-L-arginine ethyl ester (Markland et al., 1970). The esterase and thrombic activities are inhibited simultaneously by diisopropyl fluorophosphate. Injection of the enzyme causes defibrination in dogs, rabbits, and monkeys. The enzyme does not activate factors of the extrinsic system or blood coagulation factors II, V, VII, VIII, XII, and XIII (Damus et al., 1972). It does not aggregate platelets and is not inhibited by heparin. Although the enzyme does not activate components of the fibrinolytic system, it is itself fibrinolytic at very high concentration. Infusion into dogs produces well-tolerated hypofibrinogenemia is observed with high titers of nonclottable fibrinogen derivatives in the serum.

The chemical properties of the purified enzyme have been extensively studied by Markland and Damus (1971). The enzyme has esterase activity on basic amino acid esters and p-nitrophenyl esters of various N-carbobenzoxyamino acids. It exhibits no activity with a variety of N-benzoyl- or N-methylamino acid amides. The enzyme is a glycoprotein with a molecular weight of 32,700 daltons. The types of carbohydrates present in the enzyme are one fucose, two mannose, three galactose, and five glucosamine. The amino acid composition of the enzyme indicates that it is more similar to thrombin than to Arvin (Table 2). It is optimally active near pH 8 and is stable to neutral and alkaline pH; however, it loses activity upon exposure to acid pH. Both clotting and esterase activities are inhibited by DPF, indicating that the enzyme, like thrombin, is a serine esterase. A histidine residue is essential, as the enzyme is inhibited by the chloromethyl ketone of tosyl-L-lysine. Inhibition of the enzyme activity by nitration shows that tyrosine is also an essential residue. Reduction of disulfide bridges causes a loss of enzyme activity, indicating that disulfide linkages are essential.

Thrombinlike enzyme has also been isolated from the venom of a different rattlesnake, *Crotalus horridus horridus*. The molecular weight is about 19,500 (Bonilla, 1975). The enzyme is specific for the fibrinogen—fibrin conversion and does not affect other blood-clotting factors. The effects of slow and fast defibrination with purified enzyme on the hemodynamic variables were studied (Bonilla et al., 1975). Slow defibination was brought about by infusion of the enzyme; fast defibrination, by i.v. injection of the enzyme into dogs. Rapid defibrination induced significant decreases in cardiac output, stroke volume, and mean aortic arterial pressure, whereas slow defibrination had little effect.

2.5 Other Pit Viper (*Crotalidae*) Venoms

Thrombinlike enzyme isolated from the venom of *Agkistrodon acutus* from Formosa (Ouyang et al., 1971) has a molecular weight of 33,500 daltons. The amino acid composition is compared with the composition of other purified thrombinlike enzymes in Table 2. Although considerable variations within the different enzymes exist, they all have unusually high acidic amino acid residue contents. The enzyme from *A. acutus* is a glycoprotein. It does not possess hemorrhagic, local irritating, or caseinolytic activities. The toxicity is very low, as can be seen from the high value, greater than 40 μg g^{-1}, of the LD_{50} in mice. Unlike thrombin, the purified enzyme does not activate fibrin-stabilizing factor (factor XIII). The enzyme contains 8% neutral sugar, 4% sialic acid, and 1.2% hexosamine (Ouyang et al., 1972). Some amino acid residues such as —SH, histidine, and serine are considered to be essential, as the enzyme activity can be inactivated by phenylmercuric acetate, 4-chloromercuric benzoic acid, and diazonium 1-*H*-tetrazide (Ouyang and Hong, 1974).

A thrombinlike enzyme was isolated from the venom of *Trimeresurus gramineus* from Formosa (Ouyang and Yang, 1974). The molecular weight is 29,500 daltons, and the enzyme is also a glycoprotein. Of the 189 residues, 41, or 22%, are acidic amino acids. This content is also reflected in the low isoelectric point of 3.5, making the enzyme a very acidic protein. The enzyme also does not activate fibrin-stabilizing factor (factor XIII).

A similar enzyme, isolated from the venom of *T. okinavensis* from Okinawa (Anderson, 1972a), is an acidic glycoprotein with a molecular weight of 35,000 daltons. It catalyzed the hydrolysis of certain arginine esters but had very low proteolytic activity toward casein. The enzyme has a strong catalytic activity for converting fibrinogen into fibrinlike compounds. A certain factor X-activating ability persisted. It does not activate human factor XIII xymogen in plasma. Fibrinopeptide A is the main product when fibrinogen is converted to fibrin with the purified enzyme, with a small amount of fibrinopeptide B also produced (Anderson, 1972b). The enzyme does not split aromatic esters or insulin B chain. It is a serine esterase independent of metal ions. High molecular weight thrombin inhibitors do not inhibit the enzyme activity, and human plasma, which rapidly inactivates thrombin, has no effect on the venom enzyme. However, pentamidine, a low molecular weight thrombin inhibitor, is a strong inhibitor of this enzyme.

A blood-clotting factor purified from the venom of *T. flavoviridis* (Shimura et al., 1974) promotes the coagulation of whole blood, citric acid-containing blood plasma, and fibrinogen. The blood-clotting action is not antagonized by heparin or EDTA, but is enhanced by Ca(II). The factor is free of plasmin, plasminogen, and activator activities.

2.6 *Viperidae* Venoms

A coagulant enzyme partially purified from the venom of *Bitis gabonica* (Gaffney et al., 1973; Marsh and Whaler, 1974) releases fibrinopeptides A and B from fibrinogen. This activity is associated with two major protein subunits with molecular weights of 26,000

3 Effect on Fibrin

to 28,000. Like Arvin and reptilase, the partially purified enzyme defibrinates rabbits without any apparent short-term deleterious effects.

Venom of *Vipera libertina* has afibrinogenemia action (Pastorova and Kikteva, 1969). Two fractions having coagulation activity were isolated (Krylova, 1970). Unfortunately the details are not available to this author because they were published in Russian journals.

As early as 1962, Williams and Esnouf (1962) purified coagulation enzyme from *V. russellii* venom. The enzyme has amino acid esterase as well as a powerful coagulant activity. The molecular weight is about 100,000 daltons, and the optimum pH is from 7.5 to 8.5.

It was also observed that afibrinogenemia could be induced by the venoms of *Echis carinatus* (Jerushalmy, 1971; Kornalik and Pudlak, 1971) and *E. colorata* (De Vries et al., 1963). Since these workers used whole venom, it is difficult to determine the exact site of action that characterizes this *Echis* coagulant.

2.7 Fibrinogenolytic Action

Some venoms, instead of coagulating fibrinogen, digest it thus rendering the fibrinogen nonclottable. Using crude venoms of *Agkistrodon contortrix* and *A. piscivorus*, Kornalik (1971) observed that venom fibrinogenolytic protease is different from plasmin and trypsin. *Vipera aspis* venom contains an anticoagulant fraction that shows fibrinogenolytic activity (Boffa et al., 1972a). Two fibrinogenolytic enzymes, fractions 8 and 13, were isolated from the venom of *Trimeresurus mucrosquamatus* (Ouyang and Teng, 1976). Fraction 8 has a molecular weight of 26,000 and digests the β(B) chain first to yield four fragments. When fraction 8 is incubated longer with fibrinogen, the α(A) chain is also hydrolyzed by fraction 8. Fraction 13 has a molecular weight of 22,400 daltons and hydrolyzes the α(A) chain to yield two fragments.

3 EFFECT ON FIBRIN

Most studies on fibrinolysis are based on whole venom rather than on a well-defined pure component; therefore it is hard to determine the mechanism of fibrinolysis. The following are possibilities:

1. Venoms contain fibrinolytic enzymes, and fibrinolysis is due to:
 a. A specific enzyme that acts only on the fibrin molecule.
 b. A nonspecific enzyme that can digest many proteins, including fibrin.
2. Venoms do not contain fibrinolytic enzymes, but activate plasminogen. The consequence is the liberation of plasmin, which causes fibrinolytic action. If this is the case, fibrinolysis is not due to direct action of venoms but is secondary in nature.

Fibrinolytic activity can be conveniently assayed on fibrin-agar plates (Mohamed and El-Damarawy, 1974). Rosenfeld (1964) considered that the fibrinolytic effect of *Bothrops* venoms was due to the activation of plasminogen rather than the direct action of fibrin.

It is important not to confuse fibrinogenolysis with fibrinolysis; the former is digestion of the fibrinogen molecule to nonclottable fragments. When a venom has strong fibrinogenolytic activity, no blood clot is formed. Sometimes this can be mistaken as a fibrinolytic action of the venom. Venoms of *Agkistrodon piscivorus* and *A. contortrix* have no effect on plasminogen-activation. Therefore their fibrinolytic actions are due to proteases present in the venoms (Kornalik, 1966; Kornalik and Styblova, 1967; Ouyang and Huang, 1976). Unlike plasmin or trypsin, venom proteases are not inhibited by soybean trypsin inhibitor and trasylol. Kornalik (1966) reported that fibrinolytic products possess antithrombin activity.

The fibrinolytic action of *Naja naja* venom can be enhanced by some compounds, such as niflumic acid, flufenamic acid, pamoic acid, anthranilate and salicylate derivatives, in the presence of hydrazine. This effect is exerted through the pseudoglobulin fraction (Von Kaulla, 1975).

Finally, the disappearance of microfibrin clots produced by the action of Arvin on fibrinogen is due not, to fibrinolytic action of Arvin itself, but rather to locally increased activation of plasminogen and subsequent higher levels of circulating fibrinolytic activity (Silverman et al., 1973).

4. EFFECT ON FACTOR X (STEWART FACTOR)

Factor X (Stewart factor) is involved in blood coagulation in the middle phases of the coagulation scheme (see Scheme 22). MacFarlane and Barnett (1934), as well as other workers (Hawkey and Symons, 1966), observed that Russell's viper (*Viper russellii*) venom promoted blood coagulation. The promotion of blood coagulation is believed to be due to the activation of factor X in serum by a venom enzyme (MacFarlane, 1961). Factor X can also be activated by compounds other than snake venoms, such as trypsin, cathepsin c, or papain.

The activator of factor X is not restricted to the venom of *Vipera russellii*, but can also be found in other venoms such as those of *Bothrops atrox* and *B. jararaca* (Nahas et al., 1964). These venoms contain two coagulation factors: factor X activator and thrombinlike enzyme (Denson and Rousseau, 1970). Factor X activator is also present in the venoms of *Crotalus viridis helleri* and *Echis carinatus*. Venoms that do not contain factor X are those of *Agkistrodon contortrix mokeson, A. bilineatus, A. contortrix laticinctus, A. halys halys, A. halys blomhoffii, Crotalus durissus durissus, C. adamanteus, C. atrox, C. scutulatus scutulatus, Sistrurus miliarus barbouri, Trimeresurus flavoviridis*, and *Naja naja* (Denson et al., 1972).

Factor X activator was partially purified by Schiffman et al. (1969) from the venom of *Vipera russellii*. The enzyme has a molecular weight of 145,000 daltons and is calcium-dependent. Factor X activator was separated from factor V (Ac-globulin) activator. It does not interact with factor VIII (antihemophilic factor). Tosylarginine esterase activity was separated from the factor X-activating enzyme in the venom of *V. russellii* (Jackson et al., 1971). The factor X-activating enzyme is not inhibited by diisopropyl phosphorofluoridate or phenyl methyl sulfonyl fluoride, whereas the ester hydrolase is inhibited by both of these reagents. This indicates that the activation enzyme and the ester hydrolase activities are catalyzed by separate proteins.

Radcliffe and Barton (1972) activated factor X with the venom of *V. russellii* and made a physicobiochemical study of it. Activated factor X has a molecular weight of 48,000 and consists of polypeptide chains (20,000 and 30,000 daltons) that undergo reversible association–dissociation in aqueous solution. The larger component incorporated ^{32}P after inhibition of the enzymatic activity with [^{32}P]diisopropyl phosphofluoridate. It thus appears that activated factor X is a serine enzyme.

4 Effect on Factor X (Stewart Factor)

The mechanism of activation of bovine factor X by the activator from *V. russellii* venom was described by Fujikawa et al. (1972). A specific peptide bond is cleaved during the activation, resulting in the release of a glycopeptide (mol. wt. 11,000) from the amino terminal end of the heavy chain. The activation of factor X by trypsin also involves the removal of the same activation peptide as is split by the venom of *V. russellii* (Fig. 2).

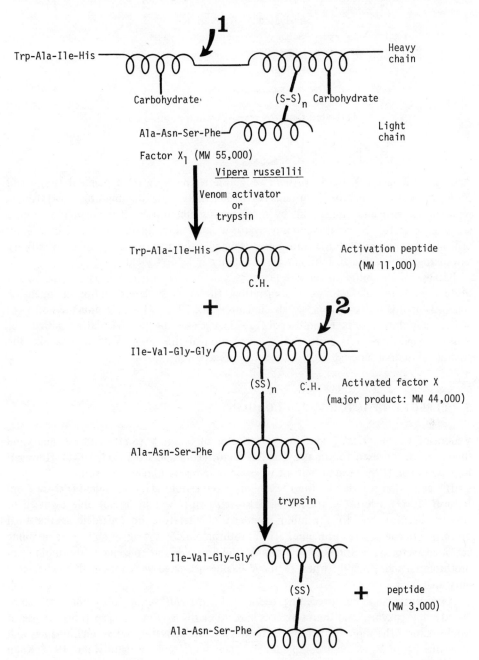

Figure 21.2 Mechanism of factor X activation by *Vipera russellii* venom activator or trypsin. (Based on Fujikawa et al., 1974.)

Once factor X is activated, it is inhibited by relatively high concentrations of diisopropyl phosphofluoride. The results suggest that factor X is converted to a specific serine protease that promotes blood coagulation.

Furie et al. (1974) observed that the activation peptide (mol. wt. 11,000) can form a complex with activated factor X. This relation is summarized in Scheme 25.

$$\text{Factor X} \longrightarrow \underset{\substack{\text{(mol. wt.} \\ 44{,}000)}}{\text{Activated factor X}} + \underset{\substack{\text{(mol. wt.} \\ 11{,}000)}}{\text{Activation peptide}}$$

$$\downarrow$$

$$\underset{(\text{mol. wt. } 55{,}000)}{\text{Activated factor X} \longrightarrow \text{Activated peptide complex}}$$

Scheme 25

Activation of factor X by *V. russellii* venom activator is Ca(II) dependent (Furie and Furie, 1974, 1975). Factor X contains two high-affinity metal binding sites. Ca(II) can be replaced by lanthanide ions; and, by marking this substitution, it is possible to form a complex of factor X–metal–venom activator without forming an activated factor X. By utilizing this property, Furie and Furie isolated venom activator protein by affinity chromatography with a 10-fold increase in activity and a 40% yield.

Because of its ability to activate factor X, venom of *V. russellii* is used clinically for the determination of factor X concentration. The factor X determination in which the venom is used is a two-stage assay (Bachmann et al., 1957). However, this method gives artifactually low results (Woolf et al., 1973) because the procoagulation activity of Russell's viper venom is destroyed by incubation of the venom with plasma plus the postmitochondrial supernatant from tissue homogenates.

5   EFFECT ON FACTOR V (Ac-GLOBULIN)

Venom of *Vipera russellii* contains an activator of factor V (Smith and Hanahan) and therefore can be used for an assay of this factor (Pool and Robinson, 1971). However, Rapaport et al. (1966) found that use of crude venom gave unreliable results.

The activation reaction by the *V. russellii* venom is not Ca(II) dependent (Prentice and Ratnoff, 1969). Factor V, treated with Russell's viper venom, is not able by itself to convert prothrombin to thrombin, and cannot therefore be regarded as the final prothrombin-converting principle. The prothrombin-converting principle, containing activated factor X, factor V, phospholipids, and Ca(II), is able to generate thrombin from prothrombin more readily when factor V has been treated with venom than when it is untreated.

As discussed in the preceding section, *V. russellii* venom also contains factor X-activating enzyme, and therefore it is hard to clarify any mechanism from the use of whole venom. The venom activating factor has a direct activating action on human as well as bovine factor V, which accompanies a change in molecular weight (Kahn, 1972; Khan and Hemker, 1972). Unactivated human factor V has a molecular weight of 410,000;

after activation the molecular weight is 110,000. This suggests that the dissociation of a tetramer occurs. For bovine factor V, the molecular weight is 400,000, which changes to 195,000 on activation; presumably a dimer instead of a monomer is produced in the case of bovine factor V activation. This was confirmed by Hanahan et al. (1972), who also observed a decrease in molecular weight from 400,000 to 205,000 when bovine factor V was activated by venom activator. The venom activator has an arginine esterase activity and can be inhibited by diisopropophosphorofluoridate (Esmon and Jackson, 1973). The enzyme accounts for one-half of the toal coagulant activity of crude Russell's viper venom. Factor V activator neither clots fibrinogen nor catalyzes the proteolysis of prothrombin. The coagulation activity of *Vipera aspis* is attributed to the specific action of factor V as well as factor X (Boffa et al., 1970).

6 EFFECT ON FACTOR III (THROMBOPLASTIN)

Factor III is a blood coagulation factor found in the extrinsic system. Any venom component, either activating or inhibiting, affecting factor III has a pronounced effect on blood coagulation. It is well known that the venom of *Vipera russellii* is procoagulant (see Section 3.4 of this chapter). Banerjee et al. (1956) noticed that addition of Ca(II) cannot induce clotting when the venom is first incubated with oxalated plasma. They considered that the venom contains antithromboplastin factor, which destroys the activity of factor III. Antithromboplastin action of action of venoms of *Agkistrodon acutus, Trimeresurus mucrosquamatus*, and *Naja naja atra* was reported by Ouyang (1957a, b). The antithromboplastin action of the last two venoms is partly responsible for their anticoagulant nature (Lee and Ouyang, 1958). *Agristodon acutus* venom, on the other hand, exhibits coagulant action at high venom concentration but is anticoagulant at low concentration. The anticoagulant action of *Naja naja* venom is also attributed to antithromboplastic activity (Kruse and Dam, 1950).

Whereas the venoms mentioned are antithromboplastic, the venom of *Cerastes cerastes* is thromboplasticlike, a property partly responsible for the coagulant nature of the venom (Mohamed et al., 1969). Venoms of *Notechis scutatus* and *Micrurus* spp. also have thromboplastinlike activity (Eagle, 1937).

MacKay et al. (1969) investigated the anticoagulative activities of three cobra venoms of *Naja melanoleuca, N. nigricollis*, and *Ophiophagus hannah*. The venoms interfered with thromboplastin formation and prolonged the thrombin clotting time. They did not interfere directly with the conversion of fibrinogen to fibrin under the action of thrombin, but increased the one-stage prothrombin time. The venoms did not appear to interfere with the activation of plasminogen by streptokinase. Similarly, the anticoagulation action of *Dendroaspis angusticeps* venom is attributed to the destruction of thromboplastin or the inhibition of thromboplastin formation (MacKay et al., 1966).

Anticoagulant fraction isolated from *Naja oxiana* venom inactivates factor IX (Christmas factor) and also acts as an antithromboplastin (Barkagen et al., 1967). The venom also suppresses the activities of factors V, VII, VIII, XI, and XII, in addition to lowering thromboplastic activity (Omarov, 1969).

Venom of *Trimeresurus okinavensis* has both coagulant and anticoagulant actions. The anticoagulant action originates from the repression of thromboplastin formation (Miyoshi, 1969).

7 EFFECT ON FACTOR II (PROTHROMBIN)

7.1 Activation

Venom of *Echis carinatus* contains prothrombin-activating principle which converts prothrombin into an active enzyme, thrombin (Kornalike, 1963; Schieck et al., 1972a). However, the thrombin activity achieved is lower than that obtained with tissue thromboplastin (Schieck et al., 1972a). When prothrombin-activating principle is injected into rats, the thrombin time is prolonged for several days. The incoagulability of blood is due to consumption of prothrombin by the venom prothrombin-activating principle. This agrees very well with the same results obtained for rabbits and monkeys (Zimmerman et al., 1971).

The prothrombin-activating principle can be separated from the fractions of hemorrhagic, caseinolytic, and fibrinogenolytic activities (Schieck et al., 1972a, b). The prothrombin-activating principle is fairly toxic and has LD_{50} values of 2.5 mg kg^{-1} by subcutaneous injection and 0.25 to 0.5 mg kg^{-1} by intravenous injection. The LD_{50} of whole venom is about 5 mg kg^{-1} by s.c. injection and 0.6 to 1.2 mg kg^{-1} by i.v. injection. Prothrombin-activating principle isolated from *E. carinatus* venom has a molecular weight of 86,000 (Kornalik et al., 1969).

Echis carinatus and *Oxyuranus scutellatus* venoms convert prothrombin to thrombin. By utilizing this property, Matsuoka et al. (1975), Denson et al. (1971), and Jimenez (1975) devised a new method for prothrombin assay in human blood using these venoms. They justified the use of *E. carinatus* venom for prothrombin (factor II) assay on the grounds that the venom can affect only factor II and does not affect other factors such as V, VII, VIII, IX, and X.

Activation of prothrombin with *Oxyuranus scutellatus* venom is not inhibited by diisopropyl fluorophosphate or soybean trypsin inhibitor, indicating that the activation enzyme is neither estrolytic nor trypsinlike (Owen and Jackson, 1973). Thrombin formed from the activation of prothrombin with *O. scutellatus* venom has multiple components (Gorman and Castaldi, 1974). The isoelectric points of these thrombin components are 7.1, 6.7, 5.6, and 6.2. The component with a p*I* of 7.1 shows coagulant and esterase activities; the one with a p*I* of 6.7 has coagulant but no esterase activity. The components with p*I*s of 5.6 and 6.2 are minor.

Notechis scutatus scutatus venom is coagulant since it activates prothrombin without affecting factor X (Eagle, 1937; Jobin and Esnouff, 1966). Therefore this venom can be used for prothrombin assay (Pirkle et al., 1972). *Naja nigricollis* venom also has prothrombinase activity (Meaume et al., 1966).

7.2 Inhibition

Venoms of *Vipera aspis* and *V. berus* are anticoagulant because they inhibit the formation of prothrombinase by forming a complex with prothrombinase that blocks the active sites of phospholipids in platelets (Boffa and Boffa, 1973; Boffa et al., 1972a, b).

8 EFFECT ON FACTOR VIII (ANTIHEMOPHILIC FACTOR)

Cases of *Bitis arietans* envenomation in human beings indicate that factor VIII levels increase significantly for the first few days (Phillips et al., 1973). Factor VIII is an antihemophilic factor present in the middle phases of the intrinsic blood coagulation system. Together with Ca(II), factor IX, and phospholipid, factor VIII activates factor X.

It is important clinically to have factor VIII free from fibrinogen for the treatment of hemophiliacs. Since *Agkistrodon rhodostoma* venom causes defibrination without affecting other blood coagulation factors, Rizza et al. (1965) used this property for the isolation of factor VIII. Green (1971) employed the same principle as Rizza et al. but used purified Arvin for the isolation of factor VIII. He was able to increase the activity 8665-fold, as compared to that of crude venom. Purified factor VIII is a conjugated protein containing 76% protein, 10% carbohydrates, and 11% lipids (Hershgold et al., 1971a). The phospholipid moiety in factor VIII is required for procoagulant activity, as treatment with phospholipase C causes the loss of the biological activity (Hershgold et al., 1971b).

9 EFFECT ON PLATELETS

Snake venoms also exert effects on platelets; in particular, they cause platelet aggregation. The venom of *Bitis nasiconris* increases platelet adhesiveness (MacKay et al., 1970). The same phenomenon was observed for *B. arietans* venom by Brink and Steytler (1974); the venom causes aggregation of platelets in the baboon (*Papio ursinus*) and in human beings. They reported that normal human platelets are extremely susceptible to *B. arietans* venoms and that the aggregation of platelets with small dosages of venom is irreversible.

In general, injury to vessels is usually followed promptly by aggregation of platelets at the point of damage and by the subsequent formation of fibrin. The formation of a hemostatic plug prevents further leakage of blood from the vessel. Injury to capillaries by *Crotalus atrox* venom is also followed by aggregation of platelets and hemostasis, as demonstrated in Fig. 3 (Ownby et al., 1974). The figure indicates that the lumen of the capillary is completely occluded by aggregates of platelets.

Adenosine-5'-diphosphate (ADP) has also been shown to induce platelet aggregation (Gaarder et al., 1961). The source of ADP for aggregation can be injured endothelial cells, erythrocytes, or the platelets themselves, and the aggregation of platelets can lead to thrombosis. The nucleotide-releasing component is different from esterolytic and coagulant fractions. In *Trimeresurus okinavensis* venom, the nucleotide-releasing activity was separated from coagulant, esterolytic, and proteolytic components (Davey and Luescher, 1967). Diisopropophosphorofluoridate inhibits the esterase and coagulative actions of the venoms of *Crotalus durissus terrificus, Trimeresurus okinavensis*, and *T. purpureomaculata*, but the compound does not reduce the platelet nucleotide-releasing activity. Boffa and Boffa (1974) and Boffa et al. (1972a, b, c) found that a component of *Vipera aspis* venom with high ADPase or 5'-nucleotidase activity had potent inhibitory action on platelet aggregation. This confirms the fact that the intact ADP molecule is responsible for platelet aggregation.

Aggregation of platelets is not restricted to snake venoms; bacterial endotoxins can also initiate it (MacKay et al., 1966; MacKay, 1973). Platelet aggregation is also known to be associated with various clinical vascular disorders, such as thrombosis, atherosclerosis, and thrombotic thrombocytopenic purpura, as well as experimentally induced reactions such as the Arthus or Shwartzman reaction (Mustard, 1967). In all of these conditions there is evidence of vascular injury.

Snake envenomation causes a decrease in platelet count. In *Crotalus ruber ruber* bite

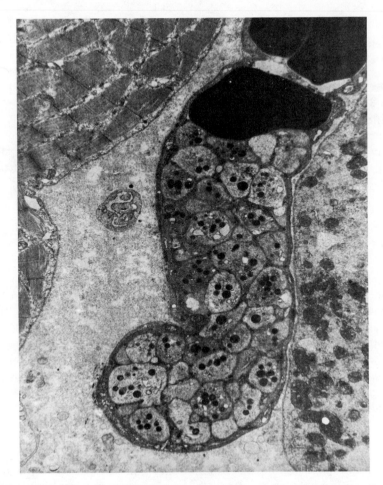

Figure 21.3 Platelet aggregation in the lumen of a capillary induced by *Crotalus atrox* venom. (Electron micrograph by Dr. Charlotte L. Ownby of author's laboratory.)

(Lyons, 1971) and in puff adder (*Bitis arietans*) bite cases in human beings low platelet counts were observed (Takahashi and Tu, 1970; Phillips et al., 1973). The results suggest that venom has a direct destructive effect on platelets *in vivo*. This view was confirmed by injecting the venom into rats, which subsequently showed a decrease in platelet count.

The component responsible for the aggregation of blood platelets was isolated from the venom of *Trimeresurus okinavensis* (Davey and Esnouf, 1969). The purified component is a glycoprotein without proteolytic, estrolytic, or coagulant activity.

A snake venom component, cobra venom factor, isolated from *Naja naja* venom, has an effect on platelets. Dodds and Pickering (1972) observed that cobra venom factor induced aggregation and release of intracellular constituents in platelet-rich plasma or suspensions of washed platelets from guinea pigs, rabbits, dogs, and human beings.

Snake venoms, especially the venom of *Crotalus durissus terrificus*, can liberate amines such as serotonin and histamine (Markwardt, 1967).

REFERENCES

Anderson, L. (1972a). Isolation of thrombin-like activity from the venom of *Trimeresurus okinavensis, Haemostasis*, **1**, 31.

Anderson, L. (1972b). Action and inhibition of the fibrinogen-clotting enzyme from the venom of *Trimeresurus okinavensis, Haemostasis*, **1**, 79.

Ashby, E. C., James, D. C. O., and Ellis, H. (1970). The effect of intraperitoneal Malayan pit-viper venom on adhesion formation and peritoneal healing, *Br. J. Surg.*, **57**, 863.

Ashford, A. and Bunn, D. R. G. (1970). The effect of Arvin on reticuloendothelial activity in rabbits, *Br. J. Pharmacol.*, **40**, 37.

Ashford, A., Ross, J. W., and Southgate, P. (1968). Pharmacology and toxicology of a defibrinating substance from Malayan pit viper venom, *Lancet*, **1**, 486.

Bachmann, F., Duckert, F., and Geiger, M. (1957). Differentiation of the factor VII complex: Studies on the Stuart–Prower factor, *Thromb. Diath. Haemorrh.*, **1**, 169.

Banerjee, R., Nath, B. B., Devi, A., and Saka, N. K. (1956). "The role of calcium and divalent metals in the coagulation of blood as induced by Russell's viper venom," in E. E. Buckley and N. Porges, Eds., *Venoms*, American Association for the Advancement of Science, Washington, D. C. pp. 118–124.

Barboriak, J. J. (1971). The effect of Arvin on plasma lipids of the rabbit, *Proc. Soc. Exp. Biol. Med.*, **136**, 313.

Barkagan, Z. S., Val'tseva, I. A., Mitel'man, L. S., Talyzin, F. F., and Yagodkin, S. I. (1967). Anticoagulatory and toxic properties of different fractions of Middle Asian cobra poison (new biogenetic anticoagulant), *Izv. Akad. Nauk SSSR, Ser. Biol.*, **116**.

Barlow, G. H., Holleman, W. H., and Lorand, L. (1970). The action of Arvin on fibrin stabilizing factor (factor XIII), *Res. Commun. Chem. Pathol. Pharamcol.*, **1**, 39.

Barlow, G. H., Lewis, L. J., Finley, R., Martin, D., and Stocker, K. (1973). Immunochemical identification of Ancrod, *Thromb. Res.*, **2**, 17.

Beaumont, K. P., Phillips, G. O., and Power, D. M. (1972). The effects of chemical inhibitors and ionizing radiation on Arvin activity, *Int. J. Radiat. Biol.*, **22**, 137.

Bell, W. R. and Regoeczi, E. (1970). Isotopic studies of therapeutic anticoagulation with a coagulating enzyme, *J. Clin. Invest.*, **49**, 1872.

Bell, W. R., Bolton, G., and Pitney, W. R. (1968a). The effect of arvin on blood coagulation factors, *Br. J. Haematol.*, **15**, 589.

Bell, W. R., Pitney, W. R., Oakley, C. M., and Goodwin, J. F. (1968b). Therapeutic defibrination in the treatment of thrombotic disease, *Lancet*, **1**, 490.

Blömback, B. (1958). Studies on the action of thrombin enzyme on bovine fibrinogen as measured by N-terminal analysis, *Ark. Kemi*, **12**, 321.

Boffa, M. C. and Boffa, G. A. (1971). Identification et séparation de différents facteurs du venin de *Vipera aspis* actifs en hémostase, *C. R. Soc. Biol.*, **165**, 2287.

Boffa, G. A. and Boffa, M. C. (1973). Chromatographie analytique et caractérisation immunochimique de la protéine anticoagulante du venin de *Vipera aspis* (inhibiteur d'aspis), *C. R. Soc. Bio.*, **167**, 654.

Boffa, M. C. and Boffa, G. A. (1974). Correlations between the enzymatic activities and the factors active on blood coagulation and platelet aggregation from the venom of *Vipera aspis, Biochim. Biophys. Acta*, **354**, 275.

Boffa, M. C., Boffa, G. A., and Josso, F. (1970). Effets du venin de *Vipera aspis* sur les facteurs plasmatiques de l'hemostase, *C. R. Acad. Sci. (Paris)*, Ser. D, **270**, 1284.

Boffa, M. C., Josso, F., and Boffa, G. A. (1972a). Action of *Vipera aspis* venom on blood clotting factors and platelets, *Thromb. Diath. Haemorrh.*, **27**, 8.

Boffa, M. C., Delori, P., and Soulier, J. P. (1972b). Anticoagulant factors from Viperidae venoms: Platelet phospholipid inhibitors, *Thromb. Diath. Haemorrh.*, **28**, 509.

Bonilla, C. A. (1975). Defibrinating enzyme from timber rattlesnake (*Crotalus horridus horridus*) venom: A potential agent for therapeutic defibrination. I. Purification and properties, *Thromb. Res.*, **6**, 151.

Bonilla, C. A., Diclementi, D., and MacCarter, D. J. (1975). Hemodynamic effects of slow and rapid defibrination with defibizyme, the thrombin-like enzyme from venom of the timber rattlesnake, *Am. Heart J.,* **90,** 43.

Bowell, R. E., Marmion, V. J., and McCarthy, C. F. (1970). Treatment of central retinal vein thrombosis with Ancrod, *Lancet,* **1,** 173.

Brink, S., and Steytler, J. G. (1974). Effects of puff-adder venom on coagulation, fibrinolysis and platelet aggregation in the baboon, *S. A. Med. Tydskr.,* **48,** 1205.

Brown, C. H., III, Bell, W. R., Shreiner, D. P., and Jackson, D. P. (1972). Effects of Arvin on blood platelets: *In vitro* and *in vivo* studies, *J. Lab. Clin. Med.,* **79,** 758.

Chan, K. E. (1969). Comparison of the antithrombotic action of the thrombinlike fraction of Malayan pit viper venom and heparin, *Cardiovasc. Res.,* **3,** 171.

Chan, K. E., Rizza, C. R., and Henderson, M. P. (1965). A study of the coagulant properties of Malayan pit-viper venom, *Br. J. Haematol.,* **11,** 646.

Cheng, H. C. and Ouyang, C. (1967). Isolation of coagulant and anticoagulant principles from the venom of *Agkistrodon acutus, Toxicon,* **4,** 235.

Clegg, J. B. and Bailey, K. (1962). The separation and isolation of the peptide chains of fibrin, *Biochim. Biophys. Acta,* **63,** 525.

Collins, J. P. (1972). Conformational changes of IRC-50 Arvin at high pH, *Biochem. J.,* **128,** 132p.

Collins, J. P. and Jones, J. G. (1972). Studies on the active site of IRC-50 Arvin, the purified coagulant enzyme from *Agkistrodon rhodostoma* venom, *Eur. J. Biochem.,* **26,** 510.

Collins, J. P., Basford, J. M., and Jones, J. G. (1971). Some catalytic properties of IRC-50 Arvin, *Biochem. J.,* **125,** 71.

Damus, P. S., Markland, F. S., Jr., Davidson, T. M., and Shanley, J. D. (1972). Purified procoagulant enzyme from the venom of the eastern diamondback rattlesnake (*Crotalus adamanteus*): *In vivo* and *in vitro* studies, *J. Lab. Clin. Med.,* **79,** 906.

Davey, M. G. and Esnouf, M. P. (1969). The isolation of a component of the venom of *Trimeresurus okinavenis* that causes the aggregation of blood platelets, *Biochem. J.,* **111,** 733.

Davey, M. G. and Luescher, E. F. (1967). Actions of thrombin and other coagulant and proteolytic enzymes on blood platelets, *Nature,* **216,** 857.

Denson, K. W. E. and Rousseau, W. E. (1970). Separation of the coagulant components of *Bothrops jararaca* venom, *Toxicon,* **8,** 15.

Denson, K. W. E., Borett, R., and Biggs, R. (1971). The specific assay of prothrombin using the Taipan snake venom, *Br. J. Haematol.,* **21,** 219.

Denson, K. W. E., Russell, F. E., Almagro, D., and Bishop, R. C. (1972). Characterization of the coagulant activity of some snake venoms, *Toxicon,* **10,** 557.

De Vries, A., Rechnic, Y., Moroz, C., and Moav, B. (1963). Prevention of *Echis colorata* venom-induced afibrinogenemia by heparin, *Toxicon,* **1,** 241.

Dodds, W. J. and Pickering, R. J. (1972). The effect of cobra venom factor on hemostasis in guinea pigs, *Blood,* **40,** 400.

Donner, L. and Houskova, J. (1967). Effect of bothropase, snake venom extract from *Bothrops jararaca,* on coagulation of blood, *Vnitr. Lek.,* **13,** 647.

Eagle, H. (1937). Coagulation of the blood by snake venoms and its physiologic significance, *J. Exp. Med.,* **65,** 613.

Egberg, N. and Nordström, S. (1969). *In vivo* effect of reptilase on fibrinogen metabolism in dogs, *Scand. J. Clin. Lab. Invest.,* **24,** 383.

Egberg, N. and Nordström, S. (1970). Effects of reptilase-induced intravascular coagulation in dogs, *Acta Physiol. Acand.,* **79,** 493

Esmon, C. T. and Jackson, C. M. (1973). Factor V activating enzyme of Russell's viper venom, *Thromb. Res.,* **2,** 509.

Esnouf, M. P. and Tunnah, G. W. (1967). The isolation and properties of the thrombin-like activity from *Agkistrodon rhodostoma* venom, *Br. J. Haematol.,* **13,** 581.

Ewart, M. R., Hatton, M. W. C., Basford, J. M., and Dodgson, K. S. (1970). The proteolytic action of Arvin on human fibrinogen, *Biochem. J.,* **118,** 603.

References

Fichman, M. and Henriques, O. B. (1962). Further studies on the purification of the blood-clotting enzyme from the venom of *Bothrops jararaca, Arch. Biochem. Biophys.*, **98**, 95.

Folk, G. E. (1960). The amino acid sequence of peptide B of co-fibrin, *Biochim. Biophys. Acta*, **44**, 383.

Folk, J. E., Gladner, J. A., and Laki, K. (1959a). The thrombin-induced formation of co-fibrin. II. Preliminary amino acid sequence studies on peptides A and B, *J. Biol. Chem.*, **234**, 67.

Folk, J. E., Gladner, J. A., and Levin, Y. (1959b). Thrombin-induced formation of co-fibrin. III. Acid degradation studies and summary of sequential evidence on peptide A, *J. Biol. Chem.*, **234**, 2317.

Follana, R., Sampol, J., and Muratore, R. (1972). Defibrinating effect of FTH50 (thrombinomimetic fraction from reptilase) in dog, *C. R. Soc. Biol.*, **166**, 1341.

Ford, P. M., Bell, W. R., Bluestone, R., Gumpel, J. M., and Webb, F. W. S. (1970). The effect of Arvin on experimental immune arthritis in rabbits, *Br. J. Exp. Pathol.*, **51**, 81.

Fujikawa, K., Legaz, M. F., and Davie, E. W. (1972). Bovine factor X_1 (Stuart factor). Mechanism of activation by a protein from Russell's viper venom, *Biochemistry*, **26**, 4892.

Funk, C., Gmür, J., Herold, R., and Straub, P. W. (1971). Reptilase -R: A new reagent in blood coagulation, *Br. J. Haematol.*, **21**, 43.

Furie, B. C. and Furie, B. (1974). Purification of proteins involved in Ca(II)-dependent protein–protein interactions (coagulant protein of Russell's viper venom), *Method Enzymol.*, **34**, 592.

Furie, B. C. and Furie, B. (1975). Interaction of lanthanide ions with bovine factor X and their use in the affinity chromatography of the venom coagulant protein of *Vipera russellii, J. Biol. Chem.*, **250**, 601.

Furie, B. C., Furie, B., Gottlieb, A. J., and Williams, W. J. (1974). Activation of bovine factor X by the venom coagulant protein of *Vipera russellii:* Complex formation of the activation fragments, *Biochim. Biophys. Acta*, **365**, 121.

Gaarder, A., Jonsen, J., Laland, S., Heilem, A., and Owren, P. A. (1961). Adenosine diphosphate in red cells as a factor in the adhesiveness of human blood platelets, *Nature*, **192**, 531.

Gaffney, P. J., Marsh, N. A., and Whaler, B. C. (1973). Coagulant enzyme from Gaboon-viper venom: Aspects of its mode of action, *Biochem. Soc. Trans.*, **1**, 1208.

Gladner, J. A., Folk, J. E., Laki, K., and Carroll, W. R. (1959). Thrombin-induced formation of co-fibrin. I. Isolation, purification, and characterization of co-fibrin, *J. Biol. Chem.*, **234**, 62.

Gorman, J. J. and Castaldi, P. A. (1974). Isolation and characterization of multiple forms of human thrombin, *Thromb. Res.*, **4**, 653.

Green, D. (1971). A simple method for the purification of factor VIII (anti-hemophilic factor) employing snake venom, *J. Lab. Clin. Med.*, **77**, 153.

Hall, G. H., Holman, H. M., and Webster, A. D. B. (1970). Anticoagulation by Ancrod for haemodialysis, *Br. Med. J.*, **4**, 591.

Hanahan, D. J., Rolfs, M. R., and Day, W. C. (1972). Observations on the factor V activator present in Russell's viper venom and its action on factor V, *Biochim. Biophys. Acta*, **286**, 205.

Harmison, C. R., Landaburu, R. H., and Seegers, W. H. (1961). Some physicochemical properties of bovine thrombin, *J. Biol. Chem.*, **236**, 1693.

Hatton, M. W. C. (1973). Studies on the coagulant enzyme from *Agkistrodon rhodostoma* venom: Isolation and some properties of the enzyme, *Biochem. J.*, **131**, 799.

Haunt, C. J. (1938). Contribution à l'étude des effets du venin de *Naja tripudians* (cobra) sur la coagulation sanguine *in vitro, Arch. Int. Physiol.*, **47**, 345.

Hawkey, C. and Symons, C. (1966). Coagulation of primate blood by Russell's viper venom, *Nature*, **210**, 141.

Henriques, O. B., Mandelbaum, F. R., and Henriques, S. B. (1959). Blood-clotting activity of the venom of *Bothrops jararaca, Nature*, **183**, 114.

Hershgold, E. J., Davidson, A. M., and Janszen, M. E. (1971a). Isolation and some chemical properties of human factor VIII (antihemophilic factor), *J. Lab. Clin. Med.*, **77**, 185.

Hershgold, E. J., Davidson, A. M., and Janszen, M. E. (1971b). Human factor VIII (antihemophilic factor): Activation and inactivation by phospholipases, *J. Lab. Clin. Med.*, **77**, 206.

Hessel, B. and Blombäck, M. (1971). The proteolytic action of the snake venom enzymes Arvin and reptilase on N-terminal chain-fragments of human fibrinogen, *FEBS Lett.*, **18**, 318.

Holleman, W. H. and Coen, L. J. (1970). Characterization of peptides released from human fibrinogen by Arvin, *Biochim. Biophys. Acta,* **200**, 587.

Holleman, W. H. and Weiss, L. J. (1976). The thrombin-like enzyme from *Bothrops atrox* snake venom: Properties of the enzyme purified by affinity chromatography of *p*-aminobenzamidine substitute agarose, *J. Biol. Chem.,* **251**, 1663.

Holt, P. J. L., Holloway, V., Raghupati, N., and Calnan, J. S. (1969). The effect of a fibrinolytic agent (Arvin) on wound healing and collagen formation, *Clin. Sci.,* **38**, 9.

Jackson, C. M., Gordon, J. G., and Hanahan, D. J. (1971). Separation of the tosyl arginine esterase activity from the factor X activating enzyme of Russell's viper venom, *Biochim. Biophys. Acta,* **252**, 255.

Jerushalmy, Z. (1971). Afibrinogemia induced by snake venoms, *Kupat-Holim Yearb.,* **1**, 52.

Jimenez, R. (1975). One-stage prothrombin assay using Taipan venom, *Bol. Med. Hosp. Infant. Mex.,* **32**, 259.

Jobin, F. and Esnouf, M. P. (1966). Coagulant activity of tiger snake (*Notechis scutatus scutatus*) venom, *Nature,* **211**, 873.

Kahn, M. J. P. (1972). Activation and modification of the molecular weight of factor V (prothrombin A) under the influence of the Russell viper, *Arch. Int. Physiol. Biochem.,* **80**, 573.

Khan, M. J. P. and Hemker, H. C. (1972). Blood coagulation factor V. V. Changes of molecular weight accompanying activation of factor V by thrombin and the procoagulant protein of Russell's viper venom, *Thromb. Diath. Haemorrh.,* **27**, 25.

Klein, M. D., Bell, W., Nejad, N., and Lown, B. (1969). The effect of Arvin upon cardiac function, *Proc. Soc. Exp. Biol. Med.,* **132**, 1123.

Kopéc, M. Z., Latao, Z. S., Stahl, M., and Wegrznowicz, Z. (1969). The effect of proteolytic enzymes on fibrin stabilizing factor, *Biochim. Biophys. Acta,* **181**, 437.

Kornalik, F. (1963). The influence of *Echis carinatus* toxin on blood coagulation *in vitro, Folia Haematol.,* **80**, 73.

Kornalik, F. (1966). Influence of snake venoms on fibrinogen conversion and fibrinolysis, *Mem. Inst. Butantan,* **33**, 179.

Kornalik, F. (1971). Fibrinolytic proteases from snake venoms, *Folia Haematol.,* **95**, 193.

Kornalik, F. and Pudlak, P. (1971). A prolonged defibrination caused by *Echis carinatus* venom, *Life Sci.,* **10**, 309.

Kornalik, F. and Stybloya, Z. (1967). Fibrinolytic proteases in snake venoms, *Experientia,* **23**, 999.

Kornalik, F., Schieck, A., and Habermann, E. (1969). Isolation, biochemical and pharmacological characterization of a prothrombin-activating principle from *Echis carinatus* venom, *Naunyn-Schmiedebergs Arch. Exp. Pathol. Pharmakol.,* **264**, 259.

Kruse, I. and Dam, H. (1950). Inactivation of thromboplastin by cobra venom, *Biochim. Biophys. Acta,* **5**, 268.

Krylova, E. S. (1970). Isolation of the coagulating factor from venom on DEAE cellulose, *Vop. Med. Khim., Biokhim. Gorm. Deistviya Fiziol. Aktiv. Veshchestv. Radiats.,* **73**.

Kwaan, H. C. and Barlow, G. H. (1971). The mechanism of action of a coagulant fraction of Malayan pit viper venom, Arvin, and of reptilase, *Thromb. Diath. Haemorrh.,* Suppl. **45**, 63.

Lee, C. Y. and Ouyang, C. (1958). Mechanism of anticoagulant action of snake venoms: A comparison of the effects of the venoms of *Naja naja atra* (cobra) and *Trimeresurus mucrosqumatus* (habu), *Proc. VIIth Int. Congr. Hematol.,* **2**, 1130.

Lewis, L. J., Martin, D. L., Buckner, S., Finley, R., Lazer, L., and Fedor, E. J. (1971). Studies on type specific immunity to the whole venom and a fraction of *Agkistrodon rhodostoma, Res. Commun. Chem. Pathol. Pharmacol.,* **2**, 649.

Lyons, W. J. (1971). Profound thrombocytopenia associated with *Crotalus ruber ruber* envenomation: A clinical case, *Toxicon,* **9**, 237.

MacFarlane, R. C. (1961). The coagulant action of Russell's viper venom, the use of antivenom in defining its reaction with a serum factor, *Br. J. Haematol.,* **7**, 496.

MacFarlane, R. C. and Barnett, B. (1934). Hemostatic possibilities of snake venom, *Lancet,* **2**, 985.

MacKay, N., Ferguson, J. C., and McNicol, G. P. (1966). The effects of the venoms of the East African green mamba (*Dendroaspis anguticeps*) on blood coagulation and platelet aggregation, *East Afr. Med. J.,* **43,** 454.

MacKay, N., Ferguson, J. C., and McNicol, G. P. (1968). Effects of three mamba venoms on the haemostatic mechanism, *Br. J. Haematol.,* **15,** 549.

MacKay, N., Ferguson, J. C., and McNicol, G. P. (1969). Effects of three cobra venoms on blood coagulation, platelet aggregation, and fibrinolysis, *J. Clin. Pathol.,* **22,** 304.

MacKay, N., Ferguson, J. C., and McNicol, G. P. (1970). Effects of the venom of the rhinoceros horned viper (*Bitis nasicornis*) on blood coagulation, platelet aggregation, and fibrinolysis, *J. Clin. Pathol.,* **23,** 789.

Mann, J. R., Breeze, G. R., Deeble, T. J., and Stuart, J. (1972). Ancrod in sickle-cell crisis, *Lancet,* **1,** 934.

Markland, F. S. and Damus, P. S. (1971). Purification and properties of thrombin-like enzyme from the venom of *Crotalus adamanteus* (eastern diamondback rattlesnake), *J. Biol. Chem.,* **246,** 6460.

Markland, F. S., Damus, P. S., Davidson, T. M., and Shanley, J. D. (1970). Thrombin enzyme from *Crotalus adamanteus, Lancet,* **1,** 1398.

Markwardt, F. (1967). Studies on the release of biogenic amines from blood platelets, *Biochem. Blood Platelets, Colloq.,* p. 105.

Marsh, N. A. and Whaler, B. C. (1974). Separation and partial characterization of a coagulant enzyme from *Bitis gabonica* venom, *Br. J. Haematol.,* **26,** 295.

Marsten, J. L., Chan, K. E., Ankeney, J. L., and Botti, R. E. (1966). Antithrombotic effect of Malayan pit viper venom on experimental thrombosis of the inferior vena cava produced by a new method, *Circ. Res.,* **19,** 514.

Matsuoka, M., Sakuragawa, N., and Takahashi, K. (1975). New method for a prothrombin assay using *Echis carinatus* venom, *Saishin Igaku (Mod. Med.),* **30,** 1253.

Mattock, P. and Esnouf, M. P. (1971). Differences in the subunit structure of human fibrin formed by the action of Arvin, reptilase and thrombin, *Nature New Biol.,* **233,** 277.

McKay, D. G., Margaretten, W., and Csavassy, I. (1966). An electron microscopic study of the effects of bacterial endotoxin on the blood vascular system, *Lab. Invest.,* **15,** 1815.

McKay, J. F. (1973). Vessel wall and thrombogenesis: Endotoxin, *Thromb. Diath. Haemorrh.,* **29,** 11.

Meaume, J., Izard, Y., and Boquet, P. (1966). A coagulating activity in *Naja nigricollis* venom, *C. R. Acad. Sci. Paris,* Ser. D, **262,** 1650.

Miyoshi, M. (1969). Influences of Okinawan pit viper venom on the mechanism of human blood coagulation, *Nippon Naika Gakkai Zasshi,* **58,** 17.

Moav, B., Moroz, C., and De Vries, A. (1963). Activation of the fibrinolytic system of the guinea pig following inoculation of *Echis colorata* venom, *Toxicon,* **1,** 109.

Moberg, A. W., Shons, A. R., Gewurz, H., Mozes, M., and Najarian, J. S. (1971). Prolongation of renal xenografts by the simultaneous sequestration of preformed antibody, inhibition of complement, coagulation and antibody synthesis, *Transplant. Proc.,* **3,** 538.

Mohamed, A. H. and El-Damarawy, N. A. (1974). Role of the fibrinolytic enzyme system in the hemostatic defects following snake envenomation, *Toxicon,* **12,** 467.

Mohamed, A. H. and Hanna, M. M. (1973). The *in vivo* anticoagulant effects of *Naja flava* venom, *Toxicon,* **11,** 419.

Mohamed, A. H., El-Serougi, M., and Khaled, L. Z. (1969). Effects of *Cerastes cerastes* venom on blood coagulation mechanism, *Toxicon,* **7,** 181.

Mustard, J. F. (1967). Recent advances in molecular pathology: A review of platelet aggregation, vascular injury and atherosclerosis, *Exp. Mol. Pathol.,* **7,** 366.

Myers, P., Rampling, M. W., and Sirs, J. A. (1973). Interaction of Arvin with erythrocyte flexibility, *J. Physiol. (London),* **230,** 51.

Nahas, L., Denson, K. W. E., and MacFarlane, R. G. (1964). A study of the coagulant action of eight snake venoms, *Thromb. Diath. Haemorr.,* **12,** 355.

Naish, P., Penn, G. B., Evans, D. J., and Peters, D. K. (1972). The effect of defibrination on nephrotoxic serum nephritis in rabbits, *Clin. Sci.,* **42,** 643.

Olsen, E. G. J. and Pitney, W. R. (1969). The effect of Arvin on experimental pulmonary embolism in the rabbit, *Br. J. Haematol.,* **17,** 525.

Omarov, S. M. (1969). Anticoagulant properties of bee venom and cobra venom, *Biol. Nauk.,* **41.**

Ouyang, C. (1957). The effects of Formosan snake venoms on blood coagulation *in vitro, J. Formosan Med. Assoc.,* **56,** 435.

Ouyang, C. and Hong, J. S. (1974). Inhibition of the thrombinlike principle of *Agkistrodon acutus* venom by group-specific enzyme inhibitors, *Toxicon,* **12,** 449.

Ouyang, C. and Huang, T. F. (1976). Purification and characterization of the fibrinolytic principle of *Agkistrodon acutus* venom, *Biochim. Biophys. Acta,* **439,** 146.

Ouyang, C. and Teng, C. M. (1972). Purification and properties of the anticoagulant principle of *Agkistrodon acutus* venom, *Biochim. Biophys. Acta,* **278,** 155.

Ouyang, C. and Teng, C. M. (1973). The effect of the purified anticoagulant principle of *Agkistrodon acutus* venom on blood coagulation, *Toxicon,* **11,** 287.

Ouyang, C. and Teng, C. M. (1976). Fibrinogenolytic enzymes of *Trimeresurus mucrosquamatus* venom, *Biochim. Biophys. Acta,* **420,** 298.

Ouyang, C. and Yang, F. Y. (1974). Purification and properties of the thrombin-like enzyme from *Trimeresurus gramineus, Biochim. Biophys. Acta,* **351,** 354.

Ouyang, C. and Yang, F. Y. (1975). Purification and properties of the anticoagulant principle of *Trimeresurus gramineus* venom, *Biochim. Biophys. Acta,* **386,** 479.

Ouyang, C., Hong, J. S., and Teng, C. M. (1971). Purification and properties of the thrombin-like principle of *Agkistrodon acutus* venom and its comparison with bovine thrombin, *Thromb. Diath. Haemorrh.,* **26,** 224.

Ouyang, C., Teng, C. M., and Hong, J. S. (1972). Purification and properties of the coagulant and anticoagulant principles of *Agkistrodon acutus, J. Formosan Med. Assoc.,* **71,** 401.

Owen, W. G. and Jackson, C. M. (1973). Activation of prothombin with *Oxyuranus scutellatus scutellatus* (Taipan snake) venom, *Thromb. Res.,* **3,** 705.

Ownby, C. L., Kainer, R. A., and Tu, A. T. (1974). Pathogenesis of hemorrhage induced by rattlesnake venom, *Am. J. Pathol.,* **76,** 401.

Pastorova, V. E. and Kikteva, L. K. (1969). Effect of antiplasmin on afibrinogenemia provoked through intravenous administration of *Vipera libertina* venom, *Byull. Eksp. Biol. Med.,* **68,** 42.

Penn, G. B., Ross, J. W., and Ashford, A. (1971). Effects of Arvin on pregnancy in the mouse and the rabbit, *Toxicol. Appl. Pharmacol.,* **20,** 460.

Phillips, L. L., Weiss, H. J., Pessar, L., and Christy, N. P. (1973). Effects of puff adder venom on the coagulation mechanism. I. *In vivo, Toxicon,* **11,** 423.

Pirkle, H., McIntosh, M., Theodor, I., and Vernon, S. (1972). Activation of prothrombin with Taipan snake venom, *Thromb. Res.,* **1,** 559.

Pitney, W. R. and Regoeczi, E. (1970). Inactivation of Arvin by plasma proteins, *Br. J. Haematol.,* **19,** 67.

Pitney, W. R., Oakley, C. M., and Goodwin, J. F. (1970). Therapeutic defibrination with Arvin, *Am. Heart J.,* **80,** 144.

Pitney, W. R., Raphael, M. J., Webb-Peploe, M. M., and Olsen, E. G. J. (1971). Treatment of experimental venous thrombosis with streptokinase and Ancrod (Arvin), *Br. J. Surg.,* **58,** 442.

Pizzo, S. V., Schwartz, M. L., and Hill, R. L., and McKee, P. A. (1972). Mechanism of Ancrod anticoagulation: A direct proteolytic action on fibrin, *J. Clin. Invest.,* **51,** 2841.

Pool, J. G. and Robinson, A. J. (1971). A change in Russell's viper venom (Stypven): Modification of factor V assay to compensate, *J. Lab. Clin. Med.,* **77,** 343.

Prentice, C. R. M. and Ratnoff, O. D. (1969). Action of Russell's viper venom on factor V and the prothrombin-converting principle, *Br. J. Haematol.,* **16,** 291.

Radcliffe, R. D. and Barton, P. G. (1972). The purification and properties of activated factor X: Bovine factor X activated with Russell's viper venom, *J. Biol. Chem.,* **247,** 7735.

Rådegran, K., Swedenborg, J., and Olsson, P. (1972). Effect of defibrinogenation and acetylsalicylic acid on the circulatory response to thrombin, *Acta Chir. Scand.,* **138,** 441.

Rahimtoola, S. H., Raphael, M. J., Pitney, W. R., Olsen, E. G. J., and Webb-Peploe, M. M. (1970).

Therapeutic defibrination and heparin therapy in the prevention and resolution of experimental venous thrombosis, *Circulation,* **42,** 729.

Rapaport, S. I., Hjort, P. F., and Patch, M. J. (1966). Rabbit factor V: Different effects of thrombin and venom, a source of error in assay, *Am. J. Physiol.,* **211,** 1477.

Regoeczi, E. and Bell, W. R. (1969). In vivo behaviour of the coagulant enzyme from *Agkistrodon rhodostoma* venom: Studies using iodine 131-labeled Arvin, *Br. J. Haematol.,* **16,** 573.

Regoeczi, E., Gergely, J., and MacFarlane, A. S. (1966). In vivo effects of *Agkistrodon rhodostoma* venom: Studies with fribinogen-I^{131}, *J. Clin. Invest.,* **45,** 1202.

Reid, H. A., Thean, P. C., Chan, K. E., and Baharom, A. R. (1963). Clinical effects of bites by Malayan pit viper (*Agkistrodon rhodostoma*), *Lancet,* **1,** 617.

Rizza, C. R., Chan, K. E., and Henderson, M. P. (1965). Separation of factor VIII (antihemophilic factor) activity from fibrinogen by means of a snake venom, *Nature,* **207,** 90.

Roitman, M. I. (1966). Blood coagulation activity of *Vipera lebetina turanica* venom and its neutralization by specific antiserum, *Vopr. Med. Khim.,* **12,** 477.

Rosenfeld, G. (1964). Fibrinolysis by snake venoms, *Sangre,* **9,** 352.

Rosenfeld, G. (1966). Fibrinolysis by snake venoms, *Proc. 3rd Int. Pharmacol. Meet.,* **6,** 119.

Ruiz Reyes, G., Pena Obando, C., and Tomayo, Perez, R. (1967). Effect of different substances upon experimental fibrinogenopenia with *Bothrops atrox* venom, *Acta Cient. Venez.,* **18,** 125.

Scheinthal, B. M., King, R. G., and Copley, A. L. (1970). The effect of temperature on the rigidity modulus of fibrin gels produced by *Bothrops atrox* venom, *Biochim. Biophys. Acta,* **214,** 260.

Schieck, A., Haberman, E., and Kornalik, F. (1972a). The prothrombin-activating principle from *Echis carinatus* venom, *Naunyn-Schmiedebergs Arch. Exp. Pathol. Pharmakol.,* **274,** 7.

Schieck, A., Kornalik, F., and Haberman, E. (1972b). Prothrombin-activating principle from *Echis carinatus* venom. I. Preparation and biochemical properties, *Naunyn-Schmiedebergs Arch. Exp. Pathol. Pharmakol.,* **272,** 402.

Schiffman, S., Theodor, I., and Rapaport, S. I. (1969). Separation from Russell's viper venom of one fraction reacting with factor X and another reacting with factor V, *Biochemistry,* **8,** 1397.

Seegers, W. H., McCoy, L., Kipper, R. K., and Murano, G. (1968). Preparation and properties of thrombin, *Arch. Biochem. Biophys.,* **128,** 194.

Sharp, A. A., Warren, B. A., Paxton, A. M., and Allington, M. J. (1968). Anticoagulant therapy with a purified fraction of Malayan pit viper venom, *Lancet,* **1,** 493.

Shiau, S. and Ouyang, C. (1965). Isolation of coagulant and anticoagulant principles from the venom of *Trimeresurus gramineus, Toxicon,* **2,** 213.

Shimura, K., Yokoyama, H., Soma, M., Ochiai, T., and Iida, Y. (1974). Blood clotting action of the venom of *Trimeresurus flavoviridis* and its applications, *Nippon Kokuka Gakkai Zasshi,* **23,** 398.

Silverman, S., Potter, E. V., and Kwaan, H. C. (1971). Effects of Arvin in mice: Immunofluorescent and histochemical studies, *Exp. Mol Pathol.,* **14,** 67.

Silverman, S., Bernik, M. B., Potter, E. V., and Kwaan, H. C. (1973). Effects of Ancrod (Arvin) in mice: Studies of plasma fibrinogen and fibrinolytic activity, *Br. J. Haematol.,* **24,** 101.

Smith, C. M. and Hanahan, D. J. (1976). The activation of factor V by factor Xa or α-chymotrypsin and comparison with thrombin and RVV-V action. An improved facto V isolation procedure, *Biochemistry,* **15,** 1830.

Soh, K. S., and Chan, K. E. (1974). Caseinolytic and esteratic activities of Malayan pit viper venom and its proteolytic and thrombinlike fractions, *Toxicon,* **12,** 151.

Stocker, K., Christ, W., and Leloup, P. (1974). Characterization of the venoms of various *Bothrops* species by immunoelectrophoresis and reactions with fibrinogen agarose, *Toxicon,* **12,** 415.

Takahashi, W. Y. and Tu, A. T. (1970). Puff adder snakebite, *J. Am. Med. Assoc.,* **211,** 1857.

Tu, A. T., Passey, R. B., and Tu, T. (1966). Proteolytic enzyme activities of snake venoms, *Toxicon,* **4,** 49.

Tuppie, A. G. G., Prentice, C. R. M., McNicol, G. P., and Douglas, A. S. (1971). *In-vitro* studies with Ancrod (Arvin), *J. Haematol.,* **20,** 217.

Tuppie, A. G. G., McNicol, G. P., and Douglas, A. S. (1972). Platelet electrophoresis: Effect of defibrination by Ancrod (Arvin), *Cardiovasc. Res.,* **5,** 101.

Turina, M., Schultz, L., Bull, B., and Braunwalt, N. S. (1972). Plasma fibrinogen recovery rate after administration of Malayan pit viper venom extracts in non-stressed and surgically stressed animals, *J. Surg. Res.,* **13,** 20.

Van der Ziel, C., Joison, J., Siso, H., Saravis, C., and Slapak, M. (1970). The effect of Arvin on blood-coagulation and patency of venovenous shunts in the dog, *Br. J. Surg.,* **57,** 856.

Von Kaulla, K. N. (1975). "*In vitro* enhancement by hydrazine and cobra venom factor of fibrinolytic activity induced by synthetic organic anions," in K. N. Von Kaulla and J. F. Davidson, Eds., *Synthetic Fibrinolytic Thrombolytic Agents,* Charles C Thomas, Springfield, SU., pp. 166–175.

Wik, K. O., Tangen, O., and McKenzie, F. N. (1972). Blood clotting activity of reptilase and bovine thrombin *in vitro:* A comparative study on seven different species, *Br. J. Haematol.,* **23,** 37.

Williams, W. J. and Esnouf, M. P. (1962). The fractionation of Russell's viper (*Vipera russellii*) venom with special reference to the coagulant protein, *Biochem. J.,* **84,** 52.

Woolf, I. L., Kipnes, R. S., and Babior, B. M. (1973). The determination of factor X in tissue homogenates: Evidence for the inactivation of Russell's viper venom by tissue homogenates in the presence of plasma, *J. Lab. Clin. Med.,* **81,** 77.

Zimmerman, H., Haberman, E., and Lasch, H. G. (1971). Der Einfluss von Gift der Sandrasselotter *Echis carinatus* aus die Hämostase, *Thromb. Diath. Haemorrh.,* **25,** 425.

22 Nerve Growth Factor

1	CHEMISTRY	361
2	IMMUNOLOGY	366
3	BIOLOGICAL EFFECTS	367
4	OCCURRENCE	369
	References	370

It is well known that snake venoms contain a potent nerve growth factor (NGF). This compound promotes the growth of the sympathetic chain ganglia *in vivo* and a halolike outgrowth of nerve fibers from embryonic or sensory ganglia cultured *in vitro*.

The neuron-stimulating effects of NGF were first observed experimentally in implantation of mouse sarcoma 180 into the body wall of 3-day chick embryos (Bueker, 1948a, b). The stimulating factor was suspected to be either a protein or nucleoprotein. To eliminate nucleoprotein stimulation, Cohen and Levi-Montaloini (1956) used a crude venom fraction containing exonuclease (phosphodiesterase) to inactivate the nucleic acids. The unexpected result was that the snake venom itself proved to be a potent source of NGF.

Nerve growth factor can also be found in the adult mouse submaxillary gland (Cohen, 1960), and more extensive studies have been conducted on mouse submaxillary NGF than on snake venom NGF because of the higher concentrations of NGF found in the mouse. In this chapter, however, the discussion will be largely limited to snake venom NGF.

1 CHEMISTRY

Initially, NGF was partially purified from the venom of *Agkistrodon piscivorus* (Cohen and Levi-Montaloini, 1956). It was shown to be a heat-labile and nondialyzable protein in whose biological activity is destroyed by trypsin, pepsin, and chymotrypsin treatment.

The partially purified NGF was 1000 times more potent than the purest sarcoma 180 fraction. Nerve growth factor purified to homogeneity from *A. piscivorus* venom (Cohen, 1959). Nerve growth-stimulating activity is not associated with many other enzymatic activities present in whole venom, such as phospholipase A_2, exonuclease (phosphodiesterase), 5'-nucleotidase, L-amino acid oxidase, DNase, diphosphopyridine nucleotidase, and ATPase. However, two activities, RNase and protease, are present in NGF, but these activities are not responsible for its biological action.

The NGF activity is stable in $0.1M$ NaOH ($26°C$, 1 hr), whereas the RNase and protease activities are completely destroyed under identical conditions. However, NGF can be present in aggregated or dissociated form. Nerve growth factor in the venom of *Agkistrodon contortrix laticinctus, Crotalus terrificus,* and *C. atrox* in aggregated form has a molecular range of 40,000 to 50,000. The isoelectric points and molecular weights of NGFs from various snake venoms and mouse salivary glands are summarized in Table 1.

Table 1 Isoelectric Points and Molecular Weights of NCF Obtained from Different Sources

Sources	PI	References	MW	References
Elapidae				
Naja naja	6.75	Angeletti, 1970b	6,000-40,000	Angeletti, 1968a,b; 1970 b
			28,000	Hogue-Angeletti, 1976
N. melanoleuca	8.0	Bailey et al, 1975	21,000	Bailey et al, 1975
N. nigricollis	9.0	Bailey et al, 1975	22,000	Bailey et al, 1975
Dendroaspis viridis	10.0	Bailey et al, 1975	34,000	Bailey et al, 1975
Viperidae				
Vipera ammodytes ammodytes	8.5	Bailey et al, 1975	34,000	Bailey et al, 1975
V. russellii	9.5-11	Banks et al, 1968	37,000	Pearce et al, 1972a,b
			38,000-42,000	Banks et al, 1968
Crotalidae				
Agkistrodon rhodostoma	10.5	Bailey et al, 1975	34,000	Bailey et al, 1975
Bothrops atrox			35,000	Glass and Banthorpe, 1975
Crotalus adamanteus			24,000-40,000	Angeletti et al, 1967
Mouse salivary gland (β-subunit)	9.2	Bailey et al, 1975	30,000	Bocchini and Angeletti, 1969

Nerve growth factor lacks the following enzyme activities:

protease (substrate, caesin), esterase (substrate, BAEE, ATEE, TAME, CBZ-proline ester, CBZ-leucine ester), dipeptidase, tripeptidase, prolidase, aminopeptidase, phospholipase A_2, lipase, ribonuclease, phosphomonoesterase, exonuclease (phosphodiesterase), 5'-nucleotidase, L-amino acid oxidease, hyaluronidase, acetylcholinesterase, and collagenase.

Purified NGF is also devoid of any lethal activity. From the chemical viewpoint, it is different from the neurotoxins of Elapidae and Hydrophiidae.

The ability to stimulate nerve growth is due to an intrinsic property of NGF and not to exogenous origins. Cohen (1959) investigated the effects of various hormones, such as insulin, glucagon, melanophore-stimulating hormone, follicle-stimulating hormone, corticotropin, gonadotropin, and thyroid-stimulating hormone, on nerve growth. None of these hormones stimulated nerve growth activity.

The carbohydrate content of *Vipera russellii* NGF is about 20%. It contains 4 residues of fucose, 8 of mannose, 9 of galactose, 14 of *N*-acetylglucosamine, and 2 of *N*-acetylneuraminic acid (Pearce et al., 1972a, b). The active fraction from *V. ammodytes ammodytes* NGF contains a large amount of carbohydrate (31.7%) composed of 3.2% mannose, 2.8% galactose, 13.8% *N*-acetylglucosamine, 4.4% *N*-acetylgalactosamine, and 7.5% sialic acid. The NGF fractions from the Elapidae venoms do not contain significant quantities of carbohydrate, the amounts being less than 2% (Bailey et al., 1975).

The NGF activity of *Vipera russellii* origin is destroyed upon treatment with pepsin but is stable against trypsin and chymotrypsin activity.

All of the four tryptophan residues in *Naja naja* NGF appear to be important to nerve growth-stimulating activity as well as to C'-fixation activity. All the tryptophan residues are readily oxidizable with *N*-bromosuccinimide, indicating positioning at the protein surface. In mouse submaxillary gland NGF, there are six trytophan residues, of which four are essential to biological activity (Angeletti, 1970a).

The NGF activity in *N. naja* venom shows a weight range of 20,000 to 40,000. Aggregation can produce an increase to as high as 120,000 daltons, wherease dissociation may cause a decrease to as low as 6000 daltons with full biological activity (Angeletti, 1968a, 1970b). The molecular weight of NGF from the venom of *Vipera russellii* is 37,000, with an axial ratio of 1 to 7 indicating a highly elongated molecule shape (Pearce et al., 1972a, b).

The NGFs isolated from the venom of *Crotalus adamanteus* and *Bothrops jararaca* (Angeletti, 1968a, b) also showed multiple molecular forms. This may be due to an association–dissociation equilibrium of the fundamental subunit, which is believed to be 13,000 daltons. Nerve growth factor isolated from the venom of *Bothrops atrox* has a molecular weight of 35,000, which is very similar to the value of 37,000 daltons for *Vipera russellii* NGF (Glass and Banthorpe, 1975). In SDS gel electrophoresis, *B. atrox* NGF has a molecular weight of 19,000, indicating that the 35,000 molecular weight protein is composed of subunits.

There are three types of NGF in the venom of *Vipera russellii* (Banks et al., 1968). The isoelectric points are very high, at p*I* 9.5 to 11. All of them contain identical *N*-terminal histidines. The sedimentation constants are 2.92, 2.83, and 2.88S, with molecular weights of 42,000, 38,000, and 39,000, respectively. The purification of NGF from the venom of *V. russellii* was described by Pearce et al. (1972a, b).

The N-terminal amino acid for the NGFs of *Vipera ammodytes ammodytes* and *Dendroaspis viridis* venoms is isoleucine. However, the N-terminal amino acid for *Agkistrodon rhodostoma* NGF is serine (Bailey et al., 1975), the same as that of NGF from mouse salivary gland (Angeletti and Bradshaw, 1971). The molecular weight of NGF from a nonvenom source such as mouse salivary gland is 30,000, as determined from S_{20} and D_{20} values (Bocchini and Angeletti, 1969).

Nerve growth factor of *Naja naja* venom has a relatively low pI of 6.75 (Angeletti, 1970b), which may be due to the presence of carbohydrates in *N. naja* NGF. The chemical and biological properties of *N. naja* NGF are remarkably similar to those of mouse submaxillary gland NGF. The two proteins have similar molecular weights, similar

Table 2 Comparison of the Amino Acid Compositions of Snake Venom Nerve Growth Factor and Mouse Submaxillary Gland NGF.

Genus Species	Naja naja		Vipera russellii	Mouse submaxillary gland
Amino Acid		Residues / Molecule		
Lysine	18	10	14	7
Histidine	8	4	6	3
Arginine	6	3	15	6
Aspartic acid	34	16	44	9
Threonine	26	13	22	11
Serine	16	8	16	9
Glutamic acid	20	10	15	8
Proline	10	5	14	2
Glycine	16	6	13	5
Alanine	12	5	14	7
Valine	18	10	33	11
Methionine	2	2	6	1
Isoleucine	12	6	12	5
Leucine	8	3	7	3
Tyrosine	4	3	7	2
Phenylalanine	8	4	12	6
Half-cystine	8	6	12	4
Tryptophan	4	3	12	2
Total residues	230	117	274	101
References	Angeletti 1970b	Hogue-Angeletti et al., 1976	Pearce et al 1972a	Bocchini 1970

Table 3 Amino Acid Sequences of Nerve Growth Factor from *Naja naja* Venom and Mouse Submaxillary Glands*

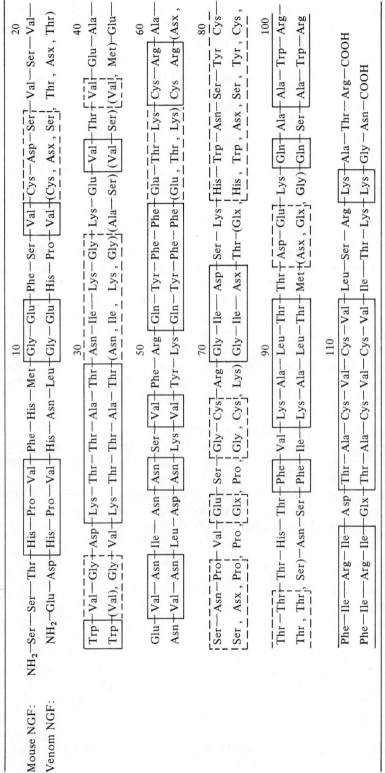

*Solid boxes indicate identical sequences. Dashed boxes indicate that the identities are positioned by homology. Table obtained from Hogue-Angeletti et al. (1976).

amino acid compositions, and comparable *in vitro* and *in vivo* activities (Angeletti, 1970b; Bocchini and Angeletti 1969). However, the NGF in isoelectric point and amino acid composition (Pearce et al., 1972a, b), as can be seen from Table 2. The NGF from *V. russellii* venom is an extremely stable protein. It retains activity in 0.1 N HCl and 0.1N NaOH, upon treatment by trypsin chymotrypsin, DNase, and RNase, and after dialysis against EDTA (Pearce et al., 1972b). However, pepsin or heating at 90° for 15 min causes complete loss of NGF activity.

Sulkowski et al. (1975) isolated a nontoxic basic protein in crystalline form from *Crotalus adamanteus* venom. This basic protein possesses NGF activity that remains constant and active through several recrystallizations. The molecular weight is 24 300, with a pI of 9.8; thus this basic protein is quite different from traditional NGF. Sulkowski et al. postulated several possibilities for the existence of the basic protein with NGF activity: (1) cocrystallization, (2) denaturation, or (3) the occurrence of two different types of NGF activity in venom.

The partial amino acid sequence of *Naja naja* NGF has been published (Hogue-Angeletti et al., 1976). Despite the many differences in immunological and chemical properties that have been reported, *N. naja* NGF is structurally similar to mouse submaxillary gland NGF, in that about 60% of the amino acid residues are identical (Server et al., 1976). This is shown in Table 3. From the amino acid sequence study, it is now known that there is a 40% difference in the amino acid sequences of the NGFs from the two sources. These structural differences may account for the results of the competition assays with $[^{125}I]$ β-NGF in which *N. naja* NGF was less effective in displacing $[^{125}I]$ β-NGF from dorsal root receptors. *Naja naja* NGF is also different in that it does not interact with the α and γ subunits of 7S NGF to form a high molecular weight complex. Nevertheless, the 60% amino acid sequence homology of the NGF proteins from mouse and cobra venom indicates that a substantial portion of their primary structures has been conserved through evolution. In particular, the regions of the NGF molecules that interact with the NGF receptors have been highly conserved.

Because of more precise studies by Hogue-Angeletti et al. (1976) and Server et al. (1976), much of the earlier work on *N. naja* NGF must be critically regarded. For instance, the data on amino acid composition and molecular weight of earlier works are quite different from the more recent reports of Hogue-Angeletti et al. (1976) and Server et al. (1976). Greater reliance should be placed on the newer data.

2 IMMUNOLOGY

Immunological data indicate a partial identity of different NGF preparations. For instance, anti-salivary gland NGF antibody cross-reacted with crude *Agkistrodon piscivorus* venom NGF (Cohen, 1960). The microcomplement fixation method demonstrated a cross-reactivity for *Crotalus adamanteus* venom NGF and mouse salivary gland NGF (Zanini et al., 1968). The antiserum to mouse salivary gland NGF inhibits the NGF activities of all snake venoms tested (*Naja naja, Crotalus admanteus, Bothrops jararaca, Vipera russellii*), and the antibodies elicited by the salivery gland NGF are able to cross-react with identical antigenic sites on the *Naja naja* NGF molecule. However, when anti-*N. naja* NGF is used for production, cross-reactivity with mouse NGF is lacking. It may be that the antibodies to *N. naja* NGF are directed toward antigenic determinants that are absent or present only in very low concentrations in mouse salivary gland NGF

3 Biological Effects

(Angeletti, 1971). Within the venom NGFs there are partial identical precipitin lines, suggesting that NGFs obtained from different venoms have some similarity but are not identical. Antiserum against *Vipera russellii* NGF cross-reacts with antisera against *Bothrops atrox* and *Agkistrodon piscivorus* NGF. This indicates that, although there is some chemical similarity among NGFs of different animal origins, venom NGF is an immunologically distinct molecular species from mouse submaxillary gland NGF (Glass and Banthorpe, 1975).

Varon (1968) has stated that NGFs from different animal sources are chemically different, although the biological effects are indistinguishable.

3 BIOLOGICAL EFFECTS

An example of the effect of *Crotalus adamanteus* venom NGF on sensory ganglia of 9-day chick embryos is illustrated in Fig. 1. The control without the addition of NGF is shown in Fig. 2.

The activity of *C. adamanteus* and *Agkistrodon piscivorus* venom NGFs is identical to that of mouse sarcoma NGF: promoting nerve fiber outgrowth from sensory and sympathetic ganglia (Levi-Montalcini and Cohen, 1956). Both the snake and the mouse NGFs increase the size and number of neurons in sympathetic ganglia and accelerate the

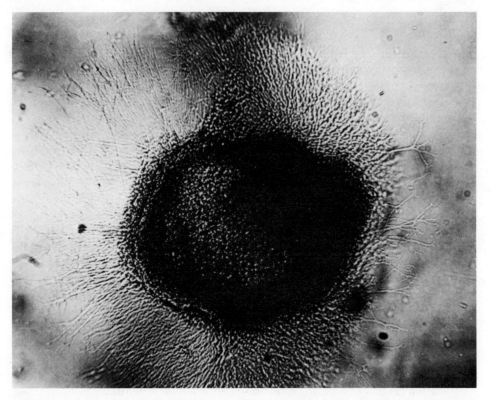

Figure 22.1 Effect of *Crotalus adamanteus* NGF on sensory ganglia of 9-day chick embryos. (Kindly supplied by Dr. Regino Perez-Polo.)

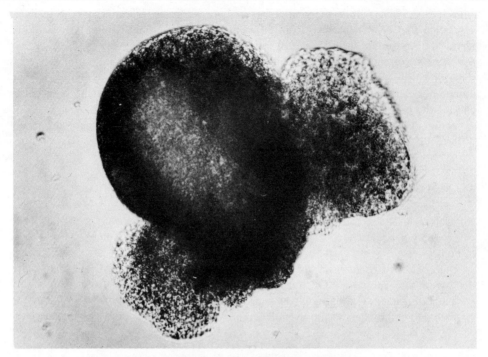

Figure 22.2   Control. (Kindly supplied by Dr. Regino Perez-Polo.)

differentiation process. These workers found that the purest fraction of snake venom is approximately 1000 times as effective as the purest fraction of mouse tumor.

Serum proteins bind to purified NGF from the venom of *Bothrops atrox* and from mouse submaxillary glands (Angeletti, 1969). This binding of NGF to serum proteins increases the biological activity of NGF. Gel filtration studies indicate that NGF forms a complex with serim proteins. This may suggest that serum proteins stabilize NGF or contain a specific activator. When ganglia cultured in the presence of NGF are transferred to medium depleted of this factor, all fiber growth ceases (Angeletti, 1969). This suggests that the role of NGF is not to promote the differentiation of primitive cells into neurons, but to activate neuronal growth. If the main effect of NGF is on the differentiation of receptive neuroblasts, pretreatment of ganglia with NGF would lead to the growth of fibers on subsequent cultures, irrespective of whether or not the media contained NGF. On the basis of this assumption, Charlwood et al. (1974) cultured sensory and sympathetic ganglia of embryonic chicks in a medium in which no growth of fibers could occur, and then transferred the ganglia to culture conditions favoring fiber growth. They examined the effect on fiber outgrowth of having NGF (*Vipera russellii* origin) in neither medium, in one medium, and in both media. They conclude that the primary effect of NGF is to maintain the viability of the neurons.

Nerve growth factor from snake venom stimulated lysine incorporation into the ganglia by 58 to 72%. Adenine incorporation into RNA was stimulated by NGF by 40 to 69%. It also stimulated the oxidation of carbon 1 of glucose by 41 to 54% (Cohen, 1959).

When the venoms of *Naja nigricollis* and *Echis coloratus* were treated with antiserum, incubated with acid for 1 hr, or heated, NGF activity was abolished (Mohamed et al., 1971).

Changes in cell surface adhesion specificity can be induced in rotating cultures of tectal cells by the addition of $10^{-7}M$ mouse submaxillary gland NGF. Mouse NGF in which two or three of the tryptophan residues have been oxidized is inactive with dorsal root ganglia, but active with tectal cells. By contrast, *Naja naja* NGF is active with dorsal root ganglia but not with tectal cells (Merrell et al., 1975). Since the concentration of mouse NGF necessary to obtain a response in tectal cells is several orders of magnitude higher than that required to induce nerve growth in dorsal root ganglia, and since the specificity of the receptor seems to be different from that found in the ganglia, it is probable that mouse NGF may be serving as an analog of another trophic factor responsible for the changes *in vivo*.

The effect of purified NGF obtained from *Vipera russellii* venom on superior cervical ganglia of neonatal mice was investigated by Banks et al. (1975). The NGF-induced hypertrophic effect arises from an increase in the rate at which the sympathetic neurons attain their mature size. The hyperplastic effect is due to an increase in the rate of production of neurons from less differentiated cells. In the developmental period, the number of neurons can exceed that found at maturity. If injection of NGF is discontinued, the excess neurons disappear; but if injection of NGF is continued to maturity, the excess number of neurons is maintained.

4 OCCURRENCE

Nerve growth factor is widely distributed in various snake venoms. Usually, Elapidae and Viperidae venoms have more potent NGF activity than do Crotalidae venoms (Cohen, 1959). The concentration of NGF in venom is very small, about 0.5% of total protein. Venoms of the following snakes have been reported to contain NGF activity:

Elapidae

Dendroaspis viridis	Shipolini et al., 1973; Bailey et al., 1975
Naja melacoluca	Bailey et al., 1975
N. naja	Cohen, 1959; Angeletti, 1968a
N. nigricollis	Mohamed et al., 1971; Bailey et al., 1975
Sependon haemachatus	Cohen, 1959

Viperidae

Bitis gabonica	Cohen, 1959
Vipera ammodytes	Cohen, 1959; Banks et al., 1968; Bailey et al., 1975
V. aspis (yellow or white)	Cohen, 1959
V. russellii	Cohen, 1959; Banks et al., 1968; Angeletti, 1968a

Crotalidae

Agkistrodon contortrix laticinctus	Angeletti, 1968a
A. piscivorus	Cohen, 1959; Banks et al., 1968
A. rhodostoma	Bailey et al., 1975
Bothrops atrox	Cohen, 1959; Angeletti, 1968a

Crotalidae *Continued*

B. jararaca	Cohen, 1959; Angeletti, 1968b
Crotalus adamanteus	Cohen, 1959; Angeletti, 1968b
C. atrox	Angeletti, 1968a
C. durissus terrificus	Angeletti, 1968a
C. horridus	Cohen, 1969

Nerve growth factor occurs in high concentrations in the submaxillary gland of the adult male mouse, in snake venoms, and in Gila monster venom (Levi-Montalcini and Angeletti, 1968; Cohen, 1959). Pearce (1973) found no NGF in the venoms of bees, scorpions, spiders, and toads.

REFERENCES

There are many review articles on NGF, such as those by Levi-Montalcini and Angeletti (1968), Angeletti et al. (1968), and Shooter and Einstein (1971). However, the Chapter in this book concerns itself primarily with NGF in snake venoms.

REFERENCES

Angeletti, R. H. (1968a). Studies on the nerve growth factor (NGF) from snake venom: Gel filtration patterns of crude venoms. *J. Chromatogr.,* **36,** 535.

Angeletti, R. H. (1968). Nerve growth factor (NGF) from snake venom: Molecular heterogeneity, *J. Chromatogr.,* **37,** 62.

Angeletti, R. H. (1969). Nerve growth factor (NGF) from snake venom and mouse submaxillary gland: Interaction with serum proteins, *Brain Res.,* **12,** 234.

Angeletti, R. H. (1970a). The role of the tryptophan residues in the activity of the nerve growth factor, *Biochim. Biophys. Acta,* **214,** 478.

Angeletti, R. H. (1970b). Nerve growth factor from cobra venom, *Proc. Natl. Acad. Sci. U.S.A.,* **65,** 668.

Angeletti, R. H. (1971). Immunological relatedness of nerve growth factors, *Brain Res.,* **25,** 424.

Angeletti, R. H. and Bradshaw, R. A. (1971). Nerve growth factor from mouse submaxillary gland: Amino acid sequence, *Proc. Natl. Acad. Sci. U.S.A.,* **68,** 2417.

Angeletti, P., Calissano, P., Chen, J. S., and Levi-Montalcini, R. (1967). Multiple molecular forms of the nerve growth factor, *Biochim. Biophys. Acta,* **147,** 180.

Angeletti, P. U., Levi-Montalcini, R., and Calissano, P. (1968). The nerve growth factor (NGF): Chemical properties and metabolic effects, *Adv. Enzymol.,* **31,** 51.

Bailey, G. S., Banks, B. E. C., Pearce, F. L., and Shipolini, R. A. (1975). A comparative study of nerve growth factors from snake venoms, *Comp. Biochem. Physiol.,* **51B,** 429.

Banks, B. E. C., Banthorpe, D. V., Berry, A. R., Davies, H. S., Doonan, S., Lamont, D. M., Shipolini, R., and Vernon, C. A. (1968). The preparation of nerve growth factors from snake venoms, *Biochem. J.,* **108,** 157.

Banks, B. E. C., Charlwood, K. A., Edwards, D. C., Vernon, C. A., and Walter, S. J. (1975). Effects of nerve growth factors from mouse salivary glands and snake venom on the sympathetic ganglia of neonatal and developing mice, *J. Physiol.,* **247,** 289.

Bocchini, V. (1970). The nerve growth factor: Amino acid composition and physiochemical properties, *Eur. J. Biochem.,* **15,** 127.

References

Bocchini, V. and Angeletti, P. U. (1969). The nerve growth factor: Purification as a 30,000 molecular-weight protein, *Proc. Natl. Acad. Sci. U.S.A.*, **64**, 787.

Bueker, E. D. (1948a). Implantation of tumors in the hind limb field of the embryonic chick and developmental response of the lumbosacral nervous system, *Anat. Rec.*, **102**, 369.

Bueker, E. D. (1948b). Hypertrophy of spinal ganglion cells in chick embryos after the substitution of mouse sarcoma 180 for the hind limb periphery *Anat. Rec.*, **100**, 735.

Charlwood, K. A., Griffith, M. J., Lamont, M. D., Vernon, C. A., and Wilcock, J. C. (1974). Effects of nerve growth factor from the venom of *Vipera russellii* on sensory and sympathetic ganglia from the embryonic chick in culture, *J. Embryol. Exp. Morphol.*, **32**, 239.

Cohen, S. (1959). Purification and metabolic effects of a nerve growth-promoting protein from snake venom, *J. Biol. Chem.*, **234**, 1129.

Cohen, S. (1960). Purification of a nerve-growth-promoting protein from the mouse salivary gland and its neurocytotoxic aniserum, *Proc. Natl. Acad. Sci. U.S.A.*, **46**, 302.

Cohen, S. and Levi-Montalcini, R. (1956). A nerve-growth-stimulating factor isolated from snake venom, *Proc. Natl. Acad. Sci. U.S.A.*, **42**, 571.

Glass, R. E. and Banthorpe, D. V. (1975). Properties of nerve growth factor from the venom of *Bothrops atrox, Biochim. Biophys. Acta*, **405**, 23.

Hogue-Angeletti, R. A., Frazier, W. A., Jacobs, J. W., Niall, H. D., and Bradshaw, R. A. (1976). Purification, characterization and partial amino acid sequence of nerve growth factor from cobra venom, *Biochemistry*, **15**, 26.

Levi-Montalcini, R. and Angeletti, P. U. (1968). Nerve growth factor, *Physiol. Rev.*, **48**, 534.

Levi-Montalcini, R. and Cohen, S. (1956). *In vitro* and *in vivo* effects of a nerve growth stimulating agent isolated from snake venom, *Proc. Natl. Acad. Sci. U.S.A.*, **42**, 695.

Merrell, R., Pulliam, M. W., Randono, L., Boyd, L. F., Bradshaw, R. A., and Glaser, L. (1975). Temporal changes in tectal cell surface specificity induced by nerve growth factor, *Proc. Natl. Acad. Sci. U.S.A.*, **72**, 4270.

Mohamed, A. H., Saleh, A. M., and Ahmed, F. (1971). Nerve growth factor from Egyptian snake venoms, *Toxicon*, **9**, 201.

Pearce, F. L. (1973). Absence of nerve growth factor in the venoms of bees, scorpions, spiders and toads, *Toxicon*, **11**, 309.

Pearce, F. L., Banks, B. E. C., Banthorpe, D. V., Berry, A. R., Davies, H. S., and Vernon, C. A. (1972a). The isolation and characterization of nerve growth factor from the venom of *Vipera russellii*, *Eur. J. Biochem.*, **29**, 417.

Pearce, F. L., Banks, B. E. C., Banthorpe, D. V., Berry, A. R., Davies, H. S., and Vernon, C. A. (1972b). "The isolation and characterization of nerve growth factor from the venom of *Vipera russellii*," in E. Zaimis, Ed., *Nerve Growth Factor and Its Antiserum*, Athlone, London, pp. 3–18.

Server, A. C., Herrup, K., Shooter, E. M., Hogue-Angeletti, R. A., Frazier, W. A., and Bradshaw, R. A. (1976). Comparison of the nerve growth factor proteins from cobra venom (*Naja naja*) and mouse submaxillary gland, *Biochemistry*, **15**, 35.

Shipolini, R. A., Bailey, G. S., Edwardson, J. A., and Banks, B. E. C. (1973). Separation and characterization of polypeptides from the venom of *Dendroaspis viridis, Eur. J. Biochem.*, **40**, 337.

Shooter, E. M. and Einstein, E. R. (1971). Proteins of the nervous system, *Ann. Rev. Biochem.*, **40**, 635.

Sulkowski, E., Kress, L. F., and Laskowski, M., Sr. (1975). Crystalline basic protein from venom of *Crotalus adamanteus, Toxicon*, **13**, 149.

Varon, S. (1968). Nerve growth factor: A selective review, *J. Pediatr. Surg.*, **3**, 120.

Zanini, A., Angeletti, P. U., and Levi-Montalcini, R. (1968). Immunochemical properties of the nerve growth factor, *Proc. Natl. Acad. Sci. U.S.A.*, **61**, 835.

23 Hemorrhage, Myonecrosis, and Nephrotoxic Action

1 HEMORRHAGE 373
   1.1 Normal Capillaries, 374
   1.2 Pathology, 375
   1.3 Biochemical Aspect, 378
       Trimeresurus spp. Venom, 378
       Agkistrodon halys blomhoffii Venom, 381
       Agkistrodon rhodostoma Venom, 381
       Vipera palestinae Venom, 381
   1.4 Antihemorrhagic Factor, 381

2 MYONECROSIS 382
   2.1 Normal Skeletal Muscle, 382
   2.2 Envenomated Muscle, 383
   2.3 Myonecrotic Toxin, 391

3 NEPHROTOXIC ACTION 391
   3.1 Ultrastructure of Kidney, 391
   3.2 Pathology of Envenomated Kidney, 392
       Crotalidae, 392
       Viperidae, 393
       Elapidae, 394
       Hydrophiidae, 396

References 396

Symptoms of snake poisoning result from the combined effects of complex protein components present in venom. Clinically, the main symptoms caused by snake envenomation can be classified into two groups, systemic and local. Venoms from snakes of Crotalidae (pit vipers) and Viperidae (vipers) have pronounced local effects on tissues. Venoms of Elapidae, such as cobras, kraits, and coral snakes, and Hydrophiidae (sea snakes) are predominantly neurotoxic and cause severe systemic effects, although they may also produce local effects to a lesser extent.

With improvements in antivenin treatment for snakebites, the majority of victims are

1 Hemorrhage

being saved from death or severe generalized symptoms. However, in many cases local tissue damage is not prevented by the use of antivenin (Stahnke, 1966) unless the latter is administered immediately (Homma and Tu, 1970).

The real problem resulting from Crotalidae envenomation is the destruction of tissues and the possible dysfunction or loss of a finger, hand, arm, or other part (McCollough and Gennaro, 1971).

1 HEMORRHAGE

Hemorrhage is a common symptom associated with local tissue damage in snake poisoning by Crotalidae and Viperidae. In severe poisoning hemorrhage can be observed in many internal organs (Tu and Homma, 1970; Tu, 1971).

Local hemorrhage is quite common after rattlesnake bites. Of a number of rattlesnake venoms tested, only Mojave rattlesnake (*Crotalus scutulutus*) venom was nonhemorrhagic (Fig. 1). All other venoms, such as those of *C. atrox, C. adamanteus, C. basiliscus, C. durissus, C. durissus terrificus, C. durissus totanacus, C. horridus horridus, C. horridus atricaudatus, C. viridis viridis,* and *Sistrurus miliarius barbouri*, were hemorrhagic (Friederich and Tu, 1971).

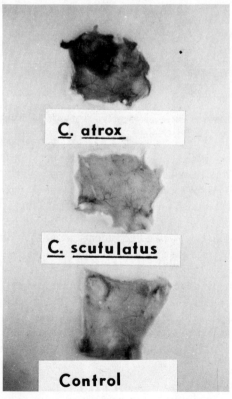

Figure 23.1 Example of hemorrhage in mouse skin caused by subcutaneous injection of (A) *Crotalus atrox* venom (½ LD_{50} dose), (B) *C. scutulatus* venom (½ LD_{50} dose), and (C), saline as a control. Most rattlesnake venoms induce hemorrhage, but *C. scutulatus* venom does not.

Figure 23.2 Photomicrograph of hemorrhage in muscle. Each dot is an erythrocyte.

An example of hemorrhage in the muscle after i.m. injection of *Crotalus atrox* (western diamondback rattlesnake) venom as examined with the light microscope is shown in Fig. 2. The onset of local hemorrhage is extremely rapid. With *C. atrox* venom, hemorrhage appears in the thigh muscles of mice within 2 min after injection (Ownby et al., 1974). Selection of tissues is important for the study of hemorrhage. For instance, cobra *Naja naja atra* and *N. naja kaouthia* venoms are nonhemorrhagic in skin or muscle (Tu et al., 1967; Stringer et al., 1971), but venoms of cobras such as *N. naja* and *N. nigricollis* can produce hemorrhage in lungs (Bonta et al., 1969; 1970). This difference is apparently due to different types of hemorrhagic toxins present in the venoms of cobras and pit vipers or to differences in the test systems. The hemorrhagic principle of cobra venom forms a complex with heparin; thus the hemorrhagic activity is abolished. However, the hemorrhagic activity of *Agkistrodon piscivorus* venom cannot be abolished by adding heparin since the latter does not form a complex with this venom.

1.1 Normal Capillaries

The escape of erythrocytes during hemorrhage can occur by at least two different routes: across a broken capillary wall or through a widened intercellular junction. In either case, the normal structure and function of capillaries are altered. Therefore a brief description

1 Hemorrhage

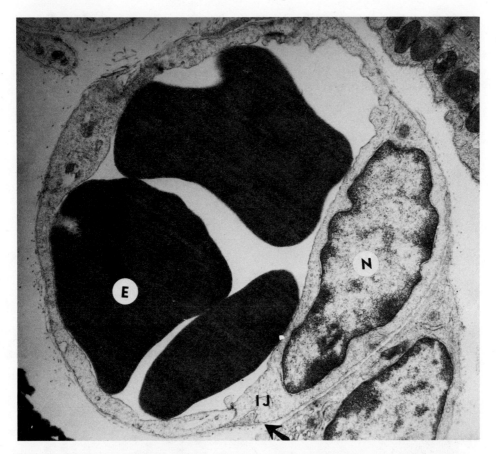

Figure 23.3 Ultrastructure of normal capillary. E: erythrocytes, IJ: intracellular tight junction, N: nucleus. (Electron micrograph by Dr. C. L. Ownby of author's laboratory.)

of capillaries in the normal condition will be helpful in understanding the hemorrhagic effect of snake venoms.

Capillaries are surrounded by a thin basement membrane. The endothelium is very thin except for the portion containing the nucleus (Fig. 3). Some capillaries are fenestrated; others are not (muscle capillaries). All of our work and most of that of other investigators was done on muscle capillaries. In Fig. 3, an erythrocyte can be seen in the lumen of a capillary. The tight junction of the endothelial cell is indicated by an arrow in Fig. 3; a higher magnification is shown in Fig. 4.

The question is, How do erythrocytes pass from the lumen to the outside of the capillaries? One possibility is that the erythrocytes escape through the tight junction (per diapedesis). The other route is through the ruptured endothelial walls (per rhexis). It is also possible that extravasation takes place by both routes simultaneously.

1.2 Pathology

Ultrastructural changes of capillaries have been investigated by several investigators using four different snake venoms. These results are briefly summarized in this section.

The effect of *Crotalus atrox* venom was studied by Ownby et al. (1974) and Ownby

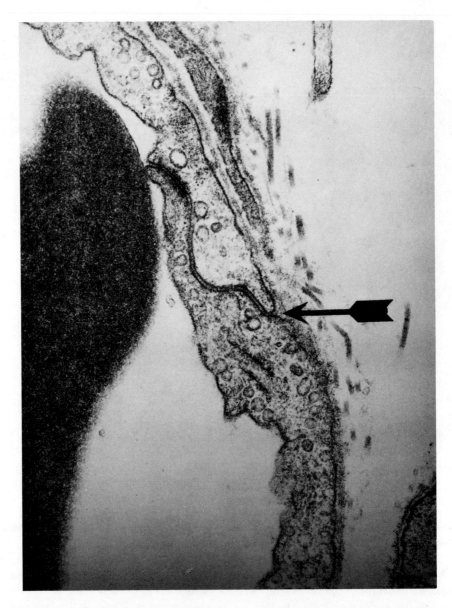

Figure 23.4 High magnification of intercellular junction (arrow). (Electron micrograph by Dr. C. L. Ownby of author's laboratory.)

(1975). Dilation of the endoplasmic reticulum of endothelial cells appeared to be the most common pathologic change initially (Fig. 5), and often the entire endothelial cell was swollen in many vessels. The endothelial cytoplasm also blebbed into the lumen. It appears that the direct lytic action on the capillary endothelium by *C. atrox* venom caused rupture of the plasma membrane, allowing the erythrocytes to escape into the connective tissue space (Fig. 6). Aggregated platelets were often present in the lumen of damaged capillaries (Fig. 3 in Chapter 21).

1 Hemorrhage

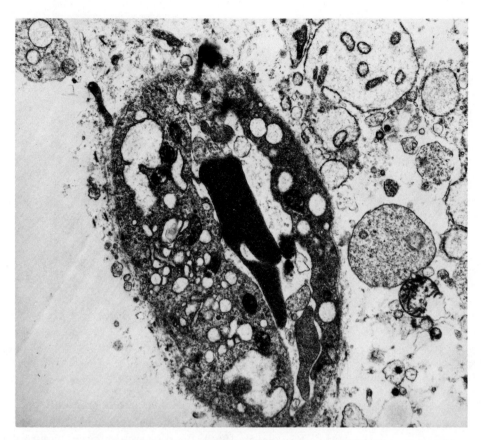

Figure 23.5 Initial stage of capillary damage due to *Crotalus atrox* envenomation. Note dilation of the endoplasmic reticulum of endothelial cells. (For details, see Ownby et al., 1974.)

It seems that *C. atrox* venom results in extravasation of erythrocytes by a direct lytic action on capillary endothelium. Intravascular hemolysis was also observed with *C. atrox* venom (Fig. 7). This is probably due to the action of venom phospholipase A_2.

McKay et al. (1970) used *Vipera palestinae* hemorrhagin to study the mechanism of hemorrhage. The results were very similar to those obtained in the work on *Crotalus atrox* venom. The hemorrhagin caused extravasation of erythrocytes by a direct lytic effect on capillary endothelium. Platelet aggregates and intravascular hemolysis were also observed.

An electron micrograph of the hemorrhagic process caused by the main hemorrhagic principle (HR1) of *Trimeresurus flavoviridis* venom indicated that the erythrocytes escaped through the tight junction (Ohsaka et al., 1971a; Tsuchiya et al., 1974, 1975). Moreover, partial disappearance of the basement membrane of the endothelial cell was observed. The effect of *Echis coloratus* venom on the brain capillaries of the mouse was investigated by electron microscopy, using horseradish peroxidase as a tracer (Sandbank et al., 1974). The envenomation resulted in breakdown of the blood–brain barrier, manifested by leakage of the peroxidase through the capillary wall. The peroxidase penetrated both the endothelial pinocytosis and through opened tight junctions between the endothelial cells. These results suggest that extravasation of erythrocytes occurs through the tight junction as well as the lysed portion of the endothelial wall.

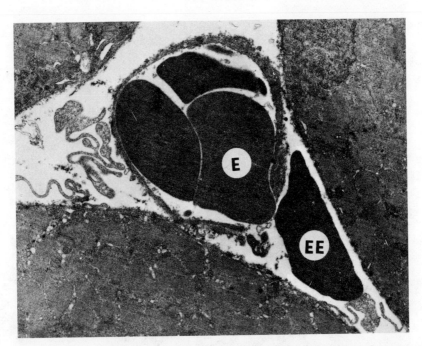

Figure 23.6 Capillary from animal injected with *Crotalus atrox* venom. The endothelium has a gap. Note extravasated erythrocyte (EE) and erythrocyte inside the capillary (E). (Electron micrograph by Dr. C. L. Ownby of author's laboratory.)

The effect of hemorrhagic principles HR1 and HR2 of *Trimeresurus flavoviridis* venom on HeLa cells, mouse embryo cells, MLg cells, and T5 cells from human embryonic fibroblasts was investigated (Yoshikura et al., 1966). The HR2 exerted the cell monolayer disrupting effect but no lytic action. The HR1 showed little of either effect. The effect of the hemorrhagic principles or venom on a culture of capillary endothelial cells has not yet been investigated.

Venoms of Crotalidae snakes damage not only capillaries but also arteries (Homma and Tu, 1971). An example of angionecrosis due to *Bothrops schlegelii* is illustrated in Fig. 8 (Tu and Homma, 1970). Hemorrhage is not restricted to the local area of the bites. Eventually venom spreads through the victim's circulatory system and causes hemorrhage in many internal organs. For instance, hemorrhage can take place in the heart, lungs, mesentery, and small intestines (Tu and Homma, 1970; Tu, 1971). An example of hemorrhage in the lungs is shown in Fig. 9.

1.3 Biochemical Aspect

Snake venoms contain a number of metals (Friederich and Tu, 1971). Among them Ca(II), Mg(II), and Zn(II) are commonly present in relatively large amounts in all venoms. When these metals were removed, hemorrhagic activity was also eliminated. This strongly suggests that hemorrhagic toxins are metalloproteins, and our recent study (unpublished data) indicates that this is indeed the case.

A number of hemorrhagic toxins have been isolated from different venoms. These will be reviewed in this section.

***Trimeresurus* spp Venom** The presence of two hemorrhagic fractions was observed in the venom of *Trimeresurus flavoviridis* (Ohsaka et al., 1960a, 1966, 1971b; Maeno et al.,

1 Hemorrhage

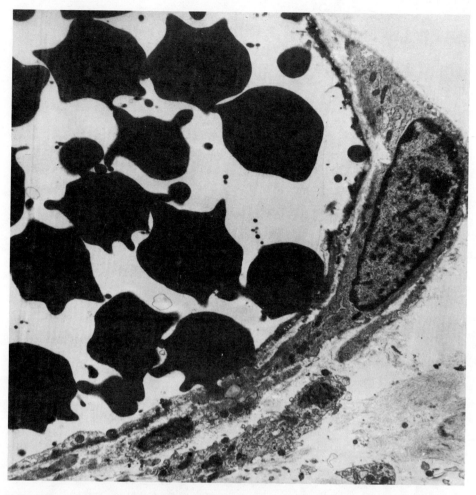

Figure 23.7 Portion of a vessel from envenomated animal. Note intravascular hemolysis. (Electron micrograph by Dr. C. L. Ownby of author's laboratory.)

1960; Omori-Satoh et al., 1967). The HR2 fraction was separated from the active lethal fraction (Ohsaka, 1960; Ohsaka et al., 1960b). Both fractions, HR1 and HR2, still contained proteolytic activity when casein was used as substrate.

When the HR1 of *T. flavoviridis* venom was purified, it had a minimum hemorrhagic dose of 0.0058 µg. It also possessed potent lethality with an LD_{50} value of 0.23 µg g^{-1} in mice. The molecular weight of HR1 is about 100,000. The hemorrhagic activity of HR1 can be eliminated with EDTA (Omori-Satoh and Ohsaka, 1970).

The HR2 of *T. flavoviridis* venom was further resolved into two fractions, HR2a and HR2b (Takahashi and Ohsaka, 1970). Neither toxin had caseinolytic activity. The hemorrhagic activity of both toxins was eliminated with EDTA.

All hemorrhagic principles, HR1, HR2a, and HR2b, can liberate peptides and carbohydrates from glomerular basement membrane preparation (Ohsaka et al., 1973; Ohsaka, 1973). Apparently, hemorrhagic toxins are very specific proteases. In an earlier study, Takahashi and Ohsaka (1970) found that hemorrhagic toxins did not hydrolyze

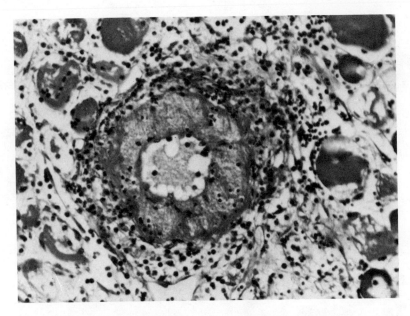

Figure 23.8 Example of angionecrosis due to *Bothrops schelegelli* venom injection. (For details, see Tu and Homma, 1970.)

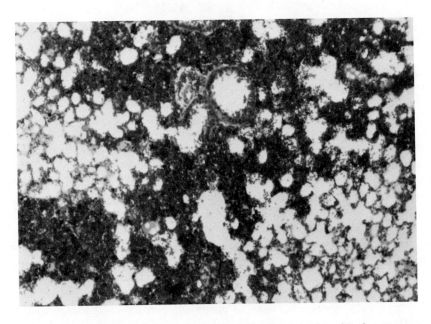

Figure 23.9 Pulmonary hemorrhage caused after 30 min by i.m. injection of *bothrops nasuta* venom into mouse. Photographed by Dr. M. Homma of author's laboratory.

1 Hemorrhage

casein. Since caseinolytic activity was measured on acid-soluble peptides after TCA precipitation, they might not have detected the very specific protease activity of hemorrhagic toxins. The basement membrane consists mainly of collagen and other proteins bound to carbohydrates. The hemorrhagic toxins (HR1, HR2a, HR2b) are probably specific proteolytic enzymes with unidentified specificity.

Trimeresurus elegans venom contains two hemorrhagic fractions, which were separated from the proteinase activity (Nozaki et al., 1974, 1975).

Agkistrodon halys blomhoffii Venom Similarly, the venom of *Agkistrodon halys blomhoffii* also contains two hemorrhagic factors, HR-I and HR-II (Omori et al., 1964; Suzuki, 1966). In this case, the lethal activity coincides with the hemorrhagic activity.

The HR-I was isolated in pure form by Oshima et al. (1972). It has a molecular weight of 80,000 to 91,000 and is an acidic glycoprotein. The minimum hemorrhagic dose is 0.0012 µg. It also retains potent lethality with an LD_{50} value of 0.36 $\mu g\ g^{-1}$ in mice. The HR-I lacks these enzyme activities; proteinase, phospholipase A_2, phosphomonoesterase, arginine ester hydrolase, bradykinin-releasing, clotting enzyme, L-amino acid oxidase, and hyaluronidase.

Agkistrodon rhodostoma Venom *Agnistrodon rhodostoma* can be fractionated into 15 components on a preparative isoelectrofocusing column (Toom et al., 1969). There are two hemorrhagic fractions; one of them is associated with arginine esterase activity. Both of them, however, separated clearly from the major lethal fraction.

Vipera palestinae Venom Hemorrhagic toxin isolated from the venom of *Vipisa palestinae* and called hemorrhagin (Grotto et al., 1967) is an acidic protein with a molecular weight of 44,000. It has gelatinase activity. The proteolytic activity can be inhibited by soybean trypsin inhibitor or DFP, whereas the hemorrhagic activity is not affected. Administration of this toxin to guinea pigs causes hemorrhage as well as hypofibrinogenemia (Grotto et al., 1969). It impairs thrombin formation, fibrinogen clottability, and platelet clot-reacting activity.

The enzyme-digested antibody that was made against hemorrhagic toxin can neutralize hemorrhagic activity (Moroz et al., 1971). This antibody fragment has a 5S unit as compared to 7S for the original immunoglobulin. Apparently the Fc fragment does not contribute to the neutralization.

Although the hemorrhagic toxin releases peptides from fibrinogen, this action does not interfere with the ability of thrombin to further split off fibrinopeptides (Grotto et al., 1972). This indicates that the bonds cleaved by the hemorrhagin in the fibrinogen are different from those susceptible to thrombin. The LD_{50} of hemorrhagic toxin in guinea pig by intracardial injection is 0.6 $\mu g\ g^{-1}$.

1.4 Antihemorrhagic Factor

A naturally occurring antihemorrhagic factor was isolated from the serum of the snake *Trimeresurus flavoviridis* (Omori-Satoh et al., 1972). The factor inhibits not only the hemorrhagic activity but also the lethal activity of the venom. The molecular weight of this factor is 70,000. The presence of an antihemorrhagic factor in the serum of the snake suggests that the snake has a mechanism for self-protection against its own venom. This explains why snakes are more resistant to their own venoms than to those of other species.

2 MYONECROSIS

Most Crotalidae venoms produce hemorrhage and myonecrosis simultaneously upon envenomation (Okonogi et al., 1964; Tu and Homma, 1970; Tu, 1971). Some snakebite victims survive, but sustain permanent tissue damage that may extend into the muscle, tendons, and cartilage (Emery and Russell, 1963; McCollough and Gennaro, 1963). The destruction of tissue, with possible dysfunction or complete loss of a body part, is an important factor in snakebite cases. A clinical example of myonecrosis in the hand due to envenomation by *Trimeresurus flavoviridis* is shown in Fig. 10. *Trimeresurus flavoviridis* is a snake of the family Crotalidae, to which American rattlesnakes, copperheads, and water moccasins also belong. They all give similar clinical symptoms in cases of severe poisoning.

Before discussing the myonecrosis induced by snake venom, it is worthwhile to review briefly the normal morphology of skeletal muscle.

2.1 Normal Skeletal Muscle

The morphology of the skeletal muscle cell is quite different from that of other types of cells. The skeletal muscle cell is extremely elongated, with several nuclei usually found at the periphery of the cell. As can be seen from a longitudinnal section, there are repeating patterns throughout (Fig. 11). The dark band in Fig. 11 is the Z band. A sarcomere extends from one Z band to another. The elongated bundles of fibers are myosin and actin filaments. The length of myosin bundles corresponds to the length of the A band (Fig. 12). When contraction takes place, the actin filaments slide toward the M band, and the light I band disappears. The thicker filaments are made up of myosin with a diameter of about 10 nm, and the thinner filaments are actins with a diameter of 6 nm. The difference in the size of myosin and actin can best be seen in a cross section of a muscle bundle, as shown in Fig. 13.

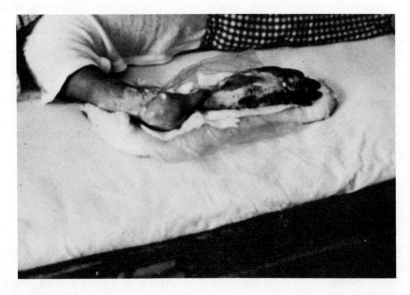

Figure 23.10 An example of myonecrosis due to envenomation by *Trimeresurus flavoviridis*. Most Crotalidae snakebites cause similar symptoms. (Photograph kindly supplied by Dr. T. Okonogi.)

2 Myonecrosis

Figure 23.11 Ultrastructure of normal muscle – longitudinal view. (Photograph by Mr. M. Stringer of author's laboratory.)

It is not the intention of the author to discuss the ultrastructure of the normal muscle cell in detail. The purpose is merely to review briefly, so that the reader can better understand the pathologic effects of snake venom.

2.2 Envenomated Muscle

An example of experimental myonecrosis in mice induced by *Crotalus atrox* (western diamondback rattlesnake) venom is shown in Fig. 14.

Elapidae venoms, including cobra venom, are usually considered neurotoxic and nonhemorrhagic; however, cobra venom does cause local myonecrosis (Reid, 1964). In

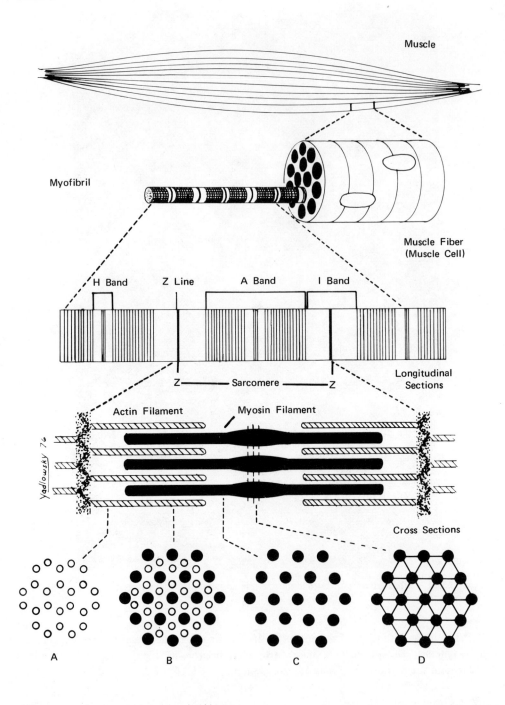

Figure 23.12 Schematic diagram of normal muscle. (Drawn by Mr. J. Yadlowsky of author's laboratory.)

2 Myonecrosis

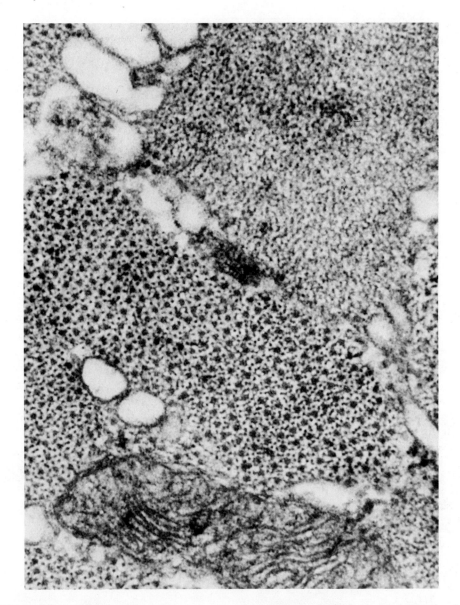

Figure 23.13 Ultrastructure of normal muscle – cross-sectional view. The big dot is myosin; the small dot, actin. (Electron micrograph by Mr. M. Stringer of author's laboratory.)

man this may easily be overlooked because the initial local reaction may be slight, and the patient may be discharged before necrosis is evident (Reid, 1968). Myonecrosis can be experimentally induced with cobra venom (Stringer et al., 1971; Fukuyama and Sawai, 1972).

Myonecrotic toxin seems to be a heat-stable factor since the venom of *Trimeresurus flavoridis* still produced a myonecrotic lesion after being heated to 100°C for 10 min (Okonogi et al., 1962; Kurashige et al., 1966). Light microscopic examination of the envenomated tissue indicated that mitochondria in the muscle disappeared.

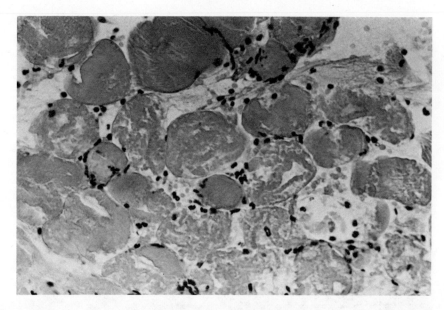

Figure 23.14 Example of myonecrosis induced by *Crotalus atrox* venom under light microscopic observation. Photograph by Dr. C. L. Ownby of author's laboratory.)

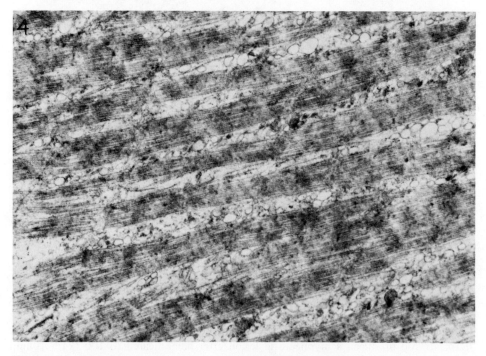

Figure 23.15 Example of myonecrosis induced by *Naja naja kaouthia* venom. Note the loss of the striated appearance of the muscle fibrils. (For details see Stringer et al., 1971.)

2 Myonecrosis

Recently, more detailed study has been done with an electron microscope. For instance, ultrastructural studies of myonecrosis induced by *Naja naja kaouthia* (Thailand cobra) venom in mice were made by Stringer et al. (1971). Degeneration of the entire muscle fiber and its constituents was observed (Fig. 15). The myofilaments coalesced to form an amorphous mass, and the sarcotubular system disappeared. The mitochondria swelled into vacuoles containing fragmented cristae. The degenerated fiber became a less electron-dense mass containing a few cellular remnants enclosed by a vestige of the plasma membrane (Fig. 16).

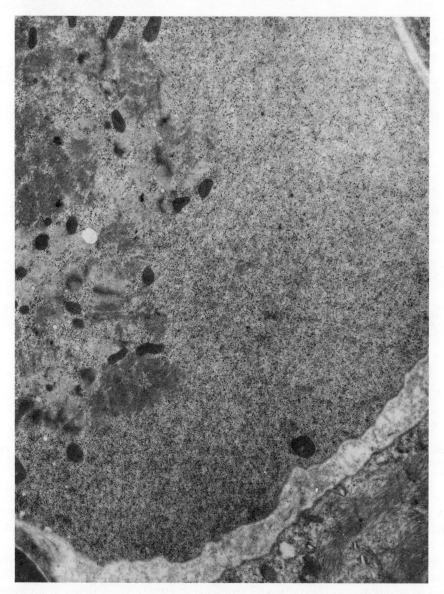

Figure 23.16 Example of myonecrosis induced by *Naja naja kaouthia* venom. Most cellular components are completely lysed. Note the increased amount of glycogen in the envenomated muscle (dark, dense dots). Electron micrograph by Mr. M. Stringer of author's laboratory.

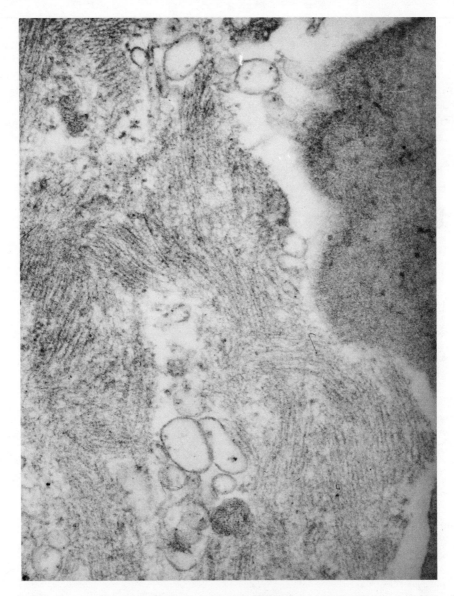

Figure 23.17 Example of myonecrosis induced by *Crotalus viridis viridis* venom. Note the disoriented meshwork of myofilaments. (For details see Stringer et al., 1972.)

A purified toxin, notexin, isolated from the venom of an Elapidae snake, *Notechis scutatus*, is myotoxic (Harris et al., 1975). It causes degenerative necrosis of the muscle, with edema and infiltration of lymphocytes, polymorphs, and macrophages.

The myonecrotic effect of Crotalidae venom is much stronger than that of cobra venoms. The myonecrotic effect of *Crotalus viridis viridis* (prairie rattlesnake) venom on mouse skeletal tissue was investigated by Stringer et al. (1972). The pathogenic process of rattlesnake venom is similar to that of cobra venom morphologically. Injured fibers contained dilated sarcoplasmic reticulum, disoriented, coagulated myofilamentous

2 Myonecrosis

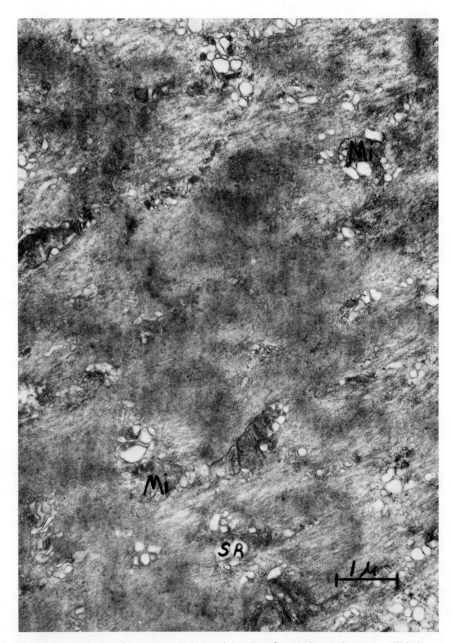

Figure 23.18 Example of myonecrosis induced by *Crotalus viridis viridis* venom. This is a more advanced pathological stage than the one shown in Fig. 17. The myofilaments have precipitated into an amorphous mass. (Electron micrograph by Mr. M. Stringer of author's laboratory.)

components, and condensed, rounded, and enlarged mitochomdria. The muscle after venom treatment is shown in Figs. 17 and 18.

Venom of a viper, *Bitis gabonica*, also damages the muscle (Mohamed et al., 1975a). It caused generalized depletion of succinic dehydrogenase activity, fasciculation, intrafibral vacuolization, and loss of transverse striations.

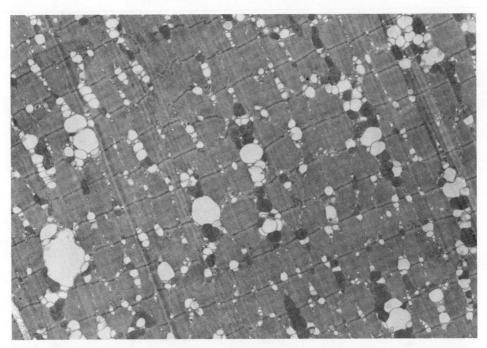

Figure 23.19 Myonecrosis induced by pure myonecrotic toxin of *Crotalus viridis viridis* venom. (Electron micrograph by Dr. C. L. Ownby of author's laboratory.)

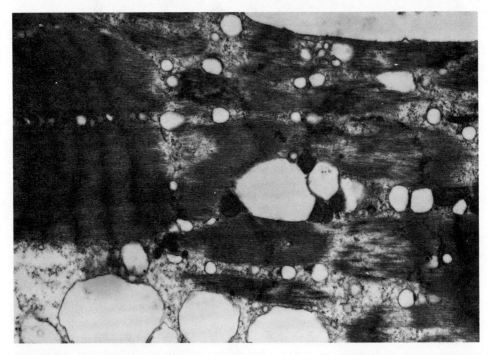

Figure 23.20 Continuation of Fig. 19, but showing a more advanced stage of necrosis. (Electron micrograph by Dr. C. L. Ownby of author's laboratory.)

Although myonecrosis is a pronounced symptom of Crotalidae envenomation, the necrosis can occur in other tissues also in cases of severe poisoning. Necrosis in the adrenocortex (Tadokoro and Kurihara, 1963), artery (Homma et al., 1967), kidneys (Schmidt et al. 1976), liver (Karimov et al., 1968), and viscera (Fayazova et al. 1969) can be observed.

2.3 Myonecrotic Toxin

Venom of *Crotalus viridis viridis* (prairie rattlesnake) contains two factors that cause myodegeneration (Ownby et al., 1976a, b). One is of high molecular weight and appeared in the void volume of a G-50 Sephadex column. The other fraction of lower molecular weight, was further fractionated until it became pure. The pure myonecrotic toxin does not induce hemorrhage in muscle, indicating that myonecrotic toxin is not identical to hemorrhagic toxin. The toxin causes extensive vacuolation (Fig. 19) and dilation of sarcoplasmic reticulum. Eventually striations and sarcomeres are disorganized (Fig. 20). These results correlated well with the myonecrosis induced by the original venom (Stringer et al., 1972). The important difference is that the pure component specifically alters skeletal muscle cells with the sarcoplasmic reticulum as the primary site of structural change.

The purified myonecrotic toxin is a basic protein with a molecular weight of 4300. It has a total of 38 amino acid residues, containing unusually high residues of lysine. The other myonecrotic toxin, of high molecular weight, has not yet been isolated.

3 NEPHROTOXIC ACTION

Snake poisoning can cause acute tubular necrosis in human victims. Clinical cases of renal failure have been reported from envenomation by sea snakes (Marsden and Reid, 1961; Reid, 1961; Furtado and Lester, 1968; Sitprija et al., 1973), *Crotalus* spp. (rattlesnakes) (Amorim and Mello, 1952, 1954; Silva et al., 1966), and Russell's viper (Sitprija et al., 1973). Before discussing the pathology of the kidney due to snake venoms, it is worthwhile to review briefly the normal kidney.

3.1 Ultrastructure of Kidney

The kidney is morphologically divided into two regions: cortex and medulla. In the cortex, glomeruli are present in which filtration of blood occurs. Also located in the cortex are proximal and distal convoluted tubules.

Mesangial cells are located in the centrolobular regions of the glomeruli (Fig. 21). The cytoplasm of the mesangial cells is much like that of the endothelial cells. The extracellular space around the mesangial cells is composed of a basement-membrane-like material called mesangial matrix. The glomerular capillaries are lined with fenestrated endothelium, which is adjacent to the glomerular basal lamina. The parietal epithelium of Bowman's capsule is composed of a simple squamous epithelium. These cells are reflected onto capillary loops at the hilar region of the glomerulus, thereby forming the visceral epithelium.

The proximal tubule cells have a number of microvilli on their luminal surfaces, while the basal surface has numerous infoldings of the plasma membrane. Abundant polyribosomes and elongated mitochondria are also present in these cells.

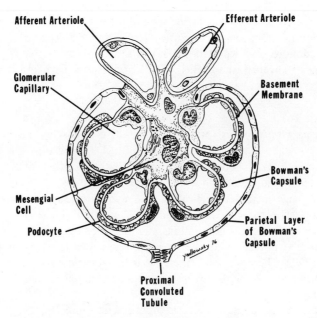

Figure 23.21 Diagram of normal kidney. Only the glomerulus portion is shown. (Drawn by Mr. J. Yadlowsky of author's laboratory.)

3.2 Pathology of Envenomated Kidney

Crotalidae Relatively few studies have been made of the venom effect on the kidney, although this is one organ susceptible to damage. In the study of Raab and Kaiser (1966a, b) a significant increase in urinary alkaline phosphatase and leucine aminopeptidase activities was observed after rats received *Agkistrodon piscivorus* venom. The increase in enzymatic activity suggested that the kidney was damaged by the venom. *Crotalus durissus terrificus* venom induces tubular casts of hemoglobin and degenerative changes in the tubular epithelium (Amorim et al., 1960).

Crotoxin causes kidney damage when injected intravenously into dogs (Hadler and Brazil, 1966). In early lesions the renal tubules are altered less than the glomeruli, which show congestion, thickness of the basement membrane, and nuclear pycnosis of some cells. In late lesions tubular damage predominates, with the proximal segment altered the most. There is intense microvascular degeneration, as well as nuclear pycnosis and necrosis of many epithelial cells. Crotoxin concentrates in the kidney, liver, lungs and spleen in that order when administered intravenously to mice.

Crotalus atrox (western diamondback rattlesnake) envenomation resulted in several ultrastructural changes of the renal corpuscles (Fig. 22) and the proximal convoluted tubules (Schmidt et al., 1976). Visceral epithelial changes included intracellular edema, blebbing, vesiculations, the formation of microvillus projections, and dilation of the endoplasmic reticulum and mitochondria (Fig. 23). Changes in the parietal epithelium were similar except that no microvillus projections were noted. Mesangiolysis was a consistent finding. Collagenous fibrils were very prominent in the lysed areas of the mesangial cells. Increased numbers of lysosome-related structures were noted in the proximal tubule cells. Most of the nuclear cisternae of the renal corpuscles and the proximal convoluted tubules were greatly dilated (Fig. 24).

3 Nephrotoxic Action

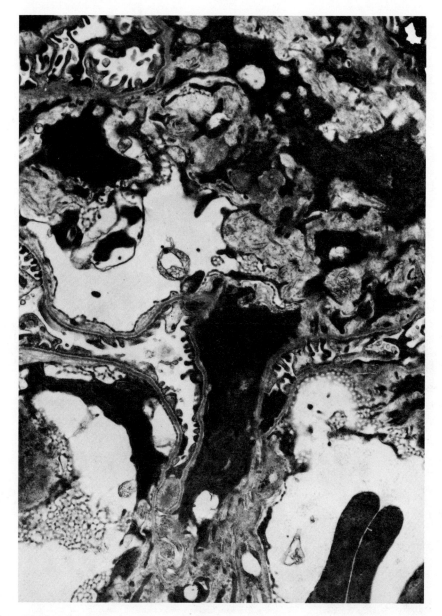

Figure 23.22 Damaged glomerulus (*Crotalus atrox* venom). (Electron micrograph by Mr. M. Schmidt of author's laboratory.)

Trimeresurus okinavenis envenomation causes kidney damage that consists of mesangiolysis (Suzuki et al., 1963).

Venom of *Crotalus viridis helleri* is toxic to human fetal kidney cell microculture; the LD_{50} is 3.5 ng per 100 cells (Jaratsch et al., 1973).

Viperidae Nephrotoxic action can also be induced by Viperidae venoms. Sant et al. (1974), Sant and Purandare (1972) and Tembe et al. (1975) reported that *Echis carinatus* venom produced lesions in kidneys of rabbits and monkeys.

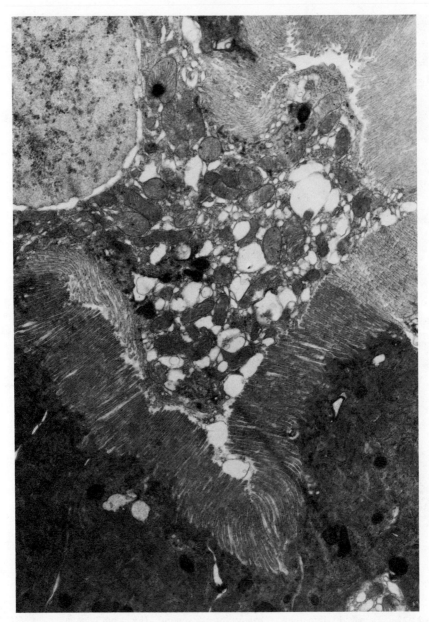

Figure 23.23 Pathologic effect of *Crotalus atrox* venom on the proximal convoluted tubule. (Electron micrograph by Mr. M. Schmidt of author's laboratory.)

Elapidae The most potent component in Elapidae venoms is neurotoxin, and the next most pontent is cardiotoxin. However, crude venoms contain more than these two components, and many Elapidae venoms cause kidney damage.

A lethal dose of *Naja haje* venom caused focal tubular degenerative lessions with intact glomeruli (Mohamed et al., 1975b). The tubular basement membrane was also intact. The brush border of the proximal convoluted tubules was destroyed only in the markedly

3 Nephrotoxic Action

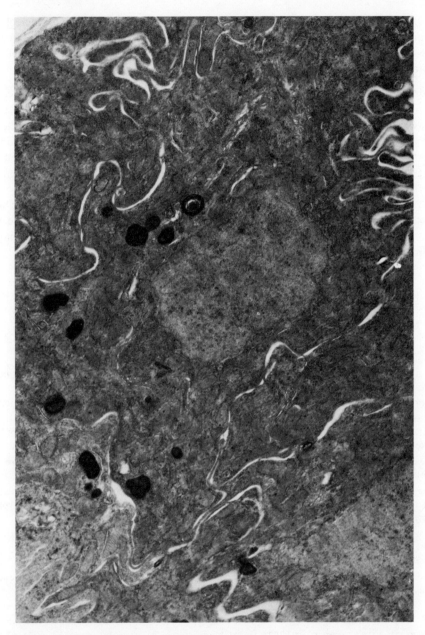

Figure 23.24 Pathologic effect of *Crotalus atrox* venom on the proximal convoluted tubule. (Electron micrograph by Mr. M. Schmidt of author's laboratory.)

degenerated tubules. Vacuolization and shading of the epithelial lining of the cells of Henle's loop and the distal convoluted tubules were observed. Multiple sublethal doses caused mainly glomerular changes with almost normal tubules. Hypertrophy of the cells of the parietal layer of Bowman's capsule appeared. The thickened empty rings of basement membrane were suggestive of endothelial damage. Distortion of some of the glomeruli with obliteration of the capsular space at certain points was observed. The cells of the proximal convoluted tubules showed double nuclei and prominent nucleoli with normal enzymatic activity of both succinic dehydrogenase and alkaline phosphatase.

Necrosis in kidney can also be produced by the venom of *Naja oxiana*, central Asian cobra.

Hydrophiidae. Envenomation by the sea snake *Laticauda semifasciata* did not result in any ultrastructural modifications of the proximal tubules (Schmidt et al., 1976). However, intracellular edema and selective swelling of organelles (endoplasmic reticulum and mitochondria) were noted in most of the visceral epithelial cells. Occasionally the plasma membranes of the light visceral epithelial cells appeared to be ruptured. The ultrastructures of the mesangial cells, the parietal epithelium, and the endothelium were unaffected.

As compared to the sea snake venom, *Crotalus* venom produces much more severe damage. This is in agreement with clinical observations that Crotalidae venoms are much more necrotic than sea snake venoms. The principal action of sea snake venom is neurotoxic; its necrotic action is rather minor.

REFERENCES

Amorim, M. F. and Mello, R. F. (1952). Neforse de nefron intermediá rio no envenenamento crotalice humano: Estudo anatomopatologico, *Mem. Inst. Butantan,* **23,** 281.

Amorim, M. F. and Mello, R. F. (1954). Intermediate nephron nephrosis from snake poisoning in man: Histopathologic study, *Am. J. Pathol.,* **30,** 479.

Amorim, M. F., Mello, R. F., and Saliba, F. (1960). Intermediate nephron nephrosis experimentally induced with *Crotalus t. terrificus* venom in the dog. *Rev. Braz. Biol.,* **20,** 359.

Bonta, I. L., Vargaftig, B. B., De Vos, C. J., and Grijsen, H. (1969). Hemorrhagic mechanisms of some snake venoms in relation to protection by estriol succinate of blood vessel damage, *Life Sci.,* **8,** 881.

Bonta, I. L., Vargaftig, B. B., Bhargava, N., and De Vos, C. J. (1970). Method for study of snake venom induced hemorrhages, *Toxicon,* **8,** 3.

Emery, J. A. and Russell, F. E. (1963). "Lethal and hemorrhagic properties of some North American snake venoms," in H. L. Keegan and Macfarlane, Eds., *Venomous and Poisonous Animals and Noxious Plants of the Pacific Region,* Pergamon, New York, pp. 409–414.

Fayazova, S. R., Brushko, Z. K., and Malikov, K. M. (1969). Histological and histochemical changes in the viscera of mice during poisoning with *Echis* venom, *Zh. Farmakol. Khimioter. Sredstva Toksikol.,* Abstr. No. 254881.

Friederich, C. and Tu, A. T. (1971). Role of metals in snake venoms for hemorrhagic, esterase and proteolytic activities, *Biochem. Pharmacol.,* **20,** 1549.

Fukuyama, T. and Sawai, Y. (1972). Local necrosis induced by cobra (*Naja naja atra*) venom, *Jap. J. Med. Sci. Biol.,* **25,** 211.

Furtado, M. A. and Lester, I. A. (1968). Myoglobinuria following snakebite, *Med. J. Aust.,* **1,** 674.

Grotto, L., Moroz, C., De Vries, A., and Goldblum, N. (1967). Isolation of *Vipera palestinae* hemorrhagin and distinction between its hemorrhagic and proteolytic activities, *Biochim. Biophys. Acta,* **133,** 356.

Grotto, L., Jerushalmy, Z., and De Vries, A. (1969). Effect of purified *Viper palestinae* hemorrhagin on blood coagulation and platelet function, *Thromb. Diath. Haemorrh.,* **22,** 482.

Grotto, L., Jerushalmy, Z., and De Vries, A. (1972). "Effects of *Vipera palaestinae* hemorrhagin on blood coagulation and platelet function," in A. De Vries and E. Kochva, Eds., *Toxins of Animal and Plant Origin,* Gordon and Breach, Vol. 2, New York, pp. 703–706.

Hadler, W. A. and Brazil, O. V. (1966). Pharmacology of crotoxin. IV. Nephrotoxicity, *Mem. Inst. Butantan,* **33,** 1001.

References

Harris, J. B., Johnson, M. A., and Karlsson, E. (1975). Pathological responses of rat skeletal muscle to a single subcutaneous injection of a toxin isolated from the venom of the Australian tiger snake, *Notechis scutatus scutatus, Clin. Exp. Pharmacol. Physiol.*, **2**, 383.

Homma, M. and Tu, A. T. (1970). Antivenin for the treatment of local tissue damage due to envenomation by Southeast Asian snakes: Ineffectiveness in the prevention of local tissue damage in mice after envenomation, *Am. J. Trop. Med. Hyg.*, **19**, 880.

Homma, M. and Tu, A. T. (1971). Morphology of local tissue damage in experimental snake envenomation, *Br. J. Exp. Pathol.*, **52**, 538.

Homma, M., Kosuge, T., Okonogi, T., Hattori, Z., and Sawai, Y. (1967). A histopathological study on arterial lesions caused by Habu (*Trimeresurus flavoviridis*) venom, *Jap. J. Exp. Med.*, **37**, 323.

Juratsch, C. E., Richters, V., and Russell, F. E. (1973). "A quantitative study of *Crotalus* venom on snake and human kidney cells in tissue culture," in E. Kaiser, Ed., *Animal and Plant Toxins*, Wilhelm Goldmann, Munich, pp. 179–184.

Karimov, Z. N., Malikov, K. M., Goncharov, V. E., and Brushko, Z. K. (1968). Toxicological and morphological characteristics of the effect of venom from the Central Asian cobra on animals, *Med. Zh. Uzb.*, **47**.

Kuraskige, S., Hara, Y., Kawakami, M., and Mitsuhashi, S. (1966). Habu (*Trimeresurus flavoviridis*) snake venom. VII. Heat-stable myolytic factor and development of its activity by addition of phospholipase A, *Jap. J. Microbiol.*, **10**, 23.

Maeno, H., Mitsuhashi, S., and Sato, R. (1960). Studies on Habu snake venom. 2c. Studies on Hβ-proteinase of Habu venom. *Jap. J. Microbiol.*, **4**, 173.

Marsden, A. T. H. and Reid, H. S. (1961). Pathology of sea snake poisoning, *Br. Med. J.*, **1**, 1290.

McCollough, N. C. and Gennaro, J. F., Jr. (1963). Summary of snake bite treatment, *J. Fla. Med. Assoc.*, **49**, 977.

McCollough, N. C. and Gennaro, J. F. (1971). "Treatment of venomous snakebite in the United States," in S. A. Minton, Ed., *Snake Venoms and Envenomation*, Marcel Dekker, New York, pp. 137–154.

McKay, D. G., Moroz, C., De Vries, A., Csavossy, I., and Cruse, V. (1970). The action of hemorrhagin and phospholipase derived from *Vipera palestinae* venom on the microcirculation, *Lab. Invest.*, **22**, 387.

Mohamed, A. H., Saleh, A. M., Ahmed, S., and El-Maghraby, M. (1975a). Histopathological and histochemical changes in skeletal muscles after *Bitis gabonica* envenomation, *Toxicon*, **13**, 165.

Mohamed, A. H., Saleh, A. M., Ahmed, S., and Beshir, S. R. (1975b). Histopathological and histochemical effects of *Naja Haje* venoms on kidney tissue of mice, *Toxicon*, **13**, 409.

Moroz, C., Hahn, J., and De Vries, A. (1971). Neutralization of *Vipera palestinae* hemorrhagin by antibody fragments, *Toxicon*, **9**, 57.

Nozaki, M., Yamakawa, M., and Hokama, Z. (1974). Purification and characterization of Sakischima habu (*Trimeresurus elegans*) venom, *Jap. J. Med. Sci. Biol.*, **27**, 83.

Nozaki, M., Yamakawa, M., and Hokama, Z. (1975). Studies on purification of Sakishima habu (*Trimeresurus elegans*) venom and preparation of test toxin for potency test of the antivenin, *Snake*, **7**, 95.

Ohsaka, A. (1960). Fractionation of habu snake venom by chromatography on CM-cellulose with special reference to biological activities, *Jap. J. Med. Sci. Biol.*, **13**, 199.

Ohsaka, A. (1973). An approach to the mechanism of hemorrhage: Hemorrhagic principles isolated from snake venom as a useful analytical tool, *Seitai No Kagaku*, **24**, 2.

Ohsaka, A., Ikezawa, H., Kondo, H., and Kondo, S. (1960a). Two hemorrhagic principles derived from habu snake venom and their difference in zone electrophoretical mobility. *Jap. J. Med. Sci. Biol.*, **13**, 73.

Ohsaka, A., Ikezawa, H., Kondo, H., Kondo, S., and Uchida, N. (1960b). Haemorrhagic activities of habu snake venom, and their relations to lethal toxicity, proteolytic activities and other pathological activities, *Br. J. Exp. Pathol.*, **41**, 478.

Ohsaka, A., Omori-Satoh, T., Kondo, H., Kondo, S., and Murata, R. (1966). Biochemical and pathological aspects of hemorrhagic principles in snake venoms with special reference to habu (*Trimeresurus flavoviridis*) venom, *Mem. Inst. Butantan*, **33**, 193.

Ohsaka, A., Ohashi, M., Tsuchiya, M., Kamisaka, Y., and Fujishiro, Y. (1971a). Action of *Trimeresurus flavoviridis* venom on the microcirculatory system of rat: dynamic aspects as revealed by cinephotomicrographic recording, *Jap. J. Med. Sci. Biol.*, **24**, 34.

Ohsaka, A., Takahashi, T., Omori-Satoh, T., and Murata, R. (1971b). "Purification and characterization of the hemorrhagic principle in the venom of *Trimeresurus flavoviridis*" in A. De Vries and E. Kochva, Eds., *Toxins of Animal and Plant Origin*, Gordon and Breach, Vol. 1, New York, pp. 369–399.

Ohsaka, A., Just, M., and Habermann, E. (1973). Action of snake venom hemorrhagic principles on isolated glomerular basement membrane, *Biochim. Biophys. Acta*, **323**, 415.

Okongi, T., Hoshi, S., Homma, M., Suto, K., Iizuka, H., and Sato, M. (1962). Local pathological changes following injection of heat-stable substances in the habu snake (*Trimeresurus flavoviridis*) venom, *Kitakanto Med. J.*, **12**, 31.

Okonogi, T., Homma, M., and Hoshi, S. (1964). Pathological studies of habu snake bite, *Gunma J. Med. Sci.*, **13**, 101.

Omori, T., Iwanaga, S., and Suzuki, T. (1964). The relationship between the hemorrhagic and lethal activities of Japanese mamushi (*Agkistrodon halys blomhoffii*) venom, *Toxicon*, **2**, 1.

Omori-Satoh, T. and Ohsaka, A. (1970). Purification and some properties hemorrhagic principle I in the venom of *Trimeresurus flavoviridis.*, *Biochim. Biophys. Acta*, **207**, 432.

Omori-Satoh, T., Ohsaka, A., Kondo, S., and Kondo, H. (1967). A simple and rapid method for separating two hemorrhagic principles in the venom of *Trimeresurus flavoviridis*, *Toxicon*, **5**, 17.

Omori-Satoh, T., Sadahiro, S., Ohsaka, A., and Murata, R. (1972). Purification and characterization of an antihemorrhagic factor in the serum of *Trimeresurus flavoviridis*, a crotalid, *Biochim. Biophys. Acta*, **285**, 414.

Oshima, G., Omori-Satoh, T., Iwanaga, S., and Suzuki, T. (1972). Studies on snake venom hemorrhagic factor I (HR-I) in the venom of *Agkistrodon halys blomhoffii:* Its purification and biological properties, *J. Biochem.*, **72**, 1483.

Ownby, C. L. (1975). *Pathogenesis and Chemical Treatment of Hemorrhage Induced by Rattlesnake Venom*, Ph.D. dissertation, Colorado State University.

Ownby, C. L., Kainer, R. A., and Tu, A. T. (1974). Pathogenesis of hemorrhage induced by rattlesnake venom, *Am. J. Pathol.*, **75**, 401.

Ownby, C L., Cameron D., and Tu, A. T. (1976a). Isolation of myotoxic component from rattlesnake venom: Electron microscopic analysis of muscle damage, *Am. J. Pathol.*, **85** 149.

Ownby, C L., Cameron, D. and Tu A. T. (1976b). "Electron microscopic analysis of myonecrosis by a pure component isolated from rattlesnake venom" in G. W Bailey ed. *34th Ann. Proc. Electron Microscopy Soc. Amer.*, p. 292.

Raab, W. and Kaiser, E. (1966a). Nephrotoxic action of snake venoms, *Mem. Inst. Butantan*, **33**, 1017.

Raab, W. and Kaiser, E. (1966b). Enzymic activity of rat urine after snake venom-induced renal damage, *Wien. Z. Inn. Med. Grenzgb.*, **47**, 327.

Reid, H. A. (1961). Myoglobinuria and sea-snakebite poisoning, *Br. Med. J.*, **1**, 1284.

Reid, H. A. (1964). Cobra bites, *Br. Med. J.*, **2**, 540.

Reid, H. A. (1968). "Symptomatology, pathology and treatment of land snake bite in India and Southeast Asia," in W. Bücherl, E. Buckley, and V. Deulofeu, Eds., *Venomous Animals and Their Venoms,* Academic, New York. pp. 611–641.

Sandbank, U., Jerushalmy, Z., Ben-David, E., and De Vries, A. (1974). Effect of *Echis coloratus* venom on brain vessels, *Toxicon*, **12**, 267.

Sant, S. M. and Purandare, N. M. (1972). Autopsy study of cases of snake bite with special reference to the renal lesions. *J. Postgrad. Med.*, **18**, 181.

Sant, S. M., Tembe, V. S., Salgaonkar, D. S., and Purandare, N. M. (1974). Lesions produced by *Echis carinatus* venom in experimental animals, *J. Postgrad. Med.*, **10**, 70.

Schmidt, M. E., Abdelbaki, Y. Z., and Tu, A. T. (1976). Nephrotoxic action of rattlesnake and sea snake venoms: An electron microscopic study, *J. Pathol.*, **118**, 75.

Silva, H. B., Brito, T., Lima, P. R., Penaa, D. O., Almeida, S. S., and Mattar, E. (1966). Acute anuric renal insufficiency due to snake, waxbee, and spider bites: Clinical and pathological observations in 3 cases, *Int. Cong. Nephrol.*, p. 273.

References

Sitprija, V., Sribhibhadh, R., Benyajati, C., and Tangchai, P. (1973). "Acute renal failure in snakebite," in A. De Vries and E. Kochva, Eds., *Toxins of Animal and Plant Origin,* Vol. 3, Gordon and Breach, New York, pp. 1013–1028.

Stahnke, H. L. (1966). *The Treatment of Venomous Bites and Stings,* Arizona State University, Tempe.

Stringer, J. M., Kainer, R. A., and Tu, A. T. (1971). Ultrastructural studies of myonecrosis induced by cobra venom in mice, *Toxicol. Appl. Pharmacol.,* **18,** 442.

Stringer, J. M., Kainer, R. A., and Tu, A. T. (1972). Myonecrosis induced by rattlesnake venom: An electron microscopic study, *Am. J. Pathol.,* **67,** 127.

Suzuki, T. (1966). Pharmacologically and biochemically active components of Japanese ophidian venoms, *Mem. Inst. Butantan,* **33,** 519.

Suzuki, Y., Churg, J., Grishman, E., Maunter, W., and Dachs, S. (1963). The mesangium of the renal glomerulus: Electron microscopic studies of pathological alterations, *Am. J. Pathol.,* **43,** 555.

Tadokoro, S. and Kurihara, N. (1963). Systemic symptoms produced by administration of habu snake (*Trimeresurus flavoviridis*) venom: Especially on acute adrenocortical necrosis, *Gunma J. Med. Sci.,* **12,** 227.

Takahashi, T. and Ohsaka, A. (1970). Purification and some properties of two hemorrhagic principles (HR2a and HR2b) in the venom of *Trimeresurus flavoviridis:* Complete separation of the principles from proteolytic activity, *Biochim. Biophys. Acta,* **207,** 65.

Tembe V. S., Sant, S. M. and Purandare, N. M. (1975) A clinico-pathologic study of snakebite cases. *J. Postgrad. Med ,* **21,** 36.

Toom, P. M., Squire, P. G., and Tu, A. T. (1969). Characterization of the enzymatic and biological activities of snake venoms by isoelectric focusing, *Biochim. Biophys. Acta,* **181,** 339.

Tsuchiya, M., Ohshio, C., Ohashi, M., Ohsaka, A., Suzuki, K., and Fujishiro, Y. (1974). Cinematographic and electron microscopic analysis of the hemorrhage induced by the main hemorrhage principle, HR1, isolated from the venom of *Trimeresurus flavoviridis, Throm. Diath. Haemorrh., Suppl.* LX, 439.

Tsuchiya, M., Oshio, C., Ohashi, M., A. Ohsaka, and Y. Fujishior (1975). Electrom microscopical study of the hemorrhage induced by the venom of *Trimeresurus flavoviridis, Bibl. Anat.,* **13,** 190.

Tu, A. T. (1971). "The mechanism of snake venom actions – rattlesnakes and other crotalids," in L. L. Simpson, Ed., *Neuropoisons, Their Pathophysiological Actions,* Vol 1: *Poisons of Animal Origins,* Plenum, New York, pp. 87–110.

Tu, A. T. and Homma, M. (1970). Toxicologic study of snake venoms from Costa Rica, *Toxicol. Appl. Pharmacol.,* **16,** 73.

Tu, A. T., Toom, P. M., and Ganthavorn, S. (1967). Hemorrhagic and proteolytic activities of Thailand snake venoms, *Biochem. Pharmacol.,* **16,** 2125.

Yoshikura, H., Ogawa, H., Ohsaka, A., and Omori-Satoh, T. (1966). Action of *Trimeresurus flavoviridis* venom and the partially purified hemorrhagic principles on animal cells cultivated in vitro, *Toxicon,* **4,** 183.

24 Autopharmacological Action

1 BRADYKININ — 400
   1.1 Bradykinin Release, 400
   1.2 Relation to Esterase Activity, 401
   1.3 Chemical Properties of Kinin-Releasing Enzyme, 402

2 BRADYKININ-POTENTIATING FACTOR AND ANGIOTENSINASE INHIBITORS — 403
   2.1 Action of Bradykinin-Potentiating Factor, 403
   2.2 Action of Angiotensinase Inhibitor, 404
   2.3 Structural Similarity, 404
   2.4 Mechanism of Action, 406

3 HISTAMINE — 407

4 OTHER AUTOPHARMACOLOGIC SUBSTANCES — 408

   References — 409

One of the very interesting properties of certain snake venoms is their autopharmacological action. Upon envenomation, substances not originally present in the venom are released from the tissues. A variety of such substances are released, including bradykinin, angiotensin, histamine, serotonin, and ATP.

1 BRADYKININ

1.1 Bradykinin Release

Bradykinin is a strong hypotensive agent because of its vasodilator action, as well as a potent substance for increasing capillary permeability. The precursor of bradykinin in the plasma globulins is called kininogen. Kallikrein or trypsin can liberate bradykinin by specific proteolysis of kininogen. Kallikrein is sometimes called kininogenase or bradykinin-releasing enzyme (EC 3.4.21.8). (See Scheme 26.)

1 Bradykinin

$$\text{Kininogen} \xrightarrow[\text{kallikrein}]{\text{Trypsin,}} \text{Bradykinin + Lysylbradykinin (kallidin)}$$

Scheme 26

Bradykinin is a nonapeptide that has the following sequence:

Arg–Pro–Pro–Gly–Phe–Ser–Pro–Phe–Arg

Kallidin has an additional lysine residue at the N-terminal.

Release of bradykinin from animal intestine, uterus, and smooth muscle by *Bothrops jararaca* was first observed by Rocha e Silva et al. (1949). In sheep and human plasma, the bradykinin-releasing activity of *Crotalus atrox* venom is more potent than that of trypsin. Bradykininogen is completely exhausted in 1 hr by a continuous infusion of 1.5 to 2 γ of venom min^{-1} kg^{-1}, and the precursor, kininogen, does not regenerate for several hours (Margolis et al., 1965). Clinically, snakebite victims of *Crotalus* experienced lower bradykininogen levels in their blood (Russell, 1965), a finding consistent with the concept that bradykinin is liberated from its precursor, bradykininogen. Actually, bovine plasma contains at least two kininogens, kininogens I and II, and venoms of snakes from Japan and Formosa release bradykinin from both types (Suzuki et al., 1967). Using the whole venom of *Bothrops jararaca*, Rothchild and Almeida (1972) observed that about two-thirds of the circulating bradykininogen in rats disappeared after a dose of 2 mg venom per gram body weight of the animal.

Although heparin does not prevent the conversion of plasma kininogen to bradykinin, it inhibits the hypercoagulative action of venom. Victims can survive the condition of massive release of bradykinin if intravascular coagulation is prevented. This evidence indicates that bradykinin-releasing activity is distinct from the lethal and blood coagulation actions of the venom.

Kinin-releasing activity (kininogenase) is not restricted to venoms of Crotalidae; Viperidae venoms also contain such activities. Henriques and Evseeva (1969) showed that venoms of snakes of the Viperidae family, such as *Vipera lebetina, V. ursini*, and *Echis carinatus*, had kininogenase activity, but such activity was absent in the venom of *Naja naja*, a member of the Elapidae family. Evseeva and Enriques (1973) found two fractions of kininogenase in the venom of *Echis carinatus*. One fraction contained a weakly active, inhibited form of kininogenase, and the other an active form of the enzyme. Venom of *Vipera russellii* also contains an inhibitor of kallikrein, plasmin, and trypsin (Takahashi et al., 1972). Its chemical properties were discussed in Chapter 10, "Enzyme Inhibitors."

1.2 Relation to Esterase Activity

The bradykinin-releasing activity of *Bothrops jararaca* venom is not related to its proteolytic activity on casein (Hamberg and Rocha e Silva, 1957a), but does correspond to its esterase activity (Hamberg and Rocha e Silva, 1957b). This was confirmed by Iwanaga et al. (1965), who also found that bradykinin release is related to hydrolytic action on arginine esters. The venom of *Agkistrodon halys blomhoffii* contains three different arginine ester-hydrolyzing activities (Sato et al., 1965), and one fraction does show bradykinin-releasing action. The other two fractions, which do not release bradykinin, show clotting and permeability-increasing activity. Altogether, 31 venoms from the Viperidae and Crotalidae families that show kininogenase (kallikrein) activity also hydrolyze BAEE (*N*-benzoyl-L-arginine ethyl ester) and TAME (*p*-toluenesulfonyl-DL-arginine methyl ester). Although purified *Bitis gabonica* kininogenase did not hydrolyze BAPA (N^2-benzoyl-DL-arginine-*p*-nitranilide), it hydrolyzed BAEE and TAME (Mebs, 1969). The purified kininogenase possessed only weak proteolytic activity.

Investigations so far show that the bradykinin-releasing action of venoms is found only in snakes of the Crotalidae and Viperidae families. Among the many venoms tested by Tu et al. (1965, 1966, 1967), only Crotalidae and Viperidae venoms hydrolyzed TAME and BAEE; none of the Elapidae venoms (with the exception of *Ophiophagus hannah*) hydrolyzed these arginine esters. Furthermore, the kininogenase (bradykinin-releasing) and arginine ester-hydrolyzing activities of many snake venoms have been investigated by Oshima et al. (1969). Crotalidae (*Agkistrodon halys blomhoffii, A. piscivorus piscivorus, A. contortrix contortrix, A. contortrix mokeson, A. acutus, Crotalus adamanteus, C. atrox, C. durissus terrificus, C. vividis, C. basiliscus, T. flavoviridis, T. okinavensis, T. mucrosquamatus, T. gramineus, Bothrops jararaca, B. atrox*) and Viperidae venoms (*Vipera russellii, V. palestinae, V. ammodytes, Echis carinatus, Causus rhombeatus, Bitis gabonica, B. arietans*) showed both activities, although the two did not necessarily parallel each other. None of the nine Elapidae venoms analyzed showed any arginine ester activity, a finding in agreement with the work of Tu et al. (1965, 1966, 1967). Moreover, Elapidae venoms showed no or very weak bradykinin-releasing activity. Therefore, on the basis of whole venom activity, there appear to be a definite relation between releasing activity and arginine ester-hydrolyzing activity. Although trypsin can release bradykinin from bradykininogen (kininogenase activity) and possesses arginine esterase activity, venom kininogenase is different from trypsin. Evidence indicates that *Vipera lebetina* kininogenase activity is not inhibited by soybean trypsin inhibitor (Evseeva and Enriques, 1969).

Bradykinin-releasing factor isolated from the venom of *Echis coloratus* has a molecular weight of 22,000 (Cohen et al., 1969) and has arginine ester-hydrolyzing activity but is devoid of proteolytic, hemorrhagic, and fibrinolytic activities (Cohen et al., 1970). The enzyme does not hydrolyze arginine amide or lysine esters. The pH optimum of kinin-releasing enzyme isolated from *Bitis gabonica* is 9.5 (Mebs, 1969). The molecular weight is 33,500, as determined by gel filtration, and the release of kinin by the enzyme is inhibited by the synthetic substrates BAEE and TAME. The enzyme isolated from the venom of *Vipera ammodytes ammodytes* has a molecular weight of 29,500 in the presence of sodium dodecyl sulfate (Bailey and Shipolini, 1976). Without detergent, the molecular weight is 33,500, as determined by gel filtration; without SDS, the molecular weight is 40,500. Here again, the release of kinin by the enzyme is inhibited by the synthetic substrates BAEE and TAME. The isoelectric point is 7.2, and the enzyme contains 10% carbohydrate. The amino acid composition is as follows: 23 Asx, 14 Thr, 14 Ser, 17 Glx, 20 Pro, 17 Gly, 16 Ala, 12 Val, 6½ Cys, 4 Met, 16 Ile, 20 Leu, 7 Try, 6 Phe, 9 His, 11 Lys, 11 Arg, and 3 Trp.

1.3 Chemical Properties of Kinin-Releasing Enzyme

The enzyme inhibitor DFP destroys arginine ester-hydrolyzing activity as well as bradykinin-releasing activity (Iwanaga et al., 1965). Trasyrol, a commercial kallikrein inhibitor, inhibits the release of bradykinin from bradykininogen. The synthetic substrate BAEE inhibits the enzyme activity, and the inhibition is dependent of the BAEE concentration. The pH optimum of kininogenase (from *Agkistrodon halys blomhoffii* venom) is 8.5, and enzyme activity is not dependent on metal ions.

Snake venom kininogenase is somewhat different from kallikrein of hog pancreatic origin (Kato and Suzuki, 1970a). Both enzymes hydrolyzed polyarginine, α-N-tosyl-derivatives of arginine, and lysine methyl esters, but did not hydrolyze α-N-tosyl-methionine ethyl ester. Polylysine can be hydrolyzed by hog kallikrein but not by *A. halys blomhoffii* kininogenase. Native kininogen II and a chemically modified substrate (reduced and S-carboxy methlated kininogen II) are hydrolyzed by venom kininogenase. However, although hog kallikrein released kallidin from the native bovine kinogen II, it did not release kinin from the modified kininogen II.

The sites of action of venom kininogenase, trypsin, plasmin, and pancreatic kallikrein on kininogen II are summarized in Scheme 27.

–Ser–Arg–Met–Lys–Arg–Pro–Pro–Gly–Phe–Ser–Pro–Phe–Arg–Ser–Val–Gln–
　　　　　　↑　　↑　　　　　　　　　　　　　　　　　　　↑
　　　　　(3)　(1)　　　　　　　　　　　　　　　　　　　　(2)

　　　　　　　　　　　　　　　　　　　　　　　　　　　　　　　　Val–Met

(1) and (2) by venom kininogenase, trypsin, plasmin
(3) and (2) by pancreatic kallikrein

Scheme 27

The enzyme isolated from *Echis coloratus* venom is thermostable with a molecular weight of 22,000 daltons (Cohen et al., 1970). It is sensitive to DFP, but insensitive to trasyol and soybean trypsin inhibitor. It hydrolyzes Phe–Lys–Ala–Ala–NH_2, Ala–Phe–Lys–Ala–NH_2 Phe–Ala–Lys–Ala–NH_2, and Ala–Ala–Lys–Ala–NH_2, but not polylysine, polyarginine, and clupein. The pH optimum of the enzyme is 8, and enzymatic activity can be seen only in the pH range of 6 to 10.

From all this evidence, we can conclude that venom bradykinin-releasing factors are similar to trypsin or pancreatic kallikrein but not identical.

2   BRADYKININ-POTENTIATING FACTOR AND ANGIOTENSINASE INHIBITORS

Certain peptides in snake venoms enhance the activity of bradykinin. These compounds are called bradykinin-potentiating factors, BPF. The BPFs were first observed in the venom of *Bothrops jararaca* by Ferreira (1965) and his co-workers (Ferreira et al., 1970a, b). The peptides are thermostable and consist of dialyzable compounds, an indication of their small molecular size.

2.1   Action of Bradykinin-Potentiating Factor

Intravenous injections of BPF in dogs and cats causes little enhancement of hypotensive action. But in the presence of bradykinin, the hypotensive action is greatly increased. Bradykinin-potentiating factor does not enhance the contraction effect elicited by histamine and acetylcholine on guinea pig ileum. The potentiating effect on BPF on bradykinin action is believed to prevent bradykinin inactivation by blood or by kininases present in smooth muscles (Ferreira and Rocha e Silva, 1965).

Peptides structurally similar to BPF have also been isolated from snake venoms, and have been shown to have angiotensinase inhibitory action (Ondetti et al., 1971). These angiotensinase inhibitors and BPF are probably identical compounds because they show chemical similarity and the same biological properties. This high degree of similarity will be discussed further in Section 2.3.

2.2 Action of Angiotensinase Inhibitor

Renin, an enzyme produced by the kidney, produces angiotensin I from angiotensinogen, a serum α_2-globulin formed by the liver. The serum enzyme angiotensinase liberates His–Leu and yields angiotensin II, a powerful hypertensive agent.

The conversion of angiotensin I to angiotensin II can be inhibited in the isolated tissues by a pentapeptide extracted from *Bothrops jararaca* venom. The pentapeptide has the structure pGlu–Lys–Trp–Ala–Pro (Aiken and Vane, 1970).

Angiotensinase inhibitor is effective in lowering blood pressure in chronic renal hypertensive rats. When Loyke (1973) injected PGlu–Asn–Trp–Pro–His–Pro–Gln–Ile–Pro–Pro subcutaneously into 18-week-old rats, the average systolic pressure before treatment was 193 mm Hg, decreasing to 176 ± 1.9 mm Hg ($p < 0.001$). Upon discontinuing the injection, the blood pressure rose to 195 mm Hg. Histologic studies by light microscopy of the liver, lungs, adrenal glands, and skin did not show any apparent toxic effect of this peptide. Angiotensinase inhibitors also possess an inhibitory action on the centrally elicited pressor effect of angiotensin II, and they are selective to central nervous system (CNS) receptor sites. Solomon et al. (1974) tested a pentapeptide, pGlu–Lys–Trp–Ala–Pro, and a nonapeptide, pGlu–Trp–Pro–Arg–Pro–Glu–Ile–Pro–Pro, for angiotensinase inhibition. Intraventicular administration of these peptides attenuated the central pressor effects elicited by both angiotensins I and II. This action appeared specific to the CNS because administration of these peptides peripherally did not antagonize angiotensin II, but did inhibit angiotensin I. The inhibitors do not significantly affect ganglionic or neural transmission in nictitating membrane studies in cats and do not alter the parasympathetic system, since there is no shift in the frequency response curve obtained by vagal stimulation. They also do not significantly alter the blood pressure responses to acetylcholine, epinephrine, or bilateral carotid occlusion.

2.3 Structural Similarity

By what mechanism does BPF enhance bradykinin activity? It is known that BPFs are strong inhibitors of angiotensinase (Ferreira et al., 1970a,b). It is also well known that certain snake venoms contain inhibitors of the angiotensin-converting enzyme, angiotensinase (Cheung and Cushman, 1973). These relations are summarized in Scheme 28.

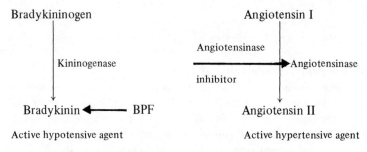

Scheme 28

The important question arises as whether BPF is identical to angiotensinase inhibitor. Upon reviewing the chemical structures of all known BPFs and the inhibitors (Table 1), one is fascinated by the extreme similarity among them. The N-terminal amino acids are pyroglutamic acid without exception; and, as can be seen from the amino sequences of the BPFs and the inhibitors, there are extremely high degrees of homology. Therefore it is

Table 1 Comparison of Peptide Structures of Bradykinin-Potentiating Factors and Angiotensinase Inhibitors

Venoms	Name	1	2A	2B	3	4	5	6	7	8	9	10	11	12	Total Residues	References
Bothrops jararaca	V-9	pGlu-	Gly-	Gly-	Trp-	Pro-	Arg-	Pro-	Gly-	Pro-	Glu-	Ile-	Pro-	Pro	13	Ondetti et al, 1971
Agkistrodon halys blomhoffii	potentiator B	pGlu-	Gly-	Leu-	Pro-	Pro-	Arg-	Pro-	Lys-	-[a]		Ile-	Pro-	Pro	11	Kato and Suzuki, 1971
A. halys blomhoffii	potentiator C	pGlu-	Gly-	Leu-	Pro-			Pro-	Lys-	Pro-	Ile-	Pro-	Pro		11	Kato and Suzuki, 1971
B. jararaca	V-2	pGlu-			Trp-	Pro-	Arg-	Pro-	Thr-	Pro-	Gln-	Ile-	Pro-	Pro	11	Ondetti et al, 1971
B. jararaca	IV	pGlu-			Trp-	Pro-	Arg-	Pro-	Thr-	Pro-	Gln-	Ile-	Pro-	Pro	11	Bodanszky et al, 1971
A. halys blomhoffii	potentiator E	pGlu-	Lys-		Trp-			Asp-	Pro-	Pro-	Val-	Ser-	Pro-	Pro	11	Kato and Suzuki, 1970b
B. jararaca	V-8	pGlu-	Ser-		Trp-	Pro-			Glu-	Pro-	Asn-	Ile-	Pro-	Pro	10	Ondetti et al, 1971
B. jararaca	V-7	pGlu-	Asn-		Trp-	Pro-	His-			Pro-	Gln-	Ile-	Pro-	Pro	10	Ondetti et al, 1971
B. jararaca	V-6-II	pGlu-	Asn-		Trp-	Pro-	Arg-			Pro-	Gln-	Ile-	Pro-	Pro	10	Ondetti et al, 1971
A. halys blomhoffii	potentiator A	pGlu-	Gly-		Arg-	Pro-		Pro-	Gly-	Pro-		Ile-	Pro-	Pro	10	Kato et al, 1973
B. jararaca	I	pGlu-	Asn-		Trp-	Pro-	His-	Pro-			Gln-	Ile-	Pro-	Pro	10	Bodanszky et al, 1971
B. jararaca	II.	pGlu-	Ser-		Trp-	Pro-	Arg-	Pro-			Gln-	Ile-	Pro-	Pro	10	Bodanszky et al, 1971
B. jararaca	V	pGlu-	Asn-		Trp-	Pro-	Arg-	Pro-	Gly-	Pro-	Asn-	Ile-	Pro-	Pro	10	Bodanszky et al, 1971
B. jararaca	V-6-I	pGlu-			Trp-	Pro-	Arg-				Gln-	Ile-	Pro-	Pro	9	Ondetti et al, 1971
B. jararaca	III	pGlu-			Trp-	Pro-	Arg-			Pro-	Gln-	Ile-	Pro-	Pro	9	Bodanszky et al, 1971
B. jararaca	VI	pGlu-	Lys-		Phe-		Ala-	Pro-							5	Bodanszky et al, 1971
B. jararaca	VII	pGlu-	Lys-		Trp-		Ala-	Pro-							5	Bodanszky et al, 1971
B. jararaca	SQ20,475 (V-3A)	pGlu-	Lys-		Trp-		Ala-	Pro-							5	Aiken and Vane, 1970 Ferreira et al, 1970 a,b
A. halys blomhoffi	peptide I	pGlu-	Lys-		Ser-										3	Okada et al, 1974
Trimeresurus mucrosquatmatus	—	pGlu-	Asn-		Trp-										3	Lo et al, 1973
A. halys blomhoffii	—	pGlu-	Glu-		Trp-										3	Kato et al, 1966
T. gramineus	—	pGlu-			Trp-	-Lys									3	Lo et al, 1973
T. gramineus	—	pGlu-	Gln-		Trp-										3	Lo et al, 1973

[a] —— No name was assigned.

very likely that these two substances are identical compounds with the same physiological activities. In Table 1, some peptides whose physiological functions have not been tested are also listed. The fact that the structures of these peptides are very similar to those of BPFs or angiotensinase inhibitors also indicates that they probably retain the same physiological activities. Finally, Ferreira et al. (1970a, b) tested the angiotensinase inhibition activity of BPF peptide V-3A, which has the structure pGlu–Lys–Trp–Ala–Pro, and found that the BPF indeed inhibited angiotensinase activity. The structure of V-3A is identical to that of the peptide tested by Aiken and Vane (1970).

2.4 Mechanism of Action

Synthetic peptides analogous to naturally occuring inhibitors present in *Bothrops jararaca* venom also inhibit angiotensinase. The following peptides:

pGlu–Lys–Trp–Ala–Pro,
pGlu–Trp–Pro–Arg–Pro–Gln–Ile–Pro–Pro,
pGlu–Ser–Trp–Pro–Gly–Pro–Asn–Ile–Pro–Pro,

inhibited the conversion of angiotensin I to angiotensin II (Bakhle, 1971). The inhibitory actions are due to the competitive binding inhibitor to the enzyme's normal substrate (angiotensin I and hippuryl–His–Leu) binding site (see Scheme 29).

Asp–Arg–Val–Tyr–Ile–His–Pro–Phe–His–Leu (angiotensin I)

| Angiotensinase
↓

Asp–Arg–Val–Tyr–Ile–His–Pro–Phe (angiotensin II)
+
His–Leu

Scheme 29

The longer peptides such as pGlu–Trp–Pro–Arg–Pro–Gln–Ile–Pro–Pro and pGlu–Asn–Trp–Pro–His–Pro–Gln–Ile–Pro–Pro are competitive inhibitors, whereas smaller peptides, for example, pGlu–Lys–Trp–Ala–Pro, are mixed competitive and non-competitive inhibitors (Cheung and Cushman, 1973). The K_i (enzyme inhibitor dissociation constant) values do not depend significantly on the substrate employed but are altered by differences in pH. The fact that the inhibitors act on angiotensinase rather than on the pressor action of angiotensin II was amply demonstrated by Engel *et al* (1972). They found that synthetic peptides similar to those of *Bothrops jararaca* venom inhibited the pressor responses induced by angiotensin I but not those induced by angiotensin II. The dual properties of BPF and angiotensin inhibitors were also demonstrated by Engel et al. (1972). They also showed that these peptides augmented the vasodepressor effect of bradykinin in rats. Bakhle (1972) proved that both synthetic BPF and the natural peptides from *Agkistrodon halys blomhoffii* venom inhibited angiotensin I-converting enzyme (angiotensinase) activity in isolated, perfused guinea pig lung or the particulate fraction of dog lung homogenates.

The identity in physiological activity and the similarity in chemical structure between BPF and angiotensinase inhibitors clearly indicate that these two substances, called by different names, are identical compounds.

3 HISTAMINES

Histamine commonly occurs in animal tissues in relatively low concentration. It is a powerful vasodilating compound and at very high concentrations may cause vascular collapse. Snake venom liberates histamine upon envenomation.

As early as 1937, Feldberg and Kellaway observed that histamine was released from perfused lung by the venoms of *Denisonia superba, Naja naja,* and *Crotalus atrox.* Histamine can also be released from many other tissues, such as dog liver (Feldberg and Kellaway, 1938) and rat diaphragm (Dutta and Narayanan, 1952, 1954; Mohamed and Zaki, 1957), thus causing an increase of histamine level in the blood (Dragstedt et al., 1938; Mohamed et al., 1971).

Crotamine is a basic protein with lethal activity isolated from the venom of the tropical rattlesnake (*Crotalus durissus terrificus*). Crotamine is known to be capable of releasing histamine from rat tissues (Gonçalves, 1956). However, venoms from certain varieties of Brazilian rattlesnakes that do not contain crotamine also have a high potency for histamine release (Rothschild, 1966). Thus crotamine is not a primary agent for the release of histamine.

Chiang et al. (1964) used crude venoms of five Formosan snakes to study the liberation of histamine from rat diaphragm preparations and found that venoms which were stronger on phospholipase A_2 activity were also stronger in histamine-releasing activity. Thus they concluded that the release of histamine was due to venom

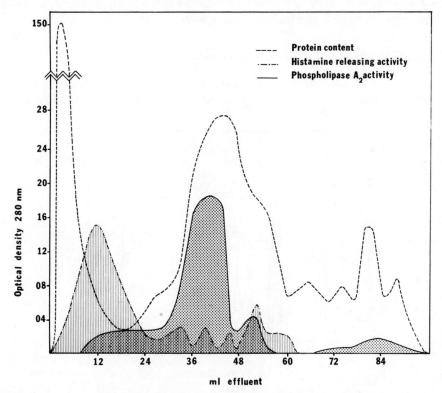

Figure 24.1. Separation of the phospholipase A_2 fraction (▓) from the histamine-releasing fraction (lll). The venom used was *Crotalus durissus terrificus*. (Reproduced from A. M. Rothschild, 1966, by permission of the Institute of Butantan and Dr. Rothschild.)

phospholipase A_2. The role of phospholipase A_2 in histamine release had been proposed as far back as 1938 by Feldberg and Kellaway. There were, however, also many items of evidence that phospholipase A_2 is not the factor for histamine release.

Later Rothschild (1966, 1967) clearly demonstrated that the histamine-releasing fraction is different from phospholipase A_2 fractions (Fig. 1). Purified phospholipase releases very little histamine from rat peritoneal cell suspensions and none in perfused guinea pig lungs (Damerau et al., 1975). However, direct lytic factor releases histamine from both systems. Histamine-releasing factor is a heat-labile (100°, 5 min), nondialyzable protein that is inactivated by trypsin (Rothschild, 1966). It is believed that the factor is a specific proteolytic enzyme.

Liberated histamine is partly responsible for local tissue reactions. Antivenin and EDTA inhibit histamine release by the venom of *Echis carinatus* (Mohamed et al., 1968, 1971).

In the fractionation of *Crotalus durissus terrificus* venom in Amberlite IRC-50, the histamine-releasing fraction is separated from the lethal and hemolytic fraction (Rothschild, 1966).

Antihistamine drugs do not reduce the toxic effects of *Naja naja* venom on rats and guinea pigs, nor is the survival time significantly affected when envenomated animals are tested with antazoline, tripelennamine, or promethazine (Dutta and Narayanan, 1954). The same results were obtained by Kaiser and Raab (1966), suggesting that histamine is not responsible for lethal action. Antiserotonin drugs do not influence the survival time of mice envenomated by *Agkistrodon piscivorous* (Kaiser and Raab, 1966). Therefore serotonin release is also a factor for lethal action of the venom.

4 OTHER AUTOPHARMACOLOGIC SUBSTANCES

Serotonin and ATP are released from platelets by the venom of *Crotalus durissus terrificus* (Markwardt et al., 1966). Serotonin-releasing factor is a protein and is clearly

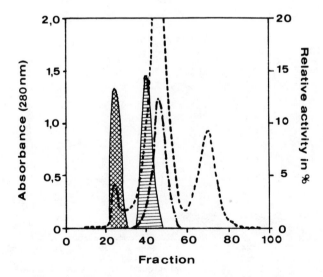

Figure 24.2. Separation of the phospholipase A_2 fraction (–•–•–) from the serotonin-releasing activity fraction (⊗) and the blood coagulation fraction (≡). (Reproduced from Markwardt et al., 1966 by permission of the author, Dr F. Markwardt and of the copyright owner, *Biochem. Z*).

separable from phospholipase A_2 and blood coagulation activity (Fig. 2). Serotonin-releasing activity is inhibited by removing Ca(II) or by the blocking of platelet glycolysis.

Lysosomes contain many different types of hydrolases. It was found by Kramar et al. (1971) that venoms of *Vipera ammodytes* and *Agkistrodon piscivorus* released cathepsin, β-glucuronidase, and acid phosphatase. Pure phospholipase A_2 or direct lytic factor (DLF) alone released relatively little acid phosphatase and β-glucuronidase but considerable amounts of cathepsin. A mixture of phospholipase A_2 and DLF greatly enhances the releasing hydrolase activity. This synergistic action of two components is also mentioned in Chapter 2, "Phospholipase A_2," Chapter 19, "Nonneurotoxic Basic Proteins, (Cardiotoxins, Cytotoxins, and Others)" and Chapter 20, "Hemolysis." Kramar et al. attributed this hydrolase release from lysosomes to the same cytolytic principles, a combination of DLF and phospholipase A_2.

REFERENCES

Aiken, J. W. and Vane, J. R. (1970). The renin–angiotensin system: Inhibition of converting enzyme in isolated tissues, *Nature,* **228,** 30.

Bailey, G. S. and Shipolini, R. A. (1976). Purification and properties of a kininogenin from the venom of *Vipera ammodytes ammodytes, Biochem. J.,* **153,** 409.

Bakhle, Y. S. (1971). Inhibition of angiotensin I converting enzyme by venom peptides, *Br. J. Pharmacol.,* **43,** 252.

Bakhle, Y. S. (1972). "Inhibition of converting enzyme by venom peptides," in J. Genest, Ed., *Hypertension,* Springer, Berlin, pp. 541–547.

Bodanszky, A., Ondetti, M. A., Ralofsky, C. A., and Bodanszky, M. (1971). Optical rotatory dispersion of the proline-rich peptides from the venom of *Bothrops jararaca, Experientia,* **27,** 1269.

Cheung, H. S. and Cushman, D. W. (1973). Inhibition of homogeneous angiotensin-converting enzyme of rabbit lung by synthetic venom peptides of *Bothrops jararaca, Biochim. Biophys. Acta,* **293,** 451.

Chiang, T. S., Ho, K. J., and Lee, C. Y. (1964). Release of histamine from the rat diaphragm preparation by Formosan snake venoms, *J. Formosan Med. Assoc.,* **63,** 127.

Cohen, I., Zur, M., Kaminsky, E., and De Vries, A. (1970). Isolation and characterization of kinin-releasing enzyme of *Echis coloratus* venom, *Biochem. Pharmacol.,* **19,** 785.

Cohen, S. (1959). Purification and metabolic effects of a nerve growth promoting protein from snake venom, *J. Biol. Chem.,* **234,** 1129.

Damerau, B., Lege, L., Oldigs, H. D., and Vogt, W. (1975). Histamine release, formation of prostaglandin-like activity (SRS-C) and mast cell degranulation by the direct lytic factor (DLF) and phospholipase A of cobra venom, *Naunyn-Schmiedebergs Arch. Exp. Pathol. Pharmakol.,* **287,** 141.

Dragstedt, C. A., Mead, F. B., and Eyer, S. W. (1938). Role of histamine in circulatory effects of rattlesnake venom (crotalin), *Proc. Soc. Exp. Biol. Med.,* **37,** 709.

Dutta, N. K. and Narayanan, K. G. A. (1952). Release of histamine from rat diaphragm by cobra venom, *Nature,* **169,** 1064.

Dutta, N. K. and Narayanan, K. G. A. (1954). Release of histamine from skeletal muscle by snake venoms, *Br. J. Pharmacol.,* **9,** 408.

Engel, S. L., Schaeffer, T. R., Gold, B. I., and Rubin, B. (1972). Inhibition of pressor effects of angiotensin I and augmentation of depressor effects of bradykinin by synthetic peptides, *Proc. Soc. Exp. Biol. Med.,* **140,** 240.

Evseeva, L. F. and Enriques, O. B. (1969). Kininogenase and kininase activity of some snake venoms, *Farmakol. Toksikol. (Moscow),* **32,** 432.

Evseeva, L. F. and Enriques, O. M. (1973). Preparation and properties of the *Echis carinatus* poison kininogenase, *Farmakol. Toksikol. (Moscow),* **36,** 462.

Feldberg, W. and Kellaway, C. H. (1937). Liberation of histamine from the perfused lung by snake venoms, *J. Physiol.,* **90,** 257.

Feldberg, W. and Kellaway, C. H. (1938). Liberation of histamine and formation of lysolecithin-like substances by cobra venom, *J. Physiol.,* **94,** 187.

Ferreira, S. H. (1965). A bradykinin-potentiating factor (BPF) present in the venom of *Bothrops jararaca, Br. J. Pharmacol.,* **24,** 163.

Ferreira, S. H. and Rocha e Silva, M. (1965). Potentiating factor from *Bothrops jararaca* venom, *Experientia,* **21,** 347.

Ferreira, S. H., Greene, L. J., Alabaster, V. A., Bakhle, Y. S., and Vane, J. R. (1970a). Activity of various fractions of bradykinin-potentiating factor against angiotensin I-converting enzyme, *Nature,* **225,** 379.

Ferreira, S., Bartelt, D. C., and Greene, L. J. (1970b). Isolation of bradykinin-potentiating peptides from *Bothrops jararaca* venom, *Biochemistry,* **9,** 2583.

Goncalves, M. (1956). "Purification and properties of crotamine," in E. E. Buckley and N. Porges, *Venoms,* American Association for the Advancement of Science, Washington D.C., pp. 261–274.

Hamberg, U. and Rocha e Silva, M. (1957a). On the release of bradykinin by trypsin and snake venoms, *Arch. Int. Pharmacodyn.,* **110,** 222.

Hamberg, U. and Rocha e Silva, M. (1957b). Release of bradykinin as related to the esterase activity of trypsin and of the venom of *Bothrops jararaca, Experientia,* **13,** 489.

Henriques, O. B. and Evseeva, L. (1969). Proteolytic, esterase, and kinin-releasing activities of some Soviet snake venoms, *Toxicon,* **6,** 205.

Iwanaga, S., Sato, T., Mizushima, Y., and Suzuki, T. (1965). Studies on snake venoms. XVII. Properties of bradykinin releasing enzyme in the venom of *Agkistrodon halys blomhoffii, J. Biochem. (Tokyo),* **58,** 123.

Kaiser, E. and Raab, W. (1966). Liberation of pharmacologically active substances from mast cells by animal venoms, *Mem. Inst. Butantan,* **33,** 461.

Kato, H. and Suzuki, T. (1970a). Substrate specificities of kinin releasing enzymes: Hog pancreatic kallikrein and snake venom kininogenase, *J. Biochem.,* **68,** 9.

Kato, H. and Suzuki, T. (1970b). Structure of bradykinin-potentiating peptide containing tryptophan from the venom of *Agkistrodon halys blomhoffii, Experientia,* **26,** 1205.

Kato, H. and Suzuki, T. (1971). Bradykinin-potentiating peptides from the venom of *Agkistrodon halys blomhoffii:* Isolation of five bradykinin potentiators and amino acid sequences of two of them, potentiators B and C, *Biochemistry,* **10,** 972.

Kato, H., Iwanaga, S., and Suzuki, T. (1966). The isolation and amino acid sequences of new pyroglutamyl peptides from snake venoms, *Experientia,* **22,** 49.

Kato, H., Suzuki, T., Okada, K., Kimura, T., and Sakakibara, S. (1973). Structure of potentiator A, one of the five bradykinin potentiating peptides from the venom of *Agkistrodon halys blomhoffii, Experientia,* **29,** 574.

Kramar, R., Lambrechter, R., and Kaiser, E. (1971). The release of acid hydrolases from lysosomes by animal venoms, *Toxicon,* **9,** 125.

Lo, K. M., Chen, S. W., and Lo, T. B. (1973). Isolation and chemical characterization of small peptides from Formosan snake venoms, *J. Chin. Biochem. Soc.,* **2,** 33.

Loyke, H. F. (1973). The effect of synthetic peptide on chronic renal hypertension in the rat, *J. Lab. Clin. Med.,* **82,** 406.

Margolis, J., Bruce, S., Starzecki, B., Horner, G. J., and Halmagyi, D. F. J. (1965). Release of bradykinin-like substance in sheep by venom of *Crotalus atrox, Aust. J. Exp. Biol. Med. Sci.,* **43,** 237.

Markwardt, F., Barthel, W., Glusa, E., Hoffmann, A., and Walsmann, P. (1966). Über ein aminfreisetzend Komponente des *Crotalus terrificus* Giftes, *Biochem. Z.,* **346,** 351.

Mebs, D. (1969). Snake venom kallikreins: Purifications and properties of kinin-releasing enzyme from the venom of the viper *Bitis gabonica, Hoppe Seylers Z. Physiol. Chem.,* **350,** 1563.

Mohamed, A. H. and Zaki, O. (1957). The *Walterinnesia* toxin as a liberator of histamine from tissues, *J. Trop. Med. Hyg.,* **60,** 275.

Mohamed, A. H., Kamel, A., and Ayobe, M. H. (1968). Effects of *Echis carinatus* venom on tissue and blood histamine and their relations to local tissue reactions and eosinophil changes, *Toxicon,* **6,** 51.

References

Mohamed, A. H., El-Serougi, M. S., and Hamed, R. M. (1971). Effects of *Naja nigricollis* venom on blood and tissue histamine, *Toxicon,* **9,** 169.

Okada, K., Nagai, S., and Kato, H. (1974). A new pyroglutamyl peptide (Pyr–Lys–Ser) isolated from the venom of *Agkistrodon halys blomhoffii, Experientia,* **30,** 459.

Ondetti, M. A., Williams, N. J., Sabo, E. F., Pluscec, J., Weaver, E. R., and Kocy, O. (1971). Angiotensin-converting enzyme inhibitors from the venom of *Bothrops jararaca:* Isolation, elucidation of structure, and synthesis, *Biochemistry,* **22,** 4033.

Oshima, G., Sato-Ohmori, T., and Suzuki, T. (1969). Proteinase, arginine ester hydrolase and a kinin releasing enzyme in snake venoms, *Toxicon,* **7,** 229.

Rocha e Silva, M., Beraldo, W. T., and Rosenfeld, G. (1949). Bradykinin, a hypotensive and smooth muscle stimulating factor released from plasma globulin by snake venoms and by trypsin, *Am. J. Physiol.,* **156,** 261.

Rothschild, A. M. (1966). Mechanism of histamine release by animal venoms, *Mem. Inst. Butantan,* **33,** 467.

Rothschild, A. M. (1967). Chromatographic separation of phospholipase A from a histamine releasing component of Brazilian rattlesnake venom *(Crotalus durissus terrificus), Experientia,* **23,** 741.

Rothschild, A. M. and Almeida, I. A. (1972). "Role of bradykinin in the fatal shock induced by *Bothrops jararaca* venom in the rat," in A. De Vries and E. Kochva, Eds., *Toxins of Animal and Plant Origin,* Gordon and Breach, Vol. 2, New York, pp. 721–728.

Russell, F. E. (1965). Bradykininogen levels following *Crotalus* envenomation, *Toxin,* **2,** 277.

Sato, T., Iwanaga, S., Mizushima, Y., and Suzuki, T. (1965). Studies on snake venoms. XV. Separation of arginine ester hydrolase of *Agkistrodon halys blomhoffii* venom into three enzymatic entities: "Bradykinin releasing," "clotting," and "permeability increasing," *J. Biochem. (Tokyo),* **57,** 380.

Solomon, T. A., Cavero, I., and Buckley, J. P. (1974). Inhibition of central pressor effects of angiotensin I and II, *J. Pharm. Sci.,* **63,** 511.

Suzuki, T., Iwanaga, S., Sato, T., Nagasawa, S., Kato, H., Yano, M., and Horiuchi, K. (1967). "Biochemical enzymes," in M. Rocha e Silva and H. S. Rothschild, Eds., *Int. Symp. Vaso-Active Polypeptides: Bradykinin Related Kinins,* pp. 27–35.

Takahashi, H., Iwanaga, J., and Suzuki, T. (1972). Isolation of a novel inhibitor of kallikrein, plasmin and trypsin from the venom of Russell's viper *(Vipera russellii), FEBS Lett.,* **27,** 207.

Tu, A. T., James, G. P., and Chua, A. (1965). Some biochemical evidence in support of the classification of venomous snakes, *Toxicon,* **3,** 5.

Tu, A. T., Chua, A., and James, G. P. (1966). Proteolytic enzyme activities in a variety of snake venoms, *Toxicol. Appl. Pharmacol.,* **8,** 218.

Tu, A. T., Toom, P. M., and Murdock, D. S. (1967). "Chemical differences in the venoms of genetically different snakes," in F. E. Russell and P. R. Saunders, Ed., *Animal Toxins,* Pergamon, New York, pp. 351–362.

25 Metabolic and Teratogenic Effects

1 METABOLIC EFFECTS 412
   1.1 Blood Glucose, 412
   1.2 Other Effects, 413

2 TERATOGENIC AND OTHER EFFECTS ON EMBRYO 415
   2.1 Chick Embryo, 415
   2.2 Mice Embryo, 416

   References 416

Snake venom is a mixture of many different pharmacologically active proteins; therefore it is understandable that snake venom has a profound effect on the metabolic processes of the host animal. Most investigators have been concerned with the structural study of venom components or with their pharmacological effects; relatively little research on the biochemical mechanism of metabolic change after envenomation has been done. In this chapter two main subjects are discussed: the metabolic and the teratogenic effects of snake venoms.

1 METABOLIC EFFECTS

1.1 Blood Glucose

The effect of snake venoms on the blood glucose content has been intensively studied; therefore this topic will be considered first. The normal glucose content in rabbit blood is 0.114%. When the venom of *Trimeresurus mucrosquamatus* was injected, the blood sugar content increased three fold (Chin, 1935). The increase in blood sugar is due to stimulation of the central nervous system, which controls carbohydrate metabolism, and also to an increase in epinephrine secretion in the suprarenal glands. Repeated injections of different venoms into rabbits eventually prevented the increase in blood glucose (Ri, 1939b). Apparently, the antibody so produced inhibited this effect (Ri, 1939f; Peng,

1 Metabolic Effects

1950). Similarly, a hyperglycemia effect was observed for the venoms of *Agkistrodon rhodostoma, Vipera russellii siamensis*, and *Bungarus fasciatus* (Rai, 1936, 1937a, b). The blood sugar began to increase 30 min after a single injection of these venoms, and hyperglycemia reached the highest value in 1 to 4 hr and then returned to normal in 10 to 12 hrs.

Since hyperglycemia did not take place in the rabbits after adrenalectomy, the role of epinephrine seemed important. Venoms of *Naja tripudians* and *N. hamadryas* did not produce hyperglycemia. Hyperglycemia was also produced by the venoms of *Trimeresurus gramineus* and *Agkistrodon acutus* but not of *Naja naja atra* and *Bungarus multicinctus* (Ri, 1939a). The increase in blood sugar occurs at the expense of liver glycogen (Chin, 1936a; Ri, 1939c). In other words, liver glycogen supplies glucose after envenomation. In severe poisoning, blood glucose is also derived from muscle glycogen.

The hyperglycemic effect of Formosan Crotalidae venoms can be prevented by antivenin (Ri, 1939e). Injection of *Trimeresurus mucrosquamatus* venom decreased the combined sugar content in the blood (Chin, 1937). The combined sugars represent the difference between the total sugar and the free glucose in the blood.

In Crotalidae envenomation, the epinephrine content in suprarenal glands decreased and the epinephrine concentration in blood increased (Chin, 1936a, b; Ri, 1939d). Therefore it is apparent that liberated epinephrine activates adenylate cyclase, which produces cyclic AMP. The cyclic AMP eventually stimulates phosphorylase, which catalyzes the breakdown of glycogen.

Hyperglycemia and a decrease in liver and muscle glycogen were also observed for the venom of *Echis carinatus* (Mohamed et al., 1963), whereas *Walterinnesia aegypta* (black snake) venom caused hypoglycemia (Mohamed and Zaki, 1959). The disappearance of the hypoglycemic effect of black snake venom after pancreatomy in dogs indicates that the venom probably produces its hypoglycemic effect in normal animals through an increase in the secretion of insulin from the β cells of the islands of Langerhans (Mohamed et al., 1965).

Mohamed et al. (1972c) found that venom of *Naja haje* produces a diabetogenic effect, which they attributed to the diminishing utilization of glucose because of the decrease in insulin activity, as well as to the stimulation of the adrenal medulla (suprarenal gland) and the release of epinephrine. Their conclusions coincided with those reported by Chin (1936a) and Ri (1939c, 1939d). Again, glycogen content in the liver and muscle decreased (Mohamed et al., 1972a). *Naja nigricollis* venom behaves differently from the venom of *N. haje* (Mohamed et al., 1972b). At the lethal dosage, former venom becomes hyperglycemic, while a sublethal dose does not affect the blood glucose level significantly.

Activity of insulin and phosphate level are also affected by snake venom poisoning. The decrease in insulin activity is due to the inhibition of insulin release (Mohamed et al., 1974a), thereby lowering the activity of glucose utilization and increasing blood phosphate levels (Mohamed et al., 1972c).

1.2 Other Effects

Venom of *Trimeresurus mucrosquamatus* induces an increase of lactic acid concentration in blood (Chin, 1940). The increase in lactic acid content is proportional to the dose received. Elevation of blood lactic acid in rattlesnake envenomation by *Crotalus viridis helleri* was also reported (Carlson et al., 1975). An increase in lactic acid level appears to be a common phenomenon in snake poisoning.

When aldolase was incubated with venoms of *Naja naja atra, Trimeresurus gramineus*, and *Agkistrodon acutus in vitro*, the enzyme was inhibited (Huang and Tung, 1953). It is believed that this effect is due to an inhibitor present in the venoms, which combined with aldolase. It is not clear, however, whether such *in vitro* inhibition has any significance in glycolysis *in vivo*.

Glycolysis can be inhibited by venom of *Notechis scutatus*, and the mechanism of inhibition was believed to be the destruction of NAD (Chain, 1937; Chain and Goldsworthy, 1938; Chain, 1939). Ghosh and Chatterjee (1948) found that succinic dehydrogenase was also inhibited in the glycolysis of pigeon brain suspension. These results indicated that there is more than one site of inhibition.

Braganca and Quastel (1953) observed that denatured *Naja naja* venom activated glycolysis, whereas the original venom caused total inhibition. They ascribed this to the NADase and ATPase actions of venoms. Glyceraldehyde-3-phosphate dehydrogenase was inhibited by *N. naja atra* venom (Yang and Tung, 1954). The component that inhibited anaerobic glycolysis of the Ehrlich ascitic cell was isolated (Brisbois et al., 1968); it did not show NADase and phospholipase A_2 activity.

Venoms of *Trimeriscurus mucrosquamatus* and *T. gramineus* both had an effect on respiratory quotient and aerobic glycolysis rate (Chin, 1943a, b). At lower dosage, they stimulated the respiratory quotient, as well as aerobic and anaerobic glycolysis rates. At higher dosage, these values decreased temporarily and then increased. The effect of *Agkistrodon acutus* venom was much weaker than the actions of the two *Trimeresurus* venoms (Chin, 1943c). Therefore the total effect of snake venoms on glycolysis is rather complex.

It is also well known that some snake venoms uncouple oxidative phosphorylation. Venoms of *Naja oxiana* and *Vipera ursini renardi* initially stimulated and later inhibited the respiration of liver mitochondria (Sakhibov et al., 1975). The inhibition may possibly be due to venom phospholipase A_2 (for a detailed discussion, see Chapter 2).

A basic protein isolated from *Crotalus adamanteus* and *C. horridus atricaudatus* venoms induced an increase in glutamic oxalacetic transaminase, hydroxybutyric dehydrogenase, and aldolase (Bonilla et al., 1971-72). The increase in these enzyme levels is believed to be due to damage inflicted on heart tissues by the basic protein.

It is well known that snake venoms increase permeability or cause disturbance of transport. Larsen and Wolff (1967) found that *Naja naja* venom protein component inhibits amino acids and 3-*O*-methyl-D-glucose transport by the small intestine, and that the uptake of *p*-amino acids and sugars by the small intestine is dependent on the integrity of the sodium pump. An effect on the sodium pump thus may be possible. However (a fact surprising to these investigators), Na(I) efflux from human erythrocytes or the inhibition of this process by ouabain is unaffected by the venom component. This may explain why the sodium pump in erythrocytes is different from that in other tissues.

The venom of *C. atrox* decreased the permeability of D-glucose-6-^{14}C in lung epithelial cells (Johnson and Bertke, 1964). It is known that glucose transport is not a simple diffusion but occurs by a carrier mechanism of active transport. These authors suggested that the change in permeability is due to the effect of the venom on phosphorylation.

Injection of *Trimeresurus flavoviridis* venom into rabbits decreased the serum phospholipid content to about half of the control value (Takaki and Sato, 1975), whereas the serum ammonia content increased.

Envenomation by *Walterinnesia aegyptea* caused a fall in the free and esterified cholesterol of both adrenals and plasma. In adrenalectomized rats, injection of venom caused hypercholesteremia with a marked increase in free cholesterol, a change that is not produced by injection of corticosterone (Zaki and Long, 1969). The same venom caused a marked eosinopenia and depletion of adrenal ascorbic acid; both effects were ascribed to hyperactivity of the suprarenal cortex, caused by the venom as a type of stress (Mohamed and Zaki, 1958). It has been reported that exposure of rats to stress is followed by a depletion of adrenal cholesterol (Sayers et al., 1944).

Vipera lebetina venom caused a fall in the nucleic acid content in the blood, spleen, and liver (Talyzin and Yurkova, 1968). Antivenin was shown to prevent this decrease.

The respiration of brain tissue was severely inhibited by incubation with the venoms of *Naja haje, N. nigricollis, Walterinnesia aegyptia, Cerastes cerastes,* and *C. vipera* (Mohamed et al., 1969). However, there was no such inhibition when skeletal muscle tissue was used. With cardiac tissue, the response was dependent on the venom used. Inhibition on cardiac tissue O_2 uptake was observed only for *Naja nigricollis* and *Walterinnesia aegyptia* venom.

The activities of succinic dehydrogenase and cholinesterase in kidney, heart, liver, spleen, and brain tissues were inhibited by the injection of *Naja naja* venom (Ahmed et al., 1974).

Catalase is an enzyme commonly present in various tissues. The effect of snake venoms on blood catalase was investigated by Takagi (1944a, b, c, d, e). Crotalidae venoms such as those of *Trimeresurus mucrosquamatus, T. gramineus,* and *Agkistrodon acutus* increased blood catalase activity (Takagi, 1944a, b, c), whereas the venoms of Elapidae such as *Naja naja atra* and *Bungarus multicinctus* decreased the activity (Takagi, 1944d, e). Snake venoms decreased the content of blood glutathione, but the mechanism of this effect was not yet clear (Fujii, 1942a, b, c, d, e, f). Both *N. naja atra* and *B. multicinctus* venoms decreased the Ca(II) and K(I) contents of blood (Chin, 1938). The venom of *Walterinessia aegypta* also decreased the blood K ion level, but the Na(I) level was increased by using a sublethal dose (Mohamed et al., 1964). Lethal doses of *W. aegypta* venom produced marked increases in blood catecholamines; a sublethal dose caused a decrease. It was suggested that a sublethal dose of *W. aegypta* venom stimulated adrenal cortical activity.

Crotalus atrox venom stimulates the activity of cyclic 3',5'-nucleotide phosphodiesterase (Cheung, 1969). A stimulatory factor is believed to be present in the venom but has not yet been isolated.

2 TERATOGENIC AND OTHER EFFECTS ON EMBRYO

Several snake venoms have been investigated for their effects on embryo tissues. Most investigators either used chick embryo or mice fetus.

2.1 Chick Embryo

It has been known for some time that cobra venom causes teratogenic effects on chick embryo (Ruch and Gabriel-Robez-Kremer, 1962, 1963). When venom is used in early developmental stages, it produces anomalies in the growth and shape of the body: coelosomy and face, limb, central nervous system, and heart malformations.

Injection of α-toxin from *Naja nigricollis* into the yolk sac at early stages of development causes an atrophy of skeletal and extrinsic ocular muscles and their innervation. In 16-day embryos treated by α-toxin, the end plates revealed by the Koelle reaction were almost completely absent. The total content and specific activities of acetylcholinesterase and choline acetyltranferase in atrophic muscles were markedly reduced (Giacobini et al., 1973).

Embryos treated with *Naja naja* venom were generally less developed than control embryos that showed normal development at the same stage of embryogenesis (Ahmed et al., 1974). The brain had not efficiently formed optic vesicles, and histological examination showed that the neural tube failed to close throughout most of its length. Extensive proliferation of the developing neuroblasts was noticed in the lumen of the neural tube of the venom-treated embryo. The proliferation of neuroblasts of the neural tube by cobra venom may be due to the presence of the so-called nerve growth-stimulating factor. The disruption of the neural tube and its failure to close properly, as well as the secondary suppression of the somites, may result from interference by the venom with the metabolic activities of the embryo, or may be direct toxic effects of the venom component.

β-Bungarotoxin inhibits the development of chick embryo. It causes muscle atrophy, a short upper beak, and a decrease in the acetylcholinesterase and succinate-cytochrome *c* reductase activities in the chick embryo (Mao et al., 1975).

2.2 Mice Embryo

Injection of *Vipera aspis* venom initially causes frequent interruptions of pregnancy (Calvert and Gabriel-Robez-Kremer, 1974). It also leads to abnormalities of the face and brain, palatal clefts, and defects of the heart and extremities. The extent and frequency of the anomalies vary according to the periods of treatment. Throughout gestation the venom is hazardous to the brain and can produce secondary teratogenic effects due to hemorrhage and necrosis.

The effects of maternal envenomation on developing fetal tissue were also studied by Mohamed et al. (1974b). The venom of *Naja nigricollis* passes through the placental barrier and injures the kidney nephrons. In the liver, it produces disorganization of lobules with cellular degeneration and congestion of the liver sinusoids. It also causes subendocardial hemorrhage, with fragmentation of the myocardium in the heart, and edema in the walls of the aorta and other large blood vessels. The endothelial linings of some vessels were ruptured, and extravasated blood was observed in the lumen of the gut and respiratory passages. Injection of the venom at early stages of pregnancy also causes deformed fetuses (Mohamed and Nawar, 1975). The skin of all envenomated fetuses appeared edematous and less wrinkled than that of the controls.

REFERENCES

Ahmed, Y. Y., Moustafa, F. A., and El-Asmar, M. F. (1974). Effect of cobra (*Naja naja*) venom on succinic dehydrogenase and cholinesterase of rat tissue, *Ind. J. Med. Res.,* **62,** 1337.

Bonilla, C. A., Fiero, M. K., and Novak, J. (1971 to 72). Serum enzyme activities following administration of purified basic proteins from rattlesnake venoms, *Chem. -Biol. Interactions,* **4,** 1.

Braganca, B. M. and Quastel, J. H. (1953). Enzyme inhibitions by snake venoms, *Biochem. J.,* **53,** 88.

Brisbois, L., Rabinovitch-Mahler, N., and Gillo, L. (1968). Isolation from cobra venom of a factor inhibiting glycolysis in Ehrlich ascites carcinoma cells, *Experientia,* **24,** 673.

Calvert, J. and Gabriel-Robez-Kremer, O. (1974). Effects on mouse gestation and embryo development of an injection of viper venom (*Vipera aspis*), *Acta Anat.,* **88,** 11.

References

Carlson, R. W., Schaeffer, R. C., Jr., Whigham, H., Michaels, S., Russell, F. E., and Weil, M. H. (1975). Rattlesnake venom shock in the rat: Development of a method, *Am. J. Physiol.*, **229**, 1668.

Chain, E. (1937). Effect of snake venoms on glycolysis and fermentation in cell-free extracts, *Quant. J. Exp. Physiol.*, **26**, 299.

Chain, E. (1939). Inhibition of dehydrogenase by snake venom, *Biochem. J.*, **33**, 407.

Chain, E. and Goldsworthy, L. (1938). Studies on the chemical nature of the antifermenting principle in black tiger snake venom, *Quant. J. Exp. Physiol.*, **27**, 375.

Cheung, W. Y. (1969). Cyclic $3',5'$-nucleotide phosphodiesterase: preparation of a partially inactive enzyme and its subsequent stimulation by snake venom, *Biochim. Biophys. Acta*, **191**, 303.

Chin, K. (1935). Über die Beeinflussung des Blutzuckers des Kaninchens durch das Gift von *Trimeresurus mucrosquamatus* Contor, *J. Formosan Med. Assoc.*, **34**, 1059.

Chin, K. (1936a). Über die Beeinflussung des Blutzuckers sowie der Glykogenmenge in Leber und Muskeln des Kaninchens durch das Gift von *Trimeresurus mucrosquamatus* Cantor, *J. Formosan Med. Assoc.*, **35**, 1646.

Chin, K. (1936b). Über die Beeinflussung des Adrenalingehaltes in Nebennieren und im Blut des Kaninchens durch das Gift von *Trimeresurus mucrosquamatus* Cantor, *J. Formosan Med. Assoc.*, **35**, 2335.

Chin, K. (1937). Über die Beeinflussung des gebundenen Blutzuckers des Kaninchens durch das Gift von *Trimeresurus mucrosquamatus* Cantor, *J. Formosan Med. Assoc.*, **36**, 606.

Chin, K. (1938). Über den Einfluss gewisser Schlangengifte auf den Ca- und K-Gehalt des Kaninchenserums, sowie dessen Beziehungen zu den Tonusanomalien des vegetativen Nervensystems. II. Studien über die Gifte formosanischer Elapidae, *J. Formosan Med. Assoc.*, **37**, 124.

Chin, K. (1940). Über den Einfluss des Giftes von *Trimeresurus mucrosquamatus* (Cantor) auf den Milchsauregehalt im Blut des Kaninchens, *J. Formosan Med. Assoc.*, **38**, 145.

Chin, K. (1943a). Einfluss der wichtigen formosanischen Schlangengifte auf die Gewebsatmung und Glykolyse von Niere Milz und Muskel der Kaninchen. I. Metteilung: Wirkung des Giftes von *Trimeresurus mucrosquamatus* Cantor, *J. Formosan Med. Assoc.*, **42**.

Chin, K. (1943b). Einfluss der wichtigen formosanischen Schlangengifte auf die Gewebsatmung und Glykolyse von Niere, Milz und Muskel der Kaninchen. II. Mitteilung: Wirkung des Giftes von *Trimeresurus gramineus* Shaw, *J. Formosan Med. Assoc.*, **42**, 53.

Chin, K. (1943c). Einfluss der wichtigen formosanischen Schlangengifte auf die Gewebsatmung und Glykolyse von Niere, Milz und Muskel der Kaninchen. III. Milleilung: Wirkung des Giftes von *Agkistrodon actus* Gunther, *J. Formosan Med. Assoc.*, **42**, 63.

Fujii, Y. (1942a). Einfluss der wichtigeren formosanischen Schlangengifte auf den Glutahionsgehalt des Kaninchenblutes. I. Mitteilung: Experimentelle Untersuchungen uber das Gift von *Trimerisurus mucroscuamatus* Cantor, *J. Formosan Med. Assoc.*, **41**, 42.

Fujii, Y. (1942b). Einfluss der wichtigeren formosanischen Schlangengifte auf den Glutathionsgehalt des Kaninchenblutes. II. Mitteilung: Experimentelle Untersuchungen uber das Gift von *Trimeresurus gramineus* Shaw, *J. Formosan Med. Assoc.*, **41**, 62.

Fujii, Y. (1942c). Einfluss der wichtigeren formosanischen Schlangengifte auf den Glutathionsgehalt des Kaninchenblutes. III. Mitteilung: Experimentelle Untersuchungen uber das gift des *Agkistrodon acutus* Gunther, *J. Formosan Med. Assoc.*, **41**, 1.

Fujii, Y. (1942d). Einfluss der wichtigeren formosanischen Schlangengifte auf den Glutathionsgehalt des Kaninchenblutes. IV. Teil: Experimentelle Untersuchungen über das Gift von *Naja naja atra* Cantor, *J. Formosan Med. Assoc.*, **41**, 25.

Fujii, Y. (1942e). Einfluss der wichtigeren Schlangengifte Formosas auf den Gluthathionsgehalt des Kaninchenblutes. V. Teil: Experimentelle Untersuchungen über das Gift von *Bungarus multitinctus* Blyth, *J. Formosan Med. Assoc.*, **41**, 16.

Fujii, Y. (1942f). Einfluss der wichtigeren Schlangengifte Formosas auf den Glutathionsgehalt des Kaninshens. VI. Teil: Über das gemeinschaftliche Verhalten bei der Gewohnung an verschiedene Schlangengifte, Betrachtet von Seiten der Veranderung des Glutatnionsgehaltes, *J. Formosan Med. Assoc.*, **41**, 1.

Ghosh, B. N. and Chatterjee, A. K. (1948). Effect of snake venoms on the oxidation of glucose and its metabolites in cell suspensions, *J. Indian Chem. Soc.*, **25**, 359.

Giacobini, G., Filogamo, G., Weber, M., Boquet, P., and Changeux, J. P. (1973). Effects of a snake α-neurotoxin on the development of innervated skeletal muscle in chick embryo, *Proc. Natl. Acad. Sci. U.S.A.,* **70,** 1708.

Huang, P. and Tung, T. C. (1953). Effect of snake venom on aldolase, *J. Formosan Med. Assoc.,* **52,** 143.

Johnson, B. D. and Bertke, E. M. (1964). Effects of *Crotalus atrox* venom on the permeability of human lung tissues to D-glucose-6-14-C, *Toxicon,* **2,** 197.

Larsen, P. R. and Wolff, J. (1967). Inhibition of accumulative transport by a protein from cobra venom, *Biochem. Pharmacol.,* **16,** 2003.

Mao, Chen, C. J., Lin, Shiau S. Y., and Lee, C. Y. (1975). Morphological and enzymatic changes of chick embryo induced by β-bungarotoxin, *J. Formosan Med. Assoc.,* **74,** 229.

Mohamed, A. H. and Nawar, N. N. Y. (1975). Dysmelia in mice after maternal *Naja nigricollis* envenomation: A case report, *Toxicon,* **13,** 475.

Mohamed, A. H. and Zaki, O. (1958). Effect of the black snake toxin on the gastrocnemius–sciatic preparation, *J. Exp. Biol.,* **35,** 20.

Mohamed, A. H. and Zaki, O. (1959). The effect of the black snake toxin on blood glucose level, *Indian J. Med. Res.,* **47,** 522.

Mohamed, A. H., El-Serougi, M., and Kamel, A. (1963). Effects of *Echis carinatus* venom on blood glucose and liver and muscle glycogen concentration, *Toxicon,* **1,** 243.

Mohamed, A. H., El-Serougi, M., and Kamel, A. (1964). Effect of *Walterinessia aegypta* venom on blood sodium, potassium and catecholamines, and urine 17-ketosteroids, *Toxicon,* **2,** 103.

Mohamed, A. H., El-Serougi, M., and Kamel, A. (1965). Effect of *Walterinnesia aegypta* venom on blood glucose in diabetic animals, *Toxicon,* **3,** 163.

Mohamed, A. H., El-Serougi, M., and Khaled, L. Z. (1969). Effect of *Cerastes cerastes* venom on blood coagulation mechanism, *Toxicon,* **7,** 181.

Mohamed, A. H., Hanna, M. M., and Selim, R. (1972a). The effects of *Naja haje* venom and its ionophoretic fractions on glucose metabolism, *Toxicon,* **10,** 1.

Mohamed, A. H., Mohamed, F. A., and El-Damarawy Nabil, A. (1972b). Diabetogenic actions of *Naja nigricollis* venom. I. Effects on glucose tolerance, plasma insulin-like activity and blood potassium, *Toxicon,* **10,** 151.

Mohamed, A. H., Mervat, A. B., and Nabil, A. E. D. (1972c). Effects of cobra (*Naja haje*) venom on blood glucose, blood phosphate and plasma insulin-like activity in dogs, *Toxicon,* **10,** 385.

Mohamed, A. H., Mervat, A. B., and El-Damarawy Nabil, A. (1974a). Effect of cobra venom (*Naja haje*) on insulin release by rat pancreas *in vitro, Toxicon,* **12,** 287.

Mohamed, A. H., Nawar, N. N. Y., and Hanna, M. M. (1974b). Some effects of *Naja nigricollis* envenomation on developing fetal tissue, *Toxicon,* **12,** 477.

Peng, M. (1950). Relation between the change of blood-sugar fluctuation and the production of immune bodies by successive injections of *Trimeresurus mucrosquamatus* venom, *J. Formosan Med. Assoc.,* **49,** 215.

Rai, K. (1936). Die Beeinflussung des Blutzuckers von Kaninchen durch die Gifte Siamesischer Schlangenarten, *J. Formosan Med. Assoc.,* **25,** 2195.

Rai, K. (1937a). Über den Einfluss der wichtigeren formosanischen Schlangengifte auf die Atmung des Kaninchens. I. Mitteilung: Die nach einer Einzeldosis der verschiedenen Schlangengifte eintretenden Veranderungen der Atmung bei Abwesenheit therapeutischer Masnahmen, *Folia Pharmacol. Jap.,* **24,** 123.

Rai, K. (1937b). The influence of the poisons of Siamese snakes on the blood-sugar of the rabbit, *Jap. J. Med. Sci.,* **10,** 163.

Ri, T. (1939a). Experimentelle Untersuchungen über die Wirkung des Giftes gewisser formonsanischer Schlangen auf den Kohlenhydrathaushalf. I. Mitteilung: Über die Beeinflussung des Zuckergehalts des Kaninchenblutes durch die Schlangengifte bei *akuter* Vergiftung, *Folia Pharmacol. Jap.,* **27,** 13.

Ri, T. (1939b). Experimentelle Untersuchungen über die Wirkung des Giftes gewisser formosanischer Schlangen auf den Kohlenhydrathaushalt. II. Mitteilung: Über die Beeinflussung des Zuckergehalts des Kaninchenblutes durch die Schlangengifte bei *Chronischer* Vergiftung, *Folia Pharmacol. Jap.,* **27,** 35.

Ri, T. (1939c). Experimentelle Untersuchungen über die Wirkung des Giftes gewisser formosanischer Schlangen auf den Kohlenhydrathaushalt. III. Mitteilung: Über die Beeinflussung des Blutzuckers sowie der Glykogenmenge in Leber und Muskeln des Kaninchens durch die Schlangengifte, *Folia Pharmacol. Jap.*, **27**, 105.

Ri, T. (1939d). Experimentelle Untersuchungen über die wirkung des Giftes gewisser formonsanischer Schlangen auf den Kohlenhydrathaushalf. IV. Metteilung: Über die Beeinflussung des Adrenalingehaltes in den Nebennieren und im Blut des Kaninchens durch die Schlangengifte, *Folia Pharmacol. Jap.*, **27**, 165.

Ri. T. (1939e). Experimentelle Untersuchungen über die Wirkung des Giftes gewisser formosanischer Schlangen auf den Kohlenhydrathaushalf. V. Mitteilung: Über den Eingluss von Heilseren auf den Blutzucker des Kaninchenblutes bei Vergiftungen mit schlangengiften, *Folia Pharmacol. Jap.*, **27**, 260.

Ri. T. (1939f). Experimentelle Untersuchungen über die Wirkung des Giftes gewisser formosanischer Schlangen auf den Kohlenhydrathaushalt. VI. Mitteilung: Über die gegenseitige Beeinflussung des Blutzuckers gewisser crotalinaer Schlangengifte nach erfolgter Gewohnung an diese Gifte, *Folia Pharmacol. Jap.*, **27**, 280.

Ruch, J. V. and Gabriel-Robez-Kremer. (1962). Action tèratogène du venin de *Naja* sur l'embryon de poulet, *C. R. Soc. Biol.*, **156**, 1508.

Ruch, J. V. and Gabriel-Robez-Kremer, O. (1963). Etude des malformations cardiaques provoquees par le venin de *Naja*, *C. R. Soc. Biol.*, **157**, 2291.

Sakhibov, D. N., Yukel'son, L. Y., Sattyev, R., and Gagel'gans, A. I. (1975). Effect of venoms of Central Asian snakes on functional parameters of mitochondria in experiments *in vivo*, *Uzb. Biol. Zh.*, **19**, 3.

Sayers, G., Sayers, M. A., Fry, E. G., White, A., and Long, C. N. H. (1944). Effects of ACTH on cholesterol esters of adrenals, *Yale J. Biol. Med.*, **16**, 360.

Takagi, T. (1944a). On the influence of the poison of Formosan snakes upon blood and organ catalase in rabbit. I. Report: On the action of poison of *Trimeresurus mucrosquamatus* (Kantor) upon catalase in rabbit. *J. Formosan Med. Assoc.*, **43**, Suppl. 3.

Takagi, T. (1944b). On the influence of the poison of Formosan snakes upon blood and organ catalase in rabbit. II. Report: On the action of poison of *Trimeresurus gramineus* (Shaw) upon catalase in rabbit, *J. Formosan Med. Assoc.*, **43**, Suppl. 3.

Takagi, T. (1944c). On the influence of the poison of Formosan snakes upon blood and organ catalase in rabbit. III. Report: On the action of poison of *Agkistrodon acutus* (Gunther) upon catalase in rabbit, *J. Formosan Med. Assoc.*, **43**, Suppl. 3.

Takagi, T. (1944d). On the influence of the poison of Formosan snakes upon blood and organ catalase in rabbit. IV. Report: On the action of poison of *Naja naja atra* (Cantor) upon catalase in rabbit, *J. Formosan Med. Assoc.*, **43**, Suppl. 3.

Takagi, T. (1944e). On the influence of the poison of Formosan snakes upon blood and organ catalase in rabbit. V. Report: On the action of poison of *Bungarus multicinctus* (Blyth) upon catalase in rabbit, *J. Formosan Med. Assoc.*, **43**, Suppl. 3.

Takaki, S. and Sato, H. (1975). Effect of *Trimeresurus flavoviridis* venom on serum ammonia and phospholipid levels, *Acta Med. Univ. Kagoshima.*, **17**, 25.

Talyzin, F. F. and Yorkova, I. B. (1968). Change in the nucleic acid level in mouse tissues caused by the venom of *Vipera lebetina*, antiserum, and propylgallate, *Tr. Univ. Druzhby Nar.*, **38**, 171.

Yang, C. C. and Tung, T. C. (1954). Cobra venom and the thiol groups of glyceraldehyde-3-phosphate dehydrogenase, *J. Formosan Med. Assoc.*, **53**, 209.

Zaki, O. A. and Long, C. (1969). Effects of adrenalectomy on the changes in cholesterol levels in rats induced by *Walterinnesia* venom, *Toxicon*, **6**, 255.

26 Immunology

1 COMPLEMENT 420

2 COBRA VENOM FACTOR 422

   2.1 History, 422
   2.2 Chemical Properties, 422
   2.3 Interaction with Serum Cofactor (C3-Proactivator), 423
   2.4 Anticomplement Action, 423
       Effect on Classical Pathway, 423
   2.5 Other Effects, 425
   2.6 Elimination, 425
   2.7 Applications, 426
   2.8 Use in Organ Transplantation, 429

3 OTHER VENOMS 429

4 POSSIBLE ROLE OF COMPLEMENT ACTIVATION IN SNAKE ENVENOMATION 000 430

   References 431

Frequently cells are injured as a result of the interaction of antigen and antibody *in vivo* through a very complicated pathway, the complement system. There is also a special pathway whereby the complement system can be activated without antigen–antibody interaction. This alternative pathway can be initiated by the compound properdin or cobra venom factor (see scheme 30).

It is not the intention of the author to review the complement system extensively; the reader should refer to a standard immunology textbook for a more comprehensive treatment of this subject. In this chapter, the complement system is described very briefly before the cobra venom factor is presented.

1 COMPLEMENT

Cobra (*Naja* spp.) venoms contain an anticomplement factor called cobra venom factor which depletes the C3 complement. The cobra venom factor is probably the most important substance found in venoms that cause important immunological activity since

1 Complement

the anticomplement action of cobra venom factor has potential use in the prevention of transplanted organ rejection. To fully understand the action of cobra venom factor, it is necessary to briefly review the complement system.

The complement system has been recognized for many years as a plasma effector pathway associated with antigen–antibody complexes. There are 11 proteins in the complement system, which make up about 10% of the globulins in normal serum of man and other vertebrates. Antigen–antibody complexes activate a rapid sequential interaction among the complement proteins which is called complement fixation. The prinicipal effect of complement fixation is on the structure and function of biological membranes. The result may be cytolysis if bacteria or erythrocytes are involved, or the activation of a noncytolytic mechanism such as the release of histamine from the mast cells, platelets, or leucocytes, enhanced phagocytosis, the contraction of smooth muscles, the aggregation and fusion of platelets, or the promotion of blood coagulation.

The components of complement C are designated C1, C2, C3, C4, C5, C6, C7, C8, and C9. Component C1 consists of three subunits, C_{1q}, C_{1r}, and C_{1s}. Intermediate complexes formed in the process of complement fixation are designated by a combination of the symbols for the individual components that make up the complex. For example, a complex composed of C1, C2, and C4 would be termed C142, the sequence of numbers indicating the order in which the individual components become part of the complex. If cells (E) and antibodies to cell surface antigens (A) are part of the complex, the notation is EAC 142. Enzyme activity associated with an activated component or complex may be indicated by placing a bar over the number, $\overline{C1}$, to differentiate it from its potentially active counterpart, C1. When a component such as C3 is fragmented in the process of activation, the fragments are designated by small letters after the notation of the component from which they originated, C3a and C3b.

The sequential action of the complement system is shown in Scheme 30.

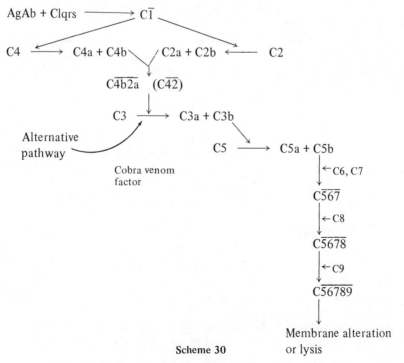

Scheme 30

Diagram indicating sequential reactions of the complement system.

2. COBRA VENOM FACTOR

2.1 History

As early as 1894, Ewing observed that snake venoms destroyed the bactericidal activity of serum. Systematic scientific study of the anticomplement activity of snake venoms started at the beginning of the twentieth century.

It is known that red blood cells can be lysed in the presence of cobra venom and human serum without the presence of red blood cell antibodies. Flexnor and Noguchi (1901) observed that the addition of blood serum was necessary for cobra venom to lyse washed red blood cells. They also observed that, if blood serum was preincubated with the venom, the lytic activity was destroyed. This was the first report associating the hemolytic activity of snake venom with a blood serum component.

As early as 1912, Ritz observed that cobra venom possessed anticomplement activity. However, elucidation of the mechanism and isolation of cobra venom factors awaited advances in immunochemistry.

Vogt and Schmit (1964) purified by gel filtration and ion-exchange chromatography a component of cobra venom that split an anaphylatoxic polypeptide from an inactive precursor. Klein and Wellensiek (1965) identified the anticomplementary activity of cobra venom as specifically involving C3. It was reported by many investigators that the administration of cobra venom factor or cobra venom to animals resulted in a decrease in the levels of circulating C3 (Nelson, 1966; Mickenberg et al., 1971; Roy, 1969).

Nelson (1966) observed that purified C3 was not inactivated when incubated with cobra venom factor *in vitro*. This suggested that the inactivation was at least a two-step reaction that involved a cofactor found in normal serum. Thus it was firmly established that cobra venom factor interfered with the C3 component of the complementary system.

2.2 Chemical Properties

Cobra venom factor (CVF) is one of the most anionic proteins in the venom of *Naja naja* and is readily separated from basic neurotoxins (Müller-Eberhard and Fjellstrom, 1971). The large molecular size of CVF allows its complete separation from other anionic proteins by passing the anionic fraction from electrophoresis through a column of Sephadex G-100. The average yield of purified CVF is 3 to 4 mg g^{-1} of crude venom, which is in agreement with the yield of 0.4% from crude venom.

Naja haje CVF is devoid of toxicity and phospholipase activity (Maillard and Zarco, 1968). and induces an almost total, long-lasting, nonaggressive *in vivo* decomplementation in guinea pigs.

Cobra venom factor is present in cobra venoms other than *Naja naja* venom; the factor was also isolated from the venom of *N. haje* (Shin et al., 1969). The molecular weight of *N. naja* CVF is 144,000, as determined from S and D combination (Müller-Eberhard and Fjellstrom, 1971). Amino acid residues per 1000 residues are, 65 Lys, 20 His, 42 Arg, 115 Asp, 66 Thr, 64 Ser, 102 Glu, 64 Pro, 60 Gly, 58 Ala, 18½ Cys, 64 Val, 15 Met, 51 Ile, 40 Tyr, 38 Phe, and 78 Leu. Cobra venom factor is a glycoprotein that contains 13% neuraminic acid, 4.2% hexosamine, and 5.7% hexose.

Immunological evidence indicates that CVF of *N. naja* venom is similar to C3, since antiserum to CVF cross-reacts with human C3. Incubation of CVF with cobra serum destroys the C3-cleaving activity of the venom factor in human serum, whereas human C3 inactivator is ineffective. Thus CVF appears to be a form of C3 (perhaps C3b); its potent action in human serum probably derives from its lack of sensitivity to human C3b inactivator (Alper and Balavitch, 1976).

2.3 Interaction with Serum Cofactor (C3-Proactivator)

A serum cofactor is necessary for cobra venom factor to react with complement (see C_3 alternative pathway in Scheme 30) (Nelson, 1966; Müller-Eberhard, 1967; Brai and Osler, 1972a, b). Evidence for complex formation between CVF and the cofactor was obtained by using ^{125}I-labeled CVF. The sedimentation rate of purified CVF is 6.75S, and that of the serum cofactor is 5S. When ^{125}I-labeled CVF was mixed with whole serum and ultracentrifuged in a sucrose density gradient, the sedimentation rate of the C3 converting activity was 8.5 to 9S, indicating that CVF and the serum cofactor had formed a complex.

The effect of the CVF–cofactor complex on C3 was studied by treating ^{125}I-labeled C3 with a mixture of CVF and serum cofactor, and then comparing the radioactivity profiles of treated and untreated labeled C3 after sucrose density ultracentrifugation. The treated sample contained low molecular weight radiolabeled material, indicating fragmentation of the C3 molecule in the activation process. The size of the fragment was indistinguishable from that induced by $C\overline{42}$.

The cofactor has been isolated from human serum and characterized (Götze and Müller-Eberhard, 1971). Its molecular weight is 80,000 daltons, and its approximate serum concentration is 100 to 200 μg ml^{-1}.

Apparently, serum cofactor forms with CVF a complex that is able to activate C3, bypassing C1, C4, and C2 (Dierich et al., 1971). The action on C3 results in activation of the terminal components and in membrane destruction, provided that suitable membrane receptors are available (Scheme 30).

This cofactor is widely distributed in serum of different species, it appears early in development, and its concentration is not affected by procedures known to greatly influence the concentration of antibody and/or of C components (Gewurz et al., 1971). Complex formation of CVF and serum cofactor is Mg(II) dependent. Purified C5 is not affected by the complex (Bitter-Suermann et al., 1972).

Cooper (1973) observed that there were two modes of interaction of cofactor with CVF. In isolated form, the cofactor and CVF formed a reversible protein–protein complex in free solution. This complex had some C3-cleaving activity.

In the presence of serum factor D, the cofactor–CVF complex was stabilized and its efficiency in cleaving C3 was greatly increased. Thus Cooper considered factor D to be an activator of serum cofactor.

2.4 Anticomplement Action

Effect on Classical Pathway Antibody-independent activation of the complement system starting with C3 can be achieved by means of cobra venom factor, which interacts with a serum cofactor. The CVF–serum cofactor complex acts C3 and activates the terminal complement sequence (see Scheme 31).

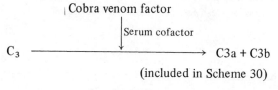

Scheme 31

Therefore once the C3 shunt is activated by CVF, erythrocytes and other cells can be lysed in the absence of specific antibody.

An activated complex of the fifth and sixth components of complement ($\overline{C56}$) can be generated by the interaction of serum with CVF (Goldman et al., 1972).

Formation of CVF–serum cofactor complex increased the 144,000 molecular weight of the former to 220,000 daltons (Müller-Eberhard et al., 1966; Müller-Eberhard and Fjellström, 1971).

To generate C3-directed enzymatic activity, CVF has to enter into a firm complex with the plasma protein, a 5S, thermolabile β-pseudoglobulin.

Cobra venom factor decreases C5 levels in whole serum from mice and human beings (Cochrane et al. 1970) and from guinea pigs (Klein and Wellenseik, 1965). Partially purified CVF's have been shown to generate anaphylatoxic fragments from C5 (Vogt and Schmidt, 1964; Jensen, 1967).

In rabbits, C6 activity was depleted to less than one-half of normal values within 16 hr (Cochrane et al., 1970). Shin et al. (1969) found that the addition of purified CVF to guinea pig serum inactivated more than 90% of C3 and considerable quantities of C5, C6, C7, C8, and C9, but had no detectable effect on C1, C4, and C2. This pattern of depletion of complement components suggested that the CVF–serum cofactor complex serves as an alternative complement-activating pathway (Nelson, 1966; Müller-Eberhard et al., 1966). Functionally, the complex is analogous to C3 covertase formed from C1, C4, and C2 by the classical pathway. The similarity of C3 convertase and CVF is evident in their ability to lyse unsensitized erythrocytes. When CVF is added to a suspension of erythrocytes in serum (Pickering et al., 1969; Ballow and Cochrane, 1969; Götze and Müller-Eberhard, 1970), or when preformed C3 convertase is added to erythrocytes suspended in a solution of purified C3-9 (Götze and Müller-Eberhard, 1970), hemolysis occurs. Both mechanisms require bivalent cations, and both hydrolyze acetyl-glycyl-L-lysine (Cooper, 1971). The physicochemical properties of the cofactor suggest that it may be related to C2, but neither C2 nor C4 could substitute for the cofactor in the reaction with CVF.

Day et al. (1972) found that starfish hemolymph contained a factor, which they called hemolymph factor, that activates vertebrate complement in the presence of CVF. This factor is capable of forming a bimolecular complex with CVF in the presence of metals. The complex is endowed with enzymatic activity presently undistinguishable from that of the C enzyme C3 convertase. It acts on C3 and initiates activation of C3-9. The most striking difference between hemolymph factor and serum cofactor (proactivator) is the former's low molecular weight of only 2000. This molecular size precludes the possibility that hemolymph factor is an enzyme. Thus Day et al. thought there must have been another activation process that had not yet been recognized.

After the completion of complement system activation induced by CVF, the latter reverts to its native state. No functional differences between native CVF and CVF reisolated from the decayed complement system could be observed. In decayed CVF–serum mixtures, properdin factor B was in its immunoelectrophoretically converted form, and the addition of fresh factor B restored full indirect lytic activity (Fearon et al., 1973a).

Drake et al. (1974) investigated the effect of CVF on C3 concentration *in vivo* to determine whether or not the total abrogation of C3 activities could be accomplished in the guinea pig. They found that circulating C3 concentration decreased upon the addition of CVF, but the abrogation of C3 could be achieved only through the combined administration of both CVF and anti-C3 (goat anti-guinea-pig C3 serum).

Intraperitoneal injection of CVF reduced C3 levels to 50% of normal after 15 hr. Injection of divided doses caused plasma C3 levels to drop to less than 5% or normal for from 1 to 4 days. Injection of as small as 20 μg per mouse of CVF induced a precipitating anti-CVF antibody response, which prevented subsequent depletion of plasma C3 by CVF (Pepys, 1975).

Purification of the plasma protein (β_2 II) reactive with antibody to β_2-glycoprotein II resulted in retention of properdin factor B activity, but loss of the ability to interact with cobra venom factor to activate the terminal complement system. The latter function was restored by the presence of two factors isolated from euglobulin and called factor D and factor E (Hunsicker et al., 1973). Factors D and E have molecular weights of 35,000 and 160,000, respectively.

Certain C derivatives, namely, $C\overline{567}$, have chemotactic activity for polymorphonuclear leucocytes. In the properdin pathway, interaction of C3b with factor B and activated factor D ($\overline{D}$) generates $C3\overline{B}$, an enzyme analogous to $C\overline{42}$ in its lability and its capacity to cleave C3 with initiation of the terminal complement sequence (Fearson et al., 1973a). Substitution of CVF for C3b yields cobra venom factor-$\overline{B}$, a convertase that functions similarly to $C3\overline{B}$ but is more stable (Fearon et al., 1973b). Ruddy et al. (1975) found that mixtures of factors B, $\overline{D}$, and either C3b or CVF, in which the corresponding enzymes $C3\overline{B}$ and CVF-$\overline{B}$ have formed, activate human polymorphonuclear leucocytes *in vitro*, as assessed by chemotaxis and cross-deactivation to structurally unrelated chemotactic factors.

2.5 Other Effects

Marney (1971) studied platelet aggregation induced by antigen–antibody reactions immunologically unrelated to the platelet membrane and demonstrated that platelet aggregation takes place in two phases. The first phase requires the first four components of complement and is completly blocked by small concentrations of cobra venom factor. It was postulated that complement-mediated membrane damage occurs during the first and the second phases and is accompanied by release of platelet ADP and endogenous materials. Dodds and Pickering (1972a) also studied the effect of CVF on platelet function and found impaired clot retraction, induced aggregation, and release of intracellular constituents.

Injection of purified CVF produced striking elevation of plasma fibrinogen and altered plasminogen levels, and significantly reduced the activities of several extrinsic and intrinsic coagulation factors (Dodds and Pickering 1972b). It seems, therefore, that there is a functional relationship between the mechanism of blood coagulation initiation and active complement. However, it is still not certain whether the effects observed are due to a direct interaction of CVF with the membrane or to an indirect complement-mediated action.

2.6 Elimination

The rate of elimination of cobra venom factor from the circulation of rabbits was investigated using ^{125}I-labeled CVF (Cochrane et al., 1970). During the first 24 hr [^{125}I] CVF was rapidly eliminated from the circulation. Over the next 5 days it was eliminated at a constant rate with a half-disappearance time of 32 hr. After 5 days the remaining [^{125}I] CVF was rapidly removed and was completely eliminated by the seventh or eighth day. The rapid removal after 5 days suggested that an immune response to CVF as an antigen may have been induced. When rabbits were injected with [^{125}I] CVF a second time 8 days after the initial injection, the levels of [^{125}I] CVF were reduced by 99% in 25 hr, supporting the view that the animals were responding by forming antibodies to CVF.

The half-life of ^{125}I-labeled CVF in mice is 24 hr (Pepys, 1975). There are some localizations of CVF in the spleen, liver, and kidneys.

2.7 Applications

Cobra venom factor has been used in the investigation of the role of complement in various immunologic reactions.

The Arthus reaction is an allergic reaction that involves edema, hemorrhage, and necrosis caused by the deposition of antigen—antibody complexes in the skin when a sensitized animal is injected intradermally with the antigen. The tissue damage can be reduced in guinea pigs by pretreating the animals with CVF (Maillard, 1968).

Passive cutaneous anaphylactic (PCA) reactions are induced by giving a series of intracutaneous injections of an antibody, followed by an i.v. injection of antigen and Evans blue. Within a few minutes blue spots appear at the sites of antigen—antibody interaction. It is thought that PCA reactions develop independently of the complement system, except for a PCA reaction in rabbits that requires neutrophils. Although PCA reactions and delayed hypersensitivity induced by O-chloromercuribenzoate in guinea pigs are not affected by the administration of CVF, the increased vascular permeability that accompanies the neutrophil-dependent PCA reaction in rabbits is prevented (Müller-Eberhard, 1967; Maillard, 1968; Cochrane et al., 1968; Henson and Cochrane, 1969; Schwartz and Naff, 1971).

The Forssman antigen is a polysaccharide and one of a group of antigens called heterophil antigens because they exist in unrelated plants or animals. These antigens are so closely related chemically that antibodies to one will cross-react with antigens to another. Intravenous injections of Forssman antiserum cause a systemic allergic reaction as a result of antigen—antibody complex formation in all tissues containing the Forssman antigen. Nelson (1966) found that, if animals were pretreated with CVF, subsequent injections of Forssman antiserum were not lethal. This is consistent with data from other studies in which animals whose complement levels were depleted by shark C4 inactivator (Jensen, 1969) and animals genetically deficient in C3 or C4 (May and Frank, 1972) were found to be resistant to the systemic Forssman reaction.

Orange et al. (1968) found that treating animals with CVF impaired their ability to generate the slow-reacting substance of anaphylaxis. In investigations of the autoimmune disease experimental allergic encephalomyelitis, CVF decreased paralysis and lowered the death rate when given immediately before or after injection of encephalitogenic protein emulsion (Pabst et al., 1971; Abrahamson, 1971).

Azar et al. (1968) studied the induction of immunologic tolerance in normal mice and in mice treated with CVF. They found that the development of immunologic tolerance to human γ-globulin was significantly inhibited in mice that had received CVF, indicating that the induction of tolerance is favored by an intact complement system.

Pepys (1972) investigated the participation of a complement in the induction of the allergic response in normal mice and in mice that had been depleted of complement, particularly C3, by pretreatment with CVF. The antibody response to sheep red blood cells and ovalbumin, antigens to which the response is thymus dependent, was significantly inhibited or delayed in animals that received CVF 0 to 2 days after injection of the antigen. In contrast, antibody responses to poly(vinylpyrrolidone) and type III pneumococcal polysaccharide, responses that are thymus independent, were unaffected by complement depletion. These results implied that complement in some way participates in T-cell function.

Erythrocytes in heterologous serum *in vitro* undergo rapid lysis because of the presence of complement-binding natural antibodies in serum. *In vivo*, heterologous erythrocytes are also rapidly removed from the circulation by the cells of the reticuloendothelial system. Prevention of the hemolysis and removal from the circulation after transfusion would be very significant. Wright et al. (1970) found that human erythrocytes in the rat survived longer if C3 was inhibited through the administration of CVF. Further improvement in erythrocyte survival was achieved when CVF was combined with either surgical splenectomy or pretreatment with ethyl palmitate.

Some immunologic reactions mediated by humeral antibody are complement dependent. This is especially true in the hyperacute form of encepholomyelitis. Experimental allergic encephalomyelitis has been classified with the delayed cellular hypersensitivities because the inflammatory infiltrate is mononuclear in nature and because the disease can be transferred passively with immunized lymphoid cells but not with serum. Levine et al. (1971) hoped the pretreatment of rats with CVF could reduce the severity of the disease but found that it is not effective in suppressing allergic encephalomyelitis. Thus it appears that the exudation of polymorphonuclear leucocytes and massive amounts of fibrin in this condition is not dependent on a complement.

Some B cells have surface receptors for C3. Levy et al. (1972) investigated whether these complement receptor lymphocytes would be susceptible to passive uptake of complement that had been activated by CVF. They found that elimination of complement receptor lymphocytes diminished the number of cells able to differentiate into plaque-forming cells in a syngeneic irradiated host, but had no effect on the population of lymphoid cells responsible for the graft-versus-host reaction.

Some erythrocytes in patients with paroxysmal nocturnal hemoglobinuria are lysed more readily by complement than are normal red cells. This susceptibility to lysis is due to a membrane defect that permits the completion of a greater proportion of initiated complement sequences. Kabakçi et al. (1972) devised a test method to detect paroxysmal nocturnal hemoglobinuria and the proportion of complement-sensitive cells, using CVF.

When *Escherichia coli* 0111:B4 endotoxic lipopolysaccharide was injected intravenously into different groups of guinea pigs, the lowest mortality rate was in normal animals, with higher rates in C4-deficient and in CVF-treated normal animals, and with the highest rate in CVF-treated, C4-deficient animals (May et al., 1972). Preincubation of lipopolysaccharide in serum from animals of similar groups indicated that some protection was afforded only by normal serum. Thus both the early and the late complement components appeared to provide protection from the lethal effects of lipopolysaccharide.

The parenteral administration of mycobacterial adjuvant into rats induces an arthropathy with features that resemble rheumatoid arthritis and Reiter's syndrome in man. This is believed to be due to cellular hypersensitivity which may involve a complement system. Kourounakis et al. (1973) investigated the role of CVF in adjuvant-induced disease and found that C9 is elevated and that CVF temporarily depletes C3 and delays the onset of arthritis. Thus they concluded that certain complement components may play a role in this hypersensitivity.

Squirrel monkeys were significantly depleted of complement by CVF of *Naja naja* (Gilbert et al., 1973). These workers also studied the effect of CVF on *E. coli* bacteremia in the monkeys. Striking neutropenia occurred rapidly in control animals, whereas the rate of occurrence of neutropenia was 20 to 30 min slower in the CVF-treated animals. This observation is consistent with a hypothesis that complement-mediated neutrophilic leucocyte function is an important host defense mechanism in gram-negative bacillary bacteremia.

In order to further study the effects of ethyl palmitate and a CVF Castro et al. (1974) used ^{51}Cr-labeled human erythrocytes, which were injected intravenously into rats. They found that erythrocyte survival was enhanced only moderately after the injection of either of these compounds alone. The combined administration of ethyl palmitate and CVF, however, resulted in marked prolongation of human erythrocyte survival consistent with a synergistic effect. This suggested that in the untreated animal complement-dependent hemolysis and reticuloendothelial phagocytosis compete for the removal of circulating heterologous erythrocytes. Both must be blocked simultaneously for maximum heterologous erythrocyte survival.

Mayer (1972) suggested that the terminal stages of complement-induced membrane damage in immune hemolysis involve attack on, or interaction of some type with, membrane lipids. Smith and Becker (1968) attributed the changes to a possible phospholipase activity associated with the terminal components of hemolytic complement. It is known that a basic peptide called direct lytic factor (DLF) opens up membranes by changing the arrangement of their structure, allowing access of phospholipase A_2 to membrane phospholipids [see Chapter 2, "Phospholipase A_2"; Chapter 19, "Nonneurotoxic Basic Proteins (Cardiotoxins, Cytotoxins, and Others)"; and Chapter 20, "Hemolysis"]. Okada and Campbell (1974) studied the role of the basic protein DLF on the complement system in sheep erythrocytes. They found that erythrocytes sensitized with antibody undergo rapid hemolysis in the presence of the eighth component of complement (C8) and DLF. However, EAC1-7 cells are not lysed by either C8 or DLF alone. Since DLF is known to have synergistic action with phospholipase A_2, the result suggests that the activated C8 may have phospholipase A_2 activity.

Guinea pigs injected with rabbit renal tubular basement membrane in Freund's adjuvant develop autoimmune renal cortical tubulointerstitial disease. Complement may be a mediator of antibody-dependent tissue injury because C3 is present along with IgG on the tubular basement membrane. In C4-deficient guinea pigs, no essential pathogenic role could be ascribed to C4, suggesting the involvement of the alternative pathway of complement activation. To determine to what extent complement C participates in the pathogenesis, Rudofsky et al. (1975) attempted to induce the disease by transfer of antitubular basement membrane autoantibodies to guinea pigs depleted of C by cobra venom factor. They found that CVF inhibited autoimmune renal tubulointerstitial disease, indicating that C plays a major role in the pathogenesis of this disease.

The effect of CVF on the hemorrhagic component in the Arthus reaction was investigated; Maillard and Zarco (1968) observed that the hemorrhagic component was strikingly reduced or abolished. Lewis and Turk (1975) administered CVF to guinea pigs 48 hr before the challenging reaction. It lowered the serum complement levels by 80 to 90%. The only effect it had on the Arthus lesion was to delay the appearance of the hemorrhage. Once it had developed, the diameter and intensity were the same as in the controls, and the development of the induration was identical. Lewis and Turk (1975) also found that the hemorrhagic component of the local Shwartzman reaction was not affected by decomplementation with CVF. Complement depletion with CVF does not prevent development of glomerular fibrin deposition, cortical necrosis, or thrombocytopenia in the generalized Shwartzman reaction (Bergstein and Michael, 1974), nor is the loss of fibrinolytic activity induced by *E. coli* endotoxin prevented by pretreatment with CVF.

Cobra venom-induced hemolysis was used to measure the C3 shunt hemolytic activity in the sera of patients with neoplastic and other diseases. Brai and Osler (1973) found that the sera of cancer patients exhibited a different C3 shunt reaction pattern than did the sera of healthy donors or those with nonneoplastic diseases. The fact that cancer sera generally manifest greater hemolytic activity after admixture with a cobra venom protein, the terminal complement components, and unsensitized guinea pig erythrocytes suggests that human neoplasia may be associated with changes in the C3 shunt complement system. Similarly, mice bearing tumors have an increased lytic capacity for unsensitized erythrocytes after interaction with a CVF (Brai and Osler, 1972b).

2.8 Use in Organ Transplantation

When an organ is transplanted, the body assumes that the foreign tissues will be harmful and antibodies are produced. Rejection of foreign tissues by the body is a serious matter. If one of the nine sequentially acting components in the complement system can be blocked, this may improve transplant survival. Thus cobra venom factor has been used as an immunosuppressor by many investigators. The purpose of this section is to review all the literature on organ transplantation in which CVF has been used.

Snyder et al. (1966) found a 13-fold increase in the survival time of CVF-treated dogs to which porcine kidneys were transplanted. Gewurz et al. (1966) also observed a significant prolongation of survival in pretreated dogs to which rabbit kidneys were transplanted. Villegas and Coppola (1968) reported a mean survival time of 7.5 days in a group of 31 untreated guinea pigs that received skin allografts. In contrast, a group of 24 allografted animals that were treated with CVF had a median survival time of 11 days.

Gewurz et al. (1967) also found that CVF-induced complement depletion prolonged the survival of rabbit renal xenografts in the dog, but had no demonstrable effect on skin allografts in the chicken or renal allografts in the dog.

Ballow et al. (1973) proposed that the basic cytotoxin impurity present in CVF is the one effective as a suppressant of graft-versus-host reaction. The basic cytotoxin from the venom of *Naja haje* has a molecular weight of 13,000 daltons. The basic cytotoxin causes lymphocyte cell death but does not kill macrophages; it inhibited local graft-versus-host reactivity. Ballow et al. found that purified CVF has less graft-versus-host activity inhibition than did the cytotoxin.

Cobra venom factor was used to study skin allografting by Glovsky et al. (1973), who found that the factor induced 2- to 3-day prolongation of allograft survival. The prolongation was attributed to alteration of complement-dependent inflammation by CVF.

3 OTHER VENOMS

A number of venoms from other species of snakes have also been analyzed for their effects on the complement system. Crude and partially purified fractions of 39 venoms from various vertebrate and invertebrate species were screened for their effects on the hemolytic complement and on components C1, C4, C2, and C3-9 (Birdsey et al., 1971). Nineteen of these were found to display anticomplement activity and were grouped into several categories:

1. Venoms of *Haemachatus haemachatus, Naja haje, N. naja, N. Melanoleuca, N. nigricollis, N. nivea*, and *Ophiophagus hannah* induce the consumption of large amounts of C3-9 to form a stable intermediate that can lyse erythrocytes.

2. Venoms of *Agkistrodon rhodostoma, Bitis arietans, Bothrops jararaca*, and *B. jararacussa* induce marked depletion of C2, C4, and C3-9.

3. Venoms of *Trimeresurus purpureomaculatus* and *T. popeorum* are selective in the consumption of C3-9, but no lysis-inducing intermediate is formed. *Lachesis muta* venom consumes a large amount of C2 in addition to C3-9.

4. *Heloderma horridum* (Gila monster) venom consumes considerable amounts of C1, C4, and C3-9.

5. Venoms of two amphibia, three arthropods, and a coelenterate do not contain anticomplementary activity.

Venoms of the following snakes have no effect on the complementary system:

Bungarus caeruleus,
B. fasciatus,
Dendraspis augusticeps,
D. polylepis,
Crotalus durissus terrificus,
Agkistrodon contortrix mokasen,
Bothrops alternata,
Trimeresurus walterinsii,
Vipera ammodytes,
V. russellii russellii,
Bitis gabonica,
B. nasicornus,
Causus rhombeatus
Laticauda colubrina

Zarco et al. (1967) reported finding anticomplementary activity in a fraction from *Agkistrodon piscivorus* venom. The crude venom reduced C4, C3, and C2 activities, a finding consistent with that of Birdsey (1971). The anticomplementary activity of a nontoxic fraction of the venom was observed with whole serum. It required serum cofactor, but a different one from that required by cobra venom factor.

Ballow and Cochrane (1969) isolated two fractions having anticomplementary action from the venom of *Naja naja*. The fraction with smaller molecular weight is identical to CVF. The fraction with a high molecular weight of 0.8 to 1.0×10^6 daltons acts on early components of the complement system but does not affect C3 or C5. It greatly inhibits the formation of EAC 14.

4 POSSIBLE ROLE OF COMPLEMENT ACTIVATION IN SNAKE ENVENOMATION

The diversity of pharmacological effects of snake venoms is evident from their complex compositions. With the exception of research to elucidate the alternative pathway to complement activation of cobra venom factor, there has been no investigation into the relationship between complement and the effects of envenomation. This is understandable in light of the fact that until recently complement fixation was thought to be associated only with antigen—antibody reaction. But now, with knowledge of alternative activation mechanisms and the production of biologically active agents by complement—noncomplement interactions, it is not unreasonable to think that the complement system may mediate some of the effects of envenomation (King, 1972).

REFERENCES

Abrahamson, H. (1971). Prevention of experimental allergic encephalomyeletis with cobra venom factor, *J. Asthma Res.,* **8,** 151.

Alper, C. A. and Balavitch, D. (1976). Cobra venom factor: Evidence for its being altered cobra C3 (the third component of complement), *Science,* **191,** 1275.

Azar, M. M., Yumis, E. J., Pickering, P. J., and Good, R. A. (1968). On the nature of immunological tolerance, *Lancet,* **1,** 1279.

Ballow, M. and Cochrane, C. G. (1969). Two anticomplementary factors in cobra venom: Hemolysis of guinea pig erythrocytes by one of them, *J. Immunol.,* **103,** 944.

Ballow, M., Day, N. K., and Good, R. A. (1973). Effect of cobra venom factor on the local GVH reaction, *J. Immunol.,* **110,** 354.

Bergstein, J. M. and Michael, A. F. (1974). Failure of cobra venom factor to prevent the generalized Shwartzman reaction and loss of renal cortical fibrinolytic activity, *Am. J. Pathol.,* **74,** 19.

Birdsey, V. (1971). Interaction of toxic venoms with the complement system, *Immunology,* **21,** 299.

Birdsey, V., Lindorfer, J., and Gewurz, H. (1971). Interaction of toxic venoms with the complement system, *Immunology,* **21,** 299.

Bitter-Suermann, D., Dierich, M., König, W., and Hadding, U. (1972). Bypass activation of the complement system starting with C3. I. Generation and function of an enzyme from a factor of guinea-pig serum and cobra venom, *Immunology,* **23,** 267.

Brai, M. and Osler, A. (1972a). Cobra venom-induced hemolysis: Activity levels in sera of patients with neoplastic and other diseases, *J. Exp. Med.,* **136,** 950.

Brai, M., and Osler, A. G. (1972b). Studies of the C3 shunt activation in cobra venom induced lysis of unsensitized erythrocytes (36623), *Proc. Soc. Exp. Biol. Med.,* **140,** 1116.

Brai, M. and Osler, A. G. (1973). Alternate complement pathway: Elevation of cobra venom-induced hemolytic activity in serums of tumor-bearing mice, *J. Immunol.,* **111,** 1598.

Castro, O., Rosen, M. W., and Finch, S. C. (1974). Mechanism of ethyl palmitate and cobra venom factor enhancement of heterologous erythrocyte survival, *Proc. Soc. Exp. Biol. Med.,* **147,** 106.

Cochrane, C. G., Müller-Eberhard, H. J., and Fjellstrom, K. E. (1968). Capacity of a cobra venom protein to inactivate the third component (C3) and to inhibit immunologic reactions, *J. Clin. Invest.,* **47,** 21a.

Cochrane, C. G., Müller-Eberhard, H. J., and Aikin, B. S. (1970). Depletion of plasma complement *in vivo* by a protein of cobra venom: Its effect on various immunologic reactions, *J. Immunol.,* **105,** 55.

Cooper, N. R. (1971). In *Proceedings of International Symposium on the Biological Activities of Complement,* Karger, New York.

Cooper, N. (1973). Formation and function of a complex of the C3 proactivator with a protein from cobra venom, *J. Exp. Med.,* **137,** 451.

Day, N., Geiger, H., Finstad, J., and Good, R. A. (1972). A starfish hemolymph factor which activates vertebrate complement in the presence of cobra venom factor, *J. Immunol.,* **109,** 164.

Dierich, M. P., Bitter-Suermann, D., König, W., and Hadding, U. (1971). Formation and function of a complement-activating enzyme generated from factors of guinea pig serum and cobra venom, *Eur. J. Immunol.,* **1,** 309.

Dodds, W. and Pickering, R. (1972a). Purified cobra venom factor: Effect on blood platelets, *Proc. Soc. Exp. Biol. Med.,* **140,** 429.

Dodds, W. J. and Pickering, R. J. (1972b). The effect of cobra venom factor on hemostasis in guinea pigs, *Blood,* **40,** 400.

Drake, W. P., Pokorney, D. R., and Mardiney, M. R., Jr. (1974). *In vivo* abrogation of serum C3 and C5 by administration of cobra venom factor and heterologous anti-C3, *J. Immunol. Method.,* **6,** 61.

Ewing, C. B. (1894). The action of rattlesnake venom upon the bactericidal power of the blood serum, *Med. Rec.,* **45,** 663.

Fearon, D. T., Austen, K. F., and Ruddy, S. (1973a). Formation of a hemolytically active cellular intermediate by the interaction between properdin factor B and D and the activated third component of complement, *J. Exp. Med.,* **138,** 1305.

Fearon, D. T., Austen, K. F., and Ruddy, S. (1973b). Serum proteins involved in decay and regeneration of cobra venom factor-dependent complement activation, *J. Immunol.,* **111,** 1730.

Flexnor, S. and Noguchi, H. (1901). Snake venom in relation to hemolysis, bacteriolysis, and toxicity, *J. Exp. Med.,* **6,** 277.

Gewurz, H., Clark, D. S., Finstad, J., Kelley, W. D., Varco, R. L., Good, R. A., and Gabrielson, A. E. (1966). Role of the complement system in graft rejections in experimental animals and man, *Ann. N.Y. Acad. Sci.,* **129,** 673.

Gewurz, H., Clark, D. S., Cooper, M. D., Varco, R. L., and Good, R. A. (1967). Effect of cobra venom induced inhibition of complement activity on allograft and xenograft rejection reactions, *Transplantation,* **5,** 1296.

Gewurz, H., Pickering, R. J., Day, N. K., and Good, R. A. (1971). Cobra venom factor-induced activation of the complement system: Developmental, experimental, and clinical considerations, *Int. Arch. Allergy,* **40,** 47.

Gilbert, D. N., Barnett, J. A., and Sanford, J. P. (1973). *Escherichia coli* bacteremia in the squirrel monkey. I. Effect of cobra venom factor treatment, *J. Clin. Invest.,* **52,** 406.

Glovsky, M. M., Ward, P. A., and Fundenberg, H. H. (1973). Role of complement in guinea pig skin allograft rejection. I. Effect of cobra venom C3 inactivator and fumaropimaric acid on rejection, *Clin. Immunol. Immunopathol.,* **1,** 165.

Goldman, J. N., Ruddy, S., and Austen, K. F. (1972). Reaction mechanisms of nascent C567 (reactive lysis). I. Reaction characteristics for production of EC567 and lysis by C8 and C9, *J. Immunol.,* **109,** 353.

Götze, O. and Müller-Eberhard, H. J. (1970). Lysis of erythrocytes by complement in the absence of antibody, *J. Exp. Med.,* **132,** 898.

Götze, O. and Müller-Eberhard, H. J. (1971). The C3-activator system: An alternate pathway of complement activation, *J. Exp. Med.,* **134,** 90.

Henson, P. M. and Cochrane, C. G. (1969). Immunological induction of increased vascular permeability, *J. Exp. Med.,* **129,** 153.

Hunsicker, L. G., Ruddy, S., and Austen, K. F., (1973). Alternate complement pathway: Factors involved in cobra venom factor (CoVF) activation of the third component of complement (C3), *J. Immunol.,* **110,** 128.

Jensen, J. (1967). Anaphylatoxin in its relation to the complement system, *Science,* **155,** 1122.

Jensen, J. A. (1969). A specific inactivator of mammalian C4 isolated from nurse shark (*Ginglymastroma cirratum*) serum, *J. Exp. Med.,* **130,** 217.

Kabakci, T., Rosse, W. F., and Logue, G. L. (1972). The lysis of paroxysmal nocturnal haemoglobinuria red cells by serum and cobra factor, *Br. J. Haematol.,* **23,** 693.

King, D. (1972). *The Interaction of Venoms and Endotoxins with Complement.* M. S. Thesis, Colorado State University, Fort Collins.

Klein, P. G. and Wellenseik, H. J. (1965). Multiple nature of the third component of guinea pig complement, *Immunology,* **8,** 590.

Kourounakis, L., Nelson, R. A., Jr., and Kopusta, M. A. (1973). The effect of a cobra venom factor on complement and adjuvant-induced disease in rats, *Arthritis Rheum.,* **16,** 71.

Levine, S., Cochrane, C. G., Carpenter, C. B., and Behan, P. O. (1971). Allergic encephalomyelitis: Effect of complement depletion with cobra venom, *Proc. Soc. Exp. Biol. Med.,* **138,** 285.

References

Levy, N. L., Scott, D. W., and Snyderman, R. (1972). Bone marrow-derived lymphoid cells (B cells): Functional depletion with cobra factor and fresh serum, *Science,* **178,** 866.

Lewis, E. and Turk, J. L. (1975). Comparison of the effect of various antisera and cobra venom factor in inflammatory reactions in guinea-pig skin. II. The Arthus reaction and the local Schwartzman reaction, *J. Pathol.,* **115,** 111.

Maillard, J. L. (1968). Decomplementization by a factor extracted from cobra venom: Effect on several immune reactions of guinea pig and rat, *Ann. Inst. Pasteur,* **114,** 756.

Maillard, J. L. and Zarco, R. M. (1968). Decomplementation pur un facteur extrait du venin de cobra: Effet sur plusieurs reactions immunes du cobaye et du rat, *Ann. Inst. Pasteur,* **114,** 756.

Marney, S. (1971). Platelet aggregation in heparinized plasma and citrated plasma: Relationship to ADP and inhibition by CVF, *J. Immunol.,* **106,** 82.

May, J. E. and Frank, M. M. (1972). Complement-mediated tissue damage: Contribution of the classical and alternate complement pathways in the Forssman reaction, *J. Immunol.,* **108,** 1517.

May, J. E., Kane, M. A., and Frank, M. M. (1972). Host defense against bacterial endotoxemia: Contribution of the early and late components of complement detoxification, *J. Immunol.,* **109,** 893.

Mayer, M. M. (1972). Mechanism of cytolysis by complement, *Proc. Natl. Acad. Sci. U.S.A.,* **69,** 2954.

Mickenberg, I. D., Snyderman, R., Root, R. K., Mergenhagen, S. E., and Wolff, S. M. (1971). The relationship of complement consumption to immune fever, *J. Immunol.,* **107,** 1466.

Müller-Eberhard, H. J. (1967). Mechanism of inactivation of the third component (C3) by cobra venom, *Fed. Proc.,* **26,** 744.

Müller-Eberhard, H. J. and Fjellstrom, K. E. (1971). Isolation of the anticomplementary protein from cobra venom and its mode of action on C3, *J. Immunol.,* **107,** 1666.

Müller-Eberhard, H. J., Nilsson, U. R., Dalmasso, A. P., Polley, M. J., and Calcott, M. A. (1966). A molecular concept of immune cytolysis, *Arch. Pathol.,* **82,** 205.

Nelson, R. A. (1966). A new concept of immunosuppression in hypersensitivity reactions and in transplantation immunity, *Surv. Ophthalmol.,* **11,** 498.

Okada, H., and Campbell, W. (1974). Immune hemolysis and a basic peptide from cobra venom, *J. Immunol.,* **113,** 1647.

Orange, R., Valentine, F., and Austen, F. (1968). Antigen-induced release of slow reacting substance of anaphylaxis (SRS-Arat) in rats prepared with homologous antibody, *J. Exp. Med.,* **127,** 767.

Pabst, H., Day, N. K., Gewurz, H., and Good, R. A. (1971). Prevention of experimental allergic encephalomyelitis with cobra venom factor, *Proc. Soc. Exp. Biol. Med.,* **136,** 555.

Pepys, M. (1972). Role of complement in induction of the allergenic response, *Nature New Biol.,* **237,** 157.

Pepys, M. B. (1975). Studies *in vivo* of cobra factor and murine C3, *Immunology,* **28,** 369.

Pickering, R. J., Walfson, M. R., Good, R. A., and Gewurz, H. (1969). Passive hemolysis by serum and cobra venom factor: A new mechanism including membrane damage by complement, *Proc. Natl. Acad. Sci. U.S.A.,* **62,** 521.

Ritz, H. (1912). Über die Wirkung des Kobragiftes auf die Komplemente. III. Mitteilung: Zugleichein Beitrag zur Kenntuisder Hamolytischen Komplemente, *Z. Immunitaetsfoirsch.,* **13,** 62.

Roy, A. C. (1969). Action of cobra and Russell's viper venom on the esterase and complement activity of blood sera, *Calcutta Med. J.,* **66,** 93.

Ruddy, S., Austen, K. F., and Goetzl, E. J. (1975). Chemotactic activity derived from interaction of factor D and B of the properdin pathway with cobra venom factor or C3b, *J. Clin. Invest.,* **55,** 587.

Rudofsky, U. H., Steblay, R. W., and Pollara, B. (1975). Inhibition of experimental autoimmune renal tubulointerstitial disease in guinea pigs by depletion of complement with cobra venom factor, *Clin. Immunol. Immunopathol.,* **3,** 396.

Schwartz, H. J. and Naff, G. B. (1971). The effect of complement depletion by cobra venom factor on delayed hypersensitivity reactions, *Proc. Soc. Exp. Biol. Med.,* **138,** 1041.

Shin, H. S., Gewurz, H., and Snydermann, R. (1969). Reaction of cobra venom factor with guinea pig complement and generation of an activity chemotactic for polymorphonuclear leukocytes, *Proc. Soc. Exp. Biol. Med.,* **131,** 203.

Smith, J. K. and Becker, E. L. (1968). Serum complement and the enzymatic degradation of erythrocyte phospholipids, *J. Immunol.,* **100,** 459.

Snyder, G. B., Ballesteros, E., Zarco, R. M., and Lynn, B. S. (1966). The prolongation of renal xenografts by C complement suppression, *Surg. Forum,* **17,** 478.

Villegas, G., and Coppola, E. (1968). Induction of prolonged survival of skin allografts in guinea pigs by complement inhibition, *Fed. Proc.,* **27,** 505.

Vogt, W. and Schmidt, G. (1964). Abtrennung des Anaphylatoxinbildenden Prinzips and Cobragift von Andern Giftkomponenten, *Experientia,* **20,** 207.

Wright, M. C., Nelson, R. A., and Finch, S. C. (1970). The effects of a cobra venom factor and ethyl palmitate on the prolongation of survival of heterologous erythrocytes, *Yale J. Biol. Med.,* **43,** 173.

Zarco, R., Schultz, D., and Vroom, D. (1967). Inactivation of guinea pig complement by *Agkistrodon piscivorus* (cottonmouth moccasin) venom, *Fed. Proc.,* **26,** 362.

27 Chemical Neutralization of Snake Venoms

Charlotte L. Ownby and Anthony T. Tu

1 CHEMICAL 436
    1.1 Steroid Compounds, 436
    1.2 Carbohydrates, 436
    1.3 Compounds Containing Sulfur, 437
    1.4 Chelating Compounds, 437
    1.5 Other Compounds, 437
        Formalin, 437
        Antihistamines, 450
        Procaine, 450
        Others, 450
    1.6 Combinations of Compounds, 450

2 RADIATION 451
    2.1 Visible Light (Photooxidation), 451
    2.2 Ultraviolet Light, 452
    2.3 Cobalt-60, 453
    2.4 X-rays, 453

References 454

At present the most effective treatment for snakebite is horse antiserum to snake venom (antivenin). This antiserum is usually produced by injecting very small amounts of venom or venoms over a long period of time. Because of the high toxicity and severe local tissue damage induced by most snake venoms, only small amounts of venom can be injected; hence the production of an antiserum with high titer is a very long and tedious procedure. There is a definite need for a technique for producing a high-titer, specific antiserum in a very short time. One approach to this problem has been to make toxoids of snake venoms that still retain their antigenicity; these toxoids or neutralized venoms are then used to produce antiserum. Attempts have been made to neutralize venom with various chemicals and various forms of irradiation. Although the purpose of most studies was to produce toxoids to serve in the production of antivenin, the aim of others was to use the toxoids obtained for the active immunization of human beings or to investigate the nature of the toxic principles involved and their mechanism of action on tissues. Many investigators have attempted to develop a chemical (nonserum) treatment for the local changes induced by snakebite, that is, hemorrhage and myonecrosis.

The purpose of this chapter is to review all these studies and to discuss the present state of our ability to neutralize snake venoms both *in vitro* (neutralization before experimental injection) and *in vivo* (neutralization after experimental injection). Many of the results of these studies are presented in tabular form in Tables 1 and 2.

1 CHEMICAL

1.1 Steroid Compounds

The antihemorrhagic action of estrogens has long been recognized, but the basis for this action is still not clear (Rona, 1963). Estriol-16,17-disodium succinate (ES) has been used to neutralize hemorrhage induced by cobra venom (see Table 1) because of its reported protective effect on blood vessels. It apparently strengthens the perivascular connective tissue by causing a change in the acid mucopolysaccharides so that the sol-gel equilibrium shifts toward the more solid gel state.

Bonta et al. (1965, 1969) tested the ability of ES to prevent hemorrhage due to *Naja naja* venom and *Agkistrodon piscivorus* venom. They found that ES could not prevent hemorrhage caused by *A. piscivorus* venom but did delay the onset and reduce the intensity of hemorrhage from *N. naja* venom (Table 1). Experiments in which heparin was also tested (Bonta et al., 1969) indicate that the protection of vessels by ES is effective only if the venom hemorrhagic action depends on a heparin-precipitable factor such as that present in *N. naja* venom. The hemorrhagic action of *A. piscivorus* venom is independent of a heparin-precipitable factor.

Adrenal cortical hormones have been tested on the basis of their role in stress reactions, since snakebite induces stress. Schöttler (1954) has shown that neither cortisone nor phenergan (antihistamine) is able to neutralize the effects of snake venoms (Table 1). In some instances the *in vivo* use of hydrocortisone seems to be of some benefit (Table 1).

1.2 Carbohydrates

Herapin is a sulfate-containing mucopolysaccharide that acts as an anticoagulant by preventing the activation of factor IX (plasma thromboplastin component, Christmas factor). It also acts with a plasma cofactor to inhibit the action of thrombin, thus preventing the coagulation of blood. Heparin is highly acidic and is known to form an irreversible complex with protamine, a highly basic protein. It is probable that heparin can form such complexes with other highly basic proteins such as those present in snake venoms. In fact, Bhargava et al. (1970) have shown that heparin forms an inactive complex with the hemorrhagic components of cobra venom, thereby neutralizing the hemorrhagic activity of this venom (Table 1).

However, heparin is unable to neutralize the hemorrhagic activity of *Agkistrodon piscivorus, Crotalus horridus horridus,* or *C. atrox* venom (see Table 1). Thus there appear to be different mechanisms of hemorrhage for venoms from these species, as compared to the venom of *Naja naja.*

Even though heparin prevented hemorrhage when mixed with cobra venom before contact with tissues, it did not prevent death, indicating that the hemorrhagic and lethal components are not the same (Bonta et al., 1970).

Esculin (in DMSO) was also tested for antihemorrhagic activity but was effective only *in vitro* (Ownby, 1975).

1.3 Compounds Containing Sulfur

Thiol compounds such as dihydrothioctic acid, thioglycolate, cysteine, glutathione, and thiourea have been used in attempts to neutralize snake venoms (see Table 1). The mechanism of antitoxic action of thiol compounds might involve either a direct effect on the venom itself or inhibition of the action of the venom on tissues. In the case of habu (*Trimeresurus flavoviridis*) venom, the activity resides in the S–S bond. Hence a low molecular compound containing an SH radical could substitute for the S–S bond and thus inactivate the venom (Kurihara and Shibata, 1971). On the other hand, thiol compounds might prevent the inhibitory action of habu venom on the activity of succinic dehydrogenase (SH enzyme). Of the thiol compounds, dihydrothioctic acid seems to be the most effective in preventing hemorrhage and death when mixed with habu venom before injection into animals (Table 1). However, it did not neutralize the toxicity of habu venom in the *in vivo* test (Table 1). Glutathione did reduce the lethality of habu venom *in vitro*, but did not reduce hemorrhage induced by *Crotalus* spp. venoms either *in vitro* or *in vivo* (see Table 1). Cysteine, thioglycolate, and thiourea did not prevent hemorrhage due to snake venom *in vitro* or *in vivo* (see Table 1).

Dimethyl sulfoxide (DMSO) has been used clinically as a solvent and to enhance the absorption of substances, since it has an outstanding ability to penetrate animal skin and mucous membranes. The results of tests of the ability of DMSO to neutralize snake venom are contradictory. Tu et al. (1970) and Ownby et al. (1975) found that DMSO neutralized the hemorrhagic and lethal activities of rattlesnake venoms *in vitro* but not *in vivo* (see Table 1). On the other hand, Tiru-Chelvam (1974) found that DMSO did not neutralize the lethal activity of *C. adamanteus* venom either *in vitro* or *in vivo* (Table 1). Therefore the effects of DMSO may vary with the species of snake, but these experiments need to be repeated.

1.4 Chelating Compounds

Venom protease activities and hemorrhagic activity can be removed by adding EDTA, a powerful chelating agent, to venoms (Goucher and Flowers, 1964; Friederich and Tu, 1971). Many purified venom proteases contain zinc and calcium (see Chapter 7). Hemorrhagic toxin itself also contains zinc and calcium ions (unpublished data); therefore the effectiveness of the chelating compound EDTA is due to the removal of metals from venom proteins. Actually, many chelating compounds can neutralize venom hemorrhagic activity. Tu et al. (1970), Ownby et al. (1975), and Ownby (1975) tested several different chelating agents *in vitro* and *in vivo* and found them all to prevent hemorrhage *in vitro* but not *in vivo* (see Table 1).

One should keep in mind, therefore, that the ability of chelating compounds to neutralize hemorrhagic activity is limited to the *in vitro* condition and is not present in *in vivo*.

1.5 Other Compounds

Formalin When formalin is mixed with snake venom before injection, the venom, regardless of its type, loses its toxicity (see Table 1). Bizzini and Raynaud (1974)

Table 1 Abilities of Various Chemicals to Neutralize Snake Venom Toxicities

Compound Tested	Venom Tested	Test System	Results	Reference
A. Steroid Compounds				
Estriol-16, 17-disodium succinate	*Naja naja*	Heart-lung prep., dogs; cpd. prior to venom	Anti-hem.	Bonta et al, 1965; 1969
Estriol-16, 17-disodium succinate	*Agkistrodon piscivorus*	Heart-lung prep., dogs; cpd. prior to venom	Not anti-hem.	Bonta et al, 1969
Cortisone	*Bothrops jararaca*	In vitro	Not anti-toxic	Schottler, 1954
Cortisone	*Crotalus terrificus*	In vitro	Not anti-toxic	Schottler, 1954
Hydrocortisone	*Crotalus adamanteus*	In vivo	Not anti-leth.	Morales et al, 1963
Hydrocortisone	*Naja naja*	In vivo	Anti-leth.	
Hydrocortisone	*Echis carinatus*	In vivo	Anti-leth.	Arora et al, 1962
Hydrocortisone	*Vipera russelli*	In vivo	Reduced leth.	Seth et al, 1971

B. Carbohydrates

Heparin	Echis carinatus	In vitro	Anti-lethal	Ahuja et al, 1946
Heparin	Naja naja	In vitro	Anti-hem.	Bonta et al, 1969; Bhargava et al, 1970
Heparin	Agkistrodon piscivorus	In vitro	Not anti-hem.	Bonta et al, 1969; Bhargava et al, 1970
Heparin	Naja naja	In vitro	Anti-hem.	Bonta et al, 1970
	Naja nigricollis		Not anti-hem.	
Heparin	Naja naja	In vivo cpd. prior to venom	Anti-hem.	Bonta et al, 1970
	Naja nigricollis			
Heparin	Crotalus horridus horridus	In vitro	Not anti-hem.	Tu et al, 1970
	Crotalidase			
Heparin		In vivo (clinical)	Anti-hem.	Raby, 1973
Heparin	Echis carinatus	In vivo (clinical)	Anti-coagulant	Christy, 1972

Table 1 *Continued*

Compound Tested	Venom Tested	Test System	Results	Reference
Heparin	Echis carinatus	In vivo (clinical)	Anti-coagulant	Weiss et al, 1973
Heparin	Crotalus atrox	In vitro	Not anti-hem.	Ownby et al, 1975; Ownby, 1975
	Crotalus atrox	In vivo	Not anti-hem.	
Esculin (in DMSO)	Crotalus atrox	In vitro	Anti-hem.	Ownby, 1975
		In vivo	Not anti-hem.	
C. Compounds Containing Sulfur				
Dihydrothioctic acid (DHTA)	Trimeresurus flavoviridis	In vitro	Anti-hem., anti-nec., anti-leth.	Sawai et al, 1963, 1969
Dihydrothioctic acid	Trimeresurus flavoviridis	In vivo	Not anti-toxic	Sawai et al, 1963
Dihydrothioctic acid	Trimeresurus flavoviridis	In vitro	Anti-hem.	Kurihara and Shibata, 1971
Cysteine	Trimeresurus flavoviridis	In vitro	Not anti-hem.	Kurihara and Shibata, 1971

Cysteine	Crotalus atrox	In vitro	Not anti-hem.	Ownby et al, 1975;
		In vivo	Not anti-hem.	Ownby, 1975
Thioglycolate	Trimeresurus flavoviridis	In vitro	Not anti-hem., reduced leth.	Kurihara and Shibata, 1971
Glutathione	Trimeresurus flavoviridis	In vitro	Reduced leth.	Yoshira, 1967
Glutathione	Naja naja	In vivo (clinical)	Reduced leth.	Watt, 1972
Glutathione	Crotalus horridus horridus	In vitro	Not anti-hem.	Tu et al, 1970
Glutathione	Crotalus atrox	In vitro	Not anti-hem.	Ownby et al., 1975
		In vivo	Not anti-hem.	Ownby, 1975
Thiourea	Crotalus horridus horridus	In vitro	Not anti-hem.	Tu et al, 1970
Dimethyl sulfoxide	Crotalus horridus horridus	In vitro	Anti-hem., anti-leth.	Tu et al, 1970
Dimethyl sulfoxide	Not given	In vivo (clinical) dogs	Reduced edema	Lee, 1972

Table 1 *Continued*

Compound Tested	Venom Tested	Test System	Results	Reference
Dimethyl sulfoxide	Naja naja	In vitro	Not anti-lethal	Tiru-Chelvam, 1974
		In vivo	"	
	Crotalus adamanteus	In vitro	"	
		In vivo	"	
Dimethyl sulfoxide	Crotalus atrox	In vitro	Anti-hem.	Ownby et al, 1975;
		In vivo	Not anti-hem.	Ownby, 1975
D. Chelating Compounds				
CaEDTA	Trimeresurus flavoviridis (Habu)	In vitro	Anti-leth. anti-nec., not anti-hem.	Sawai et al, 1961a
CaEDTA	Trimeresurus flavoviridis (Habu)	In vivo	Not anti-hem. anti-nec.	Sawai et al, 1961b
EDTA	Agkistrodon piscivorus	In vitro	Anti-hem. anti-nec.	Goucher and Flowers, 1964
EDTA	Agkistrodon piscivorus	In vivo	Anti-hem. anti-necrosis	Flowers and Goucher, 1965
	Bothrops atrox	In vivo	Anti-hem. anti-nec.	

EDTA	Crotalidae:	In vitro	Anti-hem.	Friederich and Tu, 1971
	A. acutus	"	Not anti-leth.	"
	C. atrox	"	"	"
EDTA	C. adamanteus	In vitro	Anti-hem.	Friederich and Tu, 1971
	C. basiliscus	"	Not anti-leth.	"
	C. durissus	"	"	"
	C. d. terrificus	"	"	"
	C. d. totonacus	"	"	"
	C. h. horridus	"	"	"
	C. h. atricaudatus	"	"	"
	S. milarius barbouri	"	"	"
	C. v. viridis	"	"	"
EDTA	Viperidae:	In vitro	Not anti-leth., anti-hem.	Friederich and Tu, 1971
	V. russelli siamensis	"	"	"
	B. arietans	"	"	"
	B. gabonica	"	"	"

Table 1 *Continued*

Compound Tested	Venom Tested	Test System	Results	Reference
EDTA	*C. atrox*	*In vitro*	Anti-hem.	Ownby et al, 1975; Ownby, 1975
		In vivo	Anti-hem.	Ownby et al, 1975; Ownby, 1975
DTPA	*C. horridus horridus*	*In vitro*	Anti-hem., reduced leth.	Tu et al, 1970
DTPA	*C. horridus horridus*	*In vivo*	Not anti-hem., not anti-leth.	Tu et al, 1970
DTPA	*C. atrox*	*In vitro*	Anti-hem.	Ownby et al, 1975; Ownby, 1975
		In vivo	Anti-hem.	Ownby et al, 1975; Ownby, 1975
EGTA	*C. atrox*	*In vitro*	Anti-hem.	Ownby et al, 1975 Ownby, 1975
		In vivo	Anti-hem.	Ownby et al, 1975 Ownby, 1975
N-(2-Hydroxylethyl) ethylenediaminetriacetic acid	*C. horridus horridus*	*In vitro*	Anti-hem. reduced leth.	Tu et al, 1970

N-(2-Hydroxylethyl) ethylenediaminetriacetic acid	C. horridus horridus	In vivo	Not anti-hem. not anti-leth.	Tu et al, 1970
E. Other Compounds				
Formalin	Naja flava	In vitro	Anti-toxic	Christensen, 1947
Formalin	Notechis scutatus	In vitro	Anti-toxic	Wiener, 1960
Formalin	Vipera palestinae	In vitro	Not anti-leth. anti-hem.	Moroz-Perlmutter et al, 1963
	Echis colorata	In vitro	Not anti-leth. anti-hem.	"
Formalin	Notechis scutatus	In vitro	Anti-leth. anti-hem.	Sadahiro et al, 1970
Formalin	Naja naja philippinensis	In vitro	Anti-toxic	Salafranca, 1970
Formalin	Trimeresurus flavoviridis	In vivo	Reduced leth.	Sadahiro, 1971
Formalin	Naja nigricollis	In vitro	Anti-toxic	Dumarey and Boquet, 1972

Table 1 *Continued*

Compound Tested	Venom Tested	Test System	Results	Reference
Formalin	Trimeresurus flavoviridis	In vitro	Anti-leth. anti-hem.	Kondo et al, 1973
Formalin	Trimeresurus flavoviridis	In vitro	Anti-leth. anti-hem.	Kondo and Murata, 1972
Formalin	Trimeresurus flavoviridis	In vitro	Anti-leth., anti-hem.	Sawai et al, 1972
Formalin	Naja naja	In vitro	Anti-leth.	Fukuyama and Sawai, 1973
Procaine	Bothrops jararaca	In vitro	Not anti-toxic	Schottler, 1954
Procaine	Crotalus terrificus	In vitro	Not anti-toxic	Schottler, 1954
Procaine	Crotalus atrox	In vitro	Not anti-hem.	Ownby, 1975; Ownby et al, 1975
		In vivo	Not anti-hem.	"
Epsilon Amino Caproic Acid (EACA)	Naja hannah	prior to venom	Anti-hem. anti-edema	Vick et al, 1963
	Crotalus adamanteus	"	"	

446

Ethanol, formalin	_Trimeresurus flavoviridis_	In vitro	Anti-toxic (Anti-hem. anti-leth. anti-nec.)	Sawai et al, 1972
Poly-vinyl-pyrollidone (PVP)	_Naja naja_	In vivo	increased survival time	Trethewie, 1956
Tetra-hydro-amino-acridine (THA)	_Naja naja_	In vivo	increased survival time	Trethewie, 1956
Thiabendazole (anti-fungus)	_Agkistrodon piscivorus_	In vitro	Anti-hem. anti-nec.	Stone et al, 1966
Thiabendazole (anti-fungus)	_Agkistrodon piscivorus_	In vivo	Reduced hem., reduced nec.	Stone et al, 1966
Isoxsuprine (vasodilator)	_Bothrops jararaca_	In vivo	Anti-necrosis	Rosenfeld et al, 1969a
Tetracycline	_Trimeresurus flavoviridis_	In vitro	Anti-leth. anti-nec. anti-hem.	Sawai et al, 1963

Table 1 *Continued*

Compound Tested	Venom Tested	Test System	Results	Reference
Dexamethasone	Bothrops jararaca	In vivo	Anti-necrosis	Rosenfeld et al, 1969b
Benadryl HCl - antihistamine	Crotalidae (species not given)	In vivo (clinical dogs)	Complete healing	Maier, 1951
Phenergan [10-(2-dimethyl-aminoisopropyl)-phenothiazine]-antihistamine	Bothrops jararaca	In vivo	Not anti-toxic	Schottler, 1954
Phenergan [10-(2-dimethyl-aminoisopropyl)-phenothiazine]-antihistamine	Crotalus terrificus	In vivo	Not anti-toxic	Schottler, 1954
Norepinephrine	Crotalus adamanteus	In vivo	Not anti-leth.	Morales et al, 1963

F. Combination

	Bothrops jararaca	In vivo	Red. necrosis	Rosenfeld et al, 1969c
adrenaline, hydro-cortisone and an antihistamine (pirilamine maleate)				
Empirin in DMSO	Crotalus atrox	In vitro	Anti-hem.	Ownby, 1975
		In vivo	not anti-hem.	"
Phenacetin in DMSO	Crotalus atrox	In vitro	Anti-hem.	Ownby, 1975
		In vivo	not anti-hem.	"
Rutin in DMSO	Crotalus atrox	In vitro	Anti-hem.	Ownby, 1975
		In vivo	not anti-hem.	"
Esculin in DMSO	Crotalus atrox	In vitro	Anti-hem.	Ownby, 1975
		In vivo	not anti-hem.	"

discussed the possible mechanisms by which protein toxins are detoxified by formaldehyde and reported that tyrosyl and histidyl residues participate in toxicity but lysyl residues do not. Formalin has been used to produce a toxoid of habu venom for use in active immunization as well as in antivenin production (Kondo et al., 1971; Someya et al., 1972).

Antihistamines Antihistamines have been tested because of the possible role of histamine in the systemic and local effects of snake venom (Table 1). In general, they do not appear to neutralize the effects of snake venom.

Procaine Because of the local effects of venoms, procaine HCl was tested for neutralization, but it was not effective *in vitro* or *in vivo* when used alone (see Table 1).

Others Many other compounds were also tested for neutralization of venom toxicities. Usually they were ineffective for *in vivo* neutralization.

1.6 Combinations of Compounds

The major effort to find chemicals useful in the treatment of local tissue damage induced by snake venom has been made by Tu and Ownby. In 1970, Tu et al. reported the results of tests using 28 different compounds. Of these, 3 were effective in neutralizing venom toxicity in the *in vitro* test. None was effective in the *in vivo* test; therefore, none would be of use in actual snakebite cases. Two of the compounds that reduced hemorrhage *in vitro* were chelating agents; the other was DMSO.

Ownby et al. (1975) and Ownby (1975) used combinations of various chemicals to neutralize the hemorrhagic, necrotic, and lethal activities of rattlesnake (*Crotalus atrox*) venom. The purpose of these experiments was to find a combination that would neutralize venom toxicity *in vivo* and thus be of value in the treatment of actual snakebite cases. The results of their studies are summarized in Table 2, and it can be seen that four of the combinations were effective in reducing hemorrhage in the *in vivo* test.

Table 2 Effectiveness of Various Combinations of Compounds in Reducing Hemorrhage Induced by Injection of *Crotalus atrox* Venom into Mice

Combination*	*In vitro* Test†	*In vivo* Test†
1. EDTA + procaine HCl	+	+
2. DTPA + procaine HCl	+	+
3. EGTA + procaine HCl	−	−
4. DTPA + DMSO	+	−
5. EDTA + DMSO + procaine HCl	+	−
6. DTPA + (digitoxin + DMSO)	+	+
7. DTPA + (digitoxin + DMSO) + procaine HCl	+	+

*Compounds used at their maximum tolerated dosage levels.
†Venom concentration used was varied from 20 $\mu g\ g^{-1}$ to 2.5 $\mu g\ g^{-1}$ in 0.005 ml g^{-1}; +, hemorrhage reduced; −, hemorrhage not reduced.
  Table reproduced from Ownby (1975).

Ownby et al. (1975) discussed the mechanism of action of DTPA and procaine, which constituted the most effective combination. Procaine forms a complex with DTPA and therefore increases the effectiveness of the latter *in vivo*. The increase in effectiveness is believed due to a slowdown in the diffusion of DTPA because of the formation of the following complex:

$$[(HOOCCH_2)_2-N-CH_2-CH_2]_2-\!\!-\!\!N-CH_2-COO^- + \overset{H}{\underset{\underset{\underset{\underset{O=\overset{|}{C}-O}{|}}{\overset{|}{C}}}{\overset{|}{\underset{CH_2}{|}}}}{\overset{|}{N}(C_2H_5)_2}}$$

Neither DTPA nor procaine alone is effective, but when used together they can reduce hemorrhage due to injection of *C. atrox* venom into mice or dogs (Ownby et al., 1975).

2 RADIATION

Production of antiserum for the treatment of snakebite is complicated by the high toxicity of the venom and the severe local tissue necrosis produced by some venoms. To produce antiserum of high titer and specificity without inducing death or severe tissue damage, it is necessary to detoxify the venom without decreasing its antigenicity or the protective ability of the antiserum produced. One approach to this problem has been the use of various forms of radiation, including both ionizing and nonionizing types.

2.1 Visible Light (Photooxidation)

Kocholaty (1966) investigated the effects of photooxidation in the presence of methylene blue on *Crotalus atrox* venom and found that reduction in toxicity corresponded with increase in O_2 uptake. Detoxification depended on the presence of both visible light and methylene blue. The antigenicity of the detoxified venom was not altered significantly, according to gel diffusion results, nor was the ability of the antiserum to protect mice against a lethal dose of venom. Also, proteolytic and phospholipase A_2 activities were destroyed by photooxidation. Subsequent work by Kocholaty and his co-workers showed that photooxidation with visible light in the presence of methylene blue could be used to detoxify various venoms, including those of *C. atrox, C. terrificus, Vipera russellii, Agkistrodon piscivorus, Bothrops atrox, Micrurus fulvius,* and *Naja naja* (Kocholaty and Ashley, 1966). Two of these venoms (*A. piscivorus* and *V. russelli*) were shown to retain their abilities to elicit antibodies that protected mice against the effects of untreated venom. Detoxified *M. fulvius* venom (Kocholaty et al., 1967) and *Crotalus durissus durissus* venom (Kocholaty et al., 1968a) were capable of eliciting antibodies that protected test animals against untreated homologous venom. On the other hand, antiserum produced against detoxified *B. atrox asper* venom did not protect mice against untreated venom from the same species (Kocholaty et al., 1968a). The investigators offered two possible explanations for the difference in immunological response to this venom: "the identical amino acids susceptible to photooxidation are involved as carriers of toxicity and immunogenicity; the degree of photooxidation was carried too far, resulting in a denaturation of the protein moieties responsible for the elicitation of the immunogenic response."

Detoxified cobra venom also produced antibodies that gave only poor protection to mice against untreated venom (Kocholaty et al., 1968b). This may be due to the weak antigenicity of the lethal component, since it is of low molecular weight. An alternative explanation is that the structure of the lethal (neutrotoxic) component renders it especially susceptible to photooxidation.

Huang et al. (1972, 1973) investigated the effects of photooxidation on the detoxification and immunogenicity of two Formosan snake venoms, those of *Bungarus multicinctus* and *Trimeresurus mucrosquamatus*. Both venoms were detoxified, and the loss of toxicity was found to be proportional to the rate of O_2 uptake. Antiserum produced against the detoxified venoms was similar to that produced against the untreated venoms in that the antiserum was capable of neutralizing several LD_{50} doses of the venoms. Huang et al. (1973) tested other venoms, including those of *Naja naja atra*, *Trimeresurus gramineus*, and *Agkistrodon acutus*, and found similar results. Detoxification by photooxidation appeared to be more intense against neurotoxic (*N. naja atra* and *B. multicinctus*) than hemorrhagic (*T. mucrosquamatus, T. gramineus*, and *A. acutus*) venoms. This might be explained on the basis of the characteristics of the lethal components of the two different types of venoms.

It is well known that many biological systems can be damaged by visible light if sensitized with dyes (Spikes and Glad, 1964). Molecules such as proteins can be modified by visible light when molecular oxygen and an appropriate dye are also present. Photooxidation of proteins causes destruction of specific amino acid chains, in particular, histidine and tryptophan. These primary changes lead to alteration in the specificity, enzymatic activity, antigenicity, and other properties, of protein molecules. The dye methylene blue sensitizes proteins to photooxidation by visible light. Thus, since snake venom is composed of many proteins and polypeptides, some of which are responsible for the toxicity of the venom, it is reasonable to expect photooxidation to detoxify these components. Loss of antigenicity in some instances (*Naja naja* and *Bothrops atrox asper*) but not in others (see above) is difficult to explain and awaits further experiments.

2.2 Ultraviolet Light

Macht (1935) reported that ultraviolet radiation detoxified cobra venom, that the degree of detoxification was directly related to the amount of exposure, and that ultraviolet rays of 3000 and 4000 Å were most potent in the detoxification of venom solutions. Attempts to use ultraviolet light to detoxify snake venom *in vivo* were not successful. Thus the therapeutic use of ultraviolet light in snakebite cases is not likely.

More recently Tejasen and Ottolenghi (1970) studied the effects of ultraviolet light on the toxicity, enzymatic activity, and antigenicity of cottonmouth moccasin (*Agkistrodon piscivorus*) venom. Toxicity was reduced by three to four times, as indicated by an increase in LD_{50} values for irradiated venom in mice. The reduction was directly related to the length of time the venom was exposed: threefold increase in LD_{50} value at 1 hr of ultraviolet irradiation, and fourfold increase at 3 hr of exposure. Four enzymes — phospholipase A_2, two proteinases, and phosphodiesterase — were inactivated by ultraviolet irradiation. The untreated and treated venoms were very similar in immunogenicity, as shown by immunoelectrophoresis. Irradiation of the venom reduced the amount of hemorrhage produced in rabbits but did not completely eliminate it.

2.3 Cobalt-60

Cobalt-60 has been used in attempts to detoxify cobra venoms without destroying their immunological properties. Lauhatirananda et al. (1970) irradiated cobra venom with varying doses of ^{60}Co and studied the degree of detoxification and the effect on antigenicity. They found that for aqueous samples a dose of 4.0 Mrad completely detoxified the venom, based on LD_{50} values for Swiss white mice; however, not even 60 Mrad was sufficient to detoxify dry venom. Immunodiffusion and immunoelectrophoresis indicated that the irradiated venom produced fewer precipitin bands against antivenin, and in general the higher the dose the fewer the bands formed; at doses of 3.0 and 4.0 Mrad the antigenicity of the venom was destroyed, at least for precipitating antibodies.

Salafranca (1972) studied the ability of various doses of ^{60}Co to detoxify cobra (*Naja naja philippinensis*) venom. The highest dose tested (1 Mrad) reduced the toxicity to 43% of that of the untreated venom, and the antigenicity was not decreased significantly.

Puranananda et al. (1976) studied the production of antibodies to cobra venom. Cobra venom irradiated with ^{60}Co at doses of 0.5 to 5.0 Mrad was used to produce antiserum in rabbits. This antiserum was then tested in mice for neutralization ability in agar-gel diffusion plates and for the presence of specific antibodies by immunoelectrophoresis. Antiserum produced against irradiated venom offered some protection against nonirradiated venom (3.0 Mrad, 21 LD_{50} neutralized), but not as much as antiserum produced against untreated venom (32 LD_{50} neutralized). Irradiation of venom did not reduce its ability to induce local necrosis in injected animals. Antigenicity was reduced in the irradiated venom sample; fewer precipitin bands were observed on both gel diffusion and electrophoresis. These investigators concluded that cobra venom neutralized by ^{60}Co irradiation is still capable of eliciting antibody production, and they suggest the use of 3.0 Mrad to detoxify venom to be employed in antibody production.

2.4 X-rays

The effect of X-irradiation on the necrotizing and lethal activities of cottonmouth moccasin (*Agkistrodon piscivorus*) venom was investigated by Flowers (1963). Treatment with approximately 18,000 rads reduced the necrotizing ability of this venom by 50 to 65% and the lethality by six times. Irradiation also inactivated the phospholipase activity by 95 to 100%. Even though the venom was detoxified by irradiation, the antigenicity was only slightly decreased, as indicated by precipitin patterns on gel diffusion and immunoelectrophoresis plates.

It is clear that irradiation of snake venom leads to detoxification, but the ability of detoxified venom to elicit protective antibodies varies with the venom and the type of irradiation used. Photooxidation in the presence of visible light and methylene blue can be used successfully to produce protective antibodies against most venoms tested except those of *Bothrops atrox asper* and *Naja naja* (Kocholaty and Ashley, 1966; Kocholaty et al., 1967; Kocholaty et al., 1968a, b). The results of Huang et al. (1972, 1973) agree in part with those of Kocholaty in that they found an intense detoxification of neurotoxic venoms (*N. naja atra*) due to photooxidation. However, they did not observe as great a degree of detoxification of "hemorrhagic" venoms (*T. mucrosquamatus*, etc.) as did Kocholaty (*B. atrox asper*). This discrepancy could indicate an oversimplification in the classification of venoms as neurotoxic or hemorrhagic.

Ultraviolet light is capable of detoxifying both cobra venom (Macht, 1935) and cottonmouth moccasin venom (Tejasen and Ottolenghi, 1970). The antigenicity of moccasin venom (hemorrhagic) was not changed, and that of cobra venom was not measured.

Cobalt-60 completely detoxifies cobra venom (Puranananda et al., 1976), but the immunogenicity of the detoxified venom is less than that of the untreated venom.

It appears that the antigenicity of cobra venom is more susceptible to irradiation than that of other venoms. In cobra venom irradiation probably induces some change in the secondary or tertiary structure of molecules that is responsible for their lack of antigenicity.

REFERENCES

Ahuja, M. L., Veeraraghaven, N., and Menon, I. G. K. (1946). Action of heparin on the venom of *Echis carinatus, Nature,* **158,** 878.

Arora, R. B., Wig, K. L., and Somari, P. (1962). Effectiveness of hydrocortisone and hydrocortisone-antivenene combination against *Echis carinatus* snake venom, *Arch. Int. Pharmacodyn. Ther.,* **137,** 299.

Bhargava, N., Zirinis, P., Bonta, I. L., and Vargaftig, B. B. (1970). Comparison of hemorrhagic factors of the venoms of *Naja naja, Agkistrodon piscivorus* and *Apis mellifera, Biochem. Pharmacol.,* **19,** 2405.

Bizzini, B. and Raynaud, M. (1974). La detoxication des toxines proteiques par le formal: Mecanismes supposes et nouveaux developpements, *Biochimie,* **56,** 297.

Bonta, I. L., De Vos, C. J., and Delver, A. (1965). Inhibitory effects of estriol-16, 17-disodium succinate on local haemorrhages induced by snake venom in canine heart–lung preparations, *Acta Endocrinol.,* **48,** 137.

Bonta, I. L., Vargaftig, B. B., De Vos, C. J., and Grijsen, H. (1969). Haemorrhagic mechanisms of some snake venoms in relation to protection by estriol succinate of blood vessel damage, *Life Sci.,* **8(1),** 881.

Bonta, I. L., De Vries-Krogt, K., De Vos, C. J., and Bhargava, N. (1970). Preventive effect of local heparin administration on microvascular pulmonary hemorrhages induced by cobra venom in mice, *Eur. J. Pharmacol.,* **13,** 97.

Christensen, P. A. (1947). Formol detoxification of cape cobra (*Naja flava*) venom, *S. Afr. J. Med. Sci.,* **12,** 71.

Christy, N. P. (1972). Exotic snakebite in an urban setting: heparin therapy of disseminated intravascular coagulation in a patient bitten by a saw-scale viper, *Trans. Am. Clin. Climatol. Assoc.,* **84,** 37.

Dumarey, C. and Boquet, P. (1972). Pouvoir immunogene de la toxine α du venin de *Naja nigricollis* polymerisee par l'aldehyde formique, *C. R. Acad. Sci. Paris,* **275,** 3053.

Flowers, H. H. (1963). The effects of X-irradiation on the biological activity of cottonmouth moccasin (*Ancistrodon piscivorus*) venom, *Toxicon,* **1,** 131.

Flowers, H. H. and Goucher, C. R. (1965). The effect of EDTA on the extent of tissue damage caused by the venoms of *Bothrops atrox* and *Agkistrodon piscivorus, Toxicon,* **2,** 221.

Friederich, C. and Tu, A. T. (1971). Role of metals in snake venoms for hemorrhagic, esterase and proteolytic activities, *Biochem. Pharmacol.,* **29,** 1549.

Fukuyama, T., and Sawai, Y. (1973). Experimental study on cobra venom toxoid, *Jap. J. Med. Sci. Biol.,* **26(1),** 32.

Goucher, C. and Flowers, H. (1964). The chemical modification of necrogenic and proteolytic activities of *Agkistrodon piscivorus* venom and the use of EDTA to produce a venom toxoid, *Toxicon,* **2,** 139.

Huang, C. T., Huang, J. S., Ling, K. H., and Hsieh, J. T. (1972). The effect of photooxidation on detoxification and immunogenicity of Formosan snake venoms: Photooxidative effect on venoms of *Trimeresurus mucrosquamatus* and *Bungarus multicinctus, J. Formosan Med. Assoc.,* **71,** 435.

Huang, C. T., Huang, J. S., Ling, K. H., and Lin, S. Y. (1973). Further studies on the immunogenic properties of photooxidized Formosan snake venoms, with special reference to a trial preparation of polyvalent antivenin using photooxidized multiple venom antigens: Second report, *J. Formosan Med. Assoc.,* **72,** 208.

Kocholaty, W. (1966). Detoxification of *Crotalus atrox* venom by photooxidation in the presence of methylene blue, *Toxicon,* **3,** 175.

Kocholaty, W. and Ashley, B. D. (1966). Detoxification of Russell's viper (*Vipera russellii*) and water moccasin (*Agkistrodon piscivorus*) venoms by photooxidation, *Toxicon,* **3,** 187.

Kocholaty, W. F., Ashley, B. D., and Billings, T. A. (1967). An immune serum against the North American coral snake (*Micrurus fulvius*) venom obtained by photooxidative detoxification, *Toxicon,* **5,** 43.

Kocholaty, W. F., Goetz, J. C., Ashley, B. D., Billings, T. A., and Ledford, E. B. (1968a). Immunogenic response of the venoms of fer-de-lance, *Bothrops atrox asper,* and la cascabella, *Crotalus durissus durissus,* following photooxidative detoxification, *Toxicon,* **5,** 153.

Kocholaty, W. F., Ledford, E. B., Billings, T. A., Goetz, J. C., and Ashley, B. D. (1968b). Immunization studies with *Naja naja* venom detoxified by photooxidation, *Toxicon,* **5,** 159.

Kondo, H. and Murata, R. (1972). Preparation and standardization of formal toxoid from the venom of *Trimeresurus flavoviridis* (habu), *J. Formosan Med. Assoc.,* **71,** 413.

Kondo, S., Sadahiro, S., Yamuchi, K., Kondo, H., and Murata, R. (1971). Preparation and standardization of toxoid from the venom of *Trimeresurus flavoviridis* (habu), *Jap. J. Med. Sci. Biol.,* **24,** 281.

Kondo, H., Kondo, S., Sadahiro, S., Yamauchi, K., Ohsaka, A., and Murata, R. (1973). "Preparation and immunogenicity of habu (*Trimeresurus flavoviridis*) toxoid," in A. De Vries and E. Kochva, Eds., *Toxins of Animal and Plant Origin.* Vol. 3, Gordon and Breach, New York, pp. 846–862.

Kurihara, N. and Shibata, K. (1971). Effect of thiol-compound on toxicity of habu snake (*Trimeresurus flavoviridis* Hollowell) venom, *Jap. J. Pharmacol.,* **21,** 253.

Lauhatirananda, P., Ganthavorn, S., and Hayodom, V. (1970). Radiation effects on cobra venom, *Int. At. Energy Agency,* IAEA-PL-334/10.

Lee, J. (1972). Dimethyl sulfoxide (DMSO) in experimental treatment of snakebite in dogs, *Vet. Med. Small Animal Clin.,* **67,** 404.

Macht, D. I. (1935). The effect of the ultraviolet rays on snake venoms, *Am. J. Med. Sci.,* **189,** 520.

Maier, H. K. (1951). Benadryl hydrochloride in treatment of snakebite in dogs, *Vet. Med.,* **46,** 463.

Morales, F., Bhanganada, K., and Perry, J. F. (1963). "Effect of several agents on the lethal action of two common venoms," in H. L. Keegan and W. V. Macfarlane, Eds., *Venomous and Poisonous Animals and Noxious Plants of the Pacific Region,* Macmillan, New York, pp. 385–398.

Moroz-Perlmutter, C., Goldblum, N., De Vries, A., and Gitter, S. (1963). Detoxification of snake venoms and venom fractions by formaldehyde, *Proc. Soc. Exp. Biol. Med.,* **112,** 595.

Ownby, C. L. (1975). Ph.D. dissertation, Colorado State University, Fort Collins.

Ownby, C. L., Tu, A. T., and Kainer, R. A. (1975). Effect of diethylenetriaminepentaacetic acid and procaine on hemorrhage induced by rattlesnake venom, *J. Clin. Pharmacol.,* **15,** 419.

Puranananda, C., Hayodom, V., Lauhatirananda, P., and Ganthavorn, S. (1976). Study on immune response to irradiated cobra venom in rabbits, *J. Natl. Res. Counc. Thailand,* **8,** 1.

Raby, C. (1973). Acute consumption coagulopathy following Crotalidae bites: Spectacular results with controlled heparin administration, *Nouv. Press Med. (Paris),* **2,** 2949.

Rona, G. (1963). The role of vascular mucopolysaccharides in the hemostatic action of estrogens, *Am. J. Obstet. Gynecol.,* **87,** 434.

Rosenfeld, G., de Langlada, F. G., and Kelen, E. M. A. (1969a). Experimental treatment of necrosis produced by proteolytic snake venoms. I. Action of isoxsuprine, *Rev. Inst. Med. Trop. (São Paulo,* **11(6),** 383.

Rosenfeld, G., de Langlada, F. G., and Kelen, E. M. A. (1969b). Experimental treatment of necrosis produced by proteolytic snake venoms. II. Action of dexamethasone, *Rev. Inst. Med. Trop. São Paulo,* **11(6),** 387.

Rosenfeld, G., de Langlada, F. G., and Kelen, E. M. A. (1969c). Experimental treatment of necrosis produced by proteolytic snake venoms. III. Action of a combination of antihistaminic, adrenaline, and hydrocortisone (AAC), *Rev. Inst. Med. Trop. São Paulo,* **11(6),** 390.

Sadahiro, S. (1971). Toxoids from the venom of habu. I. Detoxification of habu venom with formalin, *Jap. J. Bacteriol. (Tokyo)*, **26**, 214.

Sadahiro, S., Kondo, S., Yamauchi, K., Kondo, H., and Murata, R. (1970). Studies on immunogenicity of toxoids from habu (*Trimeresurus flavoviridis*) venom, *Jap. J. Med. Sci. Biol.*, **23**, 285.

Salafranca, E. S. (1970). "Detoxification of cobra venom and bacterial toxins for biological productions," in *Radiation Sensitivity of Toxins and Animal Poisons,"* International Atomic Energy Agency, Vienna, pp. 87–89.

Salafranca, E. S. (1972). Irradiated cobra (*Naja naja philippinensis*) venom, *Jap. J. Med. Sci. Biol.*, **25**, 206.

Sawai, Y., Makino, M., Miyasaki, S., Kawamura, Y., Mitsuhashi, S., and Okonogi, T. (1961a). Studies on the improvement of treatment of habu snake (*Trimeresurus flavoviridis*) bite. 2. Antitoxic action of monocalcium disodium ethylene diamine tetraacetate on habu venom, *Jap. J. Exp. Med.*, **31**, 267.

Sawai, Y., Makino, M., Miyasaki, S., Konto, K., Adachi, H., Mitsuhashi, S., and Okonogi, T. (1961b). Studies on the improvement of treatment of habu snake bite. 1. Studies on the improvement of habu snake antivenin, *Jap. J. Exp. Med.*, **31**, 137.

Sawai, Y., Makino, M., and Kawamura, Y. (1963). "Studies on the antitoxic action of dihydrolypoic acid (dihydrothioctic acid) and tetracycline against habu snake (*Trimeresurus flavoviridis* Hallowell) venom," in H. L. Keegan and W. V. Macfarlane, Eds., *Venomous and Poisonous Animals and Noxious Plants of the Pacific Region,* Macmillan, New York, pp. 327–335.

Sawai, Y., Kawamura, Y., Fukuyama, T., Okonogi, T., and Ebisawa, I. (1969). Studies on the improvement of treatment of habu (*Trimeresurus flavoviridis*) bites. 8. A field trial of the prophylactic inoculation of the habu venom toxoid, *Jap. J. Exp. Med.*, **39**, 197.

Sawai, Y., Chinzei, H., Kawamura, Y., and Okonogi, T. (1972). Studies on the improvement of treatment of habu (*Trimeresurus flavoviridis*) bites. 9. Studies on the immunogenicity of the purified habu venom toxoid by alcohol precipitation, *Jap. J. Exp. Med.*, **42**, 155.

Schöttler, W. H. A. (1954). Antihistamine, ACTH, cortisone, hydrocortisone and anesthetics in snake bite, *Am. J. Trop. Med. Hyg.*, **3**, 1083.

Seth, S. D. S., Arora, R. B., and Guleria, J. S. (1971). Beneficial effect of hydrocortisone and hydrocortisone-antivenene combination in the treatment of Russell's viper envenomation, *Indian J. Exp. Biol.*, **9**, 183.

Someya, S., Murata, R., Sawai, Y., Kondo, H., and Ishii, A. (1972). Active immunization of man with toxoid of habu (*Trimeresurus flavoviridis*) venom, *Jap. J. Med. Sci. Biol.*, **25**, 47.

Spikes, J. D. and Gland, B. W. (1964). Photodynamic action, *Photochem. Photobiol.*, **3**, 471.

Stone, O. J., Willis, C. J., and Mullins, J. F. (1966). Thiabendazole inhibition of venom necrosis, *J. Invest. Dermatol.*, **47**, 67.

Tejasen, P. and Ottolenghi, A. (1970). The effect of ultraviolet light on the toxicity and the enzymatic and antigenic activities of snake venom, *Toxicon*, **8**, 225.

Tiru-Chelvam, R. (1974). The effects of dimethyl sulfoxide on snake venom toxicity *in vivo* and on the efficacy of antivenins, *Curr. Therap. Res.*, **16**, 1033.

Trethewie, E. R. (1956). The effect of poly-vinylpyrollidone and tetrahydroaminoacridine on the mortality and survival time of mice injected with snake venom, *Med. J. Aust.*, **4311**, 8.

Tu, A. T., Homma, M., Hong, B., and Terrill, J. B. (1970). Neutralization of rattlesnake venom toxicities by various compounds, *J. Clin. Pharmacol.*, **10**, 323.

Vick, J. A., Blanchard, R. J., and Perry, J. F. (1963). Effects of epsilon-amino caproic acid on pulmonary vascular changes produced by snake venom, *Proc. Soc. Exp. Biol.*, **113**, 841.

Watt, D. D. (1972). Effects upon lethality of neurotoxic venoms by *in vivo* administered gluthathione, *Abstr. 1972 Gen. Mett. Int. Soc. Toxicol.*, Darmstadt, Sept. 11–13, 1972.

Weiss, H. J., Phillips L. L., Hopewell, W. S., Phillips, G., Christy, N. P., and Nitti, J. F. (1973). Heparin therapy in a patient bitten by a saw-scaled viper (*Echis carinatus*), a snake whose venom activates prothrombin, *Am. J. Med.*, **54**, 653.

Wiener, S. (1960). Active immunization of man against the venom of the Australian tiger snake (*Notechis scutatus*), *Am. J. Trop. Med. Hyg.*, **9**, 284.

Yoshira, A. (1967). Biochemical studies on habu venom with special reference to sulfhydryl radicals, *Igaku Kenky Acta. Med.*, **37**, 263.

IV OTHER VENOMS

28 Scorpion Venoms

1 CHEMISTRY 460
  1.1 Toxic Principles, 460
      Stability, 460
      Isolation and Properties, 460
      Sequences, 461
      Structure—Toxicity Relationships, 464
  1.2 Enzymes, 464
      Phospholipase A_2, 464
      Others, 464
  1.3 Nonprotein Components, 464

2 IMMUNOLOGY 465

3 TOXICOLOGY 465
  3.1 Clinical Symptoms, 465
  3.2 Yield and Toxicity, 467

4 PHARMACOLOGY 467
  4.1 Effect on Nerves, 467
      Acetylcholine Release, 467
      Catecholamine Release, 468
      Nerve Permeability, 469
  4.2 Effect on Cardiovascular System, 470
      Pathology, 470
      Physiological Effects, 471
  4.3 Hemorrhage, 474
  4.4 Effect on Skeletal Muscle, 475
  4.5 Metabolic Effect, 475
  4.6 Other Effects, 476

  References 477

Unlike snakes, all scorpions are venomous. The venom is injected by means of a stinger found at the tip of the telson, the terminal structure of the tail. Scorpions belong to the class Arachnida and order Scorpinida. The following is a brief classification of scorpions:

Family	Genus
Buthidae	*Buthus, Androctonus, Buthotus, Parabuthus, Leiurus, Isometrus, Centrurus, Centruroides, Hadrurus, Tityus, Heterometrus*
Diplocentridae	*Nebo*
Scorpionidae	*Scorpio, Opisthophthalmus, Palamnaeus, Pandinus*
Vejovidae	*Vejovis*
Chactidae	*Euscorpius*
Bothriuridae	

The most common species of scorpions in the United States are *Centruroides gertschi* and *C. sculpturatus* (Stahnke, 1966). Other species of scorpions can be found in Africa, the Middle East, India, and Central and South America. The classification of scorpions from Africa and the Middle East has been summarized by Vachon (1966).

1 CHEMISTRY

1.1 Toxic Principles

Stability. The stability of the neurotoxins of *Androctonus australis* toward denaturing agents such as temperature and variations in pH has already been noted (Rochat et al., 1967). After a solution of 1 or 2 mg of toxin in 1 ml of Tris–acetate (pH 8.6)–8M urea had been left standing for 5 hr at 20°, no loss of toxicity was observed; after 18 hr at 50° in the same solvent, 18% of the initial toxicity was still present. This stability is primarily due to the low molecular weight of the toxins and their compact secondary structures. The toxicities of the venoms of *Centruroides sculpturatus* and *Leiurus quinquestriatus* are moderately stable to heat treatment, suggesting that the toxic proteins are stable, low molecular weight polypeptides (Watt, 1964; Nitzan, 1970). Further evidence of the low molecular weight of the toxic fraction of scorpion venom is the dialyzability of the toxin (Kamon, 1965). The lethal toxins of the venoms of the scorpions *Leiurus quinquestriatus* (Nitzan et al., 1963; El-Asmar et al., 1972) and *Centruroides sculpturatus* (Watt, 1964) are dialyzable.

Isolation and Properties. Scorpion venoms, like snake venoms, consist of a mixture of many pharmacologically active proteins. Some proteins are enzymes, and others are nonenzymatic. At least 16 bands can be observed in electrophoresis for the venom of *Pandinus exitialis* (Ismail et al., 1974a).

Ultracentrifugically homogeneous toxins of *Buthus occitanus* and *Androctonus australis* were isolated. The yield of the first toxin, which has an LD_{50} level of 0.15 $\mu g\ g^{-1}$, is 7%. The other one has an LD_{50} of 0.05 $\mu g\ g^{-1}$ in mice (Miranda et al., 1964a–c). The toxins are low molecular weight, basic proteins with pI values of 8 and 9 (Miranda et al., 1964b,c). Because of the high content of ionizable groups in the neurotoxins, they tend to make nonspecific associations with other proteins (Miranda et al., 1966a). Both *Buthus* and *Androctonus* scorpions are found in North Africa. Several neurotoxins have also been purified from the venom of the South American scorpion *Tityus serrulatus* (Miranda et al., 1966b).

1 Chemistry

Rochat et al. (1967) initially isolated two neurotoxins, and later another toxin was isolated (Miranda et al., 1970), from the venom of *Androctonus australis*. *Buthus occitanus tunetanus* venom contains three neurotoxins, and *Leiurus quinquestriatus quinquestriatus* venom has five (Miranda et al., 1970). The yields (percent) of the various toxins are as follows:

Toxin	A. australis	B. occitanus	L. quinquestriatus
I	11.3	4.1	1.8
II	52.6	2.2	7.6
III	3.0	10.0	2.2
IV			4.2
V			20.6
Total %	66.9	16.3	36.4

From these values it is clear that scorpion venoms have higher toxin contents than do snake venoms. The amino acid compositions of various toxins are listed in Table 1.

The total amino acid residues of scorpion toxins are in the range of 57 to 78 residues, but most of these toxins have 62 to 66. None of the toxins contains methionine residues.

Pure toxin obtained from *Heterometrus scaber* venom has a molecular weight of 15,000 and is a glycoprotein (Nair et al., 1975). It contains 1.74% glucosamine, 0.31% sialic acid, 3.25% fucose, and 0.45% of an unidentified neutral sugar. It does not show any enzyme activities, hemolytic activity, or inhibition of succinate dehydrogenase activity, but it produces hyperglycemia in sublethal dose.

Different components in scorpion venom are responsible for the mice-lethal and larvae contraction-paralysis effects of the venom of *Androctonus australis* (Zlotkin et al., 1971a, b). The insect toxin is also different from mammal toxins in amino acid composition (Zlotkin et al., 1971c). Pure neurotoxins separated from the venom of the scorpion *A. australis* and highly toxic to mammals are inactive when tested on several arthropods. The fly larvae toxin originating from the same venom demonstrates a strong toxicity to insects but is completely inactive when applied to an arachnid or a crustacean. This toxin was isolated by Zlotkin et al. (1972a), who showed that its toxicity was destroyed by trypsin digestion. They concluded that, in addition to the toxins active on mammals and insects, the venom of *A. australis* contains another discrete protein specifically active on crustaceans. Later the protein toxic to Crustacea was isolated from the venom of *A. australis* (Zlotkin et al., 1975); it is a basic protein and contains 70 amino acid residues. This was also the case for the venoms of *Androctonus aeneas aeneas, A. amoreuxi, A. mauretanicus mauretanicus, Buthus occitanus puris, B. occitanus tunetanus*, and *Leiurus quinquestriatus* (Zlotkin et al., 1972b, c).

The insect toxin and the mammal toxin have different physiological responses. In general, the insect toxin induced an afferent transynaptic response at the sixth abdominal ganglion of the cockroach, but the mammal toxin did not affect synaptic transmission (D'Ajello et al., 1972).

Sequences. The complete sequences of toxins I and II of *Androctonus australis* venom (Rochat et al., 1970, 1972a, b) and toxins of *Centruroides sculpturatus* venoms (Babin et al., 1974, 1975) were determined and are given in Table 2. Partial sequences for the

Table 1 Amino Acid Compositions of Scorpion Neurotoxins

Genus Species Subspecies	Androctonus australis			Buthus occitanus			Centruroides sculpturatus							Leiurus quinquestriatus quinquestriatus				
Origin	Algeria			Sudan			U.S.A.							Sudan				
Name	I	II	III	I	II	III	I	II	III	IV	1	2	3	I	II	III	IV	V
Amino Acid																		
Lysine.........	6	5	6	4	5	5	9	10	8	9	8	8	8	4	4	4	6	8
Histidine......	1	2	2	1	1	1	1	1	1	1	1	0	0	3	1	1	0	0
Arginine.......	2	3	1	2	3	4	1	1	2	2	0	0	0	1	2	3	3	3
Aspartic acid..	9	8	8	9	9	9	7	5	7	7	6	5	5	10	10	9	9	10
Threonine......	2	3	0	2	1	3	5	5	4	1	3	3	3	1	2	1	3	1
Serine.........	6	2	6	2	2	2	2	6	2	4	4	5	4	2	4	3	3	3
Glutamic acid..	0	4	0	5	5	5	6	5	8	9	6	7	7	2	2	2	3	4
Proline........	6	3	6	4	3	3	4	3	3	2	4	4	4	3	4	2	3	2
Glycine........	6	7	6	5-6	6	7	10	9	9	9	7	9	9	8	5	5	7	7
Alanine........	1	3	3	4-5	5	3	0	0	3	4	3	3	3	4	2	5-6	4	3
Valine.........	5	4	6	2	2	4	1	1	3	3	1	1	1	5	3	3-4	3	2
Methionine.....	0	0	0	0	0	0	0	0	0	0	0	0	0	0	0	0	0	0
Isoleucine.....	2	1	3	3	4	1	1	1	0	0	0	0	0	4	1	2	4	2
Leucine........	4	2	4	3	3	1	5	3	6	8	4	5	5	3	2-3	2	3	2
Tyrosine.......	3	7	3	4	5	7	6	4	8	7	6	6	6	3	5	6	5	6
Phenylalanine..	1	1	1	1	1	1	2	4	3	1	2	1	1	1	0-1	1	0	2
Half-cystine...	8	8	8	8	8	8	8	6	8	8	8	8	8	8	8	8	8	8
Tryptophan.....	1	1	1	2	2	1	2	2	3	3	2	1	1	2	1	2	2	2
Total residue..	63	64	64	62	65	64	70	64	78	78	65	66	65	64	57	60	66	65
References	Rochat et al., 1970			Rochat et al., 1967			Rochat et al., 1970				McIntosh and Watt, 1972			Babin et al., 1974			Miranda et al., 1970	

toxins of *Buthus occitanus tunetanus* and *Leiurus quinquestriatus quinquesteriatus* venoms were also identified (Rochat et al., 1972a, b).

An isoneurotoxin (I′) has been found in the venom of *Androctonus australis* collected in Tunisia (area of Tozeur). It differs from neurotoxin I in the replacement of the valine residue in position 17 by an isoleucine residue. The location of disulfide bonds for toxin II of *A. australis* was identified by Kopeyan et al. (1974).

As mentioned earlier, the total number of amino acid residues in scorpion toxins is usually in the 62 to 66 range, which is similar to the values for neurotoxin Type I of Elapidae and Hydrophiidae. All scorpion toxins contain four disulfide bridges. The location of the disulfide bridges in scorpion toxins is quite different from that in snake

Table 2 Amino Acid Sequences of Scorpion Toxins

Scorpion	Toxin	Loop 1	Loop 2	Loop 3
Centruroides sculpturatus	variant 1	Lys-Glu-Gly-Tyr-Leu-Val-Lys-Lys-Ser-Asp-Gly-	Cys-Lys-Tyr-Asp-	Cys-Phe-Trp-Leu-Gly-Lys-Asn-Glu-His-Asn-Thr-
	2	Lys-Glu-Gly-Tyr-Leu-Val-Asn-Lys-Ser-Thr-Gly-	Cys-Lys-Tyr-Gly-	Cys-Leu-Leu-Gly-Lys-Asn-Glu-Gly-Asn-Lys-
	3	Lys-Glu-Gly-Tyr-Leu-Val-Lys-Lys-Ser-Asp-Gly-	Cys-Lys-Tyr-Gly-	Cys-Leu-Lys-Leu-Gly-Glu-Asn-Asp-Phe- - - -
Androctonus australis	I	Lys-Asp-Gly-Tyr-Ile-Val-Val-Glu-Lys- - -Thr-	Cys-Lys-Thr-His-Cys[a]	- - - - -Val-Pro-Pro- - - -
	II	Lys-Arg-Asp-Gly-Tyr-Ile-Val-Tyr-Pro-Asn-Asn-	Cys-Val-Tyr-His-Cys	- - - - - - - - - - - -
		Val-Lys-Asp-Gly-Tyr-Ile-Val-Asp-Asp-Val-Asn-	Cys-Thr-Tyr-Phe-Cys	- - -Gly-Arg-Asn-Ala-Tyr- - - -

Scorpion	Toxin	Loop 4	Loop 5	Loop 6
Centruroides sculpturatus	variant 1	Cys- - -Glu- - -Cys- -Lys-Ala-Lys-Asn-Gln-Gly-Gly-Ser-Gly-	-Tyr-Cys-Tyr- - -Ala- - - -Phe- - -Ala-
	2	Cys- - -Glu- - -Cys- -Lys-Ala-Lys-Asn-Gln-Gly-Gly-Ser-Tyr-Gly-	-Tyr-Cys-Tyr- - -Ala- - - -Phe- - -Ala-
	3	Cys- - -Glu- - -Cys- -Lys-Ala-Lys-Asn-Gln-Gly-Gly-Ser-Tyr-Gly-	-Tyr-Cys-Tyr- - -Ala- - - -Phe- - -Ala-
Androctonus australis	I	Cys-Asn-Arg-Glu-Cys-Lys-Trp-Lys-His-Ile-Gly-Gly-Ser-Tyr-Gly-	-Tyr-Cys-Tyr- - -Gly- - - -Phe- - -Gly-
	II	Cys-Asp-Gly-Leu-Cys- -Lys- - -Lys-Asn- - -Gly-Gly-Ser-Ser-Gly-	-Tyr-Cys-Ser-Ser-Phe-Leu-Val- -Pro-Ser-Gly-Leu-Ala-
		Cys-Asn-Glu-Glu-Cys- - -Thr-Lys-Leu-Lys-Gly-Glu-Ser- - -Gly-	-Ser-Cys-Gln-Trp-Ala-Ser-Pro-Tyr-Gly-Asn-Ala-

Scorpion	Toxin	Loop 7	Loop 8	Loop 9
Centruroides sculpturatus	variant 1	Cys-Trp-Cys-Gly-Leu-Pro-Gly-Ser-Thr-Pro-Thr-Tyr-Pro-Leu-Pro-Asn-Lys-Cys-Ser-Ser- -
	2	Cys-Trp-Cys-Gly-Leu-Pro-Gly-Ser-Thr-Pro-Thr-Tyr-Pro-Leu-Pro-Asn-Lys-Cys-Ser- - -
	3	Cys-Trp-Cys-Gly-Leu-Pro-Gly-Ser-Thr-Pro-Thr-Tyr-Pro-Leu-Pro-Asn-Lys-Cys-Ser- - -
Androctonus australis	I	Cys-Tyr-Cys-Gly-Leu-Pro-Asp-Ser-Thr-Pro-Thr-Tyr-Pro-Leu-Pro-Asn-Lys-Cys-Ser- - -
	I	Cys-Trp-Cys-Tyr-Lys-Leu-Pro-Asp-Ser-Thr-Pro-Thr-Gln-Thr-Pro-Leu-Pro-Asn-Lys-Cys-Ser-Ser- -
	II	Cys-Tyr-Cys-Tyr-Lys-Leu-Pro-Asp-His-Val-Arg-Thr-Lys-Gly- - -Pro-Gly-Arg-Cys-His- - -

[a] (- - -) indicates there is no corresponding amino acid residue.

toxins, and the two types of toxin they do not cross immunologically. The similarity between scorpion and snake toxins is not great. These structural differences also reflect the different modes of neurotoxic action of venoms of these two totally different animals.

Structure—Toxicity Relationships   Toxicity was destroyed when the toxins from the venoms of *Androctonus australis* and *Buthus occitanus* were subjected to chymotryptic and tryptic digestion (Miranda et al., 1964b). This suggests that small fragments do not have biological activity.

Toxin I of *A. australis* has three tyrosines in positions 5, 8, and 14. Iodination of the toxin resulted in iodinated tyrosine at all positions. However, the tyrosine at position 8 contained 51% of the total iodine incorporated. As there was no change in toxicity, it was concluded that Tyr-8 is not involved in the toxicity of the molecule (Rochat et al., 1972c).

1.2   Enzymes

Phospholipase A_2.   The enzyme seems to be commonly present in venoms of scorpions (Mohamed et al., 1969). Phospholipase A_2 from the venom of *Heterometrus scaber* hydrolyzed lipids from human red cell ghost but not those from intact erythrocytes (Kurup, 1966). The difference may be due to the physical state of the cell membrane phospholipids.

Others.   *Hadrurus arizonensis* venom shows acetylcholinesterase activity (Saunders and Johnson, 1970). The enzyme is also present in the venom of *Heterometrus scaber* (Nair and Kurup, 1973b).

Hyaluronidase is found in the venoms of *Scorpio maurus palmatus* (Zlotkin et al., 1972) and *Heterometrus scaber* (Nair et al., 1973).

Phosphomonoesterase and $5'$-nucleotidase are present in the venom of *Heterometrus scaber* (Nair and Kurup, 1973b), However, Russell et al. (1968) reported that *Vejovis spinigerus* venom does not contain phosphodiesterase, amylase or L-amino oxidase activity. Venoms of *Nebo hierichonticus* (Rosin, 1972) and *V. spinigerus* (Russell et al., 1968) do not contain protease. Gelatinase activity was detected in the venom of *Scorpio marus palmatus* (Zlotkin et al., 1972d). Nerve growth factor, which is commonly present in snake venom, is not found in venom of the scorpion *Palamnaeus gravimanus* (Pearce, 1973).

1.3   Non-protein Components

Scorpion venoms consist primarily of protein but also contain nonprotein materials that give strong interference in the Biuret test (Rosin, 1973). A number of free amino acids were detected in the venom of *Buthus minax* (El-Asmar et al., 1973a), and free tryptophan was found in the venom of *Heterometrus scaber* (Nair et al., 1973). Histamine, which is a hypotensive agent, is present in the venom of *Palamneus gravimanus* (Ismail et al., 1975a). 5-Hydroxytryptamine can be found in the venoms of *Leiurus quinquestriatus* (Adam and Weiss, 1956) and *Heterometrus scaber* (Nair et al., 1973). Serotonine and tryptamine were also identified in the venom of *H. scaber* (Nair et al., 1973b), the concentration of serotonin being 2.8 mg per gram of dry venom.

The venom of *H. scaber* contains significant quantities of glycosaminoglycans (Nair and Kurup, 1973a). Chondroitin sulfate A unit is present in maximum amounts, followed by fractions of chondroitin sulfate C and heparin sulfate, respectively. The fraction corresponding to chondroitin sulfate B is present in only trace amounts, and heparin is absent. Hyaluronic acid is present at a low level. The venom also contains a small amount of free hexosamine.

2 IMMUNOLOGY

Usually, the closer the phylogenetic relationship of scorpions, the more similar are the immunological properties of their venoms. There are cross-immunological reactions between the antivenin of *Hadrurus arizonensis* and the venom of *Vejovis flavis* (Potter and Northey, 1962). There is, however, no precipitation with the venoms of *Centruroides sculpturatus*, *C. gertschi*, *Androctonus australis*, *Leiurus quinquestriatus*, and *Vejovis spinigerus*. *Leiurus quinquestriatus* and *C. gertschi* antivenoms form five and two precipitation bands, respectively, with *H. arizonensis* venom.

Toxins I and II of *Androctonus australis* Hector differ immunologically from toxin I of *Buthus occitanus tunetatus* (Boquet et al., 1972). However, toxin I of *A. australis* Hector belongs to the same serology group as toxin V of *Leiurus quinquestriatus quinquestriatus*.

The antivenin made for *Buthus minax* neutralizes the lethal effect, as well as the hypertensive effect, in rats, but does not prevent the respiratory arrest (El-Asmar et al., 1973b).

Antiserum for a given scorpion venom is usually quite specific. For instance, antiserum for *Buthus quinquestriatus* shows good precipitation bands with its homologous venom, but weaker and fewer bands with heterologous venoms of *B. occitanus* and *Androctonus aeneas*, and no reaction with *Pandinus* spp. venom (Mohamed et al., 1975).

Scorpion toxins and cobra toxins are not similar immunologically (Boquet et al., 1972). This is reasonable, as they are totally different toxins.

3 TOXICOLOGY

3.1 Clinical Symptoms

Clinical studies of over 1000 randomly selected persons indicate that envenomation by the scorpion *Centruroides margaritatus* produces such symptoms as pain, local edema, and fever 1 to 20 hr after the sting (Marinkelle and Stahnke, 1965). Human volunteers receiving *Nebo hierichontictus* venom injections in the arm reported local burning pain and the formation of a small papilla, surrounded by an edematous area developing around the site of injection (Rosin, 1969b).

Watt et al. (1974) summarized the responses in man to scorpion envenomation. The constitutional symptoms include severe local pain and swelling and occasional discoloration. Sweating, pallor, restlessness, anxiety and confusion, salivation, nausea, abdominal cramps, chest pains, and headaches are also evident. There is a sensation of choking, muscle weakness, and twitching. Initial tachycardia changes to bradycardia, and initial hypertension to hypotension. There are respiratory distress and subsequent cyanosis. Death results from cardiovascular collapse and pulmonary edema. The time to death varies from less than 1 hr to several days.

The pathological findings reveal elevated urinary excretion of catecholamines and their metabolites. Serum potassium is elevated, and serum sodium is lowered. Congestion and

Table 3 Toxicity of Scorpion Venoms

Scorpion	LD$_{50}$ (μg g^{-1})	Animals	Route of Injection	Reference
Androctonus australis				
Toxin I	0.019	Mice	—	Rochat et al., 1967
Toxin II	0.010	Mice	—	Rochat et al., 1967
A. mauretanicus mauretanicus	0.170	Mice	i.v.	Cheymol et al., 1974b
Buthotus judaicus	8.46	Mice	s.c.	Nitzan (Tischler) and Shulov, 1966
Buthus occitanus	0.15	Mice	—	Miranda et al., 1964a
	1–1.2	Mice	s.c.	Nitzan (Tischler) and Shulov, 1966
Centruroides sculpturatus	1.1–1.9	Mice	i.p.	Johnson et al., 1966
	1.46	Mice	s.c.	Watt, 1964
	1.12	Mice	—	Stahnke, 1963
	0.13	Guinea pigs	—	Stahnke, 1963
	0.59	Rabbits	—	Stahnke, 1963
	1.00	Rats	i.m.	Stahnke, 1963
	0.09	Rats	—	Stahnke, 1967
	0.33	Rats	i.p.	Patterson, 1964
	2.36	Cats	—	Stahnke, 1963
	2.62	Dogs	—	Stahnke, 1963
Hadrurus arizonensis	168	Mice	i.p.	Johnson et al., 1966
Heterometrus caesar	22	Mice	i.v.	Cheymol et al., 1974b
H. scaber	0.72	Rats	i.v.	Nair et al., 1975
Leiurus quinquestriatus	0.39	Mice	—	Gotlieb, 1966
	0.34	Mice	s.c.	Nitzan (Tischler) and Shulov, 1966
Fraction 1	0.48	Mice	—	El-Asmar et al., 1972
Fraction 2	0.64	Mice	—	El-Asmar et al., 1972
Fraction 3	0.84	Mice	—	El-Asmar et al., 1972
Orthochirus innesi	2–2.67	Mice	s.c.	Nitzan (Tischler) and Shulov, 1966
Pandinus exitialis	40	Mice	i.p.	Ismail et al., 1974a
Prionurus crassicauda	0.8	Mice	s.c.	Nitzan (Tischler) and Shulov, 1966
Scorpio maurus palmatus	11.1	Mice	—	Zlotkin et al., 1972a, b
Tityus serrulatus	0.66	Mice	i.v.	Brazil et al., 1973
Vejovis spingerus	4.87	Mice	i.v.	Russell et al., 1968

hemorrhages are evident in various organs, along with pulmonary edema. The effect on the heart is focal myocardial necrosis, infiltration with monocytes and lymphocytes, and deposition of fat droplets.

Clinical experience indicated that patients stung by *Centruroides sculpturatus* and given meperdine hydrochloride (Demerol) consistently experience a more serious reaction from this venom. Assays confirmed the clinical observation that these substances act synergistically with the venoms. Therefore Stahnke and Dengler (1964) consider it unwise to use narcotics as therapeutic agents. Other narcotics also increase mortality from scorpion venom (Gotlieb, 1966).

3.2 Yield and Toxicity

The amount of venom that can be obtained from a scorpion is very small, usually less than 1 mg per specimen. There is considerable variation in the toxicities of scorpion venoms, depending on the species of animal used as subjects, the route of injection, and the species of scorpion (Table 3). It is of interest to note that the tarantula, which devours scorpions, is nontoxic to the venom of the scorpion *Centruroides limpidus tecomanus* (Wheeling and Keegan, 1972).

4 PHARMACOLOGY

Scorpion venoms affect many parts of the body. The distribution of *Leiurus quinquestriatus* venom in the rat was investigated by Ismail et al. (1974b), who found the highest radioactivity in the kidneys and the lowest radioactivity in the brain. The high content of venom in the kidneys suggests the rapid excretion rate of the venom. The venom does not appear to penetrate the blood–brain barrier, thus confirming an earlier finding (Osman et al., 1973). Relatively high radioactivity in the lungs and heart may be due to the fact that the primary sites for the action of the venom may be in these organs. Scorpion venoms have strong neurotoxic as well as other effects. Some of the more pronounced effects are discussed in this section.

4.1 Effect on Nerves

Scorpion toxins block impulse transmission at presynaptic sites of both cholinergic and adrenergic nerves (Yarom, 1970; Diniz, 1971; Moss et al., 1974). Purified toxins of *Tityus serrulatus* (Gomez and Dinitz, 1966) have been shown to cause both contraction and relaxation of isolated rat ileum. This can be the result of the release of both acetylcholine and catecholamines (Cunha Melo et al., 1970; Freire-Maia and Diniz, 1970).

Acetylcholine Release. Scorpion venoms release acetylcholine at the neuromuscular junction (Benoit and Mambrini, 1967), from postganglionic parasympathetic neurons (Diniz and Torres, 1968), from brain (Gomez et al., 1973, 1975), and from other tissues (Diniz et al., 1972; Tazieff-Depierre et al., 1974). This was demonstrated when *Androctonus australis* venom and toxin II produced spasmodic activity in isolated guinea pig ileum (Tazieff-Depierre, 1972). Tetrodotoxin antagonized the spasmogenic action of the scorpion toxin and completely inhibited the secretion of acetylcholine (Tazieff-Depierre and Andrillon, 1973).

When there is a continuous discharge of acetylcholine, normal nerve impulses cannot be transmitted to the muscle. *Heterometrus fulvipes* venom produces a tetanizing effect in isolated frog gastrocnemius muscle. This hastens the onset of fatigue, and the effect is irreversible (Venkateswarlu and Babu, 1975).

The spasmogenic action of the venom on guinea pig ileum was observed only in the presence of Ca ion, which is indispensable to the secretion of acetylcholine (Tazieff-Depierre et al., 1973).

It has been demonstrated that d-tubocurarine antigonizes the effects of *Androctonus australis* venom on frog striated muscle (Tazieff-Depierre and Nachon-Rautureau, 1975). This is another indication that the action of scorpion venom is due principally to the release of acetylcholine, a view further supported by the fact that atropine and tetrodoxine eliminated the cardiac arrhythmias due to acetylcholine secretion induced by the scorpion venom in rats (Tazieff-Depierre, 1975). The cardiotoxicity of cobra (*Naja nigricollis*) γ-toxin, however, is not decreased by atropine or enhanced by tetrodoxin. This illustrates the different mechanisms of neurotoxic action of the two toxins.

Cheymol et al. (1973) investigated the effects of three venoms of North African scorpions (*Leiurus quinquestriatus, Buthus occitanus, Androctonus australis*) and of toxins AI and AII of *A. australis* on neuromuscular action in isolated and *in situ* rat diaphragm preparations and in the rectus abdominis muscle of the frog. The venoms and toxins induced successive slow contractures, which were always preceded by a potentiation of the maximal twitch and followed by a less and less easily reversible paralysis. These contractures occurred in directly or indirectly stimulated preparations, as well as in unstimulated neuromuscular and denervated and curarized preparations. The venoms and toxins do not appear to act at the nerve ending or at end-plate receptors because they do not inhibit acetylcholine action on denervated diaphragm of the rat or rectus abdominis muscle of the frog. They cause a slow depolarization of the muscle membrane, which produces a contracture when the threshold of excitability is attained. This may result from a displacement of calcium ion or from a disturbance in membrane permeability to sodium ions, as suggested by the antagonism produced by tetrodotoxin. The effects of scorpion venoms on muscle show some similarities to the contractures induced by veratrine and by crotamine, a toxin isolated from Brazilian *Crotalus durissus terrificus* venom.

Androctonus amoreuxi venom blocks the twitch activity in both phrenic nerve-hemidiaphragm of the rat and in cat tibialis anterior preparation (Ismail et al., 1975a, b). This inhibitory activity can be prevented by the specific antivenin.

Depolarization of the neuromuscular junction with change in vesicular content usually accompanies the tissue damage produced by scorpion venom (Tafuri et al., 1971). Degeneration can also be found in the axon near the nerve terminals (Diniz et al., 1974; Böhm et al., 1974). These lesions resemble those produced by spider venom and β-bungarotoxin, indicating that the scorpion venom directly affects the nerve terminals.

Catecholamine Release. Scorpion venoms, like spider venoms, can release catecholamine as well as acetylcholine. Injection of *Tityus serrulatus* venom and its purified toxin have been shown to deplete the rat adrenal of catecholamines (Henriques et al., 1968). Catecholamine and its metabolites such as vanilmandelic acid can be detected in high content in the urine of patients severely stung by the scorpion *Orthochirus innesi* (Gueron and Weizmann, 1969).

Catecholamine depletion has also been observed in sympathetic nerves from the atrioventricular valve of rats (Rossi et al., 1974). The clinical symptoms of scorpion

4 Pharmacology

envenomation, such as hypertension, tachycardia, and pulmonary edema, are probably due to the release of catecholamines from the adrenergic nerve endings.

Leiurus quinquestriatus venom increased the serum catecholamine levels in normal and adrenalectomized rats (Moss et al., 1973). The venom also had a hyperthermic effect in rabbits (Osman et al., 1973). This is believed to be due either to noradrenalin release in the anterior hypothalamus or to a shift in the balance of Na(I) and Ca(II).

Nerve Permeability. Scorpion venoms not only depolarize the neuromuscular junction but also affect axons. When the venom of *Buthus tamulus* was applied externally to squid axon, it caused a gradual depolarization of the nerve membrane (Narahashi et al., 1972). When the venom was applied internally, however, it had no effect on the resting potential. Katz and Edwards (1972) studied the effect of *Centruroides sufussus sufussus* venom on the nerve of isolated frog sartorius nerve–muscle preparation. Single shocks applied to the nerve produced repetitive responses in both nerve and muscle. Cahalan (1975) observed the same results. Reduction of the sodium concentration in the bathing medium abolished the repetitive activity (Katz and Edwards, 1972), a finding consistent with an effect on Na(I) permeability as the mechanism of scorpion venom action. It was suggested that the site most accessible to the action of the venom is the exposed area of the axon near its termination.

Romey et al. (1975) observed that *Androctonus australis* Hector venom affects the closing of the Na(I) channel and the opening of the K(I) channel in the giant axons of crayfish and lobster nerves.

The venom of *Tityus serrulatus* and its toxin stimulated the release of acetylcholine independently of K(I) but required both Na(I) and Ca(II). The uptake of ^{24}Na(I) and of ^{45}Ca(II) by rat brain slices was significantly enhanced by the toxin (Gomez et al., 1975).

Blaustein (1975) showed that *Leiurus quinquestriatus* venom stimulated the Ca(II) uptake of rat brain. He suggested that synaptosomes may increase their permeability to Ca ion, accumulate this ion, and release neural transmitter substances, when stimulated by depolarizing agents.

It has been observed that the inward Na(I) current decays more slowly when scorpion venom is applied to the node of Ranvier. The remaining outward current consists of an unchanged leak current and a reduced K(I) current. This finding indicates that the venom mainly reduces the rate and amount of Na(I) inactivation (Koppenhöfer and Schmidt, 1968a, b).

The same conclusion was reached by Adam et al. (1966) and Adam and Weiss (1966). In the presence of venom the duration of the action potential of single myelinated nerve fibers of the frog is prolonged. Thus the main effect of the venom probably consists of an increase in the Na(I) permeability of the resting membrane and a delayed inactivation of the Na(I) permeability.

Increasing the Ca(II) concentration of the medium abolishes the incomplete Na(I) inactivation that is typical for nodes of Ranvier treated with *Leiurus quinquestriatus* venom (Schmitt and Schmidt, 1972). Similarly the maximum K(I) permeability which is reduced by the venom can be increased by raising the Ca(II) concentration.

Androctonous australis venom induces autorhythmic activity and increases the liberation of muscular Ca(II) in phrenic nerve–diaphragm preparations of guinea pig and rats, in isolated mouse diaphragm, and in the abdominal muscle of the frog (Tazieff-Depierre et al., 1968). These effects can be suppressed by an antihistamine, neoantergan; a local anesthetic, procaine; and a curarizing agent, d-tubocurarine. The decrease in autorhythmic activity results directly from suppression of Ca(II) release (Tazieff-Depierre and Czajka, 1969).

4.2 Effect on Cardiovascular System

Scorpion venoms possess potent neurotoxic activity, and strong cardiovascular responses also cannot be overlooked (Gueron et al., 1967; Patterson, 1960).

Pathology. Yarom et al. (1970) examined the heart tissue of a 2-year-old child who had died of a *Buthus quinquestriatus* sting. The postmortem examination indicated that there was myocardial degeneration. Increased calcium deposits have also been reported in the area of myofibrilar destruction (Fig. 1) in the heart of a dog envenomated with *Leiurus quinquestriatus* (Yarom and Braun, 1971b). This suggests that scorpion venom alters the myocardial cell membrane permeability to the Ca ion, an example of the direct effect of scorpion venoms on the heart.

The similarity of the histological change following scorpion venom injection to those produced by excessive noradrenaline and other adrenergic amines, especially isopro-

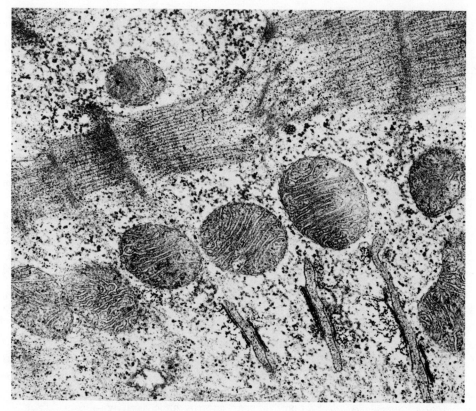

Figure 28.1 Effect of scorpion (*Leiurus quinquestriatus*) venom on the cardiac muscle. Area of intracellular damage shows typical venom-induced disruption of sarcomeres in I zones. There is much calcium deposit in the edematous sarcoplasm and in the lateral sacs, which are dilated. The mitochondria are well preserved. (Photograph kindly supplied by Dr. R. Yarom and originally published in Yarom and Braun, 1971b.)

terenol, is striking (Rohayem, 1954; Gueron et al, 1967; Yarom and Braun, 1970). The sympathomimetic nature of scorpion venom injury is further confirmed by the effect of pretreatment with adrenergic blocking agents. Gueron and Yarom (1970) examined 34 patients with severe stings from the scorpion *Buthus quinquestriatus* and observed that the electrocardiograms of some victims showed myocardial infarctionlike patterns. They also examined urinary catecholamine metabolites and found that vanylmandelic acid was elevated in 7 patients, and total free epinephrine and norpinephrine in 8. Nine of the patients died, and pathologic lesions of the myocardium were observed in 7. It is clear, therefore, that the venom's effects on the cardiovascular system are related to the level of circulating catecholamines produced by direct action on the sympathetic system.

In an electron microscopic study, Yarom and Braun (1971a) observed that *Leiurus quinquestriatus* venom had a direct pathophysiologic effect on the myocardium but did not produce muscle cell destruction. In some cases, there were scattered destruction of the I-band, edema, and lipid droplet deposition without changes in the mitochondria (Figs. 2 and 3). The venom has also been shown to cause a marked increase in contractility of isolated rat heart with a characteristic ultrastructural derangement of the I bands and Z lines (Yarom et al., 1974).

Physiological Effects. Injections of *Buthus tamulus* venom into dogs produced dramatic changes in the electrocardiogram, indicating that the venom has a pronounced cardiac effect (Rao et al., 1969).

The initial effect of *Tityus serrulatus* and *T. bahiensis* venoms is bradycardia, followed by an increase in the force and frequency of the heartbeat (Corrado et al., 1966, 1968, 1974; Diniz et al., 1966). The bradycardia is vagal mediated, since it is blocked by atropine and potentiated by prostigmine. That the latter effect is produced by the release of catecholamines is indication by the fact that it can be blocked by dichloroisoproterenol and bretylium and is absent in reserpine-treated animals.

It has also been demonstrated (Cheymol et al., 1974a) that toxins I and II of *Androctonus australis* release catecholamines from nerve endings, resulting in hypertension, peripheral vasoconstriction, lachrymation, salivation, breathing spasms, and an indirect positive inotropic effect. Subsequent ganglionic blockade (except for toxin I) produced a block of the vascular tone, leading to hypotension. Toxin II isolated from *A. australis* venom has been shown by Coraboeuf et al. (1975) to produce a marked and persistent increase in the amplitude and duration of the ventricular and atrial action potential plateaus and a positive inotropic effect in both untreated and reserpinized rats. They concluded that in the rat heart the toxin increases the plateau by slowing down the inactivation of the sodium conductance or by inducing an incomplete sodium inactivation. This effect indirectly favors the penetration of the Ca ion through the slow channel and also the development of the tonic component of cardiac contraction, both effects being responsible for the positive inotropic effect.

Androctonus mauretanicus venom also induces changes in the electrocardiogram, characterized as disorders of cardiac rhythm and conduction (Roch-Arveiller et al., 1974). Early changes were a transient sinusal bradycardia, atrioventricular block, and ectopic ventricular beats. This action was reflex, since it was inhibited by atropine, and was the result of a potent release of catecholamines by the venom. In a later stage there was a sinusal bradycardia, which resulted from direct action of the venom on the cardiac muscle with involvement of muscarinic receptors.

The cardiovascular effects of the venom of *Pandinus exitialis* appeared to be mediated

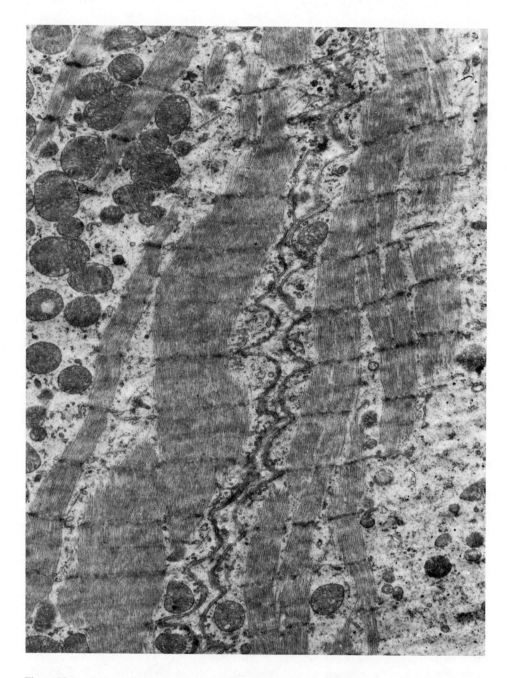

Figure 28.2 Effect of *Leiurus quinquestriatus* venom on the left-ventricle myocardium. Note the destruction of myofibrils and sarcoplasmic reticulum. (Photograph kindly supplied by Dr. R. Yarom and originally published in Yarom and Braun, 1971a. Copyright 1971 by the U.S. – Canadian Division of the International Academy of Pathology.)

4 Pharmacology

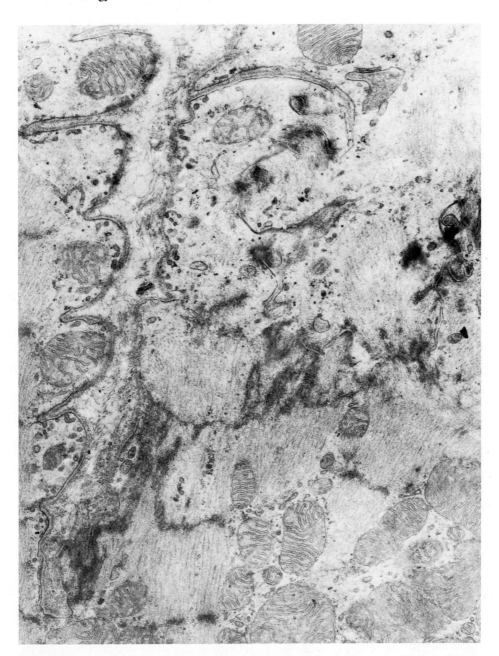

Figure 28.3 The same experiment as in Fig. 2, but showing more severe damage. (Photograph kindly supplied by Dr. R. Yarom.)

through stimulation of the autonomic nervous system, particularly the sympathetic, with release of tissue catecholamines (Ismail et al., 1972, 1974a).

Cheymol et al. (1974b) investigated the peripheral effects produced by the venoms of *Heterometrus caesar* and *Androctonus mauretanicus mauretanicus*. The marked hypertension induced by the two venoms occurred without latency and was followed by bradycardia. The mechanism of this cardiovascular action involved ganglion stimulation, followed by a ganglioplegic effect.

The venom of *Buthus minax* produced a positive inotropic effect on isolated rabbit and guinea pig heart but no alteration in rate (Ismail et al., 1973). The cardiac stimulant action was blocked by propranolol and was absent in resperpinized hearts, indicating the possibility of an indirect action of the venom, probably through the release of catecholamines. In the reserpinized hearts, and in normal hearts after propranol treatment, the venom produced bradycardia, which was blocked by atropine. The venom also produced a marked hypertensive effect in cats, dogs, rats, and guinea pigs, an action blocked by tolazoline and phenoxybenzamine. In cats and rats the hypertensive effect was preceded by brief hypotension, which was partially blocked by atropine. The venom markedly decreased the rate of flow in rat perfused hindquarter and caused a moderate increase in capillary permeability. It also stimulated rabbit intestine and guinea pig ileum, an action largely blocked by atropine. The venom increased the size of twitches in isolated, indirectly stimulated phrenic hemidiaphragm of the rat and contracted the rectus abdominis muscle of the frog. The latter effect was blocked by tubocurarine. The decrease in respiration rate was blocked by carotid sinus denervation. It may be that the venom stimulates both the sympathetic and the parasympathetic systems.

Centruroides sculpturatus venom produces different physiological responses, depending on the species of animals used. In dogs and cats, it produced hypertension, tachycardia, and delayed cardiac irregularities, whereas in rats the opposite responses, hypotension and bradycardia, occurred (Patterson, 1960, 1964).

4.3 Hemorrhage

Necropsy examinations carried out on four children and three adults who had died of scorpion (*Buthus tamulus*) venom poisoning showed congestion in all organs. Subendocardial hemorrhages were present in two, mural thrombosis of the heart in one, and massive hemorrhage of the adrenals in one and of the frontal lobe in another. There were pinpoint hemorrhages in the cerebral cortex. Histological examination showed occlusions of small blood vessels with thrombi in the heart, lungs, brain, kidneys, and adrenals (Devi et al., 1970).

Scorpion stings produce blood-stained tears and urine, intraperitoneal hemorrhage, and priapism (Balozet, 1971). This suggests that concomitant hemorrhage and thrombosis may occur. The features of the diffuse intravascular coagulation following envenomation have been described in man (Devi et al., 1970; Reddy et al., 1972). Venom of *Palamneus gravimanus* has both procoagulant and anticoagulant properties (Hamilton et al, 1974), whereas *Leiurus quinquestriatus* venom is anticoagulant with minimal coagulant activity. Neither venom has fibrinolytic activity. The procoagulant activity of *P. gravimanus* venom is due to promotion of factor X activation, while the anticoagulant fraction interferes with the action of thrombin on fibrinogen, in part by making fibrinogen less susceptible to the clot-promoting effect of thrombin.

Hemorrhage can be induced experimentally in different tissues such as the lungs and heart (El-Asmar et al., 1972; Abdel-Wahab et al., 1974), in the abdominal wall (Rosin, 1969a), in the heart (Yarom and Braun, 1971b), and in the local area around the site of injection. Local hemorrhage and necrosis can take place in mice (Rosin, 1969a) and in several vertebrates (Grasset et al., 1946). Necrosis can occur in man (Rosin, 1969a).

In addition to the hemorrhagic effect, some scorpion venoms are hemolytic (Rosin, 1972) and anticoagulant (Zlotkin et al., 1972d).

4.4 Effect on Skeletal Muscle

Scorpion venoms also have a direct effect on skeletal muscle. For instance, *Leiurus quinquestriatus* venom induces spontaneous twitches and tetanic contraction of skeletal muscle (Adam and Weiss, 1959; Adam et al., 1966). Parnas and Russell (1967) investigated the effects of venoms of five American scorpions on nerve and muscle, and concluded that they have excitatory, blocking, and lytic actions.

Short periods of exposure to *L. quinquestriatus* venom resulted in diminution and dispersal of pyroantimonate precipitates (Yarom and Meiri, 1972). Longer exposure caused a marked decrease in deposits, but after 60-min exposure to smaller doses of venom the linear precipitate recurred. The intracellular precipitate is due to ionic calcium. Yarom and Meiri suggest that *L. quinquestriatus* venom acts directly on frog muscle membrane, altering the calcium flux.

The principal effects of envenomation by *L. quinquestriatus* venom consist of hypertension, respiratory failure, and skeletal muscle stimulation (Mohamed, 1942, 1950; Adam and Weiss, 1959; Adam et al., 1966; Ismail et al., 1972). The venom also stimulates a variety of smooth muscles (Mohammed, 1950; Osman et al., 1972).

Tityus serrulatus venom also elicits spontaneous twitches, potentiates the maximum twitch, and causes a delay in its relaxation (Brazil et al., 1973). When the venom of *Heterometrus fulvipes* was injected into the cockroach, the heartbeat and spiracular movements stopped and the legs showed sagging and erratic movements without coordination. Muscle action potentials failed within 10 sec, and those of the nerve cord disappeared in 30 to 40 sec (Babu et al., 1971).

Tityus serrulatus venom injected intraperitoneally into the striated muscle of cockroaches induces varying degrees of muscle fiber degeneration. In completely degenerated muscle fibers, the plasma and basement membranes disappear (Rossi et al., 1973).

4.5 Metabolic Effect

Scorpion venoms affect many enzyme activities in various tissues. The activities of succinate dehydrogenase, lactate dehydrogenase, and acetylcholinesterase are inhibited to a greater extent in muscle homogenates than in nerve cord homogenates by the venom of *Heterometrus fulvipes* (Babu et al., 1971). A gradual drop in oxygen consumption, along with a decline in heat production, by the cockroach also takes place after venom injection.

For instance, *Leiurus quinquestriatus* venom inhibits the catalase activity of human erythrocytes (Rabie et al., 1972) and succinic dehydrogenase, lactic dehydrogenase, and acetylcholinesterase activities in the cockroach (Babu et al., 1971). *Buthus minax* venom decreases cholinesterase activity in the white matter of the brain, in the portal tract, and in the liver cells (Moustafa et al., 1974). A decrease in succinic dehydrogenase activity was observed in the kidneys, heart, and liver of chronically treated mice.

Heterometrus fulvipes venom has no effect on the succinate dehydrogenase activity of its own hepatopancreatic homogenate, whereas it has a depressive effect in brain homogenate (Selvarajan et al., 1974). In sheep the same venom increases muscle enzyme activity by 110%, but liver enzyme activity by only 34%. The differential effect may be correlated to the detoxification mechanism in the liver (Selvarajan et al., 1975).

Hyperglycemia and liver and muscle glycogenolysis were observed to occur in the rat from the venom of *Buthus quinquestriatus* (Mohamed et al., 1972). Hyperglycemia resulting from scorpion envenomation was observed earlier by many investigators

(Mohamed, 1950; Freire-Maia et al., 1959; Freire-Maia and Ferreira, 1961). These epinephrine like effects are believed to be mediated by the serotonin present in the venom. The lack of lipolytic action was attributed to inactivation of the venom serotonin by monoamine oxidase and phosphodiesterase of adipose tissue (Mohamed et al., 1972).

The venom of *Buthus minax* causes hyperglycemia in rats. Liver glycogenolysis seems to be the major factor contributing to this rise in blood glucose (El-Asmar et al., 1973a). However, Ismail et al. (1973) attributed the effect to autonomic stimulation, with a predominance of sympathetic stimulation, and release of tissue catecholamines.

Heterometrus scaber venom causes a considerable increase in the urinary excretion of valilmandelic acid, indicating that the hyperglycemia may be due to an increase in adrenaline production caused by the venom (Nair and Kurup, 1973c, d).

4.6 Other Effects

Scorpion venoms have diverse actions. They increase motility of the gastrointestinal tract, visceral hyperemia, and gastric hyperdistension (Mohamed, 1950; Patterson, 1960; Stahnke, 1965a). Patterson (1960) observed that cats and rabbits injected with venom of *Centruroides sculpturatus* developed distension of the stomach with gas, possibly carbon dioxide, which progressively extended through the pyloric sphincter into the small intestine. This was confirmed by Ismail and Osman (1973).

Centruroides sculpturatus venom causes hypertension, respiratory failure, and skeletal muscle stimulation in anesthetized animals (Patterson, 1960, 1964). The venom induces stimulation of the visceral smooth muscle. Venom added to isolated gut preparations of rat ascending colon, rabbit duodenum, and guinea pig ileum initially caused a reduction in spontaneous activity, which was followed by an increase in baseline. Later slow, rhythmic contractions developed. Atropine sulfate prevented the initial, but not the delayed, contractions. Hexamethonium chloride, a ganglionic blocking agent, failed to alter the activity of the venom in guinea pig ileum. Therefore Patterson (1962) concluded that no ganglionic stimulating action of the venom had been demonstrated.

The venom of *Leiurus quinquestriatus* produced a marked increase in the frequency and amplitude of contractions of the uterus. This is in good agreement with clinical observations on abortion induced by the venom (Osman et al., 1972). It is unlikely that the effects of the venom on the uterus are due to the presence of 5-hydroxytryptamine in the venom, since 2-bromolysergic acid diethylamide tartrate and methysergide, in concentrations effective in blocking the action of doses of 5-hydroxytryptamine far greater than those that might be found in crude venom, had no effect on the stimulant action of the venom.

Stahnke (1965a, b) reported that changes in the toxicity of *Centruroides sculpturatus* venom occurred when the recipient rats were subjected to changes in ambient temperature. Epinephrine also increased the toxicity of the venom in rats. Since epinephrine is released under stress conditions, he concluded that venom toxicity is a result of synergism between the venom and epinephrine.

Purified toxin of *Tityus serrulatus* can produce either stimulation or paralysis, depending on the dose, of the respiratory movements of rats. Apnea occurs at the expiratory phase and is not prevented by vagotomy. A shift of baseline in the abdominal respiration tracing is probably due to contraction of the skeletal muscles of the abdominal wall (Freire-Maia et al., 1970). Injection into rats of purified toxin fractions obtained from the venom of the scorpion *T. serrulatus* produced complex respiratory arrhythmias (Freire-Maia et al., 1970). The apnea recorded in intact rats after toxin injection was of long duration in comparison to that observed in denerved animals. Bilateral vagotomy abolished the apnea observed between gasping respirations (Freire-Maia et al., 1973). These workers concluded that the respiratory arrhythmias produced by scorpion toxin in rats are due mainly to stimulation of peripheral receptors and are, therefore, reflex in nature.

Pansa et al. (1973) examined the effect of *Androctonus australis* venom, its insect toxin, mammal toxin II, and the crustacean fraction on the crayfish stretch receptor organ. The crustacean fraction was able to mimic the excitatory and blocking actions of the crude venom, whereas the insect and mammal toxins were inactive. They suggested that the specificity in the action of the crustacean fraction is due to a specific affinity to the crustacean neural system.

On rat uterus the venom of *Pandinus exitialis* produced a powerful contraction which was greatly attenuated by methysergide and completely blocked by meclofenamic acid, indicating that the contraction is mediated partly by the serotonin content of the venom and partly by the release of kinins, postaglandins, and/or slow-reacting substance (Ismail et al., 1974a).

REFERENCES

Abdel-Wahab, M. F., Abdo, M. S., Megahed, Y.M., and Osman, O. H. (1974). Labelling, fractionation and toxicity of scorpion (*Leiurus quinquestriatus* H and E) venom, using iodine-125, iodine-131, and the gel-filtration technique, *Isotopenpraxis*, 10, 56.

Adam, K. R. and Weiss, C. (1956). 5-Hydroxytryptamine in scorpion venom. *Nature*, 178, 421.

Adam, K. R. and Weiss, C. (1959). Action of scorpion venom on skeletal muscle, *Br. J. Pharmacol*, 14, 334.

Adam, K. R. and Weiss, C. (1966). Pharmacology of the venoms of African scorpions, *Mem. Inst. Butantan*, 33, 603.

Adam, K. R., Schmidt, H., Stämpli, R., and Weiss, C. (1966). The effect of scorpion venom on myelinated nerve fibres of the frog, *Br. J. Pharmacol.*, 26, 666.

Babin, D. R., Watt, D. D., Goos, S. M., and Mlenjek, R. V. (1974). Amino acid sequences of neurotoxic protein variants from the venom of *Centruroides sculpturatus* Ewing, *Arch. Biochem. Biophys.*, 164, 694.

Babin, D. R., Watt, D. D., Goos, S. M., and Mlenjek, R. V. (1975). Amino acid sequence of neurotoxin I from *Centuroides sculpturatus* Ewing, *Arch. Biochem. Biophys.*, 166, 125.

Babu, S., Murali Krishna Dass, K. P., and Venkatachari, S. A. T. (1971). Effects of scorpion venom on some physiological processes in cockroach, *Toxicon*, 9, 119.

Balozet, L. (1971). "Scorpionism in the old world," in W. Bücherl and E. Buckley, Eds., *Venomous Animals and their Venoms*, Vol. III, Academic, New York, pp. 349–371.

Benoit, P. R. and Mambrini, J. (1967). Action of scorpion venom on the neuromuscular junction of frogs, *J. Physiol.*, 59, 348.

Blaustein, M. P. (1975). Effects of potassium, veratridine and scorpion venom on calcium accumulation and transmitter release by nerve terminals *in vitro*, *J. Physiol.*, 247, 617.

Böhm, G. M., Pompolo, S., Diniz, C. R., Gomez, M. V., Pimenta, A. F., and Netto, J. C. (1974). Ultrastructural alterations of mouse diaphragm nerve endings induced by purified scorpion venom, tityustoxin, *Toxicon*, 12, 509.

Boquet, P., Dumarey, C., and Ronsseray, A. (1972). Recherches immunologiques sur les toxines du venin de trois espèces de scorpions: *Androctonus australis* Hector, *Buthus occitanus tunetatus* et *Leiurus quinquestriatus*, *C. R. Acad. Sci. Paris*, 274, 1874.

Brazil, O. V., Neder, A. C., and Corrado, A. P. (1973). Effects and mechanism of action of *Tityus serrulatus* venom on skeletal muscle, *Pharmacol. Res. Commun.*, 5, 137.

Cahalan, M. D. (1975). Modification of sodium channel gating in frog myelinated nerve fibers by *Centruroides sculpturatus* scorpion venom, *J. Physiol.*, **244**, 539.

Cheymol, J., Bourillet, F., Roch-Arveiller, M., and Heckle, J. (1973). Action neuromusculaire de trois venins de scorpions Nord-Africains (*Leiurus quinquestriatus, Buthus occitanus* et *Androctonus australis*) et de deux toxines extraites de l'un d'entre eux, *Toxicon*, **11**, 277.

Cheymol, J., Bourillet, F., Roch-Arveiller, M., and Heckle, J. (1974a). Action cardiovasculaire de trois venins de scorpions Nord Africains (*Androctonus australis, Leiurus quinquestriatus, Buthus occitanus*) et de deux toxines extraites de l'un d'entre eux, *Toxicon*, **12**, 241.

Cheymol, J., Bourilet, F., Roch-Arveiller, M., and Heckle, J. (1974b). Comparative study of various peripheral effects produced by the venom of the Indian scorpion, *Heterometrus caesar,* and by the venom of the Moroccan scorpion, *Androctomus mauretanicus, C. R. Soc. Biol.*, **167**, 1574.

Coraboeuf, E., Deroubaix, E., and Tazieff-Depierre, F. (1975). Effect of toxin II isolated from scorpion venom on action potential and contraction of mammalian heart, *J. Mol. Cell. Cardiol.*, **7**, 643.

Corrado, A. P., Antonio, A., and Diniz, C. R. (1966). The mechanism of action of Brazilian scorpion venom (*Tityus serrulatus*), *Mem. Inst. Butantan,* **33**, 957.

Corrado, A. P., Antonio, A., and Diniz, C. R. (1968). Brazilian scorpion venom (*Tityus serrultatus*), an unusual sympathetic postganglionic stimulant, *J. Pharmacol. Exp. Ther.,* **164**, 253.

Corrado, A. P., neto, F. R., and Antonio, A. (1974). The mechanism of the hypertensive effect of Brazilian scorpion venom (*Tityus serrulatus* lutz e mello), *Toxicon,* **12**, 145.

Cunha Melo, J. R., Freire-Maia, L., Tafuri, W. L., and Maria, T. A. (1970). Modificações produzidas pelo toxina de escorpião na fisiologia da musculatura lisa isolada, *Abstr. 22nd Ann. Meet. Brazil. Soc. Progr. Sci.,* p. 371.

D'Ajello, V. Zlotkin, E., Miranda, F., Lissitzky, S., and Bettini, S. (1972). The effect of scorpion venom and pure toxins on the cockroach central nervous system, *Toxicon,* **10**, 399.

Devi, C. S., Reddy, C. N., Devi, S. L., Subrahmanyam, Y. R., Bhatt, H. V., Suvarnakumari, G., Murthy, D. P., and Reddy, C. R. R. R. M. (1970). Defribination syndrome due to scorpion poisoning, *Br. Med. J.,* **1**, 345.

Diniz, C. R. (1971). "Chemical and pharmacological properties of *Tityus* venoms," in W. Bücherl, and E. E. Buckley, Eds., *Venomous Animals and Their Venoms,* Vol. III, Academic, New York, pp. 311–316.

Diniz, C. R. and Torres, J. M. (1968). Release of an acetylcholine-like substance from guinea pig ileum by scorpion venom, *Toxicon,* **5**, 277.

Diniz, C. R., Gomez, M. V., Antonio, A., and Corrado, A. P. (1966). Chemical properties and biological activity of *Tityus* venom, *Mem. Inst. Butantan,* **33**, 453.

Diniz, C. R., Dai, M., Oliveira, Z. E. G., and Gomez, M. V. (1972). Release of acetylcholine from nervous tissue by a toxin of scorpion venom, *Vth Int. Congr. Pharmacol.,* p. 58.

Diniz, C. R., Piminta, A. F., Netto, J. C., Pompolo, S., Gomez, M. V., and Böhm, G. M. (1974). Effect of scorpion venom from *Tityus serrulatus* (tityustoxin) on the acetylcholine release and fine structure of the nerve terminals, *Experientia,* **30**, 1304.

El-Asmar, M. F., Ibrahim, S. A., and Rabie, F. (1972). Fractionation of scorpion (*Leiurus quinquestriatus* H and E) venom, *Toxicon,* **10**, 73.

El-Asmar, M. F., Osman, O. H., and Ismail, M. (1973a). Fractionation and lethality of venom from the scorpion *Buthus minax* (L. Koch), *Toxicon,* **11**, 3.

El-Asmar, M. F., Ismail, M., and Osman, O. H. (1973b). Immunological studies of scorpion (*Buthus minax* L. Koch) venom, *Toxicon,* **11**, 9.

Freire-Maia, L. and Diniz, C. R. (1970). Pharmacological action of a purified scorpion toxin in the rat, *Toxicon,* **8**, 132.

Freire-Maia, L. and Ferreira, M. C. (1961). Study of the mechanisms of hyperglycaemia and arterial hypertension produced by the venom of the scorpion in the dog, *Mem. Inst. Oswaldo Cruz,* **59**, 11.

Freire-Maia, L., Ferreira, M. C., and Da Silva, C. G. (1959). Hyperglycaemia in experimental scorpion poisoning of the dog, *Mem. Inst. Oswaldo Cruz,* **57**, 105.

Freire-Maia, L., Ribeiro, R. M., and Beraldo, W. T. (1970). Effects of purified scorpion toxin on respiratory movements in the rat, *Toxicon,* **8**, 307.

References

Freire-Maia, L., Azevedo, A. D., and Costa Val, V. P. (1973). Respiratory arrhythmias produced by purified scorpion toxin, *Toxicon,* **11,** 255.

Gomez, M. V. and Diniz, C. R. (1966). Separation of toxic components from Brazilian scorpion, *Tityus serrulatus,* venom, *Mem. Inst. Butantan,* **33,** 899.

Gomez, M. V., Dai, M. E. M., and Diniz, C. R. (1973). Effect of scorpion venom, tityustoxin, on the release of acetylcholine from incubated slices of rat brain, *J. Neurochem.,* **20,** 1051.

Gomez, M. V., Diniz, C. R., and Barbosa, T. S. (1975). A comparison of the effects of scorpion venom tityustoxin and ouabain on the release of acetylcholine from incubated slices of rat brain, *J. Neurochem.,* **24,** 331.

Gotlieb, A. (1966). Changes in toxic effect of the venom of the scorpion, *Leiurus quinquestriatus,* as a result of injection of atropine or morphine, *Harefuah,* **71,** 132.

Grasset, E., Schaafsma, A., and Hodgson, J. A. (1946). Studies on the venoms of South African scorpions (*Parabuthus, Hadogenes, Opisthopthalmus*) and the preparation of a specific anti-scorpion serum, *Trans. R. Soc. Trop. Med. Hyg.,* **39,** 397.

Gueron, M. and Weizmann, S. (1969). Catechol amine excretion in scorpion sting, *Isr. J. Med. Sci.,* **5,** 855.

Gueron, M. and Yarom, R. (1970). Cardiovascular manifestations of severe scorpion sting: Clinicopathologic correlations, *Chest,* **57,** 156.

Gueron, M., Stern, J., and Cohen, W. (1967). Severe myocardial damage and heart failure in scorpion sting, *Am. J. Cardiol.,* **19,** 719.

Hamilton, P. J., Ogston, D., and Douglas, A. S. (1974). Coagulant activity of the scorpion venoms *Palamneus gravimanus* and *Leiurus quinquestriatus, Toxicon,* **12,** 291.

Henriques, M. C., Gazzinelli, G., Diniz, C. R., and Gomez, M. V. (1968). Effect of the venom of the scorpion *Tityus serrulatus* on adrenal gland catecholamines, *Toxicon,* **5,** 175.

Ismail, M. and Osman, O. H. (1973). Effect of the venom from the scorpion *Leiurus quinquestriatus* (H and E) on histamine formation and inactivation in the rat, *Toxicon,* **11,** 225.

Ismail, M., Osman, O. H., Ibrahim, S. A., and El-Asmar, M. F. (1972). Cardiovascular and respiratory responses to the venom from the scorpion *Leiurus quinquestriatus E. Afr. Med. J.,* **49,** 273.

Ismail, M., Osman, O. H., and El-Asmar, M. F. (1973). Pharmacological studies of the venom from the scorpion *Buthus minax* (L. Koch), *Toxicon,* **11,** 15.

Ismail, M., Osman, O. H., Gumaa, K. A., and Karrar, M. A. (1974a). Some pharmacological studies with scorpion (*Pandinus exitialis*) venom, *Toxicon,* **12,** 75.

Ismail, M., Kertesz, G., Osman, O. H., and Sidra, M. S. (1974b). Distribution of ^{125}I-labelled scorpion (*Leiurus quinquestriatus* H and E) venom in rat tissues, *Toxicon,* **12,** 209.

Ismail, M., El-Asmar, M. F., and Osman, O. H. (1975a). Pharmacological studies with scorpion (*Palamneus gravimanus*) venom: Evidence for the presence of histamine, *Toxicon,* **13,** 49.

Ismail, M., Ghazal, A., El-Asmar, M. F., and Abdel-Rahman, A. A. (1975b). Immunological studies with scorpion (*Androctonus amoreuxi* Aud. and Sav.) venom, *Toxicon,* **13,** 405.

Johnson, B. D., Tullar, J. C., and Stahnke, H. L. (1966). Aquantitative protozoan bio-assay method for determining venom potencies, *Toxicon,* **3,** 297.

Kamon, E. (1965). Toxicity of the dialyzable fraction of the venom of the yellow scorpion, *Leiurus quinquestriatus,* to the migratory locust, *Toxicon,* **2,** 255.

Katz, N. L. and Edwards, C. (1972). The effect of scorpion venom on the neuromuscular junction of the frog, *Toxicon,* **10,** 133.

Kopeyan, C., Martinez, G., Lissitzky, S., Miranda, F., and Rochat, H. (1974). Disulfide bonds of toxin II of the scorpion *Androctonus australis* Hector, *Eur. J. Biochem.,* **47,** 483.

Koppenhöfer, E. and Schmidt, H. (1968a). Effect of scorpion venom on ionic currents of the node of Ranvier. II. Incomplete sodium inactivation, *Pfleugers Arch.,* **303,** 150.

Koppenhöfer, E. and Schmidt, H. (1968b). Incomplete sodium inactivation in nodes of Ranvier treated with scorpion venom, *Experientia,* **24,** 41.

Krup, P. A. (1966). Action of phospholipase A of the venom of *Heterometrus scaber* on human red cell ghosts and intact erythrocytes, *Naturwissenschaften,* **53,** 84.

Marinkelle, C. J. and Stahnke, H. L. (1965). Toxicological and clinical studies on *Centruroides margaritatus* (Gervais), a common scorpion in western Columbia, *J. Med. Ent.*, **2**, 197.

McIntosh, M. E. and Watt, D. D. (1972). "Purification of toxins from the North American scorpion *Centruroides sculpturatus*," in A. De Vries and E. Kochva, Eds., *Toxins of Animal and Plant Origin*, Vol. 2, Gordon and Breach, New York, pp. 529–544.

Miranda, F., Rochat, H., and Lissitzky, S. (1964a). Sur les neurotoxines de deux especes de scorpions Nord-Africains. I. Purification des neurotoxines (scorpamines) *d'Androctonus australis* (L.) et de *Buthus occitanus* (Am.), *Toxicon*, **2**, 51.

Miranda, F., Rochat, H., and Lissitzky, S. (1964b). Sur les neurotoxines de deux especes de scorpions Nord-Africains. II. Proprietes des neurotoxines (scorpamines) *d'Androctonus australis* (L.) et de *Buthus occitanus* (Am.), *Toxicon*, **2**, 113.

Miranda, F., Rochat, H., and Lissitzky, S. (1964c). Sur les neurotoxines de deux especes de scorpions Nord-Africains. III. Determinations preliminaries aux etudes de structure sur les neurotoxines (scorpamines) *d'Androctonus australis* (L.) et de *Buthus occitanus* (Am.), *Toxicon*, **2**, 123.

Miranda, F., Rochat, H., Rochat, C., and Lissitzky, S. (1966a). Complexes moléculaires présentés les neurotoxines animales. I. Neurotoxines des venins de scorpions (*Androctonus australis* Hector et *Buthus occitanus tunetanus*), *Toxicon*, **4**, 123.

Miranda, F., Rochat, H., Rochat, C., and Lissitzky, S. (1966b). Essais de purification des neurotoxines du venin d'un scorpion d'Amerique du Sud (*Tityus serrulatus* L. et M.) par des méthodes chromatographiques, *Toxicon*, **4**, 145.

Miranda, F., Kupeyan, C., Rochat, H., Rochat, C., and Lissitzky, S. (1970). Purification of animal neurotoxins: Isolation and characterization of eleven neurotoxins from the venoms of the scorpions *Androctonus australis* Hector, *Buthus occitanus tunetanus* and *Leiurus quinquestriatus quinquestriatus*, *Eur. J. Biochem.*, **16**, 514.

Mohamed, A. H. (1942). Preparation of antiscorpion serum, use of atropine and ergotoxine, *Lancet*, **2**, 364.

Mohamed, A. H. (1950). Pharmacological action of the toxin of the Egyptian scorpion, *Nature*, **166**, 734.

Mohamed, A. H., Kamel, A., and Ayobe, M. H. (1969). Studies of phospholipase A and B activities of Egyptian snake venoms and a scorpion toxin, *Toxicon*, **6**, 293.

Mohamed, A. H., Hani-Ayobe, M., Beskharoun, M. A., and El-Damarawy, N. A. (1972). Glycaemic responses to scorpion venom, *Toxicon*, **10**, 139.

Mohamed, A. H., Darwish, M. A., and Hani-Ayone, M. (1975). Immunological studies on scorpion (*B. quinquestriatus*) antivenin, *Toxicon*, **13**, 67.

Moss, J., Kazić, T., Henry, D. P., and Kopin, I. J. (1973). Scorpion venom-induced discharge of catecholamine accompanied by hypertension, *Brain Res.*, **54**, 381.

Moss, J., Toa, N. B., and Kopin, I. J. (1974). On the mechanism of scorpion toxin-induced release of norepinephrine from peripheral adrenergic neurons, *J. Pharmacol., Exp. Ther.*, **190**, 39.

Moustafa, F. A., Ahmed, Y. Y., and El-Asmar, M. F. (1974). Effect of scorpion (*Buthus minax* L. Koch) venom on succinic dehydrogenase and cholinesterase activity of mouse tissue, *Toxicon*, **12**, 237.

Nair, B. C., Nair, C., and Elliott, W. B. (1975). Action of antisera against homologous and heterologous A_2, *Toxicon*, **13**, 453.

Nair, R. B. and Kurup, P. A. (1973a). Glycosaminoglycans in the venom of the South Indian scorpion *Heterometrus scaber*, *Indian J. Biochem. Biophys.*, **10**, 133.

Nair, R. B. and Kurup, P. A. (1973b). Enzyme make-up of the venom of the South Indian scorpion, *Heterometrus scaber*, *Indian J. Biochem. Biophys.*, **10**, 230.

Nair, R. B. and Kurup, P. A. (1973c). Hyperglycemia produced by the venom of the South Indian scorpion, *Heterometrus scaber*, *Indian J. Biochem. Biophys.*, **10**, 232.

Nair, R. B. and Kurup, P. A. (1975). Venom of the South Indian scorpion *Heterometrus scaber*, *Biochim. Biophys. Acta*, **381**, 165.

Nair, R. B., Raj, R. K., and Kurup, P. A. (1973). Indole compounds of the venom of the South Indian scorpion, *Heterometrus scaber*, *Indian J. Biochem. Biophys.*, **10**, 231.

Narahashi, T., Shapiro, B. I., Deguchi, T., Scuka, M., and Wang, C. M. (1972). Effects of scorpion venom on squid axon membranes, *Am. J. Physiol.*, **222**, 850.

References

Nitzan (Tischler), M. (1970). Thermostability of the venom of the scorpion *Leiurus quinquestriatus* (H et E), *Toxicon,* **3,** 245.

Nitzan (Tischler), M. and Shulov, A. (1966). Electrophoretic patterns of the venoms of six species of Israeli scorpions, *Toxicon,* **4,** 17.

Nitzan, M., Miller-Ben-Shaul, D., and Shulov, A. (1963). Studies on the dialysable fraction of the venom of the yellow scorpion, *Leirus quinquestriatus* (H et E), *Isr. J. Exp. Med.,* **11,** 54.

Osman, O. H., Ismail, M. El-Asmar, M. F., and Ibrahim, S. A. (1972). Effect on the rat uterus of the venom from the scorpion *Leiurus quinquestriatus, Toxicon,* **10,** 363.

Osman, O. H., Ismail, M., and Wenger, T. (1973). Hyperthermic response to intraventricular injection of scorpion venom: Role of brain monoamines, *Toxicon,* **11,** 361.

Pansa, M. C., Natalizi, G. M., and Bettini, S. (1973). Effect of scorpion venom and its fractions on the crayfish stretch receptor organ, *Toxicon,* **11,** 283.

Parnas, I. and Russell, F. E. (1967). "Effects of venoms on nerve muscle and neuromuscular junction," in F. E. Russell and P. R. Saunders, Eds., *Animal Toxins,* Pergamon, Oxford, pp. 401–415.

Patterson, R. A. (1960). Physiological action of scorpion venom, *Am. J. Trop. Med. Hyg.,* **9,** 410.

Patterson, R. A. (1962). Pharmacologic action of scorpion venom on intestinal smooth muscle, *Toxicol. Appl. Pharmacol.,* **4,** 710.

Patterson, R. A. (1964). Effects of venom from the scorpion *Centruroides sculpturatus* on the rat, *Toxicon,* **2,** 167.

Pearce, F. L. (1973). Absence of nerve growth factor in the venoms of bees, scorpions, spiders, and toads, *Toxicon,* **11,** 309.

Potter, J. M. and Northey, W. T. (1962). An immunological evaluation of scorpion venoms, *Am. J. Trop. Med. Hyg.,* **11,** 712.

Rabie, F., El-Asmar, M. F., and Ibrahim, S. A. (1972). Inhibition of catalase in human erythrocytes by scorpion venom, *Toxicon,* **10,** 87.

Rao, G. P., Premalatha, B. K., Bhatt, H. V., and Haranath, P. S. R. K. (1969). Cardiac and behavioral effects of scorpion venom in experimental animals, *Indian Pediatr.,* **6,** 95.

Reddy, C. R. R. M., Suvarnakumari, G., Devi, C. S., and Reddy, C. N. (1972). Pathology of scorpion venom poisoning, *J. Trop. Med. Hyg.,* **75,** 98.

Roch-Arveiller, M., Jaillon, P., Hecklé, J., and Cheymol, G. (1974). Effets du venin de Scorpion *Androdoctonus mauretanicus* sur l'électrocardiogramme de rat, *C. R. Soc. Biol.,* **168,** 1234.

Rochat, C., Rochat, H., Miranda, F., and Lissitzky, S. (1967). Purification and some properties of the neurotoxins of *Androctonus australis* Hector, *Biochemistry,* **6,** 578.

Rochat, H., Rochat, C., Miranda, F., Lissitzky, S., and Edman, P. (1970). The amino acid sequence of neurotoxin I of *Androctonus australis* Hector, *Eur. J. Biochem.,* **17,** 262.

Rochat, H., Rochat, D., Sampieri, F., Miranda, F., and Lissitzky, S. (1972a). The amino acid sequence of neurotoxin II of *Androctonus australis* Hector, *Eur. J. Biochem.,* **28,** 381.

Rochat, H., Rochat, C., Kupeyan, C., Lissitzky, S., Miranda, F., and Edman, P. (1972b). "Structure of scorpion neurotoxins," in A. De Vries and E. Kochva, Eds., *Toxins of Animal and Plant Origin,* Vol. 2, Gordon and Breach, New York, pp. 525–528.

Rochat, C., Sampieri, F., Rochat, H., Miranda, F., and Lissitzky, S. (1972c). Iodination of neurotoxins I and II of the scorpion *Androctonus australis* Hector, *Biochimie,* **54,** 445.

Rohayem, H. (1954). Scorpion toxin and antagonisitic drugs, *J. Trop. Med.,* **56,** 150.

Romey, G., Chichoportiche, R., Lazdunski, M., Rochat, H., Miranda, F., and Lissitzky, S. (1975). Scorpion neurotoxin, a presynaptic toxin which affects both Na^+ and K^+ channels in axons, *Biochem. Biophys. Res. Commun.,* **64,** 115.

Rosin, R. (1969a). Effects of the venom of the scorpion *Nebo hierichoniticus* on white mice, other scorpions and paramecia, *Toxicon,* **7,** 71.

Rosin, R. (1969b). Sting of the scorpion *Nebo hierichonticus* in man, *Toxicon,* **7,** 75.

Rosin, R. (1972). Venom, venom effects, and poison gland of the scorpion *Nebo hierichonticus, Clienc. Cult.,* **24,** 246.

Rosin, R. (1973). Paper electrophoresis of the venom of the scorpion *Nebo hierichonticus* (Diplocentridae), *Toxicon,* **11,** 107.

Rossi, M. A., Ferreira, A. L., Paiva, S. M., and Santos, J. C. M. (1973). Myonecrosis induced by scorpion venom, *Experientia,* **29,** 1272.

Rossi, M. A., Ferreira, A. L., and Santos, J. C. M. (1974). Catecholamine-depleting effect of Brazilian scorpion (*Tityus serrulatus*) venom on adrenergic nerves of the rat atrioventricular valves, *Experientia,* **30,** 513.

Russell, F. E., Alender, C. B., and Buess, F. W. (1968). Venom of the scorpion *Vejovis spinigerus, Science,* **159,** 90.

Saunders, J., and Johnson, B. D. (1970). *Hadrurus arizonensis* venom: A new source of acetylcholinesterase, *Am. J. Trop. Med. Hyg.,* **19,** 345.

Schmitt, O. and Schmidt, H. (1972). Influence of calcium ions on the ionic currents of nodes of Ranvier treated with scorpion venom, *Pfluegers Arch.,* **333,** 51.

Selvarajan, V. R., Radhakrishna, M. C., and Swami, K. S. (1974). Influence of scorpion venom on enzyme systems of scorpion, *Heterometrus fulvipes, Curr. Sci.,* **43,** 272.

Selvarajan, V. R., Reddy, K. N., and Swami, K. S. (1975). Scorpion venom effects on succinate dehydrogenase activity of sheep tissues, *Toxicon,* **13,** 143.

Stahnke, H. L. (1963). Some pharmacological and biochemical characteristics of *Centuroides sculpturatus* Ewing scorpion venom, *Second Int. Pharmacol. Meet., Prague, Czech.,* p. 63.

Stahnke, H. L. (1965a). "Some pharmacological and biochemical characteristics of *Centuroides sculpuratus* Ewing scorpion venom," in H. W. Raudonat, Ed., *Recent Advances in Pharmacology of Toxins,* MacMillan, New York, pp. 63–70.

Stahnke, H. L. (1965b). Stress and the toxicity of venoms, *Science,* **150,** 1456.

Stahnke, H. L. (1966). Some aspects of scorpion behavior, *Bull. S. Calif. Acad. Sci.,* **65,** 65.

Stahnke, H. L. (1967). Effect of paraldehyde on scorpion and rattlesnake venom toxicity, *Southwest. Med.,* **48,** 187.

Stahnke, H. L. and Dengler, A. H. (1964). The effect of morphine and related substances on the toxicity of venoms. I. *Centrudoides sculpturatus* Ewing scorpion venom, *Am. J. Trop. Med. Hyg.,* **13,** 346.

Tafuri, W. L., Maria, T. A., Freire-Maia, L., and Cunha Melo, J. R. (1971). Effect of purified scorpion toxin on vesicular components in the myenteric plexus of the rat, *Toxicon,* **9,** 427.

Tazieff-Depierre, F. (1972). Venin de scorpion, calcium et émission d'acétylcholine par less fibres nerveuses dans l'ileon de cobaye, *C. R. Acad. Sci. Paris,* **275,** 3021.

Tazieff-Depierr, F. (1975). Cardiotoxicity in the rat of purified α-toxin isolated from *Naja nigricollis* venom and of toxins obtained from scorpion's venom, *C. R. Acad. Sci. Paris,* Ser. D, **280,** 1181.

Tazieff-Depierre, F. and Andrillon, M. P. (1973). Sécrétion d'acétylcholine provoquée par le venin de scorpion dans l'iléon de cobaye et sa suppression par la tétrodotoxine, *C. R. Acad. Sci. Paris,* **276,** 1631.

Tazieff-Depierre, F. and Czajka, M. (1969). Suppression of autorhythmic muscular activity induced by scorpion toxins by suppression of calcium release, *C. R. Acad. Sci. Paris,* Ser. D, **268,** 1228.

Tazieff-Depierre, F. and Nachon-Rautureau, C. (1975). Effect of scorpion (*Androctonus australis*) venom on a neuromuscular preparation of the frog, *C. R. Acad. Sci. Paris,* Ser. D, **280,** 1745.

Tazieff-Depierre, F., Lievremont, M., and Czajka, M. (1968). Appearance of calcium in muscle fibers due to the action of potassium chloride and toxins from scorpion (*Androctonus australis*) venom, *C. R. Acad. Sci. Paris,* Ser. D, **267,** 1477.

Tazieff-Depierre, F., Andrillon, M. P., and Goudou, D. (1973). Calcium et action spasmogène du venin de scorpion (*Androctonus australis*) sur l'iléon de cobaye, *C. R. Acad. Sci. Paris,* **276,** 2985.

Tazieff-Depierre, F., Goudou, D., and Metezeau, P. (1974). Desensitization of the guinea pig ileum to scorpion venom by variation of the ionic concentration of the external medium, *C. R. Acad. Sci. Paris,* Ser. D, **278,** 2703.

Vachon, M. (1966). Liste des scorpions connus en Egypte, Arabie, Israël, Liban, Syrie, Jordanie, Turquie, Irak, Iran, *Toxicon,* **4,** 209.

Venkateswarlu, D. and Babu, K. S. (1975). Physiological effects of scorpion venom on frog gastrocnemius muscle, *Indian J. Exp. Biol.,* **13,** 429.

Watt, D. D. (1964). Biochemical studies of the venom from the scorpion *Centuroides sculpturatus, Toxicon,* **2,** 171.

References

Watt, D. D., Babin, D. R., and Mlejnek, R. V. (1974). The protein neurotoxins in scorpion and elapid snake venoms, *J. Agr. Food Chem.,* **22,** 43.

Wheeling, C. H. and Keegan, H. L. (1972). Effects of a scorpion venom on a tarantula, *Toxicon,* **10,** 305.

Yarom, R. (1970). Scorpion venom: A tutorial review of its effects in men and experimental animals, *Clin. Toxicol.,* **3,** 561.

Yarom, R. and Braun, K. (1970). Cardiovascular effects of scorpion venom: Morphological changes in the myocardium, *Toxicon,* **8,** 41.

Yarom, R. and Braun, K. (1971a). Electron microscopic studies of the mycocardial changes produced by scorpion venom injections in dogs, *Lab. Invest.,* **24,** 21.

Yarom, R. and Braun, K. (1971b). Calcium (2+) ion changes in the myocardium following scorpion venom injections, *J. Mol. Cell. Cardiol.,* **2,** 177.

Yarom, R. and Meiri, U. (1972). Effect of scorpion venom on ultrastructure of frog sartorius muscle, *Toxicon,* **10,** 291.

Yarom, R., Gueron, M., and Braun, K. (1970). Scorpion venom cardiomyopathy, *Pathol. Microbiol.,* **35,** 114.

Yarom, R., Yallon, S., Notowitz, F., and Braun, K. (1974). Reversible myocardial damage by scorpion venom in perfused rat hearts, *Toxicon,* **12,** 347.

Zlotkin, E., Fraenkel, G., Miranda, F., and Lissitzky, S. (1971a). The effect of scorpion venom on blowfly larvae: A new method for the evaluation of scorpion venom potency, *Toxicon,* **9,** 1.

Zlotkin, E., Miranda, F., Kupeyan, C., and Lissitzky, S. (1971b). A new toxic protein in the venom of the scorpion *Androctonus australis* Hector, *Toxicon,* **9,** 9.

Zlotkin, E., Rochat, H., Kopeyan, C., Miranda, F., and Lissitzky, S. (1971c). Purification and properties of the insect toxin from the venom of the scorpion *Androctonus australis* Hector, *Biochimie,* **10,** 1073.

Zlotkin, E., Miranda, F., and Lissitzky, S. (1972a). A factor toxic to crustacean in the venom of the scorpion *Androctonus australis* Hector, *Toxicon,* **10,** 211.

Zlotkin, E., Miranda, F., and Lissitzky, S. (1972b). Proteins in scorpion venoms toxic to mammals and insects, *Toxicon,* **10,** 207.

Zlotkin, E., Fraenkel, G., Miranda, F., Lissitzky, S., and Shulov, A. (1972c). "Effect of scorpion venom on blowfly larvae," in A. De Vries and E. Kochva, Eds., *Toxins of Animal and Plant Origins,* Gordon and Breach, New York, p. 729.

Zlotkin, E. Lebovits, N., and Shulov, A. (1972d). Toxic effects of the venom of the scorpion *Scorpio maurus palmatus* (Scorpionidae), *Riv. Parassitol.,* **33,** 237.

Zlotkin, E., Menashe, M., Rochat, H., Miranda, F., and Lissitzky, S. (1975). Proteins toxic to arthropods in the venom of elapid snakes, *J. Insect Physiol.,* **21,** 1605.

29 Spider Venoms

1 BIOCHEMISTRY 485
   1.1 Toxic Protein, 485
   1.2 Enzymes, 486
       Hyaluronidase, 486
       Proteases, 486
       Other Enzymes, 487
   1.3 Nonprotein Components, 487

2 IMMUNOLOGY 488

3 TOXICOLOGY 488
   3.1 Toxicity, 488
   3.2 Clinical Symptoms, 489

4 PHARMACOLOGY 490
   4.1 Necrosis, 490
   4.2 Effect on Nerve Transmission, 490
       Cholinergic Transmission, 491
       Adrenergic and Other Effects, 493
   4.3 Effect on Cardiovascular System, 495
   4.4 Cytotoxic Action, 496

References 496

Spiders belong to the order Araneae. Systematic classification of various families is often complex, and several different classification schemes have been proposed (Cloudsley-Thompson, 1968). All spiders have fangs, and most of them have poison glands. There are only a few spiders that are dangerous to man, although many spiders can kill insects or small animals. Spiders are not insects (ants, bees, wasps); they have eight legs, whereas insects have six.

    One difficulty in spider venom research is collection of enough venom to work with. Unlike snakes, spiders yield very small amounts of venom (Table 1). Bücherl (1969) collected 12,638 specimens of *Phoneutria nigriventer*, 1964 specimens of *Loxosceles rufipes*, 2002 specimens of *L. erythrognatha*, and 219 specimens of *Lactrodectus mactans*

1 Biochemistry

Table 1 Yield of Spider Venoms

Spiders	Yield (mg/spider)	References
Lactrodectus bishopi	0.157	McCrone, 1964
L. curacaviensis	0.16	Bucherl, 1969
L. geometricus	0.097	McCrone, 1964
L. mactans mactans	0.190	McCrone, 1964
L. mactans tredecimguttatus	0.238	McCrone, 1964
L. variolus	0.254	McCrone, 1964
Loxosceles rufipes	0.13-0.27	Bucherl, 1969
L. reclusa	0.38	Morgan, 1969
Lycosa erythrognatha	0.80-2.44	Bucherl, 1969
Phoneutria nigriventer	1.60-3.15	Bucherl, 1969

for his venom research. The amounts of venom he obtained from these species were 20 g, 257 mg, 1.6 g, and 35 mg, respectively.

1 BIOCHEMISTRY

Like other venoms, spider venoms are quite complex and contain a variety of proteins and nonprotein components. However, the major biological activities reside in the protein components. For instance, the venom of the East African spider *Pterinochilus* spp. showed at least 16 components in isoelectric focusing (Perret, 1974). With a combination of gel filtration and acrylamide gel electrophoresis, a total of 26 protein bands was identified.

1.1 Toxic Protein

The toxic components of *Pterinochilus* spp. venom have a molecular weight range of 7000 to 13,000 (Perret, 1974). Necrotic and lethal toxins isolated from the venom of *Loxosceles reclusa* (Geren et al., 1973) have a molecular weight of 24,000 and a pI value of 8.3.

Necrotoxin was isolated from the venom of the Tarantula, *Dugesiella hentzi* (Lee et al., 1974). It has a molecular weight of 6700 and a pI of 10.0. The amino acid composition is as follows: 16 Lys, 2 Asp, 1 Thr, 2 Ser, 6 Glu, 4 Pro, 6 Gly, 8½ Cys, 1 Val, 5 Ile, 3 Leu, 4 Phe, and 1 Trp; There are no residues of His, Arg, Ala, Met, or Tyr.

The total amino acid residues of necrotoxin number 59, and the composition indicates that it is quite different from any known toxins of snake venoms. The purified necrotoxin is toxic to certain insects and mice, with the primary site of action on the muscle tissue in the mouse. Modification of the single tryptophan residue resulted in a loss of toxicity. Necrotoxin also increases serum creatine phosphokinase activity, another indication of muscle damage.

The principal toxic fraction of *Atrax robustus* venom is associated with a molecular weight range of 15,000 to 25,000. The toxic principle appears to be a complex of spermine (Sutherland, 1972a, b).

High molecular weight toxins were also isolated from *Loxosceles reclusa* venom (Geren et al., 1976). Both toxins 1 and 2 have isoelectric points of 8.3 and molecular weights of 34,000. Toxin 1 is a lesion-causing agent in rabbits and is lethal to mice and rabbits. Toxin 2 is not toxic to mice but is lethal to rabbits; it does not produce lesions. The toxicity of either toxin is not associated with the enzymatic activities of hyaluronidase, alkaline phosphatase, and 5'ribonucleotide phosphohydrolase, although these enzymes are found in the whole extract. It is not clear whether these lethal toxins have any neurotoxic effects.

A neurotoxin was isolated from the venom of *Latrodectus mactans tredecimguttatus*. The molecular weight of the toxin is very high and has a value of 130,000 daltons (Grasso, 1976).

1.2 Enzymes

Spider venoms contain a number of enzymes, but the presence of a particular enzyme depends on the species of spider.

Hyaluronidase This enzyme was reported in the venoms of *Phoneutria fera*, *Lycosa erythrogantha*, and *Latrodectus mactans* (McCrone, 1969). Hyaluronidase is the major constituent of *Dugesiella hentzi* venom (Schanbacher et al., 1973b). The molecular weight of venom hyaluronidase is about 39,600, and the enzyme has an isoelectric point of 6.9. The pH optimum has a rather low value of 3.5. The enzyme contains 356 amino acid residues, and its composition was determined by Schanbacher et al. (1973a, b). The amino acid composition is as follows: 48 Lys, 6 His, 9 Arg, 13½ Cys, 36 Asp, 22 Thr, 28 Ser, 32 Glu, 22 Pro, 31 Gly, 18 Ala, 17 Val, 16 Ile, 20 Leu, 13 Tyr, 15 Phe, 4 Met, and 6 Trp.

Purified hyaluronidase from *Loxosceles reclusa* venom exhibited activity against chondroitin sulfate, types A, B, and C, with optimum activity in the pH range of 5.0 to 6.6. The hyaluronidase has two components with molecular weights of 33,000 and 63,000, respectively (Wright, 1972; Wright et al., 1973).

Proteases Proteolytic enzyme activity has been detected in the venoms of *Phoneutria fera*, *Lycosa erythrogantha*, *Atrax robustus* (McCrone, 1969), and *Aphonopelma* spp. (Stahnke and Johnson, 1967). Venoms of *P. fera* and *L. erythrogantha* contain proteolytic activity toward gelatin, casein, azocoll, and denatured collagen but not toward native collagen (Kaiser and Raab, 1967).

The venom of *Pamphobetus roseus* contains three proteases when casein is used as substrate (Mebs, 1970). None of these protease fractions, however, hydrolyzed arginine esters. The enzyme activities are not inhibited by trypsin or chymotrypsin specific inhibitor. The three fractions have identical molecular weights of 10,800. Proteolytic

1 Biochemistry

enzyme activity was also detected in the venoms of *Ctenus nigriventer* and *Lycosa raptoris* (Kaiser, 1953). The three proteases isolated from *Pamphobateus roseus* venom all had molecular weights of 11,000 (Mebs, 1972). Trypsin and chymotrypsin specific inhibitors have no effect on the venom enzyme activity, indicating that spider venom protease is a different type of enzyme from its mammalian counterpart.

An absence of proteases was reported for the venoms of *Loxosceles reclusa* (Elgert et al., 1974; Wright, 1972; Wright et al., 1973) and *Latrodectus mactans* (McCrone, 1969).

Other Enzymes. Venom of *Loxosceles reclusa* (brown recluse spider) is devoid of collagenase, acetylcholinesterase, phosphodiesterase, ribonuclease, deoxyribonuclease, and phospholipase A_2 (Wright, 1972; Wright et al., 1973). Nevertheless, the venom shows esterase activities, as indicated by the hydrolysis of carbobenzoxy-L-tyrosine-*p*-nitrophenyl ester and β-naphthylacetate. Esterase was found in this venom, but lipase, catalase, acid phosphatase, alkaline phosphatase, and amylase were not detected (Elgert et al., 1974). However, Heitz and Norment (1974) reported that *L. reclusa* venom does contain an alkaline phosphatase. The enzyme activity can be destroyed by fluorescien mercuric acetate and *p*-hydroxymercuribenzoate, suggesting that a sulfhydryl group is involved in the enzyme action (Heitz and Norment, 1974).

Phosphodiesterase is present in the venoms of *Atrax robustus* and *Latrodectus mactans* (McCrone, 1969).

L-Amino acid oxidase, DNase, and RNase were not detected in the venom of *Aphonopelma* spp. (Stahnke and Johnson, 1967).

1.3 Nonprotein Components

Spider venoms are also rich in nonprotein components. Some major compounds are summarized here.

The concentration of 5-hydroxytryptamine in *Latrodectus mactans tredecimguttatus* venom is very low, 0.04 to 0.08 μg mg^{-1} of protein (Pansa et al., 1972). This compound is also present in the venoms of *Phoneutria fera* and *Lycosa erythrogantha* (McCrone, 1969). γ-Aminobutyric acid (GABA), glutamic, and aspartic acids, as well as some peptides, were identified in the venom of *Dugesiella hentzi* (Schanbacher et al., 1973a). The first compound is also present in the venoms of *Atrax robustus* and *Latrodectus mactans* (Gilbo and Coles, 1964; McCrone, 1969).

The venom of the South American bird spider also contains GABA (Fischer and Bohn, 1957). The presence of a neurohormonal transmitter in the venom may enhance its central action. Histamine was detected in the venoms of *Phoneutria fera* and *Lycosa erythrogantha* (McCrone, 1969), and the venoms of *Dugesiella hentzi* and *Aphonopelma* spp. contain ATP, ADP, and AMP (Chan et al., 1975). The ATP is the major nucleotide component and is present at a concentration of 28.1 μg μl^{-1} of *D. hentzi* venom and 56.6 μg μl^{-1} of *Aphonopelma* spp. venom; it has a synergistic toxic effect with the major toxin (necrotoxin) of *D. hentzi* venom. A nuceoside, inosine, was identified in the venom of *Loxosceles reclusa* (Geren et al. 1975), as was a guaninelike nucleotide.

Acid hydrolysis of the venom of *Atrax robustus* released spermine (Gilbo and Coles, 1964). A spermine-containing protein complex was detected in *A. robustus* venom (Sydney funnel-web spider), but it lacked toxicity (Sutherland, 1972a, b). The presence of spermine in the venom of the South American bird spider was also reported by Fischer and Bohn (1957).

2 IMMUNOLOGY

The venoms of *Latrodectus mactans mactans*, *L. variolus*, *L. bishopi*, and *L. geometricus* are very similar, and they all have essentially the same gross pathophysiological effects (McCrone, 1964; McCrone and Porter, 1965). They contain several common antigens, and an antivenin prepared against one of them will neutralize the lethal effects of all (McCrone and Netzloff, 1965). Antivenins prepared in South Africa against the venoms of *L. mactans indistinctus* and *L. geometricus* have reciprocal neutralization effects (Finlayson, 1936, 1937), and Weiner (1961) has reported that both of these South African antivenins neutralize the venom of the Australian *L. mactans hasseltii*.

Many of the venoms of closely related spider species have very similar compositions. For instance, the lethal components of *L. mactans* and *L. variolus* venoms are immunologically identical (McCrone and Hatala, 1968). Very similar results were also obtained by Smith and Micks (1968), who observed that venoms of *Loxosceles rufescens*, *L. reclusa*, and *L. laeta* had their own distinctly different immunodiffusion and immunoelectrophoretic patterns. However, they gave three identical precipitation lines common to these venoms. There is a cross-reactivity between *L. reclusa*, as tested by using human serum (Berger et al., 1973).

Venoms of *L. laeta*, *L. reclusa*, and *L. arizonica* evoke an Arthus-like reaction in the microcirculation of mouse gastroleinal mesentery (Puffer et al., 1971).

3 TOXICOLOGY

3.1 Toxicity

There is a wide range of toxicity among venoms, depending on the species of spider. Some spider venoms can be as toxic as cobra venoms, but toxicity also depends on the route of injection and on the host species. In Table 2, only the data on mice are listed.

Venom of the female *Loxosceles laeta* is more toxic than that of the male (Schenone et al., 1970). Repeated injection of the venom into rabbits produce higher antibody content, and tolerance to the toxic effect was rapidly established. This resistance lasted for at least 120 days after the last injection.

Maroli et al. (1973) investigated the LD_{50} values of *Latrodectus mactans tredecimguttatus* on amphibians, birds, insects, and mammals. The frog proved to be much more resistant to the venom than any other species tested. These results are as follows:

Animal	LD_{50} ($\mu g\ g^{-1}$)
Frog	145
Canary	4.7
Blackbird	5.9
Pigeon	0.36
Chick	2.1
Cockroach	2.7
Housefly	0.6
Guinea pig	0.08
Mouse	0.9

3 Toxicology

Table 2 Toxicities of Spider Venoms in Mice

Spiders	Route of Injection	LD_{50} ($\mu g/g$)	References
Aphonopelma sp.	—	14.4	Stahnke and Johnson, 1967
Latrodectus bishopi	i.p.	2.20	McCrone, 1964
L. geometricus	i.p.	0.43	McCrone, 1964
L. mactans mactans	i.p.	1.80	McCrone, 1964
L. mactans tredecimguttatus			
	i.p.	0.59	McCrone, 1964
	s.c.	0.68	Vicari et al, 1965
	i.v.	0.18	Vicari et al, 1965
L. variolus	i.p.	1.80	McCrone, 1964
Phoneutria fera	—	0.76	Schenberg and Pereira-Lima, 1966
Pterinochilus sp.	i.v.	8.95 (fresh)	Maretić, 1967
		8.92 (lyophilized)	Maretić, 1967

3.2 Clinical Symptoms

The symptoms of *Latrodectus mactans* envenomation were observed by McCrone and Stone (1965). The venom produces a sharp pain similar to that from a needle puncture. Usually this pain disappears rapidly, but it may persist for hours. Local muscular cramps are felt 15 min to several hours after the bite; the muscles most frequently affected are those in the thigh, shoulder, and back. Later severe pain spreads to the abdomen, and weakness and tremor are experienced. The abdominal muscles show a boardlike rigidity, respiration becomes spasmodic, and the patient is restless and anxious. This period is marked by feeble pulse, cold clammy skin, and labored breathing and speech. Convulsions, urinary retention, shock, cyanosis, neusea and vomiting, insomnia, and cold sweats also have been reported. Death may occur from the venom, depending on the victim's physical condition, his age, and the location of the bite. Adults are more resistant to the poison than are children, but individual sensitivity may enhance or depress the reaction of both age groups. Jacobs (1969) reported that envenomation by this species induced a red ring and edema around the injured area.

Loxosceles reclusa venom induces a gangrenous skin lesion (Macchiavello, 1947a, b; Atkins et al., 1958; Reed et al., 1968; Morgan 1969; Butz, 1971). On many occasions, hemolytic crisis resulting in death has been observed (Lessenden and Zimmer, 1960; Nicholson and Nicholson, 1962; Taylor and Denny, 1966). *Loxosceles reclusa* bite also causes thrombocytopenia and diffuse intravascular coagulation (Vorse et al., 1972).

Chiracanthium mildei is the most common spider in houses in the Boston area; its bite inflicts necrotizing skin lesions on human victims (Spielman and Levi, 1970). The bite of *Phidippus formosus* (jumping spider) produces pain, swelling, localized homorrhage, and an increase in skin temperature (Russell, 1970). Envenomation by *Herpyllus ecclesiasticus* causes sharp pain, pruritus, arthralgia, malaise, and nausea in human victims (Majeski and Durst, 1975).

Many other species of spiders are also implicated in bites of human beings. Russell and Waldron (1967) listed 15 species in human cases in souther California and 50 species in the United States as a whole Russell, 1973).

4 PHARMACOLOGY

Spider venoms have diverse pharmacological actions because these venoms, like snake venoms, contain many different biologically active compounds. Spider venoms have neurotoxic as well as necrotic activities. Radioactive venom distribution study indicates that spider venom has a strong affinity toward nerve tissues.

The venom of *Latrodectus mactans tredecimguttatus* was labeled with ^{32}P, and the radioactive venom distribution in guinea pig was studied by Lebez et al. (1965). A large amount of the venom remained around the site of the bite even until the death of the animal. The distribution of ^{32}P activity showed that the venom accumulates especially in the central nervous system and the peripheral nerves, and to a lesser extent in the lungs, heart, liver, and spleen.

Similar results were obtained when ^{75}Se-labeled venom of *Latrodectus mactans* was used for the study of venom distribution (Lebez et al., 1968). Most of the labeled venom remained in the snout of guinea pigs for 15 to 30 min after the bite. Less than 20% of the venom accumulated in the nervous system, with even smaller amounts occurring in the gluteal muscle, brain, spleen, and liver.

4.1 Necrosis

The bite of spiders of the genus *Loxosceles* produces local necrosis in man, guinea pig, and rabbit (Macchiavello, 1948; Micks and Smith, 1963). The lesion begins with edema and erythema and progressively develops a black eschar. The reported histopathologic changes in this lesion consist of edema and thickening of the endothelium of blood vessels, collection of inflammatory cells, vasodilation, intravascular clotting, degeneration of blood vessel walls, and hemorrhage into the dermis. The accumulation of polymorphonuclear leucocytes is especially marked, and within 3 to 5 days liquefaction and abscess formation are noted (Atkins et al., 1958; Lessenden and Zimmer, 1960; Pizzi et al., 1957; Smith and Micks, 1970).

Only one of many components of *Loxosceles reclusa* venom caused necrosis, and the effect was not prevented by the use of EDTA (Elgert et al., 1974).

4.2 Effect on Nerve Transmission

Practically all of the research to date has been done using the venom of the black widow spider, *Latrodectus mactans*, which is very common in the United States. There have also been several studies using the Australian red back spider, *L. mactans hasselti*; the results were the same as those obtained with *L. mactans*. It is known that the spider venom causes destruction of the presynaptic nerve terminals, both cholinergic and adrenergic. The venom of *L. mactans tredecimguttatus* and its fractions increase the release of acetylcholine from rat brain cortex slices, cause the disappearance of the histochemical reaction to catecholamines in adrenergic fibers of rat iris, decrease the amplitude of preganglionically evoked action potentials recorded postganglionically from rat superior cervical ganglion, and elicit the appearance of asynchronous postganglionic action potentials recorded from unstimulated ganglion (Granata et al., 1972). It has been suggested that the venom or its fractions affect neurotransmitter release from cholinergic and possibly adrenergic nerve terminals (Frontali, 1972; Frontali et al., 1973).

A purified toxin (the B_5 fraction) of *L. mactans* venom interacts irreversibly with lipid bilayer membrane to form cation selective channels (Finkelstein et al., 1976).

Cholinergic Transmission. *Latrodectus mactans* venom accelerates the spontaneous release of acetylcholine into *Torpedo* electroplax, thereby increasing the rate of discharge of miniature synaptic potentials (Walther, 1974; Palmer, 1975). The venom of *L. mactans tredecimguttatus* gave the same result (Granata et al., 1974). There was, however, no observed tissue damage in ultrastructural study.

Neri et al. (1965) investigated the effect of *L. mactans* venom and its toxic fraction on the thoracic ganglia and cercal nerve preparation of the cockroach *Periplaneta americana*. A permanent block of endogenous activity of the thoracic ganglia was observed with crude venom and its toxic fraction. A tenfold concentration was required to block the cercal nerve preparations.

Latrodectus mactans venom blocks the activity of the stretch receptor neuron of the crayfish, modifying and inhibiting the impulse frequency (Grasso and Paggi, 1967). This effect is not due to the possible presence of γ-aminobutyric acid in the venom, since picrotoxin does not prevent the venom action. It is believed that the venom is toxic at the presynaptic terminal (Palmer, 1975).

D'Ajello et al. (1969) found that the venom of *L. mactans tredecimguttatus* is capable of blocking transmission in the sixth abdominal ganglion of the cockroach, while the axonal conduction is not affected. They concluded that the venom depolarizes only the soma-dendritic region of the neuron. Thus the venom blocks the response to presynaptic stimulation, whereas the directly evoked response is unaffected. A presynaptic neurotoxic effect of *Latrodectus geometricus* venom was also observed (Del Castillo and Pumplin, 1975).

D'Ajello et al. (1971) conducted further investigations of spider venom and found that it induces the following effects:

1. A progressive decline of the resting potential of the giant neurons of the cockroach.
2. Failure of synaptic transmission through a progressive reduction of e.p.s.p. (depolarizing prepotential) in the ganglion.
3. Spontaneous repetitive firing of the giant neurons.
4. Alterations in the shape and reduction in the size of the propagated action potential.

It seems that the venom causes a massive release of transmitter, which in turn depolarizes the postsynaptic membrane. Essentially the same result was obtained by Griffiths and Smyth (1973), who concluded that both the excitatory and the inhibitory neuromuscular junctions of the cockroach are presynaptic. As indicated by postjunctional miniature excitatory and inhibitory end-plate potentials, the rate of spontaneous transmitter release increases greatly after venom application and then decreases to zero, at which time the junctions are permanently blocked.

The venom causes exhaustion of miniature end-plate activity and depletes the nerve terminal of vesicles (Longenecker et al., 1970). The venom may react with the nerve terminal membrane, thereby inducing the release of transmitter. This release of transmitter occurs independently of the presence of Ca(II) and independently of the depolarization of the terminal. Electron microscopy of the poisoned nerve-muscle junction shows a sequence of motor nerve-ending damage that culminates in disruption of the prejunctional membrane and loss of all organelles, including synaptic vesicles (Okamoto et al., 1971). The postjunctional membrane was morphologically unaffected. Thus the damage to motor nerve endings by black widow spiders was clearly demonstrated. This agrees well with the morphological changes observed by Clark et al. (1970), which indicated complete disappearance of the synaptic vesicles from the endplates.

Addition of *Latrodectus mactans hasselti* venom to guinea pig ileum increases the amplitude of the inhibitory junction potentials produced in circular muscle cells by transmural electrical stimulation (Einhorn and Hamilton, 1974). Electron microscopic examination indicates that the venom affects the nerves of the circular muscle layer of ileum.

Paggi and Rossi (1971) studied the effect of *Latrodectus* venom on isolated *in vitro* superior cervical ganglion of the rat. The venom depressed the postganglionic action potentials evoked by preganglionic stimulation, while it produced discharges of asynchronous action potentials from the unstimulated ganglion. Previous curarization effectively interfered with the production of these discharges. This can be explained by the fact that the venom enhances the release of acetylcholine at nerve endings.

When the rat cerebral cortex slices were incubated with *L. mactans tredecimguttatus* venom, the content of acetylcholine released into the medium increased twofold, with a corresponding decrease in the acetylcholine content of the tissue (Frontali et al., 1972). It was suggested that neurotransmitter depletion of cholinergic nerve terminals may help to explain the intoxication symptoms in man and other vertebrates.

Kawai et al. (1972) studied the effect of *L. mactans* venom on lobster nerve–muscle preparation and concluded that the venom has a presynaptic action at the crustacean neuromuscular junction. Both excitatory and inhibitory postsynaptic potentials (e.p.s.p. and i.p.s.p.) were first augmented and then suppressed. The frequency of miniature potentials was markedly increased by the venom. Summated postsynaptic conductance changes appeared to be responsible for the membrane depolarization and the decrease in effective membrane resistance seen in the early stages of the venom action. In the later stages both excitatory and inhibitory "giant miniature potentials" were evoked. No changes were discerned in the reversal potential of the e.p.s.p. and i.p.s.p. and in the sensitivity of the postsynaptic membrane.

Latrodectus mactans venom causes an increase in the frequency of miniature end-plate potentials and a reduction in the number of synaptic vesicles in the nerve terminal in the frog neuromuscular junction (Clark et al., 1972). Shortly after the increase in the miniature end-plate potential frequency, the presynaptic membrane of the nerve terminal has either infolded or "lifted." Examination of these infoldings or lifts reveals synaptic vesicles in various stages of fusion with the presynaptic membrane. After the supply of synaptic vesicles has been exhausted, the presynaptic membrane returns to its original position directly opposite the end-plate membrane.

A protein fraction that causes the depletion of synaptic vesicles at the neuromuscular junction was obtained from the venom of *L. mactans* (Frontali et al., 1976). It produces an increase in the frequency of miniature end-plate potentials at the frog neuromuscular junction and swelling of the nerve terminals with depletion of the synaptic vesicles. The fraction consists of four immunologically indistinguishable protein components with similar molecular weights of 130,000. The isoelectric points range from 5.2 to 5.5. The fraction contains no carbohydrate, lipolytic, or proteolytic activity.

In vitro experiments on rat superior cervical ganglion indicate that the venom of *L. mactans* blocks neurotransmission by releasing, and thus depleting, nerve-ending acetylcholine (Paggi and Toschi, 1972). Venom-mediated release of presynaptic acetylcholine persisted in a Ca(II)-free medium, whereas presynaptic acetylcholine release evoked by nerve impulses or high potassium is blocked in the absence of Ca(II).

The nerve terminals of ganglia incubated with *L. mactans* venom for 3 hrs showed characteristic changes; their axoplasm appeared electron-lucent, and only a few synaptic vesicles were encountered. The other constituents of the ganglion, namely, the nerve cells and processes, as well as the satellite cells, were not damaged (Chmouliovsky et al., 1972).

Prior treatment of neuromuscular junctions with botulinum toxin does not prevent the complete depletion of vesicles caused by *L. mactans* venom (Pumplin and Del Castillo, 1975). The creatine phosphokinase activity of ganglia homogenate was inhibited when it was incubated with *L. mactans* venom (Chmouliovsky et al., 1972). The time course of the enzyme inhibition was parallel to that of the fall of synaptic transmission. Neither the action potential nor the enzyme activity was restored by prolonged washing with physiological salt solution.

The effect of *L. mactans* venom on squid giant axon was studied by Gruener (1973), who found that axon exposed to the venom had a complete but reversible loss of excitability. The complete blockade of the action potential is a rapid event preceded by a marked shortening of the duration of the action potential. This decrease in duration has been shown to be due primarily to a large increase in the rate of membrane repolarization, concomitant with a smaller increase in the rate of membrane depolarization.

The venom of *L. mactans* has a relatively selective effect in reducing the amplitude of the second of the two spikes in the compound ganglionic action potential (Alkadhi and McIssac, 1974).

It is well known that botulinum toxin blocks the neuromuscular junction (Burgen et al., 1948). Since *L. mactans* venom increases the spontaneous output of acetylcholine quanta at the neuromuscular junction (Longenecker et al., 1970), it may counter the neurotoxic effect of botulinum poisoning. This actually occurred in a case where *L. mactans* venom treatment increased the survival time in botulinum toxin poisoning (Stern and Valjevac, 1972; Stern et al., 1975).

Adrenergic and Other Effects. Frontali (1972) demonstrated that the venom also produces catecholamine-depleting effect in iris nerve fibers. She therefore concluded that the neurotransmitter-depleting action of black widow spider venom on nerve terminals is not specific for cholinergic but extends also to adrenergic nerves in rat iris. Contraction of the nictitating membrane of the cat and midriasis could be due to the release of noradrenaline from peripheral nerve endings (Sampayo, 1944).

Frontali et al. (1973) extended their study of the effects of venom on adrenergic nerve fibres and terminals of different mammalian organs. Fluorescence observed in guinea-pig mesentery, iris, and spleen capsule disappeared when these tissues were incubated with the venom (Fig. 1). Thus the venom affects not only cholinergic nerve terminals but also various adrenergic nerve fibers and terminals of several mammalian organs.

The sympathomimetic effects caused by the venom of *Lactrodectus mactans* may

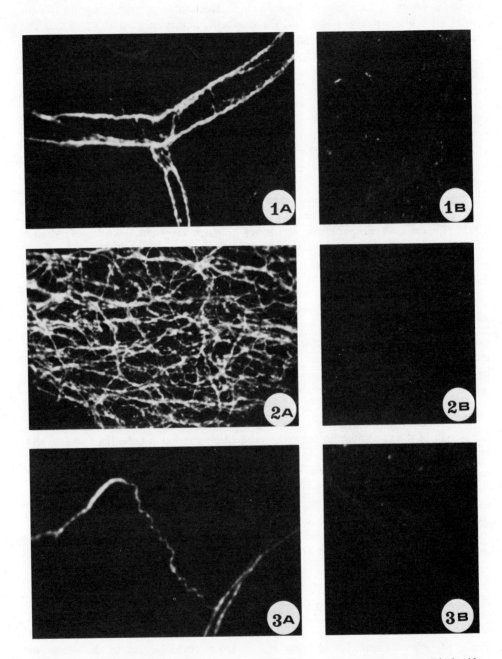

Figure 29.1 Catecholamine-depleting effect of *Latrodectus mactans tredecimguttatus* (black widow spider) venom on fibers innervating different guinea-pig tissues. (1) Mesenter. (2) Iris. (3) Spleen capsule. (A) In absence of venom. (B) In presence of venom. Note that the disappearance of fluorescence in complete or almost complete in the presence of venom. (Photographs kindly supplied by Dr. N. Frontali and originally published in Frontali et al., 1973.)

occur not only through ganglionic stimulation but through stimulation of the peripheral noradrenergic nerve terminals as well.

The venom of the Australian red back spider, *L. mactans hasselti*, causes depletion of the vesicles within the adrenergic nerve terminals of mouse vas deferens, and also increases the frequency of spontaneous excitatory junction potentials in this tissue (Einhorn and Hamilton, 1973).

Cull-Candy et al. (1973) examined the effect of the venom on a noncholinergic synapse, the insect nerve–muscle synapse where transmission is glutamatergic. An electron microscopic study indicated that locust nerve–muscle synapse had, after treatment with the venom, a paucity of synaptic vesicles, and there were numerous irregularly-shaped membrane-bound structures. The miniature discharge from these muscle fibers became very irregular, with periods of high frequency separated by periods of nearby normal frequency. The cyclic nature of these changes may have been due to the complex multiterminal innervation of insect muscle fibers, the venom acting variable delays and to variable extents at different synaptic sites.

Pretreatment of isolated spleens strips of the cat with *L. mactans* venom induced a supersensitivity to noradrenaline of similar magnitude to that produced by cocaine or surgical denervation (Pinto and Rothlin, 1974). It was concluded that the venom-induced supersensitivity was the result of an irreversible impairment of the presynaptic neural uptake of amines. This effect appears to be due to an actue *in vitro* degeneration of adrenergic nerve terminals.

In cardiac atria whose adrenergic nerve terminals were preloaded with 1-[^{3}H] noradrenaline, *L. mactans* venom produced an acceleration of tritium outflow (Pinto et al., 1974). Greater than 80% of the outflow was tritium-noradrenaline, and only a small percentage correspond to metabolites of noradrenaline. Tetrodotoxin abolished the positive chronotropic action induced by cardiac nerve stimulation but failed to antagonize the venom's rate-increasing effect. The conclusion was that the venom acted at adrenergic presynaptic sites and released noradrenaline, which subsequently activated β adrenoceptors in the guinea pig atria.

Einhorn and Hamilton (1973, 1974) further studied the effect of the venom on the purinergic nervous system of guinea pig ileum (Burnstock, 1972). The venom had an inhibitory effect on nerve impulse transmission. Electron microscopic investigation also indicated that the nerve bundle developed dilated electron-lucent varicosites containing few vesicles.

The sympathomimetic effects of *L. mactans* venom were investigated by Dagrosa et al. (1973). The venom induced contractile activity in the nictating membrane; both effects were blocked by phentolamine. Acute denervation of the nictating membrane did not affect the sympathomimetic response, but chronic denervation delayed its appearance, decreased its magnitude, and shortened its duration. The venom behaved as an agonist toward the isolated splenic capsule of the cat, its action being antagonized by blockade of the α receptors.

4.3 Effect on Cardiovascular System

As reviewed in the preceding section, spider venoms have a well-documented neurotoxic effect. The effect of venoms, in general, on the cardiovascular system have not been extensively studied. There is some evidence, however, that spider venoms influence the cardiovascular system.

Electrocardiographic tracings from four patients suffering from envenomation by

Latrodectus mactans tredecimguttatus showed sinus bradycardia, high P_2 and P_3 waves, prolongation of the QT interval, depression of the ST segment, and low to negative T waves (Maretić, 1963). In guinea pigs, supraventicular extrasystoles, alterations in atrioventricular conductivity, changes in the ST segment and T interval, prolongation of the PQ interval, and Lucciani-Wenckebach's periods were present. Tachycardia and bradycardia were also observed. Majori et al. (1972) noted that *L. mactans* venom blocked the cockroach heartbeat. The heart block may be due to impairment of the cardiac nerve ganglia function, the myocardial neuromuscular junctions, or both.

All North American spiders of the genus *Latrodectus* (*L. mactans, L. dectus, L. variolus, L. bishopi, L. geometricus*) have venoms that exert a hypertensive effect on the mammalian systemic arterial system (McCrone and Porter, 1965). *Latrodectus mactans* venom has been shown to cause dilation of capillaries and venules in mesentery of mice (Puffer and Warner, 1972). The venom of *Steatoda paykullina* (European black widow) changes the electrocardiogram pattern (Maretić et al., 1964).

Venoms of *Loxosceles laeta, L. reclusa,* and *L. arizonica* produce intestinal hemorrhage, leucocyte adhesion to vessel walls, red cell aggregation, edema, and changes in the diameter of blood vessels (Puffer et al., 1971, 1973).

4.4 Cytotoxic Action

The venom of *Latrodectus mactans tredecimguttatus* exerts a cytotoxic action on KB and Aminion cells. The fraction having cytotoxic activity is also toxic to houseflies and mice (Vicari et al., 1965). Since the original venom does not contain proteases, the cytotoxic activity is not associated with proteolytic activity.

The brown recluse spider, *Loxosceles reclusa*, native to the southeastern and central United States, can produce a necrotic lesion at the site of envenomation. This venom has a lytic action on fat and muscle tissue of the tobacco budworm lava, *Heliothis virenscens* (Norment and Vinson, 1969). Such lesions are also produced in man by intoxication with *L. aleta*, the Chilean spider (Pizzi et al., 1957; Schenone and Prats, 1961).

REFERENCES

Alkadhi, K. A. and McIsaac, R. J. (1974). Differential blockade of ganglionic transmission by extract from venom gland of black widow spider (*Latrodectus mactans*), *Toxicon,* **12,** 643.

Atkins, J. A., Wingo, C. W., Sodeman, W. A., and Wynn, J. E. (1958). Necrotic arachnidism, *Am. J. Trop. Med. Hyg.,* **7,** 165.

Berger, R. S. Millikan, L. E., and Conway, F. (1973). An *in vitro* test for *Loxosceles reclusa* spider bites, *Toxicon,* **11,** 465.

Bücherl, W. (1969). Biology and venoms of the most important South American spiders of the genera *Phoneutria, Loxosceles, Lycosa,* and *Latrodectus, Am. Zool.,* **9,** 157.

Burgen, A. S. V., Dickens, F., and Zatman, L. J. (1948). The action of botulinum toxin on the neuromuscular junction, *J. Physiol. (London),* **109,** 10.

Burnstock, G. (1972). Purinergic nerves, *Pharmacol. Rev.,* **24,** 509.

Butz, W. C. (1971). Envenomation by the brown recluse spider (Aranae, Scytodidae) and related species: A public health problem in the United States, *Clin. Toxicol.,* **4,** 515.

Chan, T. K., Geren, C. R., Howell, D. E., and Odell, G. V. (1975). Adenosine triphosphate in tarantula spider venoms and its synergistic effect with the venom toxin, *Toxicon,* **13,** 61.

Chmouliovsky, M., Dunant, Y., Graf, J., Straub, R. W., and Rufener, C. (1972). Inhibition of creatine phosphokinase activity and synaptic transmission by black widow spider venom, *Brain Res.,* **44,** 289.

References

Clark, A. W., Mauro, A., Lonenecker, H. E., and Hurlbut, W. P. (1970). Effects on the fine structure of the frog neuromuscular junction, *Nature,* 225, 703.

Clark, A. W., Hurlbut, W. P., and Mauro, A. (1972). Changes in the fine structure of the neuromuscular junction of the frog caused by black widow spider venom, *J. Cell. Biol.,* 52, 1.

Cloudsley-Thompson, J. L. (1968). *Spiders, Scorpions, Centipedes and Mites,* Pergamon, Oxford.

Cull-Candy, S. G., Neal, H., and Usherwood, P. N. R. (1973). Action of black widow spider venom on an aminergic synapse, *Nature,* 241, 353.

Dagrosa, E. E., Rothlin, R. P., Pinto, J. E. B., and Barrio, A. (1973). Sympathomimetic effects of *Latrodectus mactans* venom, *Rev. Soc. Argent. Biol.,* 48–49, 16.

D'Ajello, V., Mauro, A., and Bettini, S. (1969). Effect of the venom of the black widow spider, *Latrodectus mactans tredecimguttatus,* on evoked action potentials in the isolated nerve cord of *Periplaneta americana, Toxicon,* 7, 139.

D'Ajello, V., Magni, F., and Bettini, S. (1971). The effect of the venom of the black widow spider, *Latrodectus mactans tredecimguttatus* on the giant neurones of *Periplaneta americana, Toxicon,* 9, 103.

Del Castillo, J. and Pumplin, D. W. (1975). Discrete and discontinuous action of brown widow spider venom on presynaptic nerve terminals of frog muscle, *J. Physiol.,* 252, 491.

Einhorn, V. F. and Hamilton, R. (1973). Transmitter release by red back spider venom, *J. Pharm. Pharmacol.,* 25, 824.

Einhorn, V. F. and Hamilton, R. C. (1974). Red back spider venom and inhibitory transmission. *J. Pharm. Pharmacol.,* 26, 748.

Elgert, K. D., Ross, M. A., Campbell, B. J., and Barrett, J. T. (1974). Immunological studies of brown recluse spider venom, *Insect Immun.,* 10, 1412.

Finkelstein, A., Rubin, L. L., and Tzeng, M. C. (1976). Black widow spider venom: Effect of purified toxin on lipid bilayer membranes, *Science,* 193, 1009.

Finlayson, M. H. (1936). "Knoppie-spider" antivenene, *S. Afr. Med. J.,* 10, 735.

Finlayson, M. H. (1937), Specific antivenene in treatment of "knoppie-spider" bite, *S. Afr. Med. J.,* 11, 163.

Fischer, F. G. and Bohn, H. (1957). Die Giftsekrete der Volgelspinnen, *Liebigs Ann. Chem.,* 603, 232.

Frontali, N. (1972). Catecholamine depleting effect of black widow spider venom on iris nerve fibres, *Brain Res.,* 37, 146.

Frontali, N., Granata, F., and Parisi, P. (1972). Effects of black widow spider venom on acetylcholine release from rat cerebral cortex slices *in vitro, Biochem. Pharmacol.,* 21, 969.

Frontali, N., Granata, F., Traina, M. E., and Bellino, M. (1973). Catecholamine depleting effect of black widow spider venom on fibers innervating different guinea-pig tissues, *Experientia,* 29, 1525.

Frontali, N., Ceccarelli, B., Gorio, A., Mauro, A., Siekevits, P., Tzeng, M. C., and Hurlbut, W. P. (1976). Purification from black widow spider venom of a protein factor causing the depletion of synaptic vesicles at neuromuscular junctions, *J. Cell Biol.,* 68, 462.

Geren, C. R., Chan, T. K., Ward, B. C., Howell, D. E., Pinkston, K., and Odell, G. V. (1973). Composition and properties of extract of fiddleback (*Loxosceles reclusa*) spider venom apparatus, *Toxicon,* 11, 471.

Geren, C. R., Chan, T. K., Howell, D. E., and Odell, G. V. (1975). Partial characterization of the low molecular weight fractions of the extract of the venom apparatus of the brown recluse spider and of its hemolymph, *Toxicon,* 13, 233.

Geren, C. R., Chang, T. K., Howell, D. E., and Odell, G. V. (1976). Isolation and characterization of toxins from brown recluse spider, *Arch. Biochem. Biophys.,* 174, 90.

Gilbo, C. M. and Coles, N. W. (1964). Components of the venom of the female Sydney funnel web spider, *Atrax robustus, Aust. J. Biol. Sci.,* 17, 758.

Granata, F., Paggi, P., and Frontali, N. (1972). Effects of chromatographic fractions of black widow spider venom on *in vitro* biological systems, *Toxicon,* 10, 551.

Granata, F., Traina, M. E., Frontali, N., and Bertolini, B. (1974). Effects of black widow spider venom on acetylcholine release from *Torpedo* electric tissue slices and subcellular fractions *in vitro, Comp. Biochem. Physiol.,* 48, 1.

Grasso, A. (1976). Preparation and properties of a neurotoxin purified from the venom of black widow spider (*Latrodectus mactans tredecimguttatus*), *Biochim. Biophys. Acta*, **439**, 406.

Grasso, A. and Paggi, P. (1967). Effect of *Latrodectus mactans tredecimguttatus* venom on the crayfish stretch receptor neurone, *Toxicon*, **5**, 1.

Griffiths, D. J. G. and Smyth, T. (1973). Action of black widow spider venom at insect neuromuscular junctions, *Toxicon*, **11**, 369.

Gruener, R. (1973). Excitability blockade of the squid giant axon by the venom of *Latrodectus mactans* (black widow spider), *Toxicon*, **11**, 155.

Heitz, J. R. and Norment, B. R. (1974). Characteristics of an alkaline phosphatase activity in brown recluse venom, *Toxicon*, **12**, 181.

Jacobs, W. (1969). Possible peripheral neuritis following a black widow spider bite, *Toxicon*, **6**, 299.

Kaiser, E. (1953). The enzymatic activity of spider venom, *Mem. Inst. Butantan*, **25**, 35.

Kaiser, E. and Raab, E. (1967). Collagenolytic activity of snake and spider venoms, *Toxicon*, **4**, 251.

Kawai, N., Mauro, A., and Gundfest, H. (1972). Effect of black widow spider venom on the lobster neuromuscular junctions, *J. Gen. Physiol.*, **60**, 650.

Lebez, D., Maretić, Z., and Kristan, J. (1965). Studies on labeled animal poisons. I Distribution of P^{32}-labeled *Latrodectus tredecmguttatus* venom in the guinea pig, *Toxicon*, **2**, 251.

Lebez, D., Maretić, Z., Gubensek, F., and Kristan, J. (1968). Studies on labeled animal posions. IV. Incorporation of selenium-75 and phosphorus-32 in spider venoms, *Biol. Vestn.*, **16**, 11.

Lee, C. K., Chan, T. K., Ward, B. C., Howell, D. E., and Odell, G. V. (1974). The purification and characterization of a necrotoxin from tarantula, *Dugesiella hentzi* (Girard), venom, *Arch. Biochem. Biophys.*, **164**, 341.

Lessenden, C. M. and Zimmer, L. K. (1960). Brown spider bites, *J. Kans. Med. Soc.*, **61**, 379.

Longenecker, H. E., Hurlbut, W. P., Mauro, A., and Clark, A. W. (1970). Effects of black widow spider venom on the frog neuromuscular junction, *Nature*, **225**, 701.

Macchiavello, A. (1947a). Cutaneous arachnoidism or gangrenous spot of Chile, Puerto Rico, *J. Public Health Trop. Med.*, **22**, 425.

Macchiavello, A. (1947b). Cutaneous arachnoidism experimentally produced with the glandular poison of *Loxoscèles laeta*, Puerto Rico *J. Public Health Trop. Med.*, **23**, 266.

Macchiavello, A. (1948). Cutaneous arachnoidism experimentally produced with the glandular poison of *Loxoscèles laeta*, Puerto Rico *J. Public Health Trop. Med.*, **23**, 226.

Majeski, J. A. and Durst, G. G. (1975). Bite by the spider *Herpyllus ecclesiasticus* in South Carolina, *Toxicon*, **13**, 277.

Majori, G., Bettini, S., and Casaglia, O. (1972). Effect of black widow spider venom on the cockroach heart, *J. Insect Physiol.*, **18**, 913.

Maretić, Z. (1963). Electrocardiographic changes in man and experimental animals provoked by the venom of *Latrodectus tredecimguttatus*, *Toxicon*, **1**, 127.

Maretić, Z. (1967). "Venom of an East African orthognath spider," in F. E. Russell and P. R. Saunders, Eds., *Animal Toxins*, Pergamon, Oxford, pp. 23–28.

Maretić, Z., Levi, H. W., and Levi, L. R. (1964). The theridiid spider *Steatoda paykulliana*, poisonous to mammals, *Toxicon*, **2**, 149.

Maroli, M. Bettin, S., and Panfili, B. (1973). Toxicity of *Latrodectus mactans tredecmguttatus* venom on frog and birds, *Toxicon*, **11**, 203.

McCrone, J. D. (1964). Comparative lethality of several *Latrodectus* venoms, *Toxicon*, **2**, 201.

McCrone, J. D. (1969). Spider venoms: Biochemical aspects, *Am. Zool.*, **9**, 153.

McCrone, J. D. and Hatala, R. J. (1968). Serological relationship of the lethal components of two black widow spider venoms, *Toxicon*, **6**, 65.

McCrone, J. D. and Netzloff, M. (1965). An immunological and electrophoretical comparison of the venoms of the North American *Latrodectus* spiders, *Toxicon*, **3**, 107.

McCrone, J. D. and Porter, R. J. (1965). Hypertensive effect of *Latrodectus* venoms, *Q. J. Florida Acad. Sci.*, **27**, 307.

McCrone, J. D. and Stone, K. J. (1965). The widow spiders of Florida, *Arthropods of Florida and Neighbouring Areas*, **Vol. 2**, pp. 1–4, Florida Department of Agriculture, Gainesville, Florida.

References

Mebs, D. (1970. Proteolytic activity of spider poison, *Naturwissenschaften,* **57,** 308.

Mebs, D. (1972). "Proteolytic activity of a spider venom," in A. De Vries and E. Kochva, Eds., *Toxins of Animal and Plant Origin,* Gordon and Breach, New York, pp. 493–497.

Micks, D. W. and Smith, C. W. (1963). A compartive study of brown spider bite, *Bull. Entomol. Soc. Am.,* **9,** 174.

Morgan, P. N. (1969). Preliminary studies on venom from the brown recluse spider, *Lozosceles reclusa, Toxicon,* **6,** 161.

Neri, L. Bettini, S., and Frank, M. (1965). The effect of *Latrodectus mactans tredecimguttatus* venom on the endogenous activity of *Periplaneta americana* nerve cord, *Toxicon,* **3,** 95.

Nicholson, J. F. and Nicholson, B. H. (1962). Hemolytic anemia from spider bites (*necrotic arachnidism*), *J. Okla. State Med. Assoc.,* **55,** 234.

Norment, B. R. and Vinson, S. B. (1969). Effect of *Loxosceles reclusa gertsch* and *Mulaik* venom on *Heliothis virescens* (F.) larvae, *Toxicon,* **7,** 99.

Okamoto, M., Lonenecker, H. E., Riker, W. F., and Song, S. K. (1971). Destruction of mammalian motor nerve terminals by black widow spider venom, *Science,* **172,** 733.

Paggi, P. and Rossi, A. (1971). Effect of *Latrodectus mactans tredecimguttatus* venom on sympathetic ganglion isolated *in vitro, Toxicon,* **9,** 265.

Paggi, P. and Toschi, G. (1972). Effects of denervation and lack of calcium on the action of *Latrodectus* venom on rat sympathetic ganglion, *Life Sci.,* **11,** 413.

Palmer, M. F. (1975). Pharmacology of *Latrodectus mactans* venom, *Gen. Pharmacol.,* **6,** 325.

Pansa, M. C., Natalizi, G. M. and Bettini, S. (1972). 5-Hydroxytryptamine content of *Latrodectus mactans tredecimguttatus* venom from gland extracts, *Toxicon,* **10,** 85.

Perret, B. A. (1974). The venom of the East African spider, *Toxicon,* **12,** 303.

Pinto, J. E. B. and Rothlin, R. P. (1974). Presynaptic adrenergic supersensitivity induced by crude *Latrodectus mactans* venom, *Toxicon,* **12,** 535.

Pinto, J. E. B., Rothlin, R. P., and Dagrosa, E. E. (1974). Noradrenaline release by *Latrodectus mactans* venom in guinea-pig atria, *Toxicon,* **12,** 385.

Pizzi, T., Zacarias, J., and Schenone, F. (1967). Estudio histopatologico experimental en el enveneaminento por *Loxosceles laeta, Biologica,* **23,** 33.

Puffer, H. W. and Warner, N. E. (1972). Effect of animal toxins on the microcirculation, *Proc. West. Pharmacol. Soc.,* **15,** 27.

Puffer, H. W., Warner, N. E., Russell, F. E., and Meadows, P. (1971). Effect of toxins on the microcirculation. 1. Venom of selected species of the spider *Loxosceles, Eur. Conf. Microcirc.,* p. 142.

Puffer, H. W., Parker, J. W., Russell, F. E., and Warner, N. E. (1973). "Pathology of *Loxosceles laeta* venom poisoning in the rabbit," in Kaiser, E., Ed., *Animal and Plant Toxins,* Wilhelm Goldmann, Munich, pp. 123–127.

Pumplin, D. W. and Del Castillo, J. (1975). Release of packets of acetylcoholine and synaptic vesicles elicited by brown spider venom in frog motor nerve endings poisoned by botulinum toxin, *Life Sci.,* **17,** 137.

Reed, H. B., Hackman, R. H., and Fesmire, F. M. (1968). Variation in severity of loxoscelisms, *J. Tenn. Med. Assoc.,* **61,** 1097.

Russell, F. E. (1970). Bite by the spider *Phidippus formosus:* Case history, *Toxicon,* **8,** 193.

Russell, F. E. (1973). Venomous animal injuries, *Curr. Probl. Pediatr.,* **3,** 1.

Russell, F. E. and Waldron, W. G. (1967). Spider bites, tick bites, *Calif. Med.,* **106,** 248.

Sampayo, R. R. L. (1944). Pharmacological action of the venom of *Latrodectans mactans* and other *Latrodectus* spiders, *J. Pharmacol. Exp. Ther.,* **80,** 309.

Schanbacher, F. L., Lee, C. K., Hall, J. E., Wilson, I. B., Howell, D. E., and Odell, G. V. (1973a). Composition and properties of *Tarantula dugesiella hentzi* (Girard) venom, *Toxicon,* **11,** 21.

Schanbacher, F. L., Lee, C. K., Wilson, I. B., Howell, D. E., and Odell, G. V. (1973b). Purification and characterization of tarantula, *Dugesiella hentzi* (Girard), venom hyaluronidase, *Comp. Biochem. Physiol.,* **44B,** 389.

Schenberg, S. and Pereira Lima, F. A. (1966). Pharmacology of the polypeptides from the venom of the spided *Phoneutrai fera, Mem. Inst. Butantan,* **33,** 627.

Schenone, H. and Prats, F. (1961). Arachnidism by *Loxosceles laeta*, *Arch. Dermatol.*, **83**, 139.

Schenone, H., Courtin, L., and Knierim, F. (1970). Resistencia inducida del conejo a dosis elevadas de veneno de *Loxosceles laeta*, *Toxicon*, **8**, 285.

Smith, C. W. and Micks, D. W. (1968). A comparative study of the venom and other components of three species of *Loxosceles*, *Am. J. Trop. Med. Hyg.*, **17**, 651.

Smith, C. W. and Micks, D. W. (1970). The role of polymorphonuclear leukocytes in the lesion caused by the venom of the brown spider, *Loxosceles reclusa*, *Lab. Invest.*, **22**, 90.

Spielman, A. and Levi, H. W. (1970). Probable envenomation by *Chiracanthium mildei*, a spider found in houses, *Am. J. Trop. Med. Hyg.*, **19**, 729.

Stahnke, H. L. and Johnson, B. D. (1967). "*Aphonopelma* tarantula venom," in F. E. Russell and P. R. Saunders, Eds., *Animal Toxins,* Pergamon, Oxford, pp. 35–39.

Stern, P. and Valjevac, K. (1972). Beitrag zur Therapie der Botulinus-Intoxikation, *Arch. Toxicol.*, **28**, 302.

Stern, P., Valjevac, K., Dursum, D., and Ducic, V. (1975). Increased survival time in botulinum toxin poisoning by treatment with a venom gland extract from the black widow spider, *Toxicon*, **13**, 197.

Sutherland, S. K. (1972a). The Sydney funnel-web spider (*Atrax robustus*). 1. A review of published studies on the crude venom, *Med. J. Aust.*, **2**, 528.

Sutherland, S. K. (1972b). Sydney funnel-web spider (*Atrax robustus*). 2. Fractionation of the female venom into five distinct components, *Med. J. Aust.*, **2**, 593.

Taylor, E. H. and Denny, W. F. (1966). Hemolysis, renal failure and death, presumed secondary to the bite of a brown recluse spider, *South Med. J.*, **59**, 1209.

Vicari, G., Bettini, S., Collotti, C., and Frontali, N. (1965). Action of *Latrodectus mactans tredecimguttatus* venom and fractions on cells cultivated *in vitro*, *Toxicon*, **3**, 101.

Vorse, H., Seccareccio, R., Woodruff, K., and Benett, G. (1972). Disseminated intravascular coagulopathy following fatal brown spider bit, *J. Pediatr.*, **80**, 1035.

Walther, C. (1974). Effects of potassium, lanthanum, and black widow spider venom on miniature synaptic potentials in the *Torpedo* electroplax, *J. Comp. Physiol.*, **90**, 71.

Weiner, S. (1961). Red back spider antivenene, *Med. J. Aust.*, **43**, 1961.

Wright, R. P. (1972). Enzymic characterization of brown recluse spider venom, *Bull. Mo. Acad. Sci.,* Suppl. **2,** 93.

Wright, R. P., Elgert, K. D., Campbell, B. J., and Barrett, J. T. (1973). Hyaluronidase and esterase activities of the venom of the poisonous brown recluse spider, *Arch. Biochem. Biophys.*, **159**, 415.

30 Venoms of Bees, Hornets, and Wasps

1 BEE VENOMS 502
   1.1 Peptides, 502
       Melittin, 502
       Promelittin, 503
       Apamin, 504
       MCD-Peptide (Peptide 401), 504
       Other Peptides, 505
   1.2 Enzymes, 505
       Hyaluronidase, 505
       Phospholipase A_2, 505
       Other Enzymes, 507
       Enzyme Inhibitors, 507
   1.3 Nonpeptide Components, 508
       Histamine, 508
       Amino Acids, 508
       Carbohydrates, 508
       Lipids, 508
       Others, 508
   1.4 Immunology, 508
   1.5 Toxicology, 509
   1.6 Pharmacology, 510
       Histamine Release, 510
       Cytotoxic Action, 510
       Effect on Nervous System, 511
       Effect on Cardiovascular System, 511
       Other Actions, 512
   1.7 Metabolic Effect, 512

2 VENOMS OF HORNETS AND WASPS: GENUS *VESPA* 515
   2.1 Biochemistry, 515
   2.2 Pharmacology, 517

3 OTHER WASP VENOMS 518
   3.1 Genus *Philanthus*, 518
   3.2 Genus *Bracon* (*Microbracon*), 518
   3.3 Genus *Sceliphron*, 519
   3.4 Genus *Polistes*, 519
   3.5 Genus *Vespula* (*Paravespura*), 519

References 520

Insects, bees, hornets, wasps, and yellow jackets are all closely related members of the order Hymenoptera. Classification of these species is complex and beyond the scope of this book. However the following very brief outline is given primarily to facilitate understanding of the taxonomic differences between bees, hornets, wasps, and yellow jackets.

Superfamily	Family	Genus
Ichneumonidea	Braconidae	*Bracon (Microbracon)*
Apoidea	Apidae	*Apis*
Vespoidea	Vespidae	*Vespa*
		Vespula (Paravespura)
		Dolichovespula
		Polistes
	Bombidae	*Bombus*
Sphecoidea	Sphecidae	*Philanthus*
		Sceliphron
Formicoidea	Formicidae	*Paraponera*
		Myrmecia
		Pseudomyrmex
		Solenopsis
		Myrmicaria
		Myrmica
		Iridomyrmex
		Dolichoderus
		Tapinoma

For information on the biology of bees, wasps, and ants, the reader is referred to the article by Maschwitz and Kloft (1971).

The high mortality rate from bee, hornet, and wasp stings is attributed to anaphylactic shock resulting from hypersensitivity to venom polypeptides. The biochemistry and pharmacology of bee and wasp venoms have been well reviewed by Habermann (1972) and O'Connor and Peck (1975). Venoms from these insects are not identical. O'Connor et al. (1964a, b) investigated the electrophoretic patterns of venoms of honeybees (*Apis mellifera*), wasps (*Polistes apachus*), yellow hornets (*Vespula arenaria*), yellow jackets (*Vespula pennsylvanica*), and bumblebees (*Bombus huntii* and *B. occidentalis*) and found that the venom proteins of these insects are similar but show some differences.

1 BEE VENOMS

1.1 Peptides

Melittin   Melittin is the primary constituent of *Apis mellifera* venom and has important biological actions. It causes the lysis of erythrocytes (Habermann and Kowallek, 1970) and the release of marker ions from liposomes (Sessa et al., 1969) and of histamine from mast cells (Rothschild, 1965). Melittin also interacts with phosphatidylcholine to form a rather tight complex that can be isolated from free melittin by Sephadex chromatography (Mollay and Kreil, 1973).

Melittin has been observed in two forms, melittin I and II, and both are believed to be present in venom. The structure of melittin I is as follows (Habermann and Jentsch, 1967; Jentsch, 1968, 1969):

```
 1                  5                   10                      15
H–Gly–Ile–Gly–Ala–Val–Leu–Lys–Val–Leu–Thr–Thr–Gly–Leu–Pro–Ala–Leu–Ile–
       20                  25
Ser–Trp–Ile–Lys–Arg–Lys–Arg–Gln–Gln–NH₂
```

Melittin II has 27 amino acid residues and shows a slight difference in sequence after residue 21 (Schröder et al., 1971):

```
                              21             25
........................Ser–Arg–Lys–Lys–Arg–Gln–Gln–NH₂
```

Melittins appear to be species specific. Consequently, melittins obtained from the venoms of *A. dorsata A. florea,* and *A. mellifera* are different. However, melittin isolated from the venom of *A. cerana* is identical to that from *A. mellifera* (Kreil, 1973a, 1975). This is reasonable since these two species are closely related biologically and can cross-fertilize. The melittin differences are as follows:

	Position of Amino Acid Residue					
	5	10	15	22	25	26
A. mellifera } *A. cerana*	Val	Thr	Ala	Arg	Gln	Gln
A. dorsata	Ile	Ser	Ala	Arg	Gln	Glu
A. florea	Ile	Ala	Thr	Asn	Lys	Gln

Both melittin I and melittin II have been synthesized chemically, and both produce the full hemolytic and toxicologic activities of natural melittin (Schröder et al., 1971; Habermann and Zeuner, 1971).

Franklin and Baer (1974) observed that melittin can interact with a serum component to form an insoluble complex. Dirks and Sternburg (1972) also found that venoms of *A. mellifera* and of *Polistes exclamans, P. annularis, P. metricus,* and *Vespula maculata* were capable of complexing with normal rabbit serum. The venom component responsible for the complexing is very likely to be melittin, as Franklin and Baer (1974) observed. For this reason, observation of venom immunoprecipitation may result in erroneous conclusions.

Promelittin Promelittin is a precursor molecule that can be isolated in the bee venom gland. It is less potent than melittin in biological activity, and can be converted to the active toxin only after removal from the gland's ribosomes. The existence of this precursor was demonstrated in 1971 by Kreil and Bachmayer, who detected the entire amino acid sequence of melittin in promelittin. Furthermore, at the amino terminal, promelittin appears heterogeneous (Kreil, 1973b). The major type of promelittin contains eight more amino acids than does melittin. Other species of different chain lengths are present in varying amounts, but the structure of the main component is as follows:

Glu–Pro–Glu–Pro–Asp–Pro–Glu–Ala–melittin (1-26)

Apamin   Apamin, another major peptide in bee venom, was first isolated by Habermann and Reiz (1965). This is a neurotoxic peptide (Spoerri et al., 1973, 1975) with the following structure (Haux et al., 1967):

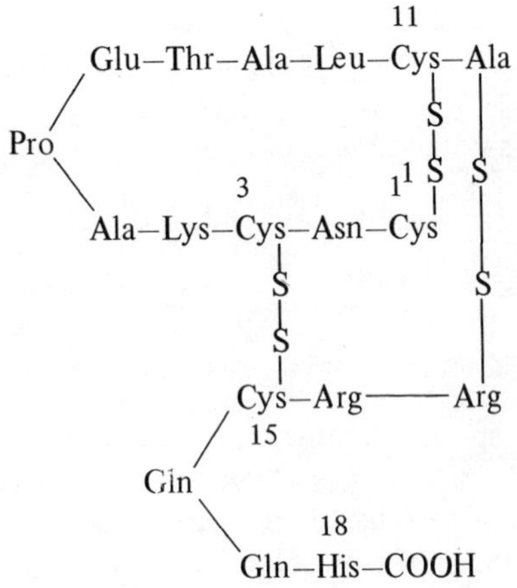

```
                          11
         Glu—Thr—Ala—Leu—Cys—Ala
        /                    S
     Pro                     |
        \        3         1¹ S   S
         Ala—Lys—Cys—Asn—Cys
                     S            S
                     |            |
                     S            |
                     |            |
                     Cys—Arg——————Arg
                    / 15
                  Gln
                     \   18
                      Gln—His—COOH
```

In order to study the role of the disulfide bridge in biological activity, apamin was first converted to the tetra-S-sulfonate derivatives by oxidative sulfitolysis. After reduction of the derivatives by mercaptoethanol, followed by careful air oxidation, the disulfide bridges could be constituted. The biological activities of the reoxidation products were restored to about 50 to 60% of the original values (Hartter and Weber, 1975).

The naturally occurring bee venom apamin has been synthesized by the solid-phase method (Van Rietschoten et al., 1975). The synthetic peptide has chemical and toxicological properties identical to those of native apamin.

MCD-Peptide (Peptide 401).   This is the third major peptide in the venom of *Apis mellifera*. It is highly effective in degranulating mast cells and hence is called MCD-peptide, mast cell degranulating peptide (Fredholm, 1966; Breithaupt and Haberman, 1968). It has also been called peptide 401 (Gauldie et al., 1976). The sequence (Von Haux, 1969; Vernon et al., 1969) and the position of the disulfide bonds (Billingham et al., 1973) have been elucidated.

The structure of MCD-peptide is as follows:

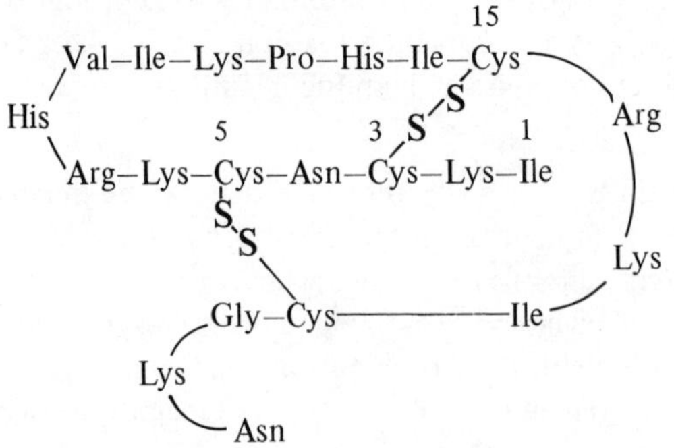

```
                           15
       Val—Ile—Lys—Pro—His—Ile—Cys
      /                       S
    His                    S
      \      5        3  S     1      Arg
       Arg—Lys—Cys—Asn—Cys—Lys—Ile
                  S
                  S                    Lys
             Gly—Cys————————Ile
            /
          Lys
             \
              Asn
```

The relative positions of the disulfide bridges in MCD-peptide are very similar to those in apamin.

1 Bee Venoms

Other Peptides. *Apis mellifera* venom also contains other peptides, such as secapin, tertiapin, and melittin F. The amino acid compositions of these peptides are shown in Table 1, together with those of apamine, melittin, and MCD-peptide. These peptides are indeed minor ones, as the amounts present in the venom are very small. Gauldie et al. (1976) determined the percentages of peptide present in whole venom as follows: 45% melittin, 3% apamine, 2 to 3% MCD-peptide, 1% secapin, < 1% tertiapin, and < 1% melittin F. All of the three minor peptides are highly basic and migrate to the cathode upon electrophoresis at pH 5.6.

Procamine, a polypeptide found in the venom of *A. mellifera*, contains a histamine residue at the C terminal. This was the first histamine-containing peptide to be isolated from a natural source (Peck and O'Connor, 1974). The structures of procamine are as follows:

Ala–Gly–Pro–Gln–Histamine

Ala–Gly–Gln–Gly–Histamine

Lowy et al. (1971) isolated a very interesting peptide called minimine, which has a molecular weight of 6000. Its amino acid composition is listed in Table 1. Minimine has a unique physiological activity on *Drosophila melanogaster* larvae. Larvae that receive a sublethal dose of minimine produce miniature flies as small as one-fourth the normal size. When these reproduce, their progeny are of normal size at the larval and adult stages.

1.2 Enzymes

Bee venom contains a number of enzymes, although not as many as snake venoms.

Hyaluronidase. Hyaluronidase was detected in the venom of *Apis mellifera*, but it has not been characterized extensively (Favilli, 1956; Franklin and Baer, 1975).

Hyaluronidase is not the highest molecular weight component of bee venom: there is another protein with as yet unknown biological activity (Ivanov et al., 1972).

Hyaluronidase isolated from bee venom has arginine as the N-terminal group, and only traces of alanine and threonine were detected. No free SH groups were found in the enzyme (Krusteva et al., 1973). Four moles of mannose was associated with each mole of the enzyme.

Phospholipase A_2. At least two phospholipase A_2 enzymes were found in *Apis mellifera* venom (Jentsch and Dielenberg, 1972). Phospholipase A_2 isolated from this venom has a molecular weight of 18,500 (Munjal and Elliott, 1971b), which is very similar to the corresponding values for the enzyme isolated from various snake venoms (see Chapter 2). The structure of the enzyme as determined by Shipolini et al. (1971, 1974a, b) is shown in Fig. 1. In contrast to previous reports, the molecular weight deduced from the sequenced structure is 14,629.

The phospholipase A_2 of bee venom is highly basic with a pI of 10.5. The enzyme hydrolyzes long-chain phosphatidylphospholipids most rapidly, and it shows product activation (Moores and Lawrence, 1972). The bee venom enzyme is also activated by fatty acid anions such as palmitate and oleate (Lawrence and Moores, 1975).

Two constituents of bee venoms, namely, phospholipase A_2 and melittin, show a marked synergism toward cell membranes, therefore a mixture of the two readily lyses erythrocytes (Vogt et al., 1970). Melittin also stimulates enzymatic hydrolyis of egg

Table 1 Amino Acid Compositions of Peptides in the Venom of *Apis mellifera*

Amino Acid	Apamine	Melittin	MCD-Peptide	Secapin	Tertiapin	Melittin F	Minimine
Lysine	1	3	5	2	4	2	4
Histidine	1	0	2	0	1	0	2
Arginine	2	2	2	3	1	2	2
Aspartic acid	1	0	2	2	2	0	6
Threonine	1	2	0	0	0	2	4
Serine	0	1	0	1	0	1	4
Glutamic acid	3	2	0	0	0	2	3
Proline	1	1	1	5	1	1	2
Glycine	0	3	1	1	1	1	4
Alanine	3	2	0	0	1	1	3
Valine	0	2	1	3	0	1	2
Methionine	0	0	0	0	1	0	1
Isoleucine	0	3	4	3	2	2	1
Leucine	1	4	0	0	1	3	3
Tyrosine	0	0	0	1	0	0	2
Phenylalanine	0	0	0	1	0	0	2
Half-cystine	4	0	4	2	4	0	5
Tryptophan	0	1	0	0	1	1	1
Total residues	18	26	22	24	20	19	51

1 Bee Venoms

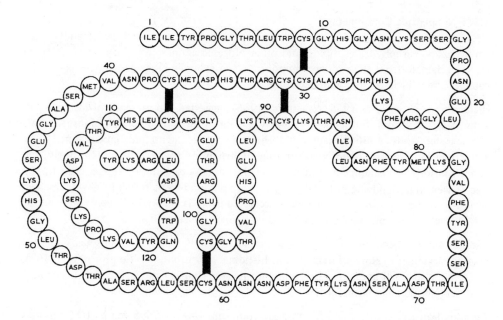

Figure 30.1 Amino acid sequence of phospholipase A_2 isolated from *Apis mellifera* venom. The dark lines indicate the positions of disulfide bridges. (Reproduced from Shipolini et al., 1974b.)

lecithin by venom phospholipase A_2 (Mollay and Kreil, 1974). The direct lytic factor obtained from *Naja naja* venom and bacterial toxin polymyxin B can replace melittin in this respect. The venom enzyme can hydrolyze phospholipids, but a faster rate can be obtained when phospholipoproteins are used (Slotta et al., 1971). *Apis mellifera* venom inhibits mitochondrial oxidative phosphorylation by the liberation of fatty acid and lysolecithin due to phospholipase A_2 action on mitochondrial membranes (Vázquez-Cotón and Elliott, 1966). Mitochondrial phospholipids decrease by 31% after the injection of *A. mellifera* venom, probably because of the activity of venom phospholipase A_2 (Yurkiewicz, 1968).

The optical properties of phospholipase A_2 show no change from 10° to 90°C, as determined by the ORD method (Nair et al., 1976). However, enzyme activity decreases above 60°C. The explanation may be that the slight physical change incurred on heating does not produce significant changes in the optical properties of the enzyme.

Other Enzymes. In addition to hyaluronidase, *Apis mellifera* venom was reported to contain esterase and acid and alkaline phosphatase (Benton, 1967). Hexosaminidase activity was detected in the venom, but the activity in the venom sac was much stronger (Hsiang and Elliott, 1975).

Acetylcholinesterase is absent in bee venom (Neumann and Habermann, 1956).

Enzyme Inhibitors. Protease inhibitor is present in the venom of *Apis mellifera* (Shkenderov, 1973). The inhibitor is heat stable and resistant to pepsin digestion, the high stability being conferred by the low molecular weight of the polypeptide. When a bee stings man or animal, venom is exposed to the victim's proteolytic enzymes. The protease inhibitor probably protects venom components such as hyaluronidase, phospholipase A_2, and toxic peptides from hydrolysis by these proteases.

1.3 Nonpeptide Components

Bee venoms also contain a number of nonpeptide components such as histamine, free amino acids, carbohydrates, lipids, and other biogenic amines. Although the main toxic effects originate from the protein components, the venom's biogenic amines may be the cause of the pain associated with bee stings.

Histamine. As early as 1935, Nagamitu (1935) suggested the presence of histamine in *Apis mellifera* venom. Subsequently, this was demonstrated by a number of investigators (Tetsch and Wolff, 1936; Schacter and Thain, 1954; Neumann and Habermann, 1954). Reinert (1936) reported the concentration of histamine to be 1.5%. Interestingly, the concentrations of histamine and histidine in the venom correlate to the age of the honey bee (Owen and Braidwood, 1974). These compounds are not found in newly emerged worker bees but appear in the venom of week-old bees. The amounts present increase for 3 to 4 weeks, reach maxima, and then decline again in 6-week-old bees.

Amino Acids. *Apis mellifera* venoms contain a number of free amino acids such as alanine, arginine, cystine, glutamic acid, histidine, and proline (O'Connor et al., 1967). There are also small amounts of γ-aminobutyric acid (0.04%) and β-aminoisobutyric acid (0.02%) (Nelson and O'Connor, 1968).

Carbohydrates. Glucose and fructose are found in *Apis mellifera* venom (O'Connor et al., 1967). The total sugar content is about 2% by dry weight.

Lipids. Five or six lipids, probably phospholipids, are present in *Apis mellifera* venom. The total lipid content in the venom is about 5% by dry weight (O'Connor et al., 1967).

Others. Dopamine and noradrenaline are present in the venom of *Apis mellifera* (Owen, 1971). The amounts of dopamine and noradrenaline in single honey bee venom reservoirs are 775 and 115 ng, respectively. The presence of 5-hydroxytryptamine in epithelial cells of the venom gland of *A. mellifica* was confirmed by Grzcki and Czerny (1972). Also, acetylcholine is present in most Hymenoptera venoms, including bee venom (Neumann and Habermann, 1956).

1.4 Immunology

Many Hymenoptera sting victims are killed by anaphylactic shock. Therefore an understanding of the immunological properties of bee venom is important.

Bee venom contains a specific antigen that is not found in the body of the bee (Schulman et al., 1966). Four to five antigens are present in *A. mellifera* venom (Munjal and Elliott, 1971a). Two of the antigens are common to the venoms of the honeybee, wasp, yellow hornet, black hornet, and yellow jacket (Foubert and Stier, 1958; Langlois et al., 1965). Immunologically, melittin and phospholipase A_2 are totally different antigens (Shepherd et al., 1974). Apamin, the second major component in bee venom, is nonantigenic (Munjal and Elliott, 1971a). Whereas phospholipase A_2 of snake venom origin has low antigenicity (Högberg and Uvnäs, 1957; Keller and Schwarz, 1963; Middleton and Phillips, 1964; Munjal and Shivpuri, 1968), phospholipase A_2 of bee venom provides good antigenicity (Munjal and Elliott, 1971a). Antibodies to venom phospholipase A_2 have been found in beekeepers (Mohammed and El Karemi, 1961).

Actually, the antigenic properties of bee venom are derived mainly from large molecular weight polypeptides. In bee venom, the high molecular weight components include hyaluronidase, MCD-peptide, proteinase inhibitors, and phospholipase A_2 (Shkenderov, 1974). Melittin, apamine, MCD-peptide, and protease inhibitors did not

1 Bee Venoms

show any anaphylactic properties after sensitization with or without Freud's complete adjuvant.

Busse et al. (1975) were able to desensitize a person who has high risk of future bee stings. Bee venom extracted into a coca solution was administered parenterally in increasing daily doses until the equivalent of one venom sac was given per day, and this was continued for a month. Anaphylaxis did not occur after subsequent bee-sting challenges. With this treatment, there was an increase in the level of IgG-blocking antibody and a decrease in IgE titer. Rothenbacher and Benton (1972) also suggested a weekly injection of bee venom for desensitization.

The allergenic activities of hyaluronidase, phospholipase A_2, melittin, and apamin were investigated by measuring the amount of histamine released upon injection of the antigen (King et al., 1976). Hyaluronidase and phospholipase A_2 are 2 and 8 times, respectively, more active than whole venom. Melittin is about one-tenth as active as venom, and apamin is inactive as an allergen. Chemical modification of amino groups and carboxymethylation of phospholipase A_2 decrease its allergenic activities. These results indicate that the antigenic determinants of phospholipase A_2 depend on the charge, the primary structure, and the conformation of the molecule.

1.5 Toxicology

The LD_{50} values of *A. mellifera* venom and its components have been investigated by a number of workers and are listed in Table 2. Apamine and melittin are comparable to the original venom in toxicity. Cardiopep is a very weak toxic compound. Surprisingly, bee venom phospholipase A_2 is quite toxic. Melittin is a toxic peptide, and its toxicity is greatly enhanced by another venom component, phospholipase A_2. Melittin is very similar to cobra (*Naja naja*) cardiotoxin and direct lytic factor (DLF). Melittin does have a cardiotoxic effect but is much less potent than DLF (Slotta and Vick, 1969).

Table 2 Toxicities of *Apis mellifera* Whole Venom and Its Components as Tested in Mice

Sample	LD_{50} (μg/g)	Route of Injection	Reference
Venom	4.0	—	Tetsch and Wolff, 1936
	0.15	Intracerebral	Derevici et al, 1970
	12.5	i.p.	Derevici et al, 1970
	3.5	—	Vick et al, 1974
Apamine	15.0	—	Vick et al, 1974
	4.0	i.v.	Habermann and Cheng-Raude, 1975
	5.0	—	Hartter and Weber, 1975
Cardiopep	15.0	—	Vick et al, 1974
Melittin	4.0	i.v.	Jentsch and Habermann, 1967
Phospholipase A_2	0.37	i.p.	Slotta et al, 1971

1.6 Pharmacology

Histamine Release. As previously mentioned, bee venom has histamine-releasing properties in animal tissues. Two principles are responsible for this effect (Feldberg and Kellaway, 1937; Fredholm and Haegermark, 1967, 1969): phospholipase A_2, and a "basic peptide" (Fredholm and Haegermark, 1969). Rothschild (1965) demonstrated that phospholipase A_2 released only small amounts of histamine from rat peritoneal fluid cell suspension, whereas melittin released histamine readily. We now know that the action of phospholipase A_2 is indirect. The hydrolysis product, lysophosphotidylcholine, is responsible for the histamine release. Melittin releases histamine from tissues by direct action, which correlates with the observation that melittin itself has a strong mast cell degranulation activity.

Cytotoxic Action. It has been known for some time that bee venom has hemolytic action. The two hemolytic principles in bee venom are phospholipase A_2 which gives indirect action, and melittin, which is a basic peptide and provides direct lytic action.

The relationship between structure and hemolytic action was extensively studied by Schröder et al. (1971), using synthetic peptides. As can be seen from Table 3, the fragments of melittin do not have hemolytic activity, although some retain surface activity. Melittins contain hydrophobic groups in the N-terminal region with hydrophilic residues in the C-terminal region. This is typical of cationic surface-active compounds like

Table 3 Hemolytic Activity and Action on Surface Tension of Aqueous Solution of Melittins and Related Peptides

Peptide	Hemolytic Activity %	Surface Activity %
1————————————————26 (C terminal / N-terminal)	100	100
1—————————————————27	100	100
7——————————————27	3	70
7——————————(23)-(24)——(27)	6	120
15——————————26	1	5
15———————(21)-(23)——26	5	100
18—————————————27	Inactive	Inactive
18——————(23)-(24)——27		65
18———(21)-(23)——26	1	70
1————————20	Inactive	110
7——————20	Inactive	90
1————14	Inactive	8
4————14	Inactive	50
7————14	Inactive	Inactive

Table reproduced from Schröder et al. (1971). A circled number indicates the blocking of the ϵ-amino group of lysine.

detergents. However, experiments indicate that hemolytic activity cannot be solely explained by the cationic surface activity of a compound. Bee venom contains another surface-active component (Shipman and Cole, 1968), but it is different from the hemolytic component. Therefore Mitchell et al. (1971) suggested that the toxicity of melittin is due to membrane disruption.

Heparin inhibits the hemolytic action of bee venom (Sergeeva, 1974). This is believed to be due to the complex formation of heparin with the hemolytic component of bee venom. Cole and Shipman (1969) incubated mouse bone marrow cells with *Apis mellifera* venom or its components, and then assayed for the ability to form hemopoietic splenic colonies when transfused into lethally X-irradiated recipient mice. Colony-forming ability was annulled or decreased by venom, melittin, and phospholipase A_2. These workers therefore suggested that the cellular toxicity is mediated via effects at the cell surface.

Degranulation of mast cells is the earliest observable response to the subcutaneous injection of bee venom (Higginbotham and Karnella, 1970). During the mast cell reaction, the granules exhibit a marked swelling and a decrease in electron density (Bloom and Haegermark, 1967). Because heparin and bee venom form a complex, and heparin is a major constituent of the mast cell granules, secretion of heparin in response to bee sting may be a protective mechanism of the venom recipient (Saelinger and Higgenbotham, 1974). There are three components in bee venom that cause mastocytolytic action: phospholipase A_2, melittin, and MCD-peptide. Of these, phospholipase A_2 is an indirect mastocytolytic agent (Habermann and Breithaupt, 1968). The mechanism of lytic action of *Escherichia coli* cell membrane was studied by electron spin resonance spectroscopy (Williams and Bell, 1972), using spin labels. The results suggest that melittin participates in a hydrophobic interaction with the hydrocarbon region of the bilayer and acts as a disrupter of matrix ordering.

Effect on Nervous System. Unlike many toxins of snakes, spiders, and scorpions, which are peripherally neurotoxic, bee venom is centrally neurotoxic. Because of the relatively large size of the toxins of snakes, spiders, and scorpions, they do not pass the blood–brain barrier easily and thus are not neurotoxic to the central nervous system. However, because of the small molecular size of apamin, bee venom is toxic to the CNS (Habermann and Reiz, 1965). The central neurotoxicity of apamin gives a much stronger response when injected intraventicularly or intralumbarly than by the intravenous or subcutaneous route (Habermann and Cheng-Raude, 1975). The CNS effects of melittin were also reported by Vyatachannikov and Sinka (1973). However, Ishay et al. (1975b) considered that the depression in the electroencephalogram was due to decreases in systemic blood pressure and to serious disturbances evident in the electrocardiogram, rather than to direct effects on the CNS. Bee venom increases the afferent impulse of the sinus nerve and blood pressure and stimulates respiration through action on the chemoreceptors of the carotid glomus (Korneva and Orlov, 1969). The venom induces damage in the cultured cerebral cortex of the mouse. This damage includes disordered Nissi bodies, widened and curved cisternae of the endoplasmic reticulum, disrupted mitochondria, vacuolated cytoplasm, and infolded nuclei.

Effect on Cardiovascular System. Bee venom decreases the blood pressure and the rate of cardiac contractions in anesthetized cats. Vagotomy eliminates the bradycardiac effect and reduces the intensity of its hypotensive action. Cardiac receptors apparently participate directly in the mechanism of bee venom toxic action (Korneva, 1972). *Apis mellifera* venom contains a cardioactive compound that has been called cardiopep by

Vick et al. (1974). When this compound is injected into the coronary circulation of perfused heart, it produces a 50% increase in rate and a 150% increase in force of contraction, but no change in coronary vascular resistance. Cardiopep is relatively nontoxic with an LD_{50} of 15 $\mu g\, g^{-1}$. The compound is a potent β-adrenergiclike stimulant not entirely blocked by propranolol and possesses antiarrhythmic properties. Cardiopep comprises 0.7% of whole bee venom and appears to be different from melittin, a major venom component.

Melittin itself has a cardiotoxic action. It causes bradycardia, arrhythmia, and atrioventricular block. When injected into cat aorta, it irreversibly stopped the heartbeat (Krylov, 1973). Melittin and apamin cause a decrease in blood pressure, narrowing of the pulse pressure, and increases in the plasma cortisol levels and in circulating epinephrine and norepinephrine. The heart rate decreases sharply without cardiac arrhythmias (Vick and Brooks, 1972). Apamin increases the heart rate and contractile force without changing the coronary vascular resistance when injected into the coronary circulation of perfused monkey and dog hearts (Vick et al., 1972a).

Venom phospholipase A_2 decreased the arterial blood pressure and heart rate, suspended respiration, and caused the death of dogs in 10 min (Vick and Brook, 1972; Vick and Shipman, 1972). Venom phospholipase A_2, as well as melittin, also increased capillary permeability (Kireeva, 1970).

The action of venom is similar to that of aldosterone or vasopressine in increasing the Na(I) transport of toad bladder without affecting the conductance of the membrane (Marumo et al., 1968).

Other Actions Bee venom has a diverse action in addition to the effects already mentioned. For instance, an antibacterial property was reported by Fennel et al. (1968). *Apis Mellifera* venom and melittin are effective against a penicillin-resistant strain of *Staphylococcus aureus*.

Radioprotective activity was also reported. Shipman and Cole (1967) found that the resistance of mice to X-irradiation increases greatly after subcutaneous injection of bee venom. Similar protective action was reported by Ginsberg et al. (1968) and Kanno et al. (1970), with mellitin responsible for this protection.

Melittin also binds to components of skin to form a tight complex (Shipman and Cole, 1972). This complex formation appears to be at least somewhat specific, as melittin does not form a complex with albumin.

Bee venom has an antialkylating activity and is effective against poisoning with bis(2-chloroethyl)methylamine-HCl when given prophylactically (Rauen et al., 1972). Melittin and apamin are not responsible for this action.

The venom induces hemorrhage on the lung surface of dogs when applied topically, while heparin forms an inactive complex with the vasculotoxic compenents of *A. mellifera* venom (Bonta et al., 1972).

1.7 Metabolic Effect

Although *Apis mellifera* venom does not change the physical properties of fibrinogen, it increases the fibrinolytic activity of rat blood *in vivo* (Omarov, 1970).

Bee venom also increases the secretion of corticosterone in rats (Couch and Benton, 1972; Alfano et al., 1973) and monkeys (Vick et al., 1972c, d). This may be the reason for its antiarthritic action in experimental arthritis (Neumann and Stracke, 1951).

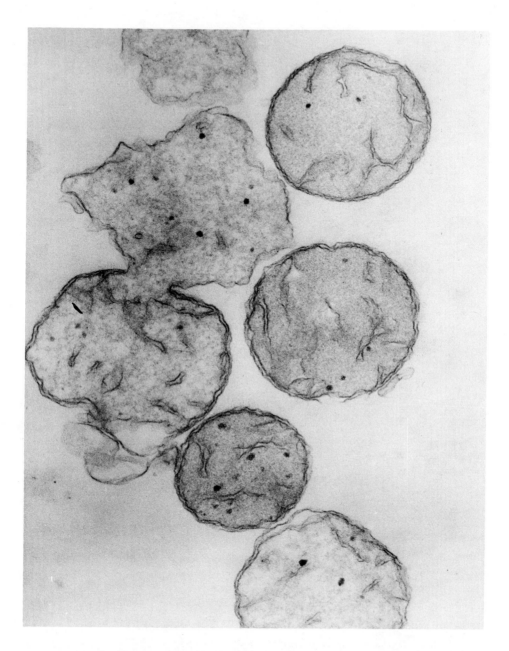

Figure 30.2 Mitochondria treated with melittin concentration that caused loss of phosphate acceptor control. Note lysis of the outer membrane and continuity of the inner membrane after pure melittin treatment at a concentration of 23.8 $\mu g\ mg^{-1}$ mitochondrial protein. (Photograph kindly supplied by Drs. D. Munjal and W. B. Elliott and originally published in Olson et al., 1974.) Permission was granted by the copyright owner, Pergamon Press.

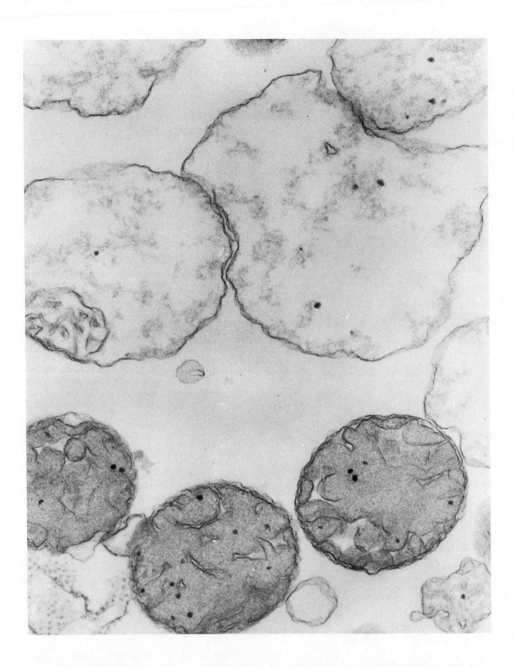

Figure 30.3 Mitochondria treated with melittin concentration that totally inhibited respiration. Pure melittin at a concentration of 242 $\mu g\ mg^{-1}$ mitochondrial protein causes releases of inner membrane vesicles from the outer membrane. (Photograph kindly supplied by Drs. D. Munjal and W. B. Elliott and originally published in Olson et al., 1974.) Permission was granted by the copyright owner, Pergamon Press.

Cortisol elevation in plasma suggests that there is a generalized stimulation of the adrenal gland (Vick and Shipman, 1972; Vick et al., 1972b).

The bee venom component peptide 401 has anti-inflammatory activity, as measured by an inhibition of the increase in vascular permeability. Five other peptides do not have such activity, implying that anti-inflammatory activity is specific for peptide 401 (Banks et al., 1976). The anti-inflammatory peptide was also indentified as peptide 401 or MCD-peptide by Billingham et al. (1973). Thus bee venom can be used in anti-inflammatory therapy (Lorenzetti et al., 1972). However, this effect has also been attributed to the action of melittin by Vick et al. (1972c).

Bee venom inhibits mitochondrial oxidative phosphorylation in a manner similar to that of snake venoms (Vázquez-Colón and Elliott, 1966). This action is attributed to phospholipase A_2 in the venom. In addition, the venom also increases the osmotic fragility of erythrocytes (Hort and Herz, 1968a, b), an effect that is also probably due to the action of phospholipase A_2. Melittin also has an inhibitory effect on mitochondrial respiration (Olson et al., 1974). Electron micrographs of treated mitochondria show outer membrane lysis with preservation of inner membrane structure (Figs. 2 and 3). Melittin also inhibits acetylcholinesterase (Mitchell et al., 1971). The inhibition is a noncompetitive type and probably forms a complex that may be one contributory factor for the toxic effect on *Drosophila* larvae.

Blood glucose level is increased when bee venom is subcutaneously injected into rabbits. The venom may stimulate adrenaline secretion in the animals and thereby increase glyconeogenesis (Artemov et al., 1972).

2 VENOMS OF HORNETS AND WASPS: GENUS: VESPA

The oriental hornet (*Vespa orientalis*) is widely distributed in many countries of the Middle East, the Mediterranean basin and southeast Asia. It is a social insect and forms annual colonies.

Vespa orientalis sting may result in the death of animals and human subjects because of nephrotoxic, hepatotoxic, and allergic involvement (Jonas and Shugar, 1964; Schaller, 1964). Severe muscle necrosis developed in patients who sustained multiple stings from the hornet *V. affinis* (Shilkin et al., 1972). Often death is due to anaphylactic shock, and the record of wasp sting deaths has been described in detail (O'Connor et al., 1964b).

2.1 Biochemistry

Vespa venom is a colorless liquid and contains 27% solid materials, of which 76% are protein (Fischel et al., 1972). Dialysis does not diminish the lethality or hemolytic activity, suggesting that the toxic principles are high molecular weight substances. This is in significant contrast to bee (*Apis mellifera*) venom, in which the toxic principles are dialyzable. By electrophoresis, six to seven fractions are detectable. Four hemolytic fractions are present in *Vespa orientalis* venom.

Four components are responsible for antibody production (Ishay et al., 1972). The biological activities of two components have been identified; one component has hemolytic activity, and the other has lysozyme activity. *Paravespula germanica* venom reacted with the antiserum for *V. orientalis* to give two precipitin arcs, suggesting partial cross-immunity. In addition, immunodiffusion studies indicate that venoms of honeybees (*Apis mellifera*), paper wasps (*Polistes apachus*), yellow hornets (*Vespula arenaria*), and yellow jackets (*V. pennsylvanica*) contain some common antigens and others specific to each venom (O'Connor and Erickson, 1965).

Although *Vespa orientalis* releases aliphatic and cyclic ketones as alarming and poisoning substances for various colony members of the Oriental hornet, none of the ketones has been detected in the venom (Saslavasky et al., 1973).

The venom sac extract induces hyperglycemia in cats (Ishay, 1975). The hyperglycemic factor is probably a protein or a protein-bound substance, as it is not dialyzable and is inactivated by heating at 100°C.

The venom also has anticoagulant activity (Joshua and Ishay, 1975). It inhibits *in vitro* formation of thromboplastin and thrombin by human plasma and inactivates previously formed tissue thromboplastin. It also has a strong fibrinogenolytic and fibrinolytic activity when tested on purified human fibrinogen. The fibrinolysis is inhibited by soybean trypsin inhibitor, as well as by human serum.

The venom of *V. orientalis* contains proteases, hyaluronidase, histamine-releasing factor, and hemolytic activity (Edery et al., 1972). Nonprotein components include histamine, acetylcholine, 5-hydroxytryptamine, kinins, adrenaline, noradrenaline, and dopamine. The venom hemolyzed erythrocytes of man, guinea pig, rabbit, cat, mouse, and rat, but not of sheep, ox, horse, or camel. The hemolytic effect was enhanced by the addition of egg yolk phospholipids. This hemolytic activity is probably due to venom phospholipase A_2 and basic proteins. Sublytic doses of whole venom caused potassium leakage from susceptible erythrocytes and increased their osmotic fragility (Joshua and Ishay, 1973).

Venom of other hornets have similar compositions. For instance, *V. vulgaris* (European hornet, common wasp) contains acetylcholine, 5-hydroxytryptamine, histamine, and kinin (Jaques and Schachter, 1954; Bhoola et al., 1961). Kinin is believed to be the major pain-producing factor of this insect sting. Hyaluronidase activity was detected in the venoms of *V. orientalis* and *V. crabro*, as well as of *Paravespula germanica, P. vulgaris, Dolichovespula saxonica, D. media*, and *Polistes gallicus* (Allalouf et al., 1972).

Free amino acids in the venom sac of *Vespa orientalis* were extensively investigated by Ikan and Ishay (1973). The sac contains the following:

Amino Acid	Amount (μ mole) per gram of Venom Sac
Alanine	5.6
Arginine	1.5
Asparagine	0.2
Glutamine	0.2
Glutamic acid	1.9
Glycine	0.9
Leucine	0.3
Proline	5.5
Serine	0.9
Taurine	1.4
Threonine	0.8
Tyrosine	0.3
Valine	0.1

Histine, isoleucine, and ornithine are present only in traces.

2 Venoms of Hornets and Wasps: Genus *Vespa*

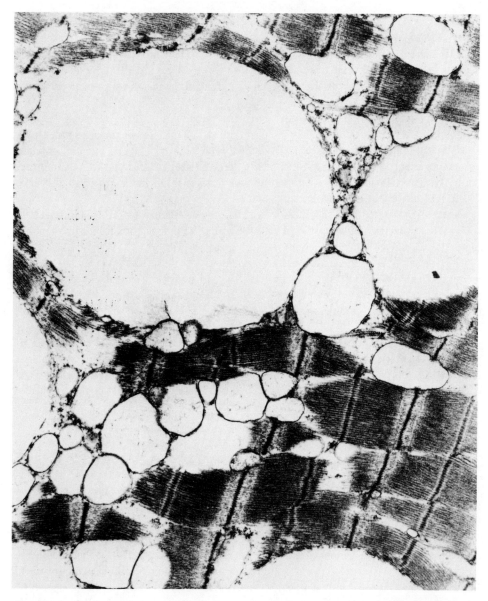

Figure 30.4 Muscle damage caused by the venom of *Vespa orientalis*. The T tubules and terminal cisternae are dilated to such an extent that it is no longer possible to recognize the triad unit. (Photograph kindly supplied by Dr. J. Ishay and originally published in Ishay et al., 1975a.) Permission was granted by the copyright owner, Pergamon Press.

2.2 Pharmacology

The LD_{50} of *Vespa orientalis* venom in mice by i.v. injection is $2.5\ \mu g\ g^{-1}$ (Edery et al., 1972). The venom induces contraction of isolated smooth muscle preparations and bronchioconstriction in anesthetized guinea pigs, and, when injected intradermally into rats and rabbits, increases the permeability of microcirculation vessels. There is a morphological change in mitochondria in the muscle of guinea pig after envenomation

(Sandbank et al., 1971). The cristae collapse and agglutinate in the middle or adhere to the inner membrane of the mitochondrion. Clumped cristae dissolve into an amorphous granular material. Further study by Ishay et al. (1975a) indicates that most of the muscle filaments are divided by enlarged vacuoles into strips of myofilaments (Fig. 4). The T tubules and the terminal cisterns are greatly extended, giving the whole myofilament system a disrupted appearance.

The venom has a suppressive effect on the development of insect larvae and on the metamorphosis of the tadpole (Barr-Nea and Ishay, 1975). It also has a skin-bleaching effect in the tadpole. Histological analysis of the tail showed that the pigment granules normally present in the melanophores of the dermis disappeared from the melanophores of the treated animals. The venom may induce hormonal inbalance, which may be one reason for impairment of development.

Vespa orientalis venom also causes kidney damage by tubular necrosis (Sanbank et al., 1973). The proximal tubular epithelial cells showed an abundance of autophagic vacuoles, suggesting disintegration of mitochondria.

The venom affects the cardiovascular dynamics of dogs (Kaplinsky et al., 1974). Respiration increased, and mean aortic pressure and peripheral vascular resistance fell, while cardiac output was increased by 60%. High doses produced arrhythmias, shock, and death. Part of the cardiovascular effect may be due to a venom component, 5-hydroxytryptamine.

Central neurotoxic effects, manifested by isoelectric EEG and by depression of respiratory centers, appeared before cardiovascular manifestations when venom was injected into the vertebral artery of cats (Ishay et al., 1974). However, no pathological changes were observed in the cortex or cerebellum. Therefore the functional changes due to venom cannot be recognized by histology of CNS tissue.

3 OTHER WASP VENOMS

3.1 Genus *Philanthus*

Philanthus triangulum F. (the digger wasp, the bee wolf wasp) is a wasp that preys on honeybee workers. The active component of the venom has a molecular weight of less than 700 (Visser and Spanjer, 1969). The bees are paralyzed by the wasp venom, which has an inhibitory action on neuromuscular transmission (Rathmayer, 1962a, b). Piek (1966) found that the venom affected the muscle action potential, whereas the nerve action potential was not affected, and that the venom blocks both excitatory and inhibitory transmission, with a concurrent decrease in amplitude of the postsynaptic potentials. Piek and Njio (1975) studied the effect of the venom on the flight muscle of the honeybee worker by recording miniature excitatory postsynaptic potentials. The venom decreased the frequency by a factor of 3.85, a good indication of a presynaptic effect. However, ultrastructural investigation indicated that there was no difference between paralyzed and control muscles in the number and distribution of presynaptic vesicles. Also, no difference in other structures was observed. Therefore the venom prevents the release of transmitter substance without altering the visible ultrastructure in nerve terminals.

3.2 Genus *Bracon (Microbracon)*

Microbracon is an exophagous parasite of hosts who lead a hidden life (Salt, 1931). These hosts frequently are larvae of Lepidoptera, belonging to the families Gelechiidae,

Tortricidae, Pyralididae, Pyraustididae, or Noctuidae. Usually the hosts are permanently paralyzed. *Microbracon hebector* venom has paralytic action on *Galleria* larvae, *Philosamia* larvae, and *Musca* adults (Van der Meer et al., 1965). The venom is toxic to insects, but the host susceptibility depends on the species. A high susceptibility to this paralyzing venom is restricted to the order Lepidoptera. A few species of Hemiptera, Diptera and Hymenoptera can also be paralyzed but only if high doses of the venom are injected (Drenth, 1974).

The venom specifically blocks the neuromuscular transmission in the somatic muscles (Piek and Simon Thomas, 1969; Piek et al., 1974). It especially affects the frequency of the miniature excitatory postsynaptic potentials and therefore is assumed to act at a presynaptic site. Walther and Rathmayer (1974) suggested that the venom has specific affinity for glutaminergic synapses.

The action of *Microbracon gelechiae* is very similar to that of *M. hebector* (Piek et al., 1974). The venom of *M. gelechiae* also affects the frequency of the spontaneous miniature excitatory postsynaptic potentials, again indicating the presynaptic nature of the venom effect. Electron micrographs of the synaptic region in muscle fibers, however, do not show any difference between normal and paralyzed animals. The active principle of the venom is estimated to have a molecular weight of 60,000.

3.3 Genus *Sceliphron*

Sceliphron caementarium is the mud-dauber wasp. None of the protein components of this venom possesses antigenic character, in common with many honeybee, wasp, yellow jacket, or hornet venoms (Rosenbrook and O'Connor, 1964a). These authors further identified the components of the venom and found 17 nonprotein constituents, including 3 free amino acids and phosphatidylcholine substances (Rosenbrook and O'Connor, 1964b). They suggested that the activity of the venom is due to the combined effects of several components.

3.4 Genus *Polistes*

Venom of this genus contains two minor kinins and one major. The structure of the major kinin is as follows (Pisano, 1966a, b).

Pyroglu—Thr—Asn—Lys—Lys—Lys—Leu—Arg—Gly—Arg—Pro—Pro—Gly—Phe—Ser—Pro—

Phe—Arg

Hyaluronidase was detected in the venom of *Polistes gallicus* (Allalouf et al., 1972). It hydrolyzed both hyaluronic acid and chondroitin sulfate.

3.5 Genus *Vespula (Paravespura)*

The major component of *Paravespura germanica* venom is a basic protein, and it is similar to the major component found in *Vespa orientalis* (Fischl et al., 1972). Both venoms possess marked hemolytic activity. The LD_{50} for mice is 117.5 $\mu g\, g^{-1}$ (Ishay et al., 1973). The *P. germanica* venom shows protease and hyaluronidase activity and causes hemolysis. It also affects the neuromuscular junction in cat striated muscle but does not affect the muscle cell membrane.

The yellow jacket (*Vespula maculifrons*) venom sac contains two vasoactive polypeptides, vespulakinins 1 and 2 (Yoshida et al., 1976). Vespulakinin 1 is a heptadecapeptide, vespulakinin 2, a pentadecapeptide. They are both highly basic and contain the nonapeptide bradykinin at their carboxyl termini. Most unique is the presence of carbohydrates. Vespulakinins are the first reported naturally occurring glycopeptide derivatives of bradykinin and the first reported vasoactive glycopeptides. The structure of vespulakinin 1 is as follows:

```
    Carbohydrate (1)         Carbohydrate (2)
           |                        |
    Thr—Ala—Thr————————Thr—Arg—Arg—Arg—Gly—Bradykinin
```

Vespulakinin 2 lacks the amino terminal Thr—Ala, but has the same carbohydrate composition as vespulakinin 1.

REFERENCES

Alfano, J. A., Elliott, W. B., and Brownie, A. C. (1973). The effect of bee venom on serum corticosterone levels and adrenal mitochondrial cytochrome P-450 in intact and hypophysectomized rats, *Toxicon,* **11,** 101.

Allalouf, D., Ber, A., and Ishay, J. (1972). Hyaluronidase activity of extracts of venom sacs of a number of Vespinae (Hymenoptera), *Comp. Biochem. Physiol.,* **B, 43,** 119.

Artemov, N. M., Kireeva, V. F., and Gudenko, N. A. (1972). Effects of bee venom on the blood sugar level, *Uch. Zap. Gor'k. Gos. Univ.,* **5.**

Banks, B. E. C., Rumjanek, F. D., Sinclair, N. M., and Vernon, C. A. (1976). Possible therapeutic use of a peptide from bee venom, *Bull. Inst. Pasteur,* Paris, **74,** 137.

Barr-Neal, L. and Ishay, J. (1975). Effect of the venom sac content of the Oriental hornet (*Vespa orientalis*) on the metamorphosis of the toad tadpole (*Bufo viridis*), *Experientia,* **31,** 212.

Benton, A. W. (1967). Esterases and phosphatases of honey bee venom, *J. Apicult. Res.,* **6,** 91.

Bhoola, K. D., Calle, J. D., and Schachter, M. (1961). Identification of acetylcholine 5-hydroxytryptamine, histamine, and a new kinin in hornet venom (*V. Crabro*), *J. Physiol.,* **159,** 167.

Billingham, M. E. J., Morley, J., Hanson, J. M., Shipolini, R. A., and Vernon, C. A. (1973). Anti-inflammatory peptide from bee venom, *Nature,* **245,** 163.

Bloom, G. D. and Haegermark, O. (1967). Studies on morphological changes and histamine release induced by bee venom, n-decylamine and hypotonic solutions in rat peritoneal mast cells, *Acta Physiol. Scand.,* **71,** 257.

Bonta, I. L., Bhargava, N., and Vargaftig, B. B. (1972). "Dissociation between hemorrhagic, enzymatic and lethal activity of some snake venoms and of bee venom as studied in a new model," in A. de Vries and E. Kochva, Eds., *Toxins of Animal and Plant Origin,* Vol. 2, Gordon and Breach, New York, pp. 707–718.

Breithaupt, H. and Habermann, E. (1968). MCD (mast cell-degranulating) peptide from bee venom: Isolation and biochemical and pharmacological properties, *Naunyn-Schmiedebergs Arch. Exp. Pathol. Pharmakol.,* **261,** 252.

Busse, W. W., Reed, C. E., Lichtenstein, L. M., and Reisman, R. E. (1975). Immunotherapy in bee-sting anaphylaxis: Use of honey-bee venom, *J. Am. Med. Assoc.,* **231,** 1154.

Cole, L. J. and Shipman, W. H. (1969). Chromatographic fractions of bee venom: Cytotoxicity for mouse bone marrow stem cells, *Am. J. Physiol.,* **217,** 965.

Couch, T. L. and Benton, A. W. (1972). The effect of the venom of the honey bee, *Apis mellifera L.,* on the adrenocortical response of the adult male rat, *Toxicon,* **10,** 55.

Derevici, A., Filotti, A., and Toaxan, E. (1970). Immunogenic properties of bee venom, *Bull. Apicult. Inform. Doc. Sci. Tech.,* **10,** 195.

Dirks, T. F., Jr. and Sternburg, J. G. (1972). Non-immunochemical complexing between hymenopteran venoms and rabbit sera, *Toxicon,* **10,** 381.

Drenth, D. (1974). Susceptibility of different species of insects to an extract of the venom gland of the wasp *Microbracon hebetor* (Say), *Toxicon,* **12,** 189.

References

Edery, H., Ishay, J., Lass, I., and Gitter, S. (1972). Pharmacological activity of Oriental hornet (*Vespa orientalis*) venom, *Toxicon,* **10,** 13.

Favilli, G. (1956). "Occurrence of spreading factors and some properties of hyaluronidases in animal parasites and venoms," in E. E. Buckley and N. Porges, Eds., *Venoms,* American Association for the Advancement of Science, Washington, D.C., pp. 281–289.

Feldberg, W. and Kellaway, C. H. (1937). Liberation of histamine and its role in the symptomatology of bee venom poisoning, *Aust. J. Exp. Biol. Med. Sci.,* **15,** 461.

Fennell, J. F., Shipman, W. H., and Cole, L. J. (1968). Antibacterial action of melittin, a polypeptide from bee venom, *Proc. Soc. Exp. Biol. Med.,* **127,** 707.

Fischl, J., Ishay, J., Goldberg, S., and Gitter, S. (1972). Investigation of protein fractions and haemolytic properties of wasp venom, *Acta Pharmacol. Toxicol.,* **31,** 65.

Foubert, E. L., Jr. and Stier, R. A. (1958). Antigenic relationships between honeybees, wasps, yellow hornets, black hornets, and yellow jackets, *J. Allergy,* **29,** 13.

Franklin, R. M. and Baer, H. (1974). Immune and nonimmune gel precipitates produced by honey bee venom and its components, *Proc. Soc. Exp. Biol. Med.,* **147,** 585.

Franklin, R. and Baer, H. (1975). Comparison of honeybee venoms and their components from various sources, *J. Allergy Clin. Immunol.,* **55,** 285.

Fredholm, B. (1966). Studies on a mast cell degranulating factor in bee venom, *Biochem. Pharmacol.,* **15,** 2037.

Fredholm, B. and Haegermark, O. (1967). Histamine release from rat mast cell granules induced by bee venom fractions, *Acta Physiol. Scand.,* **71,** 357.

Fredholm, B. and Haegermark, O. (1969). Studies on the histamine releasing effect of bee venom fractions and compound 48/80 on skin and lung tissue of the rat, *Acta Physiol. Scand.,* **76,** 288.

Gauldie, J., Hanson, J. M., Rumjanek, F. D., Shipolini, R. A. and Vernon, C. A. (1976). The peptide components of bee venom, *Eur. J. Biochem.,* **61,** 369.

Ginsberg, N. J., Dauer, M., and Slotta, K. H. (1968). Melittin used as a protective agent against X-irradiation, *Nature,* **220,** 1334.

Grzycki, S. and Czerny, K. (1972). Cytochemical studies on the poison gland of honeybee sting, *Acta Anat.,* **82,** 91.

Habermann, E. (1972). Bee and wasp venoms: The biochemistry and pharmacology of their peptides and enzymes are reviewed, *Science,* **177,** 314.

Habermann, E. and Breithaupt, H. (1968). MCL-peptide, a selectively mastocytolytic factor isolated from bee venom, *Naunyn-Schmiedebergs Arch. Exp. Pathol. Pharmakol.,* **260,** 127.

Habermann, E. and Cheng-Raude, D. (1975). Central neurotoxicity of apamin, crotamin, phospholipase A and α-amanitin, *Toxicon,* **13,** 465.

Habermann, E. and Jentsch, J. (1967). Sequential analysis of melittin from tryptic and peptic fragments, *Biochem. Z.,* **348,** 37.

Habermann, E. and Kowallek, H. (1970). Modifikationen der Aminogruppen und des Tryptophans im Melittin als Mittel zur Erkennung von Struktur-Wirkungs-Beziehungen, *Hoppe-Seylers Z. Physiol. Chem.,* **351,** 884.

Habermann, E. and Reiz, K. G. (1965). A new method for the separation of the components of bee venom, especially of the centrally active peptide apamine, *Biochem. Z.,* **343,** 451.

Habermann, E. and Zeuner, G. (1971). Comparative studies of native and synthetic melittins, *Naunyn-Schmiedebergs Arch. Exp. Pathol. Pharmakol.,* **270,** 1.

Hartter, P. and Weber, U. (1975). Basische Peptide des Bienengifts. I. Isolierung, Reduktion und Reoxidation von Apamin and MCD-Peptid, *Hoppe-Seylers Z. Physiol. Chem.,* **356,** 693.

Haux, P., Sawerthal, H., and Habermann, E. (1967). Sequenzanalyse des Bienengift-Neurotoxins (Apamin) aus seinen tryptischen und chymotryptischen Spaltstucken, *Hoppe-Seylers Z. Physiol. Chem.,* **348,** 737.

Higginbotham, R. D. and Karnella, S. (1970). The significance of the mast cell response to bee venom, *J. Immunol.,* **106,** 233.

Högberg, B. and Uvnas, B. (1957). The mechanism of the disruption of mast cells produced by compound 48/80, *Acta Physiol. Scand.,* **41,** 345.

Hort, I. and Herz, A. (1968a). Effect of the venom of the bee *Apis mellifera* on osmotic fragility of cattle RBC, *Toxicon,* **5,** 181.

Hort, I. and Hertz, A. (1968b). The influence of bee venom on the osmotic fragility of human red blood cells, *Experientia,* **24,** 254.

Hsiang, H. K. and Elliott, W. B. (1975). Differences in honey bee (*Apis mellifera*) venom obtained by venom sac extraction and electrical milking, *Toxicon,* **13,** 145.

Ikan, R. and Ishay, J. (1973). Free amino acids in the haemolymph and the venom of the Oriental hornet, *Vespa orientalis, Comp. Biochem. Physiol.,* **44B,** 949.

Ishay, J. (1975). Hyperglycemia produced by *Vespa orientalis* venom sac extract, *Toxicon,* **13,** 221.

Ishay, J., Fischl, J., and Gitter, S. (1972). Investigation of wasp venom: Antigenic relationship, *Acta Pharmacol. Toxicol.,* **31,** 71.

Ishay, J., Nadler, E. Z., and Gitter, S. (1973). Pharmacological activity of *Paravespura germanica* wasp venom, *Acta Pharmacol. Toxicol.,* **33,** 157.

Ishay, J., Lass, Y., Ben-Schachar, D., Gitter, S., and Sandbank, U. (1974). The effects of hornet venom sac extract on the electrical activity of the cat brain, *Toxicon,* **12,** 159.

Ishay, J., Lass, Y., and Sandbank, U. (1975a). A lesion of muscle transverse tubular system by Oriental hornet (*Vespa orientalis*) venom: Electron microscopic and histological study, *Toxicon,* **13,** 57.

Ishay, J., Ben-Shachar, D., Elazar, Z., and Kaplinsky, E. (1975b). Effects of mellitin on the central nervous system, *Toxicon,* **13,** 277.

Ivanov, C., Shkenderov, S., and Krusteva, M. A. (1972). Isolation and purification of hyaluronidase from bee venom, *Dokl. Bolg. Akad. Nauk,* **25,** 229.

Jaques, R. and Schachter, M. (1954). The presence of histamine, 5-hydroxytryptamine and a potent, slow contracting substance in wasp venom, *Br. J. Pharmacol.,* **9,** 53.

Jentsch, J. (1968). Amino acid sequence of melittin. I. Edman degradation of melittin to the 20th position, *Z. Naturforsch,* B, **23,** 1613.

Jentsch, J. (1969). Amino acid sequence of melittin. II. Preferred cleavage of valine, leucine, and isoleucine bonds by α-protease from *Crotalus atrox* venom, *Z. Naturforsch.,* B, **24,** 415.

Jentsch, J. and Dielenberg, D. (1972). Phospholipase A (EC 3.1.1.4) from bee venom. II. At least two phospholipases A in bee venom, *Justus Liebigs Ann. Chem.,* **757,** 187.

Jentsch, J. and Habermann, E. (1967). Structure of melittin, the toxic main peptide from bee venom, *Peptides, Proc. Eur. Peptide Symp.,* pp. 263–270.

Jonas, W. and Shugar, M. (1964). Severe hepatic and renal damage following wasp stings, *Dapim Refuiim (Folia Med.),* **22,** 3.

Joshua, H. and Ishay, J. (1973). Hemolytic properties of the Oriental hornet venom, *Acta Pharmacol. Toxicol.,* **33,** 42.

Joshua, H. and Ishay, J. (1975). The anti-coagulant properties of an extract from the venom sac of the Oriental hornet, *Toxicon,* **13,** 11.

Kanno, I., Ito, Y., and Okuyama, S. (1970). Radioprotection by bee venom, *J. Jap. Med. Radiat.,* **29,** 30.

Kaplinsky, E., Ishay, J., and Gitter, S. (1974). Oriental hornet venom: Effects on cardiovascular dynamics, *Toxicon,* **12,** 69.

Keller, R. and Schwarz, M. (1963). Versuch zu Charakterisierung der anaphylactischen Reakton isoliertier Gewebemast zellen der Albinoratte mittels spezifischer Antiseren, *Pathol. Microbiol.,* **26,** 100.

King, T. P., Sobotka, A. K., Kochoumain, L. and Lichtenstein, L. M. (1976). Allergens of honey bee venom, *Arch. Biochem. Biophys.,* **172,** 661.

Kireeva, V. F. (1970). Capillary permeability changes resulting from the effect of bee venom, *Uch. Zp Zap. Gor'k. Gos. Univ.,* **101,** 113.

Korneva, N. V. (1972). Mechanism of change of blood pressure and frequency of heartbeat caused by bee venom, *Uch. Zap. Gor'k. Gos. Univ.,* **56.**

Korneva, N. V. and Orlov, B. N. (1969). Effect of bee venom on cartoid chemoreceptors, *Biol. Nauk.,* **2,** 42.

Kreil, G. (1973a). Structure of melittin isolated from two species of honeybees, *FEBS Lett,* **33,** 241.

Kreil, G. (1973b). Biosynthesis of melittin, a toxic peptide from bee venom: Amino acid sequence of the precursor, *Eur. J. Biochem.,* **33,** 558.

Kreil, G. (1975). The structure of *Apis dorsata* melittin: Phylogenetic relationships between honeybees as deduced from sequence data, *FEBS Lett.,* **54,** 100.

Kreil, G. and Bachmayer, H. (1971). Biosynthesis of melittin, a toxic peptide from bee venom: Detection of a possible precursor, *Eur. J. Biochem.,* **20,** 344.

Kristeva, M., Mesrob, B., Ivanov, C., and Shkendervo, S. (1973). Partial characterization of hyaluronidase from bee venom, *Dokl. Bolg. Akad. Nauk,* **26,** 917.

Krylov, V. N. (1973). Effect of bee venom on the heart, *Mater. Povolzh Konf. Fiziol., Uchastiem Biokhim., Farmakol. Morfol.,* **1,** 86.

Langlois, C., Shulman, S., and Arbesman, C. E. (1965). The allergic response to stinging insects. III. The specific venom sac antigens, *J. Allergy,* **36,** 109.

Lawrence, A. J. and Moores, G. R. (1975). Activation of bee venom phospholipase A_2 by fatty acids, aliphatic anhydrides and glutaraldehyde, *FEBS Lett.,* **49,** 287.

Lorenzetti, O. J., Fortenberry, B., and Busby, E. (1972). Influence of bee venom in the adjuvant-induced arthritic rat mode, *Res. Commun. Chem. Pathol. Pharmacol.,* **4,** 339.

Lowy, P. H., Sarmiento, L., and Mitchel, H. K. (1971). Polypeptides minimine and melittin from bee venom: Effects on *Drosophila, Arch. Biochem. Biophys.,* **145,** 338.

Marumo, F., Sasaoka, T., Asano, Y., and Endou, H. (1968). The increasing effect of bee venom on the sodium transport of the toad bladder, *Proc. Jap. Acad.,* **44,** 569.

Maschwitz, U. W. J. and Kloft, W. (1971). "Morphology and function of the venom apparatus of insects – bees, wasps, ants, and caterpillars," in W. Bücherl and E. E. Buckley, Eds., *Venomous Animals and Their Venoms,* Vol. 3, Academic, New York, pp. 1–60.

Middleton, E. and Phillips, G. B. (1964). Distribution and properties of anaphylactic and venom induced slow reacting substance and histamine in guinea pigs, *J. Immunol.,* **93,** 220.

Mitchell, H. K., Lowy, P. H., Sarmiento, L., and Dickson, L. (1971). Melittin: Toxicity to *Drosophila* and inhibition of acetylcholinesterase, *Arch. Biochem. Biophys.,* **145,** 344.

Mohamed, A. H. and El Karemi, M. M. (1961). Immunity of bee-keepers to some components of bee venom: Phospholipase A antibodies, *Nature,* **189,** 837.

Mollay, C. and Kreil, G. (1973). Fluorometric measurements on the interaction of melittin with lecithin, *Biochim. Biophys. Acta,* **316,** 196.

Mollay, C. and Kreil, G. (1974). Enhancement of bee venom phospholipase A_2 activity by melittin, direct lytic factor from cobra venom and polymyxin B, *FEBS Lett.,* **46,** 141.

Moores, G. R. and Lawrence, A. J. (1972). Conductiometric assay of phospholipids and phospholipase A, *FEBS Lett.,* **28,** 201.

Munjal, D. and Elliott, W. B. (1971a). Studies of antigenic fractions in honeybee (*Apis mellifera*) venom, *Toxicon,* **9,** 229.

Munjal, D. and Elliott, W. B. (1971b). A simple method for the isolation of a phospholipase A from honeybee (*Apis mellifera*) venom, *Toxicon,* **9,** 403.

Munjal, D. and Shivpuri, D. N. (1968). "Preliminary studies on snake venom phospholipase A and its effect on tissues," in *Aspects of Allergy and Applied Immunology,* New Height Publishers, Delhi, p. 57.

Nagamitu, G. (1935). Beitrage zur physiologischen Wirkung des Histamins: Über das Gift der Honigbienen, *Okayama Igakkai Zasshi,* **47,** 3005.

Nair, C., Hermans, J., Munjal, D., and Elliott, W. B. (1976). Temperature stability of phospholipase A, activity. I. Bee (*Apis mellifera*) venom phospholipase A_2, *Toxicon,* **14,** 35.

Nelson, D. A. and O'Connor, R. (1968). The venom of the honeybee (*Apis mellifera*): Free amino acids and peptides, *Can. J. Biochem.,* **46,** 1221.

Neumann, W. and Habermann, E. (1954). Beiträge zur Charakterisierung der Wirkstoffe des Bienengiftes, *Naunyn-Schmiedebergs Arch. Exp. Pathol. Pharmakol.,* **222,** 367.

Neumann, W. and Habermann, E. (1956). "Paper electrophoresis separation of pharmacologically and biochemically active components of bee and snake venoms," in: E. E. Buckley and N. Porges, Eds., *Venoms,* American Association for the Advancement of Science, Washington, D.C., pp. 171–174.

Neumann, W. and Stracke, A. (1951). Untersuchen mit Bienegift und Histamin an der Formaldehydearthritis der Ratte, *Naunyn-Schmiedebergs Arch. Exp. Pathol. Pharmakol.,* **213,** 8.

O'Connor, R. and Erickson, R. (1965). Hymenoptera antigens: An immunological comparison of venoms, venom sac extracts and whole insect extracts, *Ann. Allergy,* **23,** 151.

O'Connor, R. and Peck, M. L. (1975). Venoms of Apidae, *Intern. Encyclo. Pharmacol.,* Spring Verlag.

O'Connor, R., Rosenbrook, W., Jr., and Erickson, R. (1964a). Disc electrophoresis of Hymenoptera venoms and body proteins, *Science,* **145,** 1320.

O'Connor, R., Stier, R. A., Rosenbrook, W., Jr., and Erickson, R. W. (1964b). Death from "wasp" sting, *Ann. Allergy,* **22,** 385.

O'Connor, R., Henderson, G., Nelson, D., Parker, R., and Peck, M. L. (1967). The venom of the honeybee (*Apis mellifera*): General character, *Animal Toxins,* Pergamon, New York, p. 17.

Olson, F. C., Munjal, D., and Malviya, A. N. (1974). Structural and respiratory effects of melittin (*Apis mellifera*) on rat liver mitochondria, *Toxicon,* **12,** 419.

Omarov, Sh. M. (1970). Effects of bee venom and cobra venom on fibrinogen and fibrinolytic activity in blood, *Uch. Zap. Gor'k. Gos. Univ.,* **101,** 122.

Owen, M. D. (1971). Insect venoms: Identification of dopamine and noradrenaline in wasp and bee stings, *Experientia,* **27,** 544.

Owen, M. D. and Braidwood, J. L. (1974). A quantitative and temporal study of histamine and histidine in honey bee (*Apis mellifera* L.) venom, *Can. J. Zool.,* **52,** 387.

Peck, M. L. and O'Connor, R. (1974). Procamine and other basic peptides in the venom of the honeybee (*Apis mellifera*), *J. Agr. Food Chem.,* **22,** 51.

Piek, T. (1966). Site of action of the venom of the digger wasp, *Philanthus triangulum* F., on the fast neuromuscular system of the locust, *Toxicon,* **4,** 191.

Piek, T. and Njio, K. D. (1975). Neuromuscular block in honeybees by the venom of the bee wolf wasp (*Philanthus triangulum* F.), *Toxicon,* **13,** 199.

Piek, T. and Simon Thomas, R. T. S. (1969). Paralyzing venoms of solitary wasps, *Comp. Biochem. Physiol.,* **30,** 13.

Piek, T., Spanjer, W., Njio, K. D., Veenendaal, R. L., and Mantel, P. (1974). Paralysis caused by the venom of the wasp *Microbracon gelechiae, J. Insect Physiol.,* **20,** 2307.

Pisano, J. J. (1966a). Structure of the major kinin in wasp venom, *Int. Symp. Vaso-Active Polypeptides: Bradykinin Related Kinins,* Ribeirao Preto, Brazil, p. 35.

Pisano, J. J. (1966b). Wasp kinin, *Mem. Inst. Butantan,* **33,** 441.

Rathmayer, W. (1962a). Paralysis caused by the digger wasp *Philanthus, Nature,* **196,** 1148.

Rathmayer, W. (1962b). Das Paralysierungs problem beim Bienewolf *Philanthus triangulum, Z. Vergl. Physiol.,* **45,** 143.

Rauen, H. M., Schriewer, H., and Ferie, F. (1972). Alkylans alkylandum reactions. 10. Antialkylating activity of bee venom, melittin, and apamin, *Arzneim-Forsch,* **22,** 1921.

Reinert, M. (1936). *Zur Kenntnis des Bienengiftes Festschrift,* Emil Barell, Basel.

Rosenbrook, W., Jr., and O'Connor, R. (1964a). The venom of the mud-dauber wasp. II. *Sceliphron caementarium:* Protein content, *Can. J. Biochem.,* **42,** 1005.

Rosenbrook, W. and O'Connor, R. (1964b). The venom of the mud-dauber wasp. III. *Sceliphron caementarium:* General character, *Can. J. Biochem.,* **42,** 1567.

Rothenbacher, H. and Benton, A. W. (1972). Pathologic features in mice hyposensitized to bee venom, *Am. J. Vet. Res.,* **33,** 1867.

Rothschild, A. M. (1965). Histamine release by bee venom phospholipase A and mellitin in the rat, *Br. J. Pharmacol.,* **25,** 59.

Saelinger, C. and Higginbotham, R. D. (1974). Significance of heparin in the local response to bee sting, *Tex. Rept. Biol. Med.,* **32,** 553.

Salt, G. (1931). Parasites of the wheat-germ sawfly, *Cephus pygmeus* Linneus, in England, *Bull. Entomol. Res.,* **22,** 479.

References

Sandbank, U., Ishay, J., and Gitter, S. (1971). Mitochondrial changes in the guinea pig muscle after envenomation with *Vespa orientalis* venom, *Experientia,* **27,** 303.

Sandbank, U., Ishay, J., and Gitter, S. (1973). Kidney changes in mice due to Oriental hornet (*Vespa orientalis*) venom: Histological and electron microscopical study, *Acta Pharmacol. Toxicol.,* **32,** 442.

Saslavasky, H., Ishay, J., and Ikan, R. (1973). Alarm substances as toxicants of the Oriental hornet, *Vespa orientalis, Life Sci.,* **12,** 135.

Schachter, M. and Thain, E. M. (1954). Chemical and pharmacological properties of the potent, slow contracting substance (kinin) in wasp venom, *Br. J. Pharmacol.,* **9,** 352.

Schaller, H. (1964). Anaphylakyischer Schock und schwerer Hirnschaden als Komplikation eines Hymenopterenstiches, *Schweiz. Arch. Neurol. Neurochir. Psychiatr.,* **94,** 92.

Schröder, E., Lübke, K., Lehmann, M., and Beetz, I. (1971). Haemolytic activity and action on the surface tension of aqueous solutions of synthetic melittins and their derivative, *Experientia,* **27,** 764.

Sergeeva, L. I. (1974). Heparin-induced inhibition of the hemolytic activity of bee venom, *Uch. Gor'k. Gos. Univ.,* **175,** 130.

Sessa, G., Freer, J. G., Colacicco, G., and Weismann, G. (1969). Interaction of a lytic polypeptide, melittin, with lipid membrane systems, *J. Biol. Chem.,* **244,** 3575.

Shepherd, G. W., Elliott, W. B., and Arbesman, C. E. (1974). Fractionation of bee venom. I. Preparation and characterization of four antigenic components, *Prep. Biochem.,* **4,** 71.

Shilkin, K. B., Chen, B. T. M., and Khoo, O. T. (1972). Rhabdomyolysis caused by hornet venom, *Br. Med. J.,* **1,** 156.

Shipman, W. H. and Cole, L. J. (1967). Increased resistance of mice to X-irradiation after the injection of bee venom, *Nature,* **215,** 311.

Shipman, W. H. and Cole, L. J. (1968). Surfactant bee venom fraction: Separation on a newly devised constant-flow-rate chromatographic column and detection by changes in effluent drop volume, *U. S. Clearinghouse Fed. Sci. Tech. Inform.,* AD-677597.

Shipman, W. H. and Cole, L. J. (1972). Complex formation between bee venom melittin and extract of mouse skin detected by Sephadex gel filtration, *Experientia,* **28,** 171.

Shipolini, R. A., Callewaert, G. L., Cottrell, R. C., and Vernon, C. A. (1971). The primary sequence of phospholipase A from bee venom, *FEBS Lett.,* **17,** 39.

Shipolini, R. A., Callewaert, G. L., Cottrell, R. C., and Vernon, C. A. (1974a). The amino-acid sequence and carbohydrate content of phospholipase A_2 from bee venom, *Eur. J. Biochem.,* **48,** 465.

Shipolini, R. A., Doonan, S., and Vernon, C. A. (1974b). The disulphide bridges of phospholipase A_2 from bee venom, *Eur. J. Biochem.,* **48,** 477.

Shkenderov, S. (1973). Protease inhibitor in bee venom: Identification, partial purification, and some properties, *FEBS Lett.,* **33,** 343.

Shkenderov, S. (1974). Anaphylactogenic properties of bee venom and its fractions, *Toxicon,* **12,** 529.

Shulman, S., Bigelsen, F., Lang, R., and Arbesman, C. (1966). Allergic response to stinging insects: Biochemical and immunologic studies of bee venom and other bee body preparations, *J. Immunol.,* **96,** 29.

Slotta, K. H. and Vick, J. A. (1969). Identification of the direct lytic factor from cobra venom as cardiotoxin, *Toxicon,* **6,** 167.

Slotta, K. H., Vick, J. A., and Ginsberg, N. J. (1971). "Enzymatic and toxic activity of phospholipase A," in A. De Vries and E. Kochva, Eds., *Toxins of Animal and Plant Origin,* Vol. I, Gordon and Breach, New York, 401–418.

Spoerri, P. E., Jentsch, J., and Glees, P. (1973). Apamin from bee venom: Effects of the neurotoxin on cultures of the embryonic mouse cortex, *Neurobiology,* **3,** 207.

Spoerri, P. E., Jentsch, J., and Glees, P. (1975). Apamin from bee venom. II. Effects of the neurotoxin on subcellular particles of neural cultures, *FEBS Lett.,* **53,** 143.

Tetsch, C. and Wolff, K. (1936). Untersuchungen über Analysen zwischen Bienen und Schlangen (*Crotalus*)-Gift, *Biochem. Z.,* **288,** 126.

Van der Meer, C., Drenth, D., Nijhof, J. K., Piek, T., and Simon, R. T. (1965). Paralyzing animal poisons, *U. S. Dept. Comm.,* AD624456.

Van Rietschoten, J., Granier, C., Rochat, H., Lissitzky, S., and Miranda, F. (1975). Synthesis of apamin, a neurotoxic peptide from bee venom, *Eur. J. Biochem.,* **56,** 35.

Vázquez-Colón, L. and Elliott, W. B. (1966). On the response of rat liver mitochondria to treatment with bee venom, *Toxicon,* **4,** 61.

Vernon, C. A., Hanson, J. M., and Brimblecombe, R. W. (1969). Brit. Pat. No. 1314823.

Vick, J. A. and Brooks, R. B., Jr. (1972). Pharmacological studies of the major fractions of bee venom, *Am. Bee. J.,* **112,** 288.

Vick, J. A. and Shipman, W. H. (1972). Effects of whole bee venom and its fractions (apamin and melittin) on plasma cortisol levels in the dog, *Toxicon,* **10,** 377.

Vick, J. A., Shipman, W. H., Brooks, R. B., Jr., and Hassett, C. C. (1972a). The β-adrenergic and anti-arrhythmic effects of apamin, a component of bee venom, *Am. Bee J.,* **112,** 339.

Vick, J. A., Brooks, R. B., Jr., and Shipman, W. H. (1972b). Therapeutic applications of bee venom and its components in the dog, *Am. Bee J.,* **112,** 414.

Vick, J. A., Mehlman, B., Brooks, R., Phillips, S. J., and Shipman, W. (1972c). Effect of bee venom and melittin on plasma cortisol in the unanesthetized monkey, *Toxicon,* **10,** 581.

Vick, J. A., Mehlman, B., Brooks, R. B., Philips, S. J., and Shipman, W. (1972d). Effect of bee venom and melittin on the plasma cortisol level in unanesthetized monkeys, *Tr. Mezhdunar. Simp. Primen. Prod. Pchelovod, Med. Vet.,* **49.**

Vick, J. A., Shipman, W. H., and Brooks, R., Jr. (1974). Beta adrenergic and anti-arrhythmic effects of cardiopep, a newly isolated substance from whole bee venom, *Toxicon,* **12,** 139.

Visser, B. J. and Spanjer, W. (1969). Biochemical study of two paralysing insect venoms, *Acta Physiol. Pharmacol. Néerl.,* **15,** 107.

Vogt, W., Patzer, R., Lege, L., Oldigs, H. D., and Willie, G. (1970). Synergism between phospholipase A and various peptides and SH-reagents in causing hemolysis, *Naunyn-Schmiedebergs Arch. Exp. Pathol. Pharmakol.,* **265,** 442.

Von-Haux, P. (1969). Die Aminosäurensequenz von MCD-Peptid, einem spezifisch Mastzellen-degranulierenden Peptid aus Bienengift, *Hoppe-Seylers Z. Physiol. Chem.,* **350,** 536.

Vyatchannikov, N. K. and Sinka, A. Y. (1973). Effect of melittin, the major constituent of bee venom, on the central nervous system, *Farmakol. Toksikol.,* **36,** 625.

Walther, C. and Rathmayer, W. (1974). Effect of *Microbracon* venom on excitatory neuromuscular transmission in insects, *J. Comp. Physiol.,* **89,** 23.

Williams, J. C. and Bell, R. M. (1972). Membrane matrix disruption by melittin, *Biochim. Biophys. Acta,* **228,** 255.

Yoshida, H., Geller, R. G., and Pisano, J. J. (1976). Vespulakinins: new carbohydrate containing bradykinin derivatives, *Biochemistry,* **15,** 61.

Yurkiewicz, W. J. (1968). Phospholipids of the greater wax moth, *Galleria mellonella,* as affected by bee venom, *Proc. Pa. Acad. Sci.,* **42,** 20.

31 Ant Venoms

1	BULLDOG ANT VENOMS 1.1 Toxicology, 527 1.2 Biochemistry, 528	527
2	FIRE ANT VENOMS	528
3	OTHER ANT VENOMS	529
	References	529

Ants belong to the order Hymenoptera, to which bees and wasps also belong, and the family Formicidae. There are nine subfamilies of ants. Relatively little characterization has been done on ant venoms.

1 BULLDOG ANT VENOMS

The Australian bulldog ant, *Myrmecia pyriformis* (*M. forficata*), is found in south Australia. Upon envenomation, it produces a sting similar to, but more painful than, that of a bee.

1.1 Toxicology

The venom is contained in an abdominal sac and is injected into the victim by means of a retractable stinging organ at the tip of the abdomen.

The local reaction is typified by a wheal and erythema, which are immediate in onset and are followed by edema and tenderness or itching that sometimes persists for several days (Lewis and De La Lande, 1967). The venom contains a muscle-stimulating component that produces a slow, persistent contraction of rat uterus and of mepyramine-treated guinea pig ileum, and a prolonged hypotensive response in anesthetized cats (De La Lande et al., 1963a, b, 1965).

The LD_{50} in mice is 2 to 5 $\mu g \ g^{-1}$ by intraperitoneal injection.

1.2 Biochemistry

The venom is a rich source of histamine, which comprises 1 to 3% by dry weight (De La Lande et al., 1963a, 1965). As mentioned above, it also contains a smooth muscle-stimulating substance. Hyaluronidase and a direct hemolytic component are also present in the venom (Cavill et al., 1964). In addition to containing endogeneous histamine, the venom is capable of releasing histamine upon envenomation (Thomas and Lewis, 1965; De La Lande and Lewis, 1966). However, the venom does not possess bradykinin-releasing activity (Lewis and De La Lande, 1967). The venom possesses phospholipase A_2 activity (Lewis et al., 1968). On incubation with liver phosphatidylcholine, a large amount of unsaturated fatty acids (oleic, linoleic, and arachidonic acids) is released, indicating that the site of hydrolysis is at the β position.

The venom of *Myrmecia gulosa* inhibits the respiration of insect mitochondria; the site of inhibition is between cytochromes b and c in the terminal electron transfer system (Ewen and Ilse, 1970). The inhibitor is a small molecular weight basic protein. Inhibition of mitochondrial respiration can be achieved by using either succinate or α-glycerophosphate as substrate. Therefore the inhibition is not at the substrate level.

A smooth muscle-stimulating component was isolated in pure form from the venom of *M. pyriformis* by Wanstall and De La Lande (1974). The component is free from enzyme activity but has red cell-lysing and histamine-releasing activity and resembles the major peptide from bee venom, melittin.

2 FIRE ANT VENOMS

In the United States there are two common subspecies of *Solenopsis saevissima* that are important to public health: *S. saevissima saevissima* and *S. saevissima richteri*, believed to have arrived here from South America in 1919. They are usually found in the southeastern part of the United States. In addition, *Solenopsis geminata* commonly occurs in Texas and Florida, and *S. xyloni* in the southeastern United States. The biology of venomous ants of the genus *Solenopsis* has been described in detail by San Martin (1971).

The venom of *S. saevissima richteri* inhibits Na(I), K(I)-ATPase, and oligomycin-sensitive Mg(II)-ATPase (Koch and Desaiah, 1975). It was observed that fire ant abdomen homogenates and whole ant homogenates are essentially devoid of ATPase activities. It is suggested by Koch and Desaiah that fire ant venom is responsible for the lack of observable ATPase activity in the homogenates.

It is rather interesting that *S. saevissima*, *S. geminata*, and *S. xyloni* venoms contain only a small amount of protein. The major components are various 2-methyl-6-*n*-alkyl(or alkenyl) piperidines (MacConnell et al., 1971, 1976; Brand et al., 1972) having the following structures:

I.	R = $(CH_2)_{10}$Me	II.	R = $(CH_2)_{10}$Me
III.	R = $(CH_2)_{12}$Me	IV.	R = $(CH_2)_{12}$Me
V.	R = $(CH_2)_{14}$Me	VI.	R = $(CH_2)_{14}$Me
VII.	R = *cis*-$(CH_2)_3$CH:CH$(CH_2)_7$Me	VIII.	R = *cis*-$(CH_2)_3$CH:CH$(CH_2)_7$Me
IX.	R = *cis*-$(CH_2)_5$CH:CH$(CH_2)_7$Me	X.	R = *cis*-$(CH_2)_5$CH:CH$(CH_2)_7$Me

3 Other Ant Venoms

Gas chromatography and mass spectra were used to identify *cis*-2-methyl-6-undecylpiperidine (I), *trans*-2-methyl-6-undecylpiperidine (II), *cis*-2-methyl-6-tridecylpiperidine (III), *trans*-2-methyl-6-tridecylpiperidine (IV), *cis*-2-methyl-6-pentadecylpiperidine (V), *trans*-2-methyl-6-pentadecylpiperidine (VI), *cis*-2-methyl-6-(*cis*-4-tridecen-1-yl)piperidine (VII), *trans*-2-methyl-6-(*cis*-4-tridecen-1-yl)piperidine (VIII), *cis*-2-methyl-6-(*cis*-6-pentadecen-1-yl) piperidine (IX), and *trans*-2-methyl-6-(*cis*-6-pentadecen-1-yl)piperidine (X). The venom of the red *S. saevissima* contained all 10 compounds plus another that was not identified, with the trans forms predominating and the cis forms present as traces. Compound X was present in the largest amount, with lower and about equal amounts of IV and VIII. The venom of the black form of *S. saevissima* had mainly I, II, III, IV, VII, and VIII; only traces of IX and X were found, and the amount of VIII was much greater than that of IV. Obviously, the black form is not just a color variation. Venoms of *S. xyloni* and *S. geminata* contained I, II, III, and VII. Compounds I and II were the major components, III and VII were present only in traces, and none of the C_{15} side-chain compounds was found.

The venom of *S. xyloni* contained a compound not found in the other venoms; degradation and mass spectra showed it to be 2-methyl-6-undecyl-$\Delta^{1,2}$-piperideine. The structure of this compound is as follows:

$$CH_3 \overset{}{\diagup} N \diagdown (CH_2)_{10}-CH_3$$

It may be a precursor or an intermediate in the metabolism of the other components.

The ratio of *cis*-2-methyl-6-*n*-undecylpiperidine (cis C_{11}) to *trans*-2-methyl-6-*n*-undecylpiperidine (trans C_{11}) in the venom of individual workers and soldiers of *S. geminata* and in individual alate queens of *S. geminata*, *S. xyloni*, *S. invicta*, and *S. richteri* was estimated by Brand et al. (1973). A considerable variation in this ratio was demonstrated between individuals of a particular caste within a species. In spite of considerable individual variation, the various species nevertheless exhibit a fair degree of control of the biosynthetic system for these 2,6-disubstituted piperidine alkaloids. Results obtained on one colony of *S. geminata* soldiers suggested that these individuals may have arisen from at least two fertile queens in the nest. Venoms and synthetic venom alkaloids had no significant effects on phrenic nerve–diaphragm or mammalian cardiovascular survey preparations (Buffkin and Russell, 1974).

3 OTHER ANT VENOMS

Venom of the desert ant, *Pogonmyrmex barbatus*, causes piloerection and sweating (Williams and Williams, 1964). The LD_{50} in mice is $1.29\,\mu g\,g^{-1}$ by intraperitoneal injection (Williams and Williams, 1965). The death of envenomated animals was preceded by a characteristic twisting and stretching of the abdomen, followed first by respiratory distress and ataxia and then by extreme lethargy with labored, slowed breathing.

On the basis of ultraviolet absorption, the venom appears to be proteinaceous.

REFERENCES

Brand, J. M., Blum, M. S., Fales, H. M., and McConnell, J. G. (1972). Fire ant venoms: Comparative analyses of alkaloidal components, *Toxicon,* **10,** 259.

Brand, J. M., Blum, M. S., and Barlin, M. R. (1973). Fire ant venoms: Intraspecific and interspecific variation among castes and individuals, *Toxicon,* **11,** 325.

Buffkin, D. C. and Russell, F. E. (1974). A study of the venom of the imported fire ant: Physiopharmacology, chemistry, and therapeutics, *Proc. West. Pharmacol. Soc.,* **17,** 223.

Cavill, G. W. K., Robertson, P. L., and Whitfield, F. B. (1964). Venom and venom apparatus of the bull ant, *Myrmecia gulosa, Science,* **146,** 79.

De La Lande, I. S. and Lewis, J. C. (1966). Constituents of the venom of the Australian bull ant, *Myrmecia pyriformis, Mem. Inst. Butantan,* **33,** 951.

De La Lande, I. S., Thomas, D. W., and Tyler, M. J. (1963a). Preliminary analysis of the venom of the bulldog ant *(Myrmecia forficata), Biochem. Pharmacol.* (Conf. Issue), **12,** 187.

De La Lande, I. S., Thomas, D. W., and Tyler, M. J. (1963b). "Pharmacological analysis of the venom of the 'bulldog' ant *Myrmecia forficata,*" *Recent Advances in the Pharmacology of Toxins,* Pergamon, New York, pp. 71–75.

De La Lande, I. S., Thomas, D. W., and Tyler, M. J. (1965). Pharmacological analysis of the venom of the bulldog ant *(Myrmecia forficata), Proc. Second Int. Pharmacol. Meet., Prague,* **9,** 71.

Ewen, L. M. and Ilse, D. (1970). An inhibitor of mitochondrial respiration in venom of the Australian bulldog ant, *Myrmecia gulosa, J. Insect. Physiol.,* **16,** 1531.

Koch, R. B. and Desaiah, D. (1975). Sensitivity of ATPase activities to fire ant venom and abdomen preparations, *Life Sci.,* **17,** 1315.

Lewis, J. C. and De La Lande, I. S. (1967). Pharmacological and enzymic constituents of the venom of an Australian 'bulldog' ant, *Myrmecia pyriformis, Toxicon,* **4,** *225.*

Lewis, J. C., Day, A. J., and De La Lande, I. S. (1968). Phospholipase A in the venom of the Australian bulldog ant, *Myrmecia pyriformis, Toxicon,* **6,** 109.

MacConnell, J. G., Blum, M. S., and Fales, H. M. (1971). The chemistry of fire ant venom, *Tetrahedron,* **26,** 1129.

MacConnell, J. G., Blum, M. S., Buren, W. F., Williams, R. W., and Fales, H. M. (1976). Fire ant venoms: Chemotaxonomic correlations with alkaloidal compositions, *Toxicon,* **14,** 69.

San Martin, P. R. (1971). "The venomous ants of the genus *Solenopsis,*" in W. Bücherl and E. E. Buckley, Eds., *Venomous Animals and Their Venoms,* Vol. 3, Academic, New York, pp. 95–102.

Thomas, D. W. and Lewis, J. C. (1965). Histamine release by the ant, *Aust. J. Exp. Biol. Med. Sci.,* **43,** 275.

Wanstall, J. C. and De La Lande, I. S. (1974). Fractionation of bulldog ant venom, *Toxicon,* **12,** 649.

Williams, M. W. and Williams, C. S. (1964). Collection and toxicity studies of ant venom, *Soc. Exp. Biol. Med.,* **116,** 161.

Williams, M. W. and Williams, C. S. (1965). Toxicity of ant venom, further studies of the venom from *Pogonomyrmex barbatus, Proc. Soc. Exp. Biol. Med.,* **119,** 344.

32 Gila Monster Venoms

 1 TOXICOLOGY 532

 2 CHEMISTRY 533

   References 533

The Gila monster is a lizard belonging to the order Squamata, the same order to which the poisonous snakes belong. This is the only lizard that produces a venom. There are two species of *Heloderma: H. suspectum* and *H. horridum. Heloderma suspectum* is composed of two subspecies: *H. suspectum cinctum* and *H. suspectum suspectum* (Fig. 1).

Figure 32.1 *Heloderma suspectum suspectum* (Gila monster), captured in southern Utah and kept in captivity at Utah State University. (Photograph by the author.)

Heloderma horridum is generally acknowledged to be composed of three subspecies: *H. horridum exasperatum*, *H. horridum horridum*, and *H. horridum alvarezi*.

The Gila monster is restricted to North America, ranging from west-central Chiapas, Mexico on the headwaters of the Rio de Chiapa on Atlantic drainage, and northwards along the Pacific coastal areas and inland to the extreme southwest corner of Utah and the southern tip of Nevada.

1 TOXICOLOGY

The LD_{50} of *Heloderma suspectum* has been determined by a number of investigators, with the following results:

Animal Used	Route of Injection	$LD_{50} (\mu g\ g^{-1})$	Reference
Mice	i.v.	2.0	Tu and Murdock, 1967
Mice	–	0.8	Mebs and Raudonat, 1967
Mice	i.p.	3.0	Patterson, 1967a
Mice	s.c.	4.0	Stýblová and Kornalik, 1967
Mice	i.v.	0.4	Stýblová and Kornalik, 1967
Rat	s.c.	14.0	Stahnke et al., 1970
Rat	Intracardial	1.4	Patterson, 1967a

Heloderma horridum venom gave an LD_{50} value of $0.8\ \mu g\ g^{-1}$ body weight in mice (Mebs and Raudonat, 1967).

As far back as 1897, Santesson extracted the venom from *H. suspectum* and injected it into frogs, white mice, and rabbits. In frogs, he observed a curarelike action on the respiratory center. Dyspnea occurred in mice and preceded respiratory failure. Loeb et al. (1913) also observed that the main cause of death was respiratory failure. In addition, they reported that the venom caused hemorrhage in the gastrointestinal tract and congestion and edema in the lungs.

Dogs and rats envenomated with *H. suspectum* venom developed hypotension, tachycardia, and ventilatory difficulties (Patterson, 1967a). The venom had the immediate effect of causing a reduction in carotid blood flow. This was followed by hypotension and changes in intrathoracic and postcaval blood pressures. The venom contains a relatively heat-stable, noncholinergic, and nonhistamine smooth muscle-stimulating factor (Patterson, 1967a, b). The venom did not have significant effect on the blood coagulation system in rabbit and cat (Patterson and Lee, 1969).

The fraction having hemorrhagic effect also possesses arginine ester-hydrolyzing activity (Mebs, 1972). Hemorrhage occurs in the intestines, kidneys, and lungs. Tubular necrosis is seen affecting the distal as well as the proximal convoluted tubules. The lumina of the tubules and the space of Bowman's capsules are filled with a granular or hyaline substance.

2 CHEMISTRY

Gila monster venom consists primarily of proteins, and the toxic principles are in nondialyzable fractions. Both *Heloderma suspectum* and *H. horridum* venoms contain hyaluronidase, phospholipase A_2, kinin-releasing activity, and slight proteolytic activity at pH 8.5 (Mebs and Raudonat, 1966; Mebs, 1968). They also contain sodium, potassium, calcium, and magnesium and traces of aluminium, zinc, copper, iron, and silicon (Mebs and Raudonat, 1967). No hemolytic activity against washed human erythrocytes was detected in either venom.

Heloderma suspectum suspectum venom shows arginine esterase as well as peptidases (Tu and Murdock, 1967; Murdock, 1967). The venom hydrolyzed Ala–Gly, Ala–Phe, Ala–Val, Gly–Gly, Gly–Leu, Gly–Tyr, Gly–Val, Leu–Gly, Leu–Phe, Phe–Gly, Gly–Gly–Ala, Gly–Phe–Phe, and Leu–Gly–Phe. The dipeptides not hydrolyzed were Gly–Asp, Gly–Glu, and Leu–Val.

A kallikrein isolated from the venom of *H. suspectum* (Mebs, 1969a, b) releases kinin from plasmaglobulin and hydrolyzes arginine esters. The esterase and kinin-releasing activities are completely blocked by DFP, and the release of kinin by the enzyme is inhibited by BAEE and TAME. The molecular weight is approximately 76,000.

L-Amino acid oxidase, phospholipase A_2, fibrinolytic activity, hyaluronidase (spreading factor), phosphodiesterase, and proteolytic activities have been detected in the venom of *H. suspectum* (Stýblová and Kornalik, 1967). Serotonin is also present in this venom (Zarafonetis and Kalas, 1960).

REFERENCES

Loeb, L., Alsberg, C. L., Cooke, E., Corson-White, E. P., Heisher, M. S., Fox, H., Githens, T. S., Leopold, S., Meyers, M. K., Rehfuss, M. E., Rivas, D., and Tuttle, L. (1913). The venom of *Heloderma*, *Carnegie Inst. Publ.*, 177, Washington, D.C.

Mebs, D. (1968). Some studies on the biochemistry of the venom gland of *Heloderma horridum*, *Toxicon*, 5, 225.

Mebs, D. (1969a). Isolierung und Eigenschaften eines Kallikreins aus dem Gift der Krustenechse *Heloderma suspectum*, *Hoppe-Seylers Z. Physiol. Chem.*, 350, 821.

Mebs, D. (1969b). Purification and properties of a kinin liberating enzyme from the venom of *Heloderma suspectum*, *Naunyn-Schmiedebergs Arch. Exp. Pathol. Pharmakol.*, 264, 280.

Mebs, D. (1972). "Biochemistry of *Heloderma* venom," in A. De Vries and E. Kochva, Eds., *Toxins of Animal and Plant Origin*, Vol. 2, Gordon and Breach, New York, pp. 499–514.

Mebs, D. and Raudonat, H. W. (1966). Biochemical investigations on *Heloderma* venom, *Mem. Inst. Butantan*, 33, 907.

Mebs, D. and Raudonat, H. W. (1967). Biochemistry of venom of the Gila monsters *Heloderma suspectum* and *Heloderma horridum*, *Naturwissenschaften*, 54, 494.

Murdock, D. S. (1967). *A Biochemical Investigation of Selected Venoms: Cobra and Gila Monster*, M.S. thesis, Utah State University.

Patterson, R. A. (1967a). Some physiological effects caused by venom from the Gila monster, *Heloderma suspectum*, *Toxicon*, 5, 5.

Patterson, R. A. (1967b). Smooth muscle stimulating action of venom from the Gila monster, *Heloderma suspectum*, *Toxicon*, 5, 11.

Patterson, R. A. and Lee, I. S. (1969). Effects of *Heloderma suspectum* venom on blood coagulation, *Toxicon*, 7, 321.

Santesson, C. G. (1897). Über das Gift von *Heloderma suspectum,* einer giftigen eidechse, *Nord. Med. Ark.,* **30,** 1.

Stahnke, H. L., Heffron, W. A., and Lewis, D. L. (1970). Bite of the Gila monster, *Rocky Mt. Med. J.,* **67,** 25.

Stýblová, Z. and Kornalik, F. (1967). Enzymatic properties of *Heloderma suspectum* venom, *Toxicon,* **5,** 139.

Tu, A. T. and Murdock, D. S. (1967). Protein nature and some enzymatic properties of the lizard *Heloderma suspectum suspectum* (Gila monster) venom, *Comp. Biochem. Physiol.,* **22,** 389.

Zarafonetis, C. J. D. and Kalas, J. P. (1960). Serotonin, catecholamines, and amine oxidase activity in the venoms of certain reptiles, *Am. J. Med. Sci.,* **24,** 764.

Appendix Common Names of Poisonous Snakes

For scientific research, it is best to use scientific names. Common names are sometimes convenient in that they are easier to pronounce, but can lead to confusion because one species may have many different common names. For clarification, the common names of some poisonous snakes are listed here. No attempt has been made to include the names of all snakes.

Hydrophiidae (Sea Snakes)

Aipysurus laevis	Olive-brown sea snake
Astrotia stokesii	Stokes's sea snake
Enhydrina schistosa	Beaked sea snake, common sea snake
Hydrophis caerulescens	Many-toothed sea snake
H. cyanocinctus	Annulated sea snake
H. fasciatus	Banded small-headed sea snake
H. lapemoides	Persian Gulf sea snake
H. mamillaris	Bombay sea snake
H. ornatus	Reef sea snake
H. spiralis	Yellow sea snake
Lapemis curtus	Short sea snake
L. hardwickii	Hardwicke's sea snake
Laticauda colubrina	Yellow-lipped sea snake
Microcephalophis gracilis	Graceful small-headed sea snake, common small-headed sea snake
M. cantoris	Cantor's small-headed sea snake
Pelamis platurus	Pelagic sea snake, yellow-bellied sea snake
Praescutata viperina	Viperine sea snake

Elapidae (Elapids)

Acanthophis antaricus	Death adder
Boulengerina spp.	Water cobras
B. annulata	Banded water cobra
Bungarus spp.	Kraits
B. candidus	Common krait
B. fasciatus	Banded krait
B. multicinctus	Formosan banded krait
Calliophis spp.	Oriental coral snakes
Dendraspis (Dendroaspis) spp.	Mambas
D. angusticeps	Green mamba
D. jamesoni	Jameson's mamba
D. polylepis	Black mamba
D. viridis	Western green mamba
Hemachatus hemachatus (*Haemachatus haemachatus*)	Ringhals
Leptomicrurus spp.	Slender coral snakes
Maticora spp.	Long-glanded coral snakes
Micruroides spp.	Arizona coral snakes
Micrurus spp.	American coral snakes
M. annellatus	Annelated coral snake
M. diastema	Atlantic coral snake
M. distans	Broad-banded coral snake
M. frontalis	Southern coral snake
M. fulvius	Eastern coral snake
M. hemprichii	Hemprich's coral snake
M. miapartitus	Black-ringed coral snake
M. nigrocinctus	Black-banded coral snake
M. spixii	Amazonian coral snake
M. surinamensis	Surinam coral snake
Naja spp.	Cobras
N. goldii	West African cobra
N. haje (N. haie)	Egyptian cobra, asp
N. melanoleuca	Black cobra, forest cobra
N. naja (N. naia)	Indian cobra, common cobra, spectacled cobra
N. naja atra	Formosan cobra, Taiwan cobra, Chinese cobra
N. naja kaouthia (*N. naja siamensis*)	Thailand cobra
N. naja miolepis	Borneo cobra
N. naja philippinensis	Philippine cobra
N. naja sputatrix	Malay cobra
N. nigricollis	Spitting cobra, black-necked cobra
N. nivea	Cape cobra, yellow cobra
N. oxiana (N. naja oxiana)	Central Asian cobra, oxus cobra

Appendix

Notechis scutatus	Tiger snake
Ophiophagus hannah (*Naja hannah*)	King cobra
Oxyuranus scutellatus	Taipan
Pseudechis australis	Australian Mulga snake
P. papuanus	Papuan black snake
P. porphyriacus	Red-bellied black snake, Australian black snake
Walterinnesia aegyptia	Desert black snake

Viperidae (Viperids, Vipers)

Bitis arietans	Puff adder
B. atropos	Berg adder
B. caudalis	Horned puff adder
B. gabonica	Gaboon viper
B. inornata	Cape puff adder
B. nasicornis	River Jack
Causus spp.	Night adders
C. rhombeatus	Rhombic night adder
Cerastes cerastes	African desert horned viper, Sand viper
C. cornuta	Egyptian sand adder
C. vipera	Sahara sand viper
Echis carinatus	Saw-scaled viper, carpet viper
Pseudocerastes persicus	Persian horned viper
Vipera ammodytes	Horn viper, European viper, Sand natter, long-nosed viper
V. aspis	European asp, asp viper
V. berus	Marasso pulustre, common viper, common adder, Kreuzotter
V. lebetina	Levantine viper
V. maurifanica	Sahara rock viper
V. russellii	Russell's viper
V. russellii formosensis	Formosan Russell's viper
V. orsinii	Orsini's viper, meadow viper
V. superciliaris	African lowland viper
V. xanthina	Near East viper

Crotalidae (Croralids, Pit Vipers)

Agkistrodon acutus	Hundred pace snake
A. bilineatus	Cantil, tropical moccasin
A. caliginosus	Chosen-mamushi
A. contortrix	Copperhead
A. halys	Pallas' viper
A. halys blomhoffii	Mamushi
A. hypnale	Hump-nosed
A. mokasen	Highland moccasin

A. piscivorous	Cottonmouth, water moccasin
A. rhodostoma	Malayan pit viper
A. saxatilis	Sangaku-mamushi
Bothrops alternatus	Urutu
B. atrox	Barba amarilla, fer-de-lance
B. bilineatus	Amazon tree viper
B. caribbaeus	St. Lucia serpent
B. godmani	Godman's viper
B. itapetiningae	Cotiarinha
B. jararaca	Jararaca
B. jararacussu	Jararacussu
B. lanceolatus	Fer-de-lance
B. lansbergii	Lansberg's hog-nosed viper
B. nasuta	Hog-nosed viper
B. neuwiedi	Jararaca pintada
B. nigroviridis marchi	Black-spotted palm viper
B. nummifer (*B. nummifera*)	Jumping viper, timba
B. schlegelii	Eyelash viper, horned palm viper
Crotalus adamanteus	Eastern diamondback rattlesnake
C. atrox	Western diamondback rattlesnake
C. basiliscus	Mexican west-coast rattlesnake
C. cerastes	Sidewinder, horned rattlesnake
C. confluentus	Prairie rattlesnake
C. durissus	Cascabel
C. durissus terrificus	South American rattlesnake, Tropical rattlesnake
C. horridus	Timber rattlesnake, banded rattlesnake
C. lepidus	Green rattlesnake
C. mitchellii	White rattlesnake, bleached rattlesnake
C. molossus	Black-tailed rattlesnake
C. ruber	Red diamond rattlesnake
C. scutulatus	Mojave rattlesnake, Mojave diamond rattlesnake
C. tigris	Tiger rattlesnake
C. unicolor	Aruba rattlesnake
C. viridis lutosus	Great Basin rattlesnake
C. viridis oreganus	Pacific rattlesnake
Lachesis mutus	Bushmaster
Sistrurus catenatus	Massasauga rattlesnake
S. miliarius	Pigmy rattlesnake
S. ravus	Mexican pigmy rattlesnake
Trimeresurus albolabris	White-lipped tree viper
T. elegans	Sakishima habu

Appendix

T. flavoviridis — Habu
T. gramineus — Green pit viper
T. monticola — Chinese mountain pit viper
T. mucrosquamatus — Formosan habu
T. okinavensis — Hime habu
T. popeorum — Pope's tree viper, green pit viper
T. purpureomaculatus — Mangrove viper
T. stejnegeri — Chinese green tree viper, Formosan bamboo viper
T. tokarensis — Tokara habu
T. wagleri — Wagler's pit viper

Colubridae (Colubrids, Rear Fang Snakes)

Dispholidus typus — Boomslang
Thelotornis kirtlandii — Bird snake

Index by Subjects

Acetylcholinesterase, 97–102
  chemistry, 97–100
  inhibitor, 100
  occurrence, 100–102
  receptors in, 240–243, 251
Acetylcholinesterase inhibitor, 140
Acetylcholine receptor:
  binding with snake neurotoxins, 243–251
  localization, 252–253
  properties, 240–243
Acid phosphatase, 78–83
Alkaline phosphatase, 78–83
Amino acid composition:
  bee venom components, 506
  scorpion, neurotoxins, 462
  snake venoms:
    acetylcholinesterase, 242
    acetylcholine receptor, 242
    anticoagulant factor, 332
    Arvin, 336
    bradykinin-releasing factor, 402
    cardiotoxins, 304, 305
    cobra venom factor, 422
    cytotoxins, 308, 309
    elapid neurotoxins, 264, 265, 267
    nerve growth factor, 422
    phospholipase A_2, 28, 29
    pit viper venom components, 215, 220
    proteinase inhibitor, 144
    proteolytic enzymes, 108, 109
    sea snake neurotoxins, 171, 172
    thrombin-like enzymes, 341
    viper neurotoxins, 171, 172
  spider venoms:
    hyaluronidase, 486
    neurotoxin, 485

L-Amino acid oxidase, 85–95
  absorption spectra, 86
  biological significance, 93
  enzyme mechanism, 89–93
  fluorescence spectra, 86
  isozymes, 89
  occurrence, 93–95
  preparation of α-keto acids, 90
  prosthetic group, 15, 16, 87, 89, 92
Amino acids, free:
  bee venoms in, 508
  scorpion venoms in, 464
  snake venoms in, 9, 10
  spider venoms in, 487
  wasp venoms in, 516
Amino acid sequence
  bee venoms:
    apamin, 504
    MCD peptide (peptide 401), 504
    melittin, 503
    phospholipase A_2, 507
    promelittin, 503
  scorpion neurotoxins, 461–463
  snake venoms:
    angiotensin, 406
    angiotensinase inhibitor, 405
    bradykinin, 401
    bradykinin potentiating peptides, 9, 10, 405
    cardiotoxins, 307, 308, 310–312
    crotamine, 216
    cytotoxins, 307, 308, 310–312
    direct hemolytic factor, 324
    fibrinopeptides, 334, 335
    nerve growth factor, 365
    neurotoxins, postsynaptic, 266–271

neurotoxins, presynaptic, 293, 294
phospholipase A_2, 30–32
proline-rich peptide, 9, 10
proteinase inhibitor, 142, 143
wasp venom kinin, 519
Amylase, 134
Angiotensinase inhibitor, 140, 141, 406
Antibacterial action, 512
Anticomplement factor, 420–421
Antihistamines, effect of, 450
Ant venoms, 527–529
bulldog ant, 527, 528
fire ant, 528, 529
others, 529
Apamin, 504, 506
Arginine ester hydrolase, Gila monster venom, 532
Arginine ester hydrolase, snake venoms:
Arvin, 335
blood coagulation, 129, 332, 343
brodykinin release, 129, 401, 402
isolation, 128, 129
separation from proteases, 105, 106, 128
Arginine residue modification, 276
Arvin (Ancrod), 14, 333–340
chemistry, 334–337
clinical trial, 340
microlot, 337, 338
physiological effect, 338–340
ATP:
hydrolysis, 72, 73
release, 408
Autopharmocological action, 400–409
bradykinin-potentiating factor, 403–406
bradykinin release, 400–403
histamine release, 407
other substances, 408

Basic nonneurotoxic proteins:
cardiotoxins, 302–306
cobramine A, 316
cobramine B., 284, 316
cytotoxins, 306–316
other basic proteins, 317
peptide 4.9.6, 317
similarity to neurotoxins, 291–293
Bee venoms:
chemistry, 502–508
classification, 502
enzymes, 505–507
immunology, 508, 509
metabolic effect, 512–515
nonpeptide components, 508
pharmacology, 510–512
protease inhibitor, 507
toxicology, 509
Biogenic amines, 16, 464, 487, 508, 516, 528, 533

Blood coagulation, 329–352
anticoagulation, 330–333
antihemophilic factor, 350, 351
factor II, effect on, 350
factor III, effect on, 349
factor V, effect on, 348, 349
factor VIII, effect on, 350, 351
factor X, effect on, 346–348
fibrin, effect on, 345, 346
fibrinogen, effect on, 345
pit viper venoms, 344
procoagulation, 330, 333–350
prothrombin, effect on, 350
rattlesnake venoms, 343, 344
thrombin-like enzyme, 333–345
viper venoms, 344, 345
Bradykinin:
bradykinin-potentiating factor, 9, 10, 403–406
chemistry, 400–403
Gila monster venom, 533
esterase, relation to, 401, 402
snake venoms in, 16, 401–406
Bungarotoxin:
alpha-, 185–187, 240–245, 247–251
beta-, 185–187, 189, 293
history, 184, 185

Carbohydrates:
bee venoms in, 508
scorpion venoms in, 461, 464, 465
snake venoms in, 12–14
L-amino acid oxidase, 87
Arvin, 337
blood coagulation enzymes, 343, 344, 347, 351
cobra venom factor, 422
nerve growth factor, 363
proteolytic enzymes, 110, 111
reptilase, 341
Carboxyl group modification, 278
Cardiolipin, hydrolysis of, 41
Cardiotoxins, 223, 225–227, 302–306
amino acid composition, 220, 304, 305
biochemistry, 219–222, 291, 292, 303–306
pharmacological activity, 225–227, 302, 303
Chelating compounds, effect of, 437, 442–444
Chemical modification, 272–279
Chemical neutralization, 435–454
Circular dichroism (CD), 43, 282, 283
Cobra venoms, 179–184
Cobra venom factor (CVF), 420–431
Cobramine A, 316
Cobramine B., 284, 316
Collagenase, 120–122
Colubridae venoms, 234, 235

Index by Subjects

Conformation:
   neurotoxins, 279–291
   phospholipase A_2, 43
Convulxin, 217
Coral snake venoms, 190–192
Crotalidae venoms, 221–229
Crotamine, 215, 216
Crotoxin, 213–215
Cytotoxins, 207, 208, 271, 306–316

Distribution of poisonous snakes, 1–3
Distribution of venoms, 236–239
Disulfide bonds, 277, 278, 288–290
DNase, 73, 74

Elastase, 122, 123
Enzyme inhibitors
   acetylcholinesterase, 140, 403–406
   angiotensinase, 140, 141
   bee venoms in, 507
   phospholipase A_2, 139, 140
   proteinase, 141–145, 507
Enzymes
   in ant venoms, 528
   in bee venoms, 505, 507
   in Gila monster venoms, 533
   in scorpion venoms, 464
   in snake venoms, 23–136
   in spider venoms, 486, 487
   in wasp venoms, 516
Exonuclease, 64–73

Fatty acids, 15, 32–39, 42, 43
Fibrinogenolytic action, 345
Fibrinolytic action, 345, 346

Gila monster venom, 531–533
Glucose metabolism, effect on, 412, 413
Glycoproteins, see Carbohydrates

Hemorrhage, 372–381
Hemolysis, 321–326
Hexosamidase, 507
Histamine
   in ant venoms, 528
   in bee venoms, 508
   in hornet venoms, 516
   in scorpion venoms, 464
   in snake venoms, 16
   in spider venoms, 487
Histamine release, 407, 408, 510
Histidine residue mofification, 278
Hornet venoms, 515–518
Hyaluronidase
   in ant venoms, 528
   in bee venoms, 505
   in Gila monster venom, 533

   in hornet venoms, 516
   in scorpion venoms, 464
   in snake venoms, 132–134
   in spider venoms, 486
   in wasp venoms, 516, 519

Isoelectric points
   L-amino acid oxidase, 89
   anticoagulant, 332
   arginine ester hydrolase, 129
   crotoxin, 213
   Mojave toxin, 219
   nerve growth factor, 362
   phospholipase A_2, 44, 45
   pit viper venom, 166
   sea snake neurotoxins, 166
   sea snake venom, 166
Isozymes
   L-amino acid oxidase, 89
   phospholipase A_2, 26, 44–46

Krait venoms, 184–190

Lactate dehydrogenase, 136
LD_{50}, see Toxicity
Lipids, 14–16, 508
Lysine residue modification, 276, 277

MCD-peptide (peptide 401), 504, 506
Melitin, 502, 503
Melitin F, 505, 506
Membranes, effect on, 46–56, 325, 326
Metabolic effect, 412–415
Metals
   L-amino acid oxidase, effect on, 59
   cobra venom factor, effect on, 423
   Gila monster venom in, 533
   hemorrhagic toxins in, 378
   $5'$-nucleotidase, effect on, 81
   phosphodiesterase, effect on, 70, 71
   phospholipase A_2, effect on, 38
   phospholipase A_2 inhibitor, effect on, 140
   phosphomonoesterase, effect on, 79
   proteolytic enzymes in, 110
   snake venoms in, 6–9
Minimine, 505, 506
Mitochondria, effect on, 47–51, 184, 190, 208, 389, 414, 470, 513, 514, 515, 518, 528
Myonecrosis, 55, 152, 193, 222, 382–391, 470–473

NAD nucleosidase, 134, 135
Nephrotoxic action, 152, 207, 217, 391–396, 518
Nerve growth factor, 317, 361–370
Nervous system, effect on

bee venoms, 511
Gila monster venom, 532
hornet venoms, 518
scorpion venoms, 467–470
snake venoms
  phospholipase A_2, 48–54
  neurotoxins, 160, 161, 179, 182, 184, 185, 189, 191, 193, 194, 202, 203, 206, 207, 217, 219, 222, 225, 226, 228, 229, 237, 240–253
spider venoms, 490–495
wasp venoms, 518, 519
Neurotoxins, bees, 502–504, 506, 511
Neurotoxins, scorpions, 460–464, 467–470
Neurotoxins, snakes
  basic nonneurotoxic proteins, similarity to, 291–293
  circular dichroism, 282, 283
  conformation, 279–291
  disulfide linkages, 279–281
  fluorescence, 282
  isolation, 262–266
  laser Raman spectroscopy, 283–291
  optical rotatory dispersion, 282, 283
  postsynaptic, 262
  presynaptic, 258, 259
  Raman spectroscopy, 283–291
  receptor binding, 240–253
  reconstitution, 214
  relation to phospholipase A_2, 214, 293
  sequences of amino acids, 266–271, 293, 294
  size and toxicity, 279
  stability, 281
  structure-toxicity relationship, 272–279
  type I, 266
  type II, 266
  X-ray diffraction, 281
Neurotoxins, spiders, 486, 490–495
Nucleotides, 10, 11

Occurrence of
  acetylcholinesterase, 101, 102
  L-amino acid oxidase, 93–95
  biogenic amines, 16, 464, 487, 508, 516, 528, 533
  carbohydrates, 12–14, 461, 464, 465, 508
  cobra venom factors, 430
  exonuclease, 71, 72
  lipids, 14–16, 508
  metals, 6–8
  nerve growth factor, 369, 370
  nucleotides, 10, 11
  nucleosides, 10, 11
  phospholiesterase, 71, 72
  phospholipase A_2, 57, 464, 505, 507, 516, 533
  phosphomonoesterase, 80, 82, 83
  proteolytic enzymes, 117–120, 122
Optical rotatory dispersion, 282, 283, 507
Organ transplantation, 429

Phosphatase, 78–83
Phosphodiesterase
  endopolynucleotidase, 73, 74
  exonuclease, 64–73, 487
  hydrolysis of ATP, 72, 73, 528
  isolation, 70, 71
  occurrence, 71, 72
Phospholipase A_2
  bee venoms, 505, 507
  Gila monster venom, 533
  hornet venoms, 516
  scorpion venoms, 464
  snake venoms, 23–57
Phospholipase A_2 inhibotor, 139, 140
Phosphomono esterase, 78–83
Pipecolic acid, 10
Pit vipers, see Crotalidae venoms
Plateletets, effect on, 351, 352
Promelittin, 503
Proteinase inhibitors, 141–145, 507
Proteolytic enzymes, 104–123, 486, 487, 533

Radiation effect, 451–454
Rattlesnake venoms, 211–227
Reptilase, 341–343
RNase, 73, 74

Scorpion venoms, 459–477
Sea snake venoms, 151–174
Secapin, 505, 506
Serotonin, 16
Sequences, see Amino acid sequence
Spermine, 16
Spider venoms, 484–496
Steroid compounds, effect of, 436, 438

Teratogenic effect, 415, 416
Tertiapin, 505, 506
Toxicity
  ant venoms, 527, 529
  bee venoms, 509, 512
  Gila monster venoms, 532
  scorpion venoms, 466
  snake venoms
    L-amino acid oxidase, 93
    anticoagulation factor, 333
    basic proteins, nonneurotoxic, 317
    cardiotoxins, 302
    Crotalidae venoms, 223–225
    elapidae venoms, 186–189, 194
    hemorrhagic toxin, 379, 381
    Hydrophiidae venoms, 154–157
    5′-nucleotidase, 82

Index by Subjects

    phospholipase A_2, 56, 57
    pit viper venoms, 223–223
    proteinase inhibitor, 145
    proteolytic enzymes, 116
    sea snake venoms, 154–157
    Viperidae venoms, 204, 205
  spider venoms, 488, 489
  wasp venoms, 517
Transaminase, 136
Tryptophan residue modification, 39, 272–275
Tyrosine residue modification, 275, 276, 464

Vespulakinins, 519, 520

Viperidae venoms, 201–208

Wasps, 518–520

X-ray effect, 453, 454

Yield of venoms
  coral snakes, 190, 191
  Crotalidae, 152
  Elapidae, 152
  scorpion, 467
  sea snakes, 152, 153
  Viperidae, 152

Index by Species

Sometimes different names have appeared in published papers for the same venoms. In this index, the names used by the original investigators are used. Entries appear in the order that they are discussed in the text.

Snakes
Hydrophiidae (Sea snakes)
Acalytophis peronii
  yield, 153
Aipysurus eydouxii
  toxicity, 154
  yield, 153
A. laevis
  biochemistry, 172, 269
  immunology, 158
  toxicity, 154, 156
  yield, 153
Astrotia stokesii
  immunology, 158
  toxicity, 154, 156
Enhydrina schistosa
  biochemistry, 166, 167, 171
  cytotoxic action, 315
  immunology, 157–159, 235
  isoelectric point s, 166
  mitochondria, effect on, 184
  neurotoxins, 171, 274, 284, 293
  pathology, 163
  pharmacology, 161
  phosphodiesterase, 71
  phosphomonoesterase, 80
  proteolytic enzymes, 117
  Raman spectroscopy, 284, 286, 288
  toxicity, 154–156
  yield, 153
Hydrophis belcheri
  toxicity, 154
  yield, 153
H. cyanocinctus
  biochemistry, 167, 172
  immunology, 157–159
  pharmacology, 163
  toxicity, 154, 156, 281
  yield, 153
H. elegans
  immunology, 158
  toxicity, 154, 156
  yield, 153
H. klossi
  toxicity, 154
H. major
  immunology, 158
  toxicity, 156
H. melanososa
  toxicity, 154
H. ornatus
  yield, 153
H. spiralis
  toxicity, 154, 156
  immunology, 157–159
H. torquatus diadema
  yield, 153
Kerilia jerdoni
  immunology, 157
  toxicity, 154
Lapemis hardwickii
  biochemistry, 170, 171
  cytotoxic action, 315
  immunology, 157–159

neurotoxins, 170, 171, 272–274, 284, 288
pharmacology, 163
Raman spectroscopy, 284, 286, 288
toxicity, 154–156
yield, 153
Laticauda colubrina
biochemistry, 172
toxicity, 154
L. laticaudata
biochemistry, 167, 172
immunology, 157, 158
pharmacology, 161
toxicity, 154, 281
L. semifasciata
biochemistry, 166–169, 172, 274, 276, 278, 281, 286, 288, 289, 292
cytotoxic action, 315
hemolysis, 324
immunology, 157–160
nephrotoxic action, 396
neurotoxins, 157, 160, 168, 169, 172
pathology, 163–165
pharmacology, 160, 161
phospholipase A_2, 24, 28, 34, 39, 44–46, 52, 55–57
phosphomonoesterase, 78, 80
Raman spectroscopy, 286, 288, 289
toxicity, 154–157, 281
Microcephalus gracilis
immunology, 157
Pelamis platurus
biochemistry, 167, 171, 173, 278, 280, 284, 291
carbohydrate, 13
cytotoxic action, 313, 315
immunology, 157–159
neurotoxins, 157, 162, 171, 173
pharmacology, 162, 163
Raman spectroscopy, 284–286, 291
toxicity, 155, 157
yield, 153
Praescutata viperina
immunology, 157, 158
toxicity, 155
yield, 153

Elapidae (Elapids)
Acanthophis antarcticus
acetylcholinesterase, 101
hyaluronidase, 133
nucleoside, 11
pharmacology, 193
toxicity, 186
Bungarus caeruleus
acetylcholinesterase, 101
L-amino acid oxidase, 93
esterase, 130
toxicity, 186

B. fasciatus
acetylcholine receptor, binding to, 251
acetylcholinesterase, 98–101
L-amino acid oxidase, 93
cardiotoxin, 303, 304
metabolic effect, 413
NADase, 134, 135
pharmacology, 190
proteolytic enzyme, 118, 119
phosphodiesterase, 71
phosphomonoesterase, 82
toxicity, 184, 187
yield, 153
B. multicinctus
acetylcholine receptor, binding to, 240, 243–245, 247–253
acetylcholinesterase, 97, 101
L-amino acid oxidase, 93
chemical neutralization, 452
chemistry, 264
distribution of venom, 236, 237
hyaluronidase, 133
metabolic effect, 413, 415
NADase, 134, 135
pharmacology, 184, 185, 189, 258, 259, 261
phospholipase A_2, 47, 56
phosphomonoesterase, 80, 82
toxicity, 184, 186, 228
Demansia textilis
acetylcholinesterase, 101
esterase, 130
lactate dehydrogenase, 136
Dendraspis (Dendroaspis) Spp.
pharmacology, 194, 195
D. angusticeps
acetylcholinesterase, 101
basic protein, nonneurotoxic, 317
blood coagulation, 331, 349
esterase, 130
hemolysis, 323
nucleoside, 11
phosphodiesterase, 71
phosphomonoesterase, 82
proteinase inhibitor, 141
toxicity, 187
D. jamesoni
anticoagulation, 331
esterase, 130
lactate dehydrogenase, 136
toxicity, 187
D. polylepis
anticoagulation, 331
esterase, 130
lactate dehydrogenase, 136
proteinase inhibitor, 141–144
toxicity, 187
D. viridis

Index by Species

basic protein, non-neurotoxic, 317
immunology, 235
nerve growth factor, 362, 364, 369
neurotoxin, 270
toxicity, 187
Denisonia superba
acetylcholinesterase, 101
histamine release, 407
hyaluronidase, 133
nucleotide, 11
toxicity, 187
D. textilis
L-amino acid oxidase, 93
Elaps corallinus
acetylcholinesterase, 101
Hemachatus hemachatus (Haemachatus haemachatus)
anticomplement action, 430
direct hemolytic factor, 304, 316, 323, 324
phosphodiesterase, 65, 71
phospholipase A_2, 24, 50
phosphomonoesterase, 81, 82
proteinase inhibitor, 141, 144
proteolytic enzyme, 118
toxicity, 187
Micrurus spp.
blood coagulation, effect on, 349
M. alleni
immunology, 192
M. carinicauda dumerilii
immunology, 192
toxicity, 188
yield, 191
M. frontalis frontalis
immunology, 192
pharmacology, 191
toxicity, 188
M. fulvius fulvius
L-amino acid oxidase, 94
biochemistry, 192
chemical neutralization, 451
pharmacology, 191
toxicity, 188
yield, 191
M. mipartitus anomalus
immunology, 192
M. mipartitus herwigii
immunology, 192
toxicity, 188
M. nigrocinctus
toxicity, 188
Naja bungarus
acetylcholinesterase, 101
N. flava
acetylcholinesterase, 101
blood coagulation, effect on, 331
chemical neutralization, 445
phospholipase A_2, 56
toxicity, 188
N. haje
acetylcholinesterase, 101
anticomplement factor, 430
biochemistry, 274, 275, 277, 283
cobra venom factor, 422, 429
esterase, 130
hyaluronidase, 133
immunology, 235
lactate dehydrogenase, 136
metabolic effect, 413, 415
nephrotoxic action, 394
phosphodiesterase, 71
phosphomonoesterase, 78, 80, 82
proteinase inhibitor, 141
proteolytic enzyme, 118
toxicity, 188, 281
transaminase, 136
N. haje annulifiera
cytotoxin, 308, 310–312
neurotoxin, 292
N. hamadryas
metabolic effect, 413
N. melanoleuca
acetylcholinesterase, 101
L-amino acid oxidase, 94
basic protein, 292
blood coagulation, effect on, 331, 349
carbohydrates, 12
cytotoxins, 308, 310–312, 315
hemolysis, 323
immunology, 235
lactate dehydrogenase, 136
nerve growth factor, 362, 369
phospholipase A_2, 24, 28, 39, 43, 293, 294
phosphomonoesterase, 78, 80
proteolytic enzyme, 118
toxicity, 188
N. mossambica mossambica
anticomplement action, 430
cytotoxin, 308, 310–312
pharmacology, 184
N. naja
acetylcholine receptor, binding to, 244, 245, 249
blood coagulation, effect on, 346, 349
carbohydrates, 12
cardiotoxin, 302–304, 310–312
chemical neutralization, 436, 438, 439, 441, 442, 447, 451–453
cobra venom factor, 422–430
distribution of venom, 237, 238
hemolysis, 316, 322–325
hemorrhage, 374
histamine release, 407, 408
immunology, 182–184, 235
lactate dehydrogenase, 136

metabolic effect, 414, 415
metals, 6, 8
nerve growth factor, 362–366, 369
neurotoxins, 264, 271, 274
pharmacology, 180–183, 225
phospholipase A_2, 24, 28, 41, 43, 44, 47, 49, 52, 56, 57
phospholipase A_2 inhibitor, 139
phosphodiesterase, 71
phosphomonoesterase, 80, 82
proteolytic enzyme, 118, 119
teratogenic effect, 416
toxicity, 155, 188, 281

N. naja atra
acetylcholinesterase inhibitor, 140
blood coagulation, effect on, 349
cardiotoxins, 302–304, 310–312
chemical neutralization, 452, 453
cytotoxic action, 314–316
distribution of venom, 237
enzyme inhibition, 414
esterase, 130
hyaluronidase, 133
immunology, 182, 183
inorganic constituents, 8
lactate dehydrogenase, 136
metabolic effect, 413–415
metals, 6, 8
NADase, 134
neurotoxins, 264, 271, 274–276, 278–283
pharmacology, 180–182, 184
phosphodiesterase, 71
phospholipase A_2, 24, 28, 33, 39, 47, 48, 56
proteolytic enzyme, 118
toxicity, 155, 188, 228, 281

N. naja ceylonicus
cytotoxin, 309

N. naja philippinensis
chemical neutralization, 445, 453
cytotoxic action, 315
immunology, 183
neurotoxins, 265, 282, 283
toxicity, 188

N. naja samarensis
cytotoxins, 309
phosphomonoesterase, 78, 80
proteolytic enzyme, 119

N. naja siamensis (N. naja kaouthia)
acetylcholine receptor, binding to, 244, 246, 247, 249
L-amino acid oxidase, 94
cardiotoxin, 303, 309
immunology, 182, 183
myonecrosis, 181, 386, 387
neurotoxins, 264, 265, 276, 277, 279
phosphodiesterase, 71

phospholipase A_2, 52, 54
phosphomonoesterase, 80, 82
toxicity, 188
yield, 153

N. nigricollis
acetylcholine receptor, binding to, 242–244, 246, 249, 252
acetylcholinesterase, 101, 102
acetylcholinesterase inhibitor, 140
anticomplement action, 430
blood coagulation, effect on, 330, 331, 349, 350
cardiotoxin, 180, 304, 310–312, 316
chemical neutralization, 439, 445
hemolysis, 323
hemorrhage, 374
hyaluronidase, 133
immunology, 183, 235
lactate dehydrogenase, 136
metabolic effect, 413, 415
nerve growth factor, 362, 368, 369
phosphodiesterase, 71
phospholipase A_2, 24, 28, 44, 46, 56
phosphomonoesterase, 80, 82
proteolytic enzyme, 119
teratogenic effect, 413, 415
toxicity, 188

N. nigricollis mossambica
toxicity, 188

N. nivea
acetylcholinesterase, 101
anticomplement action, 430
biochemistry, 278
esterase, 130
lactate dehydrogenase, 136
pharmacology, 181
phosphodiesterase, 71
phosphomonoesterase, 82
proteinase inhibitor, 141
proteolytic enzyme, 119
toxicity, 189

N. oxiana (N. naja oxiana)
acetylcholinesterase, 97, 98, 101
blood coagulation, effect on, 349
cytotoxin, 309–312
hemolysis, 323, 324
hyaluronidase, 133
immunology, 182
metabolic effect, 414
nephrotoxic action, 396
neurotoxins, 178, 179, 265, 282
pharmacology, 182
phosphodiesterase, 71, 74
phospholipase A_2 inhibitor, 140
phosphomonoesterase, 80, 82
proteolytic enzyme, 119
toxicity, 188

N. tripudians

Index by Species

acetylcholinesterase, 101
blood coagulation, effect on, 331
hyaluronidase, 133
Notechis scutatus
  acetylcholinesterase, 101
  blood coagulation, effect on, 349, 350
  chemical neutralization, 445
  hyaluronidase, 133
  immunology, 158
  metabolic effect, 414
  neurotoxins, 190, 193, 262, 263, 267, 293, 294
  nucleosides, 11
  nucleotides, 11
Ophiophagus hannah (Naja hannah)
  L-amino acid oxidase, 94
  anticomplement action, 430
  arginine ester hydrolase, 401
  blood coagulation, effect on, 331, 349
  chemical neutralization, 446
  cytotoxic action, 315
  immunology, 182
  neurotoxin, 265
  phosphodiesterase, 71
  phosphomonoesterase, 78–80, 82
  proteolytic enzyme, 118
  toxicity, 189
  yield, 153
Oxyuranus scutellatus
  biochemistry, 192
  blood coagulation, effect on, 350
  toxicity, 189
O. scutellatus canni
  biochemistry, 192
Pseudechis australis
  acetylcholinesterase, 101
  toxicity, 189
P. collettii
  phosphodiesterase, 71
  phosphomonoesterase, 82
  proteolytic enzyme, 118
P. purphyriacus
  acetylcholinesterase, 101
  toxicity, 189
Pseudonaja textilis
  toxicity, 189
Sepedon haemachates
  acetylcholinesterase, 101
  carbohydrates, 12
  nerve growth factor, 369
Tropidechis carinatus
  enzymes, 193
Walterinessia aegyptia
  acetylcholinesterase, 101
  hyaluronidase, 133
  metabolic effect, 413–415
  pharmacology, 193, 194
  toxicity, 189

transaminase, 136

Viperidae (Vipers)

Bitis arietans (B. lachesis)
  acetylcholinesterase, 101
  anticomplement action, 430
  arginine ester hydrolase, 138, 402
  blood coagulation, effect on, 331, 350
  bradykinin release, 402
  chemical neutralization, 443
  cytotoxic action, 207, 315
  distribution of venom, 237
  hemolysis, 323
  metals, 6, 8
  nucleoside, 11
  pharmacology, 207
  phospholipase A_2, 24, 29, 35
  platelets, effect on, 351, 352
  proteolytic enzymes, 106–108, 111, 116, 118
  toxicity, 155, 204
B. gabonica
  acetylcholinesterase, 101
  arginine ester hydrolase, 402
  blood coagulation, effect on, 344
  bradykinin release, 402
  carbohydrates, 12–14
  chemical neutralization, 443
  hemolysis, 323
  metals, 6
  myonecroses, 389
  nerve growth factor, 369
  pharmacology, 207
  phosphodiesterase, 72
  phospholipase A_2, 24, 28, 32, 35
  phosphomonoesterase, 82
  toxicity, 155, 204
B. gabonica rhinoceros
  cytotoxic action, 315
B. nasicornis
  blood coagulation, effect on, 331
  collagenase, 120
  platelets, effect on, 351
  proteolytic enzymes, 118, 120
Causus rhombeatus
  arginine ester hydrolase, 128, 402
  bradykinin release, 402
  NADase, 135
  proteolytic enzymes, 118
Cerastes cerastes
  blood coagulation, effect, on, 349
  metabolic effect, 415
  pharmacology, 207
  toxicity, 204
C. cornutus
  acetylcholinesterase, 102
  hyaluronidase, 133
  toxicity, 204
  transaminase, 136

C. vipera
  acetylcholinesterase, 102
  hyaluronidase, 133
  metabolic effect, 415
  toxicity, 204
  transaminase, 136

Echis carinatus
  acetylcholinesterase, 102
  arginine ester hydrolase, 128, 402
  blood coagulation, effect on, 330, 346, 350
  bradykinin release, 401, 402
  carbohydrates, 12, 14
  chemical neutralization, 438–440
  hyaluronidase, 133
  immunology, 235
  metabolic effect, 413
  metals, 8
  nephrotoxic action, 393
  pharmacology, 207
  proteolytic enzymes, 118, 119
  toxicity, 204
  transaminase, 136

E. coloratus (E. colorata)
  acetylcholinesterase, 102
  blood coagulation, effect on, 330
  bradykinin release, 402, 403
  chemical neutralization, 445
  cytotoxic action, 308
  hemolysis, 322
  hemorrhage, 377
  nerve growth factor, 368
  toxicity, 204
  transaminase, 136

Vipera ammodytes
  acetylcholinesterase, 102
  amino acids, free, 9
  L-amino acid oxidase, 16, 85–88, 93, 94
  arginine ester hydrolase, 402
  basic protein, 317
  bradykinin release, 402
  distribution of venom, 238
  hemolysis, 323
  hyaluronidase, 133
  nerve growth factor, 362–364, 369
  neurotoxin, 202, 203
  pharmacology, 206, 207
  phospholipase A_2, 24, 43
  proteolytic enzyme, 118
  riboflavin, 16
  toxicity, 205

V. aspis
  acetylcholinesterase, 102
  L-amino acid oxidase, 94
  blood coagulation, effect on, 330, 345, 350
  collagenase, 120
  cytotoxin, 309
  hyaluronidase, 133
  immunology, 235
  nerve growth factor, 369
  pharmacology, 207
  platelets, effect on, 351
  proteolytic enzyme, 118, 120
  teratogenic effect, 416
  toxicity, 155, 205

V. aspis hugyi
  acetylcholinesterase, 102

V. berus
  blood coagulation, effect on, 350
  carbohydrates, 12
  immunology, 235
  pharmacology, 207
  phospholipase A_2, 24, 29, 35, 44
  toxicity, 205

V. latifi
  toxicity, 205

V. lebetina
  acetylcholinesterase, 102
  L-amino acid oxidase, 94
  arginine ester hydrolase, 128
  blood coagulation, effect on, 330, 345
  bradykinin release, 401, 402
  carbohydrates, 12
  hyaluronidase, 133
  metabolic effect, 415
  phosphodiesterase, 70
  phosphomonoesterase, 81, 82
  proteolytic enzyme, 118
  toxicity, 205

V. lebetina turanica
  L-amino acid oxidase, 87, 94

V. palestinae
  L-amino acid oxidase, 87, 89, 94
  arginine ester hydrolase, 402
  bradykinin release, 402
  cytotoxic action, 207, 208
  hemolysis, 325
  hemorrhage, 377, 381
  metals, 8
  neurotoxin, 202, 203
  pharmacology, 206
  phospholipase A_2, 25, 29, 43, 45, 48, 51, 56, 57
  proteolytic enzyme, 118
  toxicity, 205
  venom injected, amount of, 238

V. persica
  toxicity, 205

V. russellii (V. russelli)
  acetylcholinesterase, 102
  arginine ester hydrolase, 402
  blood coagulation, effect on, 345–349
  bradykinin release, 402
  chemical neutralization, 438, 451
  cytotoxic action, 315
  hemolysis, 325

hyaluronidase, 132, 133
immunology, 206
lethal toxin, 203
metals, 8
nerve growth factor, 362–364, 366–369
pharmacology, 207
phosphodiesterase, 72, 73
phospholipase A_2, 25, 44, 56
phosphomonoesterase, 80–82
proteinase inhibitor, 141–144, 401
toxicity, 205

V. russellii formosensis
L-amino acid oxidase, 94
hyaluronidase, 133
pharmacology, 206
phosphomonoesterase, 80, 82
proteolytic enzyme, 118
toxicity, 205

V. russellii siamensis
blood glucose, effect on, 413
chemical neutralization, 443
metals, 6
toxicity, 155, 205
yield, 153

V. ursini
L-amino acid oxidase, 94
arginine ester hydrolase, 128
bradykinin release, 401
mitochondria, effect on, 208
proteolytic enzyme, 118

V. ursini renardi
metabolic effect, 414

V. xanthina palaestinae
toxicity, 205

Crotalidae (Crotalids, Pit vipers)

Agkistrodon acutus
arginine ester hydrolase, 402
blood coagulation, effect on, 330–333, 341, 344, 349
bradykinin release, 402
chemical neutralization, 443, 452
enzyme inhibition, 414
hyaluronidase, 133
immunology, 229
metabolic effect, 413–415
NADase, 134
peptides, 9
pharmacology, 225, 228
phosphodiesterase, 72
phospholipase A_2, 56
phosphomonoesterase, 78–80, 83
toxicity, 156, 223, 228

A. bilineatus
L-amino acid oxidase, 94
blood coagulation, effect on, 346
phosphodiesterase, 72
proteolytic enzymes, 117
toxicity, 223

A. caliginosus
L-amino acid oxidase, 87, 89, 94

A. controtrix
arginine ester hydrolase, 402
blood coagulation, effect on, 345, 346
bradykinin release, 402
carbohydrates, 12
collagenase, 120
NADase, 134
proteolytic enzyme, 117, 120
serotonin, 16
toxicity, 223

A. contortrix laticinctus
arginine ester hydrolase, 128
blood coagulation, effect on, 346
cytotoxic action, 315
metals, 6
NADase, 135
nerve growth factor, 362, 369
pathology, 228, 229
toxicity, 223

A. contartrix mokasen (A. mokeson, A. contortrix mokeson, A. mokason)
L-amino acid oxidase, 94
arginine ester hydrolase, 402
blood coagulation, effect on, 346
bradykinin release, 402
NADase, 134
proteolytic enzymes, 117

A. halys (A. halys halys)
L-amino acid oxidase, 94
arginine ester hydrolase, 128
blood coagulation, effect on, 346
hemolysis, 322
proteolytic enzyme, 117

A. halys blomhoffii (A. halys blomhoffi)
L-amino acid oxidase, 94
angiotensinase inhibitor, 406
arginine ester hydrolase, 128, 401, 402
blood coagulation, effect on, 346
bradykinin potentiating factor, 405, 406
bradykinin release, 401–403
hemolysis, 325
hemorrhage, 381
NADase, 134
peptides, 10
phosphodiesterase, 70, 72
phospholipase A_2 25, 29, 32, 43, 44
phosphomonoesterase, 80–83
toxicity, 223

A. piscivorus (A. piscivorus piscivorus)
anticomplement action, 430

arginine ester hydrolase, 402
blood coagulation, effect on, 345, 346
bradykinin release, 402
carbohydrates, 12, 14
chemical neutralization, 436, 439, 442, 447, 451–453
components, 227
cytotoxic action, 315
distribution of venom, 237
hemorrhage, 374
hyaluronidase, 133
NADase, 134
nephrotoxic action, 392
nerve growth factor, 361, 362, 366, 367, 369
pharmacology, 229
phosphodiesterase, 72
phospholipase A_2 25, 39, 47, 48, 50, 52
phosphomonoesterase, 80, 83

A. piscivorus leucostoma
arginine ester hydrolase, 128
collagenase, 120
proteolytic enzymes, 106–108, 110, 112, 116, 117, 120

A. rhodostoma
L-amino acid oxidase, 94
anticomplement action, 430
arginine ester hydrolase, 128, 129
Arvin (Ancrod), 333, 336–343, 346
blood coagulation, effect on, 331, 342, 351
carbohydrates, 14
hemorrhage, 381
isoelectric points, 166
metabolic effect, 413
nerve growth factor, 362, 364, 369
phosphodiesterase, 72
proteolytic enzyme, 117
toxicity, 223, 228
yield, 153

Bothrops alternatus (B. alternata)
acetylcholinesterase, 101
phospholipase A_2, 44
toxicity, 223

B. atrox
arginine ester hydrolase, 402
blood coagulation, effect on, 342, 343, 346
bradykinin release, 402
chemical neutralization, 442, 451
distribution of venom, 238, 239
nerve growth factor, 362, 363, 367–369
phosphodiesterase, 70–72, 74
phospholipase A_2 inhibitor, 140
phosphomonoesterase, 79–81, 83
platelets, effect on, 352
proteolytic enzymes, 117
toxicity, 156, 223

B. atrox asper
chemical neutralization, 451–453
phosphodiesterase, 72
proteolytic enzyme, 117

B. atrox marajoensis
reptilase, 341

B. atrox moojeni
blood coagulation, effect on, 341

B. itapetiningae
L-amino acid oxidase, 93

B. jararaca
angiotensinase inhibitor, 140, 141, 403, 405, 406
anticomplement action, 430
arginine ester hydrolase, 129, 402
blood coagulation, effect on, 331, 341, 342, 346
bradykinin potentiating factor, 403, 405, 406
bradykinin release, 401, 402
chemical neutralization, 438, 446–449
nerve growth factor, 363, 366, 370
peptides, 9
phosphodiesterase, 72
phospholipase A_2, 25, 44
phospholipase A_2 inhibitor, 140
phosphomonoesterase, 83
proteolytic enzymes, 107, 110, 112, 117
reptilase, 341
toxicity, 223

B. jararacussu
anticomplement action, 430
blood coagulation, effect on, 342
phospholipase A_2, 25
phospholipase A_2 inhibitor, 140
toxicity, 223

B. nasuta
distribution of venom, 238
hemorrhage, 380
toxicity, 156, 223

B. neuwiedii
phospholipase A_2, 25, 36, 44, 47
phospholipase A_2 inhibitor, 140
toxicity, 223

B. neummifera (B. nummifer)
distribution of venom, 238
toxicity, 156, 223

B. picadoi
distribution of venom, 238
toxicity, 156, 223

B. schlegelii
distribution of venom, 238
hemorrhage, 378, 380
toxicity, 156, 223

B. venezuelae
proteolytic enzymes, 117

Crotalus adamanteus
arginine ester hydrolase, 402

Index by Species

ATP, hydrolysis of, 73
basic proteins, 219, 220, 317
blood coagulation, effect on, 341, 346
bradykinin release, 402
chemical neutralization, 438, 442, 443, 446, 448
distribution of venom, 237, 239
elastase, 122
hemolysis, 324, 325
hemorrhage, 373
hyaluronidase, 134
metals, 7
metabolic effect, 414
nerve growth factor, 362, 363, 366, 367, 370
pharmacology, 222, 225
phosphodiesterase, 65–68, 70, 72–74
phospholipase A_2, 25–27, 29, 33
phosphomonoesterase, 80, 81, 83
toxicity, 156, 223

C. atrox
L-amino acid oxidase, 94
arginine ester hydrolase, 402
blood coagulation, effect on, 346
bradykinin release, 401, 402
chemical neutralization, 440–444, 446, 449–451
cytotoxic action, 313–315
distribution of venom, 237
hemolysis, 324
hemorrhage, 373–378
histamine release, 407
hyaluronidase, 134
immunology, 227
metabolic effect, 414, 415
myonecrosis, 383, 386
nephrotoxic action, 392–395
nerve growth factor, 370
pharmacology, 225
phosphodiesterase, 72
phospholipase A_2, 25, 29, 37, 39, 40, 43
phosphomonoesterase, 80, 83
platelets, effect on, 351
toxicity, 156, 224

C. bascilliscus
arginine ester hydrolase, 402
bradykinin release, 402
chemical neutralization, 443
hemorrhage, 373
metals, 7
proteolytic enzymes, 117
toxicity, 224

C. cerastes
toxicity, 224

C. durissus (C. durissus durissus)
L-amino acid oxidase, 94
blood coagulation, effect on, 346
chemical neutralization, 443, 451
hemorrhage, 373
metals, 7
phosphodiesterase, 72
proteolytic enzyme, 117

C. durissus cumanensis
L-amino acid oxidase, 94
phosphodiesterase, 72
phosphomonoesterase, 83

C. durissus terrificus (C. terrificus terrificus)
L-amino acid oxidase, 90, 91, 94
arginine ester hydrolase, 402
autopharmacological action, 352, 402, 407, 408
chemical neutralization, 438, 443, 446, 448, 451
chemistry, 213–217, 238
cytotoxic action, 315
hemolysis, 325
hemorrhage, 373
hyaluronidase, 134
metals, 7
nephrotoxid action, 392
nerve growth factor, 362, 370
pharmacology, 217–219
phosphodiesterase, 65, 72
phospholipase A_2, 25, 44, 56
phosphomonoesterase, 83
platelets, effect on, 351
toxicity, 224

C. durissus totonacus
chemical neutralization, 443
hemorrhage, 373
metals, 7

C. horridus (C. horridus horridus)
L-amino acid oxidase, 94
blood coagulation, effect on, 341, 343
chemical neutralization, 439, 441, 443–445
hemorrhage, 373
metals, 7
nerve growth factor, 370
phosphodiesterase, 72
proteolytic enzymes, 117
toxicity, 156, 224

C. horridus atricaudatus
cytotoxic action, 315
hemorrhage, 373
metals, 7
metabolic effect, 414

C. ruber
platelets, effect on, 351
toxicity, 224

C. scutulatus
L-amino acid oxidase, 94
blood coagulation, effect on, 346
cardiotoxin, 306
chemistry, 219–222
electrophoresis, 212

isoelectric focusing, 212, 219
pharmacology, 222, 225–227
phosphodiesterase, 72
proteolytic enzyme, 117
Raman spectroscopy, 220–222
toxicity, 224

C. viridis (C. viridus viridis)
L-amino acid oxidase, 94
arginine ester hydrolase, 402
bradykinin release, 402
chemical neutralization, 443
cytotoxic action, 313
hemorrhage, 373
metals, 7
myonecrosis, 388–391
phosphodiesterase, 72
proteolytic enzyme, 117
toxicity, 224

C. viridis helleri
L-amino acid oxidase, 94
blood coagulation, effect on, 346
metabolic effect, 413
pharmacology, 225
phosphodiesterase, 72
proteolytic enzyme, 117
toxicity, 224

C. viridis lutosus
toxicity, 224

Lachesis muta (L. mutus)
anticomplement action, 430
phosphodiesterase, 72
phosphomonoesterase, 82, 83
toxicity, 224

Sistrurus miliarius barbouri
L-amino acid oxidase, 94
blood coagulation, effect on, 346
chemical neutralization, 443
hemorrhage, 373
metals, 7
phosphodiesterase, 72
proteolytic enzyme, 117
serotonin, 16

Trimeresurus albolabris
toxicity, 225

T. elegans
amino acids, free, 9
hemorrhage, 381
metals, 9
phosphodiesterase, 72
phosphomonoesterase, 80

T. flavoviridis
antihemorrhagic factor, 381
arginine ester hydrolase, 402
blood coagulation, effect on, 344, 346
bradykinin release, 402
chemical neutralization, 437, 438, 440–442, 445–447
cytotoxin, 309

hemorrhage, 377–379
immunology, 229
metabolic effect, 414
metals, 9
myonecrosis, 382, 385
pharmacology, 229
phosphodiesterase, 72
phospholipase A_2, 41
phospholipase A_2 inhibitor, 140
phosphomonoesterase, 80, 83
riboflavin, 16
toxicity, 225

T. gramineus
L-amino acid oxidase, 94
arginine ester hydrolase, 402
blood coagulation, effect on, 330, 332, 333, 341, 344
bradykinin potentiating factor, 405
bradykinin release, 402
chemical neutralization, 452
enzyme inhibition, 414
hyaluronidase, 134
immunology, 229
metabolic effect, 413–415
NADase, 134, 135
peptides, 9
pharmacology, 225, 228
phosphodiesterase, 72
phosphomonoesterase, 80, 83
toxicity, 225, 228

T. mucrosquamatus
amino acids, free, 9
arginine ester hydrolase, 402
blood coagulation, effect on, 345, 349
bradykinin potentiating factor, 405
bradykinin release, 402
chemical neutralization, 452, 453
hyaluronidase, 134
metabolic effect, 412–415
peptides, 9
pharmacology, 225, 228
phosphodiesterase, 72
phosphomonoesterase, 80, 83
toxicity, 225, 228

T. okinavensis
L-amino acid oxidase, 94
arginine ester hydrolase, 402
blood coagulation, effect on, 344, 349
bradykinin release, 402
immunology, 229
nephrotoxic action, 393
nucleotide release, 351
phosphomonoesterase, 80, 83
platelets, effect on, 352
toxicity, 225

T. popeorum
anticomplement action, 430

T. purpureomaculata

Index by Species

   anticomplement action, 430
   platelets, effect on, 351
 T. wagleri
   toxicity, 225
Colubridae (Colubrids)
 Atractaspis microlepidota
   immunology, 235
 Boiga cyanea
   immunology, 235
 Dispholidus typus
   case report, 234
   immunology, 235
   toxicity, 234
   yield, 234
 Enhydris bocourti
   immunology, 235
 Homalopsis buccata
   immunology, 235
 Leptodiera annulata
   phosphodiesterase, 72
   phospholipase A_2, 57
   phosphomonoesterase, 82
   proteolytic enzyme, 117, 118, 234
 Thelotornis kirtlandi
   toxicity, 234
Scorpions
 Androctonus aeneas (A. aeneas aeneas)
   Crustacean toxin, 461
   immunology, 465
   toxic principle, 461
 A. amoreuxi
   Crustacean toxin, 461
   neurotoxic effect, 468
 A. australis
   cardiovascular effect, 471
   immunology, 465
   isolation, 460, 461
   neurotoxins, 462, 467–469, 477
   sequence of amino acids, 461, 463
   structure-toxicity relationship, 464
   toxic principle, 460
   toxicity, 460
 A. mauretanicus
   cardiovascular effect, 471, 473
   toxic principle, 461
   toxicity, 466
 Buthus judaicus
   toxicity, 466
 B. minax
   amino acids, free, 464
   immunology, 465
   metabolic effect, 475, 476
 B. occitanus
   amino acid composition, 462
   immunology, 465
   isolation, 460, 461
   neurotoxic effect, 468
   structure-toxicity relationship, 464

   toxicity, 460
 B. occitanus puris
   toxic principle, 461
 B. occitanus tunetanus
   immunology, 465
   sequence, 462
   toxic principle, 461, 462
 B. tamulus
   cardiovascular effect, 471
   hemorrhage, 474
   neurotoxic effect, 469
 Centruroides gertschi
   immunology, 465
 C. limpidus tecomanus
   toxicity, 467
 C. margaritatus
   clinical symptoms, 465
 C. sculpturatus
   amino acid composition, 462
   cardiovascular effect, 474
   clinical symptoms, 467
   immunology, 465
   physiological effect, 476
   sequence of amino acids, 461, 463
   toxicity, 460
 C. sufussus sufussus
   neurotoxic effect, 469
 Hadrurus arizonensis
   acetylcholinesterase, 464
   immunology, 465
   toxicity, 466
 Heterometrus caesar
   cardiovascular effect, 473
   toxicity, 466
 H. fulvipes
   metabolic effect, 475
   myotoxic effect, 475
   neurotoxic effect, 467
 H. scaber
   enzymes, 464
   metabolic effect, 476
   non-protein components, 464
   toxic principle, 461
   toxicity, 466
 Leiurus quinquestriatus (Buthus quinquestriatus)
   blood coagulation, effect on, 474
   cardiotoxic effect, 470–473
   distribution of venom, 467
   enzyme inhibition, 475
   immunology, 465
   myotoxic effect, 475
   neurotoxins, 461, 462, 468
   non-protein components, 464
   physiological effect, 476
   toxicity, 460
 Nebo hierichontictus
   clinical symptoms, 465

Orthochirus innesi
  neurotoxic effect, 468
  toxicity, 466
Pelamneus gravimanus
  blood coagulation, effect on, 474
  histamine, 464
Pandinus exitialis
  cardiovascular effect, 471
  immunology, 465
  physiological effect, 477
  toxicity, 466
Prionurus crassicauda
  toxicity, 466
Scorpio maurus palmatus
  enzymes, 464
  toxicity, 464
Tityus bahiensis
  cardiovascular effect, 471
T. serrulatus
  cardiovascular effect, 471
  isolation, 460
  myotoxic effect, 475
  neurotoxic effect, 467–469
  physiological effect, 476
  toxicity, 466
Vejovis flavis
  immunology, 465
V. spinigerus
  immunology, 465
  toxicity, 466

Spiders

Aphonopelma spp.
  nonprotein component, 487
  proteases, 486
  toxicity, 489
Atrax robustus
  enzymes, 486, 487
  nonprotein components, 487
  toxic components, 486
Chiracanthium mildei
  clinical symptoms, 489, 490
Ctenus nigriventer
  enzyme, 487
Dugesiella hentzi
  necrotoxin, 485, 486
  nonprotein components, 487
Herpyllus ecclesiasticus
  clinical symptoms, 490
Latrodectus bishopi
  cardiovascular effect, 496
  immunology, 488
  toxicity, 489
  yield of venom, 485
L. curacaviensis
  yield of venom, 485
L. dectus
  cardiovascular effect, 496
L. geometricus
  cardiovascular effect, 496
  immunology, 488
  neurotoxic effect, 491
  yield of venom, 496
L. mactans
  clinical symptoms, 489
  cardiovascular effect, 496
  distribution of venom, 490
  enzymes, 486, 487
  immunology, 488
  neurotoxic effect, 490–493, 495
  nonprotein components, 487
  toxicity, 489
  yield of venom, 485
L. mactans hasseltii
  immunology, 488
  neurotoxic effect, 490, 492, 495
L. mactans indistinctus
  immunology, 488
L. mactans tredecimguttatus
  cardiovascular effect, 496
  cytotoxic effect, 496
  distribution of venom, 490
  neurotoxic effect, 490, 491, 494
  neurotoxin, 486
  nonprotein components, 487
  toxicity, 488, 489
  yield of venom, 485
L. variolus
  cardiovascular effect, 496
  immunology, 488
  toxicity, 489
  yield of venom, 485
Loxosceles aleta
  necrosis, 496
L. arizonica
  cardiovascular effect, 496
  immunology, 488
L. laeta
  cardiovascular effect, 496
  immunology, 488
  toxicity, 488
L. reclusa
  cardiovascular effect, 496
  clinical symptoms, 489
  enzymes, 486
  immunology, 488
  necrosis, 490, 496
  nonprotein component, 487
  toxic component, 486
  yield of venom, 485
L. refescens
  immunology, 488
L. rufipes
  yield of venom, 484, 485
Lycosa erythrogantha
  enzymes, 486
  nonprotein components, 487

Index by Species

yield of venom, 484, 485
L. raptoris
  enzyme, 487
Pamphobetus roseus
  enzyme, 486
Phidippus formosus
  clinical symptoms, 490
Phoneutria fera
  enzymes, 486
  nonprotein components, 487
  toxicity, 489
P. nigriventer
  yield of venom, 484, 485
Pterinochilus spp.
  isoelectric focusing pattern, 485
  toxic component, 485
Steatoda paykullina
  cardiovascular effect, 496

Hymenoptera
Bees (Apidae)
Apis cerana
  melittin, 503
A. dorsata
  melittin, 503
A. florea
  melittin, 503
A. mellifera
  amines, biogenic, 508
  amino acid composition, 506
  amino acids, free, 508
  antibiotic action, 512
  antiinflammatory action, 515
  apamin, 504, 506, 599, 512
  carbohydrates, 508
  cardiopep, 509
  cardiovascular effect, 511
  cytotoxic action, 510
  enzymes, other than phospholipase A_2, 505, 507
  hemolytic activity, 510
  hemorrhage, 512
  histamine, 508
  histamine release, 510
  immunology, 508, 515
  lipids, 508
  MCD-peptide (peptide 401), 504, 506
  melittin, 502, 506, 509, 510, 512, 515
  melittin F, 505, 506
  metabolic effect, 512
  minimine, 505, 506
  mitochondria, effect on, 507, 513–515
  phospholipase A_2, 505, 507, 509, 510, 515
  procamine, 505
  promelittin, 503
  protease inhibitor, 507
  secapin, 505, 506
  surface tension, effect on, 510

tertiapin, 505, 506
toxicity, 509

Hornets and Wasps
Dolichovespula media
  hyaluronidase, 516
D. saxonica
  hyaluronidase, 516
Microbracon (Bracon) hebector, 519
M. gelechiae, 519
Paravespula: see Vespula
Philanthus triangulum, 518
Polistes annularis
  melittin, 503
P. apachus
  electrophoretic pattern, 502
  immunology, 515
P. exclamans
  melittin, 503
P. gallicus
  hyaluronidase, 516
  kinin, 519
P. metricus
  melittin, 503
Sceliphron caementarium, 519
Vespa spp.
  composition, 515
V. affinis
  clinical symptoms, 515
V. orientalis
  amino acid composition, 516
  clinical symptoms, 515
  composition, 516
  enzymes, 516
  hemolysis, 515
  immunology, 515
  pharmacology, 517, 518
  toxicity, 517
Vespula (Paravespula) arenaria
  immunology, 515
V. crabro
  hyaluronidase, 516
V. germanica
  hemolysis, 519
  hyaluronidase, 516, 519
  immunology, 515
  major component, 519
  protease, 519
  toxicity, 519
V. maculata
  melittin, 503
V. maculifrons
  vespulakinins, 519, 520
V. pennsylvanica
  immunology, 516
V. vulgaris
  composition, 516
  hyaluronidase, 516

Ants
Myrmecia gulosa
  electron transport inhibition, 528
M. pyriformis (M. forficata)
  histamine, 528
  histamine release, 528
  hyaluronidase, 528
  phospholipase A_2, 528
  smooth muscle stimulation, 528
  toxicity, 527
Pogomyrmex barbatus, 529
Solenopsis germinata
  composition, 528, 529
S. invicta
  composition, 529
S. saevissima
  composition, 528, 529
S. saevissima richteri
  ATPase inhibition, 528
  composition, 529
S. xyloni
  composition, 528, 529

Gila Monster
Heloderma horridum
  anticomplement action, 430
  toxicity, 532
H. suspectum
  autopharmacological action, 533
  enzymes, 533
  pharmacology, 532
  serotonin, 533
  toxicity, 532